Pipeline Integrity Management Under Geohazard Conditions

Edited by

Mamdouh M. Salama (ConocoPhillips)

Yong-Yi Wang (CRES)

Don West (Golder)

Alexander McKenzie-Johnson (Geosyntec)

Azam B A-Rahman (Petronas)

Guiyi Wu (TWI)

Jens Petter Tronskar (DNVGL)

Jim Hart (SSD Inc)

Bernt J. Leira (NTNU)

ASME

SETTING THE STANDARD

© 2020, The American Society of Mechanical Engineers, 2 Park Avenue, New York, NY 10016, USA (www.asme.org)

All rights reserved. Printed in the United States of America. Except as permitted under the United States Copyright Act of 1976, no part of this publication may be reproduced or distributed in any form or by any means, or stored in a database or retrieval system, without the prior written permission of the publisher.

INFORMATION CONTAINED IN THIS WORK HAS BEEN OBTAINED BY THE AMERICAN SOCIETY OF MECHANICAL ENGINEERS FROM SOURCES BELIEVED TO BE RELIABLE. HOWEVER, NEITHER ASME NOR ITS AUTHORS OR EDITORS GUARANTEE THE ACCURACY OR COMPLETENESS OF ANY INFORMATION PUBLISHED IN THIS WORK. NEITHER ASME NOR ITS AUTHORS AND EDITORS SHALL BE RESPONSIBLE FOR ANY ERRORS, OMISSIONS, OR DAMAGES ARISING OUT OF THE USE OF THIS INFORMATION. THE WORK IS PUBLISHED WITH THE UNDERSTANDING THAT ASME AND ITS AUTHORS AND EDITORS ARE SUPPLYING INFORMATION BUT ARE NOT ATTEMPTING TO RENDER ENGINEERING OR OTHER PROFESSIONAL SERVICES. IF SUCH ENGINEERING OR PROFESSIONAL SERVICES ARE REQUIRED, THE ASSISTANCE OF AN APPROPRIATE PROFESSIONAL SHOULD BE SOUGHT.

ASME shall not be responsible for statements or opinions advanced in papers or . . . printed in its publications (B7.1.3). Statement from the Bylaws.

For authorization to photocopy material for internal or personal use under those circumstances not falling within the fair use provisions of the Copyright Act, contact the Copyright Clearance Center (CCC), 222 Rosewood Drive, Danvers, MA 01923, tel: 978-750-8400, www.copyright.com.

Requests for special permission or bulk reproduction should be addressed to the ASME Publishing Department, or submitted online at: https://www.asme.org/publications-submissions/journals/information-for-authors/journal-guidelines/rights-and-permissions

ISBN: 978-07918-6199-8
Book No.: 861998

CONTENTS

INTRODUCTION

While the traditional pipeline design mainly focuses on pressure containment by limiting the hoop stress to certain percentage of the specified minimum yield strength (SMYS), pipelines that are subject to ground movement may experience high longitudinal stress exceeding the SMYS and strains exceeding yield strain and approaching 2% or more. If longitudinal stress/strain is not properly considered during the design and construction phases, premature failure can occur. Failures have occurred in many countries including the US, Canada China, Malaysia, etc. These failures occurred in both liquid and gas pipelines due to soil movement (landslides, subsidence, heavy rain etc.). The failure is manifested by either rupture of the pipe mainly at the girth weld due to the tensile longitudinal stress or buckling/wrinkling of the pipe due to the compressive stress. Unfortunately, existing codes lack the depth to allow the design against these failures. As example, girth welds that are made to comply with industry standards such as API 1104 are not guaranteed to have high resistance to longitudinal strains. In addition, ASME codes such as B31.8 specifies that the maximum permitted longitudinal strain is to 2% without providing any direction on how to ensure that the pipe can achieve this strain capacity. Worse yet, some analysts or company have taken this 2% strain value as their pipelines' strain capacity without any justification.

During the last decade extensive effort was devoted by the pipeline industry to manage ground movement hazards and design against them to ensure the integrity of pipelines. In an effort to assemble all the recent advances in a single document, ASME organized in 2019 as part of its conference series on Asset Integrity Management (AIM) a specialty conference on "Pipeline Integrity Management under Geohazard conditions (PIMG)". Key industry subject matter experts who are actively working to address the subject were invited to prepare 42 peer referred papers to the conference. The papers compile the results of extensive research as well as assemble pipeline operators' experiences in tackling geohazard challenges for both new and vintage pipelines. In addition to the papers of the conference the editors of the book prepared an introduction to each section that includes summary review of the different papers in the section. The papers are presented in 10 sections addressing all aspects of geohazard integrity management.

The first section highlights the geohazard impact on pipeline integrity. The next four sections focused on the geohazard demand and its management by addressing geohazard characterization, monitoring, management and mitigation. The following 4 sections focused on the strain capacity by addressing strain capacity predictions, management and monitoring. The last chapter tied both strain demand and capacity through structural reliability and risk assessment protocols.

The information on this book is not only intended for use by pipeline designers and operators but it is expected to also be used by regulators and standard writing organizations. It is therefore, the intention of ASME to update this book on regular bases as new data, case studies and advancement of the state of practice become available. The update is expected to be done by organizing future conferences on the same topic every 3 years.

The Editors

Copyright © 2019 by ASME

SECTION 1
GEOHAZARD IMPACT

INTRODUCTION

By Mamdouh M. Salama
Mamdouh.m.salama@conocophillips.com

Pipelines subjected to geohazard require special treatment during the design phase. The design process starts with identifying the geohazard mechanisms in the pipeline ROW and these may include landslides, discontinuous permafrost, seismic, hydro-mechanical/erosion, ground subsidence, shallow ground water, collapsible and expansive soils, etc. If geohazards cannot be avoided, the strain demand is predicted based on geotechnical models and soil/pipe interaction. Based on the strain demand the pipeline designer can specify the pipeline material, geometry and construction practices to ensure that the strain capacity of the pipe exceeds the strain demand. Figure 1 presents the key components of efficient strain-based design (SBD) strategy for pipelines.

Improper consideration of geohazards in route selection, design, and construction of pipelines can lead to premature failure. These failures manifest themselves by rupture of the girth weld or buckling (wrinkling) of the pipe as shown in Figure 2. The US regulators PHMSA has reported that between 2009 and 2018 the number of pipeline failures due to loads induced by natural forces such as earth movement (landslides, subsidence, etc.) and heavy rain/floods, was 273. In 2018, the number of failures was 13 for gas pipelines and 14 for liquid pipelines. PHMSA records only include significant incidents that results in one of the following: fatality or injury requiring in-patient hospitalization, $50,000 or more cost, measured in 1984 dollars, release of 5 or more bbls of highly volatile liquid (HVL), release of 50 or more bbls of non-HVL liquid, and release resulting in an unintentional fire or explosion.

Paper PIMG 2019-1041 by Mekalled and Paviglianiti provides additional information on geohazard pipeline incidents that have been reported by both the Canadian and US regulators. In addition, the paper highlighted the importance of proper soil-pipeline interaction analysis in predicting the strain values and of soil monitoringat locations of potential hazards.

Paper PIMG2019-1028 by Hongyuan, et al presents the failure analysis of a pipeline in China due to geohazard loads. Thepipe was 32-inch spiral welded APIL 5L X 80 and the cracks were initiated near the extrados. The failure analysis indicated that the pipe failed ata strain of 0.57%

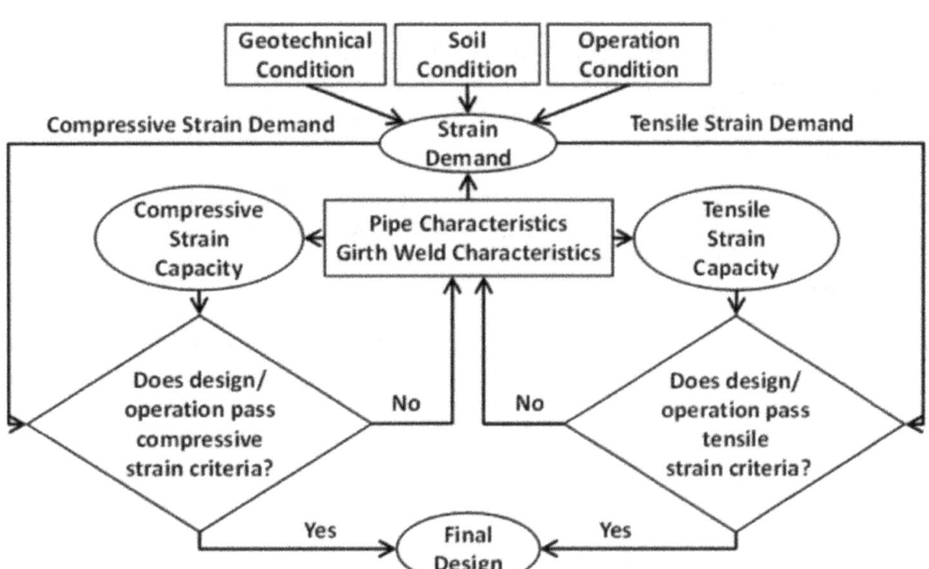

Figure 1. Key components of pipeline design to account for geohazards

Copyright © 2019 by ASME

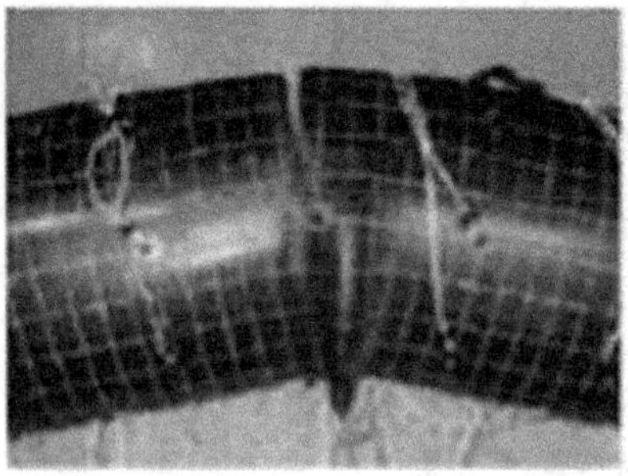

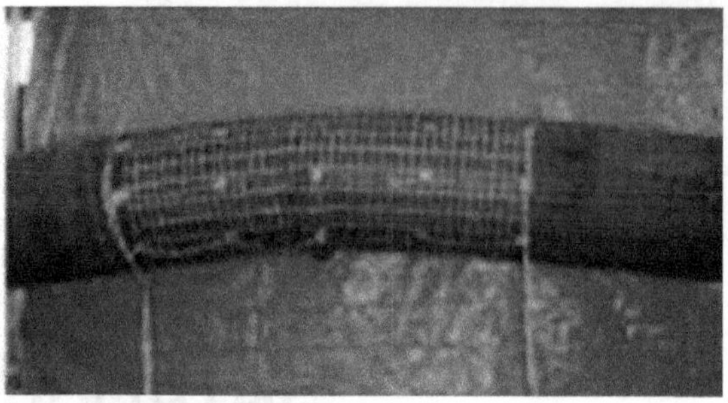

Figure 2. Illustration of rupture and bucking due to high tensile and compressive longitudinal strain

In order to ensure against these failures due to natural forces, PHMSA published several Codes of Federal Regulation such as 192.317 and 192.613. CFR 192.317 requires the pipeline operator to take all practicable steps to protect the pipeline from washouts, floods, unstable soil, landslides, or other hazards that may cause the pipeline to move or to sustain abnormal loads. CFR 192.613 addresses surveillance requirements that may be accomplished by the following.

1) Periodic visual inspection of pipeline facilities to identify items such as changes in topography, and pipeline exposure or movement.
2) Periodic review and analysis of records, such as the following: 1) Patrols, and 2) Facility failure investigations.

Paper PIMG2019-1067 by Wang who has been working extensively with both PRCI and PHMSA to develop guidelines for proper design and management of pipelines subjected to geo hazard condition, provides an overview of the process to design a pipeline to have sufficient longitudinal strain capacity to withstand geohazard loads. In addition, the paper presented a list of remaining gaps in the current technology. These gaps are currently being addresses through several projects and they will be addressed in future editions of this book.

While the focus of this book is on pipeline integrity subjected to geohazard conditions, i.e. failure due to natural forces, other failure mechanism should also be addressed to ensure the overall integrity of pipelines. Table 1 provides PHMSA records of high consequence incidents in the US for the period 2009 to 2018.

Table 1. PHMSA record of Pipeline incidents from 2009 to 2018

Incident Cause	Hazardous Transmission	Gas Transmission
Corrosion	806	201
Excavation Damage	153	150
Incorrect Operation	555	61
Material/Welding	2068	482
Natural Forces	176	97
Other causes	197	141
Total	**3956**	**1132**

Although the focus on this book is to address pipeline subjected to longitudinal strain, one cannot overlook the hoop stress and strain not only with respect to the pressure induced hoop stress which is the addressed as paper of the normal pipeline design using Barlow's equation but also regarding the effect of longitudinal strain on the hoop strain in the intrados of a bend. The paper by Zhu (AIMPIMG2019-1006) focused on evaluating the existing methods for predicting the burstpressure of pipelines, without defects and with corrosion defects. For the defect free pipe, the paper compared the prediction of four strength models, namely Tresca, Von-Mises, Zhu-Leis and flow stress with test results for pipes with different strain hardening.The paper also evaluatedseveral corrosionassessment models including B31G, B31G modified, RSERENG, LPC, PCORRC, modified PCORRC etc.

Chapter 1 Impact of soil movement on Pipelines (AIMPIMG2019-1041)
BahaaMekalled and Joe Paviglianiti, National Energy Board, Canada

Chapter 2 Failure analysis of a Stress-based pipeline under plastic strain (AIMPIMG2019-1028)
Chen Hongyuan, Feng Hui, Chi Qiang, and HuoChunyong, Tubular Goods Research Institute of China National Petroleum Corporation, China

Chapter 3 Strain-Based Design and Assessment - Concepts and Gaps (AIMPIMG2019-1067)
Yong-Yi Wang, Center for Reliable Energy Systems (CRES), USA

Chapter 4 Burst Failure Models and Their Predictions of Buried Pipelines (AIMPIMG2019-1006)
Xian-Kui Zhu, EWI, USA

Copyright © 2019 by ASME

AIMPIMG2019-1041

Chapter 1

IMPACT OF GEOHAZARDS ON PIPELINES AND REGULATORY POWERS AND RESTRICTIONS IN CANADA AND THE UNITED STATES

Bahaa Mekalled P. Eng
National Energy Board
Calgary, Alberta, Canada

Joe Paviglianiti P. Eng.
National Energy Board
Calgary, Alberta, Canada

ABSTRACT

Several buried pipeline incidents have previously been triggered by the surrounding soil behavior in reaction to groundwater fluctuation events or creep movement that happen over an extended period of time. Depending on the soil type, the water cycle and the season, what was previously in a stable environment might become more vulnerable to slope movement and slides. Soil movement triggers stresses in the pipes whether soil failures occurs or not. The impact of soil movement on pipelines are considered in this paper based on the existing information in public domain. The tools available to the regulator to assess the need for intervention when soil movement is suspected are also be discussed. This paper examines incidents that occurred in Canada and the United States of America (USA).

Note: *the views, judgments, opinions, statements and recommendations expressed in this paper do not necessarily reflect those of the National Energy Board (NEB or Board), its Chair or Members, nor is the Board obliged to adopt any of them.*

INTRODUCTION

An examination of publicly available oil and gas pipeline incident data issued by North American federal regulators has identified reporting requirements for geohazard threats. The Canadian Federal Regulator, the National Energy Board (NEB) [1] has an incident reporting category that enables identification of incidents related to geohazards that may have not led to a failure (i.e. leak or rupture). The United States Pipeline and Hazardous Materials Safety Administration (PHMSA) Office of Pipeline Safety (OPS) [2] requires reporting of accidents which impact people or the environment such as a release of product or monetary impact. Due to the authors' familiarity with the NEB data and because they include a wider range of geohazard events compared to the PHMSA data, a more detailed analysis was performed on this data.

The NEB geohazard related incident data was analyzed and an obvious increase in incidents was observed. However; this may be partially attributed to a change in clarification on incident definition scope and reporting requirements. These incidents might have not all resulted in product release, but they may have a significant impact on the current operating conditions compared to the pipeline's original design and may result in the need for changes to the pipeline operating conditions.

In addition, the current North American oil and gas pipeline regulations the NEB Onshore Pipeline Regulations (OPR) [3], and the PHMSA 49 CFR Part 195 [4] for Hazardous Liquids and 49 CFR Part 192[5] for Natural Gas do not address geohazard threats in explicit detail. The standards either referenced by the regulations or used by industry in North America, such as the Canadian Standards Association (CSA) Z662 Oil and Gas Pipelines Systems [6] or the American Society of Mechanical Engineers (ASME) liquid and gas pipeline standards ASME B31.4 [7] and ASME B31.8 [8] respectively, also do not provide significant detail on how to address geohazards. Another potential gap is that the standards and regulations do not provide details or guidance on the how to address and assess the possible interaction of the geohazard threat with other threats such as cracking, corrosion, or girth welds.

This paper examines the geohazard related regulatory and standard requirements and associated geohazard incident reporting requirements. In addition, the paper considers the increase in geohazard related incidents to identify the reason for the increase. Geohazard incidents are evaluated and discussed from the perspective of soil displacement related to erosion, washouts, slope stability, settlement, frost heave, and landslides and the associated challenges associated with the absence of explicit regulatory regulations or guidance. This lack of regulatory guidance requires competent geohazard professional judgment to assess whether the pipeline is at risk due to the soil

Copyright © 2019 by ASME

displacement event that has taken place. In addition, the interaction between the condition of the pipeline (i.e. corrosion, cracking) and the soil displacement may result in a combined effect that could lead to a pipeline incident.

Since most of the readily publicly available geohazard incident data is from Canadian regulatory sources this paper focuses on the associated pipeline geohazard issues encountered and reported in Canada.

NOMENCLATURE

ASME:	American Society of Mechanical Engineers
BC	British Columbia
CFR	Code of Federal Regulations
CSA:	Canadian Standards Association
CSA-Z662:	Oil and gas pipeline systems standard
CSCC	Circumferential Stress Corrosion Cracking
EA	Engineering Assessment
ERG	Event Reporting Guidelines
HCA	High Consequence Area
mi.	miles
NEB	National Energy Board
NTSB	National Transportation Safety Board
OPR	Onshore Pipeline Regulation
PHMSA	Pipeline and Hazardous Materials Safety Administration.
PRCI	Pipeline Research Council International
ROW	Pipeline Right of Way
SCC	Stress Corrosion Cracking
TSB	Canadian Transportation Safety Board

SUMMARY OF REGULATORY AND STANDARDS REQUIRMENTS RELATED TO GEOHAZARD

Geohazard Design and Assessment Requirements

The regulatory requirements imposed on North American pipelines regulated by the Federal regulators do not provide explicit detailed guidance on how to prevent and assess potential geohazard threats. In addition, the North American oil and gas pipeline standards do not provide guidance on how to address geohazard issues and when to intervene and react to the geohazard threat. There is also very little guidance directing companies to identify how to assess the interaction of the geohazard threat with other integrity threats such as circumferential stress corrosion cracking or metal loss.

For example, PHMSA for natural gas lines through CFR Part 192.317 [5] requires companies to consider the following:

"The operator must take all practicable steps to protect each transmission line or main from washouts, floods, unstable soil, landslides, or other hazards that may cause the pipeline to move or to sustain abnormal loads."

The relevant ASME B31.8 [8] natural gas pipeline standard does not provide much more guidance but states in Section 841.1.10:

"(a) When pipelines and mains must be installed where they will be subject to natural hazards, such as washouts, floods, unstable soil, landslides, earthquake related events (such as surface faulting, soil liquefaction, and soil and slope instability characteristics), or other conditions that may cause serious movement of, or abnormal loads on, the pipeline, reasonable precautions shall be taken to protect the pipeline, such as increasing the wall thickness, constructing revetments, preventing erosion, and installing anchors."

The Canadian NEB regulations are even less explicit, OPR [3] Section 40 states:

"A company shall develop, implement and maintain an integrity management program that anticipates, prevents, manages and mitigates conditions that could adversely affect safety or the environment during the design, construction, operation, maintenance or abandonment of a pipeline."

The NEB regulation OPR [3] incorporates by reference all of the CSA Z662 [6] oil and gas pipeline standard. Similar to the ASME standard this standard does not provide much detail on how to design a pipeline to prevent geohazards threats and how to assess the geohazards. CSA-Z662 [6] Clause 4.2.4 states:

"The stress design requirements in this standard are specifically limited to design conditions for operating pressure, thermal expansion ranges, temperature differential, and sustained force and wind loadings. However loadings not addressed are those related to slope movements, fault movements, frost heave, and vortex shedding."

In addition, Clause 10.3 of CSA-Z662 [6] identifies that the pipeline company's integrity management system shall include procedures to identify conditions that can lead to failure and to eliminate or manage these potential failure conditions. As a follow-up, the standard states that if a company becomes aware of potential failure conditions, it shall conduct an Engineering Assessment (EA) to determine what part is susceptible to failure and if it is suitable for continued service. CSA identifies unstable slopes as one of the examples of conditions that can lead to failures.

Geohazard Incident Reporting Requirements

PHMSA in CFR Part 195.55 [4] requires companies to report the following associated with geohazard.

"(a) Except as provided in paragraph (b) of this section, each operator shall report in accordance with §195.56 the existence of any of the following safety-related conditions involving pipelines in service:

(1) ...

(2) Unintended movement or abnormal loading of a pipeline by environmental causes, such as an earthquake, landslide, or flood that impairs its serviceability.

(3) Any material defect or physical damage that impairs the serviceability of a pipeline."

Copyright © 2019 by ASME

However, as stated below PHMSA further defines accidents by the quantity of the spill, fatality, injury, events that result in emergency shutdown, or significance of event. A review of the PHMSA data is unable to determine whether companies are reporting geohazard events, such as pipe exposures without pipe movement, or damage and washouts if they do not meet the following definitions of accident for gas and oil pipelines respectively:

> *"Gas Transmission Onshore "Significant Incident" are those including any of the following conditions:*
> *(1) Fatality or injury requiring in-patient hospitalization;*
> *(2) $50,000 or more in total costs, measured in 1984 dollars*
> *(3) Highly volatile liquid releases of 5 barrels or more or other liquid releases of 50 barrels or more; and*
> *(4) Liquid releases resulting in an unintentional fire or explosion. Gas distribution incidents caused by a nearby fire or explosion that impacted the pipeline system are excluded from this definition."*

"Crude Oil/ Refined Petroleum/ Biofuel pipeline accidents Impacting People or the Environment (IPE) – if either criterion 1 or 2 below is met, the accident counts as IPE:
1. Regardless of Location of Accident (B10):
Fatality; or
Injury requiring in-patient hospitalization; or
Ignition; or
Explosion; or
Evacuation; or
Wildlife impact; or
Water contamination (ocean, groundwater, drinking); or
Public/Non-Operator Private Property Damage
2. For Location of Accident":
Unintentional Release Volume greater than or equal to 5 gallons AND HCA; or
Unintentional Release Volume (greater than or equal to 5 barrels AND HCA; or
Water contamination (surface); or
Soil contamination.""

The Canadian federal Regulator, the NEB, requires reporting of incidents as specified in OPR [3]. This regulation defines incidents as an occurrence that results in:

"(a) the death of or serious injury to a person;
(b) a significant adverse effect on the environment;
(c) an unintended fire or explosion;
(d) an unintended or uncontained release of LVP hydrocarbons in excess of 1.5 m3;
(e) an unintended or uncontrolled release of gas or HVP hydrocarbons;
(f) the operation of a pipeline beyond its design limits as determined under CSA Z662 or CSA Z276 or any operating limits imposed by the Board."

In 2014 the NEB provided reporting expectations in its Event Reporting Guidelines [9] (ERG) which were implemented in 2015. Of interest is the guidelines and expectations for the above incident reporting criteria (f) which relate to geohazard reporting. The Guidelines stated in section 3.1.3:

"For the purposes of notification under section 52 of the OPR, the Board employs the following definition for an "operation beyond design limits":
any operation of the pipeline beyond the criteria for which the pipeline was designed and/or the operation of the pipeline beyond criteria imposed to mitigate a condition on the pipeline. This includes conditions that have been demonstrated by an engineering assessment to provide continued fitness for service of the pipeline (e.g., pressure restrictions).
Events that fall under this definition include, but are not limited to:
- *exposure of the pipeline to conditions where significant vibration or excess stress can occur;*
- *slope movements that affect a pipeline;*
- *exposures of pipelines in waterbodies (e.g. rivers, wetlands); and ..."*

To provide additional clarity to the ERG [10] they were revised in 2017 and implemented in 2018 by revising the geohazard related guidance for the OPR incident reporting criteria (f) which changed to the following:

"For the purposes of notification under section 52 of the OPR, the Board employs the following definition for an "operation beyond design limits...
- *slope movements that exceed what was predicted at the design stage or were not predicted in the design stage;*
- *unintended exposures of pipelines including in waterbodies (e.g. rivers, wetlands) and on land; and ..."*

The 2015 reporting change, outlined in the Guidelines, resulted in the ability for the NEB to capture and analyze more comprehensive geohazard-specific data. This also caused an increase in geohazard incident, since previous geohazard related incidents were only associated with the OPR incident reporting requirements (a) to (e) that resulted in consequences such as a release or injury.

The NEB, falls under the oversight of the Canadian Transportation Safety Board [11] (TSB). The TSB also requires companies under NEB jurisdiction to report incidents directly to the TSB. The TSB defines pipeline geohazard occurrences as reportable if a geotechnical, hydrotechnical or environmental activity poses a threat to the safe operation of the pipeline.

HISTORICAL INCIDENTS AND DATA RELATED TO SOIL DISPLACEMENT IN CANADA

The NEB only regulates approximately 73,000 km. (45,360 mi.) of a total of 840,000 km (522,063 mi.) of pipelines in Canada. The oil and gas pipelines the NEB regulates are predominately large diameter transmission pipelines that cross provincial or international boundaries. This compares to PHMSA who regulate over 711,180 km [12] (442,002 mi.) where the onshore gas transmission pipelines constituted 478,767 km (297,556 mi) and the liquid pipelines constituted 232,413 km (144,446 mi). This means PHMSA regulates approximately 10 times the amount of oil and gas transmission pipelines than the NEB. This scale of pipeline mileage is for comparison purposes

Copyright © 2019 by ASME

to be able to realize the extent of the issues in the US in comparison to the data presented in Canada. Figure 1 shows a map of PHMSA regulated pipelines and Figure 2 shows a map of the NEB regulated pipelines. These figures make a good visual comparison of the extent of the pipeline mileages in the 2 countries. Note that the PHMSA figure includes all of PHMSA regulated pipelines interstate and intrastate lines.

Figure 1: Map of PHMSA Regulated Pipelines;

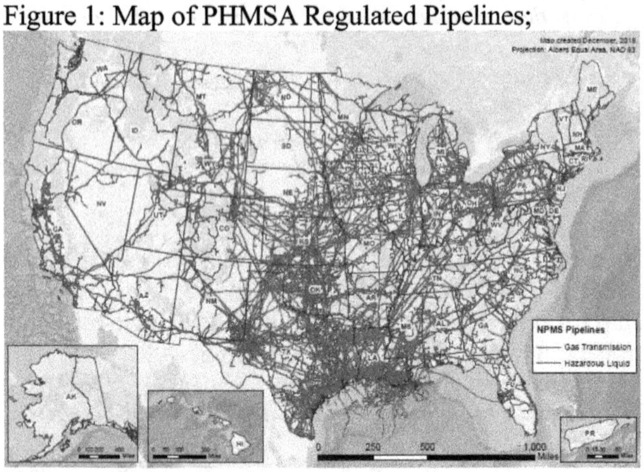

Note:

1. Map shows all of the PHMSA regulated pipelines including the transmission and distribution lines including intrastate and interstate pipelines
2. https://www.npms.phmsa.dot.gov/Documents/NPMS_Pipelines_Map.pdf

Figure 2: Map of NEB Regulated Pipelines

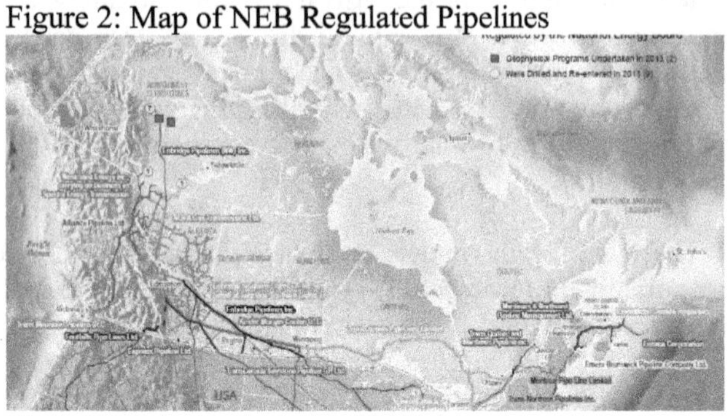

Note:
https://www.neb-one.gc.ca/sftnvrnmnt/sft/dshbrd/rgltdpplns-eng.pdf

PHMSA [13] and [14] published incident data for the time period 2010 to 2017. The total reported geohazard related incidents under the PHMSA Natural Force data category for this time period is 37 incidents and that's including incidents due to earth movements and heavy rains/floods and excludes lightning and other natural force damage.

Since the NEB revised its reporting geohazard requirements in 2015 a comparison of the NEB data with the PHMSA data would mean looking at the 2015 to 2017 accident data. 2018 data will not be considered as a part of this comparison since TSB is yet to publish data for 2018 and the NEB incidents are yet to be finalized and closed out. The total reported geohazard related accidents under the PHMSA Natural Force data category for the years 2015 to 2017 were 18, Table 1. This included 4 incidents for the category Earth Movement and 14 for the Heavy Rain/Flood category for gas and oil transmission pipelines combined.

The TSB [15] has also published geohazard incident data from the period 2010 to 2017 under the category Geotechnical/Hydrotechncial activity, and reported a total of 22 incidents for this period. For the 2015 to 2017 period there were a total of 19 reported incidents in this category. And, the NEB reported incidents from 2015 to 2017 was 29 incidents related to the category Geotechnical/Hydrotechnical.

The following discussion is focused on the NEB geohazard data in Canada because the NEB Guidelines require a more comprehensive reporting of geohazard incidents that do not only result in failures or damage to property. This will allow a more detailed analysis of pipe to soil interactions and geohazard interactions with other integrity hazards such as corrosion.

Between 2010 and 2017, approximately 37 incidents related to geohazard issues were reported to the NEB. During that time period some of the reported incidents were due to frost heave, other incidents were due to buoyancy, settlement, slope movement and landslides and exposure in watercourses. Note that, incidents of massive soil displacement such as landslides and beaver dam breach were also reported even though they did not result in pipe failure or product release. It should be noted that between 2013 and 2015 and prior to the 2016 change in reporting regulation, the NEB received 12 geohazard related incidents out of which 7 were frost heave related. This indicates that only 40% of the incidents reported were related to soil displacement. After the change in reporting requirements in 2016 the NEB received, 2 incidents due to slope movement and landslide, 22 incidents due to erosion and exposure near a water, and only one incident related to frost heave. The spike in reporting that occurred after 2016 as shown in Table 1 is attributed to the change made in the reporting requirements.

Table 1. Number of Incidents reported in the US and Canada between 2010 and 2017

Year	TSB	NEB	PHMSA[3]
2010	0	0	0
2011	0	0	9
2012	1	0	4
2013	1	2	3
2014	1	6	3
2015	0	4	9
2016	3	7	5
2017	16	18	4
2018	N/A	39	N/A

Copyright © 2019 by ASME

Total Reported Incidents (2015-2017)	19	29	18

Note: 1. N/A: data not available at the time of writing the paper
2. Numbers are based on published data, may sometimes vary slightly based on incident reporting analysis
3. PHMSA [14]

Another change in reporting requirement was implemented in 2018 for the NEB. That change in reporting requirement resulted in another spike in incident reporting. In 2018 alone 39 incidents related to Geohazards were reported to the NEB compared to 18 in 2017 and 7 in 2016 and 0 between 1995 and 2012.

If we extrapolate that there is 10 times more pipeline mileage in the US, then the number of similar incidents that should have been reported in the US should be significantly more than that currently reported under the current PHMSA regulations. For example, based on just linear interpolation for the 39 incidents reported to the NEB in 2018 there should have been 390 incidents reported in the US. This calculation is just for the purpose of comparison. Other factors such as geology, terrain and weather related conditions play a significant role in the occurrence of natural hazards incidents.

Table 2. Nature and number of pipeline geohazards related incidents reported to the NEB between 2013 and 2018

Year	Frost Heave	Geotechnical Slope Movement/landslides	Hydrotechnical Exposure in Waterbody
2013	1	0	1
2014	4	2	0
2015	2	1	1
2016	0	0	7
2017	1	2	15
2018	0	2	37

Based on the data presented in Table 2, it is evident that before the change in the reporting requirements that the NEB implemented in 2015, the predominant reported incident types were associated with frost heave. However, after the change in the reporting requirements, there was an increase in the incidents reported and a change in the type of incidents. In 2016, there was 7 hydrotechnical incidents reported. This trend continued to 2017 when the second adjustment was implemented to the reporting requirements. In 2017, the NEB received 15 hydrotechnical, 2 geotechnical, and one frost heave incidents. Furthermore, in 2018, the NEB received more than 4 times the incidents that were received in 2016 and more than 30 times the incidents received prior to 2015.

The hydrotechnical incidents that were reported to the NEB in 2016 were mainly related to pipeline exposures in waterbodies such as rivers, creeks, swamps and muskeg. The reason for the exposure was attributed to heavy rainfall and snowmelt events that resulted in washout events. One of the exposure reported in 2016 had a 3 m free span, the remainder of the reported exposures had no free span. This trend continued to 2017 where 15 hydrotechnical related incidents/exposures were reported. These exposures ranged from 0.1 m to 80 m with and without free span. The reported free span ranged from 1 m to 40 m. The percentage of the pipeline that reported free span was only 10%. Incidents reported in 2018 are still being processed and companies are still filling related information at the time of writing this paper, therefore this data is not being considered.

The Geotechnical incidents related data reported in 2016 was nil, while only 2 incidents were reported in 2017. One of the incidents reported in 2017 was due to a beaver dam breach while the other one was due to slope movement detected by slope monitoring process.

With regards to the reported NEB incidents and company remediation and monitoring plans after the incident is reported, a general trend is noticed. The company addresses the site depending on the circumstances and the intensity of the incident. Monitoring plans are in general put in place to ensure the stability of the site with and without the implementation of remediation. Pipeline EAs are normally carried out to ensure that the pipeline is suitable for return to service. Several factors are considered in an EA including the condition of the exposed pipe and its defects. Some of the key issues that should be considered, and are not always, are corrosion, dents, and coating conditions. Often the pipeline company performs an EA prior to returning the pipeline to service if it was deemed to stop operation after the incident. Pipeline defects such as corrosion and stress corrosion cracking should be linked to the soil displacement event and pipeline exposure that was reported in order to determine whether the pipeline is safe to return to service.

EXAMPLES OF SOME OF THE REPORTED INCIDENTS

In general, incidents that involve pipeline rupture and or triggered pipeline shut down were related to slope stability issues and soil displacement. For example:

1. This incident that occurred in 2015 resulted in a pipe rupture without any signs of slope failure. The slope movement was slow and steady in this case. The stresses induced by the slope movement were enough to push the pipeline to failure even though the pipeline had few defects.
2. Another case was reported to the NEB where the company elected to shut their pipeline and run a temporary replacement surface pipe because the observed soil displacement threatened the pipeline safety. This temporary surface pipe arrangement was put in place until permanent mitigation takes place. The first incident resulted in a rupture due to creep slope movement that was not originally detected.

The second case did not result in any pipe rupture, rather the slope movement was detected early and alternative measure were taken. The soil displacement problem was detected through

Copyright © 2019 by ASME

regular monitoring program. Therefore, an incident that could have resulted in product release was prevented.

Incidents related to peak rainfall events that resulted in flooding and washout of soil along the pipeline RoW were also reported. These events are a result of a combination of weather and soil condition. The pipeline may become exposed and sometimes suspended when heavy rainfall and snowmelt events take place due to loss of cover and supporting soil. Four of the incidents reported in 2016 were related to heavy rainfall events in mountainous terrain. These exposures resulted in the pipeline being shut down until remediation was applied.

Incidents related to erosion near water bodies have also been reported. These exposures vary in severity and location. A few exposures were near river banks and at the water front, others consisted of total pipe exposure over the river bed and banks. These exposures vary in length across the water body and may or may not include free spans. Heavy rainfall and snow melt leading to higher water levels in the water body trigger erosion of the river banks and degradation of the river bed.

Of all of the reported incidents only the ones with potential stress corrosion cracking and or pipeline wrinkle or buckle have resulted in a rupture and product release. The combined effect of pipeline defect and soil displacement is a critical factor to evaluate the safe operation of a pipeline. A pipe that is free of SCC may be able to survive through natural process such as erosion, slope movement, soil displacement and frost heave, but one that has a defect of some sort may not be able to survive the natural hazard.

Porter et al [16], 2016 has compiled data from different resources and reported the number of geohazards related incidents in different countries that included, Australia, Canada, Western Europe, United States and South America, and concluded that failures due to geohazards occur often. In addition, several pipeline failures might have been misdiagnosed as being a result of design and installation defects, stress corrosion, temperature fluctuations and extremes, or unknown causes. Furthermore, Porter et al. has concluded that caution should be exercised when using the databases as a source of pipeline vulnerability and failure due to geohazards as many of the incident might have been misdiagnosed.

NEB INCIDENT REVIEW AND CLOSE OUT PROCEDURE

Upon incident reporting to the NEB, a review process takes place to ensure that the company is following up with the pipeline conditions. The company has to demonstrate to the NEB that:

1. The pipeline is safe to operate through the established process and an EA;
2. The causes of the incident and the contributing factors have been identified and investigated and that the proper corrective actions have been implemented; and
3. The company has to demonstrate that they have implemented proper mitigation and/or monitoring plans, in order to ensure the continued safe pipeline operation.

The NEB has also identified through the incident review, inspection, and audit processes that:

1. Companies do not always consider interacting defects. They may consider the geohazard but not the combined effect of a pipeline defect that might impair the safety of the pipeline. If a pipeline has SCC it might not have the same tolerance for soil movement as a pipeline free of SCC. In the presence of CSCC the cracks are perpendicular to the soil movement and therefore they are more liable to open up and result in a pipeline incident.
2. The impact of free debris (gravel, boulders, trees, etc.) that might be present in the vicinity of the exposed pipeline is often neglected;
3. After the company determines that the pipeline is safe to operate the pipeline remediation is put on hold depending on the hazards rating. Often the pipeline at that location is put on a monitoring plan until the company is able to implement the mitigation.
4. Certain locations in high consequence areas are dealt with as a priority case and mitigation is often implemented as soon as the engineering investigation is done and warranty a mitigation plan.

TYPE OF SOIL DISPLACMENTS AND PIPELINES

Buried pipelines cross different terrains to transport various energy related products from one area of a continent or country to another. Terrains such as mountains, water bodies, prairies, desert, valleys and tundra are often assessed during the early pipeline design stages for route selection. Slopes as well as various waterbody crossings are present along most of these terrains and are identified and rated with different severity as geohazards during early pipeline stages design. Pipeline built in the previous era were designed with little consideration to geohazards, therefore some of these pipeline cross areas of relatively steeper slopes and are sitting at shallow burial depth within the waterbodies than pipelines that are being currently built. Current pipeline design aims at reducing the footprints of the pipelines within geo-hazardous terrain. Whether new or old pipelines cross slopes of different intensity, inclination or height, it depends on the area, ROW and terrain being crossed. The aging pipelines have several deficiencies from the condition of the pipeline to the terrain that they were placed in without due consideration to the existing geohazards.

In addition, it should be made clear that most of the naturally occurring slopes are stable and may experience little to no movement if conditions along these slopes remain unchanged. However, upon ROW clearing and installation of buried pipeline certain aspects along these slopes are manipulated and changed. Slope stability issues are triggered by several factors that result in decrease in soil strength due to various phenomenon such as

Copyright © 2019 by ASME

increase in groundwater elevation, pore pressure and seismic events. Proper design of these critical areas and a better system to provide advanced warning of critical weather events will aid in maintaining better stability and may enhance pipeline reliability.

Several types of soil displacement/slope instability may be encountered along a pipeline ROW. This may happen as:

1. Soil erosion, which may occur as a result of heavy rainfall and snow melt events and consists of upper soil displacement and loss of cover and sometimes pipeline free span. These events are sudden and occur as a result of extreme weather conditions. The pipeline may sometimes become suspended when these events take place. The severity of this case depends on the extent and the length of the suspended pipeline portion. Pipeline companies normally shut the pipeline down as soon as this is discovered and perform an EA
2. Creep movement that may take several months or years to be detected and may never result in slope failure, however, may still result in pipeline rupture. These cases are of particular interest since conditions at these locations may go undetected for several years before a pipeline rupture occurs. Pipe defects such as CSCC and buckling may increase the risk to the pipeline at these locations.
3. Rapid landslides where soil is displaced in a few minutes/seconds typically result in loss of cover and consequently pipeline exposure and possibly free span. When the pipeline is exposed, any defects such as CSCC, SCC, coating damage, or buckles will make the pipe more vulnerable to failure than when initially buried and free of defects.
4. Deep seated failure consists of slope failure from the crest to the toe of the slope. The state of the pipeline when this occurs depends on the length of the slope and free span, the pipe layout along the slope and the pipeline defects.
5. Landslides may be encountered even in relatively flat terrain due to changes to the soil properties and loading conditions. Soil may become loose and lose its strength. When this happens soil becomes liquefiable. This condition is especially common in marine clay where the soil will become liquid and flow resulting in pipeline suspension and loss of cover. The most critical issue when this happens is pipeline suspension and free span.

SOIL-PIPE INTERACTION

When a pipeline is buried along a slope any soil displacement along the slope is felt by the pipeline. A pipeline is in particular able to deform in tension well beyond its effective yield strength without rupture, however when a pipeline is in compression (buckling and wrinkling), it can become in a more critical state. Different models have been developed in an attempt to assess the stability of pipelines subjected to slope movement and one of the critical factors to applying those models is that the pipeline must be free of any defects.

Soil-pipe interaction studies are becoming a common tool that companies resort to in order to predict the impact of soil displacement and behavior changes due to temperature and groundwater fluctuation on pipelines. Several factors play a role in this analysis, these factors include: soil type, soil properties and characteristics, soil to pipe interface, direction of soil displacement relative to the pipeline axis, site layout, site conditions, and pipe characteristics.

Seismic movement also changes soil strength and modifies soil properties, however these are not discussed here as they are special events and have special characteristics.

Nasser [17] listed the different types of modeling techniques used to study soil pipe interaction such as centrifuge modeling and numerical modeling. The paper made a comparison of the two techniques. This approach should be further explored and used to try and predict locations where a pipeline might be experiencing stresses that may become critical to the pipeline operation especially due to additional factors such as pipe defects.

STRESS CORROSION CRACKING AND PIPELINE DEFECTS

SCC in pipelines is initiated when small cracks develop on the exterior surface of a pipeline. The cracks initiate as distinct anomalies, however with time they become denser and they connect into what is commonly referred to as colonies. When these colonies grow, cracks propagate and lead to pipeline failure. Certain pipeline types and environmental conditions need to be met for the cracks to form and propagate. Most importantly surrounding soil conditions and pipeline material lead the way to the formation of SCC. These conditions whether they relate to soil or pipe may result in a potent formula that will cause the pipeline to corrode faster than anticipated.

In buried pipelines the surrounding environment consists of soil. Soil characteristics such as soil type, drainage conditions, pH, temperatures, electrical conductivity, and moisture conditions will play a role in SCC formation. Other pipeline related factors that may assist in the formation of SCC include pipe manufacturing process, steel type and grade, steel composition, coating type and operating temperature. Prevention methods such as pipe coating and cathodic protection are used to prevent the formation of SCC.

CONCLUSION

Based on the incident data available at the time of this research, it is apparent that geohazards play a significant role during the operational life of a pipeline and may result in pipeline safety issues. In addition, proper soil pipeline interaction analyses should be performed to ensure that any soil displacement does not result in pipeline failure. Soil monitoring at locations that are identified as having potential geohazards should be recommended and implemented based on the identified risk level. The type of monitoring and frequency should be relevant to the type of the identified geohazard and

Copyright © 2019 by ASME

associated risk. Currently, the available regulation in Canada have a gap in addressing the issue of geohazards and related consequences in the case of their occurrence. The Regulator does not currently have any standards to follow to ensure that geohazards are dealt with appropriately.

Reporting requirements for the US and Canada are different resulting in a difference in the number of reported incidents. Relative to the mileage of existing pipelines in the US and Canada, it appears that more incidents are reported in Canada than the US. For 45,360 miles in Canada a total of 29 incidents were reported compared to 18 incidents reported in the US for 442,000 miles between 2015 and 2017. This may be an indicator that the operating companies in the US may not be reporting as many of the incidents occurring along the RoW in the US due to gap in the US reporting requirements.

PRCI [18] 2009 provides recommendations to assess new and existing gas and liquid pipelines where there is a potential for soil displacement; however, the Regulator is unable to enforce these tools as they are not a part of the regulatory system. There appears to be a missing link on how to adopt these guidelines and be able to use them as regulation, and standards that may guide companies and make them better equipped to deal with geohazards. Furthermore, the combined effect of pipeline imperfections and geohazards does not appear to have been examined as a potential combined threat, companies currently rely on EA to examine those.

It is important to identify geohazards along existing, new, and proposed pipelines in order to reduce incidents and enhance public confidence, environment protection, and public safety, and give the regulating bodies more leverage to regulate. First step has already taken place by reporting and publishing the incident data as per the NEB [19]. Analyzing this data and recognizing what cases are more vulnerable is the next step. Change is difficult to bring about and the need for it is not easily identified, nevertheless it is very important. Several books, guidelines, papers, and conferences have been held related to geohazards yet the Regulator struggles to identify what the company should be obliged and committed to do when those hazards are identified. Steps should be developed to identify expectation of the operating company and put some guidance into the steps to perform and move forward in order to make both the Regulator and the company capable of identifying and remediating geohazards and pipeline combined and separate effects. This will promote public and environmental safety. Whether this is achieved through ISO, PHMSA, or CSA, or adopted by the regulating agencies directly, this step will be beneficial on many levels.

ACKNOWLEDGMENTS

The authors wish to thank our director Ken Nnamoko and professional leader Iain Colquhoun for their encouragement and support to write this paper and attend the conference.

REFERENCES

[1] National Energy Board (NEB); http://www.neb-one.gc.ca/index-eng.html

[2] Pipeline and Hazardous Materials Safety Administration, (PHMSA); https://www.phmsa.dot.gov/

[3] National Energy Board. "Onshore Pipeline Regulations SOR/99-294." Minister of Justice Canada. (2018). http://laws-lois.justice.gc.ca/eng/regulations/SOR-99-294/index.html

[4] Code of Federal Regulation. "Pipeline Safety Regulations, Title 49 CFR Part 195, Transportation of Hazardous Liquids by Pipeline." U.S. Government Publishing Office, Washington, DC, United States. (2018) https://www.govinfo.gov/content/pkg/CFR-2018-title49-vol3/pdf/CFR-2018-title49-vol3.pdf

[5] Code of Federal Regulation. "Pipeline Safety Regulations Title 49 CFR Part 192 – Transportation of Natural Gas and Other Gas by pipeline, Minimum Federal Safety Standards" U.S. Government Publishing Office, Washington, DC, United States. (2018) https://www.govinfo.gov/content/pkg/CFR-2018-title49-vol3/pdf/CFR-2018-title49-vol3.pdf

[6] Standard Council of Canada. "National Standard of Canada, CAN/CSA-Z662-15 Oil and Gas Pipeline Systems." CSA Group, Toronto, Ontario, Canada (2015).

[7] American Society of Mechanical Engineers. "Pipeline Transportation Systems for Liquid Hydrocarbons and Other Liquids, ASME Code for Pressure Piping, B31, ASME B31.4" New York, NY, United States. (2002)

[8] American Society of Mechanical Engineers. "Gas Transmission and Piping Systems, ASME B31.8, ASME Code for Pressure Piping, B31." New York, NY, USA. (2004).

[9] National Energy Board. "Event Reporting Guidelines." Calgary, Canada (2014).

[10] National Energy Board. "Event Reporting Guidelines." Calgary, Canada (2017).

[11] Transportation Safety Board of Canada. "Transportation Safety Board Regulations, Section 4(1)". Minister of Justice, Canada, (2018). https://laws-lois.justice.gc.ca/eng/regulations/SOR-2014-37/page-3.html#h-6.

[12] Pipeline and Hazardous Materials Safety Administration. US Department of Transportation, Washington, DC, United States (2018). https://www.phmsa.dot.gov/faqs/gathering-pipelines-faqs

[13] Pipeline and Hazardous Materials Safety Administration. US Department of Transportation, Washington, DC, United States. (2019). https://www.phmsa.dot.gov/data-and-statistics/pipeline/pipeline-incident-20-year-trends

Copyright © 2019 by ASME

[14] Pipeline and Hazardous Materials Safety Administration. "National Pipeline Performance measures." US Department of Transportation, Washington, DC, United States (2018). https://www.phmsa.dot.gov/data-and-statistics/pipeline/national-pipeline-performance-measures.
[15] Transportation Safety Board of Canada. "Statistical Summary: Pipeline Occurrences." Gatineau, Quebec (2018). http://www.tsb.gc.ca/eng/stats/pipeline/2017/ssep-sspo-2017.
[16] Porter, Michael. Ferris, Gerald. Leir, Mark. Leach, Miguel, and Haderspock, Mario. "Updated Estimates of Frequencies of Pipeline Failures Caused by Geohazards." Proceedings of the 2016 11th international pipeline Conference. IPC2016-64085. Calgary, Alberta, Canada, September 26-30, 2016. ,
[17] Daiyan, Nasser. Kenny, Shawn. Phillips, Ryan, and Pospescu, Radu. "Investigative pipeline-soil interaction under axial lateral relative movements in sand." Canadian Geotechnical Journal. Vol 48 No 11. (2011), DOI: 10.1139/t11-061
[18] Pipeline Research Council International. "Guidelines for Constructing Natural gas and Liquid Hydrocarbon Pipelines through Areas Prone to Landslide." Technical Toolboxes, Houston, Texas. (2008).
[19] National Energy Board. "Incident Data." Safety and Environment. Government of Canada. (2018). http://www.neb-one.gc.ca/sftnvrnmnt/sft/dshbrd/mp/dt-eng.html

Copyright © 2019 by ASME

AIM-PIMG2019-1028

Chapter 2

FAILURE ANALYSIS OF A STRESS-BASED PIPELINE UNDER PLASTIC STRAIN

Chen Hongyuan, Feng Hui, Chi Qiang, Huo Chunyong,
State key Laboratory for Performance and Structure Safety of Petroleum Tubular Goods and Equipment Materials,
Tubular Goods Research Institute of China National Petroleum Corporation, Xi'an, China

ABSTRACT

Strain based design concept have been extensively used for pipelines experience expectable plastic strain for both installation and service, meanwhile most transmission pipelines are designed by hoop stress induced by internal pressure, and maximum longitudinal stress which is typically limited by 90% specific minimum yield strength of pipe. However, some stress-based designed pipelines have experienced unforeseen plastic strain in past decade that resulted in failure. Strain criteria including the strain demand from ground movement and strain capacity of the girth weld are often used in strain based design. In the paper, the failure causes of rupture of a girth weld of 32 inches API 5L X80 SAWH gas pipeline are present for demonstration of the failure analysis.

Keywords: Failure analysis, pipeline, plastic strain, girth weld

INTRODUCTION

Because of the high efficiency and low cost, pipeline is considered as the safest and most economical form of oil & natural gas transmission. The hydrocarbon transmission pipeline network is constructed to transfer oil & gas from field to refineries distribution and purchasers. The pipelines are usually designed by stress criteria because of high internal pressure and hoop stress to ensure pressure containment. The axial loadings are also limited for the stability of the pipeline. A typical limit is 90% SMYS [1]. As for the pipelines under plastic strain, an alternative approach is to limit the axial strain, and strain based design concepts have been developed and codified. In past years, some stress-based designed pipelines experienced plastic strain caused by exterior load from environment and failed during service. It seems the strain-based criteria can be used for failure analysis in these cases because the strain is always easy to quantify.

In the paper, the failure causes of rupture of a girth weld of 32 inches API 5L X80 SAWH gas pipeline are present. The pipeline were joint by FCAW welding process with V-shape bevel. It has served for many years until broken after a landslide of roadbed because of continues rainfall. The cracks location and the status of the pipeline are schematically shown in Figure 1 (a) and (b). The displacement of the pipeline is quantified by 3-dimensional measurement. Then the critical strain and stress can be quantified by FEA under the displacement-controlled load input. The mechanical test and analysis are also executed for requiring the failure cause.

FAILURE ANALYSIS

Two sample pipes of the ruptured pipeline with a length of 12 meters were extracted from the breakpoint for both sides. The segments pose a helical weld and also a circumferential weld for each end.

Figure 2 illustrates appearance of fracture surface for one sample. It is clear that the crack is initiated near extrados after visual inspection on fracture surface.

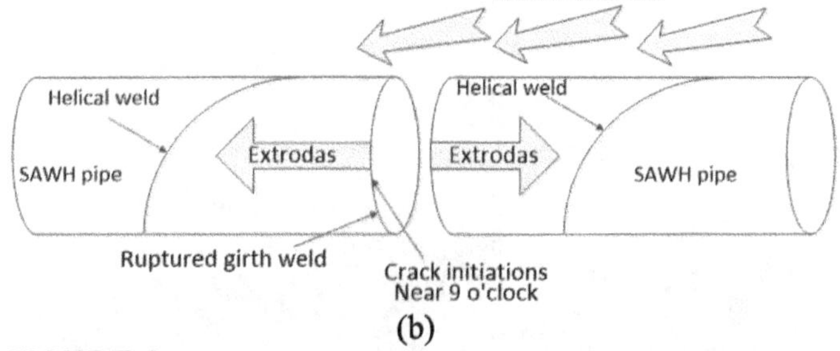

FIGURE 1: CRACK LOCATION ON PIPELINE SAMPLE.

Copyright © 2019 by ASME

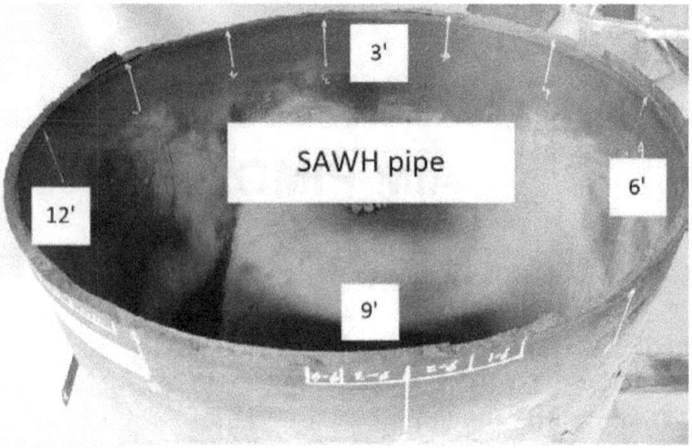

FIGURE 2: APPEARANCE OF FRACTURE SURFACE

2.1 Marco Appearance of the Pipeline

The whole fracture surface of the girth weld was distinguished to the initiation zone, brittle failure zone of weld, ductile failure zone of weld, ductile failure zone of body by macro appearance.

The initiation zones are located in the 9 o'clock and 10:30 of the fracture surface which is the extrados of the bend under the pressure from the landslide mass. As shown in figure 3, the zones marked"1", "2" , "3"and"4". The texture is converged in the zones.

Brittle failure zone of weld is the fracture surface with brittle morphology and along weld metal. There are 3 brittle failure zones in the whole circle located in the 2 o'clock to 4 o'clock, 8 o'clock and 11 o'clock separately.

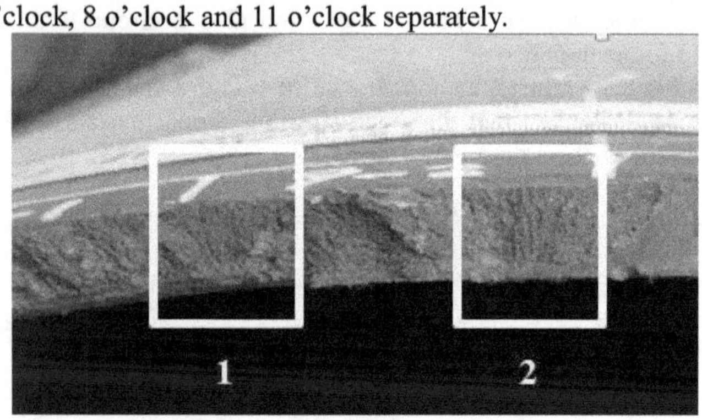

(a) Initiation zones at 9 o'clock

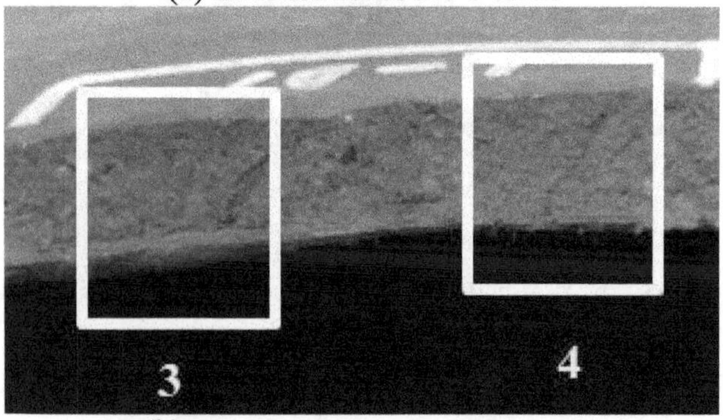

(b) Initiation zones near 10 o'clock

FIGURE 3: FRACTURE INITIATION ZONES

Ductile failure zone is the fracture zone with ductile surface for weld metal and pipe body. The ductile failure zone of weld is located in 5 o'clock while the ductile failure zone of pipe body is located in 5 o'clock to 8 o'clock.

2.2 Geometric Measurement and Reconstruction

Before sampling tests, the pipes are measured by Laser Tracker to obtain 3-D profile of pipe surface. The CAD model can be reconstructed from the point cloud to quantify the deformation which is the criteria of strain or stress analysis. The point cloud and the reconstructed CAD model are shown in figure 4 and figure 5 separately. Figure 5 shows that the deflection at the pipe center is 178.8mm. It means that the critical strain/stress can be evaluated by deforming the pipeline under a displacement-controlled load to reach a same deflection.

Thickness measurements were carried out in different displacement (10mm, 20mm, 40mm, 80mm, as shown in figure 6) from the failure girth weld for whole circumference by an ultrasonic thickness meter. As shown in figure 7, the readings were between 11.5 and 14 mm. The thickness reduction is remarkable in the zone related to ductile tear zone while the thickness is almost same as original in the zones related to brittle fracture zone and initiation zone, as shown in figure 8.

FIGURE 4: POINT CLOUD FOR 3-D PROFILE OF PIPE SURFACE

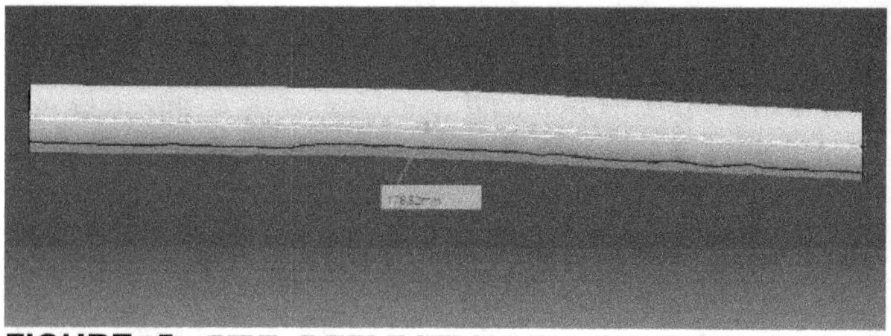

FIGURE 5: PIPE DEFLECTION IN RECONSTRUCTED CAD MODEL

Copyright © 2019 by ASME

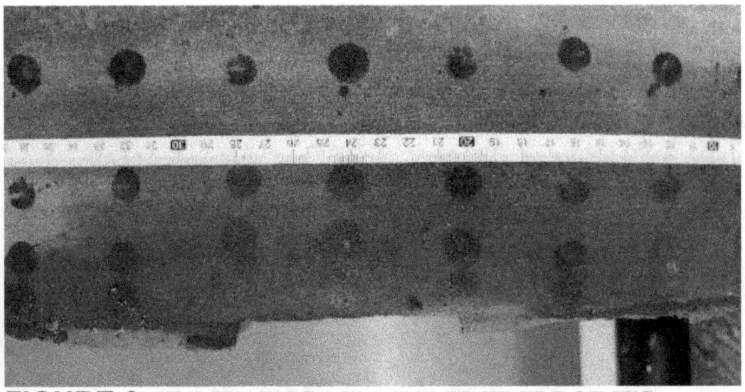

FIGURE 6: THICKNESS MEASUREMENT POINTS

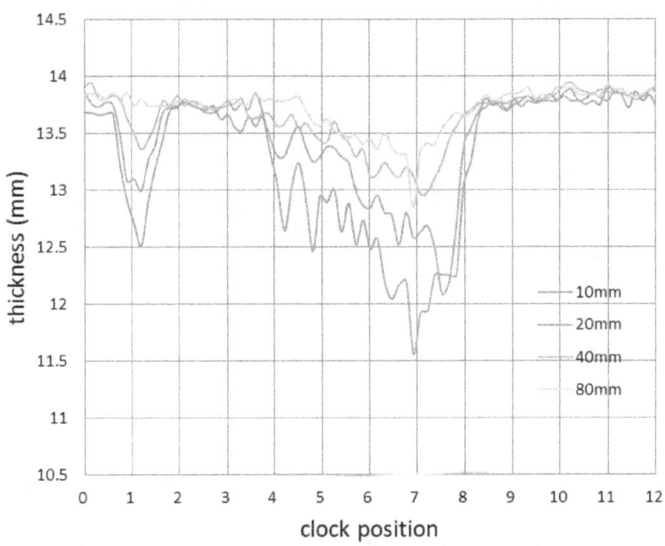

FIGURE 7: THICKNESS MEASUREMENT RESULTS

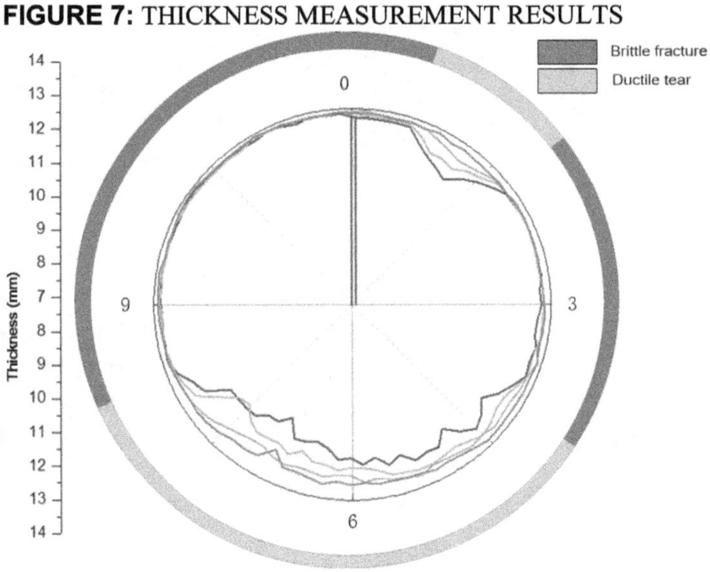

FIGURE 8: THICKNESS REDUCTION vs. CLOCK POSOTION

2.3 Mechanical Properties

2.4.1. Tensile test

Test specimens parallel and perpendicular to the pipe axis were prepared on the pipe body in accordance to test method ASTM E8[2]. The specimens for all-weld metal are also prepared on the adjacent girth weld. Strip specimens are used for longitudinal pipe body and weld joint, while the round bar specimens are used for transverse pipe body and all-weld metal tensile specimens, as shown in table 1. The sizes of specimens are also shown in the table.

The results of tensile test and the accepted values are shown in Table 1. Values of tensile strength and yield strength are located in the range of standard API 5L X80. The DIC is also executed during the tensile test of weld joint then the axial strain for the pipe body (point 1 and point 2) and weld root (point 3) are illustrated. The test results from pipe body and all-weld metal, and the DIC contour of weld joint tensile test show a slight under-match of weld yield strength, as table 1 and figure 9 shown. Although the tensile strength result shows that the girth weld overmatch the pipe body, the figure 9 (b)~(d) show that strain concentration occurs at the root during the tensile. It is also the cause that the weld root has a high incidence for crack initiation. The investigation for yield strength under match result in strain concentration on root for V-shape weld joint would not mentioned here due to limit space.

2.4.2 Impact test

Charpy impact test was carried out in accordance to ASTM E23[3] test method. Charpy impact test specimens were prepared for near girth weld and the temperature was held at −10 °C. Table 2 summarizes the results of Charpy impact test. The minimum value and the average value of the absorbed energy are 80 J and 60 J in accordance to pipeline specification. All values obtained were in the range of standard and specification.

Table1 Mechanical Properties of the pipeline

Sample No.	Position, orientation	Specimens size (mm)	Yield Strength (MPa)	Ultimate strength (MPa)
1	Body,longitudinal	38.1×50	599	644
2	Body,transverse	Φ6.25×25	626	649
3	Girth weld joint	38.1×50	/	725
4	All-weld metal	Φ5×25	592	661
API 5L X80 (PSL2)	/	/	555	625

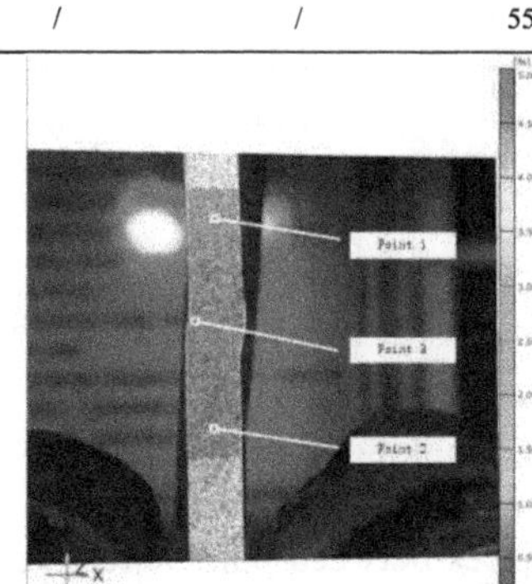

(a)

Copyright © 2019 by ASME

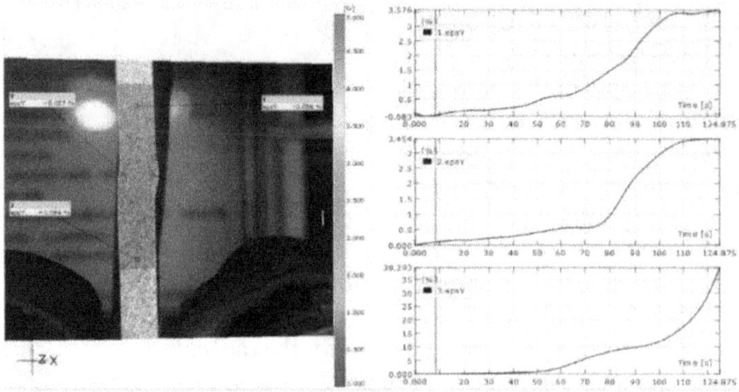

(b)

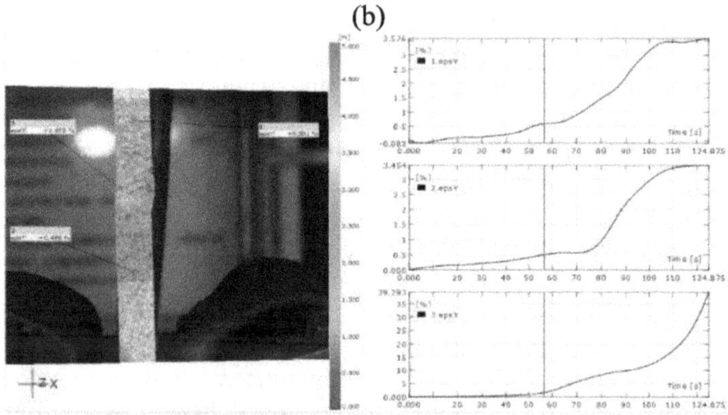

(c)

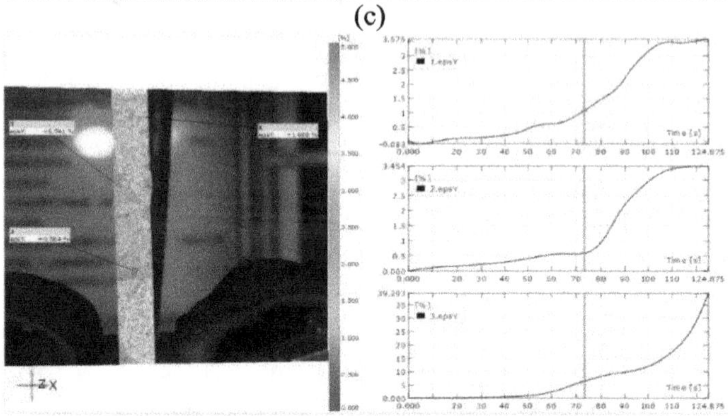

(d)

FIGURE 9: STRAIN CONTOUR OF THE WELD JOINT ALONG THE AXIS

Table2 Charpy impact test results

Sample No.	Position,	Specimens size (mm)	Absorbed energy (J)	Absorbed energy mean value (J)
1	Weld	10×10	143	
2	Weld	10×10	123	141
3	Weld	10×10	158	
4	HAZ	10×10	68	
5	HAZ	10×10	207	110
6	HAZ	10×10	55	
WPS	/	/	60	80

2.4.3 Hardness measurement mapping

Vickers hardness measurement mapping is used to characterize the strength distribution of the girth weld joint. The test force is 0.5kg while the spaces between points are 0.3mm, as shown in figure 10 (a). The test results imply that the ultimate strength of pipe body is higher than the weld (especially the root), as figure 10 (b) shown. Although the ultimate strength of weld metal is slightly higher that pipe body, the obvious under match is present at the root which may cause strain concentration for V-shape weld joint.

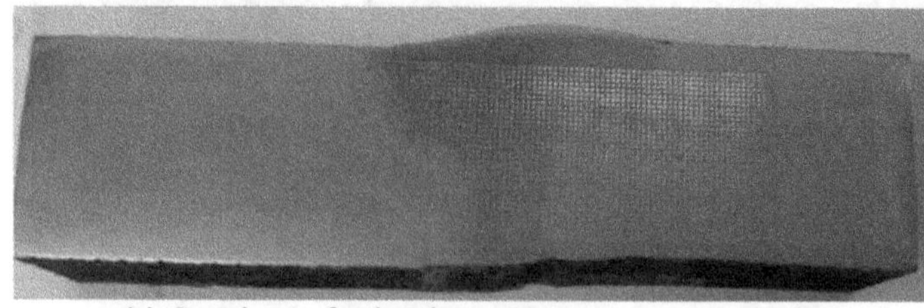

(a) Specimen for hardness measurement mapping

(b) Hardness measurement mapping

FIGURE 10: HARDNESS MEASUREMENT MAPPING

(a) Morphology and EDS analysis for initiation zone 1 and 2

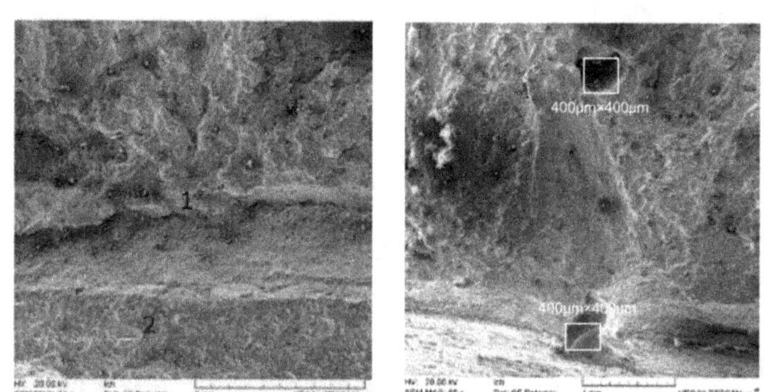

(b) Morphology and EDS analysis for initiation zone 3 and 4

FIGURE 11: MORPHOLOGY AND EDS ANALYSIS FOR INITIATION ZONES

Copyright © 2019 by ASME

2.5. Fracture surface investigation

The specimens of fracture surface were cut into small parts and cleaned by alcohol and ultrasonic. In order to investigation of the fractured surface, scanning electron microscopy (SEM) equipped with EDS was used to survey the initiation zones marked 1~4 in figure 3 and figure 4.

The test results of SEM in figure 11 shows that there is no large size inclusions in the initiation zone 1, 2, and 3, while a 400μm-size weld defect is observed in zone 4, which do not exceed the specification.

3. FINITE ELEMENT ANALYSIS

3.1 FEA model

The FEA model was constructed on the basis of pipeline size around failure point. The figure 12 illustrates the 50 meters of the related pipeline including deformed pipeline of 26 meters.

There are 100 quadratic pipe32H elements (i.e., quadratic pipe element of 3-nodes and 2 integration points, and H indicates hybrid element) and 201 nodes for characterizing the pipeline. The original point locates on the -20m point, and positive for x axis is the right side.

The pipe element was considered as thick wall cylinder within 24 nodes on cross profile which locate in the inner surface, outer surface and middle point between them, as shown in figure 12. The material parameter includes the stiffness, strength, density, expansion factor.

The soil constraints are simulated by an annular tube structure. The thickness of the annular tube and the pipeline are set to 0.5m and the real thickness. The ITT (i.e., Interaction between Tube-and-Tube) pipe-pipe friction contact elements are used between the pipeline and the annular tube to simulate the soil-pipe constraints. Figure 13 illustrates the nodes between the pipeline and the annular tube, and the contact conditions. The inner radius of annular tube and the outer radius of the pipeline are presented by a and b separately. The numbers present the serial number of nodes. The n is the normal direction while the t is tangential direction.

Figure 14 shows the FEA model of the pipeline and the annular tube. The properties of annular tube and soil are defined by experience as a phenomenological analysis.

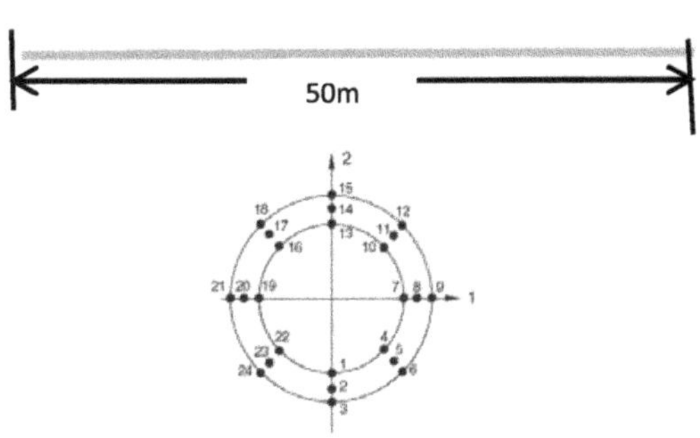

FIGURE 12: ILLUSTRATION OF THE PIPELINE MODEL

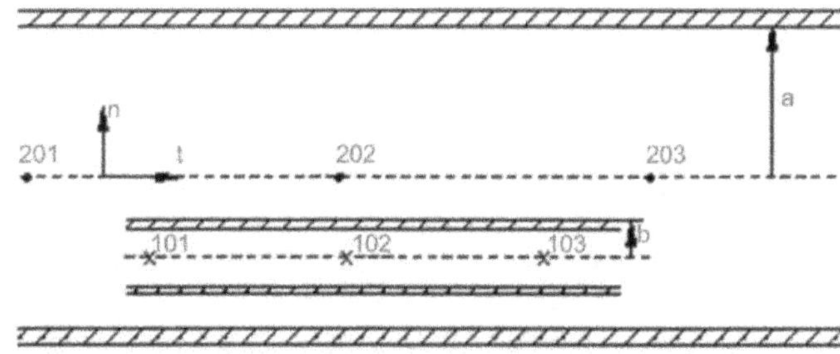

FIGURE 13: ILLUSTRATION OF THE NODES AND CONTACT CONDITIONS

FIGURE 14: FEA model of the pipeline and the annular tube

3.2 Load and the displacement

The displacement constraint is imposed on the pipeline of 26 meters adjacent to the breakpoint to quantify the critical strain and stress of rupture as shown in figure 15. The displacement-controlled load makes the soil which presented by the annular tube deformed and produced the reversed force. The remote pipeline do not have bending strain, it was introduced for remitting the end constraint from the both sides. On the basis of the model above, a displacement-controlled load is imposed on the pipeline, when the pipeline deformed just reached the deflection as the sample pipe (178.8mm at the center point, as figure 15 shown), the strain and the stress of the breakpoint on axis is 0.57% and 639MPa. It means that the critical strain and the critical stress of the pipeline under the displacement result from landslide are 0.57% and 639MPa separately.

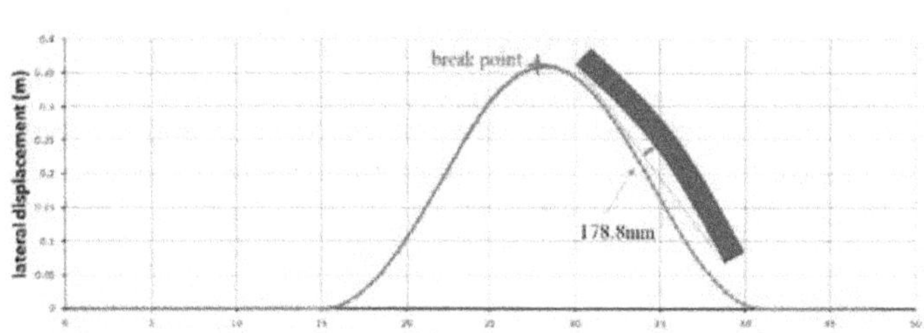

FIGURE 15: DISPLACEMENT CONTROLLED LOAD IN FEA MODEL

Copyright © 2019 by ASME

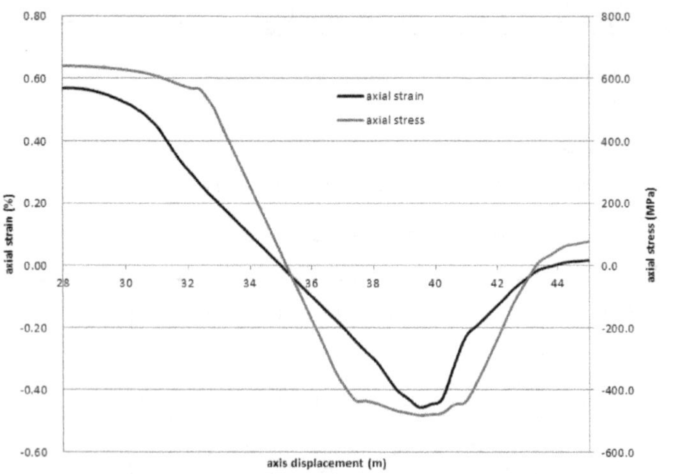

FIGURE 16: CRITICAL STRAIN AND CRITICAL STRESS

4. RESULTS AND DISCUSSION

Results of experiments and related evidences discussed, revealed that the girth weld has suffered displacement-controlled load result from the landslide and failed. The critical strain and stress when failed have been calculated by FEA under the support from the measure and test. The evidences are as follows:

1. Both of the results of tensile tests with DIC and the hardness measurement mapping show that the girth weld is under matched with the pipe body on yield strength, even the ultimate strength for root is also under match.
2. By laser geometric measurement and reconstruction the CAD model, the plastic strain of the failed pipeline has been determined as the 178mm deflection on residual deformation.
3. The macro morphology implies the fracture initiated on the extrados of the girth weld (tensile side of the bending) and propagated by brittle facture and ductile tearing ways.
4. The SEMs show that there is no oversize flaw in the initiation zones. In the meantime the absorbed energy of Chapy impact is also higher than critical value of standard. It means that the external load may cause the failure.
5. By bending the pipeline to the state of the failed sample pipe in FEA, it can conclude that the critical strain and critical stress of the pipeline is 0.57% and 639MPa separately which higher than the elastic limit and the strength limit of weld joint.

5. CONCLUSION

The cause of the failure discussed in this paper was external load from landslide on girth weld of pipeline which ultimately resulted in rupture of the structure. There are various contributing factors to the weld failure, including:

1. External load that exceed the normal service condition which should be axial stress lower than 0.9 YS or axial strain lower than 0.5%;
2. Under match of the girth weld especially for the root, although the tensile strength of tensile test is higher than the standard limit.
3. Additionally, when definite geological hazard occurred, the strain/displacement criteria are effective to the pipeline failure analysis by using the CAD reconstruction and FEA.

6. ACKNOWLEDGEMENTS

Sincere acknowledgements should be presented to State Key Laboratory for Performance and Structure Safety of Petroleum Tubular Goods and Equipment Materials for the finance support (National Key R&D Program of China 2016YFC0803201, CNPC project 2017D-5006-12) and their agreement of the publication of the paper.

7. REFERENCES

[1]ASME, B31.8, Gas Transmission and Distribution Piping Systems, (2016)

[2] ASTM, E8, Standard test methods for tension testing of metallic materials, Annual Book of ASTM Standards, (2001)

[3] ASTM, E23, Standard Test Methods for Notched Bar Impact Testing of Metallic Materials, (2007)

Copyright © 2019 by ASME

AIMPIMG-1067

Chapter 3

STRAIN-BASED DESIGN AND ASSESSMENT – CONCEPTS AND GAPS

Yong-Yi Wang
Center for Reliable Energy Systems
Dublin, OH, USA
ywang@cres-americas.com

ABSTRACT

There have been significant advances in the development of the strain-based design (SBD) methodology from the mid-2000s to the early 2010s in anticipation of new pipeline construction in northern regions of North America. Many of the expected pipelines have not been built due to changes in producing regions. SBD has been used on smaller scales on a number of onshore projects. At the same time, strain-based assessment (SBA) has gained wider use for vintage and new in-service pipelines that were designed using traditional stress-based design.

This paper provides an overview of conceptual understandings and practical implementations of SBDA. Key steps in the strain-based design of new pipelines are covered in detail. Key concepts relevant to SBDA, such as loading modes, failure modes under tensile loading, compressive strain capacity relevant to different loading modes, and safety factors, are explained. Similarly, key steps for strain-based assessment of in-service pipelines are introduced. Features relevant to the assessment of in-service pipelines, such as the determination of strain demand, understanding key drivers to tensile strain capacity by vintage of the pipelines, welding processes and inspection practice, and detection and characterization of girth weld flaws, are highlighted.

The last part of the paper covers the gaps in the current technology and practice. These gaps must be understood to apply SBDA properly. The execution of SBDA requires collaboration among multiple functional areas of a company from an operational viewpoint. From a technical perspective, SBDA requires multi-disciplinary approaches that keep practical constraints in mind.

The focus of this paper is high-level concepts and relationships among key components in SBDA. The space limit of this paper does not permit exhaustive reviews of all relevant work carried out by many organizations. Such reviews can be found in publications by this and other authors.

KEYWORDS

Strain-based design, strain-based assessment, strain demand, tensile strain capacity, compressive strain capacity, loading modes, failure modes under tension, gaps in SBDA

NOMENCLATURE

Strain demand (SD): the strain imposed on a pipeline by its operational and environmental conditions

Strain capacity (SC): the highest strain level a pipe segment can sustain without negative consequences

Tensile strain capacity (TSC): the strain capacity in tension

Compressive strain capacity (CSC): the strain capacity in compression

Compressive strain capacity by maximum load (CSC_{ML}): the CSC by maximum load (bending moment or compressive load). This is the CSC commonly defined in current codes and standards, such as CSA Z662

Post-maximum-load compressive strain capacity (CSC_{PML}): This is the CSC beyond the maximum load

Safety factor (SF): In the context of SBDA, the concept of safety is applied to separate the strain capacity and strain demand by a sufficient margin. An appropriate safety factor may be set with the considerations of the uncertainties in strain demand, strain capacity, and risk tolerance level at the site of interest. The safety factor is implemented by setting of strain demand limit.

Strain demand limit (SDL): the limit the strain demand is allowed to reach. SDL is obtained by one of two ways: (a)

Copyright © 2019 by ASME

multiplying the strain capacity by a safety factor that is less than 1.0, or (b) subtracting a finite number from the strain capacity.

1. INTRODUCTION

There have been significant advances in the development of the strain-based design (SBD) methodology from the mid-2000s to the early 2010s in anticipation of new pipeline construction in the northern regions of North America [1]. The expected construction did not occur in part due to the shift in producing regions driven by shale production. SBD has been used on smaller scales on a number of onshore projects. At the same time, strain-based assessment (SBA) has gained wider use for vintage and new in-service pipelines that were designed with traditional stress-based design. Strain-based design and assessment (SBDA) is still a maturing field. Proper execution of SBDA requires collaboration among multiple functional areas of a company from an operational viewpoint. From a technical perspective, SBDA requires multi-disciplinary approaches while keeping practical constraints in mind.

This paper provides an overview of conceptual understandings and practical implementations of SBDA. Gaps in the current technology are reviewed so proper actions can be taken.

1.1 Stress- vs. Strain-Based Design

Most pipelines in service today were designed with a primary focus on pressure containment through limiting the hoop stress to a certain percentage of the pipe's strength, such as the specified minimum yield stress (SMYS). Such a design is often referred to as stress-based design. Traditional pipeline design methods have some parts of their roots in the design of pressure vessels. One of the major differences between pressure vessels and pipelines is the external loads imposed by the environment. Most pipelines are subjected to some level of longitudinal loads imposed by their environment. The magnitude of these loads is often unknown and not easily controllable. In some cases, such loads could cause pipe segments to deform beyond the yield condition of the materials. The longitudinal loads on most pressure vessels are known to a greater precision and would almost never cause yielding of materials by design.

Strain-based design (SBD) refers to pipeline design methodologies with a specific goal of maintaining pipeline service and integrity under large longitudinal strains (typically defined as longitudinal strains greater than 0.5%). SBD provides systematic and quantitative procedures to account for the effects of high longitudinal strain. SBD is complementary to the traditional stress-based design, not a replacement of the traditional design.

1.2 Strain-Based Assessment

The principles and procedures developed for SBD can be applied to the maintenance of existing pipelines in areas with ground movement [2]. The process of using SBD principles and procedures to assess existing pipelines is sometimes termed SBA or SBDA, i.e., strain-based assessment or strain-based design and assessment.

1.3 Limit States in SBDA

SBDA encompasses at least two limit states: tensile rupture and compressive buckling. Tensile rupture is an ultimate limit state which is related to the breach of the pressure boundary. The compressive buckling could be either a service limit state or an ultimate limit state. Each limit state is defined by strain demand (SD) and strain capacity (SC). The strain demand is the strain imposed on a pipeline by its operational and environmental conditions. The strain capacity is the highest strain level a pipe segment can sustain without negative consequences. A negative consequence is not necessarily a leak or rupture and it can depend on the specific field conditions. For instance, it can be an operational limit, such as preventing the passage of inspection pigs, or a condition that may lead to future integrity concerns, such as coating damage, fatigue after the formation of wrinkles, etc.

2. SBD FOR NEW PIPELINES

2.1 Overall Process

A few key steps in SBD are:

- *Determine the nature of the strain demand.*
- *Estimate levels of strain demand.*
- *Determine target strain capacity.*
- *Assess options to achieve the target strain capacity.* The requirements for linepipe specifications, welding procedure qualification, flaw acceptance, and other necessary conditions to ensure the achievement of the target strain capacity are developed in this step.
- *Develop processes to meet material specifications and qualification requirements.*
- *Develop and implement field quality assurance and monitoring processes.*

A safe design is achieved when the strain demand is less than the strain demand limit . If a safe design is not achieved initially, a reduction of strain demand, an increase of strain demand limit, or both may be explored to achieve a satisfactory design. The steps listed above may be repeated if the necessary conditions for a safe and cost-effective design and post-construction maintenance cannot be met. When feasible, reduction of strain demand in the form of a route selection or special trenching and pipe support should be the first choice for new pipeline construction. Tailored project-specific SBD procedures may be exercised depending on conditions specific to a project.

Copyright © 2019 by ASME

2.2 Further Definition of Key Steps

2.2.1 Strain Demand

2.2.1.1 Loading Mode

Loadings on pipelines are classified as "primary" or "secondary" loadings. Primary loads such as internal pressure result from an equilibrium requirement and cannot be relieved by yielding or deformation of the pipe. Secondary loads such as those induced by restrained thermal expansion/contraction result from a compatibility requirement and are both self-limiting and can be relieved by yielding or deformation. Both primary and secondary loads cause stress and strain in pipes.

Analogous to primary and secondary loads are the concepts of load- and displacement-controlled loading. A loading is considered load-controlled if the magnitude of such a loading is not affected by the amount of deformation or displacement. Examples of a load-controlled loading are dead-weight loading, soil load on a span, and internal pressure. A loading is considered displacement-controlled if the amount of deformation is not affected by the load-carrying capacity of the component/structure being subjected to the deformation. Examples of displacement loading are pipe bending by a mandrel and reeling-on a pipe string in spool-based installation. Most loadings on pipelines are a combination of displacement- and load-controlled loadings. Certain types of ground movement are primarily displacement-controlled.

Understanding the nature of a loading is important. When a pipe segment is under a load-controlled loading, the reserve load capacity is limited after the pipe segment reaches its yield condition. In contrast, if the pipe segment is under a displacement-controlled loading, there can be a considerable reserve in deformation capacity after the pipe segment reaches its yield condition. Therefore, SBDA provides greater benefits for displacement-controlled loading than load-controlled loading.

Many other features of loading are important for SBDA. For instance, a slow ground moving event can be monitored, and mitigation may be applied before a critical state is reached. A fast event, such as a seismic event, does not permit intervention during the occurrence of the event; in this case, having an adequate design could be more important. More on this subject is discussed in a companion paper [3].

2.2.1.2 Estimating Levels of Strain Demand

This process may start from the estimated amount of ground or soil movement, which may come from historical data, ground, or aerial surveys. Such movements are imparted on a pipeline through pipe-soil interaction. Models of various complexities may be used to estimate strains imposed on the pipeline due to the ground movement [4,5,6,7]. Having appropriate soil properties in the model input is a key in the accurate prediction of strain demand using pipe-soil interaction models.

2.2.2 Target Strain Capacity

The target strain capacity may be set by dividing the strain demand by a safety factor that is less than one or adding a finite number (another way to add a safety factor) to the strain demand. This process effectively sets the strain demand limit equal to the strain demand.

2.2.3 Options to Achieve Target Strain Capacity

2.2.3.1 Tensile Strain Design

(1) Factors Affecting Tensile Strain Capacity

Parameters that have been identified as having a dominant influence on TSC are:

- Materials' strain hardening capacity,
- Weld strength mismatch,
- Extent and level of HAZ softening,
- Girth weld profile,
    - Cap reinforcement
    - High-low misalignment
    - Bevel angle
- Girth weld flaws,
    - Type
    - Location
    - Dimensions
- Toughness, and
- Internal pressure.

(2) Design Scenarios for Tensile Strain Capacity

The TSC of a pipeline is often controlled by the TSC of its girth welds unless special measures are taken, as the welds may contain welding imperfections and have deteriorative material properties in comparison to the base pipe material. However, girth welds do not have to be the weakest link.

Possible failure modes of a pipe segment under longitudinal tensile loading can be conceptually organized by four different categories as shown in Figure 1 and Figure 2 [8]. These failure modes are the result of collective observations of experimental tests of varying scales, from small laboratory size specimens, e.g., CTOD, SENT, and cross-weld tensile specimens, to curved wide plates and full-scale pipe tests. These categories are given as a function of normalized flaw size, as weld flaws have been traditionally considered a key driver for weld integrity. In the context of Figure 1, the parameters affecting TSC, with the exception of flaw size, are held constant. The categorization of the failure modes is intended to facilitate the rationalization of a material's tensile strain response. The boundaries between these modes can overlap and are not absolute.

(a) *Category I Failure: Failure in the Pipe Body*

When there is a small flaw in an overmatching weld, the crack driving force imparted on the flaw increases upon the initial application of remote straining. The driving force can become saturated as the flaw is effectively protected by the

Copyright © 2019 by ASME

overmatching weld metal. Almost all of the subsequent applied longitudinal displacement acts to only deform the pipe body, leading to a failure in the pipe body. The crack driving force is never large enough to cause a failure in the flawed plane.

(b) *Category II Failure: Flaw Failure with Limited Flaw Growth until the Final Point of Failure*

When the flaws are shallow and long or moderately deep and short, there is a dynamic equilibrium between the crack-driving force needed to grow the flaw and the straining of the pipe body. There is limited flaw growth when the applied load approaches the maximum load. Substantial straining can occur in the pipe body before the attainment of the maximum load. This dynamic equilibrium between the straining of the flawed plane and the pipe body means that the eventual failure location is highly dependent on the strain hardening behavior of the materials around the flawed plane and the pipe body. Although the final failure is in the flawed plane, a high strain capacity approaching the pipe uniform strain can be achieved.

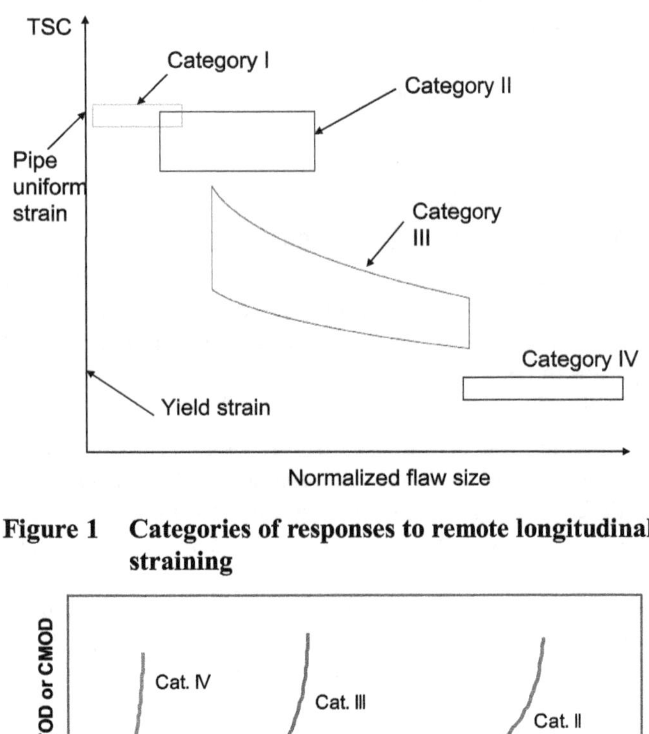

Figure 1 Categories of responses to remote longitudinal straining

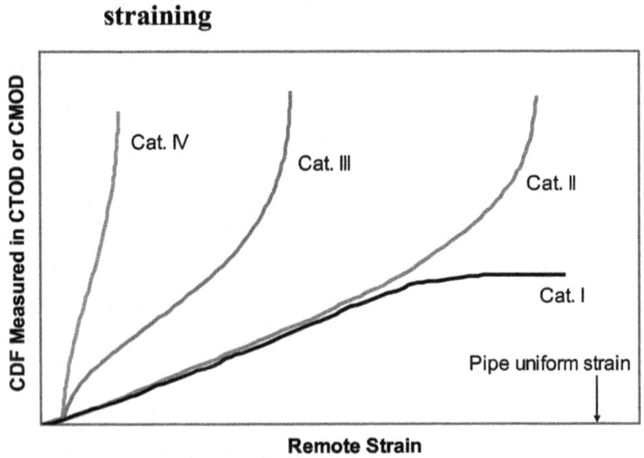

Figure 2 Categories of responses expressed in crack driving force (CDF) vs. remote strain relations

(c) *Category III Flaw Failure with Finite Flaw Growth*

If the increase in the crack-driving force is gradual after gross section yielding of the pipe body, flaw growth occurs gradually with the remote straining. The flaw may fail by (1) local instability, i.e., the flaw popping-through the wall, but the overall load capacity of the specimen may still increase, or (2) global instability if the specimen is load-controlled, i.e., the flaw rapidly propagates in the circumferential direction. Many large-scale tests are performed in displacement-controlled mode and the tests are stopped when the overall maximum load is reached. In displacement-controlled mode, the attainment of maximum load does not necessarily lead to instantaneous overall instability.

(d) *Category IV Flaw Failure in Elastic Strain Range*

If the flaw is large, or there is gross weld strength undermatching, or there is a large magnitude of high-low misalignment, or some combination of those conditions, the growth of the crack-driving force may be unbounded. The remotely applied displacement is almost entirely taken up by the flawed plane. The failure strain under this scenario is near or less than the yield strain.

Categories I and II are desirable for high strain capacity. This implies either limiting the deformation to the pipe body or ensuring minimum flaw growth. This can be achieved by having a consistent and high degree of weld strength overmatching, a limited extent and degree of heat-affected zone (HAZ) softening, a tight flaw acceptance level, and a favorable weld profile to prevent strain localization in the weld region. Good weld profiles could include a minimal level of high low misalignment and sufficiently wide weld cap reinforcement that provides structural reinforcement of the weld and prevents plastic shearing along the softened HAZ.

Tensile strain models typically address Categories II and III behavior. TSC may be estimated using available models with a range of possible material, dimensional, and operational parameters [9]. What-if scenarios may be exercised to determine the best approach to achieve the target strain capacity. The associated requirements for material properties, construction practices, and post-construction maintenance must be considered when working through the what-if scenarios.

Category IV behavior should be avoided. This behavior is possible even in the absence of weld flaws when there is a high level of weld strength undermatching and HAZ softening [10].

2.2.3.2 Compressive Strain Design

(1) Determine Relevant Limit States

CSC is often defined as the strain corresponding to the point of the maximum bending moment in a lateral bending test, CSC_{ML}. In most cases, pipes have a very minimal amount of bulging or wrinkling at this point of loading (i.e., maximum moment). In a displacement-controlled loading scenario,

Copyright © 2019 by ASME

reaching the CSC_{ML} does not negatively affect the pipeline service. The ovalization of the pipe's cross section is small and does not impede the passage of inspection pigs. An alternative CSC corresponding to the post-maximum-load behavior, i.e., CSC_{PML}, can be more appropriate. If the loading is load-controlled, the appropriate CSC is CSC_{ML}. On the other hand, if the loading is mostly displacement-controlled, the appropriate CSC is CSC_{PLM}.

If the loading is load-controlled, the pipe may collapse immediately after the maximum bending moment is reached, possibly forming severe wrinkles. Large and severe wrinkles can lead to ruptures/leaks from high local tensile strains in either the hoop or longitudinal direction. If the wrinkles survive the initial formation, the long-term integrity of the wrinkles may be affected by possible fatigue damage and/or coating/corrosion related concerns. Knowledge of the buckling process and the nature and magnitude of subsequent loading is necessary to make the right decisions about the need and timing of mitigation measures.

(2) Factors Affecting CSC_{ML}

The following factors are known to affect CSC_{ML}:

- Pipe D/t ratio,
- Pipe strain hardening behavior, sometimes represented by Y/T ratio or yield strength
- Stress-strain curves of the pipes
- Internal pressure or external overpressure,
- Geometric imperfections or features, and
- Loading mode.

(3) Models for CSC_{ML}

A review of models for CSC_{ML} is given in Reference [11].

(4) Approach to CSC_{PML}

There are no well-established procedures to estimate CSC_{PML}. One example of justifying the strain limit of 2.0 in ASME B31.4 and B31.8 is given in Reference [12].

2.2.3.3 Output from the Evaluation of Target Strain Capacity

Requirements for the parameters affecting the target strain capacity must be clearly specified. These requirements may include

- Linepipe specifications,
- Girth weld qualification, and
- Girth weld flaw acceptance criteria

2.2.3.4 Safety Factors for Tensile and Compressive Strain Designs

(1) Safety Factor for TSC in Tension

Curved wide plate and full-scale experimental tests have shown that the measured tensile strain capacity can differ by a factor of two under "identical" material, geometric, and flaw conditions [13]. Assuming the predicted TSC from tensile strain models is in the middle of the low- and high-end capacity from the experimental measurement, the low-end strain capacity would be 1/1.5 of the predicted value. In order to have a predicted strain capacity equal to the lower-end strain capacity, a safety factor of 1/1.5 should be applied to the predicted value. Consequently, the appropriate safety factor is 0.67 (=1/1.5). PHMSA's SBD special permit conditions have a safety factor of 0.60 [14]. CSA Z662 Annex C on limit state design has a resistance factor that is equivalent to a safety factor of 0.70. Here, a safety factor of 0.67 is recommended.

In some case-specific strain capacity analysis, such as the process defined in the Level 4 procedures of the PRCI-CRES tensile strain models [9], most of the input parameters are assumed to be near-worst-case scenarios. This process leads to a predicted TSC on the lower bound of the distribution of possible TSCs. Using a safety factor of 0.67 on such a TSC can be overly conservative. A safety factor greater than 0.67 may be used to determine the strain demand limit when the predicted strain capacity is biased towards lower-bound values.

(2) Safety Factor for CSC in Compression

CSA Z662 Annex C sets the safety factor for compressive strain capacity to 0.80. PHMSA's SBD SP conditions also use a safety factor of 0.80. Therefore, a safety factor of 0.8 may be applied to SDL to obtain CSC_{ML}.

2.2.4 Develop Processes to Meet the Material Specifications and Qualification Requirements

The material qualification processes must be accompanied by well-defined test procedures that cover all aspects of testing, from specimen position in pipes/welds to post-test data analysis and reporting.

2.2.5 Develop and Implement Field Quality Assurance and Monitoring process

Creating consistently good welds that meet the requirements in the welding procedure qualification is crucial to ensure the target strain capacity can be achieved. Field conditions can vary greatly. A quality assurance program is necessary. One possible reference is PHMSA's special permit conditions for SBD [14].

3. SBA OF IN-SERVICE PIPELINES

3.1 Overall Process

Strain-based assessment (SBA) is used for the integrity assessment of in-service pipelines experiencing high longitudinal strains. The same fundamental procedures used in SBD can be applied to SBA.

For pipelines that are not specifically designed to sustain high longitudinal strains, there can be considerable challenges in performing SBA. Some of the key elements in SBA are:

(1) Determination/estimation of strain demand,
(2) Detection and sizing of girth weld flaws, particularly planar flaws, and

Copyright © 2019 by ASME

(3) Determination/estimation of girth weld properties, particularly toughness properties.

3.2 Determination of Key Elements

3.2.1 Determination/Estimation of Strain Demand

In addition to the pipe-soil interaction models described in Section 2.2.1.2, there are methods available for estimating strain demand of an in-service pipeline.

(1) *IMU (Inertial Measurement Unit).* IMU provides the centerline position of a pipeline. The "bending strain" in an IMU report is computed from the centerline profiles of the pipes by assuming pipes are straight when no strain is present. Therefore, bending strains include strains from all events that cause the pipe profile to change from a straight profile, including construction bends and all subsequent profile changes. The computed strains originating from construction bends are not relevant for the assessment of the pipe integrity. Therefore, the contribution of those bends to the reported bending strain must be subtracted for integrity assessment. For in-service pipelines that do not have the original pipe profile at the time of construction, determining the bending strains that are most relevant to integrity assessment is possible for segments of pipes that are believed to be straight at the time of construction. Locations with hot and cold bends can be identified by IMU based on their characteristic bending strain profiles. However, determining the magnitudes of the strains caused by external loads at those locations can be a challenge. For in-service pipelines, successive IMU runs are often used to detect locations with pipe movements. IMU tools cannot detect uniform tensile or compressive strains which do not cause changes in pipe profiles. In most cases, the bending strains are reported when the values are greater than a threshold value (e.g., 0.15% strain).

(2) *Direct measurement.* Strain gages have been widely used in locations with expected high strains. Fiber optic cables can measure strains over long segments of pipelines.

Each of those strain estimation/determination methods offers certain advantages and disadvantages. For instance, the strains measured by strain gages are considered to be highly reliable. However, strain gages can only monitor isolated locations. Pipe and soil interaction models can be used for long segments of pipelines. The strains computed from such models are highly dependent on the assumed soil properties. Site-specific soil properties are generally not available and can change over time or even from season to season as the water content in the soil can affect soil properties.

3.2.2 Determination/Estimation of Strain Capacity

Key factors affecting the tensile strain capacity of girth welds are described in Section 2.2.3.1. In comparison to new pipelines being considered for SBD, in-service pipelines present additional challenges in determining the appropriate values for the key factors.

Girth weld properties are affected by the linepipe and welding processes. Girth welds may be divided into at least three groups for understanding their respective properties.

Group 1: Vintage girth welds made to relatively high-carbon normalized steels with typical carbon content of 0.2% or higher and before US federal regulations which effectively led to virtually all girth welds being inspected during construction.

Group 2: Newer pipelines constructed with stress-based design principles using microalloyed TMCP steels, and

Group 3: Newer pipelines constructed with SBD principles using microalloyed TMCP steels.

Very few pipes fall into Group 3. If a pipeline or pipe segment falls into Group 3, the weld properties, including mechanical properties and flaw characteristics, should be known and on file.

Welds in Group 2 are typically viewed as having good toughness. Any weld imperfections remaining after construction should be small as virtually all girth welds have been inspected during construction. Other than the recent girth weld incidents attributable to high levels of weld strength undermatching and HAZ softening, this group of welds generally have a reasonable strain capacity. However, the strain capacity of individual welds can vary greatly.

Welds in Group 1 could contain volumetric and planar imperfections depending on the vintage of construction, welding methods, and inspection practice. Volumetric imperfections are typically not a structural integrity concern unless they are linked up with planar flaws. One of the biggest threats to Group 1 welds is the occasional existence of hydrogen cracks. On the positive side, welds in Group 1 could have wide weld caps. The cap reinforcement and weld strength overmatching can reduce the negative impact of weld flaws. A systematic testing and evaluation of the mechanical properties and flaw characteristics of vintage girth welds demonstrated that many vintage girth welds have good strain capacity [15].

In addition to the uncertainties in the mechanical properties of the welds, the detection and sizing of girth weld flaws, particularly planar flaws, of in-service pipelines can be a challenge. Reliable and proven ILI tools (either free-flowing or tethered) do not exist. In-ditch inspection methods can be effective but are not necessarily 100% reliable or accurate.

Copyright © 2019 by ASME

Tensile strain models targeted for vintage girth welds are being developed by CRES for PRCI.

4. GAPS IN SBDA

4.1 Sufficiency of Standards in Ensuring Good Girth Weld Strain Capacity

There is a somewhat prevailing misunderstanding about the longitudinal stresses or strains a girth weld can take if the girth welds are qualified and inspected by field welding standards such as API 1104. In reality, girth welds made to comply with API 1104 are not guaranteed to have a high resistance to longitudinal stresses or strains. An analysis of the basic requirements in girth weld procedure qualification and experimental tests of completed welds show that the strain capacity of API 1104-compliant girth welds can vary from as low as 0.2% to more than 2%. For welds made before the qualification and inspection criteria came into force, one of the major contributors to girth welds' low tolerance to longitudinal stresses or strains is large flaws that occasionally remain after construction. These flaws can become unstable when additional stresses or strains are imposed on the welds after the pipelines have been in satisfactory service for decades.

4.2 Consistent Measure of Strains

There are no consistent and uniform measures of strains for either strain demand or strain capacity. This can lead to different values of reported strains for the same event. Using strain demand and strain capacity measured and reported in different ways can increase the uncertainty in the assessment outcome.

- A fixed gage length, e.g., 10 feet, is often used when computing the "bending strain" from IMU runs.
- When pipe and soil interaction models are used, the peak strain from pipe or beam elements are taken as strain demand. The concept of gage length in such models is seldom discussed.
- Strains measured by strain gages is a local strain in a small area of the material.
- The compressive strain capacity (CSC_{ML}) reported in most experimental tests is measured over a gage length of one or two pipe diameters.
- TSC reported by TSC models could be nominal strains remote from a girth weld or strains across a girth weld.
- The strain capacity reported in curved wide plate or full-scale tests could be either a nominal strain away from a girth weld or an averaged strain across a girth weld.

4.3 Interaction of Strain Capacity and Strain Demand

The amount of strain imposed on the pipeline is the result of the interaction between the pipe and surrounding soil. The strain computed from the pipe and soil interaction models is affected by the pipe properties and the representation of the pipe in the models. For instance, internal pressure affects the stiffness of the pipelines, and therefore, the response of the pipelines to the ground movements. The same internal pressure affects the strain capacity. Therefore, the strain demand is intrinsically related to the strain capacity of the pipe. Higher internal pressure can reduce the strain demand due to the stiffening effects of the internal pressure and reduce the TSC.

4.4 Flaw Interaction Rule

Flaw interaction rules developed for strain-based assessment, such as those in API 1104 Annex A, CSA Z662 Annex K, and BS 7910, are not conservative for SBDA. Flaws start to interact at a far great distance under high longitudinal strain conditions than low strain conditions prior to material yielding [16].

4.5 Significance of Post-Buckling Behavior

If the loading is load-controlled, a pipe may collapse immediately after the maximum bending moment is reached (CSC_{ML}), possibly forming severe wrinkles. Large and severe wrinkles can lead to rupture/leak from high local tensile strains in either hoop or longitudinal direction. If the wrinkles survive the initial formation, the long-term integrity of the wrinkles may be affected by possible fatigue damage and/or coating/corrosion related concerns. The knowledge of the buckling process and the nature and magnitude of subsequent loading is necessary to make right decisions about the need and timing of mitigation measures.

If the loading is displacement-controlled, reaching CSC_{ML} is typically not an ultimate limit state (associated with leak or rupture) or even a service limit state (e.g., passage of inline inspection tool). The magnitude of wrinkle or buckle at the point of CSC_{ML} can be small. In some cases, design against CSC_{ML} can be more limiting than design against TSC. The relevance of CSC_{ML} versus CSC_{PML} depends on site conditions, i.e., interaction of pipe and surrounding environment, and plans for monitoring and mitigation when CSC_{ML} is reached. More work is needed in this area.

4.6 Roles of Large-Scale Testing

Typical large-scale testing for TSC includes curved wide plates and full-scale tension tests. There are no uniformly accepted test procedures. Specimen dimensions, gage placement and data acquisition, test execution procedures, loading conditions and sequence, and post-test data processing all affect the reported TSC.

5. CONCLUDING REMARKS

Key steps and processes for SBDA are provided for the design and integrity management of pipelines subjected to high longitudinal strains. SBDA is a complex process that requires fresh thinking and approaches to material specification, design, welding, field construction, and post-construction integrity management.

Copyright © 2019 by ASME

Significant progress has been made in the last decade on the development of SBDA framework and methodology. While SBDA is gaining wider acceptance, it is still a maturing field. New approaches, different from the customary practice, are necessary to ensure pipeline integrity. While SBDA offers opportunities for safe and cost-effective design and integrity management, it must be recognized that pipelines expected to experience high longitudinal strains need to be designed, constructed, and maintained with greater precision than traditional pipelines with hoop-stress-dominated loading. As longitudinal strains are typically associated with ground movement hazards, input from geotechnical experts in the design and maintenance of pipelines is critical to minimize the impact of hazards. Careful consideration must be given to the nature of the loads over the entire service life of a pipeline. The overall industry experience in SBDA is limited. It is recommended that SBDA procedures, processes, and best practice guidelines be continuously updated as the industry gains practical experience and new technologies are developed.

ACKNOWLEDGMENT

A number of organizations and individuals have contributed to the work referenced here. In particular, the support of PRCI and US DOT PHMSA is gratefully acknowledged. The author is grateful for the support of Dr. Dunji Yu, Dr. Dan Jia, and Mr. Alex Wang in preparing this paper.

REFERENCES

1 Wang, Y.-Y., "Strain-Based Design and Assessment and 0.8 Design Factor", DOT PHMSA Research and Development Forum, July 18-19, 2012 https://primis.phmsa.dot.gov/rd/mtgs/071812/YongYiWang.pdf

2 "Management of Ground Movement Hazards for Pipelines," a joint industry project led by CRES, final report, 2017.

3 Wang, Y.-Y., Yu, D., "Management of In-Service Pipelines in Areas of Ground Movement," Proceedings of the 1st AIM-PIMG Conference, AIMPIMG-1065, Houston, TX, March 25-28, 2019.

4 Zhang, F., Liu, M., and Wang, Y.-Y., Yu Z., and Tong, L., 2012, "Strain demand in areas of mine subsidence", *Proceedings of the 9th International Pipeline Conference*, Calgary, Alberta, Canada, September 24-28.

5 Zhang, F., Liu, M., Wang, Y.-Y., Yu Z., and Tong, L., 2012, "Integrity of pipeline in area of mine subsidence", *Proceedings of the 9th International Pipeline Conference*, Calgary, Alberta, Canada, September 24-28.

6 Liu, M., Wang, Y.-Y., and Yu, Z., "Response of Pipelines under Fault Crossing," Proceedings of the 18th International Offshore and Polar Engineering Conference (ISOPE 2008), Vancouver, Canada, July 6-11, 2008.

7 Zhang, F. and Wang, Y.-Y., "Assess the Seismic Wave Influence on Buried Pipelines with ASCE Soil-Spring Model," Proceedings of the ASME 2013 33rd International Conference on Ocean, Offshore and Arctic Engineering, San Francisco, CA, June 8-13, 2014.

8 Wang, Y.-Y., Liu, M., McColskey, D., Weeks, T., and Horsley, D., "Broad Perspectives of Girth Weld Tensile Strain Response," Proceedings of the 8th International Pipeline Conference, Paper No. IPC2010-31369, September 27 – October 1, 2010, Calgary, Alberta, Canada.

9 Wang, Y.-Y., Liu, B., and Wang, B., "Tensile Strain Models and Their Applications," Proceedings of the 1st AIM-PIMG Conference, AIMPIMG-1066, Houston, TX, March 25-28, 2019

10 Wang, Y.-Y., Jia, D., Rapp, S., Johnson, D., "Low Strain Capacity Girth Welds of Newly Constructed Pipelines and Mitigative Approaches," Proceedings of the 1st AIM-PIMG Conference, AIMPIMG-1064, Houston, TX, March 25-28, 2019.

11 Liu, M., Wang, Y.-Y., Zhang, F., Wu, X., and Nanney, S., "Refined Compressive Strain Capacity Models," Proceedings of the 6th Pipeline Technology Conference, Ostend, Belgium, October 7-9, 2013

12 Liu, M., Wang, Y.-Y., Sen, M., and Song, P., "Integrity Assessment of Post-Peak-Moment Wrinkles," Proceedings of the 11th International Pipeline Conference, Paper No. IPC2016-64654, September 26-September 30, 2016, Calgary, Alberta, Canada.

13 Wang, Y.-Y., Liu, M., Long, X., Stephens, M., Petersen, R., and Gordon, R., 2011, "Validation & Documentation of Tensile Strain Limit Design Models for Pipelines," US DOT Contract No. DTPH56-06-T000014, final report http://primis.phmsa.dot.gov/matrix/PrjHome.rdm?prj=200.

14 "Strain-Based Design – Possible Conditions," US DOT PHMSA, February 2015.

15 Kotian, K. and Wang, Y.-Y., "Mechanical Properties of Vintage Girth Welds," Proceedings of the 11th International Pipeline Conference, Paper No. IPC2016-64420, September 26-September 30, 2016, Calgary, Alberta, Canada.

16 Wang, Y.-Y., Liu, M., Long, X., Stephens, M., Petersen, R., and Gordon, R., 2011, "Validation & Documentation of Tensile Strain Limit Design Models for Pipelines," US DOT Contract No. DTPH56-06-T000014, final report, http://primis.phmsa.dot.gov/matrix/PrjHome.rdm?prj=200.

Copyright © 2019 by ASME

AIM-PIMG2019-1006

Chapter 4

BURST FAILURE MODELS AND PREDICTIONS OF BURIED PIPELINES

Xian-Kui Zhu
EWI
1250 Arthur E Adams Drive
Columbus, Ohio 43221, USA

ABSTRACT

The oil and gas transmission pipelines are made of carbon steels and usually buried underground. They are susceptible to external corrosion that is a big threat to the integrity of aging pipelines. Different corrosion assessment methods have been developed for managing the aging pipeline integrity. This paper reviews and evaluates the existing burst failure models and their predictions for buried aging pipelines, and both pipes and elbows are considered. The review is first given to the burst failure prediction models for defect-free pipes, including strength solutions and flow solutions of the burst pressure as well as their experimental validations. Followed are evaluations on the corrosion assessment methods that are categorized as three-generation models for corroded line pipes from low to high strength steels. After then, the burst pressure models are briefly discussed for the pipeline elbows with and without corrosion defects. On this basis, the better burst failure models are recommended, and the major challenges are discussed.

NOMENCLATURE

A_0	area of axial cross section with a length L
A_d	area of a complex corrosion defect profile
ASSY	average shear stress yielding criterion
d	defect depth
d/t	defect depth to wall thickness ratio
D	outside diameter of a pipe
D_m	mean diameter of a pipe
FEA	finite element analysis
GrB	Grade B
L	defect length
L_f	Lorenz factor
LPC	line pipe corrosion
M	bulging factor
n	strain hardening exponent of a power-law material
P_0	Tresca strength failure pressure
P_b	burst pressure of pipes
P_{b0}	burst pressure of pipes for infinite long defects
P_{f0}	flow stress-based failure pressure
P_{M0}	von Mises strength failure pressure
P_{ZL0}	Zhu-Leis strength failure pressure
P_T	Tresca flow burst pressure
P_M	von Mises flow burst pressure
P_{ZL}	Zhu-Leis flow burst pressure
PCORRC	pipeline corrosion criterion
r	radius of a pipe
R_b	radius of an elbow
SMYS	specified minimum yield stress
t	wall thickness of a pipe
UTS or T	ultimate tensile stress
Y	yield stress
Y/T	yield to tensile stress ratio
ZL	Zhu-Leis criterion
ASME	American Society of Mechanical Engineers
DNV	Det Norske Veritas
DOT	Department of Transportation
EWI	Edison Welding Institute
PHMSA	pipeline and hazardous materials safety administration
PRCI	Pipeline Research Council International
σ_{aa}	axial stress in a pipe
$\sigma_{\theta\theta}$	hoop stress in a pipe
σ_f	flow stress of material
σ_{ys}	yield stress (Y) of material
σ_{uts}	ultimate tensile stress (T or UTS) of material

Copyright © 2019 by ASME

1. INTRODUCTION

The pipeline system is an important nation's infrastructure and used worldwide to transport oil or natural gas over long distances to meet the increasing demands of energy. Historical practice showed that pipeline is an effective and safe vehicle used for oil and gas transportation. However, the pipelines are made of carbon steels and buried underground per the regulation requirements. Because steel pipes are highly susceptible to external corrosion due to underground water or soil moisture, external corrosion can weaken the pipeline strength, and threatens the pipeline integrity. In addition to external corrosion attack, the buried pipelines under high pressure may fail occasionally due to geohazards, natural forces, mechanical damage, steel making and weld quality, stress corrosion cracking, and other anomalies in a pipeline.

Pipeline and Hazardous Materials Safety Administration (PHMSA) of the U.S. Department of Transportation (DOT) reports the U.S. pipeline incidents and the statistic trends over the 20 years at its website [1] and updates the database regularly. Based on the PHMSA pipeline failure database, Dai et al. [2] statistically analyzed the recent six-year significant incidents from 2010 to 2015 that contain 432 crude oil pipeline failures and 238 gas transmission pipeline failures. Figures 1(a) and 1(b) show the failure causes that were categorized for oil pipelines and gas pipelines, respectively. As evident in those figures, the top three failure causes are corrosion, pipe/weld material failure, and equipment failure for the oil pipeline failure, while those failure causes become to the pipe/weld material failure, excavation damage, and corrosion for the gas pipelines. As a result, corrosion is one of the top three failure causes for either oil or gas pipelines, and so corrosion assessment is critical to buried pipeline integrity.

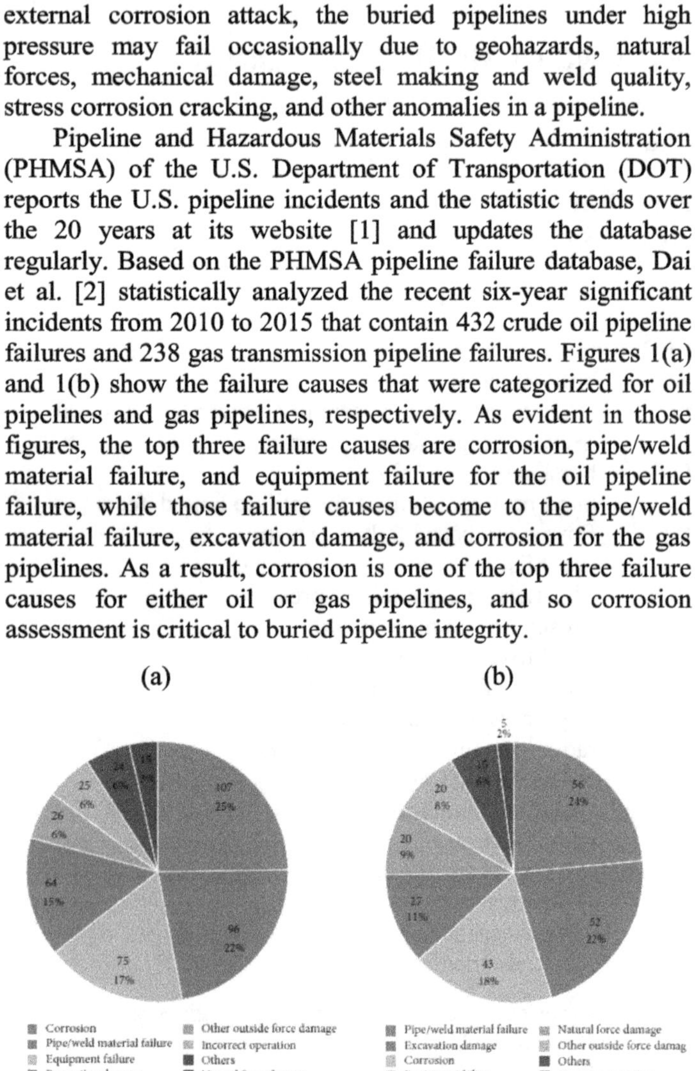

Figure 1. Failure causes of (a) oil and (b) gas pipelines

Unprotected buried steel pipelines are highly susceptible to corrosion, and without proper corrosion protection every steel pipeline will eventually deteriorate. Corrosion can reduce the pipe wall thickness, weaken the pipeline strength and degrade its loading-carrying capacity. To prevent steel pipelines from external corrosion, coating and cathodic protection are required by regulation. Usually, buried pipeline management must cope with other extreme geological conditions in a range of geohazards, such as ground shaking in seismically active regions, land sliding, ground subsidence and settlement, geological fault displacements, and freeze-thaw displacements [3-4]. This work only considers corrosion effect on buried pipelines subject to internal pressure. In such cases, overburden pressure on a buried pipeline can be neglected since it is insignificant in comparison to internal pressure [5-6].

Before corrosion assessment methods were developed, many analytical, empirical, or numerical methods have been developed for defect-free pipes. As reviewed by Zhu [7], Christopher et al. [8] evaluated many available burst pressure prediction models for both thin-wall and thick-wall pressure vessels. Law and Bowie [9] assessed the burst pressure prediction models for high-strength line pipes. More recently, Zhu and Leis [10] evaluated a series of burst pressure prediction models for a variety of pipeline steels in terms of the strength theories and the plastic flow theories, and the most accurate model was identified for defect-free pipes.

For pipes with blunt defects, Miller [11] presented a good review of early investigations on the limit load of flawed pipes or pressure vessels. Cosham et al. [12] reviewed the available corrosion assessment methods in predicting burst pressure for corroded pipelines. In the oil and gas industry, ASME B31G [13] and Modified B31G [14] remain the primary corrosion assessment methods used in the daily integrity assessment of aging pipelines, even though those methods are known to be overly conservative. Recently, Zhu [15] briefly reviewed the corrosion assessment methods and numerical approaches used to predict burst pressure of line pipes with metal loss defects, and Leis et al. [16] presented their new numerical results to minimize B31G model uncertainty in the corrosion assessment.

A real pipeline typically contains straight pipes and curved pipes that are called elbows, as shown in Figure 2. An elbow is a pipe fitting and used to accommodate the direction change of the pipeline during construction to meet its route design, particularly in mountainous terrains. In contrast to straight pipes, elbows usually have significant flexibility, higher stresses and strains, and very noticeable ovalization. Thus, the integrity assessment of elbows is more important than pipes, but unfortunately its study is very limited in public literature.

Figure 2. Pipes and elbows in a pipeline

Copyright © 2019 by ASME

This paper delivers a critical review and evaluation of existing assessment methods and their failure predictions for line pipes and elbows with and without corrosion defects. This includes the strength and flow solutions of burst pressure for defect-free pipes, three-generation corrosion models for corroded pipes, and simple models for pipeline elbows, their limitations and experimental or numerical validations. Different models are graphically compared or validated with burst test data. Major technical challenges are discussed.

2. BRIEF REVIEW OF BURST MODELS FOR PIPES

The methods and burst prediction models for defect-free pipes are the basis to develop corrosion assessment models for corroded pipes and is first discussed. A recent review on the burst models, challenges and innovative technology for defect-free pipes was presented by the present author [15]. The results are briefly reviewed below.

2.1. Strength Models and Their Validations

For a thin-wall defect-free pipe, four strength solutions [17] of burst pressure were obtained in terms of the ultimate tensile stress (UTS)-based Tresca, von Mises, Zhu-Leis (ZL) criteria and the flow stress-based criterion.

(1) Tresca strength solution:

$$P_0 = \frac{2t}{\bar{D}} \sigma_{uts} \tag{1}$$

(2) von Mises strength solution:

$$P_{M0} = \frac{4t}{\sqrt{3} D_m} \sigma_{uts} \tag{2}$$

(3) Zhu-Leis strength solution:

$$P_{ZL0} = \frac{1}{2}\left(1 + \frac{2}{\sqrt{3}}\right) \frac{2t}{\bar{D}} \sigma_{uts} \tag{3}$$

(4) Flow stress-based failure pressure:

$$P_{f0} = \frac{2t}{D} \sigma_{flow} \tag{4}$$

where D is the outside diameter, D_m is the mean diameter, and t is the wall thickness. $\sigma_f = (\sigma_{ys} + \sigma_{uts})/2$, σ_{ys} is the yield stress and σ_{uts} is the UTS. Zhu-Leis strength solution in Eq. (3) was determined using the average shear stress criterion (ASSC) proposed by Zhu and Leis [17, 18]. Equations (1) to (3) show that Zhu-Leis strength solution is equal to the average of Tresca and Mises strength solutions.

To evaluate the accuracy of those four strength models, experimental data for more than 100 full-scale tests of thin-wall cylindrical pressure vessels made of various carbon steels were compared with the burst pressure predictions from those strength criteria, as shown in Fig. 3. All materials used in the full-scale tests are ductile steels with a wide range of strain hardening exponent n from 0.02 to 0.18 that cover low to high-strength pipeline steels. Experimental details and references cited in Fig. 3 were described in Zhu and Leis [18].

In Fig. 3, points are experimental data, and lines are predictive results. The points within the three circles from right to left represent burst test data for low, medium, and high-strength carbon steels. Four dash lines starting from the top denote burst pressure predictions by von Mises, Zhu-Leis, Tresca, and flow stress criteria. It is observed that (1) von Mises criterion overestimates burst pressure for all steels, (2) Zhu-Leis criterion well predicts burst pressure for high-strength steels, (3) Tresca criterion predicts burst pressure for medium-strength steels, (4) the flow stress criterion predicts the most conservative results for all steels, and (5) except for the flow stress criterion, the other three strength criteria cannot consider the effect of strain hardening rate on the burst pressure data. This promoted to develop flow theory-based solutions.

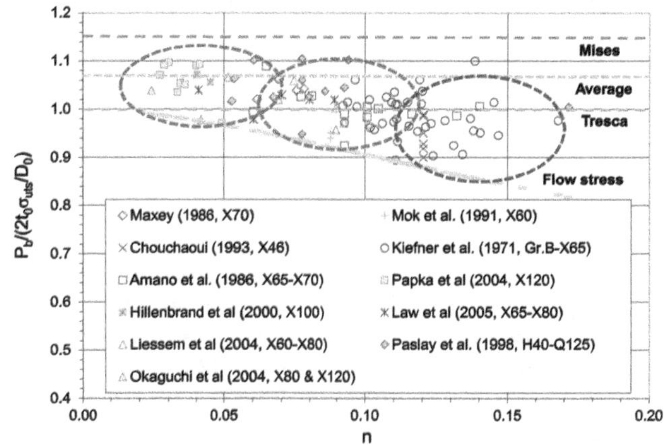

Figure 3. Comparison of experimental data of burst pressure and predictions by von Mises, Zhu-Leis, Tresca, and flow stress criteria for various pipeline steels [18]

2.2. Flow Models and Experimental Validation

The von Mises and Tresca theories are two classic plastic flow theories but determine different solutions of burst pressure for a same pipe with a difference more than 15%. To fill the technical gap, this author [17-18] proposed a new multi-axial flow theory of plasticity, i.e. average shear stress flow theory that was often referred to as Zhu-Leis theory in literature. For a power-law hardening material, three flow solutions of burst pressure were determined for defect-free pipes in terms of Tresca, Zhu-Leis (ZL) and von Mises criteria as:

$$P_T = \left(\frac{1}{2}\right)^{n+1} \frac{4t}{D_m} \sigma_{uts} \tag{5}$$

$$P_{ZL} = \left(\frac{2+\sqrt{3}}{4\sqrt{3}}\right)^{n+1} \frac{4t}{D_m} \sigma_{uts} \tag{6}$$

$$P_M = \left(\frac{1}{\sqrt{3}}\right)^{n+1} \frac{4t}{D_m} \sigma_{uts} \tag{7}$$

where n is the strain hardening exponent of a pure power-law material.

The experimental data of burst pressure shown Fig. 3 are reused to validate the flow theories. Figure 4 compares the experimental data with Tresca, Zhu-Leis and von Mises flow

Copyright © 2019 by ASME

solutions in Eqs. (5), (6) and (7). This figure shows that (1) all test data and predictions of burst pressure are functions of strain hardening exponent n in a similar trend, (2) von Mises flow solution is an upper bound of burst data, (3) Tresca flow solution is a lower bound of burst data, and (4) Zhu-Leis flow solution matches best with the burst pressure data on average. Thus, Zhu-Leis flow theory filled the technical gap between Tresca and von Mises flow theories.

Using other full-scale burst test databases, Zimmermann et al [19], Zhou and Huang [20] and Bony et al. [21-22] evaluated available burst prediction models and drew the same conclusion that the Zhu-Leis flow solution is the most accurate one for predicting burst pressure of defect-free pipes.

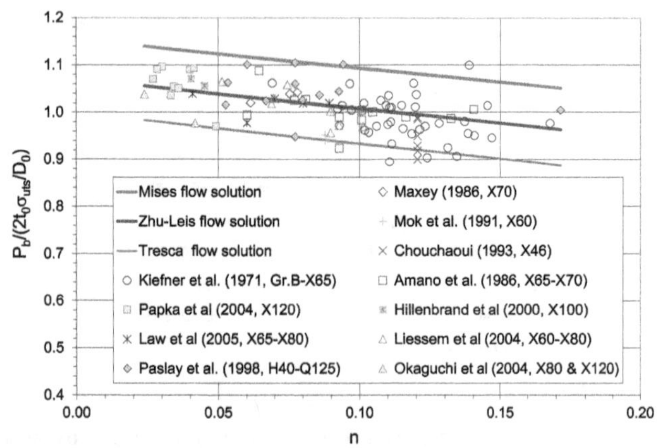

Figure 4. Comparisons of three flow solutions and experimental data for various carbon steels [18]

2.3. Numerical Validation of Zhu-Leis Flow Solution

It is well-known that all commercial finite element analysis (FEA) software codes, such as ABAQUS [23], adopt the von Mises flow theory and determine the von Mises-based burst pressure that is overestimated for defect-free pipes. Recently, Zhu and Leis [24] obtained a Mises stress-based failure criterion in reference to the Zhu-Leis flow solution and von Mises flow theory. Figure 5 shows the FEA results of von Mises effective stress versus the internal pressure normalized by ZL solution that were obtained in Reference [24] for X65 defect-free pipe using ABAQUS. Also included in this figure are two theoretical solutions determined using the von Mises flow theory and Zhu-Leis flow theory, respectively.

Figure 5 reveals that (1) von Mises effective stress at the von Mises burst failure is higher at about 18% than that at the corresponding Zhu-Leis burst pressure, (2) the von Mises burst pressure is higher at about 8% than the Zhu-Leis burst pressure, and (3) the proposed Mises stress-based material failure criterion can determine a more accurate numerical result of burst pressure for the defect-free pipe. As a result, the proposed material failure criterion can be used in ABAQUS or any other FEA software to determine a more accurate burst pressure of a defect-free pipe or pressure vessel at plastic collapse that was defined by the Zhu-Leis flow theory.

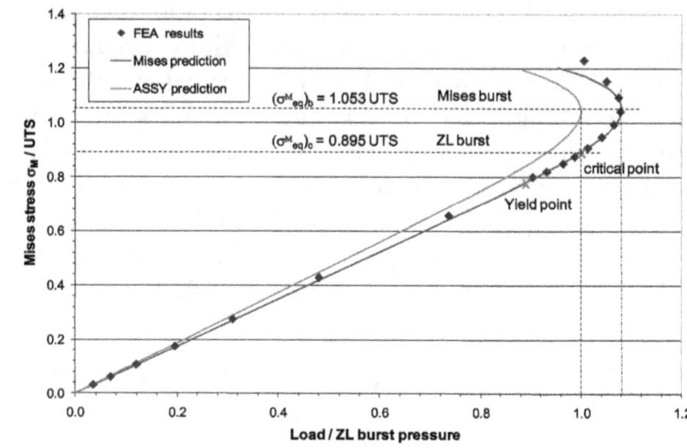

Figure 5. Variation of Mises stress with pressure for X65 steel [18]

Recently, four FEA calculations were performed at EWI on a X65 pipe of 30"-diameter and 0.5"-wall thickness, where the yield stress is fixed at 72 ksi, but the stain hardening exponent n varies from 0.05, 0.1, 0.15 to 0.2. Using the material failure criterion and the ABAQUS RIKS model, the FEA results of Zhu-Leis and von Mises failure pressures were determined and compared with the theoretical solutions, as shown in Figure 6. It shows that the FEA results agree very well with theoretical solutions over the large range of strain hardening rate for both flow theories. Thus, the Zhu-Leis flow solution of burst pressure is numerically validated here, in addition to the experimental validation in Fig. 4.

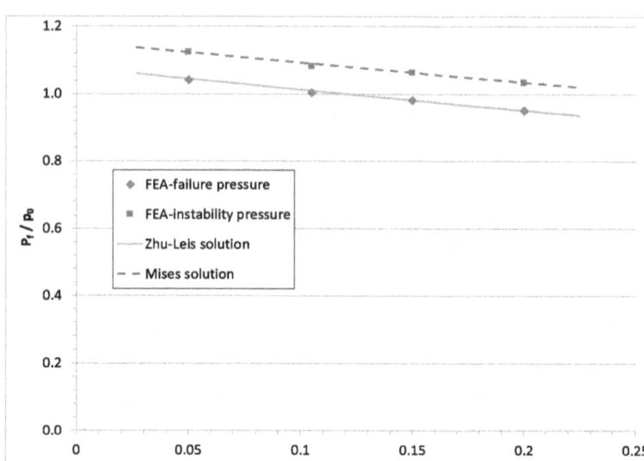

Figure 6. Comparison of FEA results and theoretical solutions for X65 over a large range of strain hardening exponent n

3. CORROSION ASSESSMENT MODELS

Actual corrosion defects in an aging pipeline can be very complicated in size, shape, location and orientation. Only longitudinally oriented blunt defects are considered, and soil pressure is ignored here. This section reviews three-generation corrosion models. The first one was empirically developed in the 1960s to 1970s, the second one was numerically developed in the 1990s to 2000s, and the third one remains ongoing.

Copyright © 2019 by ASME

3.1. The First-Generation Models (1970s – 1980s)

3.1.1. ASME B31G model

Based on full-scale burst data obtained in 1960s for steels up to X65, Maxey et al. [25-26] developed a semi-empirical model (i.e., NG-18 equation) for estimating the remaining strength of pipes with cracks in the early 1970s. On this basis, a corrosion assessment method ASME B31G was developed in 1981 and updated in 1991 [13] and 2009 [14]. For a short defect, burst pressure of a corroded pipeline is estimated by:

$$p_b = \frac{2t\sigma_f}{D}\left[\frac{1-(2/3)(d/t)}{1-(2/3)(d/t)/M}\right], \text{ for } L \le \sqrt{20Dt},\ d/t \le 0.8 \quad (8)$$

For a long corrosion defect, failure pressure of a corroded pipeline is determined by:

$$p_b = \frac{2t\sigma_f}{D}\left[1-\frac{d}{t}\right], \quad \text{for } L > \sqrt{20Dt},\ d/t \le 0.8 \quad (9)$$

where $\sigma_f = 1.1\text{SMYS}$ is the flow stress, SMYS is the specified minimum yield stress, d is the defect depth, L is the total defect length, and M is the bulging factor and expressed as:

$$M = \sqrt{1+0.8\left(\frac{L}{\sqrt{Dt}}\right)^2} \quad (10)$$

ASME B31B model was developed and calibrated with the vintage pipeline grades up to X65, and thus may be inadequate to use for modern pipeline steels like X80 or X100.

3.1.2. Modified B31G (0.85 dL) model

Practical application showed that ASME B31G method can be overly conservative, and thus a modified version of ASME B31G was proposed by Kiefner and Vieth [27] in the late 1980s in the associated RSTRENG software as:

$$p_b = \frac{2t\sigma_f}{D}\left[\frac{1-0.85(d/t)}{1-0.85(d/t)/M}\right] \quad (11)$$

where the flow stress was redefined as $\sigma_f = SMYS + 69\,(MPa)$, d/t≤0.8, and the bulging factor was replaced by:

$$M = \sqrt{1+0.6275\left(\frac{L}{\sqrt{Dt}}\right)^2 - 0.003375\left(\frac{L}{\sqrt{Dt}}\right)^4}, \quad \text{if } L \le \sqrt{50Dt} \quad (12a)$$

$$M = 3.3 + 0.032\left(\frac{L}{\sqrt{Dt}}\right)^2, \quad \text{if } L > \sqrt{50Dt} \quad (12b)$$

Note that the 0.85 factor used in Eq. (11) replaces the 2/3 factor used in Eq. (8). This modified B31G model was accepted by ASME in the version of B31G-2009 [14]. Ma et al. [28] compared their different versions and discussed differences.

3.1.3. RSTRENG (effective area method)

To better approximate a real corrosion area with a river bottom profile, an effective area method was proposed by Kiefner and Vieth [27] to estimate the remaining strength:

$$P_b = \frac{2t\sigma_f}{D}\left[\frac{1-A_d/A_0}{1-A_d/A_0M}\right] \quad (13)$$

where the flow stress and the bulging factor are the same as defined in Modified B31G, A_d denotes the effective area of a complex corrosion defect profile, and A_0=tL is the axial cross-section area with the maximum defect length L. This method allows to determine more accurate corroded area using the discrete approach, and thus to determine more accurate burst pressure of a real corrosion defect. A personal-computer code called RSTRENG was developed to facilitate the analysis of a real corroded area via the effective area method.

For a machined rectangular defect, Eq. (13) becomes as:

$$P_b = \frac{2t\sigma_f}{D}\left[\frac{1-(d/t)}{1-(d/t)/M}\right] \quad (14)$$

If a defect is long enough, Equation (14) reduces to Eq. (9).

3.2. The Second-Generation Models (1990s)

3.2.1. LPC model

In the middle 1990s, based on full-scale burst test data and a large amount of elastic-plastic FEA results for corroded pipes up to X65, Fu et al. [29-30] at British Gas developed a line pipe corrosion (LPC) criterion to determine burst pressure:

$$P_b = \frac{2t\sigma_{uts}}{D-t}\left[\frac{1-d/t}{1-d/Qt}\right] \quad (15)$$

where σ_{uts} is the UTS of pipeline steels, and the parameter Q was curve-fit from the FEA results as:

$$Q = \sqrt{1+0.31\left(\frac{L}{\sqrt{Dt}}\right)^2} \quad (16)$$

The LPC criterion was adapted and incorporated into DNV RP-F-101 [31] in 2004 with 1.05 UTS to replace 1.0 UTS.

3.2.2. PCORRC model

In the late 1990s, based on experimental data and elastic FEA results of corrosion defects, Leis et al. [32-33] at Battelle developed the following PCORRC criterion:

$$P_b = \frac{2t\sigma_{uts}}{D}\left[1-\frac{d}{t}\left(1-\exp(-0.157\frac{L}{\sqrt{D(t-d)/2}})\right)\right] \quad (17)$$

For a very long corrosion defect, Equations (15) and (17) reduce to the simple equation:

$$p_{b0} = \frac{2t\sigma_{uts}}{D}\left[1-\frac{d}{t}\right] \quad (18)$$

As a result, the PCORRC criterion is identical to the LPC criterion for long corrosion defects.

Copyright © 2019 by ASME

3.3. The Third-Generation Models (2000s to now)

3.3.1. Modified PCORRC models

Since 2000s, further experimental and numerical studies of corrosion models have been continued [34-45], with a focus on high-strength pipeline grades X70, X80 and X100. To consider the effect of strain hardening rate, Zhu and Leis [39] adopted the von Mises flow solution in Eq. (7) as the reference stress, and the PCORRC criterion in Eq. (17) is modified as:

$$p_d = \frac{4}{\left(\sqrt{3}\right)^{1+0.239\left(\frac{1}{Y/T}-1\right)^{0.596}}} \frac{t}{D}\sigma_{uts} \times \left(1-\frac{d}{t}\left(1-\exp\left(-0.157\frac{L}{\sqrt{D(t-d)/2}}\right)\right)\right) \quad (19)$$

Because von Mises flow solution is nonconservative, this modified PCORRC model may be nonconservative too.

Ma et al. [42] further modified the modified PCORRC in Eq. (19). Based on FEA results for X70 and X80, these authors obtained a new defect geometry function and a modified PCORRC equation for high-strength pipeline steels:

$$p_d = \frac{4}{\left(\sqrt{3}\right)^{1+n}} \frac{t}{D}\sigma_{uts} \times \left(1-\frac{d}{t}\left(1-0.7501\exp\left(\frac{-0.4174L}{\sqrt{Dt}}\right)\left(1-\frac{d}{t}\right)^{-0.1151}\right)\right) \quad (20)$$

More recently, based on full-scale burst tests and the FEA calculations for X70 corroded pipes, Yeom et al. [45] obtained another modified PCORRC criterion in a conservative form:

$$p_d = 0.9\sigma_{uts}\frac{2t}{D} \times \left(1-\frac{d}{t}\left(1-\exp\left(-0.224\frac{L}{\sqrt{D(t-d)/2}}\right)\right)\right) \quad (21)$$

Equation (21) did not consider the effect of strain hardening rate and reduces to $P_b=0.9\sigma_{uts}(2t/D)$ for defect-free pipes. Thus, this modified PCORRC model may be too conservative.

3.3.2. Analytic Model for long corrosion defects

For long corrosion defects, the LPC and PCORRC models reduce to Eq. (18) that is equivalent to the Tresca strength solution with a factor of (1-d/t) for the thinner wall. So motivated, Zhu [43] simply extended the flow solutions in Eqs (5) to (7) for defect-free pipes to long corrosion defects (L>D):

(1) Tresca flow solution for a long corrosion defect:

$$P_T = \left(\frac{1}{2}\right)^{n+1} \frac{4t\sigma_{uts}}{D}\left[1-\frac{d}{t}\right] \quad (22)$$

(2) Zhu-Leis flow solution for a long corrosion defect:

$$P_{ZL} = \left(\frac{2+\sqrt{3}}{4\sqrt{3}}\right)^{n+1} \frac{4t\sigma_{uts}}{D}\left[1-\frac{d}{t}\right] \quad (23)$$

(3) Mises flow solution for a long corrosion defect:

$$P_M = \left(\frac{1}{\sqrt{3}}\right)^{n+1} \frac{4t\sigma_{uts}}{D}\left[1-\frac{d}{t}\right] \quad (24)$$

As similar to the Zhu-Leis solution in Eq. (6), Reference [37] showed that the Zhu-Leis solution in Eq. (23) is a better prediction of burst pressure for a long defect. Thus, Equation (23) can be regarded as a third-generation assessment criterion for very long defects, but not for finite defect lengths.

3.3.2. New Corrosion Models

Based on Zhu-Leis flow solution in Eq. (23) for a long defect, burst pressure for a corrosion defect with finite sizes was assumed by Zhu [15] as the following function:

$$P_b = \left(\frac{2+\sqrt{3}}{4\sqrt{3}}\right)^{n+1} \frac{4t\sigma_{uts}}{D}\left[1-\frac{d}{t}f\left(\frac{L}{\sqrt{Dt}}\right)\right] \quad (25)$$

where $f(L/\sqrt{Dt})$ is a function of defect length and determined from the FEA calculation. For X65 steel, FEA results of burst pressure were obtained in [15], and the *f* function in Eq. (25) were curve-fit from the FEA results. Thus, Eq. (25) becomes:

$$P_b = \left(\frac{2+\sqrt{3}}{4\sqrt{3}}\right)^{n+1} \frac{4t\sigma_{uts}}{D}\left[1-\frac{d}{t}\left(1-1/\left(1+0.1385\frac{L}{\sqrt{Dt}}+0.1357\left(\frac{L}{\sqrt{Dt}}\right)^2\right)\right)\right] \quad (26)$$

In addition, using the Zhu-Leis solution in Eq. (6) and the defect geometry function of PCORRC in Eq. (17), another new burst pressure model is obtained in this work to consider the strain hardening effect on the corrosion defect failure:

$$p_d = \left(\frac{2+\sqrt{3}}{4\sqrt{3}}\right)^{n+1} \frac{4t\sigma_{uts}}{D} \times \left(1-\frac{d}{t}\left(1-\exp\left(-0.157\frac{L}{\sqrt{D(t-d)/2}}\right)\right)\right) \quad (27)$$

All above-noted three-generation corrosion models are compared analytically or evaluated using burst test data next.

3.4. Evaluations of Corrosion Assessment Methods

3.4.1. Evaluation of the first and second-generation models

The first and second-generation models are frequently used in the integrity assessment of aging pipelines. To understand the difference of the different models, the PHMSA in the U.S. DOT sponsored a large project [46] in the mid-2000s to thoroughly evaluate the corrosion models. A large experimental database that contain 313 full-scale vessel and ring expansion test data of real and machined metal loss defects was collected for line pipes, and the experimental tests were carried out by a number of organizations over the last fifty years in the oil and gas industry. The metal loss defects were axially isolated, only internal pressure was applied with known pipe geometry, defect sizes, and material properties. Actual material properties were used in the calculations of failure pressure.

In the final Report [46], it was concluded that (1) the RSTRENG model is the most accurate one with the least scatter, and most data points fall into a relatively small scatter; (2) the B31G model is the most conservative one with a large scatter range; (3) the Mod B31G is less conservative than the B31G with a reduced scatter range; and (4) the LPC and PCORRC models predict comparable results with scatter

Copyright © 2019 by ASME

ranges similar to the Modified B31G results. Note that these two models can determine more accurate burst pressure of defects with uniform bottom, but not for actual corrosion defects with non-uniform or river-bottom depths.

Because of the large scatters, the pipeline industry has not accepted either the UTS-based LPC or PCORRC model for predicting the remaining strength of actual corrosion defects, and continues to use the B31G, Modified B31G, or RSTRENG model in their daily integrity assessment of real corrosion defects in service.

3.4.2. Comparison of the second-generation models

Because both LPC and PCORRC models use the UTS, their comparison is easier. Figure 7 compares the remaining strength predicted by the LPC and PCORRC models for four machined defect depths of d/t=0.2, 0.4, 0.6 and 0.8 with defect lengths up to $L/(Dt)^{0.5}=20$, where the remaining strength, $Rs=P_b/P_0$, is the ratio of predicted burst pressure of corrosion defects divided by the Tresca strength solution in Eq. (1) for defect-free pipe. It's observed that (1) both models predict the very similar results for various defect sizes with slight conservatism for PCORRC, and they are comparable, and (2) both predictions of remaining strength become independent of defect length for all defect depths if their lengths $L/(Dt)^{0.5}>10$ and approach to those predicted by the Zhu-Leis flow solution in Eq. (23) for long corrosion defects.

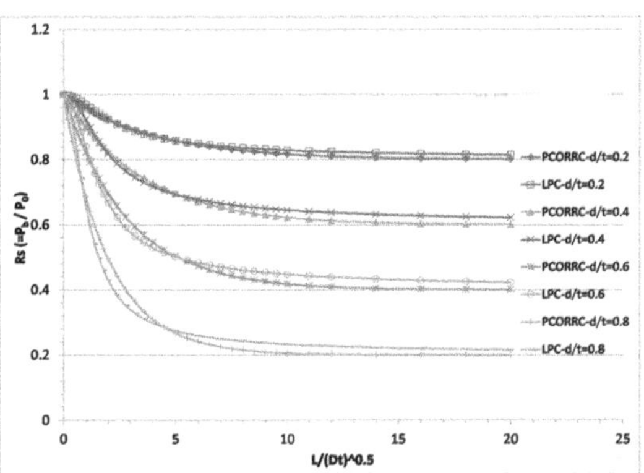

Figure 7. Comparison of residual strength by PCORRC and LPC

3.4.3. Validation of the long corrosion models

To evaluate the long defect assessment models in Eqs (22) to (24), a set of burst pressure tests for real long corrosion defects conducted by Souza et al [37-38] are employed here. Two pipe specimens (TS6.1 and TS6.2) are grade X42, and the other four pipe specimens (TS02, TS04, TS05, TS06 and TS10) are grade X46. All real corrosion defects are irregular in shape and orientation, and the defect length is about 3.65D for the X42 pipes and about 6.0D for the X46 pipes. As such, those real corrosion defects are very long and adequate to be used for validating above-noted long corrosion assessment models. With the averaged defect depths (or the maximum defect depths), their burst pressures can be predicted by models.

Five corrosion models in Equations (8), (11), (17), (22) and (23) are used to predict the burst pressure for the six real corroded pipes in reference to B31G, Modified B31G, PCORRC, von Mises flow solution and Zhu-Leis solution for long defects, respectively. Figure 8 shows the five predictions of burst pressure that are normalized by the measured burst data for each pipe. This figure shows that (1) von Mises flow solution overly predicts the burst data except for test pipe TS06, and its mean error is 21%; (2) Zhu-Leis flow solution underestimates the burst data for the four X46 pipes, but overestimates the burst data for other two X42 pipes because their mean depth is much smaller than the maximum depth, and the mean error is -9%; (3) the PCORRC predictions are similar to the Zhu-Leis solutions with a mean error of -8%; and (4) B31G and Modified B31G predictions are conservative for all seven pipes with a mean error of -39% and -29%, respectively. These results confirm that the von Mises solution is non-conservative, B31G and Modified B31G results are overly conservative, and Zhu-Leis solution and PCORRC prediction are better with a fair degree of conservatism.

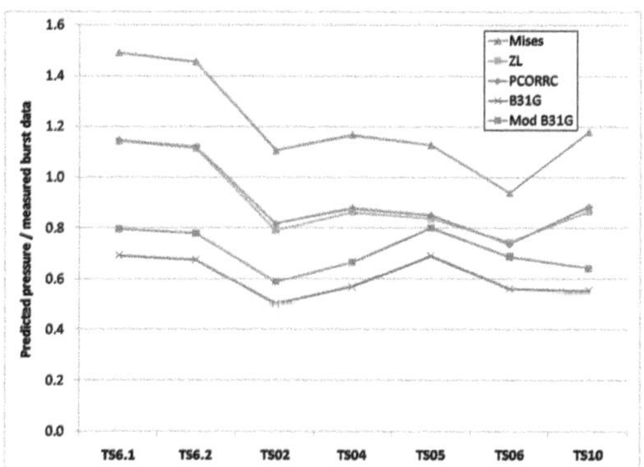

Figure 8. Comparison of burst pressure predictions for the X42 and X46 pipe specimens with very long axial defects [37]

3.4.4. Experimental validations of all three-generation models

Kim et al. [34] carried out a set of full-scale experimental tests for corrosion defects in a Korean X65 pipeline steel, and reported six burst test data for machined defects with a fixed uniform depth of d/t=0.5 and six lengths of L=50, 100, 200, 300, 600, and 900 mm. The burst pressures are predicted using the corrosion models discussed above, and compared with the test data, as shown in Figs 9(a) to 9(c).

Figure 9(a) compares the burst test data with predictions from three first-generation models (B31G, Mod B31G, and RSTRENG) and two second-generation models (LPC and PCORRC). It shows that both LPC and PCORRC predictions match well the test data with small conservatism. The Mod B31G predictions are reasonably good over the full range of defect length, while the B31G predictions are conservative, particularly for the long defect of $L>\sqrt{(20Dt)}$. All RSTRENG predictions underestimate the test data. This is anticipated because the RSTRENG was developed for use of the actual area of corrosion defects with non-uniform depths.

Copyright © 2019 by ASME

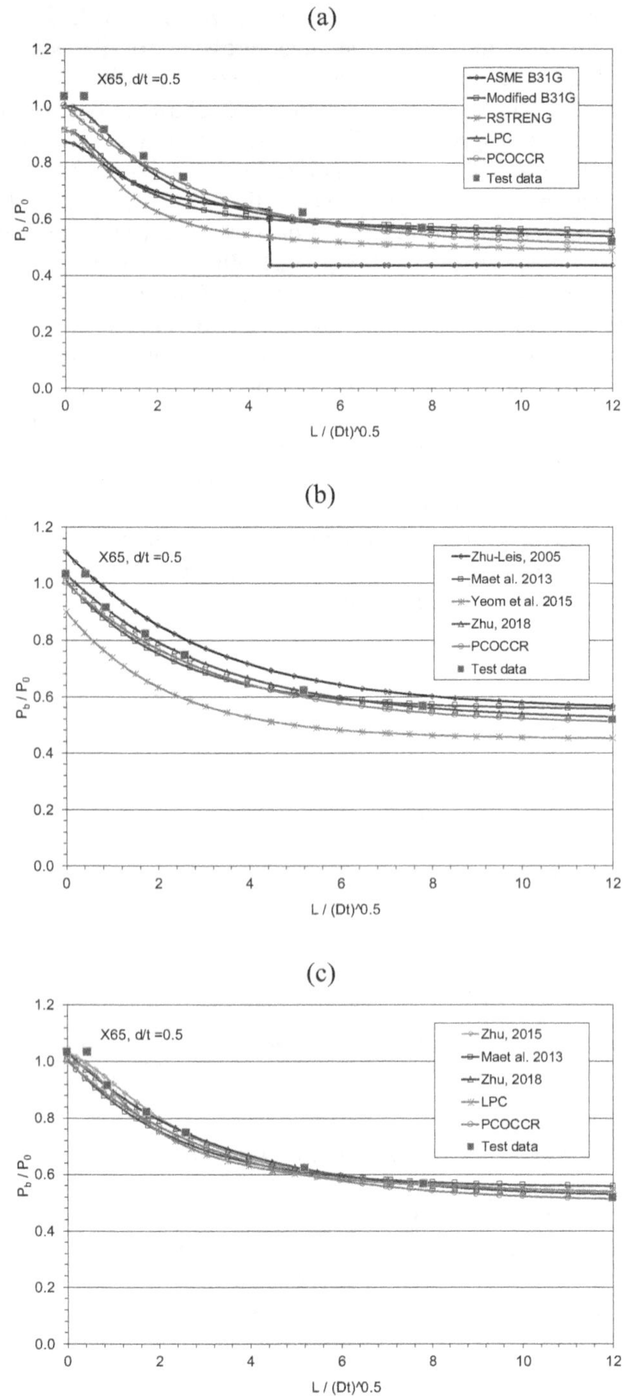

Figure 9. Comparison of full-scale test data and burst pressure predictions for X65 pipe with a deep defects d/t=0.5: (a) first and second-generation models, (b) PCORRC and its modified models, and (c) second and third generation models.

Figure 9(b) compares the burst test data with predictions by the original PCORRC model in Eq. (17) and its modified models of Zhu 2005 [39] in Eq. (19), Ma et al. 2013 [42] in Eq. (20), Yeom et al. 2015 [45] in Eq. (21) and the newly-proposed Zhu 2018 in Eq. (27). It shows that the Zhu 2005 model overestimates the burst data in an error less than 7%, while the Yeom et al. 2015 model overly underestimates the burst data for about 20%. The other two modified models of Ma et al. 2013 and Zhu 2018 predict comparable results as the original PCORRC model for X65, and match well the burst test data. As a result, Zhu 2005 and Yeom et al. 2015 models should not be used in practical applications. The other two modified and original PCORRC models may be used in applications of corrosion defects with a flat bottom.

Figure 9(c) compares the burst test data with predictions by the second-generation models of LPC and PCORRC and the third-generation models of Ma et al. 2013 [42] in Eq. (20), Zhu 2015 [15] in Eq. (26) and Zhu 2018 in Eq. (27). It shows those five models predict comparable results for those defects, and all predictions match well the test data for this X65 corroded pipe. Among them, improved PCORRC model proposed in this work (Zhu 2018) predicts the best result of burst pressures of the defects in the X65 pipe.

To further evaluate their differences of those five corrosion models for other defect depths, more full-scale burst data are needed for different pipeline grades and different defect sizes. This is beyond the scope of the present work.

3.4.5. Effect of steel grade on corrosion assessment models

Due to lack of burst test data, the steel grade effect on the corrosion assessment models is analytically investigated here. For this purpose, two representative grades, two typical flat defect depths and five corrosion models are selected. Two grades are a low grade of GrB and a high grade of X80. Two defect depths are a shallow defect depth of d/t=0.3 and a deep defect depth of d/t=0.8. The five selected corrosion models are ASME B31G, Mod B31G, PCORRC and two modified PCORRC models by Ma et al. 2013 and Zhu 2018.

Figure 10(a) and 10(b) compares those five models' predictions of burst pressure for GrB pipes with a shallow defect of d/t=0.3 and a deep defect of d/t=0.8, respectively. As evident in Fig. 10(a), the flow stress-based models are overly conservative in comparison to the UTS-based models. Three UTS-based models' predictions are comparable. Zhu 2018 model is slightly conservative than the PCORRC model at a difference of about 6%, while Ma et al. 2013 model predict the nearly same results as Zhu 2018 model for short defects and the similar results as PCORRC for very long defects. Figure 10(b) shows that the conservatism of B31G and Mod B31G is reduced for the deep defect of d/t=0.8 than that for the shallow defect of d/t=0.3, the Mod B31G predictions can be larger than those by the three UTS-based models. While the three UTS-based models still predict comparable results, the difference of burst pressures by Zhu 2018 and PCORRC models is reduced slightly, and the Ma et al. 2013 model is larger than the other two UTS-based models. As a result, Zhu 2018 modified PCORRC model could be a better model that can consider the effect of strain hardening rate and predict more accurate burst pressures of shallow and deep defects in GrB pipes. Off course, further full-scale experimental test data are needed to verify the prediction models and the implication.

Copyright © 2019 by ASME

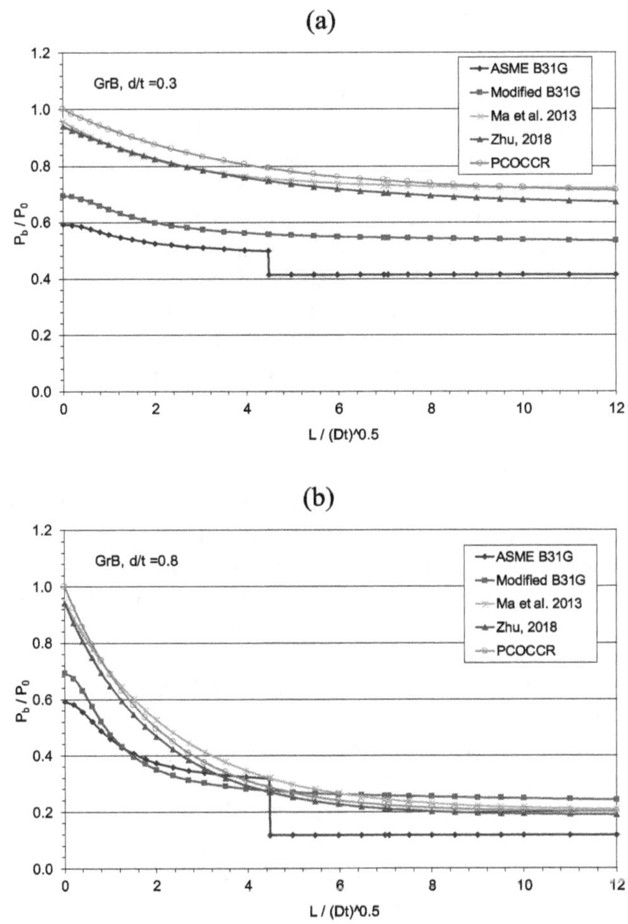

Figure 10. Comparison of five corrosion models as marked for GrB steel and for (a) d/t=0.3, and (b) d/t=0.8

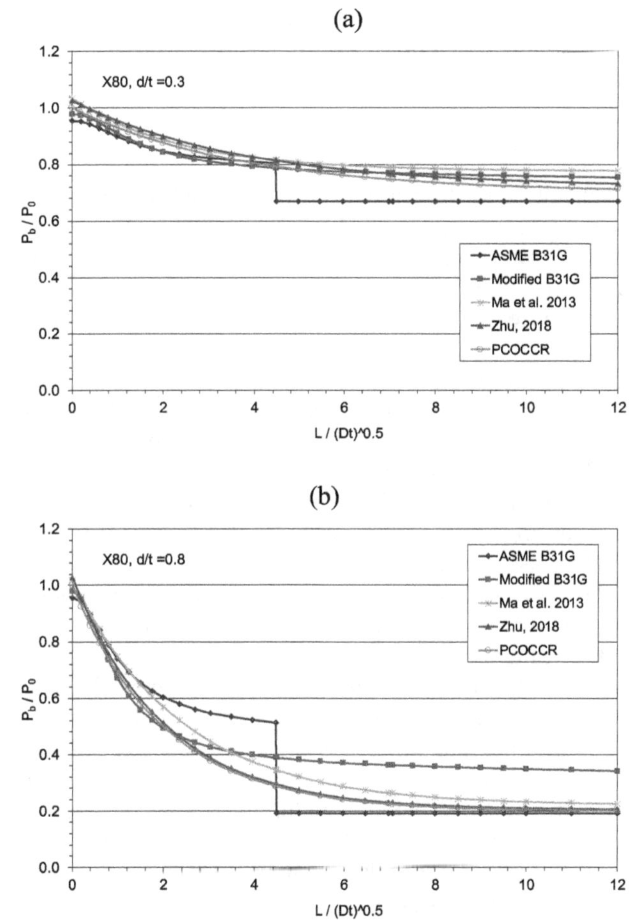

Figure 11. Comparison of five corrosion models as marked for X80 steel and for (a) d/t=0.3, and (b) d/t=0.8

Figure 11(a) and 11(b) compares those five models' predictions of burst pressure for X80 pipes with a shallow defect of d/t=0.3 and a deep defect of d/t=0.8, respectively. As shown in Fig. 11(a), all models predict comparable results, except for B31G for long defects. Because of strain hardening effect, Zhu 2018 model predicts slightly higher results than the PCORRC model at a difference of about 3%, while Ma et al. 2013 model predict the largest for long defects. Figure 11(b) shows that both B31G and Mod B31G can overly estimate burst pressure the deep defect of d/t=0.8 in X80 pipes except for very short defects for both models and long defects for B31G model. While the three UTS-based models still predict comparable results, the burst pressure predictions by Zhu 2018 and PCORRC models are almost identical to each other, but the Ma et al. 2013 model predicts much larger results than other two UTS-based models at an error of about 20%. This implies that the Ma et al. 2013 model may be non-conservative for deep defects in X80 pipes, which is consistent with the implication for GrB pipes with deep defects. Thus, Zhu 2018 modified PCORRC model would be a better model to consider the effect of strain hardening rate and to predict more accurate burst pressures of shallow and deep defects in X80 pipes. This implication needs verification of full-scale test data.

4. Burst Assessment Methods for Elbows

4.1. Bust pressure models for defect-free elbows

In a PRCI report [47] in the early 1990s, the burst pressure failure predictions for pipeline elbows were studied based on a simple elastic theory. The maximum elastic hoop and axial stresses occur at the intrados of the elbow and are expressed as:

$$\sigma_{\theta\theta} = \frac{Pr}{t} L_f, \ \sigma_{aa} = \frac{Pr}{2t} \tag{28}$$

where $L_f = (R_b/r\text{-}0.5)/(R_b/r\text{-}1.0)$ is called as Lorenz factor, R_b is the radius of the elbow, r is the pipe radius, and t is the wall thickness of the pipe and the elbow.

Miller [11] gave a lower bound limit pressure of an elbow based on the Tresca yield criterion:

$$P_{T0} = \frac{t\sigma_{uts}}{r\,L_f} \tag{29}$$

Kim and Oh [48] extended it to the von Mises criterion using FEA results and gave an upper bound limit pressure as:

$$P_{M0} = \frac{2}{\sqrt{3}} \frac{t\sigma_{uts}}{r\,L_f} \tag{30}$$

Copyright © 2019 by ASME

Similarly, using the Zhu-Leis yield criterion, a medium strength solution of burst pressure for a defect-free elbow is obtained as:

$$P_{ZL0} = \frac{1}{2}\left(1 + \frac{2}{\sqrt{3}}\right)\frac{t}{rL_f}\sigma_{uts} \quad (31)$$

Bubenik and Rosenfeld [47] collected a set of burst pressure data of full-scale tests for defect-free elbows for different carbon steels in a range of Y/T from 0.54 to 0.69. From Eqs (29) to (31), the burst pressure of those elbows are predicted based on the Tresca, Mises and Zhu-Leis strength criterion. Figure 12 compares the burst pressure test data with the three predictions for those tested elbows without defects. It is observed that except for one case of Y/T=0.68, (1) all Tresca predictions are conservative, particularly for steels with Y/T < 0.61; (2) the Zhu-Leis predictions reduce the conservatism and improve the Tresca prediction for Y/T<0.61; (3) von Mises predictions are fine for steels with Y/T<0.61, and (4) both Zhu-Leis and Mises solutions are nonconservative for steels with Y/T>0.61. Overall, the scatters of each prediction model for elbows are very large, and multiple factors can contribute to those big scatters. The key factor is that burst test data collected in Report [47] were primarily for small-diameter elbows, but not the large-diameter pipeline elbows. Another important factor is that the wall thickness of an elbow at its intrados is typically large than the wall thickness at other locations of the elbow. On the other hand, all prediction models used the nominal wall thickness of the pipe or elbow, and thus the predictions are generally less than the experimental data of burst pressure for the elbows with Y/T<1.

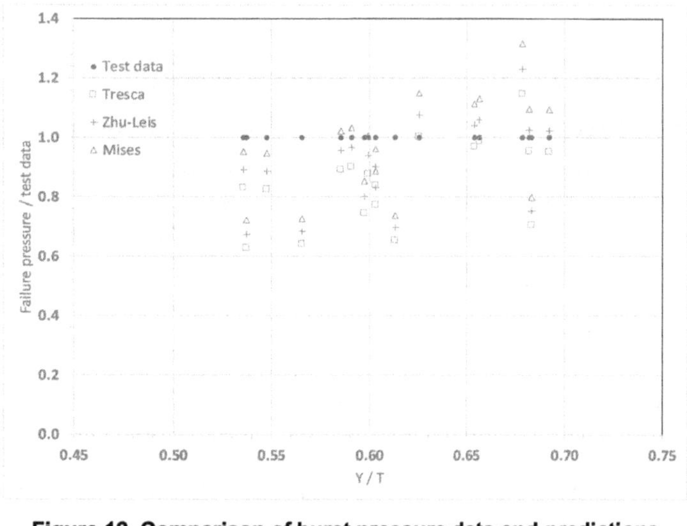

Figure 12. Comparison of burst pressure data and predictions

Further theoretical and experimental studies are needed to determine more accurate burst pressure and to consider strain hardening or Y/T effect on burst pressure predictions of pipeline elbows.

4.2. Corrosion models for corroded elbows

In the PRCI report [47], the corrosion models for corroded elbows were determined simply using the Lorenz factor to divide Eqs (8), (11), and (13) for B31G, Mod B31G and RSTRENG models. As a result, the reduced burst pressure of a corroded elbow was determined in comparison to a corroded straight pipe with the same pipe geometry and defect sizes. For a series of corroded elbows under internal pressure, the results in Report [47] showed that such predicted burst pressures are conservative in comparison to experimental data. Further analytical and experimental investigations are needed to determine a more accurate burst pressure for corroded elbows.

6. Conclusions

This paper presented a technical review and evaluation of existing methods for predicting burst failure of buried aging pipelines, including straight pipes and curved elbows with and without metal loss defects. This included the strength and flow solutions of burst pressure for defect-free vessels, and three-generation corrosion models for pressure vessels with corrosion defects, their limitations and experimental validation. Several new corrosion assessment models were proposed in this work for pipes with corrosion defects. The burst prediction models for pipeline elbows were briefly discussed due to limited studies. The results indicated that further theoretical and experimental studies are needed to improve the corrosion assessment models for predicting more accurate burst pressure of straight pipes and curved elbows with corrosion defects to ensure the pipeline integrity.

5. References

[1] US DOT PHMSA, Pipeline Incident 20 Year Trends, www.phmsa.dot.gov/data-and-statistics/pipeline, 2018.

[2] Dai L, et al., Analysis and comparison of long-distance pipeline failures, *Journal of Petroleum Engineering*, Vol. 2017, pp. 3174636.

[3] Nyman DJ, Lee EM, Audibert JME, Mitigating geohazards for international pipeline projects: challenges and lessons learned, *Proceedings of the 7th International Pipeline Conferences*, Sept 29- Oct 3, 2008, Calgary, Canada.

[4] Porter M, et al., Updated estimates of frequencies of pipeline failures caused by geohazards, *Proceedings of the 11th International Pipeline Conferences*, September 26-30, 2016, Calgary, Canada

[5] Peng L-C, Stress analysis methods for underground pipe lines, *Pipe Line Industry*, Vol. 47 (5), 1978: 67-71.

[6] American Lifeline Alliance, *Guidelines for the Design of Buried Steel Pipe*, July 2001.

[7] Zhu XK. Strength criteria versus plastic flow criteria used in pressure vessel design and analysis, *Journal of Pressure Vessel Technology*, Nol. 138, 2016, Paper ID: 041402.

Copyright © 2019 by ASME

[8] Christopher T, Sarma BS, Potti PK, Rao BN, and Sankar K, A comparative study on failure pressure estimation of unflawed cylindrical vessels, *International Journal of Pressure Vessels and Piping*, Vol. 79, 2002: 53-66.

[9] Law M and Bowie G., Prediction of failure strain and burst pressure in high yield-to-tensile strength ratio line pipes, *International Journal of Pressure Vessels Piping*, Vol. 84, 2007: 487-492.

[10] Zhu XK and Leis BN, Evaluation of burst pressure prediction models for line pipes, *International Journal of Pressure Vessels and Piping*, Vol. 89, 2012: 85-97.

[11] Miller AG, 1988, Review of limit loads of structures containing defects, *International Journal of Pressure Vessels and Piping*, Vol. 32, pp. 197-327.

[12] Cosham A, Hopkins P, and Macdonald, KA, Best practice for the assessment of defects in pipelines, *Engineering Failure Analysis*, Vol. 14, 2007: 1245-1265.

[13] American Society of Mechanical Engineers, *Manual for Determining the Remaining Strength of Corroded Pipelines*, ASME B31G-1991, New York.

[14] American Society of Mechanical Engineers, *Manual for Determining the Remaining Strength of Corroded Pipelines*, ASME B31G-2009, New York.

[15] Zhu XK, A new material failure criterion for numerical simulation of burst pressure of corrosion defects in pipelines, *Proceedings of the ASME pressure vessels and Piping Conference*, July 19-23, 2015, Boston, Massachusetts, USA, Paper ID: PVP2015-45713.

[16] Leis B, Zhu XK, Orth F, Aguiar D, and Perry L. Minimize model uncertainty in current corrosion assessment criteria, *Proceedings of the 21th PRCI-APGA-EPRG Joint Technical Meeting on Pipeline Research*, May 1-5, 2017, Colorado Springs, Colorado.

[17] Zhu XK and Leis BN, Accurate prediction of burst pressure for line pipes, *Journal of Pipeline Integrity*, Vol. 4, 2004: 195-206.

[18] Zhu XK and Leis BN, Average shear stress yield criterion and its application to plastic collapse analysis of pipelines, *International Journal of Pressure Vessels Piping*, Vol. 83, 2006: 663-671.

[19] Zimmermann S, Marewski U, and Hohler S, Burst pressure of flawless pipes, *3R International*, Special Edition, Vol.46, 2007:28-33.

[20] Zhou W and Huang T, Model error assessment of burst capacity models for defect-free pipes, *Proceedings of International Pipeline Conference*, September 25-28, 2012, Calgary, Canada.

[21] Bony M, Alamilla JL, Vai R, and Flores E, Failure pressure in corroded pipelines based on equivalent solutions for undamaged pipe, *Journal of Pressure Vessel Technology*, Vol. 132, 2010, ID: 051001.

[22] Alamilla JL, Sosa E, Sanchez-Magana CA, Andarde-Valencia R, and Contreras A, Failure analysis and mechanical performance of an oil pipeline, *Materials and Design*, Vol. 50, 2013: 766-773.

[23] ABAQUS Standard Version 2016, *Analysis User's Manual*, Dassault Systémes Simulia Corporation, Providence, RI, 2016.

[24] Zhu XK and Leis BN, Theoretical and numerical predictions of burst pressure of pipelines, *Journal of Pressure Vessel Technology*, Vol. 129, 2007: 644-652.

[25] Maxey WA, Kiefner JF, Eiber RJ, and Duffy AR, Ductile fracture initiation, propagation and arrest in cylindrical vessels, *Fracture Toughness*, ASTM STP 514, Part II, 1972, pp. 347-362.

[26] Kiefner JF, Maxey WA, Eiber RJ, and Duffy AR, Failure stress levels of flaws in pressurized cylinders, *Progress in Flaw Growth and Fracture Toughness Testing, ASTM STP 536*, American Society for Testing and Materials, 1973, pp. 461-481.

[27] Kiefner JF and Vieth PH, *A Modified Criterion for Evaluating the Remaining Strength of Corroded Pipe*, Final report on Project PR 3-805 to the Pipeline Research Committee of the American Gas Association, 1989.

[28] Ma et al, Analysis on the latest assessment criteria of ASME B31G-2009 for the remaining strength of corroded pipelines, *Journal of Failure Analysis and Prevention*, Vol. 11, 2011: 666-667.

[29] Fu B and Kirkwood MG, Determination of failure pressure of corroded linepipes using nonlinear finite element method, *Proceedings of the 2nd International Pipeline Technology Conference*, Vol.II, 1995, pp. 1-9.

[30] Fu B and Batte AD, New methods for assessing the remaining strength of corroded pipelines, *EPRG/PRCI 12th Biennial Joint Technical Meeting on Pipeline Research*, Groningen, The Netherlands, 1999, Paper 28.

[31] Det Norske Veritas (DNV), *Recommended Practice DNV-RP-F101 – Corroded Pipelines,* 2004.

[32] Leis BN and Stephens DR, An alternative approach to assess the integrity of corroded line pipes – Part I: current status and Part II: alternative criterion, *Proceedings of the Seventh International Offshore and Polar Engineering Conference*, May 25-30, 1997, Honolulu, Hawaii, USA.

[33] Stephens DR, Leis BN, Kure MD, and Rudland DL, *Development of an Alternative Failure criterion for Residual Strength of Corrosion Defects in Moderate-to-High Toughness Pipe*, PRCI Project Report PR3-9509, 1999.

[34] Kim WS, Kim YP, Kho YT, Choi JB, Full scale burst test and finite element analysis on corroded gas pipeline, *Proceedings of the fourth International Pipeline Conference*, Calgary, 2002, Alberta, Canada.

Copyright © 2019 by ASME

[35] Choi JB, Goo BK, Kim JC, Kim YJ and Kim WS, Development of limit load solutions for corroded gas pipelines, *International Journal of Pressure Vessels and Piping*, Vol. 80, 2003: 121-128.

[36] Benjamin, AC and Andrade EQ, Modified method for the assessment of the remaining strength of corroded pipelines, *Proceedings of 2003 Rio Pipeline Conference*, 2003, IBP413-03.

[37] Souza RD, Benjamin AC, Freire JLF, Vieira RD, Diniz JLC, Burst tests on pipeline containing long real corrosion defects," *Proceedings of International Pipeline Conference*, Calgary, Alberta, Oct 4-8, 2004.

[38] Souza RD, Benjamin AC, Vieira RD, Freire JLF, Castro JTP, Rupture tests of pipeline segments containing long real corrosion defects, *Experimental Technique*, Vol. 31, 2007: 46-51.

[39] Zhu XK and Leis BN, Influence of yield-to-tensile strength ratio on failure assessment of corroded pipelines. *Journal of Pressure Vessel Technology,* Vol. 127, 2005: 436-442.

[40] Chiodo MSG and Ruggieri C, Failure assessments of corroded pipelines with axial defects using stress-based criteria: numerical studies and verification analyses, *International Journal of Pressure Vessels and Piping*, Vol. 86, 2009: 164-176.

[41] Besel M, Zimmerman S, et al. Corrosion assessment method validation for high-grade line pipe, *Proceeding of the 8th International Pipeline Conference*, 2010, Calgary, Canada.

[42] Ma B, Shuai J, Liu D, Xu K, Assessment on failure pressure of high strength pipeline with corrosion defects, *Engineering Failure Analysis*, Vol. 32, 2013: 209-219.

[43] Zhu XK, Corrosion assessment methods for pipelines with long blunt defects, *Journal of Pipeline Engineering*, Vol. 14, 2015: 111-120.

[44] Chen Y, et al., Failure analysis of high strength pipeline with single and multiple corrosions, *Materials & Design*, Vol.67, 2015: 552-557.

[45] Yeom KJ, Lee YK, OH KH, and Kim WS. Integrity assessment of a corroded API X70 pipe with a single defect by burst pressure analysis, *Engineering Failure Analysis*, Vol. 57, 2015: 553-561.

[46] Chauhan V and Brister J, *A Review of Methods for Assessing the Remaining Strength of Corroded Pipelines*, US DOT Final Report 153A, November 2009.

[47] Bubenik TA and Rosenfeld MJ, *Assessing the Strength of Corroded Elbows*, PRCI NG-18 Report No. 206, Battelle Memorial Institute, May 1993.

[48] Kim YJ and Oh CK, Limit loads for pipe bends under combined pressure and in-plane bending based on finite element limit analysis, *International Journal of Pressure Vessels and Piping*, Vol. 83, 2006: 148-153.

Copyright © 2019 by ASME

SECTION 2
GEOHAZARD CHARACTERIZATION

INTRODUCTION

By Don West
(Don_West@Golder.com)

Geohazard characterization, in its basic form and definition for pipelines refers to the development of a conceptual/physical/numerical model of the geologic process, mechanism or geologic feature whose presence and activity may be adversely affecting a pipeline and thus resulting in a potential threat to the pipeline. The geologic model of the geohazard contains elements that describe the type, location, nature, extent, behavior and age or level of activity of the geologic process, all of which may contribute to a potential hazard or threat to a pipeline.

Knowledge of these geologic model elements, individually and in combination, is critical to geohazard characterization as it relates to pipeline threat assessment, threat mitigation planning and design, and long-term monitoring for pipeline operations. The details to which geohazard characterization, and geologic model development, are taken is dependent on the nature and objectives of the particular geohazard assessment being addressed. If, for example, a regional-scale geohazard and threat assessment is being considered, then geohazard characterization and the supporting geologic model development may be general and more conceptual. If site-specific hazard or threat mitigation is being considered, then hazard characterization and model development will be more detailed, comprehensive and complete.

Knowing the type of the geologic hazard (process) is important because not only do the different general types of potential hazards (e.g., landslide, seismic, subsidence, erosion, etc.) affect and threaten a pipeline differently, there are significant as well as subtle differences within hazard types that may result in differences regarding the potential threat. Therefore, it is critical to identify and characterize the hazard type. For example, there are at least five types of landslide movement (fall, topple, slide, flow, spread), and each type has distinctive movement characteristics that could affect a pipeline differently. If there is a combination of movement types in a single landslide complex, then it also would affect a pipeline differently than if there was a single landslide movement type. Thus, correct identification and characterization of the type of hazard (individual or in combination) is important to geohazard characterization overall.

The location of the hazard or threat relative to the pipeline is also critical for geohazard characterization. If the hazard is a landslide, does the pipeline traverse the landslide axially, perpendicularly, or obliquely, or is the pipeline located in an area subject to upslope landside retrogression? Knowing the location and orientation of the hazard relative to the pipeline (and visa versa) helps in under-standing the severity of the threat, and provides input to develop-ment of mitigation options, and long-term operational decisions.

The nature, extent (including geometry) and behavior of a particular geologic hazard relates to how the associated geologic process manifests itself in the natural environment. For example, an earth flow may be deep-seated and engage a buried pipeline, or it could be shallow and flow over a buried pipeline. Thus, understanding the vertical extent of the earth flow relative to the burial depth of the pipeline is important to assessing the risk of future or continued landslide movement to the pipeline, and this affects the need for and nature of hazard or threat mitigation.

Finally, an understanding of the age, or level of activity of a geologic process (hazard) is critical to assessing the risk or threat to a pipeline. A landslide, for example, could be active, dormant or relict. An active landslide could pose a current threat to a pipeline, but a dormant landslide could be re-activated by human activities (e.g., pipeline construction) or adverse weather events, while a relict landslide could become re-activated by significant climate change. The level of activity of tectonic surface faults, such as the recurrence interval for surface faulting events, also relates directly to the threat to a pipeline if the faults cross the alignment.

In summary, we see that geohazard characterization may take many forms and be detailed and comprehensive, or general, depending on the nature and objectives of the geohazard assessment being considered. There is, however, a need to understand and develop a model of the geologic process that describes the geologic hazard or threat to the pipeline. Within that model are important basic elements which describe, characterize and constrain the geologic process that may be a hazard; e.g., hazard type, location, nature, extent, behavior and level of activity, and which are critical to assessing the geohazard, and correctly evaluating the risk or threat to the pipeline. Correctly understanding and evaluating geohazard threats leads to more complete and appropriate decision making regarding hazard mitigation and management.

None of the six papers in Chapter 2 directly addresses geohazard characterization, however, each one makes use of some level of hazard identification, characterization and model development to support their thesis subjects and outcomes of their individual papers. The basic content and nature of each of the six papers, and

Copyright © 2019 by ASME

their relationships to geohazard characterization are discussed in the following paragraphs.

The first paper is "Ground Movement Hazards (Landslides, Subsidence) and Pipelines – An Overview and Primer," by Donald West of Golder Associates Inc. In this paper, the author seeks to provide an overview of landslide and subsidence hazards so that pipeline owners, operators and engineers use consistent and technically appropriate terminologies for these potential hazards to reduce the potential for misunderstandings that may come from the use of colloquial terms and descriptions for the hazards. In providing the overview, the author also describes a technical framework and terminology that could be used to help characterize landslide and subsidence hazards.

The second paper is "Guidelines for Managing Geohazards Affecting the Engineering and Construction of New Oil and Natural Gas Pipelines," by Rodolfo Sancio (Geosyntec Consultants), Anthony Rice (Geosyntec Consultants), Jean Audibert (Independent Consultant), David Morgan (Independent Consultant) and Jordan Rattray (Geosyntec Consultants). The authors describe a geohazard management guidance document, developed for PRCI, that incorporates management guidance, apparently for new-build projects, into a single volume of four parts: 1) the common types of geohazards affecting pipelines, 2) the typical processes used in practice to identify and characterize the geohazards throughout the various project phases, 3) the techniques used to mitigate the geohazards, and 4) the approaches and methods recommended to project teams of owners, operators, engineers,and contractors to successfully mitigate project risks. In discussing Part 2, the authors note that it is important for pipeline owners and designers to know early on in project planning about potential geohazards including their types, locations, and strategies that may be used to reduce the risks to the pipeline and the public. Because detailed characterization of geohazards is time-consuming and costly, and therefore impractical, the authors outline a phased approach to hazard characterization, that begins at the desktop level, proceeds through a field reconnaissance, and culminates in site-specific field characterization.

In the third paper, "Quantifying Surface Fault Rupture Hazard Potential at the Fault/Pipeline Intersection," Alan Hull, Eric Cannon and John Hennessey (Golder Associates Inc.), provide three case histories of fault rupture hazard assessments for pipelines, and discussions of the types of assessments that may be employed to assess fault rupture hazards (e.g., simple deterministic, simplified probabilistic, and logic-tree-based deterministic). The authors emphasize that because surface fault rupture is a complex geologic process and causes permanent ground displacements, pipeline owners and operators need to understand the location of surface faults, the amounts and directions of potential surface rupture, the width of the rupture zone, the likelihood for future rupture/creep, and the associated uncertainties in all these fault parameters. They note that the fault parameters are a critical input to any modeling of pipeline performance expected during surface fault displacement, and the post-earthquake integrity of the pipeline system, and this implies that appropriate geohazard characterization is essential.

The fourth paper is "Advanced Pipeline Geohazard Simulation: Evaluation of Pipeline Response to Lateral Slope Movements" by AbdelfettahFredj andAaron Dinovitzer ofBMT. The authors present a summary paper on the development and application of numerical modeling to study the behavior of buried pipelines interacting with soil movements caused by geohazard events. Their numerical modeling required characterization of a number of geotechnical, pipe and operational parameters including; slope, soil types, width of soil movement, ground movement pattern, pipe geometry, material grade, operating pressure and temperature (axial load).

In the fifth paper, "On Estimating Pipeline Reliability at Slope Crossings," Millan Sen and SherifHassanien (Enbridge), present three probabilistic approaches to evaluate pipeline reliability when crossing slopes where ground movement may occur. The models used in each of the probabilistic approaches required characterization and development of a number of parameters including slope movement history, slope angle, expected soil strength, triggering factors, type and nature of site mitigations, results of slope monitoring, length of pipe across the slope, crossing angle and pipe diameter.

The sixth paper is "A Comprehensive Study on the Effects of Soil Movement on Pipelines due to Manmade Hazards" by Abhimanyu Kumar, N. O. Akinci, N. Simpson and X. Dai of WS Atkins. Their paper presents the results of case histories and finite element modeling to study the effects of offshore rig operations on subsea pipelines caused by soil movement as a result of the operations. To carry out the modeling, the sub-bottom soil stratigraphy was characterized in terms of type and thickness.

Copyright © 2019 by ASME

AIM-PIMG2019-1008

Chapter 5

GROUND MOVEMENT HAZARDS (LANDSLIDES, SUBSIDENCE) AND PIPELINES: AN OVERVIEW

Donald O. West[1]
Golder Associates Inc.
Redmond, WA

ABSTRACT

Because of the linear and often long nature of pipeline systems, they cross varied terrain and geologic conditions susceptible to ground movement hazards, such as landslides and subsidence. Landslide and subsidence hazards can pose significant risks to pipelines, to public health and safety, and to the environment. Over time, a large number of releases/ruptures, or pipe damage can result from the effects of landslide or subsidence movement, if the effects are left unaddressed. The timely and systematic identification, mapping, characterization and threat classification of potential ground movement hazards are important and allow pipeline owners/operators to understand the nature, spatial distribution, magnitude, extent and level of activity of the hazards, as well as their potential impacts or risks to the pipeline systems. With this information, informed hazard mitigation and informed risk management and operational decisions can be implemented.

Ground movement hazards that could affect a pipeline system are the result of naturally-occurring, active geologic processes. The geologic processes occur in the absence of a pipeline system, and only become potential hazards if they affect, or potentially affect the integrity of an existing pipeline, or the routing, design and construction of a proposed pipeline. Although landslide and subsidence processes (i.e., hazards) are naturally-occurring; they can also be triggered or exacerbated by human activity or intervention.

Whether a particular pipeline is vulnerable (at risk) or is exposed to the effects of ground movement hazards depends on several factors, including:

- The location, type, nature, age, activity, rate and extent of the hazard relative to the pipeline and its orientation.
- The pipeline's exposure to the hazard, and the likelihood of the consequences of pipeline damage or rupture if the hazard is active.
- The type, age, material, design parameters, and construction methods and practices of the pipeline, and the operational and maintenance environment, processes and procedures of the pipeline.

Development of a basic knowledge and understanding of the types, natures, and behaviors of ground movement hazards that may impact pipeline systems, in particular landslides and subsidence, is essential for pipeline owners, operators and pipeline engineers. This paper seeks to provide that basic understanding.

INTRODUCTION

Pipeline systems, such as midstream and downstream, are often long and extensively distributed allowing them to cross a variety of topographic terrains and geologic conditions. The topography may range from flat plains to rugged and steep hills and mountains, while the geologic conditions may vary from stable to unstable soils or bedrock. Such variability in topography and geologic conditions may mean that a pipeline system is exposed to multiple threats from ground movement hazards, including landslides and ground subsidence.

In this paper, we address and provide an overview of common subsidence hazards and landslide hazards, but with a particular focus on landslides, by answering the question of what a landslide is, describing landslide types and mechanics, reviewing landslide classifications and movement characteristics, as well as the common features and terminology used to describe landslides, and also describing common triggers/causes of landslides. We will also address the common impacts that landslide movement can have on pipelines. However, we start the discussion with a general overview of common subsidence processes and their potential hazards.

COMMON SUBSIDENCE PROCESSES/HAZARDS

Land subsidence is a gradual settling or sudden sinking of the Earth's surface resulting from the subsurface movement of earth materials (rock and soil) [1,2]. More than 17,000 mi^2 (44,000 km^2) in 45 states of the United States have been affected

[1] Contact author: dwest@golder.com

Copyright © 2019 by ASME

by surface subsidence [1,2]. Ground subsidence contributes to permanent inundation of coastal lands because sea level encroaches into the settled areas; it aggravates watercourse flooding, alters topographic gradients, disrupts stream channel flows, and ruptures the ground surface [2]. These changes to the ground surface can adversely affect buried pipeline systems.

The primary subsidence hazards that could affect a pipeline system include both naturally-occurring processes, and processes that are caused or triggered by human activity. The primary types of subsidence of the ground surface that could be hazards to, and affect pipeline systems include [3]:

- Naturally-occurring processes:
    - Subsurface dissolution of bedrock
    - Thawing of permafrost
    - Soil piping (subsurface erosion)
- Human-caused/triggered processes:
    - Fluid withdrawal and aquifer system compaction
    - Collapse of underground mine workings
    - Drainage of organic soil
    - Hydrocompaction of soil
    - Localized settlement of pipe trench and pipe construction backfill

Of the naturally-occurring subsidence hazards, the dissolution of bedrock appears to be the most widespread (at least in the United States) when addressing potential threats to pipelines. Additionally, this natural process may be influenced by human activity commonly associated with changes to surface and subsurface water flow.

The dissolution typically occurs in soluble carbonate sedimentary bedrock (e.g., limestone, dolomite), or in soluble evaporite sedimentary bedrock (e.g., gypsum, salt) as a result of rainfall on the ground surface, infiltration and subsurface flow of groundwater [4,1,5,6,7,8,9]. The dissolution of bedrock in the subsurface forms small to large voids or caverns that settle or collapse, producing subsidence features at the ground surface which are termed karst topography or terrain [9,10]. The prominent surface subsidence features in karst terrain are sinkholes, which are closed depressions in the ground that have no natural external surface drainage [9,11]. According to [9], about 20 percent of the United States is underlain by karst terrain.

There are essentially three types of sinkholes produced from the dissolution of bedrock [11]: dissolution sinkholes, cover-subsidence sinkholes (Figure 1), and cover-collapse sinkholes (Figure 2). These sinkhole types are generally characterized on the basis of the presence and thickness of soil cover over the soluble bedrock, their relative size, the rate at which the sinkholes develop, and the severity of the subsidence effects of the sinkhole at the ground surface [9,11].

Dissolution sinkholes form where there is no, or only a thin soil cover (overburden) over carbonate or evaporite bedrock. The voids underlying dissolution sinkholes are generally small and may be filled with soil. Dissolution sinkholes may not affect pipeline systems, but they can provide clues to the presence of karst terrain, and possibly more significant sinkhole issues.

Cover-subsidence sinkholes involve thicker amounts of overburden over the soluble bedrock, and the soil is typically granular and permeable (e.g., sand) [11]. Cover-subsidence sinkholes tend to form gradually over time [9]. Figure 1 illustrates the surface expression of a large cover-subsidence sinkhole, formed above a dissolution void in gypsum in the Andes of Argentina. Cover-subsidence sinkholes may have initiating features that can be identified early on and mitigated by pipeline operators.

Cover-collapse sinkholes can form abruptly, over a period of hours to a few days, and result in significant damage to surface and near surface infrastructure [11,9,5]. Figure 2 shows the surface effects from a sudden cover-collapse sinkhole in a Florida neighborhood. Because cover-collapse sinkholes can occur rapidly with no apparent warning, they may not be able to be mitigated.

Figure 1. Cover-subsidence sinkhole formed from dissolution of evaporite bedrock (gypsum), Argentina (2011 photo; D. West)

Figure 2. Cover-collapse sinkhole formed from a limestone dissolution cavern, Florida (1981 photo; AP)

Surface subsidence due to the collapse of active and abandoned underground mine openings appears to represent the

Copyright © 2019 by ASME

greatest threat to pipelines from human-caused/triggered subsidence, because of their prevalence and geographic extent in North America.

Active coal mining in North America, and elsewhere, tends to employ two basic extraction methods which are room-and-pillar and longwall [12,13,14,15]. Longwall mining is termed a high-extraction process where all of the coal is removed along a working face, resulting in total collapse, and surface subsidence is expected, planned for, and can be accommodated and mitigated by pipeline operators. Alternatively, room-and-pillar mining is a low-extraction process where only about 40 to 50 percent of the coal is removed (in modern mines), and subsidence is not expected or planned [14,15]. Bauer [15], however, also notes that retreat mining is sometimes employed where about 80 to 90 percent of the coal is removed from a room-and-pillar mine, which may result in surface subsidence.

When an underground mine opening collapses, the collapsed opening may migrate to the surface inducing tensile and compressive stresses, and forming subsidence features, as illustrated on Figures 3 and 4. Figure 5 shows the effects of compressive stresses above an active longwall coal mine on a blown-down, abandoned natural gas pipeline.

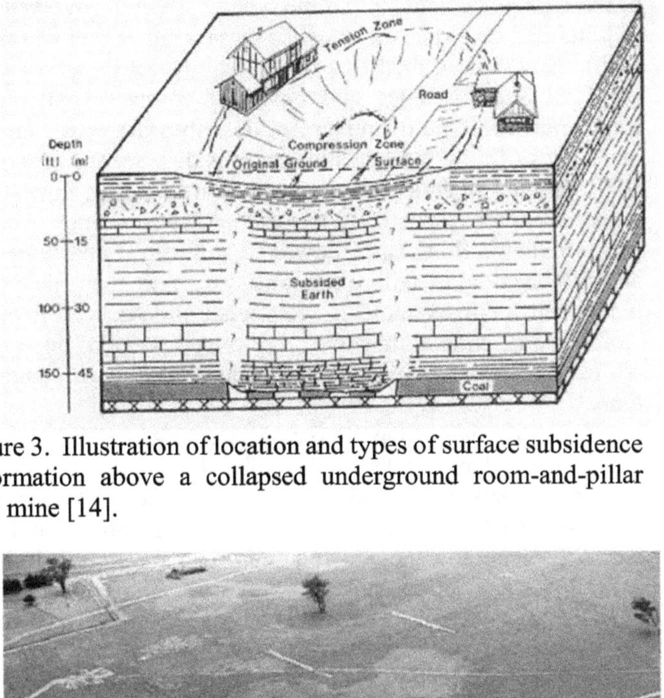

Figure 3. Illustration of location and types of surface subsidence deformation above a collapsed underground room-and-pillar coal mine [14].

Figure 4. Surface effects from subsurface block caving above a mine in Australia [16]

Figure 5. Uplifted pipe and ground breakout due to underground longwall mining [3].

POTENTIAL LANDSLIDE HAZARDS

In this section, we address landslide types, terminology, classification, characteristics and triggers/causes. The purpose is to provide pipeline owners/operators and pipeline engineers with a consistent and common understanding of what potential landslide hazards may be to pipelines.

What Is A Landslide?

Landslides are complex natural geologic phenomena. The complexity comes from the fact that there are numerous types of landslides and combinations of landslides, varied and complex geomorphic and geologic environments in which landslides occur, numerous causes or triggers of landslides, and varied sizes and behaviors of landslides, once triggered (e.g., [17, 18]). The variability and complexity of landslides in general contribute to the difficulty in understanding their types, natures, location, extents, age and timing of movement.

Geologists, engineers, and other professionals often rely on unique and slightly differing definitions of landslides, and this diversity in definitions reflects the complex nature of the many disciplines associated with studying landslide phenomena. For our purposes, "landslide" is a general term used to describe the downslope movement of soil, rock, and organic materials under the effects of gravity, and also describe the landform that results from such movement. A landslide occurs when the downslope component of forces (driving forces) acting on the slope exceeds the resistance of the materials underlying the slope (resisting forces). Driving forces can be increased by changes to slope geometry (e.g., erosion or grading of the toe), or by increased loading on the slope (e.g., placing fill on the slope or earthquake shaking). Resisting forces can be reduced by mechanical and chemical weathering of the material underlying the slope (which weaken the material), and by increased/raised groundwater/pore water pressure levels in the slope.

Landslides may be shallow (at or near the ground surface; i.e., only a few feet thick) or they may be deep-seated (several

Copyright © 2019 by ASME

tens to hundreds of feet thick), and may occur extremely rapidly, in seconds, minutes or hours, and may be episodic, or as extremely slow, on-going processes occurring over months, years and centuries. Temporal physical changes to the environment such as intense precipitation events, earthquakes, undercutting and erosion by streams, and/or activities of humans can initiate or trigger landslides. The stability of a slope is directly related to the:

- Material properties of the soil and rock underlaying the slope,
- State of stress and stress history of the slope,
- Surface and groundwater/pore water conditions within the slope and along rupture surfaces, and
- Slope geometry.

Landslide Classification

Landslides have colloquially been termed "slips," "slumps," "mudslides," or "slope anomalies" by pipeline operators. However, these terms do not fully describe the variable natures, mechanics and behaviors of the landslides that can occur along pipelines, possibly leading an incomplete understanding of the potential impacts that landslides may have on pipelines. Thus, a more descriptive and consistent landslide classification is desirable.

Varying classifications of landslides are associated with specific mechanics of slope movement and the properties and characteristics of the types of movement. Regardless of the exact definition used or the type of landslide under discussion, understanding the basic parts of a typical landslide is instructive, and allows numerous parties to communicate effectively regarding landslides. Figure 6 shows the physical positions and the most common terms used to describe the unique parts of a landslide.

Landslides initiate in the source area (sometimes called the zone of depletion) and move downslope to deposit in the toe area (sometimes called the zone of accumulation). Landslide movement in the source area displaces originally intact ground along a surface of rupture, while deposition in the accumulation zone occurs over intact terrain along a surface of separation [17, 18].

The upper limit of the source area is the headscarp (also called the main scarp) of the landslide, and the lateral limits are defined by the right and left flanks. The main body of the landslide almost always exhibits disturbed hummocky topography which stands out from more uniform, undisturbed adjacent terrain, as illustrated on Figure 6.

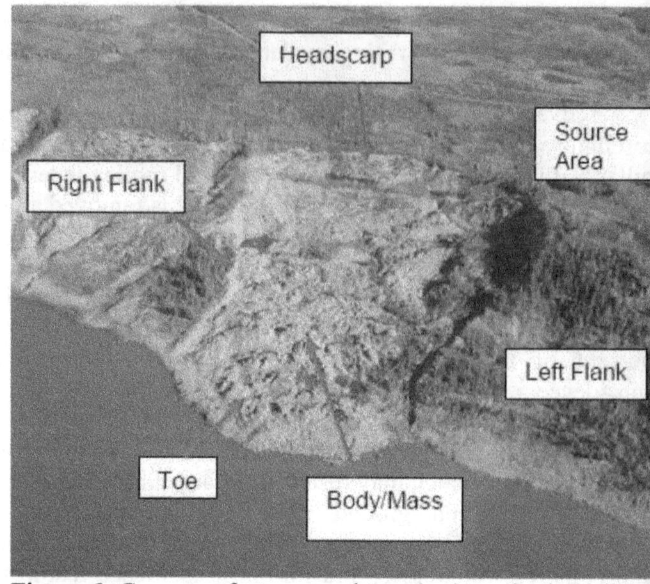

Figure 6. Common features and terminology for landslides (2006 photo; D. West)

Landslides can be classified into different types on the basis of the type of movement and the type of material involved [18, 17]. In brief, material in a landslide mass is either ***rock*** or ***soil*** (or both). Soil is described as ***earth*** if mainly composed of sand-sized or finer particles and ***debris*** if composed of coarser fragments. The type of movement describes the actual internal mechanics of how the landslide mass is displaced: ***fall***, ***topple***, ***slide***, ***spread***, or ***flow***. Thus, landslides are described using two terms that refer respectively to material and movement (that is, rockfall, debris flow, earth flow, and so forth). Landslides may also form a complex failure encompassing more than one type of movement. For example, a "rotational slide-debris flow," or "translational slide-earth flow." For the purposes of this report, we treat "type of movement" as synonymous with "landslide type."

Based on more than 30 years evaluating landslide hazards along more than 90,000 miles of natural gas transmission pipeline systems in the United States, Canada and Mexico, the landslides that seem to commonly affect pipelines in North America, generally include rotational and translational slides, earth flows, and debris flows. General descriptions of each of the landslide types are discussed below, based primarily on [18].

Rotational Slide: A landslide on which the surface of rupture is curved upward (spoon-shaped) and the slide movement is more or less rotational about an axis that is parallel to the contour of the slope (Figures 7 and 8). The displaced mass may, under certain circumstances, move as a relatively coherent mass along the rupture surface with little internal deformation. The head of the displaced material may move almost vertically downward, and the upper surface of the displaced material may tilt backwards toward the headscarp. The velocity of rotational slides may vary from extremely slow (less than 1 foot every 5 years) to moderately fast (5 feet per month), to rapid (e.g., 6 feet per hour). Figure 8 shows a rotational landslide that displaced a

Copyright © 2019 by ASME

road and engulfed a neighborhood. Rotational slides are the type of landslide that may colloquially be known as a "slip" or "slump."

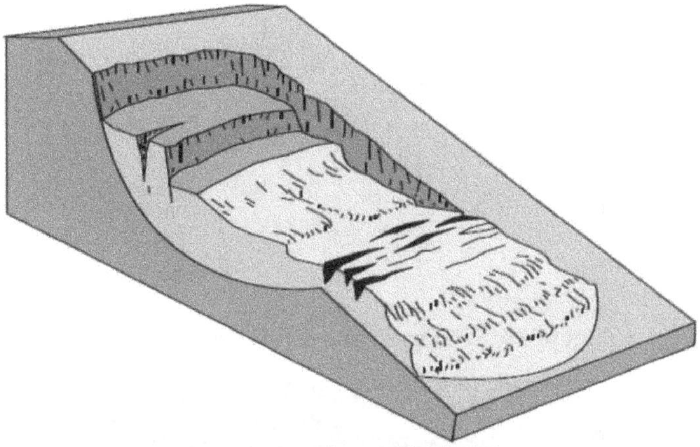

Figure 7. Schematic of a rotational slide [18]

Figure 8. Example of a rotational slide, California (1997 photo; R.L. Schuster)

Translational Slide: The mass in a translational landslide moves out, or down and outward, along a relatively planar surface with little rotational movement or backward tilting (Figures 9 and 10). This type of slide may progress over considerable distances if the surface of rupture is sufficiently inclined, in contrast to rotational slides, which tend to quickly restore the slide equilibrium. The material in the slide may range from loose, unconsolidated soils to extensive slabs of rock, or both. Translational slides commonly move or rupture along geologic discontinuities such as faults, joints, bedding surfaces, or the contact between rock and soil. In northern, or arctic environments the slide may also move along the permafrost layer. Movement velocities may initially be slow (5 feet per month) but many are moderate in velocity (5 feet per day) to extremely rapid (e.g., >10 miles per hour). With increased velocity, the landslide mass of translational slides may disintegrate and develop into a debris flow, as shown on Figure 10, where a translational slide, which initiated within dipping volcanic lava flows, locally transitioned to a debris flow, and flowed out over an alluvial plain.

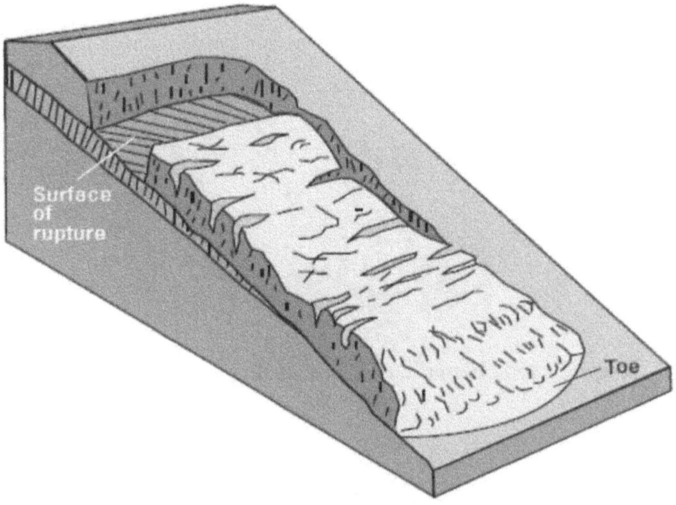

Figure 9. Schematic of a translational slide [18]

Figure 10. Example of a translational slide transitioned to a debris flow, Washington (1977 photo; D. West).

Earth Flow: Earth flows (Figures 11 and 12) can occur/initiate on gentle to moderate slopes, generally in fine-grained soil, commonly clay or silt, but also in very weathered, clay-bearing bedrock. The mass in an earth flow moves as a plastic or viscous flow with strong internal deformation. Susceptible marine clay (quick clay) when disturbed is very vulnerable and may lose most shear strength with a change in its natural moisture content and suddenly liquefy, potentially inundating large areas and flowing for several kilometers. The earth flow source or initiation zone commonly increases through headscarp retrogression, upslope of the original zone of disturbance. Slides or lateral spreads may also evolve downslope into earth flows. The velocity of earth flows can range from very slow (creep) to extremely rapid (e.g., >10 miles per hour).

Copyright © 2019 by ASME

H

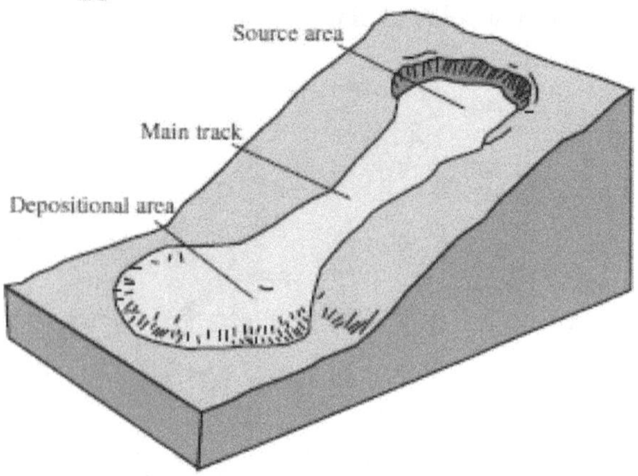

Figure 11. Schematic of an earth flow [19]

Figure 12. Example of an earth flow, Washington (1995 photo; D. West).

Debris Flow: A form of rapid mass movement in which loose soil, rock and sometimes organic matter combine with water to form a slurry that flows downslope (Figures 13 and 14). They have been informally and inappropriately called "mudslides" because of the large quantity of fine material that may be present in the flow. Occasionally, as a rotational or translational slide gains velocity and the internal mass loses cohesion or gains water, it may evolve into a debris flow. Dry flows can sometimes occur in cohesionless sand (sand flows). Debris flows can be deadly because they can be rapid (e.g., 6 feet per hour) to extremely rapid (e.g., 35 mph) and may occur without warning.

Figure 13. Schematic of a debris flow [18]

Figure 14. Example of a debris flow, Washington (2006 photo; D. West).

Landslide Causes/Triggers

Landslides are most commonly caused or triggered through natural processes but can also be triggered by human activities. Natural causes can be divided into three main groups [17, 18]. These three groups are: ***geological, morphological***, and ***physical*** causes.

Common geological causes/triggers of landslides include weak soil and rock materials, weathered soil and rock, and contrast(s) in groundwater permeability in slopes on which landslides occur. Common morphological causes or triggers include fluvial (stream) erosion of a slope toe and vegetation removal by fire or drought (which increases groundwater infiltration and runoff). Prolonged exceptional precipitation, intense precipitation, elevated groundwater and earthquake shaking make up common physical triggers of landslides.

Copyright © 2019 by ASME

Although landslide processes (i.e., hazards) are naturally-occurring, they can also be triggered or exacerbated by human activity. For example, the construction and operation of a road or a pipeline on a steep slope, or through a landslide can trigger landslide movement on ground that would have otherwise been stable. Another example is irrigation on or above a slope that raises the local groundwater table and groundwater pore pressures, which can lead to landslide movement. The construction of the pipeline itself may alter the surface water flow environment (through right-of-way grading and water bar placement and orientation) and near-surface ground water regime (construction of pipe ditch and trench plugs/breakers) that may have an adverse effect on ground stability.

COMMON LANDSLIDE EFFECTS/IMPACTS ON PIPELINES

Whether a pipeline is vulnerable, or at risk to the effects of landslide movement depends on many factors including [3]:

- Pipe burial depth (including any above-ground pipe)
- Pipe diameter, wall thickness, grade of steel, specification/quality of welds, and coating type
- Presence of interacting anomalies
- Dimensions and geometry of the pipe trench, and the trench backfill material properties (e.g., grain size, grain/particle shape, consistency/density, moisture content and strength)
- Landslide type, dimensions, material properties, moisture content, and amount and velocity of movement
- Landslide depth relative to pipe burial depth
- Pipeline crossing location on or adjacent to landslide, and
- Pipeline orientation relative to the direction of landslide movement and landslide mass deformation (axial, transverse, oblique).

Experience with landslides and pipelines in North America indicates that there are several common ground movement scenarios for a landslide to potentially damage or rupture a pipeline [3]. The three apparently most commonly encountered general pipeline/landslide interaction scenarios are illustrated in Figure 15. The landslide movement may be parallel to the orientation of the pipeline (axial), it may be oblique to the pipeline, or it may be perpendicular (transverse). In the three scenarios depicted on Figure 15, the pipeline is engaged by the landslide; i.e., the pipeline is within the moving landslide mass.

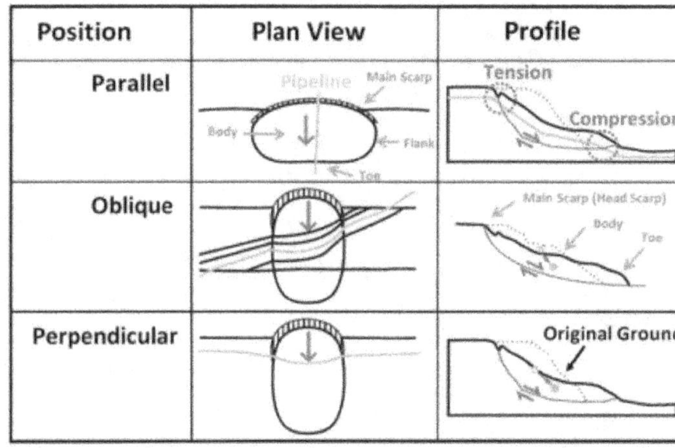

Figure 15. Common pipeline/landslide interaction scenarios (modified from [3])

In the parallel movement scenario, discrete sliding or slow flow would impart tensile stresses on the pipeline at or near the top of the slope, while compressive stresses would be induced at the toe of the landslide, and in this case, the base of the slope. Internal movement in the landslide mass may induce variable tensile and compressive stresses on the pipe. Where the landslide occurs at the depth of burial of the pipeline, landslide and backfill material may be dragged across and along the pipe possibly damaging the coating.

Where the pipeline is oriented obliquely across the landslide and discrete sliding or slow flow occur, landslide-induced loading on the pipeline would generally depend on the skew angle of the oblique movement relative to the pipeline. The pipeline may experience lateral shear stresses where it traverses the lateral limits of the landslide, and tensile and compressive stresses within the internal landslide mass.

In the perpendicular, or transverse scenario, discrete sliding or slow flow movement typically puts the pipe in bending or shear stresses at the slide flanks, and, depending on the nature of the movement, may put the pipe variably in tensile or compressive stresses in the body of the landslide as the landslide mass extends downslope. Figure 16 shows the approximately 4 feet of lateral deformation/deflection of a pipeline centerline (identified from field wand locates) by the transverse or perpendicular movement of a landslide across the pipe alignment.

Copyright © 2019 by ASME

FIGURE 16. Lateral deflection of a pipeline centerline by transverse landslide movement (2012 photo; D. West)

SUMMARY

Subsidence and landslide ground movement hazards can affect the integrity of pipeline systems up to the point of release or rupture. To manage the risk associated with these ground movement hazards, it is critical for pipeline owners, operators and engineers to understand the types, natures and behaviors of landslide and subsidence movements to effectively develop mitigation approaches, strategies and methods in order to assure that pipeline integrity is maintained.

ACKNOWLEDGEMENTS

This paper has benefitted greatly from the many years of geologic hazard investigations for, and various discussions with Kent Ryan (formerly with Williams Northwest Pipeline), Bill Liu (currently with TransCanada), and Alex Mckenzie-Johnson (currently with Kinder Morgan). In addition, the many discussions with sponsors, participants and authors of the Joint Industry Project (JIP) on "Management of Ground Movement Hazards for Pipelines" (completed in 2017), have been especially insightful and helpful. Furthermore, my geologic hazard colleagues at Golder, in particular, John deLaChapelle and Bailey Theriault, have been my greatly appreciated sounding boards for all things geologic as they relate to pipelines.

REFERENCES

[1] Galloway, D.L., D.R. Jones, and S.E. Ingebritsen, eds. *Land Subsidence in the United States*, U.S. Geological Survey Circular 1182, 1999.

[2] Galloway, D.L., G.W. Bawden, S.A. Leake, and D.G. Honegger. *Chapter 2, Land Subsidence Hazards*, in Baum, R.L., D.L. Galloway, and E.L. Harp, eds., Landslide and Land Subsidence Hazards to Pipelines: U.S. Geological Survey Open-File Report 2008-1164, p. 33-106, 2008.

[3] Wang, Y.-Y., B. Wang, K. Kotian, D. West, D. Dewar, W. Webster, S. Rapp, J. Hart, A. Mckenzie-Johnson, and T. Fore. *Management of Ground Movement Hazards for Pipelines*, CRES Project No. CRES-2012-M03-01, Submitted to a Group of Sponsors, 2017.

[4] Ege, J.R., 1984. *Mechanisms of surface subsidence resulting from solution extraction of salt*, in Holzer, T.L., ed., Man-induced land subsidence: Geological Society of America Reviews in Engineering Geology, p. 203-221, 1984.

[5] U.S. Geological Survey (USGS). *Land Subsidence in the United States*, Fact Sheet-165-00, December 2000.

[6] U.S. Geological Survey (USGS). *Sinkholes,* Fact Sheet 2007-3060, July 2007.

[7] U.S. Geological Survey (USGS). *The Science of Sinkholes, Science Features*, http://www.usgs.gov/blogs/features/usgs_top_story/the-science-of-sinkholes/?from=texlink, last modified September 14, 2011.

[8] U.S. Geological Survey (USGS). *Sinkholes: The USGS Water Science School*, http://ga.water.usgs.gov/edu/sinkholes.html, last modified 08-Mar-2013.

[9] Kentucky Geological Survey. *Karst Is a Landscape*, University of Kentucky, http://www.uky.edu/KGS/water/general/karst/karst_landscape.htm, last modified September 21, 2006.

[10] Zhou, W. and B.F. Beck. *Management and mitigation of sinkholes on karst lands: an overview of practical applications,* Environmental Geology, v. 55, p. 837-851, 2008.

[11] Nelson, S.A. *Subsidence: Dissolution & Human Related Causes,* EENS 3050, Natural Disasters: Tulane University, 7/9/2012.

[12] Gray, R.E. and R.W. Bruhn. *Coal mine subsidence – eastern United States,* in Holzer, T.L., ed., Man-induced land subsidence: Geological Society of America Reviews in Engineering Geology, p. 123-150, 1984.

[13] Bell, F.G. and D.D. Genske. *The Influence of Subsidence Attributable to Coal Mining on Environment, Development and Restoration: Some Examples from Western Europe and South Africa,* Environmental & Engineering Geoscience, v. VII, n. 1, p. 81-99, 2001.

[14] Bauer, R.A. *Mine Subsidence in Illinois: Facts for Homeowners*, Illinois State Geological Survey, Circular 569, 28 p, 2006.

[15] Bauer, R.A. *Planned Coal Mine Subsidence in Illinois: A Public Information Booklet*, Illinois Department of Natural Resources, Circular 573, 2008.

[16] Mattox, A., D. Coil, N. Hoagland, and B. Higman. *Block Caving,* ttp://www.groundtruthtrekking.org/Issues/MetalsMining/block-caving-underground-mining-method.html, last modified 21 March 2013.

[17] Cruden, D.M. and D.J. Varnes. *Chapter 3, Landslide Types and Processes*, in Turner, K.A., and Schuster,

Copyright © 2019 by ASME

R.L., eds., Landslides – Investigation and Mitigation, Special Report 247: Transportation Research Board, National Research Council, National Academy Press, p. 36-75, 1996.

[18] Highland, L.M and P. Bobrowsky. *The Landslide Handbook – A Guide to Understanding Landslides*, U.S. Geological Survey Circular 1325, 129 p, 2008.

[19] U.S. Geological Survey. *Landslide Types and Processes*, Fact Sheet 2004-3072, July 2004.

Copyright © 2019 by ASME

AIM-PIMG2019-1040

Chapter 6

GUIDELINES FOR MANAGING GEOHAZARDS AFFECTING THE ENGINEERING AND CONSTRUCTION OF NEW OIL AND NATURAL GAS PIPELINES

Rodolfo B. Sancio
Geosyntec Consultants, Inc.
Houston, Texas, USA

Anthony H. Rice
Geosyntec Consultants, Inc.
Seattle, Washington, USA

Jean Audibert
Independent Consultant
Kerrville. Texas. USA

David Morgan
Independent Consultant
Workington, Cumbria, UK

Jordan Rattray
Geosyntec Consultants, Inc.
Oakland, California, USA

ABSTRACT

Pipelines are inextricably exposed to geohazards given that they are long linear facilities that encounter variable terrain and geologic conditions, including watercourses. As such, reconnaissance, feasibility, and engineering design studies for new or existing pipelines need to include analyses that identify, characterize, and mitigate the potential long-term effects of geohazards on pipeline integrity. Also, there is a growing recognition that potential short-term effects of geohazards during construction, need to be addressed. Developing a pipeline project without considering geohazards early in the project cycle and mitigating them, as appropriate, is likely to lead to greater construction costs, contractor claims, schedule delays, and ultimately higher operational costs.

The technical literature includes numerous publications describing approaches, methodologies, and recommendations, for identifying, characterizing, and mitigating geohazards, as well as case histories of pipelines affected by geohazards and how their effects were mitigated. However, these publications are not comprehensive, separately provide information for identification, characterization, and mitigation, do not include all geohazards affecting pipelines, or may not have been developed specifically for pipelines.

Under the auspices of the Pipeline Research Council International, Inc. (PRCI), a guidance document has been prepared by the authors to efficiently incorporate geohazard management guidance into a single volume addressing: i) the types of geohazards affecting pipelines, ii) the typical process used in practice to identify and characterize the geohazards throughout the various project stages, iii) the techniques used to mitigate the geohazards, and iv) the approaches and methods recommended to project teams of owners, operators, engineers, and contractors to successfully mitigate project risks. The guidance document is intended for a wide audience that includes technical and nontechnical managers at pipeline owning and operating companies, geoscience and engineering teams involved in studies and assessments of pipeline integrity hazards, and construction staff, regulators, lenders, and other stakeholders involved in executing pipeline projects. This paper presents an abridged version of the PRCI guidance document [1] with updates.

INTRODUCTION

A geohazard is defined as a naturally occurring or human-triggered geologic process of a given intensity that has caused, or may result in, damage to the operation of a pipeline or associated facility; may impede the operation of a pipeline or associated facility; or may negatively affect the land, structures, and health and safety of landowners and other stakeholders who reside near the pipeline. Pipelines are inextricably exposed to geohazards given that they are long linear facilities that encounter variable terrain and geologic conditions, including watercourses. As such, reconnaissance, feasibility, and engineering design studies for new pipelines must include analyses that identify, characterize, and mitigate the potential effects of geohazards on pipeline integrity.

Developing a pipeline project without considering, studying, inventorying, and mitigating geohazards early in the project cycle has significant potential to increase risks to the public, heighten construction risks to workers, lead to greater construction costs, contractor claims, schedule delays, and ultimately higher operational costs. Avoiding a geohazard during design and construction can be orders of magnitude less costly than mitigation during operation.

Copyright © 2019 by ASME

The following sections will review and discuss i) the types of geohazards affecting pipelines, ii) the typical process used in practice to identify and characterize the geohazards throughout the various project stages, iii) the techniques used to mitigate the geohazards, and iv) the approaches and methods recommended to project teams of owners, operators, engineers, and contractors to successfully execute projects.

LANDSLIDES

The term "landslide" is used to describe the downslope movement of soil, rock, and organic material under the effects of gravity (Figure 1) and the landform that results from such movement [2]. Alternate terms such as "mass movement," "slope instability," "unstable area," or "slope failure" are also commonly used. However, we recommend against using the term "slope failure" during geohazard studies because "failure" can be misconstrued to imply that a pipeline system is unable to perform satisfactorily in the presence of a landslide; whereas, depending on circumstances, a landslide may not adversely affect a pipeline, or the landslide can be avoided entirely through pipeline routing decisions.

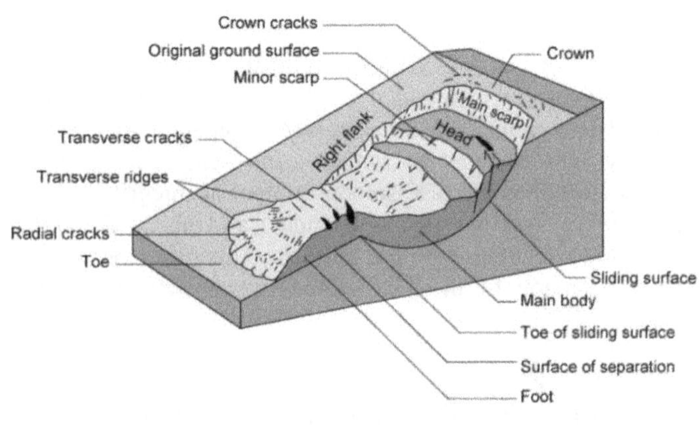

Figure 1: Isometric Illustration of a Rotational Landslide [2]

The landslide mass may be composed of soil, rock, or both (i.e., earth). The type of movement may comprise slides, spreads, flows, falls, and topples.

Sloping ground is subjected to destabilizing gravitational forces. The ground will remain static as long as the destabilizing forces remain smaller than the resisting forces, which derive from the shear strength of the earth mass. A landslide is triggered when the destabilizing forces become larger than the resisting forces, which occurs when the destabilizing forces increase or the resisting forces decrease.

There are three main contributors to natural landslide susceptibility: i) slope inclination, ii) soil/rock strength, and iii) surface water and groundwater conditions. Other important triggers for landslides include inertial forces caused by earthquakes and the reduction of soil shear strength due to seismically induced liquefaction. Therefore, a proper landslide susceptibility analysis should take into consideration potential triggering factors, such as ground motions and changes in surface water and groundwater, slope inclination, and strength of the slope materials.

Landslides can present a hazard to pipelines and associated facilities in the form of pipe-induced compressive or tensile stresses and buckling that can ultimately result in pipeline rupture or impact operation of pipelines and associated facilities [3]. Figure 2 [4] presents three sketches showing the locations of pipeline bending, tension, and compression when a pipeline is subjected to ground displacement caused by landslides.

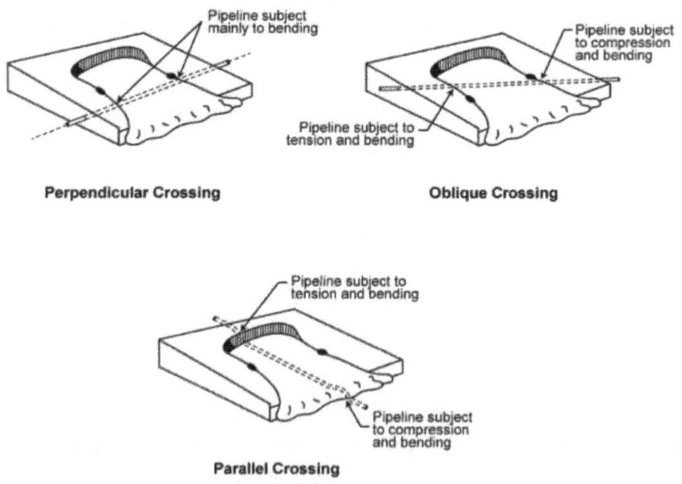

Figure 2. Principal modes of soil-pipeline interaction at landslides (modified from [4])

There is uncertainty in the process of identifying and characterizing landslides, particularly during desktop studies and even after conducting field observations. Therefore, describing a potential issue as a failure a priori, could have unintended effects on routing and geohazard management. Similarly, early liberal use of the term "landslide" can lead to the perception that a pipeline route is more hazardous than it is.

Landslides vary widely in type and size, with small, surficial landslides presenting less of a threat to pipelines than larger deep-seated landslides. The severity of a landslide hazard to a pipeline depends on the location of the pipeline with respect to the unstable area, amount of permanent ground displacement, relative direction of movement, soil type, strength and density of the earth materials, and the design details for the pipeline, such as pipe diameter, wall thickness, type of weld, and weld stress-strain characteristics.

In addition to the direct effects that landslides can have on the integrity of the pipeline, landslides emanating from the pipeline right-of-way (ROW) due, for example, to uncontrolled side casting and instability of spoil storage areas, may affect the surrounding community and, even if naturally occurring, may be perceived as the responsibility of owner/operator.

Copyright © 2019 by ASME

It is in the interest of the owner and operators to recognize the requirement for spoil disposal, both temporary and permanent, during the early phases of the project to ensure locations can be selected based on minimal geohazard risk. Transferring this requirement to the contractor may result in the selection being schedule driven with the potential for residual risks being carried forward into the construction phase or the operation phase.

WATERCOURSE MIGRATION

Watercourses (streams) are morphologically active and continually evolving in response to imbalances in sediment supply, changes in watershed hydrology, and local hydraulic conditions.

Changes in land use within a watershed over the design life of the pipeline or during short-term events such as wildfires can lead to changes in the equilibrium of streams in the watershed, which can lead to changes in its past patterns of vertical and/or horizontal migration or can cause migration where none was present previously. For example, urbanization and development within a watershed can result in reduced infiltration of precipitation which can lead to greater runoff volumes and more rapid flow rates. Deforestation within a watershed due to timber harvesting, farming or following a wildfire can also create changes in stream equilibrium. Loss of vegetation can cause increased surface erosion during precipitation events and greater runoff volumes because of a decreased evapotranspiration from natural vegetation.

Pipelines are either buried beneath the stream bed or elevated over the stream channel. When the pipeline becomes exposed within the stream, it is subject to loss of integrity that may be caused by such things as hydrodynamic loading, debris impact, fatigue caused by vortex induced vibrations, loss of support caused by stream bed erosion, and loss of cathodic protection or increased corrosion [5].

Mitigating these hazards by burying the pipeline beneath or by elevating it above the floodplain must consider potential channel changes over the design life of the pipeline, including long-term vertical migration (e.g., scour, degradation, and aggradation), short-term bed mobility, and horizontal migration (e.g., bank erosion, encroachment, avulsion, and meander cutoffs).

For example, Figure 3 (modified from [6]), shows the migration of an active channel from 1930 to 2000. These channel migration changes (horizontal and vertical) can occur over long timeframes due to natural channel evolution or over shorter timeframes in response to episodic events such as floods and mass wasting events (i.e., debris flows) or anthropogenic activities (e.g., infrastructure construction (including pipelines), flow diversions, channel confinement, gravel/aggregate mining).

In high-relief areas, landslides that develop outside of the ROW may temporarily dam watercourses. When these dams breach, the resulting hydrograph may be higher than what is predicted solely from extreme rainfall events and may cause the stream to migrate and affect the pipeline at a crossing some distance downstream. As such, crossing design in high-relief areas should consider the combined effects of landslides and stream migration resulting from dam break flooding.

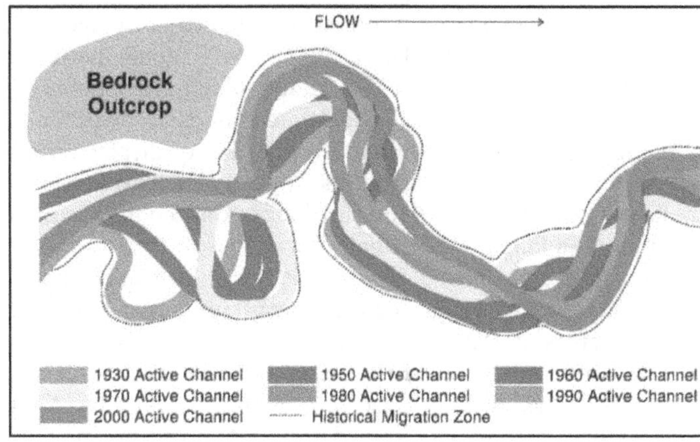

Figure 3. Evolution of an active stream channel between 1930 and 2000 (modified from [6])

SUBSIDENCE

Subsidence is a time-dependent process, either natural or human induced, in which there is a lowering of the ground surface over time in response to the removal of gas, liquid, or solid matter.

Factors that can cause or trigger subsidence include the following:

- Mining: removal of solid matter from the subsurface, which produces a void that may collapse and cause downward movement of the ground surface.
- Sinkholes: are spherical, ellipsoidal, or tubular shaped depressions that develop either naturally in relatively soluble evaporite (salt and gypsum) deposits and carbonate (limestone and dolomite) rocks or anthropogenically through processes that accelerate the dissolution of these deposits.
- Subsurface Fluid Withdrawal: removing fluids (water, oil, gas) from the sub-surface causes a pressure reduction in the fluid, which causes an increase in the vertical stress, which in turn is accompanied by a reduction in volume that manifests itself at ground surface as downward vertical displacements.
- Permafrost Thawing: climatic or anthropogenic causes, such as the heat released by the contents of the pipeline, can lead to increases in temperature that thaw the permafrost layer and cause volume reduction and subsequent downward vertical displacement of the ground surface.

Although subsidence generally refers to a downward movement of the ground surface, lateral ground movement can also accompany subsidence. These movements can result in deformation of a pipeline that crosses the affected area and potentially lead to overstressing and failure of the pipeline [3]. In many cases, the vertical movement is spread out over a large enough length of pipeline that the imposed bending stress is relatively minor, however the horizontal movement can drive a buried pipeline beyond its longitudinal (axial and bending) strain capacity.

Copyright © 2019 by ASME

ACTIVE FAULTS

Active faults are fractures of the earth's crust defined by planes or rupture zones that are considered active for purposes of pipeline design because they have exhibited displacement within the Holocene, or the last 11,000 years. Tectonic faults are the most common type of active faults, which are capable of large displacements over very short time periods. Growth faults are active faults that exhibit much slower rates of displacement, and yet, can impose displacements beyond tolerable levels to pipelines, particularly when the pipeline is expected to remain in operation for a long period of time.

Active faults induce permanent ground displacements (PGD) and represent a significant geohazard for buried pipelines. Pipelines crossing fault zones must be allowed to deform longitudinally and in flexure to accommodate ground surface offsets [7]. The sketch on Figure 4 illustrates how fault displacements can subject buried pipes to a combination of bending and shear.

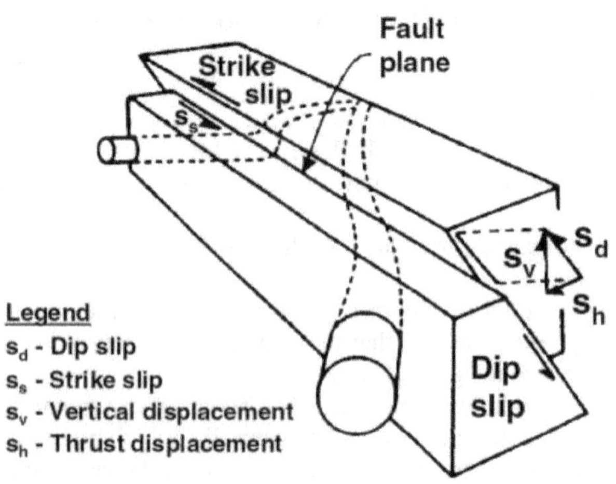

Figure 4. Principal modes of soil-pipeline interaction at fault crossings (modified from [7]).

Many of the world's most active faults have been studied (e.g., [8], and interactive maps of faults are available online (e.g., Quaternary faults (may have moved in the past 2.6 million years) in the US [9] and seismogenic faults in Europe [10]). These Quaternary and seismogenic faults are not necessarily active as defined above for pipeline design, and any such faults crossed by a pipeline will need to be further studied to evaluate if mitigation is required. Likewise, a focused study of a ROW may reveal other active faults that had not been identified in previous evaluations.

EXPANSIVE AND COLLAPSIBLE SOILS

Expansive soils are those that swell or contract at constant applied loading following an increase or decrease in soil water content. Their swell or shrinkage potential derives from volume changes associated with the adsorption or release of water by the constituent clay minerals. In a natural environment, swelling and shrinking occur in response to cycles of wet and dry weather (rainfall and evaporation).

Only the upper portion of the soil profile, described as the "active zone," is affected by swelling or shrinking due to changes in water content. Pipelines buried within the active zone will cyclically move upward and downward following periods of soil wetting and drying. These displacements will induce stresses at locations where differential swell/shrinkage occurs, particularly at above ground connections. These stresses and displacements are typically ignored by modern day designers of welded steel pipelines, but some cast iron gas pipelines installed in the late 19th century and early 20th century are suspected to have ruptured due to such stresses [11].

Collapsible soils cover significant areas around the world and are typically present in arid regions (e.g., [12]). They are unsaturated soils that can withstand relatively high vertical stress without showing significant change in volume. These soils are susceptible to a large and sudden reduction in volume upon wetting and may undergo considerable reduction of shear strength and deformation modulus. This process is also described as hydrocompaction and can induce displacement and stresses on pipelines at locations where differential vertical movement occurs.

SOIL LIQUEFACTION

Soil liquefaction is a term used to describe the temporary reduction of the shear strength of soil that is typically caused by seismically induced ground motions. Liquefaction also reduces the volume of the soil, and this reduction may manifest itself as vertical settlement. Three factors are required for liquefaction to develop: i) the presence of susceptible soils such as sand and other granular materials in a relatively loose state, ii) soil saturation, and iii) sufficiently large cyclic ground motions (typically from an earthquake but could also be caused by construction equipment or blasting operations).

A reduction in soil shear strength may lead to lateral spreads or flow slides, typically on shallow slopes near streams, and flotation or sinking of the pipeline. Liquefaction is not discussed any further in this paper as it is not considered a geohazard on its own, rather it is a trigger for landslides, flotation, or sinking.

GEOHAZARD MANAGEMENT

Overview

Each geohazard that can affect a pipeline has a likelihood of occurrence (i.e., an associated probability of occurrence). Occurrence of the geohazard (with its associated stresses and displacements on the pipeline) does not necessarily mean that the pipeline will be irreversibly damaged or lose its ability to contain the product it is conveying, because some geohazards will not impose significant stresses or displacements on the pipeline. Moreover, when the geohazard cannot be avoided, it may be possible to mitigate through engineering design.

The conditional susceptibility to loss of pipeline integrity upon occurrence of a geohazard is defined as vulnerability and is probabilistically quantified with a value between 0 and 1,

Copyright © 2019 by ASME

where 1 denotes certainty that the pipeline will lose containment upon development of the stresses or displacements caused by the geohazard. Risk is therefore measured as the product of i) the probability of occurrence of the geohazard, ii) the vulnerability, and iii) the consequence(s) of loss of containment. Consequences are typically expressed as volume of product lost, monetary losses and reputation damage to the owner/operator, loss of life, and financial costs for spill cleanup. Financial costs for spill cleanup include legal fees and reimbursement for environmental, property, and personal damages. Consequences are also uncertain; they too can be expressed in probabilistic terms.

For onshore oil and gas pipeline projects, risk represents the level of uncertainty with respect to unplanned events or unforeseen circumstances such as ground conditions and exposure to geohazards. Geohazard-driven risk is therefore one of the major considerations while developing a pipeline project.

Accurately quantitatively estimating the geohazard likelihood (i.e., probability of occurrence) is a complex process that requires characterizing the triggering mechanism and site-specific data, which are typically not available over the large area associated with a pipeline ROW. Therefore, typical practice involves using semiquantitative or qualitative hazard assessments that rely on the expert judgment of the geologists and engineers who are evaluating the geohazards. These assessments are done within a framework that considers factors that affect hazard triggering such as for landslides, subsurface lithology and stratigraphy, slope inclination, precipitation frequency and intensity, groundwater conditions, and seismicity. The output of these hazard analyses is expressed on a relative hazard scale that can be as simple as low hazard, moderate hazard, and high hazard, or more complex with an annual probability of occurrence. However simple, this hazard categorization is very important when incorporating geohazard assessment into pipeline routing decisions. For example, a route that requires traversing terrain with a higher percentage of moderate to high hazard areas is likely to be costlier than a route that traverses a lower percentage of moderate to high hazard areas.

Geohazard management refers to the process of identifying, characterizing, and mitigating geohazard-driven risks. Given the three components of risk (hazard, vulnerability, and consequence), this section provides a recommended process for identifying and characterizing geohazards for new pipelines, while the next section describes mitigating the geohazards.

Geohazard Identification and Characterization

Pipelines traverse tens to thousands of miles through many different types of terrain. The cumulative risks presented by geohazards within this differing terrain coupled with unforeseen ground conditions form a significant project risk that needs to be effectively managed while developing and executing the project to ensure the residual risks, to both the public and for operations, have been reduced to the maximum extent possible. It is therefore very important for the pipeline representatives (owners and designers alike) to know as early as possible during project planning what type of geohazards may exist, where they are located, and what strategy will be used to reduce the risk they pose to the public and the pipeline.

At project inception, the potential route corridor for a planned pipeline can be as much as 10 to 100 or more miles wide. Implementing detailed geohazard analyses throughout such a large area is not cost effective. Therefore, geohazard studies need to be conducted through a phased approach that starts at a coarse scale and progressively decreases as route certainty is developed and initial geohazard characterization efforts identify those areas of the corridor requiring more detailed evaluation.

There is no consistent nomenclature to describe the various phases of pipeline geohazard evaluations. However, many US-based pipeline owners and operators and their consultants have developed a standard of practice that involves a three-phase approach that starts in the conceptual design phase consistent with the steps needed to finalize a pipeline route soon after the start of the FEED phase of a project:

- Phase 1 – Desktop Study
- Phase 2 – Reconnaissance Level Field Study
- Phase 3 – Detailed Field Characterization

The reader should refer to the PRCI report [1] for detailed discussions of the various steps involved in each of the three phases. Only a brief description of each Phase will be given below.

Phase 1 – Desktop Study: The purpose is to identify and delineate, at a regional or local scale, hazard-prone areas along a potential pipeline route or corridor. This is done through the integrated evaluation of geology, geomorphology, hydrology, and geotechnics by SMEs.

Phase 2 – Reconnaissance Level Field Study: The objective is to reduce uncertainties in geologic and geomorphic characterization by collecting additional data from reconnaissance-level ground-truthing, particularly for landslide hazards, active faults, stream crossing hazards, and subsidence hazards thus verifying the hazard analysis conducted in Phase 1 through a site-specific reconnaissance-level field study.

Phase 3 – Detailed Characterization: Typically, after the route has been finalized to a single option, the geohazard project team conducts site-specific intrusive investigations to obtain geotechnical data that can be used to further characterize the ground conditions and develop improved ground models or mitigation designs for the site.

Geohazard Identification throughout Project Stages

Geohazard identification is an on-going process throughout the three Project Stages typically used by pipeline owners and operators during execution of the design and construction of a project:

1. Pre-FEED (front-end engineering design) or concept
2. FEED
3. EPC (engineering, procurement, and construction)

Routing studies and adjustments are conducted during all the project stages; therefore, geohazard studies should be conducted in parallel. Phase 1 geohazard studies should begin in pre-FEED because the earlier geohazards are assessed, the more likely it will be to find the lowest cost, readily constructible route

Copyright © 2019 by ASME

(e.g., [13], [14], [15]). Most of the Phase 2 work should be conducted during Pre-FEED to be able to find the apparent best route by the end of this stage of the project. Initiation of geohazard studies upon project conception reduces the potential for construction cost overruns due to unforeseen conditions and can also avoid significant operational costs.

Failure to conduct geohazard studies early in the project cycle or delaying the start of geohazard studies until after defining the pipeline route will likely lead to greater project risk, substantial project costs, and heightened operational risks that could have been averted otherwise. In some jurisdictions, geohazard studies are required for permitting by regulating authorities and should therefore be conducted early in the project.

Some Phase 2 studies may also be considered during pre-FEED, but these might be budget-constrained given that some pipeline project proponents (i.e., owners and their partners) may only be able to allocate limited resources during initial project stages given the uncertainty of the project viability. As stated in [15], this reluctance to commissioning adequate geohazard studies *"...almost always leads to difficulties in the EPC stage, particularly deviations in construction scope emanating from incomplete engineering during FEED."*

Phase 1 and Phase 2 studies should be completed by the end of FEED. Some Phase 3 studies should begin during FEED and be substantially completed prior to or during the early stages of the engineering component of EPC contract. Identifying, characterizing, and mitigating geohazards prior to construction can avoid costly delays and change orders, but due to the nature of pipeline construction, the need for additional geohazard mitigation could be identified during construction, and contractual risk allocation (discussed below) should take this possibility into account.

Typical Deliverables of a Geohazard Study

Geohazard studies are commonly conducted by teams of external consultants who have expertise in the subject matter. Therefore, documenting geohazard management in a manner that conveys the process of identifying, characterizing, and mitigating the hazards is critical for effective communication between the geohazard study team and the project management team, as well as with other public and private stakeholders such as project partners, regulatory bodies, non-governmental organizations, and the public.

The following are the critical components of a report documenting a geohazard study:

- Geohazard inventory: a GIS based catalog or database of the geohazards identified along a pipeline route together with information associated with the hazard
- Recommendations for geohazard mitigation: as part of the detailed design, geohazard mitigation packages should be developed that include the construction drawings and specifications the contractor will follow in implementing the mitigation measures.
- Recommendations for geohazard monitoring: to help reduce the risk by allowing an owner/operator to respond with corrective action when the instruments report threshold performance values, prior to an incident occurring.
- Geohazard risk register: a live document that the management team should maintain and review regularly throughout the pipeline project, including the development, design, construction, and operation phases.

GEOHAZARD MITIGATION

Overview

As previously described, geohazard management includes applying processes to identify, characterize, and mitigate risk. Geohazard-driven risk is mitigated by:

- avoiding the geohazard through route selection (the preferred option),
- decreasing the likelihood of hazard occurrence, or
- reducing the vulnerability of the pipeline to the hazard.

However, proper geohazard mitigation prior to construction is affected by scheduling, priority given by the owner/operator, competence of the geohazard management team, and basis for design.

Geohazard Risk Mitigation Techniques

If the geohazard cannot be completely avoided through route selection, there are two simple and straightforward categories for mitigating geohazard effects:

- Mitigation of the Geohazard involves implementing activities during the project that reduce the potential loads that the geohazard could impose on the pipeline through PGDs. For example, landslide mitigation may include stabilization, with its ancillary effect of decreasing PGD, by reducing the driving forces or increasing the resisting forces. Similarly, at pipeline watercourse crossings, the natural forces and processes that generate lateral and vertical migration and cause exposure of buried pipelines may be of watershed scale and thus involve areas larger than a pipeline owner/operator can influence during pipeline design and construction. Therefore, mitigation of the geohazard at watercourse crossings involves implementing measures that reduce the likelihood of localized vertical or lateral migration of the stream.
- Reduction of Pipeline Vulnerability involves mitigating activities that recognize that the geohazard with its associated PGD and subsequent loads imposed on the pipeline cannot be adequately mitigated and that ground displacement will occur during the design life of the pipeline. Therefore, the goal is to reduce the effect the geohazard and its associated PGD have on the integrity of the pipeline. Reduction of the pipeline vulnerability is placed into five main categories:
    - Avoidance
    - Isolation
    - Trench modification
    - Increase of pipeline resistance
    - Implementation of a robust monitoring and maintenance plan

Copyright © 2019 by ASME

A geohazard assessment and risk mitigation management process may involve activities in one or more categories. For example, at active faults, the design team with fault specialists may design a special trench, select specific line pipe material, over match welds, and reduce the weld defect criteria to accommodate the anticipated displacements should the fault rupture.

Geohazard mitigation techniques are described in detail in the PRCI guidance document [1].

Issues Affecting Geohazard Mitigation Throughout Project Stages

There are numerous examples of pipelines routed through geohazard-prone areas that have been operating without major interruptions because the owner/operator instituted a geohazard management program from concept through operations that effectively identified, characterized, and mitigated the geohazard associated risk. Conversely, there are examples of pipelines that were installed with insufficient studies to manage geohazards that have experienced numerous incidents, which have led to service interruptions and spills.

Some of the most common issues that affect geohazard-induced pipeline risk that should be foreseen and avoided include the following:

Scheduling of Geohazard Studies: Commercial viability of many pipeline projects requires development under very compressed schedules, which can lead to insufficient time to conduct adequate geohazard management studies prior to construction. As stated in [7], project scheduling is the root cause of the majority of geohazard mitigation issues that arise on pipeline projects.

Importance Given by Owner/Operator to Geohazard Identification and Mitigation: Properly conducted geohazard management studies can lead to considerable cost savings if geohazards are identified and mitigated early in the project. However, it is sometimes difficult for project managers to consider the value of the geohazard studies because geohazards that are avoided are not included in a cost-benefit analysis or considered cost savings.

Team of Subject Matter Experts: When evaluating the merits of a company vying to conduct geohazard studies, the owners/operators should have an experienced internal team that can competently evaluate the geohazards team proposed for the work. The right team of SMEs can provide significant value to the project, whereas the wrong team of SMEs may fail to properly identify or characterize a geohazard and fail to properly mitigate it.

Basis of Design: This is a document that discusses the pipeline performance requirements, risk tolerance and expectations of the owner and regulators. The level to which geohazards will be mitigated is defined in the basis of design, as not all owners are willing to accept the same risk.

Recommended Process for Construction Handover to Operations

Companies operating pipelines in North American jurisdictions are required to prepare and annually update procedural manuals for operations, maintenance, and emergency response. It is good practice to design and implement a robust geohazard monitoring and maintenance plan to ensure geohazard mitigation continues following construction handover and throughout the operating life of the pipeline. These plans should include routine and focused monitoring of high-risk locations to provide early warnings of potential problems before threats to integrity of the pipeline develop. Special attention should be paid during the first few years following completion of the pipeline as this is the period when the pipeline trench and reinstated ROW are consolidating, vegetation is not fully established, and installed erosion control measures may need to be adjusted or maintained based on actual conditions.

A well-prepared GIS-based operations, maintenance, and emergency response procedural manual will facilitate a relatively seamless handover from the pipeline construction team to the pipeline operations and maintenance team.

CONTRACTUAL MANAGEMENT OF GEOHAZARD-DRIVEN RISK

Background

Around the globe, onshore oil and gas pipeline projects have become larger, more technically challenging, and increasingly difficult to execute due to remote locations, higher environmental standards and socioeconomic benefit expectations, more difficult terrain, and more demanding performance criteria and safety and security specifications.

These factors combine to increase both the quantity and magnitude of risks. Therefore, the need to properly identify, allocate, and manage these risks becomes ever more important to ensuring the success of such challenging projects for the benefit of all stakeholders [16], [17] and [18].

Risk Perception

Often, the perception of risk will be influenced by past experiences and value systems, both personal and organizational. A pipeline owner may be focused on long-term geohazard risks associated with pipeline integrity while overlooking short-term geohazard risks associated with pipeline construction. Hence, a contractor in a competitive tendering situation may feel that it is being asked to assume risks over which it has no control. On the other hand, the investor/owner may consider that those risks have been properly allocated to the contractor as the party best able to manage them. Moreover, third parties such as the project financier may seek to allocate maximum risk to the contractor for the benefit of cash flow, insisting on an allocation of risk even more narrow than that which might otherwise have been negotiated between experienced onshore pipeline industry participants. There is a growing recognition that geohazard assessments need to consider both long-term risks associated

Copyright © 2019 by ASME

with pipeline integrity and short-term risks associated with pipeline construction [19].

Project Risk

The primary project-level risks can generally be categorized as technical, operational, quality, and health and safety risks. Due to their linear nature, any pipeline construction project involves risk, and there is no possibility to eliminate all the risks associated with a project. It is important to understand those risks early in the project, control the risks allocated to each party, and properly manage those risks.

During the typical life cycle of an oil and gas development project, decisions are taken at different project phases. Therefore, definitive processes are required to systematically and appropriately identify, assess, control, and reduce risk during all phases of a project. Clearly, the risks which each party consider the most significant to them will guide their focus on risk management, the allocation of those risks, and their choice of disciplines called upon to inform their decision-making process.

Geohazard-Driven Risk

The effectiveness of geohazard mitigation is uncertain, and appropriate design solutions will typically rely on historical data and qualitative risk assessment to determine the most effective solution. Consequently, avoidance is the preferred mitigation wherever possible.

A project's ability to effectively manage geohazard risks decreases as project development progresses. It is important to reduce the geohazard risk to the maximum extent possible prior to entering the construction phase of any pipeline project.

The cost and schedule implications of mitigating the geohazard during the construction phase will be significantly higher than during the design phase, and may become prohibitive due to external influences such as delays in obtaining permits, land access, etc.

Therefore, early identification is essential to understand the risks and ensure any mitigation measures are integrated into the overall project execution schedule.

Contractual Risk Allocation

A variety of factors distinguish pipeline construction contracts from most other type of contracts; these include the length of the project, its complexity, diversity of terrain, size, and the fact that the price agreed to and amount of work completed may change due to unforeseen conditions as construction proceeds.

However, owners/operators often press for lump sum contracts coupled with early completion dates. It is the combination of high complexity and tight deadlines that results in onshore oil and gas pipeline projects being exposed to a wide array of risks.

Under normal competitive contracting conditions, the owner/operator (the party normally having stronger bargaining power) may want to transfer as much risk as can be leveraged to the contractor. In turn, the contactor will attempt to push back, but inevitably the result is often that the owner/operator manages to transfer the risk to the EPC contractor, and the EPC contractor then attempts to transfer its allocation of risks down the contractual chain to its subcontractors and suppliers, where the entities can neither manage the risk nor mitigate the financial consequences [20].

Current Risk Allocation Approach

The options for dealing with geohazard risks with both potentially positive and negative outcomes are similar, although the interpretation and implications are clearly different. The following are opportunities for mitigating geohazard risks:

- Risk elimination by avoidance (not proceeding or proceeding on a different alignment)
- Risk reduction (by undertaking further surveys, desktop ground truthing, geotechnical site investigations)
- Risk retention (self-insurance, bearing a large deductible, internal management of risk)
- Risk transference (by legal, contractual and insurance)

These mitigation strategies, particularly risk transference, are given effect contractually by using contractual exclusions, limitations of liability, indemnity clauses, guarantees, performance bonds, and insertion of a risk premium.

While organizations like IPLOCA [17,18] advocate a balanced and equitable allocation of risks between the parties, inequality in bargaining power between contractors and large groups of owners/operators coupled with the desire of contractors to secure new projects in a competitive market, means that risks (especially geohazard risks) are not always allocated to the party best able to manage them. In addition, due to competitive forces, the ability of a contractor to insist upon an appropriate risk premium in exchange for having taken on that risk is also curtailed.

While the owner/operator may consider an underpriced lump-sum bid as a benefit, the contractor may already see the opportunity for change orders and scope increases during the execution of the work. The natural consequences arising from such opportunistic behavior (from both parties) are disputes, delays, and additional costs, which are seldom to any party's advantage and always to the detriment of the project.

Benefits of Realistic Risk Allocation

The pipeline construction industry is beginning to recognize the basic principles of appropriate risk allocation, which can be summarized as follows:

- Risks belong with those parties who are best able to evaluate, control, bear the cost, and benefit from, the assumption of risks.
- Every risk has an associated and unavoidable cost which must be assumed somewhere in the process.
- Many risks and liabilities, especially short-term construction related geohazard risks, are best shared.

Survey data suggest that inappropriate allocation of risk results in nearly a 20% contingency or risk premium being included within bids [21]. In principle, a significant portion of

Copyright © 2019 by ASME

these increased costs could be saved by observing the principles of realistic risk allocation.

Risk allocation in any contract affects cost, time, quality, and the potential for disputes, delays, and claims. In addition, contractual misallocation of risk is known to be a leading cause of many construction disputes.

The importance of selecting the correct risk allocation strategy cannot be overemphasized, especially when a pipeline project has significant geohazard exposure.

Adopting a realistic risk allocation between the parties is a fundamental factor for moving away from the adversarial climate that impacts many onshore oil and gas pipeline projects and leads to many of the substantial cost and schedule overruns. Inappropriate risk allocation on a pipeline construction project is not cost effective and directly affects the chances of project success.

Recommended Strategies for Managing Geohazard-Driven Risk in Construction Contracts

Overview

Onshore pipeline projects frequently extend over hundreds to thousands of miles and are likely to encounter terrain and associated ground conditions differing from those foreseen during the preliminary design and tendering phase. Therefore, the high likelihood of change(s) is inherent in onshore oil and gas pipeline projects. This factor alone requires that parties involved analyze potential short-term and long-term geohazard-driven risks at a very early stage of the tendering process and ensure that the contractual baseline is set to ensure that anticipated risks are realistically allocated to the party best able to manage them.

It is essential that the contract recognizes and makes allowance for the specific aspects of onshore oil and gas pipeline projects as well as the unique nature of the type of terrain traversed by these projects.

An efficient way of ensuring a fit-for-purpose solution is to make provisions for reimbursable scopes of work to cover defined geohazard risks and engineering implementation of identified geohazard mitigations.

Geohazard Risk Register

Preparation of a geohazard risk register by the owners/operators during the early phase of a pipeline project is fundamental to facilitate effective management of geohazard-driven risks and minimize the potential cost and schedule implications for the project. The risk register should be maintained throughout development of the project and made available during the tendering process to facilitate allocation of the risks to the party(ies) best able to manage and mitigate the impact(s) of geohazards during project design and execution. During construction the geohazard risk register should be updated as some risks are mitigated and as new risks become evident along the ROW.

Early Contractor Involvement

When developing the contracting strategy for a new onshore pipeline project, the owners/operators should identify and assess the core competencies of each of the perspective contractors and adjust the contracting strategy to match the contractor's core competency. Adopting this type of process will ensure that geohazard risks are appropriately allocated and reflect both the current status of geohazard mitigation and the scope of the project. Engaging an EPC contractor during the early phases of a pipeline project allows the contractor to participate in developing the design and identifying the risks and to provide constructability input into the process. This involvement ultimately allows geohazard-driven risk mitigation to be integrated into the construction schedule.

RECOMMENDATIONS FOR PIPELINE OWNERS AND OPERATORS

Geohazards are identifiable and therefore avoidable. Developing a pipeline project without considering geohazards early in the project cycle has significant potential to increase risks to the public, heighten construction risks to workers, lead to greater construction costs, contractor claims, schedule delays, and ultimately higher operational costs.

Phase 1 studies should be conducted essentially as soon as other planning and permitting activities begin. Phase 2 and 3 studies should follow as appropriate. Phase 1 studies are the minimum level of due diligence that should be conducted by owners/operators.

Collection of LIDAR should not be considered optional for Phase 1 studies. LIDAR is not only very useful for identifying and delineating geohazards, but can also be used for other evaluations such as for wetland delineation, routing, archaeology, etc.

To effectively manage geohazards, pipeline owners should:

- Select a team of qualified subject matter experts that have experience in the area of interest and with geohazard studies for pipelines.
- Convey to the SMEs, through a basis for design, the owner's/operator's performance requirements and risk tolerance, including government-mandated performance requirements.
- Document all geohazards in a geohazard-driven risk register.
- Develop contractual conditions that will lead to mitigation of geohazards that appear or develop during construction because mitigating a geohazard during construction can be orders of magnitude less costly than mitigation during operation.

ACKNOWLEDGEMENTS

This project would not have been possible without the oversight, guidance, and resolve of Mr. Alexander McKenzie-Johnson, of Kinder Morgan, Inc., as well as others in the Design Materials and Construction Committee members of PRCI. In addition, the authors greatly appreciate the assistance of several Geosyntec reviewers during the development of this guidance

Copyright © 2019 by ASME

document. Their comments and suggestions were valuable in refining the content and organization of these guidelines.

REFERENCES

[1] Sancio, R., Rice, A., Audibert, J., Morgan, D. and Rattray, J. 2018. *Guidelines for Management of Geohazards Affecting the Engineering and Construction of New Oil and Natural Gas Pipelines*, Contract PR-616-164506, Contractor Project Number: TXG0022, Prepared for the Design Materials and Construction Committee of Pipeline Research Council International, Inc. by Geosyntec Consultants, Inc.

[2] Highland, L.M., and P. Bobrowsky. 2008. *The Landslide Handbook – A Guide to Understanding Landslides*. Reston, VA, United States Geological Survey Circular 1325, 129 pp.

[3] Baum, R.L., D.L. Galloway, and E.L. Harp. 2008. *Landslide and Land Subsidence Hazards to Pipelines.* United States Geological Survey Open File Report, 2008-1164, prepared in cooperation with U.S. Dept. of Transportation, Pipeline Research Council International, and DGH Consulting Inc.

[4] O'Rourke, T.D. 1998. *"An Overview of Geotechnical and Lifeline Earthquake Engineering."* Proceedings of Geotechnical Earthquake Engineering and Soil Dynamics Conference Geotechnical Special Publication No. 75, ASCE. Vol. 2, 1392–1426.

[5] CEPA. 2014. *Pipeline Watercourse Management, Recommended Practices*. First Edition, December. Canadian Energy Pipeline Association, 70 pp.

[6] Rapp, C. F., and T.B. Abbe. 2003. *A Framework for Delineating Channel Migration Zones.* Washington State Department of Transportation: Ecology Final Draft Publication #03-06-027. 135 pp.

[7] Nyman, D.J., E.M. Lee, and J.M.E. Audibert. 2008. *"Mitigating Geohazards for International Pipeline Projects: Challenges and Lessons Learned."* Proceedings of IPC2008, 7th International Pipeline Conference. Paper IPC2008-64405.

[8] Yeates, R. 2012. *Active Faults of the World.* Cambridge University Press. 621 pp.

[9] U.S. Geological Survey and New Mexico Bureau of Mines and Mineral Resources. 2017. *"Quaternary fault and fold database for the United States"* accessible from USGS web site: https://earthquake.usgs.gov/hazards/qfaults/.

[10] Basili R., V. Kastelic, M.B. Demircioglu, D. Garcia Moreno, E.S. Nemser, P. Petricca, S.P. Sboras, G.M. Besana-Ostman, J. Cabral, T. Camelbeeck, R. Caputo, L. Danciu, H. Domac, J. Fonseca, J. García-Mayordomo, D. Giardini, B. Glavatovic, L. Gulen, Y. Ince, S. Pavlides, K. Sesetyan, G. Tarabusi, M.M. Tiberti, M. Utkucu, G. Valensise, K. Vanneste, S. Vilanova, and J. Wössner. 2013. The European Database of Seismogenic Faults (EDSF) compiled in the framework of the Project SHARE. http://diss.rm.ingv.it/share-edsf/.

[11] Ng, P.C.F. 2013. "*Discussion: Response of a plastic pipe buried in expansive clay*" Proceedings of the Institution of Civil Engineers – Geotechnical Engineering, Vol. 166 Issue 3, June 2013, pp. 328–330.

[12] Gaaver, K.E. 2012. "Geotechnical properties of Egyptian collapsible soils." Alexandria Engineering Journal 51(3):205–210.

[13] Fookes, P.G. 1997. "Geology for Engineers: The Geological Model, Prediction and Performance." *Quarterly Journal of Engineering Geology* 30(4):293–424.

[14] Brunsden, D. 2005. "Geo-hazards and Pipeline Engineering." *Proceedings of the International Conference on: Terrain and Geohazard Challenges facing Onshore Oil and Gas Pipelines*. Thomas Telford, London. pp. 9–27.

[15] Nyman, D.J., W.J. Hall, and L.S. Cluff. 2006. "Observations and Lessons Learned from Three Decades of Seismic Hazard Mitigation for Major Oil and Gas Pipeline Projects." *Proceedings of the 8th U.S. National Conference on Earthquake Engineering*. April 18–22. Paper 1016.

[16] INGAA. 2016. *Mitigation of Land Movement in Steep and Rugged Terrain for Pipeline Projects: Lessons Learned from Constructing Pipelines in West Virginia*. Interstate Natural Gas Association of America. Prepared by Golder Associates, Inc. Final Report No. 2015-03, April 2016. 134 pp.

[17] IPLOCA. 2009. *Onshore Pipelines, The Road to Success*. First edition. September.

[18] IPLOCA. 2013. *Onshore Pipelines, The Road to Success*. Third edition. September.

[19] Read, R.S., 2018 '*Pipeline Geohazard Assessment: Bridging the Gap Between Integrity Management and Construction Safety Contexts."* Proceedings of IPC2018, 12th International Pipeline Conference. Paper IPC2018-78225

[20] Nielsen, K.R. 2006. Experienced Based Recommendation on Risk Allocation for Both Owners and Contractors.

[21] Construction Industry Institute. 2006. *Equitable Risk Allocation, Research Summary* 210-1, September.

Copyright © 2019 by ASME

AIM-PIMG2019-1044

Chapter 7

QUANTIFYING SURFACE FAULT RUPTURE HAZARD AT THE ACTIVE FAULT–BURIED PIPELINE INTERSECTION

Alan Hull[1], Eric Cannon, John Hennessey
Golder Associates Inc.
Lake Oswego, Oregon USA

ABSTRACT

Surface faults are planar structures where near-surface crustal strain is accommodated by horizontal and vertical displacement. Major surface faults, such as the San Andreas fault, typically rupture in association with large earthquakes with sudden (coseismic) vertical and horizontal displacements ranging from a few millimeters to more than 15 m. Displacement is rarely confined to a single plane, with total coseismic displacement distributed across zones from 10s to 1,000s of meters wide. The recurrence interval for coseismic surface displacement on a single fault can be as short and 20 to 30 years, and longer than 10,000 years.

Not all faults capable of surface fault displacement rupture in association with large earthquakes. Growth faults, such as those developing in the fluvial-deltaic environments of the Gulf Coast of the United States, form linear and curvilinear escarpments at the ground surface. Growth faults show slow creep episodes with creep at rates ranging from a few to 10 mm/year. These growth faults are aseismic, regional-scale slide planes.

Because surface fault rupture causes permanent ground displacements, pipeline owners and operators need to understand the location of surface faults, the amounts and directions of potential surface rupture, the width of the rupture zone, the likelihood for future rupture/creep, and the associated uncertainties. These fault parameters are a critical input to any modeling of pipeline performance expected during surface fault displacement, and the post-earthquake integrity of the pipeline system. Logic-tree-based deterministic and simplified probabilistic methods can incorporate uncertainties associated with the fault data. Future challenges are to reduce geological uncertainties and incorporate them into the engineering analyses of pipeline performance.

INTRODUCTION

The connection between large earthquakes and rupture at surface faults was not clearly made until the late 19th century when post-earthquake field observations recorded meter-scale horizontal and vertical offsets of roads, fences and other man-made features in New Zealand and Japan [1]. The effects of coseismic fault rupture on urban infrastructure were starkly revealed from investigations following the 1906 San Francisco earthquake and the associated surface rupture along the northern section of the San Andreas fault in California, USA. The post-earthquake investigations revealed how coseismic surface rupture had displaced and severely damaged the water supply dams, pipelines and other critical San Francisco water supply infrastructure.

Development and expansion of the North American transcontinental pipeline system from the mid-20th century increased its exposure to the potential effects of earthquake-related hazards, including surface fault rupture. By the 1970s, detailed mapping of seismically active faults, particularly in California, started to reveal the hazard that surface fault rupture posed to the national oil and gas pipeline distribution network.

An early recognition of the potential for fault rupture to impact the operation of pipelines was incorporated into the design and construction of the 1,870-km-long Trans-Alaska Pipeline System in Alaska in the 1970s [2]. The potential for surface rupture along the Denali fault was recognized, and the expected fault displacement quantified [3] from detailed geological investigations. An innovative, above-ground surface fault rupture mitigation system was constructed and successfully tested during the 2002 Denali earthquake [4] when coseismic rupture of 2.3 m horizontal and 0.8 m vertical extended through the pipeline without substantial damage (Figure 1).

[1] Contact author: ahull@golder.com

Copyright © 2019 by ASME

FIGURE 1: POST-2002 DENALI EARTHQUAKE ALIGNMENT OF THE TRANS-ALASKA PIPELINE (PHOTO M. MERTZ)

By the end of the 1990s and early 2000s, the locations of active faults [5] revealed that seismically active faults were common throughout much of the western USA. While the hazard to pipeline systems from surface fault rupture is typically localized, the response of buried pipelines to fault rupture was beginning to be incorporated into industry guidelines and engineering practice [6], [7]. Implementation of guidelines over the last 15 years has required many large- and small-scale investigations of surface faults where they intersect points along the 275,000 km of buried crude oil pipelines and 2,464,00 km of buried gas pipelines that extended across the USA in 2010 [8]. Investigations of the surface fault rupture hazard to buried pipelines have resulted in both the development of mitigation design options, and the advancement of the methods used to quantify the locations and amounts of possible surface fault rupture.

This paper describes three examples of recent geological studies at surface fault-pipeline crossings that illustrate the type of geological information required to support engineering analyses of pipeline performance during fault rupture. These examples are presented in the context of the nature of surface fault rupture and the existing pipeline industry guidelines used for the seismic design of buried pipeline systems. Following these examples, we briefly describe some recent trends in the advancement of the analytical methods used to quantify fault displacement at the pipeline-fault intersection, including suggestions on how these methods can be developed to serve better the needs of pipeline owners and integrity managers.

SURFACE FAULT RUPTURE

A fault is a planar structural failure surface within the Earth's crust. Faults can have both vertical and horizontal displacements, although one type of movement is usually dominant. The cumulative movement on any fault is a function of its area (length and depth) and the total strain that has accumulated and been released at the fault. For example, faults with larger cumulative displacements (100s of meters to 100s of kilometers) have moved repeatedly along the same plane, often over millions of years.

Where a fault ruptures the Earth's surface, it disrupts the topography, subsurface stratigraphy, existing geologic structure, and anthropogenic features, where present. Because faults are planar features, and because the outcrop of a fault at the Earth's surface describes a line (i.e., a fault trace), fault rupture of the Earth's surface occurs along a line that can be straight, curved, or "stepped," depending on scale and number of surface ruptures that have occurred along the fault. Surface displacement (rupture) on a fault is nearly always associated with large earthquakes (i.e., coseismic).

To identify the potential for fault surface rupture, the important faults are those that are accumulating strain in the present-day tectonic strain field. Empirical studies indicate that only the larger faults rupture to the ground surface, and it is these larger faults that can present a significant hazard to buried pipelines. For example, most surface fault ruptures occur in tectonically active areas, have single-event horizontal and/or vertical surface displacements that range from about 100 mm to 20 m, and are associated with a moderate to large earthquake (moment magnitude [**M**] ≥6).

FAULT RUPTURE INFORMATION FOR PIPELINE ANALYSIS

In their analysis of the response of buried pipelines subjected to earthquake effects, O'Rourke and Liu [6] drew upon available research and post-earthquake investigations from 12 global earthquakes to show that permanent ground displacements (PGD) pose a significant hazard to buried pipelines. They described various analytical methods and results from modeling a range of fault displacements across narrow fault rupture zone widths. Most of the analyses were for strike-slip fault (horizontal displacements), and they noted the paucity of studies that modeled vertical displacements from normal and reverse faults at that time.

In developing guidelines for the seismic design of buried oil and gas pipelines, Honegger and Nyman [7] recommended that the analysis and potential mitigations of the surface rupture hazard to buried pipelines be restricted to those faults with demonstrated displacement in the Holocene Epoch (i.e., last 11,700 years), or what are generally known in the western USA as "active" faults.

They noted that experienced geologists should be consulted to develop estimates of expected fault displacements, although simplified estimates could be made based on available empirical equations (e.g., Wells and Coppersmith [9]) that can provide fault displacement estimates from information on fault rupture

Copyright © 2019 by ASME

length, fault rupture depth, average fault slip rate, and earthquake magnitude. Honegger and Nyman [7] noted that there is significant variability and uncertainty associated with the displacement estimates from geological data.

From a geologist's perspective, the tasks needed to acquire robust fault displacement estimates for buried pipeline analyses include:

- Compilation and desktop review of available geological maps, active fault maps and associated reports, published geological research; and available consultant reports that describe the nature and activity of the fault(s) near fault-pipeline crossing locations
- Acquisition and review of available information on historical earthquakes and past surface fault rupture on the fault or faults that intersects the pipeline
- Review of available geomorphic evidence for locations of past surface fault ruptures based on historical aerial and ground photographs, high-resolution digital elevation models, and bathymetric data surrounding the fault-pipeline crossing locations
- Development of a geographic information systems (GIS) database of relevant data
- Field investigations near the fault-pipeline crossing sites, including detailed subsurface trenching investigations to gather additional information on fault location and/or past cosesimic surface displacements .

Analyses of the data gathered from the geological research and field investigations permits geologists to more reliably estimate the key required fault parameters of:

- Fault location and width of deformation zone
- Fault dip
- Fault zone complexity (i.e. multiple traces or single trace)
- Sense and amount of fault displacement
- Average fault slip and/or creep rate

A major goal for undertaking this level of geological analysis is to understand better the variability in the geological parameters. This understanding is required because of the complex nature of surface fault rupture, and to quantify the uncertainties in the geological estimates of the fault parameters used for any pipe strain analyses.

The following examples illustrate the geological analyses undertaken to define surface fault rupture parameters at pipeline-fault crossings in three different tectonic settings. The examples are from a high-slip plate boundary fault (San Andreas fault), a low-slip reverse fault within the wider plate boundary zone (Southern Whidbey Island fault zone), and episodic "growth faults" in the Gulf Coast of Texas.

SAN ANDREAS FAULT EXAMPLE

The San Andreas fault (SAF) is the main active fault within the San Andreas Fault System that accommodates relative motion between the Pacific and North American plates in the San Francisco Bay region (Figure 2). About 41 mm/yr. of right-lateral (dextral) shear occurs across the San Andreas Fault System at the latitude of San Francisco Bay, representing about 75% to 80% of the total present-day relative plate motion. The SAF trace is expressed geomorphically as a distinct broad, linear trough within the Coastal Range foothills of the San Francisco Peninsula. This linear trough was dammed in the 1870s by the Spring Valley Water Company to create water reservoirs, including San Andreas Lake and the Upper and Lower Crystal Springs reservoirs to service the growing city of San Francisco. Water from these lakes was conveyed through a network of steel pipelines that extended along and across the SAF trace.

The 1906 San Francisco earthquake was accompanied by a SAF surface rupture length of about 477 km (Figure 1) [10]. The rupture extended through the Upper and Lower Crystal Springs reservoirs embankment (Figure 3) that is now part of California State Route 92 (SR 92), and the present-day location of a transmission gas pipeline. In the region surrounding the Upper and Lower Crystal Springs reservoirs, Lawson [11] summarized reports of the SAF horizontal and vertical offsets observed in 1906. The observations were based largely on measurements of displacements of man-made features such as fences, roads, pipelines, dams, and tunnels (Figure 4). From San Andreas Lake to southeast of the Upper Crystal Springs Reservoir, horizontal offsets ranged from about 2.44 m to 3.96 m [11], [12]. Measured offsets were located across one or more ground ruptures that in places were part of a deformation zone up to 610 m wide [11].

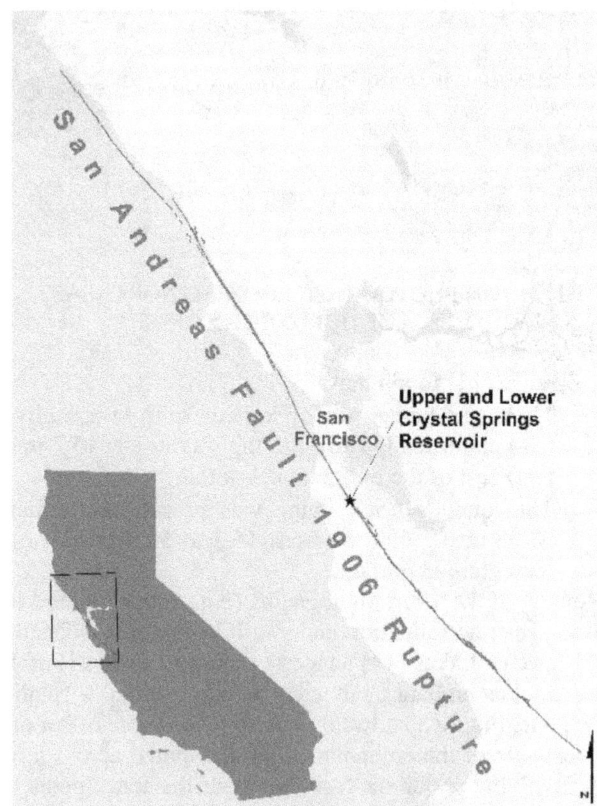

FIGURE 2: LOCATION AND EXTENT OF THE 1906 SAN ANDREAS SURFACE FAULT RUPTURE. FAULT LENGTH AFTER [10]

Copyright © 2019 by ASME

In the early 1920s, the brick-lined outlet tunnel between the Upper and Lower Crystal Springs reservoirs was reoccupied and repaired following its blockage from fault rupture accompanying the 1906 earthquake [13]. The outlet tunnel is a horseshoe-shaped, brick-lined tunnel constructed through a serpentinite ridge that formed the northeast abutment of the Crystal Springs Upper Dam (Figure 3, 5). The 236-m-long tunnel (1.83 m high by 1.68 m wide) met a 27.4-m-deep regulating gate shaft northeast of the dam [13] (Figure 3).

Lawrence [13] described the reconstruction of the former outlet tunnel undertaken in 1924. Because the brick lining of the outlet tunnel was intact except near the 1906 SAF rupture, the outlet tunnel reconstruction provided information on the location and extent of the 1906 SAF rupture directly northwest of the embankment and SR 92.

FIGURE 3: 1906 PHOTOGRAPH LOOKING NORTHEAST ALONG SR 92 TO 2.8 M OF HORIZONTAL OFFSET

Lawrence [13] found that:

- The outlet tunnel was offset 1.68 m right-laterally and 0.30 m vertically (down to the southwest) 40.2 m (132 feet) east of the outlet tunnel portal.
- The tunnel's brick lining was broken and collapsed within a 6.1 m zone from 46.3 to 52.4 m east of the outlet tunnel portal.

Review of the 1906 photographs (e.g., Figure 3) and high-resolution seismic reflection and swath bathymetry beneath the Crystal Springs Lakes [14] suggests that most of the 1906 SAF rupture on the embankment crest occurs within a relatively narrow zone. Figure 3 shows that the zone of fence offset on the northwest side of the embankment is distributed across a fence length involving about six fence posts. If the fence posts were constructed at typical 6- or 8-ft centers (1.8 to 2.4 m), then the width of the fault displacement on the crest of the embankment in 1906 was about 11 to 14.6 m. Based on the length of deflected fences along the embankment shown on Figure 3, the width of the 1906 fault rupture deformation zone was about 9.1 m on the embankment crest.

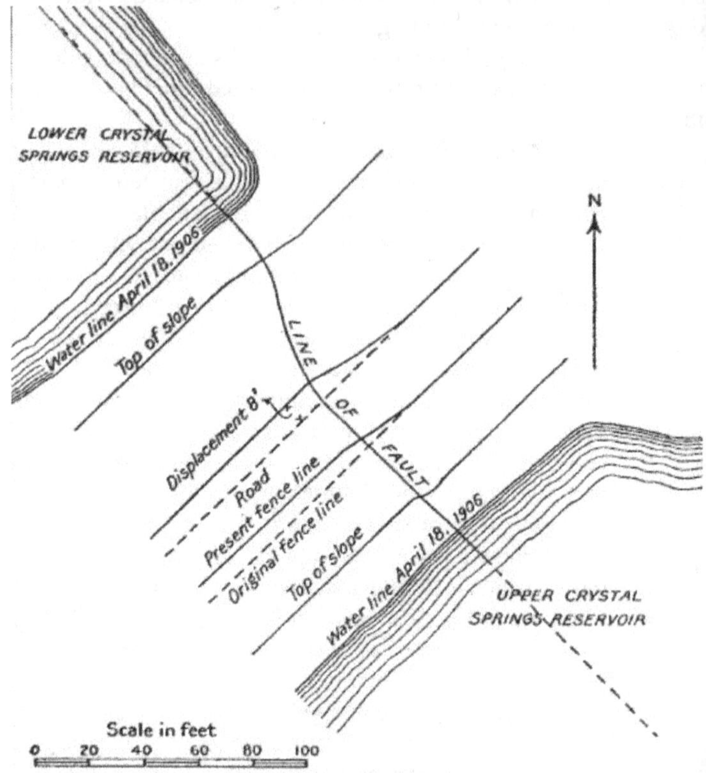

FIGURE 4: OFFSETS MEASURED ALONG THE CREST OF UPPER AND LOWER CRYSTAL SPRINGS RESERVOIR EMBANKMENT IN 1906 [11]

The alignment of the fences to the northeast and southwest of the main offset seen in the 1906 photos indicates that little or no secondary or off-fault deformation occurred in the 1906 rupture at this location. For example, in Figure 3, the fence on the northeast side of the embankment extends 244 m northeast to the ridge near the outlet tunnel shaft cover structure. This fence indicates no offset or misalignment from movement along secondary traces northeast of the main SAF rupture.

Fault mapping shows that the strike of the SAF at and to the northwest of the embankment has an azimuth of 325° (N35°W). Confirmation of the 1906 rupture location from review of historical photographs [11] and tunnel offset measurements [13] indicates that a fault strike azimuth of 324° ±2° is a reasonable estimate for the SAF orientation at this location. Similarly, 1906 measurements of coseismic fault rupture width in 1906 [11], [15], from reoccupation of the outlet tunnel [13], and from our review of historical photographs indicate a fault rupture width of 6.1 to 14.6 m, suggesting a total fault location uncertainty width of about 15 m.

Figure 5 shows an interpretation of the lake bottom bathymetry [14], offset measured from 1906 photographs and maps [11], [12], and the observations from the 1924 engineering works [13]. The interpretation indicates that the 1906 SAF

Copyright © 2019 by ASME

surface rupture at the Upper and Lower Crystal Springs reservoirs was:

- A single, northwest-striking (N36°W) fault located within the Franciscan Complex rocks at the northeastern end of Upper and Lower Crystal Springs embankment
- A right-lateral displacement of 2.68 ±0.18 m when accounting for uncertainties associated with fault strike measurement and the offset tunnel alignment measured in the 1920s
- Within a zone of about 6.1 to 10 m wide in rock, but increasing to 11 to 14.6 m wide at the embankment crest
- Not associated with secondary displacement based on post-earthquake observations of fences aligned nearly perpendicular to the fault.

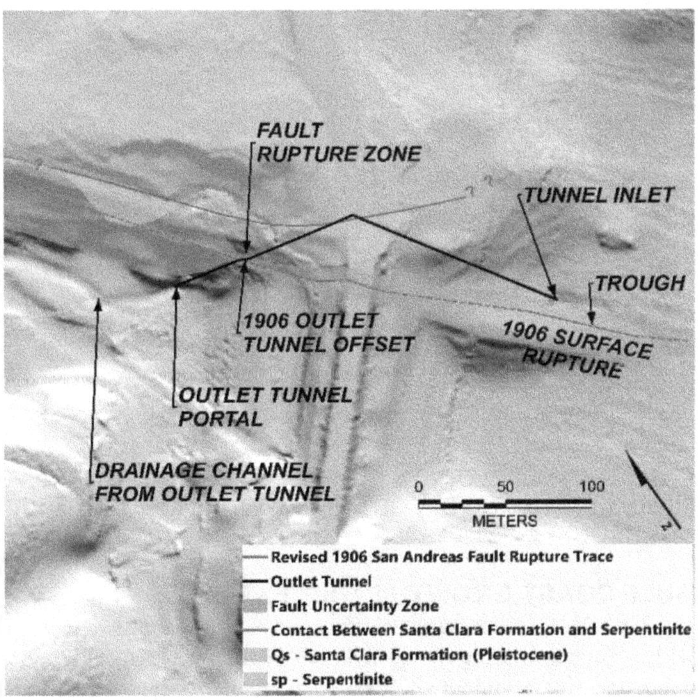

FIGURE 5: SAN ANDREAS FAULT AT THE UPPER AND LOWER CRYSTAL SPRINGS RESERVOIR EMBANKMENT WITH FAULT LOCATION UNCERTAINTIES

This analysis of the SAF at SR 92 shows how detailed evaluation of post-earthquake observations, analysis of historical records of engineering works, and the interpretation of modern high-resolution elevation data can all be used to constrain the amount, orientation, and width of fault rupture at the fault-pipeline intersection. The geological analyses are used to develop estimates for the surface fault rupture parameters, including uncertainties, needed to model buried pipeline performance during cosesimic displacement.

SOUTHERN WHIDBEY ISLAND FAULT ZONE EXAMPLE

The Southern Whidbey Island fault zone (SWIFZ) is one of several faults within the western margin of the North American continental tectonic plate. In the Puget Lowland, the northwest-southeast striking SWIFZ accommodates crustal deformation north of Seattle, Washington. The U.S. Geological Survey Quaternary Fault and Fold Database [16] shows the SWIFZ three, northwest-striking oblique slip (reverse, left-lateral) faults ranging up to approximately 159 km in length. The faults dip at 45° to 80° to the northeast. The most recent prehistoric deformation is classified as less than 15,000 years ago [16]. Average fault slip rates on the SWIFZ are estimated at between 0.2 and 1 mm/yr.

Much of the information on the location and potential activity of the SWIFZ comes from a multi-faceted investigation of the SWIFZ [17]. Analysis and interpretation of aeromagnetic geophysical surveys identified a series of northwest-trending lineaments interpreted from aeromagnetic anomaly contrasts. The aeromagnetic anomaly contrast lineaments have the same general trend as the SWIFZ on Whidbey Island, and on this basis, the SWIFZ has been extended southeastward from Whidbey Island across Possession Sound and into the Puget Sound mainland (Figure 6). Regional, high-resolution elevation data were used to identify possible fault scarps and lineaments and Holocene-active traces of the SWIFZ [17].

Detailed information on the location, nature, and type of deformation on the SWIFZ comes from interpretation of paleoseismic trenches (Figure 6) excavated across topographic scarps aligned with aeromagnetic anomaly contrast lineaments [18] [19] [20]. Evidence from the paleoseismic trenches suggests at least four surface-rupturing earthquakes occurred between about 16,400 years ago to less than 2,700 years ago. The past earthquakes were inferred from localized zones of intra- and inter-formation shears or faults, over-steepened sedimentary contacts, and near-surface zones of liquefied sediment exposed in the trenches. This type evidence for fault displacement suggests that past surface rupture on these backthrusts of the SWIFZ did not rupture a single fault plane that extends to the ground surface. Rather, surface fault rupture on this part of the SWIFZ results in folding, warping and uplift (to the northeast) rather than a discrete fault scarp.

Analysis of the available data for the location of possible strands of the SWIFZ near the pipeline crossing (Figure 6) indicated that only one possible strand of the SWIFZ intersects the buried pipeline. This strand occurs along an unnamed aeromagnetic lineament identified as "less certain" [17]. Analysis of available high-resolution topography and field mapping confirmed that a generally northwest-trending topographic lineament coincides with the magnetic anomaly lineament.

Analyses of the origin and location of SWIFZ strands surrounding the pipeline-SWIFZ intersection found no definitive evidence for the existence and past surface fault rupture of the fault strand mapped to intersect the pipeline. There is a reasonably well defined magnetic anomaly lineament that may

Copyright © 2019 by ASME

be a strand of the SWIFZ, but a clear topographic lineament is not apparent. Nevertheless, nearby studies show a strong correlation between mapped, northwest-trending magnetic anomaly lineaments and repeated surface displacement of SWIFZ strands (Figure 6). For the purposes of assessing the potential surface fault rupture hazard to the pipeline, the magnetic anomaly lineament is assumed to be a strand of the SWIFZ capable of future surface fault rupture. Fault parameters estimated for the pipeline-SWIFZ crossing are: 1) fault strike of 295°±5°; 2) fault dip of 60°±15° to the northeast; 3) fault displacement of 1.49±0.49 m; and 4) a fault location uncertainty width of 100.6 m (Figure 7).

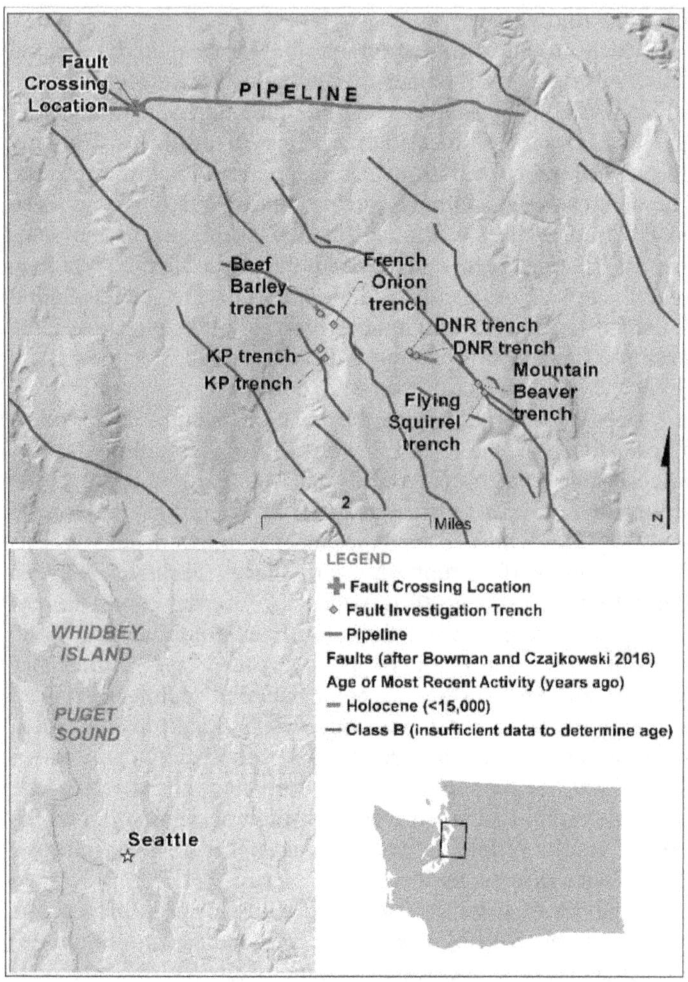

FIGURE 6: LOCATION OF INFERRED TRACES OF THE SOUTHERN WHIDBEY FAULT ZONE AND FAULT INVESTIGATION TRENCHES EAST OF PUGET SOUND, WASHINGTON

Analysis of the SWIFZ pipeline intersection showed that there are major uncertainties associated with the identification of the fault trace, confirming its location, and confirming its recent surface activity. Faults were identified based principally on geophysical anomaly contrasts that on closer examination were associated with weak topographic lineaments. The link between geophysical anomaly and topographic lineament was not well developed at the pipeline-fault crossing area. For the purposes of pipeline performance, however, coseismic surface faulting displacement was assumed.

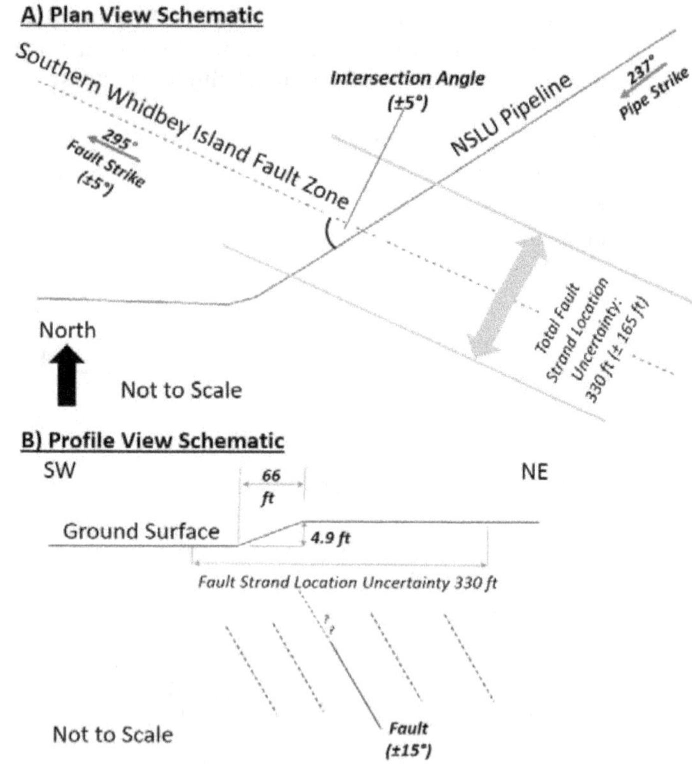

FIGURE 7: DEVELOPMENT OF FAULT PARAMETERS FOR PIPELINE CROSSING OF SOUTHERN WHIDBEY FAULT ZONE NORTH OF SEATTLE, WASHINGTON (1 FT = 0.3048 M)

GULF COAST GROWTH FAULT EXAMPLE

Growth faults are extensional (normal-slip) faults that form within accumulating sedimentary sequences as they are deposited (i.e., syndepositional). Growth faults typically occur along continental margins with high rates of sediment supply and deposition, such as the Gulf Coast of the United States. Movement along growth faults results from unconsolidated sediments moving under gravitational forces, and thus growth faults are generally considered non-tectonic (i.e., not associated with large earthquakes).

The surface traces of growth faults are well known in the Gulf Coast region of Texas and Louisiana, with hundreds of active faults mapped in the southern portions of these states [21], [22], [23]. About 80% of surface faults along the Gulf Coast occur in radial patterns formed around columns of salt (salt domes or diapirs). Surface faults have relatively short mapped fault lengths measuring between 0.8 to 4.8 km [23]. The remainder of faults represent 4.8 km- to 18 km-long regional growth faults that trend approximately parallel to the coastline [23]. Movement along growth faults has caused damage to

Copyright © 2019 by ASME

engineered structures, including at least one oil pipeline rupture [22]. One example, the Long Point fault in west Houston, measures about 16-km long and effects about 240 buildings [23]. Movement along growth faults can be either naturally occurring or triggered by human activities such as groundwater withdrawal or petroleum extraction [24], [25].

Movement along growth faults is aseismic [23], meaning that movement is not associated with an observable/recordable earthquake. Individual movement events for growth faults in the Houston area are typically in the sub-millimeter to millimeter range [23]. Average slip rates of active growth faults in the Houston area are estimated at between 5 to 20 mm/year. The hazard associated with growth faulting generally decreases with greater distance inland (north) from the coast, as the geologic environment transitions from depositional to erosional.

One example of active growth faults is the Wilcox-Frio fault zones (WFFZ) that trend along and parallel to the Gulf Coast and extend to approximately 130 km inland near the Houston area (Figure 8) These WFFZ generally comprise a series of southward dipping *en echelon* normal-slip faults rooted in salt formations. Growth fault formation developed as a result of sedimentary loading and southward migration of salt formations that began in the Paleocene Epoch (~56 to 66 million years ago) and continued to grow through the Miocene Epoch (~5 to 23 million years ago) [26], [27]. Most of the faults within the WFFZ are associated with natural geologic processes such as salt movement. Many faults show little to no evidence of surface movement in the Quaternary Period (i.e., last 2.6 million years). Ground subsidence caused by fluid withdrawal, however, may be locally accelerating or reactivating some growth faults because many actively growing fault scarps in the Houston area have developed within the last century [27], [28].

One possible, southeast-facing growth fault scarp that intersects a local buried gas pipeline measures about 1 m high and is at least 8.8 km long. The mapped trend of this growth fault follows that of other nearby mapped growth fault traces [21], as well as that of growth faults along the Gulf Coast that trend generally parallel to the coastline (Figure 8). Furthermore, the possible down-dropped portion of the fault also agrees with mapped growth faults in the region that have a down-dropped side of the fault toward the coast/southeast. Based on our review of high-resolution digital elevation models, aerial imagery, and field observations, the growth fault can be assigned the following surface displacement parameters for pipeline strain analyses:

- Fault strike of 45° (northeast)
- Dip of 60 to 85° to 132° (southeast)
- Pipeline intersection angle at 58°
- Fault scarp slope angle (5 to 10°)

The large number of growth faults extending across the Gulf Coast region of Texas and Louisiana that contains many buried oil and gas pipelines suggests that growth faults pose a substantial hazard to the operation of a major section of the USA oil and gas production and pipeline distribution systems. Because the growth faults are generally aseismic and move in small, creep-like episodes, their surface fault rupture potential is not been well characterized, perhaps because the scale of features is too large, and/or because growth fault morphology can be subtle because of their relatively slow movement rates. Nevertheless, our studies reveal that characterization of these features can be improved to support pipeline strain analyses for pipeline system integrity assessment.

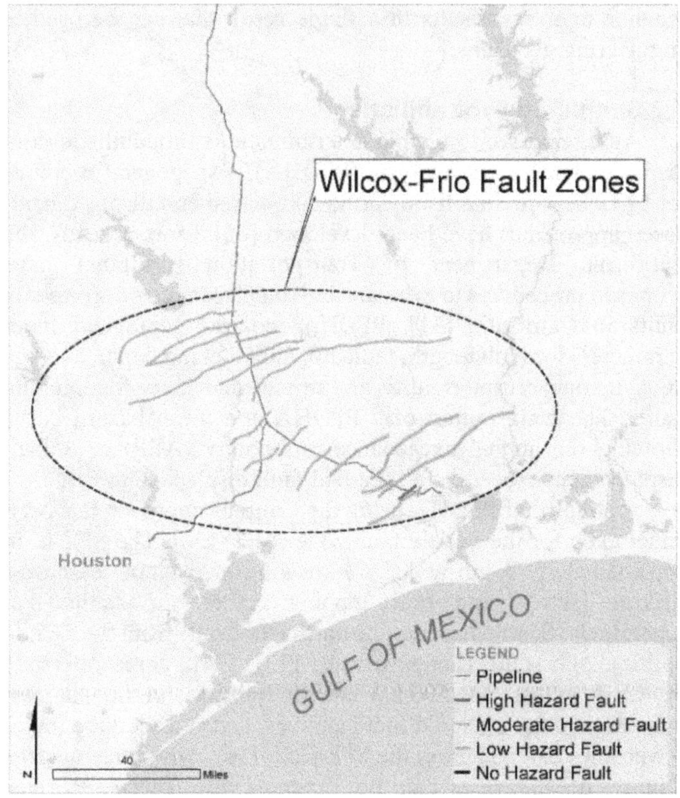

FIGURE 8: LOCATION OF GROWTH FAULTS WITHIN THE WILCOX-FRIO FAULT ZONES NORTHEAST OF HOUSTON, TEXAS

FAULT DISPLACEMENT HAZARD ASSESSMENT

The three examples above illustrate how geological information is acquired and interpreted to develop fault parameter estimates for pipeline strain analysis. The geological parameters are, however, associated with uncertainties, some of which arise from the inherent variability in the fault rupture process (i.e., every rupture is unique) and some because of the uncertainties associated with the availability and resolution of the available geological data. In order to provide robust fault parameters to pipeline performance modelers, geologists are developing approaches to quantify the uncertainties within the geological estimates of surface fault rupture. These approaches are described below.

1.1 Simple Deterministic

Deterministic analyses have long been the industry practice for both earthquake ground motion and fault rupture parameter estimates. In the simple deterministic approach, geologists review the available data; and using their experience and

Copyright © 2019 by ASME

professional judgement, develop single-value estimates for fault parameters required for pipe strain analysis. While the geologist is aware of the uncertainties of some or all of the input data, these uncertainties are either not formally quantified or only mean or median values are provided from statistical relations used in the analyses. The approach is simple and practical because the pipeline analyses results in a single result that can be used in engineering decisions.

1.2 Simplified Probabilistic

An alternative to simple deterministic is probabilistic fault displacement hazard analysis (PFDHA). Two general methods for PFDHA—probabilistic-earthquake-based and displacement-based approaches have been developed [30]. More recently, the California Department of Transportation (Caltrans) have proposed procedures to evaluate fault displacement on strike-slip faults in California [31]. PFDHA requires geological input parameters for fault length, fault dip, fault locking depth, average fault slip rate, crustal rigidity, and an aseismic factor for creeping faults. The main output of a PFDHA is a mean hazard curve showing the annual exceedance probability (AEP) of a fault displacement exceeding a specified fault displacement value.

A simplified PFDHA using the Caltrans approach has been undertaken for the Seattle fault zone (SFZ) [32]. The SFZ is an approximately 5-km-wide, 70-km-long series of east-west striking, south- and north-dipping reverse-slip faults that separate the Seattle basin to the north (footwall) from the Seattle uplift in the south (hanging wall) [33], [34]. Three interstate freeways, at least one 500 kW electric transmission line, at least three natural gas or liquid fuel pipelines, and at least three major sewer lines extend across the SFZ [35], [36]. Any future surface rupture of the SFZ can be expected to generate surface displacement across these lifelines, including the buried oil and gas pipelines.

Figure 9 shows the mean hazard curve for SFZ displacement [32]. Since the estimated characteristic **M**7.3 earthquake has a recurrence interval of about 3,300 years, the fault displacement for return periods less than about 3,300 years is negligible. The estimated mean fault displacement for 10,000-year return period is close to 1.6 m. A parametric study of the major PFDHA inputs showed that the surface fault rupture hazard is most sensitive to the average fault slip rate and dip angle, and is much less sensitive to the locking depth and total fault length. For example, an increase in dip angle from 25° to 55° results in a 4% and 19% increase in fault displacement at 5,000-year and 10,000-year return periods, respectively. By comparison, the estimated fault displacement only increases by 1% and 2% at 5,000-year and 10,000-year return periods, respectively, when the fault length increases from 69 to 75 km.

The Caltrans procedures have several assumptions that limit their wider applicability. For example, they assume that the surface fault rupture at any point is the average displacement occurring on the fault, even though historic surface fault ruptures show asymmetric slip distributions. Another assumption is that the SFZ is the only contributor to the fault displacement hazard at the site. That is, all fault slip occurs on the SFZ and the slip is from earthquakes occurring on the specific fault rather than triggered by large slip on other crustal faults. The off-fault displacement from other crustal faults is assumed to be negligible compared to the primary fault displacement. We recognize that in some pipeline fault crossing design applications it will be necessary to estimate the amount and distribution of off-fault deformation as well as slip on the primary rupture surface.

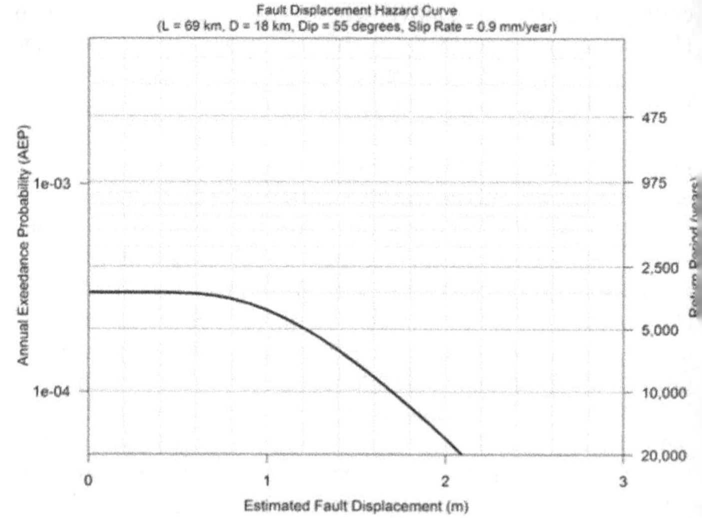

FIGURE 9: MEAN FAULT DISPLACEMENT HAZARD CURVE FOR THE SEATTLE FAULT ZONE USING CALTRANS PROCEDURES [32]

1.3 Logic-Tree-Based Deterministic

Thompson et al. [29] describe a logic-tree methodology that that incorporates uncertainties in fault displacement hazard estimates at a point along a fault. They argue that the logic-tree-based methodology is more useful that the simple deterministic approach or a probabilistic approach that underestimates the epistemic uncertainty or provides only a mean hazard curve.

The main advantage of the logic-tree-based approach is its ability to capture a full range of uncertainties associated with the inputs into the calculation of the fault displacement values used in pipeline analyses. Using the earthquake magnitude approach, they develop logic-trees to quantify and weight uncertainties associated with fault length, fault width, different empirical fault rupture length magnitude scaling relations, and relations that capture additional epistemic uncertainty and aleatory variability associated with the fault rupture process.

The output of their analyses are exceedance curves that show the probability of exceedance *vs* displacement (Figure 9). These curves graphically show a weighted mean estimate of fault displacement at the pipeline-fault intersection. Variability is shown by the fractiles curves that capture the aleatory variability associated with the mean fault displacement estimate (Figure 9). Analysis of the sensitivity of the results from a worked example in Figure 9 shows that the greatest uncertainties arise from inclusion of additional epistemic uncertainty, the rupture length of the maximum credible earthquake estimated on the fault, and

Copyright © 2019 by ASME

the standard deviations from the empirical fault length *vs* earthquake magnitude scaling relations [29].

A major advantage of the value in the logic-tree-based approach is that [29]:

- Geologists must document and justify their judgements for their preferred fault parameters, and must consider the uncertainties associated with their preferred values
- Results are more reliable and technically defensible, particularly when used at critical locations or when substantial investment in pipeline mitigation is proposed
- Sensitivity analyses can be used to direct investment into reducing the main uncertainties, rather than gathering more data on parameters that may not result in improved results
- Capturing uncertainties allows an easy conversion to the presentation of results as simplified probabilistic fault displacement hazard curves.

The logic-tree deterministic approach represents a significant advancement from simple deterministic and the simplified probabilistic described below. By using the approach to estimate fault displacement values at a point along a pipeline, geologists provide engineers with a clearer understanding of the uncertainties associated with mean or preferred values. Unfortunately, at this stage these uncertainties remain high, such that 84th-percentile displacement estimates can be more than two times median estimates [29].

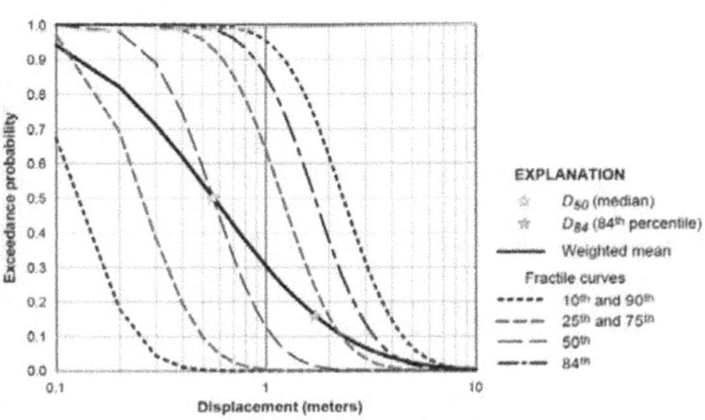

FIGURE 10: EXAMPLE EXCEEDANCE CURVES FOR A WEIGHTED MEAN MODEL OF SURFACE FAULT DISPLACEMENT AT APPOINT USING THE LOGIC-TREE-BASED APPROACH. FIGURE FROM [29]

CONCLUSIONS

The assessment of the surface rupture hazard at a point along a fault where it intersects a buried pipeline requires substantial information on the nature, location and width of possible fault rupture. The three examples outlined here illustrate that careful research of existing geological and tectonic geomorphological information can reveal substantial detail on fault displacement parameters, even in environments away from high-slip faults with short (100s of years) recurrence intervals for surface rupture. Despite the availability of data, however, the uncertainties associated with the fault information remain high, and result in a wide range of possibilities for the amount and direction of surface fault rupture to be considered in modelling of pipeline performance.

Fortunately, new methods are being developed to incorporate uncertainties in deterministic analyses (e.g., [29]) and probabilistic fault displacement analyses so that geologists can document defensible judgements of fault parameters for input in engineering analyses. As these new methods are refined and tested, it will be important that the engineering methods that use these fault parameters also develop methods to incorporate engineering uncertainties, resulting in improved quantification of pipeline system performance.

ACKNOWLEDGEMENTS

Special thanks to Clay Johnson, Golder, for his support on the San Andreas fault study. Rob Givler and Steve Thompson at Lettis Consultants International; and Chris Madugo at Pacific Gas and Electric were part of many discussions on quantifying fault parameters for pipeline performance analysis. Doug Honegger helped educate the geologists on finite element analysis of coseismic fault displacement at pipelines, and how engineers deal with geological and engineering uncertainties.

REFERENCES

[1] Yeats, R., Sieh, K., Allen, C. "*Earthquake Geology.*" Oxford University Press. (1997).

[2] Honegger, D., Nyman, D., Johnson, E., Cluff, L. Sorenson, S. "Trans-Alaska Pipeline System performance in the 2002 Denali Fault, Alaska, earthquake." *Earthquake Spectra* 20 No. 3 (2004): pp. 707-738.

[3] Woodward-Lundgren and Associates. "Summary report basis for pipeline design for active-fault crossings for the Trans-Alaska Pipeline System." Appendix A-3 (1974).

[4] Newmark, N., Hall, W. "Pipeline design to resist large fault displacement." *Proceedings of the U.S. National Conference on Earthquake Engineering*: University of Michigan, Ann Arbor, Michigan, 1975.

[5] U.S. Geological Survey. Quaternary fault and fold database for the United States: http//earthquake.usgs.gov/hazards/qfaults: (2006).

[6] O'Rourke, M., Liu, X. "*Response of buried pipelines subject to earthquake effects.*" MCEER Monograph Series (1999).

[7] Honegger, D., Nyman, D. "Guidelines for the seismic design and assessment of natural gas and liquid hydrocarbon pipelines." Pipeline Research Council International Inc., Contract PR-268-9823, 2004.

[8] O'Rourke, M., Liu, X. "*Seismic design of buried and offshore pipelines.*" MCEER Monograph MCEER-12-MN04, (2004).

[9] Wells D., Coppersmith K. "New Empirical Relationships among Magnitude, Rupture Length, Rupture Width, Rupture Area, and Surface Displacement." *Bulletin of the Seismological Society of America* 84(4) (1994): pp. 974-1002.

Copyright © 2019 by ASME

[10] Zoback, M.. The 1906 earthquake and a century of progress in understanding earthquakes and their hazards. *GSA Today*: 16, no. 4/5, (2006) doi: 10.1130/GSAT01604.1.

[11] Lawson, A. C., chairman. The California earthquake of April 18, 1906, Report of the State earthquake investigation commission: Washington, D.C., Carnegie Institution of Washington, Publication 87, (1908).

[12] Schussler, H. "*The water supply of San Francisco, California, before, during and after the earthquake of April 18, 1906, and the subsequent conflagration*." New York, Martin Brown Press (1906).

[13] Lawrence, W. B. "A bit of earthquake damage." *San Francisco Water* 3(4) (1924): pp.13–16.

[14] Finlayson, D. Triezenberg, P., Hart, P. "Geophysical Surveys of the San Andreas and Crystal Springs Reservoir System including Seismic-Reflection Profiles and Swath Bathymetry, San Mateo County, California" U.S. Geological Survey Open-File Report 2010-1030 (2010).

[15] Taber, S. Local effects of the California earthquake of 1906. *Journal of Geology* 14(9) (1906): pp. 303–315.

[16] Johnson, S., Blakely, R., Brocher, T., Haller, K., Barnett, E., Sherrod, B., Kelsey, H., and Lidke, D., compilers. "Fault number 572, Southern Whidbey Island fault zone, in Quaternary fault and fold database of the United States" 2016. U.S. Geological Survey website, https://earthquakes.usgs.gov/hazards/qfaults, accessed 07/18/2017

[17] Blakely, R., Sherrod, B., Wells, R., Weaver, C., McCormack, D., Troost, K., and Haugerud, R. "The Cottage Lake Aeromagnetic Lineament: A Possible Onshore Extension of the Southern Whidbey Island Fault, Washington". U.S. Geological Survey Open-File Report 2004-1204 (2004).

[18] Sherrod, B., Barnett, E., and Kelsey, H. "Excavation logs of two trenches across a strand of the southern Whidbey Island fault zone near Grace, Washington." U.S. Geological Survey Open-File Report 2005-1013 (2005a).

[19] Sherrod, B., Blakely, R., Weaver, C., Kelsey, H., Barnett, E., and Wells, R. "Holocene Fault Scarps and Shallow Magnetic Anomalies along the Southern Whidbey Island Fault Zone Near Woodinville, Washington." U.S. Geological Survey Open-File Report 2005-1136 (2005b).

[20] Sherrod, B., Blakely, R., Weaver, C., Kelsey, H., Barnett, E., Liberty, L., Meagher, K., Pape, K. "Finding concealed active faults: Extending the southern Whidbey Island fault across the Puget Lowland, Washington." *Journal of Geophysical Research Solid Earth* 113, B05313 (2008).

[21] Ewing, T. "Tectonic Map of Texas (Four Quadrants)". Bureau of Economic Geology, University of Texas at Austin. Scale 1:750,000 (1990).

[22] Verbeek, E., Ratzlaff, K., Clanton, U. "Faults in parts of north-central and western Houston metropolitan area, Texas." U.S. Geological Survey No. 1136 (1979).

[23] Norman, C. "Ground Subsidence and Active Faults in the Houston Metropolitan Area." *Coastal Subsidence, Sea Level and the Future of the Gulf Coast*. HGS/ECH Conference, Field Trip Guide (2005)

[24] Engelkemeir, R., Khan, S. "LiDAR Mapping of Faults in Houston, Texas, USA." *Geosphere* 4 (1) (2008): pp 170-182; doi: 10.1130-GES00096.1.

[25] Huang, J., Van Nieuwenhuise, D., Khan, S. "Integrated Fault and Hazard Analysis in Downtown Houston, Texas." *Gulf Coast Association of Geological Societies (GCAGS) Journal* 4 (2015): pp. 147-154.

[26] Ewing, T. "Structural Styles of the Wilcox and Frio Growth-Fault Trends in Texas: Constraints on Geopressured Reservoirs." Bureau of Economic Geology, University of Texas at Austin. Report of Investigations No. 1541986 1986.

[27] Ewing, T. "The Tectonic Framework of Texas". Text to Accompany "*The Tectonic Map of Texas*". Bureau of Economic Geology, University of Texas at Austin (1991).

[28] Verbeek, E., Clanton, U. "Map Showing Surface Faults in the Southeastern Houston Metropolitan Area, Texas." U.S. Geological Survey Open-File Report 78-797 (1978).

[29] Thompson, S., Madugo, C., Lewandowski, N., Lindvall, S., Ingemansson, B., Ketabdar, M. "Fault displacement hazard analysis methods and strategies for pipelines." *11th US National Conference on Earthquake Engineering Integrating Science, Engineering & Policy*, Los Angeles, California June 25-29, 2018.

[30] Youngs R., Arabasz W., Anderson R., Ramelli A., Ake J., Slemmons D., McCalpin J. "A methodology for probabilistic fault displacement hazard analysis (PFDHA)." *Earthquake Spectra* 19(1) 2003: pp. 191-219.

[31] Shantz T. "Caltrans Procedures for Calculation of Fault Rupture Hazard." http://www.dot.ca.gov/newtech/structures/peer_lifeline_program/docs/Caltrans_Procedures_for_Fault_Rupture_Hazard_Calculation.pdf, accessed 03/10/2015.

[32] Li, F., Hsiao, E., Lifton, Z., Hull, A. "A simplified probabilistic fault displacement hazard analysis procedure: Application to the Seattle Fault Zone, Washington, USA." *6th International Conference on Earthquake Geotechnical Engineering*, Christchurch, New Zealand, 1-4 November 2015.

[33] Johnson S., Dadisman S., Childs J., Stanley W. "Active tectonics of the Seattle fault and central Puget Sound, Washington—Implications for earthquake hazards." *Geological Society of America Bulletin* 111(7) (1999): pp. 1042-1053.

[34] Liberty L., Pratt T. "Structure of the eastern Seattle fault zone, Washington State: New insights from seismic reflection data." *Bulletin of the Seismological Society of America* 98(4) (2008): pp. 1681-1695.

[35] Haugerud R., Ballantyne D., Weaver C., Meagher K., Barnett E. "Lifelines and earthquake hazards in the greater Seattle area." U.S. Geological Survey website, http://geomaps.wr.usgs.gov/pacnw/lifeline/ index.html, accessed 03/10/2015.

[36] Ballantyne D, Bartoletti S, Chang S, Graff B, MacRae G, Meszaros J, Weaver C. "Scenario for a Magnitude 6.7 Earthquake on the Seattle Fault." Earthquake Engineering Research Institute, Oakland, California, https://www.eeri.org/wpcontent/uploads/2011/05/seattscen_full_book.pdf, accessed 03/10/2015.

Copyright © 2019 by ASME

AIMPIMG2019-1073

Chapter 8

ADVANCED PIPELINE GEOHAZARD SIMULATION: EVALUATION OF PIPELINE RESPONSE TO LATERAL SLOPE MOVEMENTS

Dr. Abdelfettah Fredj
BMT
Ottawa, Ontario, Canada

Aaron Dinovitzer
BMT
Ottawa, Ontario, Canada

ABSTRACT

While pipeline materials are generally ductile and thus resistant to elevated strains, US Department of Transportation (DOT) statistics suggest that between 1991 and 2000, around 30 pipeline failures were attributed to natural forces. Natural forces are defined in general causal categories including subsidence, landslide, mudslide, frost heave and washout. To enhance pipeline safety and serviceability through better understanding of geotechnical hazards, BMT with Pipeline Research Council International (PRCI), the US DOT and pipeline operating company support, has developed an advanced pipe-soil interaction design and assessment support tool to provide guidance on the pipe deformations and stress states generated due to geohazard events. Risk mitigating measures or modified designs based upon the information provided by a tool of this type can be implemented to preclude throughput interruption, environmental contamination, and damage to neighboring assets and potential harm to the public.

Pipe-soil interaction numerical simulation requires a three-dimensional analysis process considering large deformations, non-linear material behavior and soil constitutive models. This requirement is difficult to meet considering Winkler (i.e., soil modeled using springs) and Lagrangian soil models. BMT has made use of the LS-DYNA Multi-Material Arbitrary Lagrangian Eulerian (MM-ALE), Smooth Particle Hydrodynamic (SPH) and Discrete Element Methods (DEM) capabilities for modeling the pipe-soil interaction problems inherent in geohazard assessment. The development and validation of this tool has considered both the soil constitutive model and the pipe response and deformation modes to permit consideration of a range of pipe limits states including pipe yield, wrinkling/buckling, and girth weld defect assessment. The application of the tool in engineering assessments and mitigation measure development has considered amongst other events, slope movement along and transverse to the pipeline, subsidence, excavation process effects, wrinkle bend integrity management and strain relief scenarios.

This paper will provide an overview of the constituent elements and application of the BMT pipe-soil interaction modeling tool for geohazard assessment. The LS-DYNA- based pipe-soil interaction model will be used to demonstrate the treatment of slope movements colinear and perpendicular to the pipeline axis. These sample applications illustrate the information that can be derived from this type of modelling tool and how remedial action plans can be developed and remedial actions defined.

INTRODUCTION

This paper presents the summary of work involved in the development and application of numerical modeling tools to investigate the mechanical behavior of buried pipelines interacting which implied active soil movements. Detailed pipe-soil interaction analyses are completed for commonly observed modes of ground movement. The objective of this research is to develop an engineering tool to define the effects of operational and geotechnical loads on pipeline systems to support decision making regarding threat severity or repair scheduling. This tool could be incorporated in strain-based design and assessment to facilitate the consideration of complex loading scenarios inducing significant flexural loads, including pipeline subsidence and ground movements inducing lateral pipeline movements. The results of this project define the local nominal strain state that can be used to assess localized anomalies / defects (e.g., corrosion, cracks, dents, weld faults, gouges, etc.) or the potential for the formation of a wrinkle or buckle.

The advancement of these tools and their initial application are supported by the US Department of Transportation, Pipeline and Hazardous Materials Safety Administration (PHMSA) and the Pipeline Research Council International.

Copyright © 2019 by ASME

NOMENCLATURE

ALA	American Lifeline Alliance
ASCE	American Society of Civil Engineers
BS	British Standard
CSA	Canadian Standards Association
DEM	Discrete element method
DOT	US Department of Transportation
MM-ALE	Multi-material Arbitrary Lagrangian Eulerian
NPS	Nominal pipe size
PHMSA	Pipeline and Hazardous Materials Safety Administration
PRCI	Pipeline Research Council International
SMYS	Specified minimum yield strength
SPH	Smooth particle hydrodynamic
δ	Displacement
D	Diameter
n	Soil displacement function parameter
t	Thickness
Te	Temperature differential (installation – operating)
W	Ground displacement width

MODEL AND VALIDATION

Soil-pipe interaction numerical simulation requires large deformation analysis capability; this requirement is difficult to meet for Lagrangian-based codes. LS-DYNA [1] offers the possibility to use continuum Multi-material Arbitrary Lagrangian Eulerian (MM-ALE) finite element model, continuum Smooth Particle Hydrodynamic (SPH) and Discrete Element Method (DEM) capabilities for modeling materials that undergo large deformations.

The LS-DYNA [1] implementation of these high energy methods (ALE, SPH and DEM) to represent soils was found to be efficient in assessing geotechnical hazards. Pipe-soil interaction using advanced modeling techniques (ALE, SPH and DEM) within the non-linear explicit code, LS-DYNA [1], are validated and calibrated by the author based on available published full-scale experimental data including pipelines subjected to lateral and axial soil movements. The models are further calibrated for nominal soil types to replicate the pipe-soil load displacement properties outlined in ASCE (ALA) [2] recommendations. The model description and validation results are presented in previous papers [3, 4, 5, 6, 7 and 8].

The SPH, ALE and DEM simulation processes continue to be developed by the author. As such, further validation and engineering applications have been completed and documented. The model has proven useful in the design of slope remediation and maintenance activities for pipelines. The simulated pipeline behaviors have been used to assess interactions with active slope movement and evaluate pipeline strain demand and strain limit capacity [3, 4, 5, 6, 7 and 8]. These engineering applications have further demonstrated the ability and utility of the pipe-soil interaction tool to predict pipe response, support engineering decision making and contribute to geotechnical hazard mitigation.

In this paper, the soil-pipe interaction under soil movement is investigated with attention to both the soil constitutive model and the use of the LS-DYNA DEM.

The method does not require element meshing and can handle extreme material distortion, permitting the simulation of 3D, large soil displacements interacting with a shell element representation of the pipeline. In the DEM method, the soil material is treated as particles that have their masses smoothed in space, and the pipe is modeled using shell element with five (5) integration points through the thickness. The coupling between the DEM grid points (soil) and the pipe is accomplished through a contact definition. As the method incorporates mechanical contact between particles, in the case of wet soil (particles), the liquid surface tension that may cause an attraction force between adjacent particles may be considered. The model was developed and validated for the assessment of pipe-soil interaction modeling of geotechnical events.

The model is a coupled 3D continuum model that can consider the effects of layered soils, trench geometries, operating conditions, and pipe materials stress-strain behavior including differences in tensile and compressive material behaviors.

MODELING OF LANDSLIDE HAZARD

The sample applications of the DEM pipe-soil interaction geotechnical hazard assessment method were employed to investigate the strains developed in pipelines subjected to lateral soil movement. This paper presents some examples of the DEM modelling work completed as part of larger exploration of pipe response to geotechnical hazards [8]. The primary focus of this work was a nominal pipe size (NPS) 24 (610 mm) pipe system, however, other pipe sizes were considered. The objective of the work was to complete a sensitivity study to define the problem parameters and the pipe strains developed in lateral soil movement events and identify trends. The analyses considered 90 lateral ground movement scenarios.

This work was focussed on ground movements involving deformable soils. Care should be taken in applying these results to movement across a pipeline surrounded by rock or very stiff soils that will not deform and will cause extremely localized pipe deformations. The scenarios considered in the larger sensitivity study [8] included the following:

- Diameters NPS 12 (324 mm), NPS 18 (457 mm), NPS 24 (610 mm) and NPS 30 (762 mm);
- Three pipe wall thicknesses for each pipe;
- Ground displacement. Direction perpendicular to the pipe axis;
- Ground displacement width (W): 10, 20, 30, 40 50 and 100 m wide;
- Two soil conditions (properties consistent with sand and sand/clay till);
- Two slope angles of 0 and 16 degrees; and

Copyright © 2019 by ASME

- Pipe burial depths of 1.2 m (4 ft), 1.5 m (5 ft) and 1.8 m (6 ft) to top pipe (the effect of burial depth is not described in this paper).

The assumptions for the analyses included in the sensitivity study considering pipeline response to soil displacement pattern were as follows:

- 72%, 50% and 36% SMYS internal pressure;
- Thermal Loads (installation - max operating temperature) considering 0 °C, 45 °C and 65°C;
- Generic stress strain curves for Grade X52 (SMYS =359 MPa; UTS =455 MPa , Y/T =0.79) and X70 (SMYS = 483 MPa, UTS =565 MPa, Y/T =0.85) were considered as shown in Figure 1; these curves were developed based on the minimum specified material properties outlined in CSA Z245[9]; and
- Two ground displacement patterns - The lateral ground deformation was initiated using a cosine function raised to the power "n" similar to Suzuki et al. [48] and M. O'Rourke [47] function (with values of n=0.1 to 2): n value of 2 and 0.1 were selected to represent specially distributed and abrupt transverse ground displacement.

$$y(x) = \delta\left(\cos\frac{\pi x}{W}\right)^n$$

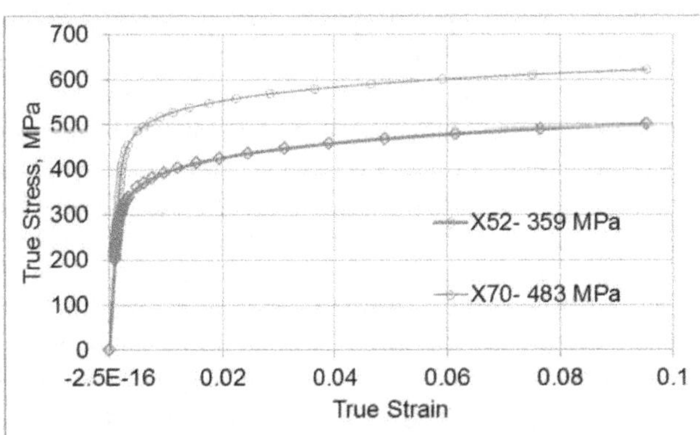

Figure 1. True Stress – Strain Curves for Grade X52 and X70 (359 and 483 MPa) Materials

SENSITIVITY STUDIES

The following section presents a summary of some results used to define the problem parameter and the pipe strains developed in lateral soil movement events and identify trends. The full results of this research are presented in DOT report [8]

Effect of Ground Displacement Width

A sensitivity study has been completed to understand the effect of ground movement width on pipe deformations or strains. The results indicate that for the six (6) widths considered, the strains in the pipe are affected by the width of the soil movement and there is a critical movement width that promotes the largest tensile and compressive strain in the pipe for any given value of ground displacement.

The critical width, defined as the soil movement width that produces the highest pipe strain, is a function of the pipe stiffness (geometry), soil movement pattern and pipeline operating pressure. The results indicate that:

- For 30-inch outer diameter pipe with wall thickness of 7.92 mm (0.312 inch) (D/t=96) Grade X70 and operating at 72% SMYS the critical width is 30 m considering a soil displacement parameter n equal to 2;
- For 24-inch outer diameter pipe with wall thickness 9.52 mm (0.375 inch) wall thickness (D/t =64), Grade X52 and operating at 72% SMYS the critical width is 20 m and 10 m considering soil movement pattern parameter n equal to 2 and 0.1, respectively;
- For 18-inch outer diameter pipe with wall thickness of 6.35 mm (0.25) (D/t =72), Grade X52 and operating at 72% SMYS the critical width is 10 m considering soil parameter n equal to 2;
- For 12-inch outer diameter pipe with wall thickness of 4.57 mm (0.18 in); (D/t =71) Grade X52 and operating at 72% SMYS the critical width is 10 m.

This work was focussed on ground movements involving deformable soils. Care should be taken in applying these results to ground movement across a pipeline surrounded by rock or very stiff soils that will not deform and will cause extremely localized pipe deformations where this critical width concept may not apply. It would be conservative to consider the maximum strain for all ground movement widths greater than the critical width.

Effect of Ground Movement Pattern

In order to evaluate the effect soil movement pattern, the simulation using cosine function raised to power 2 was repeated using cosine function raised to the power 0.1. Figure 2 presents the alternative soil movement patterns that were simulated.

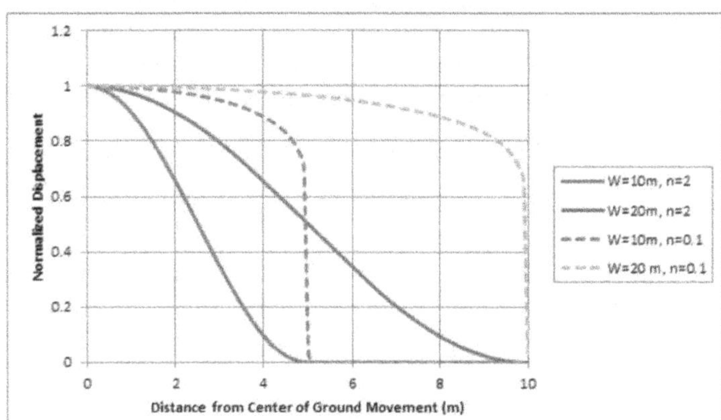

Figure 2: Soil Movement Patterns, Width W=10m, and 20 m

Figures 3 and 4 illustrate the simulated displaced position of the pipe and the soil patterns for both loading scenario considering ground movement width of 20 m. These figures are for X52 grade NPS 24 (610mm) pipe with t =9.5 mm (3/8 in) and D/t=64. The strains (Figure 3) for n = 0.1 are maximized at a displacement width of 10 m, the critical width, whereas lower strains (lower curvature) are developed for n = 0.1 for the soil displacement width of 20 m. The different soil colours in

Copyright © 2019 by ASME

Figure 3 and Figure 4 are used to denote different geometric zones in support of the model creation.

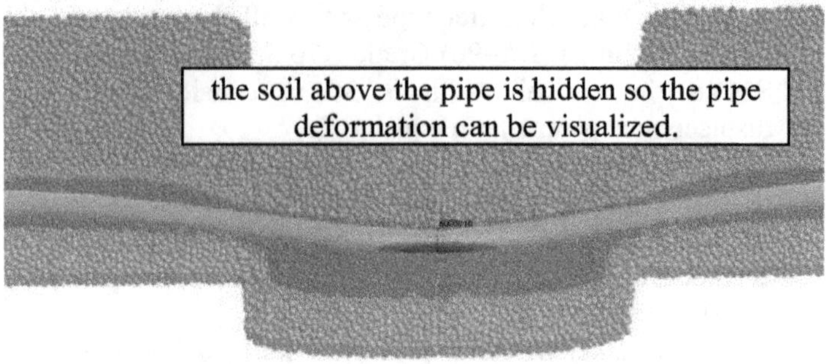

Figure 3: Pipe Deformation and Axial Strains for given Soil Movement of 2 m and W=10 m (Top View), n=0.1

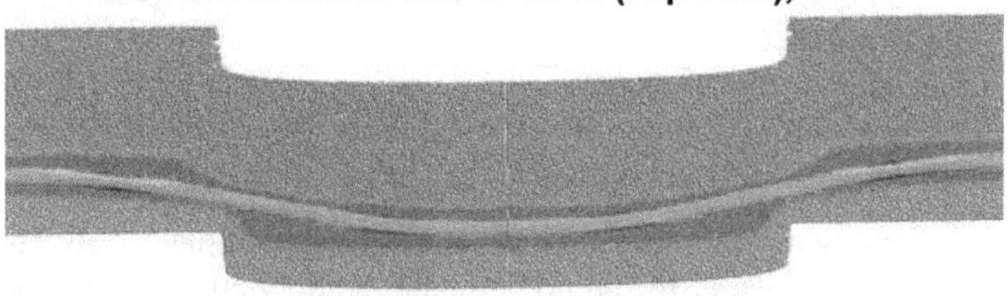

Figure 4: Pipe Deformation and Axial Strains for given Soil Movement of 2 m and W=20 m (Top View), n=0.1

Figures 5 and 6 present the pipe maximum tensile and compressive strains versus the ground movement for displacement widths of W = 10 m and 20 m. These figures are for X52 grade NPS 24 (610mm) pipe with t = 9.5 mm (3/8 in).

The results show that the critical ground movement depends on the soil movement function (pattern). From these results it is noted that:

- The width of 20 m (66 ft) results in the largest tensile and compressive strain in the pipe for any given value of ground displacement considering soil parameter n equal to 2.
- The width of 10 m (33 ft) results in the largest tensile and compressive strain in the pipe for any given value of ground displacement considering soil parameter n equal to 0.1.

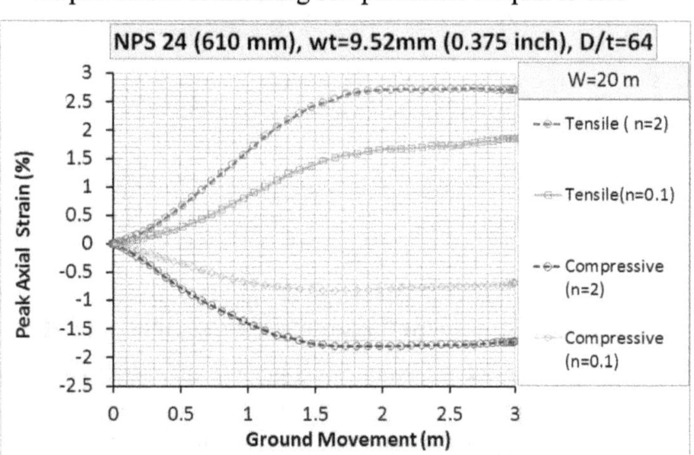

Figure 5: NPS 24 (D/t=64) – Maximum Axial Strain vs Ground Movement for the Displacement. Width of 20 m (66 ft)

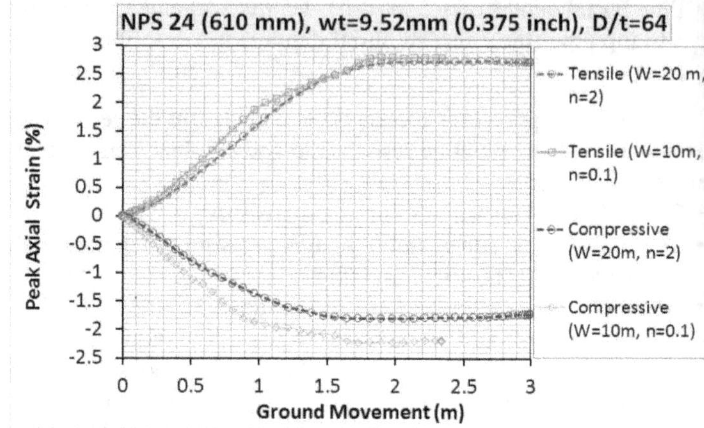

Figure 6: NPS 24 (D/t=64) - Maximum Axial Strain vs Ground Movement for 20 m (66 ft) and 10 m (33ft) Critical Widths

Effect of Pipe Wall Thickness And D/T

The structural performance of an NPS 24 pipeline Grade X52 (359 MPa) subjected to landslide was evaluated using a combination of three pipe wall thicknesses including: 9.52 mm (0.375 in), 6.35 mm (0.25 in) and 12.7 mm (0.5 in); corresponding to D/t ratios of 64, 96 and 48, respectively.

A comparison of the pipe strain accumulation results for the three pipe wall thicknesses is presented for ground displacement perpendicular to pipeline axis considering the critical ground displacement width of 20 m, in Figure 7. The results indicate that a pipeline with a higher D/t ratio (lower pipe wall thickness) results in higher computed tensile and compressive strains. The magnitudes of the tensile strains are noted to be higher than the compressive strains. The strains ranged from +3.25% and -2.77% for 6.35 mm (0.25 inch) wall thickness to +2.7% and -1.72% for 9.53 mm (0.375 inch) wall thickness while for 12.7 mm (0.5 inch) the corresponding strain range is +2.38% and -1.58%. There is very little difference between the computed magnitudes of the compressive strain for pipes with a D/t ratio of 64 and 48. The difference is about 0.14%; however, the maximum difference in compressive strain is higher than 1% for the pipe with D/t equal to 64 and 96.

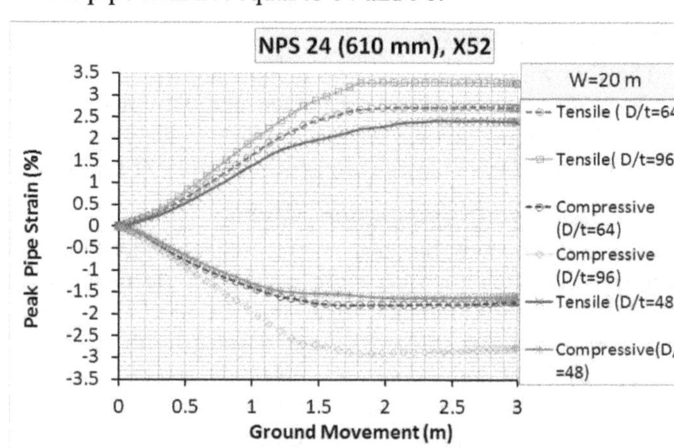

Figure 7: NPS 24 – Maximum Axial Strain vs Ground Movement for the Critical Widths of 20 m (66 ft)

Copyright © 2019 by ASME

Figure 8 shows the NPS 24 (609 mm) pipe maximum tensile strains as a function of the maximum soil displacement for various ground displacement widths. The maximum strains are developed for the critical width (20 m).

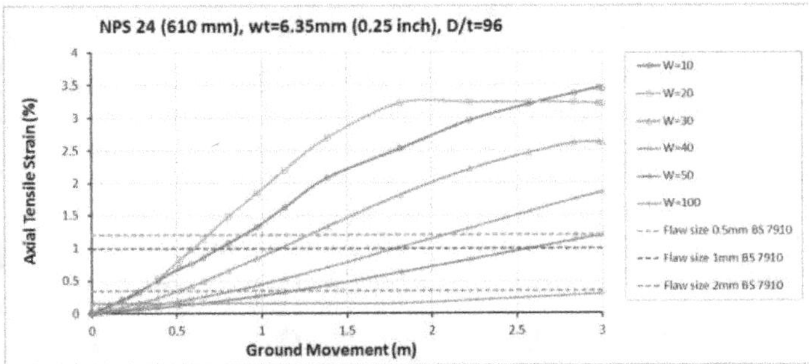

Figure 8: NPS 24 (D/t=96) - Maximum Pipe Tensile Strain vs Ground Movement Width

Figures 9 and 10 present the maximum local and nominal compressive strains in the longitudinal direction of the pipeline versus the maximum ground displacement for various ground displacement widths. The nominal compressive strain is defined as the average compressive strain measured over a gauge length of 2D (where D is the pipe outside diameter) and is comparable to limit strain criteria in codes and standards.

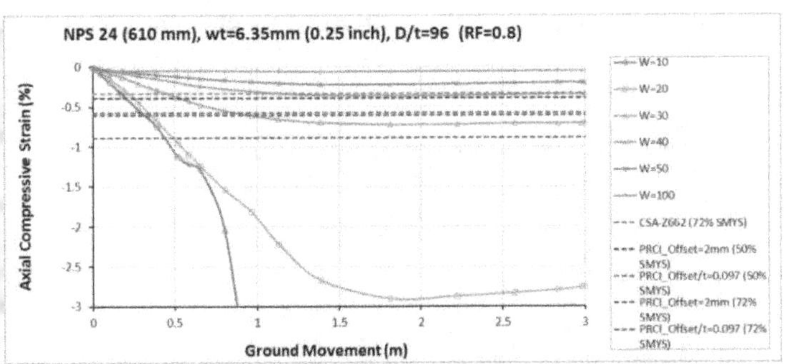

Figure 9: NPS 24 (D/t=96) - Maximum Local Pipe Compressive Strain vs Ground Movement Width

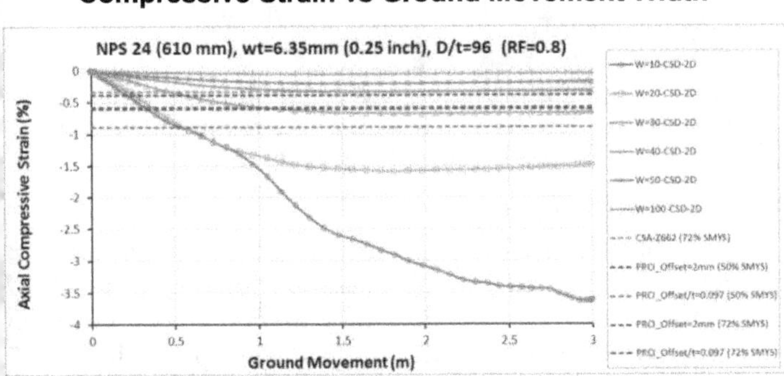

Figure 10: NPS 24 (D/t=96) - Maximum Nominal Pipe Compressive Strains vs Ground Movement Width

The results in Figures 9 and 10 are for NPS 24 (609 mm) with a wall thickness of 6.35 mm (0.25 inch) and D/t ratio of 96. The results illustrate the importance of the width of ground movement and how the critical width is a function of pipe stiffness (D/t ratio). The critical width is between 10 and 20 m. The width of 20 m (66 ft) results in the largest tensile strain in the pipe for any given value of ground displacement similar to the 9.52 mm (0.375) wall (D/t =64). However, the width of 10 m (33 ft) results in the largest compressive strain in the pipe for any given value of ground displacement. A comparison of the strain results of the two wall thicknesses 6.35 mm (0.25 inch) and 9.52 (0.375 inch) is provided in Figure 11 for W=10 m. The pipe with 6.35 mm (0.25 inch) wall exceeds the compressive strain limit and a buckle/wrinkle forms at 1.5% local compressive strain or 1.12% nominal compressive strain for soil displacement of 0.7 m.

Figure 12 shows the soil and pipe deformation (wrinkle formation) for soil displacements up to 3 m. The figurealso shows the total axial strain distribution (blue color is compressive, and red is tensile strain). In Figure 12 the soil above the pipe is hidden so the pipe deformation can be visualized.

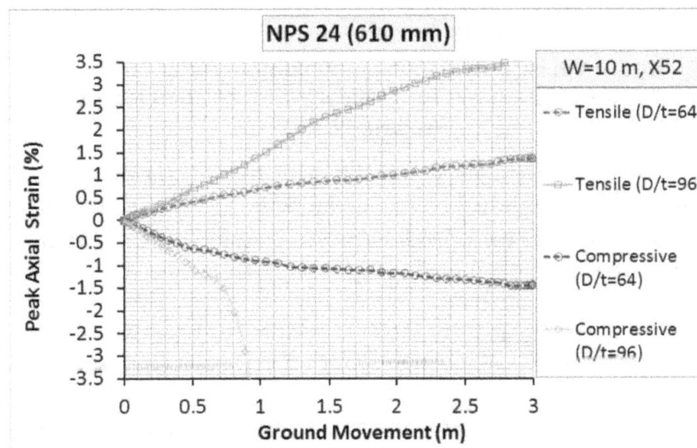

Figure 11: NPS 24 (Grade X52) – Maximum Axial Strain vs Ground Movement for the Critical Width of 20 m (33 ft)

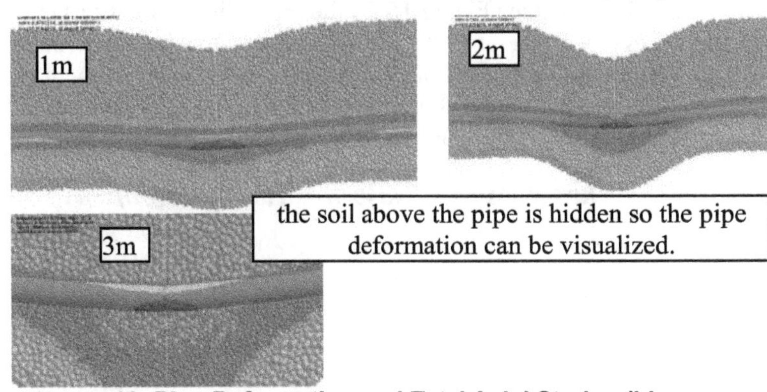

Figure 12: Pipe Deformation and Total Axial Strains (blue color is compressive, and red is tensile strain) for Given Soil Movement of 1 m, 2 m and 3 m (Top View)

Effect of Operating Pressure

Comparison of the pipe responses for three operating pressures of 4 MPa, 5.6 MPa and 8.1 MPa are considered to illustrate the impact of the pipeline operating at an internal pressure that produces a pipe wall stress of 36%, 50% and 72% of its SMYS. The results of the analyses for critical width of 20 m are provided in Figure 13 and Figure 14. These figures are

Copyright © 2019 by ASME

for grade X52 NPS 24 (610mm) pipe with t =9.5 mm (3/8 in) and D/t=64. Figure 13 presents the pipe maximum tensile and compressive strains versus maximum ground displacement for the critical displacement width of 20 m (66 ft).

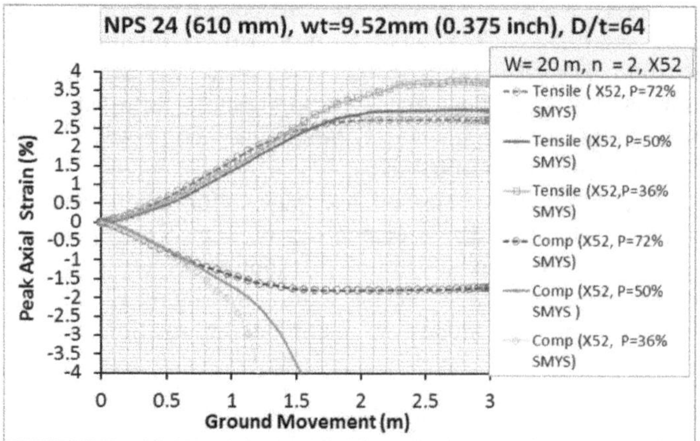

Figure 13: NPS 24 (D/t=64) – Maximum Axial Strain vs Ground Movement for Critical Width of 20 m (66 ft) – Pressure Effect

Figure 14 presents the sum of tensile (positive) and compressive (negative) strains in the pipe. The results indicated that reduction in the operating pressure results in and increase in the maximum computed compressive and tensile strains. The results also indicate that:

- Pipes operating at 72% SMYS do not exceed the compressive strain limit for the applied ground movement of 3.0 m. (see Figure 15 for the pipe soil deformation).
- Pipes operating at 50% SMYS exceed the buckle/wrinkle compressive strain limit (0.65%) for a soil displacement of 1.4 m. The wrinkle amplitude is 8 mm for a soil movement of 2.25 m. (see Figure 16 for the pipe soil deformation).
- Pipes operating at 36% SMYS exceeds the buckle/wrinkle compressive strain limit (0.55%) for a soil displacement of 1.0 m. The wrinkle amplitude is 28 mm for a soil movement of 2.25 m. (see Figure 17 for the pipe soil deformation).

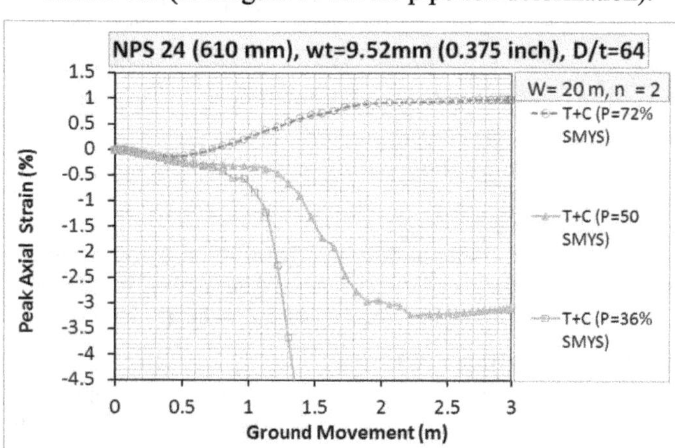

Figure 14: NPS 24 (D/t=64) – Sum of the Maximum Tensile and Compressive Strain vs Ground Movement for the Critical Widths of 20 m (66 ft) – Pressure Effect

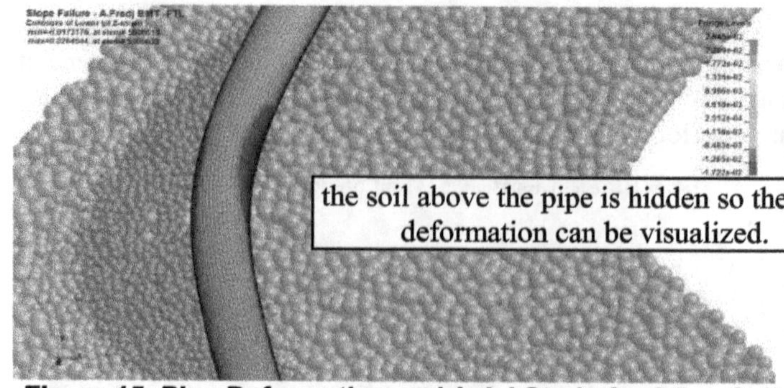

Figure 15: Pipe Deformation and Axial Strain for 2.5 m Soil Movement (Top View) – 0.72 SMYS Pressure (No Wrinkle)

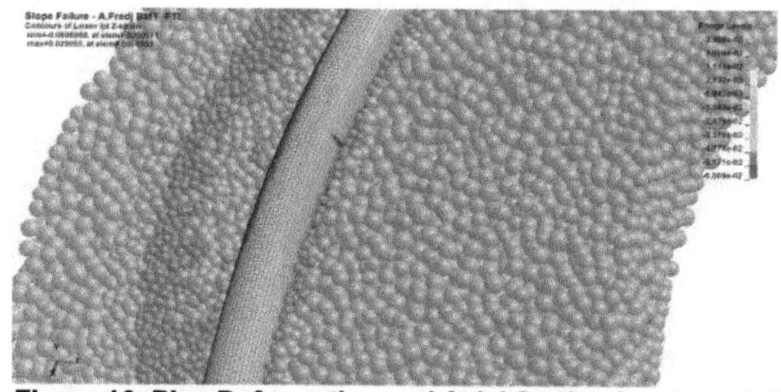

Figure 16: Pipe Deformation and Axial Strains for given Soil Movement of 2.5 m (Top View) – 0.50 SMYS Pressure (Wrinkle observed, Amplitude = 8 mm)

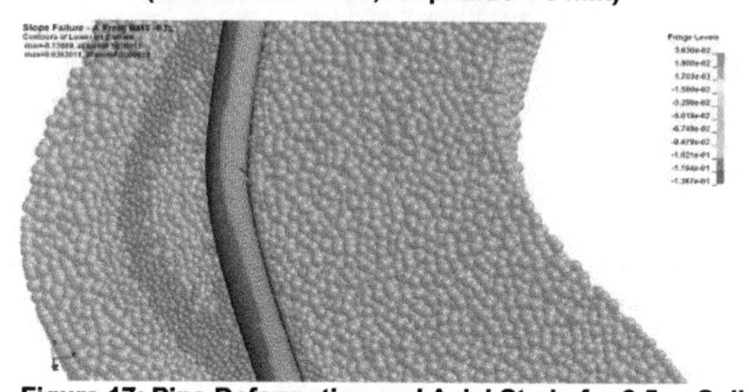

Figure 17: Pipe Deformation and Axial Strain for 2.5 m Soil Movement (Top View) – 0.36 SMYS Pressure (Wrinkle observed, Amplitude = 28 mm)

Effect of Temperature

Comparisons between the results of three temperature differentials (0, 45 and 65 °C) analyzed for critical width of 20 m (base case) are provided in Figures 18 and 19. Both figures are for grade X52 NPS 24 (610 mm) pipe with t =9.5 mm (3/8 in) [D/t=64].

Figure 18 presents the maximum tensile and compressive strains in the pipe versus the maximum ground displacement for critical ground displacement width of 20 m (66 ft). Figure 18 also presents the sum of tensile (positive) and compressive (negative) strains in the pipe. The results indicated that the calculated compressive strains are slightly higher at ground displacement less than 0.75 m for cases with a temperature

Copyright © 2019 by ASME

differential; this is mainly due to compressive strain induced by thermal loading. At large ground displacement, tensile strains are higher than compressive strain. This differential in peak tensile and compressive strains indicates that there is a significant uniform axial strain in the pipe due to larger catenary effect.

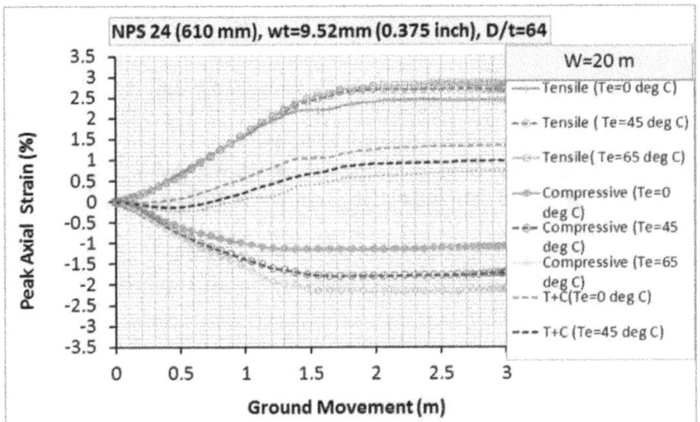

Figure 18: NPS 24 (D/t=64) – Maximum Axial Strain vs Ground Movement for Critical Width of 20 m (66 ft) – Temperature Effect

Reduction in the differential temperature results in reduction in the maximum computed compressive strains. The maximum computed tensile strain is not significantly affected. The computed tensile strain for case with no temperature is more than twice the computed compressive strain. Also, for large soil displacement the computed compressive strain for case with temperature differential is more than twice the computed compressive strain for cases with no thermal load.

Figure 19 shows maximum pipe displacement versus the maximum ground displacement. The maximum pipe displacement follows the ground displacement up to certain ground displacement than the pipe displacement increases slowly with ground displacement, which results in slowly increase in pipe strains. The results indicated that a higher operating temperature differential results in larger pipe displacement.

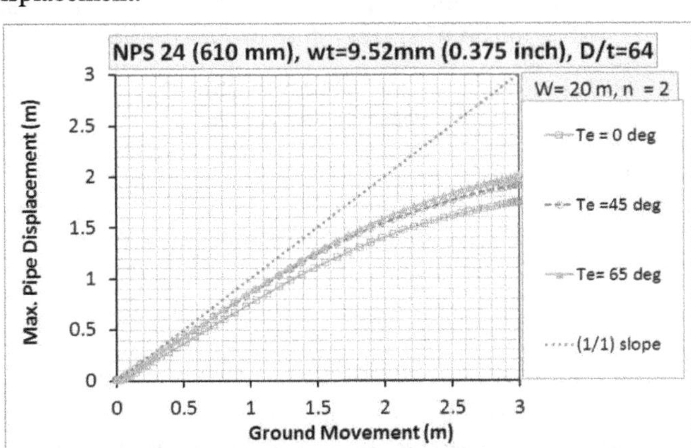

Figure 19: NPS 24 (D/t=64) – Maximum Pipe Displacement vs Ground Movement for the Critical Widths of 20 m (66 ft) –Temperature Effect, Grade X52

STRAIN DEMAND ENVELOPES

The strain demand determined from the analysis was first compared with the calculated strain capacity defined from the guidance; BS 7910 for tensile strain, and CSA Z662 and PRCI for compressive strain [3, 8]. While there are many existing models for calculating the compressive and tensile strain capacity in codes and standards, it was decided to develop a simple a strain-based design methodology that can be easily adapted to any code and standards, which is the purpose of this section.

The tensile and compressive strain demand envelopes along the longitudinal direction of the pipeline are presented in Figures 20 through 33. These figures show the maximum soil displacement versus ground width displacement as a series of 'envelopes' for a given pipe tensile and compressive strain demands of 0.25%, 0.5%, 0.75%, 1%, 1.25%, 1.5%, 1.75% and 2%. The safe pipeline operating goal of pipeline subjected to lateral ground displacement is to stay within the normal operating strain range defined by codes and standards.

Detailed results are presented in Figures 20 through 33 for a range of pipe grades, operating pressures and pipe D/t values. The results illustrate the differences in behavior across pipe, material pressure and geometry combinations supporting evaluation of a range of conditions and can be used to compare various design or assessment scenarios.

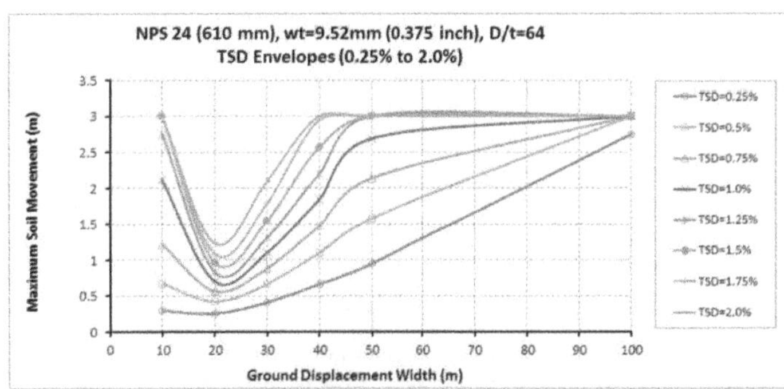

Figure 20: Tensile Strain Demand vs Width (W) – Grade X52 NPS 24 (D/t=64) operating at 72% SMYS

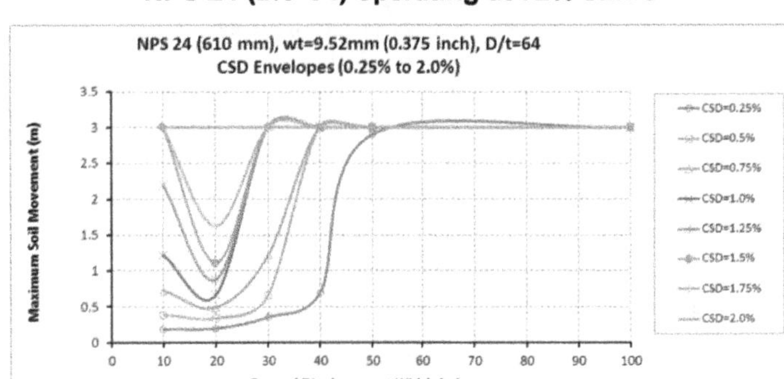

Figure 21: Compressive Strain Demand vs Width (W) – Grade X52 NPS 24 (D/t=64) Operating at 72% SMYS

Copyright © 2019 by ASME

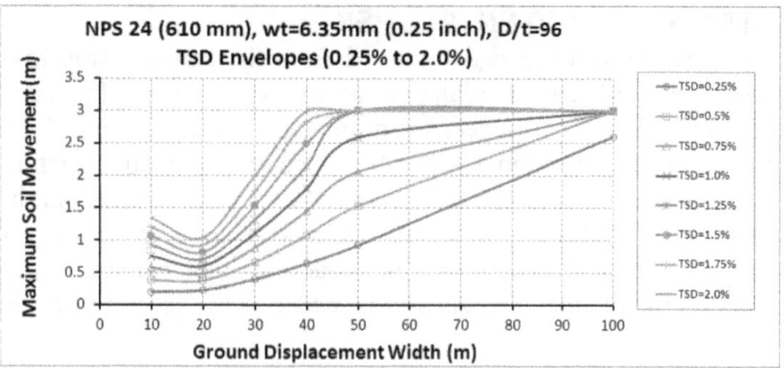

Figure 22: Tensile Strain Demand vs Width (W) – Grade X52 NPS 24 (D/t =96) Operating at 72% SMYS

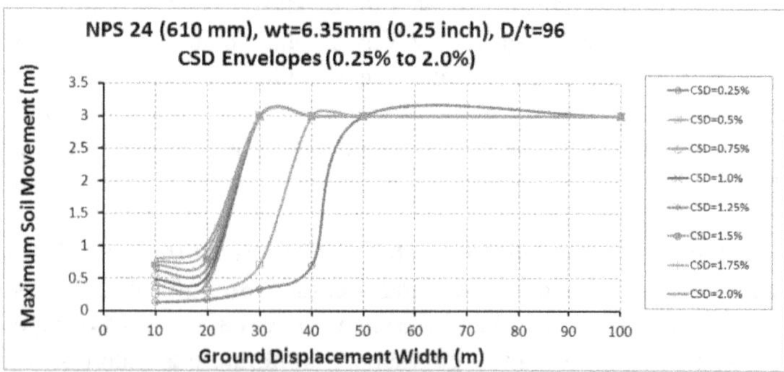

Figure 23: Compressive Strain Demand vs Width (W) – Grade X52 NPS 24 (D/t =96) Operating at 72% SMYS

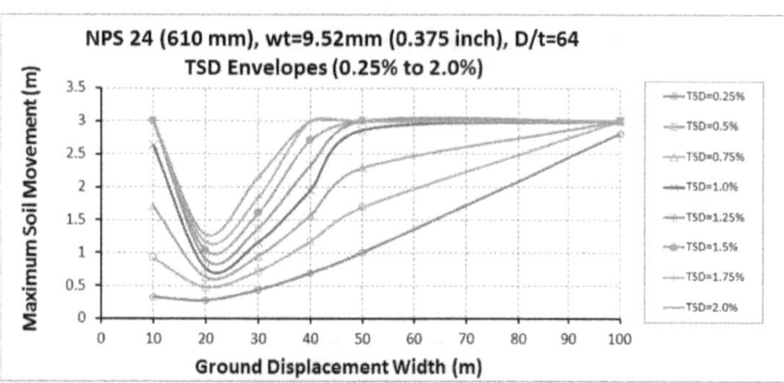

Figure 24: Tensile Strain Demand vs Width (W) – Grade X52 NPS 24 (D/t =64) Operating at 36% SMYS

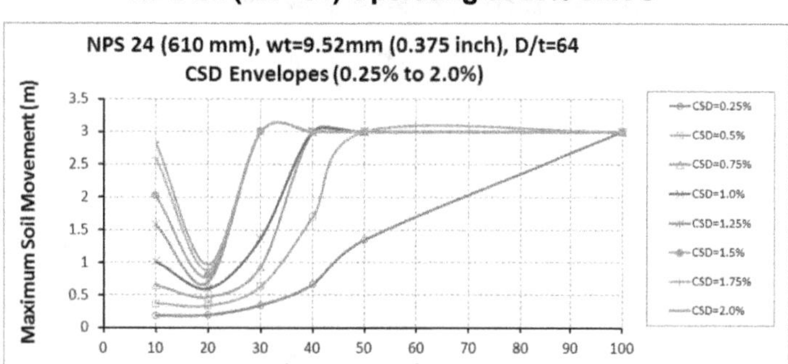

Figure 25: Compressive Strain Demand vs Width (W) – Grade X52 NPS 24 (D/t =64) Operating at 36% SMYS

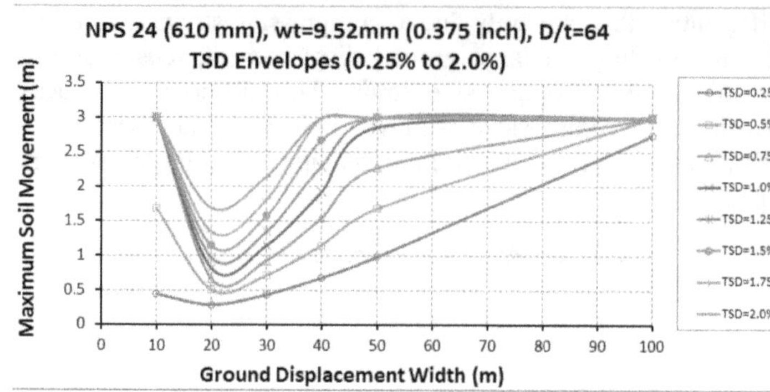

Figure 26: Tensile Strain Demand vs Width (W) - Grade X70 – NPS 24 (D/t =64) Operating at 72% SMYS

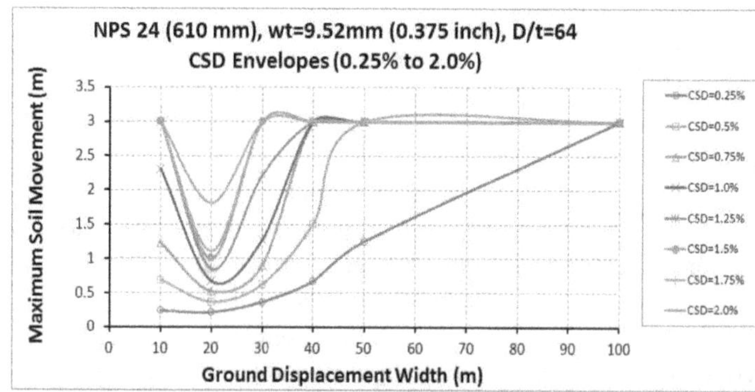

Figure 27: Compressive Strain Demand vs Width (W) - Grade X70 NPS 24 (D/t =64) Operating at 72% SMYS

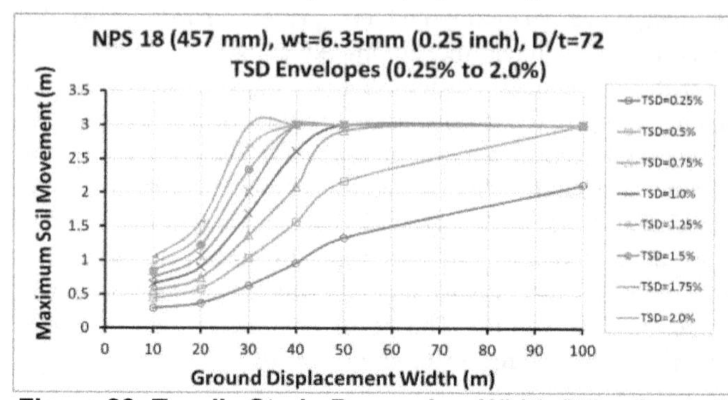

Figure 28: Tensile Strain Demand vs Width (W) – Grade X52 NPS 18 (D/t =72) Operating at 72% SMYS

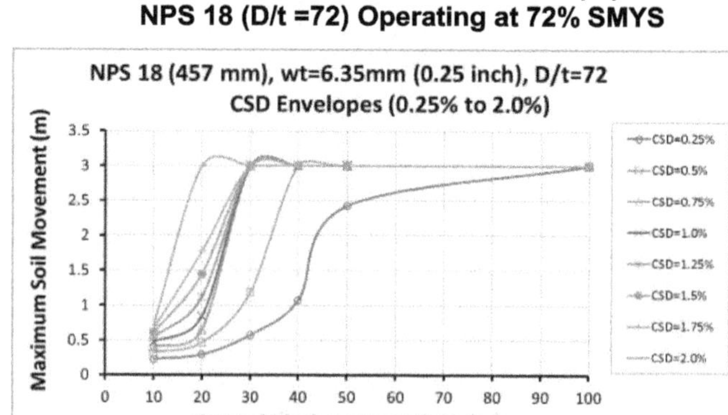

Figure 29: Compressive Strain Demand vs Width (W) – Grade X52 NPS 18 (D/t =72) Operating at 72% SMYS

Copyright © 2019 by ASME

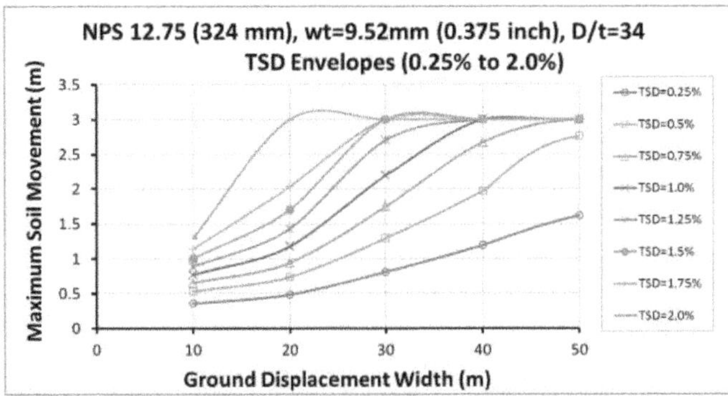

Figure 30: Tensile Strain Demand vs Width (W) – Grade X52 NPS 12 (D/t =34) Operating at 72% SMYS

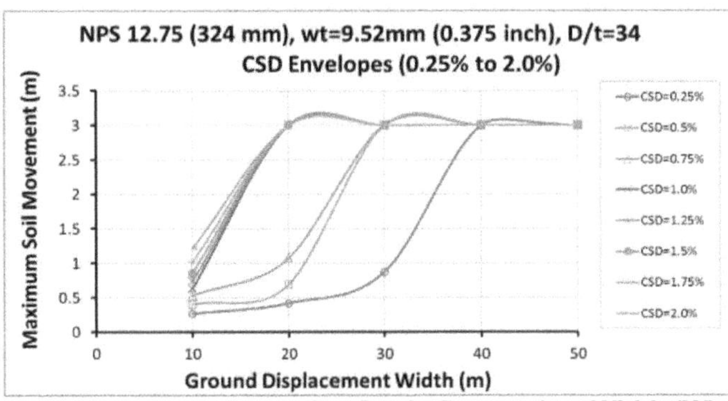

Figure 31: Compressive Strain Demand vs Width (W) - Grade X52 NPS 12 (D/t =34) Operating at 72% SMYS

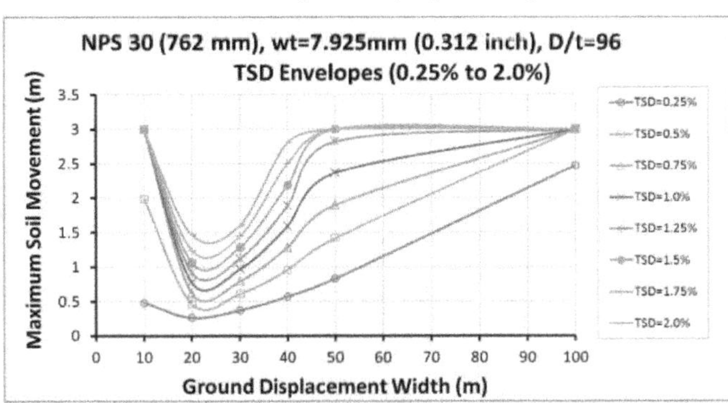

Figure 32: Tensile Strain Demand vs Width (W) – Grade X70 NPS 30 (D/t =96) Operating at 72% SMYS

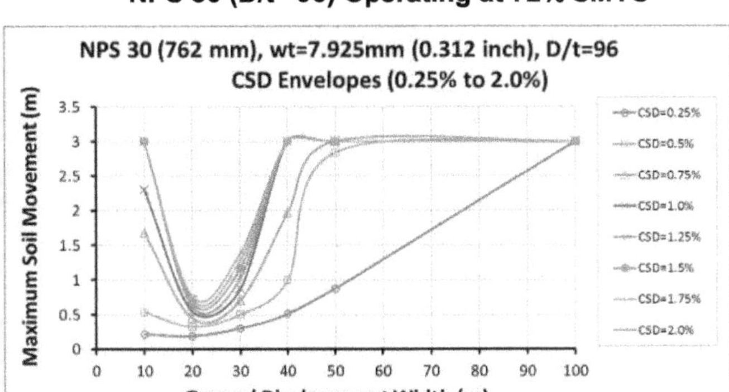

Figure 33: Compressive Strain Demand vs Width (W) – Grade X70 NPS 30 (D/t =96) Operating at 72% SMYS

CONCLUSION

The current paper provides a sample application of the DEM pipe-soil interaction geotechnical hazard assessment method employed to investigate the strains developed in pipelines subjected to lateral soil movement. The primary focus was an NPS 24 (610 mm) pipe system; however, the analysis also considered 12-inch (305 mm), 18-inch (457 mm) and 30-inch (762 mm) diameter pipes. The use of the simulated pipe responses to soil movement to consider the onset of tensile and compressive limit states in a strain-based assessment was demonstrated and the sensitivity of the pipe response to a number or geotechnical hazard scenario factors were presented, including:

- Pipe geometry
- Soil types
- Operating Pressure
- Ground movement pattern
- Material Grade
- Widths of soil movement
- Slope
- Temperature (axial load)

The analysis results presented in this paper and in related work [8] illustrate that not all parameters investigated had the same magnitude of impact. The effect of the parameters considered in these sensitivity studies indicate that the parameters can be ranked based upon their impact on strain demand as follows:

Higher Importance	Medium Importance	Lower Importance
- Magnitude of soil movement - Width of soil movement - Soil stiffness	- Pipe geometry (D/t) - Operating pressure - Temperature differential	- Soil to pipe friction - Material grade

The tensile and compressive strain demand envelopes along the longitudinal direction of the pipeline were developed. The maximum soil displacement versus ground width displacement as a series of 'envelopes' was developed for a given pipe tensile and compressive strain demands of 0.25%, 0.5%, 0.75%, 1%, 1.25%, 1.5%, 1.75% and 2%. The safe pipeline operating goal of pipeline subjected to lateral ground displacement is to stay within the normal operating strain range defined by codes and standards.

The strain for critical width lateral movement ground movement events (strain demand or applied strain) may be used conservatively to evaluate the response of pipeline system with similar geotechnical hazards. The allowable strain (strain capacity or resistance) should be evaluated for each pipe geometry, material and operating condition.

The observations made in this paper are related to lateral ground movements involving deformable soils. Care should be taken in applying these results to ground movement across a pipeline surrounded by rock or very stiff soils that will not deform and will cause extremely localized pipe deformations.

Copyright © 2019 by ASME

In future, it would be desirable to consider ground movement along the pipe axis or other geotechnical hazard scenarios to provide a more complete reference data set for the geotechnical hazard assessment. The results developed in this project could be extended to consider any arbitrary direction of soil movement relative to the pipeline.

ACKNOWLEDGMENTS

The Authors gratefully acknowledge the technical direction and sponsorship of the US Department of Transportation, Pipeline and Hazardous Materials Safety Administration, the Pipeline Research Council International (PRCI) and individual companies in supporting research related to the information presented in this paper.

REFERENCES

[1] LS-DYNA *"Keywords User's Manual", version 971.1*, Livermore Software and Technology Corporation, Livermore, CA. May 2014.

[2] ALA 2001, American Lifelines Alliance (ALA): Guidelines for the Design of Buried Pipeline, July 2001.

[3] A. Fredj and A. Dinovitzer: Application of Discrete-Elements Method to Evaluate Pipeline Response to Slope Movement -IPC2016 -64508, September 26-30, 2016 Calgary, Alberta, Canada.

[4] Fredj, A., Dinovitzer, A., and Zhou, J., 2008a, A 3-Dimentional Continuum ALE Model for soil-Pipe Interaction, Proceedings of International Pipeline Conference, Alberta, IPC2008-64624.

[5] Fredj, A., and Dinovitzer, A., 2010a, Three-Dimensional Response of Buried Pipelines Subjected to Large Soil Deformation Effects- Part I: 3D Continuum Modeling Using ALE and SPH Formulations, Proceedings of International Pipeline Conference, Alberta, IPC2010-31516.

[6] Fredj, A., and Dinovitzer, A., 2010b, Three-Dimensional Response of Buried Pipelines Subjected to Large Soil Deformation Effects- Part II: Effects of the Soil Restraint on the Response of Pipe/Soil Systems, Proceedings of International Pipeline Conference, Alberta, IPC2010-31517.

[7] Fredj, A., Dinovitzer, A., "Simulation of the Response of Buried Pipelines to Slope Movement Using 3D Continuum Modelling", IPC2012-31516.

[8] Fredj, A., "Definition of Geotechnical and Operational Load Effects on Pipeline Anomalies – Final Report", US Department of Transportation, PHMSA sponsored research contract DTPH56-14-H-00008, 2017.

[9] CSA Z662 –Standards Oil and Gas Pipeline Systems, 2011.

Copyright © 2019 by ASME

AIMPIMG2019-1034

Chapter 9

ON ESTIMATING PIPELINES RELIABILITY AT SLOPE CROSSINGS

Millan Sen
Enbridge
Edmonton, Alberta, Canada

Sherif Hassanien
Enbridge
Edmonton, Alberta, Canada

ABSTRACT

Transmission pipelines often traverse land slopes along the right-of-way; whereby some of these slopes might have a potential for instability related movements. In order to analyze the risk of failure of pipelines due to slope movement it is beneficial to establish probabilistic approaches that can quantify the likelihood of failure at each site given both aleatory and epistemic uncertainties. Estimation of such likelihood would support comparison of a pipelines' safety to reliability targets, validate whether slope management programs are acceptable, and support selection of mitigation alternatives. There is currently a gap in pipeline integrity literature in terms of available probabilistic approaches to analyze, assess, and manage pipelines crossing potentially moving slopes.

Three probabilistic approaches are presented herein: (1) A qualitative ranking analysis of slope hazards (QuRASH) which adopts site scores based on readily available slope crossing characteristics; (2) A semi-quantitative analysis of slope hazards (SQuASH) which is a reliability-based explicit limit state approach; and (3) A fully quantitative analysis of slope hazards (QuASH), which utilizes finite element modelling to establish a probabilistic strain demand for evaluating the pipeline's probability of failure through implicit limit state functions. The resulting slope threat prioritizations from these probabilistic approaches compared favorably with those slopes evaluated by experts to exhibit elevated threats.

The reliability approaches were applied to support decision making at a pipeline slope crossing that required mitigation. Various remediation/integrity mitigation designs were proposed by the project team, each with varying benefits. Determination of the probability of failure associated with each mitigation option/design enabled quantification of associated risks. Such information was considered by the decision makers for selection of the most favorable mitigation option.

INTRODUCTION

Pipelines in transmission pipeline networks often traverse land slopes along the right-of-way; especially near water crossings. While the vast majority of these slopes are stable, some might have a potential for instability related movements. Accordingly, pipelines subjected to these movements are susceptible to strain overload which may cause loss of containment in terms of buckling and/or tensile elongation failure modes. In order to analyze the risk of failure of pipelines due to slope movement it is beneficial to establish probabilistic approaches that can quantify the likelihood of failure at each site given both aleatory and epistemic uncertainties. Estimation of such likelihood would support prioritization of integrity mitigation actions and confirm pipelines' safety. There is a gap in pipeline literature in terms of available probabilistic approaches to analyze, assess, and support management of the slope movement integrity threat. Three probabilistic approaches are presented herein: (1) A qualitative ranking analysis of slope hazards (QuRASH); (2) A semi-quantitative analysis of slope hazards (SQuASH); and (3) A fully quantitative analysis of slope hazards (QuASH). Of these three approaches reported herein the QuRASH method is reported in fulsome, while the SQuASH and QuASH methods are presented conceptually due to their complexity.

For the QuRASH approach, qualitative crossing characteristic scores are assigned to each site. The focus of the approach is on the likelihood of slope movement initiating given the vast majority of slope crossings are stable; however the likelihood of pipeline failure should movement occur is also considered. Based on the resulting qualitative score for each site, the probability of failure (*PoF*) was established through application of historical-based failure rates. Select site scores were assigned a likelihood of failure based on historical events, and all slopes were assigned a *PoF* based on their relative scores to these sites.

Copyright © 2019 by ASME

For the SQuASH approach, a semi-quantitative methodology for establishing the *PoF* at slope crossings was developed. The probability of movement at each slope was directly assigned by geotechnical engineers through guidelines that relate slope characteristics with likelihood of creep instability. The probability of sudden mass landslide occurring at each site was also established. Subsequently the strain demand that would occur should these movements initiate was predicted through the use of a closed form equation. This strain demand was compared to the compressive and tensile pipeline strain capacities. Finally the *PoF* was the conditional probability of the likelihood of slope movements initiating multiplied by the likelihood of the strain demand exceeding the strain capacity should this occur.

For the QuASH approach, a fully quantitative methodology for establishing the *PoF* at slope crossings was developed. This approach is usually applied to slopes with measurable deep seated movements. The strain demand is determined using finite element analysis in consideration of the pipeline parameters, slope characteristics, and operating conditions. The strain capacity is determined through closed form equation. Distributions are applied to the strain demand through either application of previously reported FEA variability or development of a response surface; and distribution is applied to the strain capacity through consideration of input uncertainties. The *PoF* is established as the likelihood of the FEA strain demand exceeding the calculated strain capacity.

These probabilistic approaches have been applied to a large simulated sample of slope crossings that exhibit diverse characteristics representing North America geotechnical profiles. The resulting probabilities of failures were directly compared to those predicted based on expert judgement, and the top 50 probabilistic sites compared favorably to those evaluated to exhibit elevated threats by slope management program managers.

The reliability approaches were applied to support decision making at a simulated pipeline slope crossing that required mitigation. Various remediation/integrity mitigation designs were proposed by the pipeline project team, each with varying benefits. Determination of the *PoF* associated with each mitigation option/design enabled quantification of associated risks. Such information was considered by the decision maker for selection of the most favorable mitigation option.

INTEGRITY MANAGEMENT OF SLOPE CROSSINGS

When assessing the threat from geotechnical slope instability on crossing pipelines, expert based assessment methods are currently predominantly employed. For example: site observations are conducted at slope crossing sites by geotechnical engineers, indications of slope instabilities are reported, on-site monitoring may be installed, and additional assessments may be conducted. Integrity program managers would subsequently execute mitigation projects based on these evaluations. The entire management approach is predominantly based on the initial visual observations conducted by the geotechnical engineer. However, there is a potential that every high threat site would not be identified through this current approach for example: if a site was overgrown with vegetation during an inspection and the next inspection is scheduled long into the future; a site has not exhibited recent movements but has experienced ancient movements that could potentially become reactivated; or if a site exhibits limited likelihood for instability but the pipelines would likely fail in the event of minor movements. Accordingly, there is benefit in developing a check to the expert-based assessments conducted by geotechnical program managers. Ideally this check should be independent and take a completely different approach to maximize redundancy potential. Integrity program managers may then consider these alternate assessments and re-evaluate their management plans as warranted. A recent advanced integrity management system has been presented and employed by pipeline operators; namely, the pipeline integrity core process as shown in Figure 1 [6], which utilizes a plan-do-check-act scheme. Herein, the proposed reliability approaches serve the check-reliability targets box.

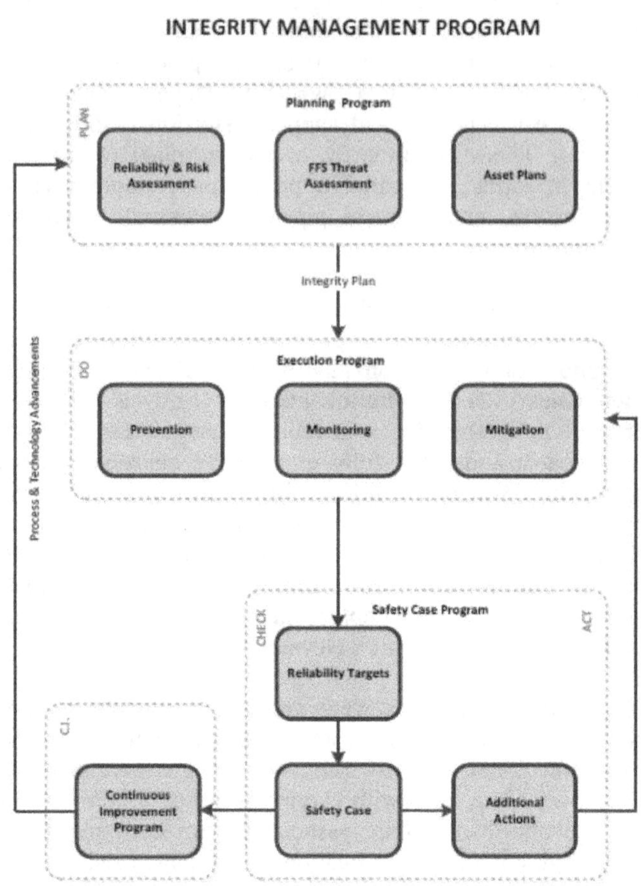

Figure 1 – Pipeline Integrity Core Process

Copyright © 2019 by ASME

RELIABILITY AT SLOPE CROSSINGS

The check-reliability targets step shown in Figure 1 consists of analysis of the *PoF* of the pipeline at a slope crossing and comparing it to a permissible probability of failure (PoF_p) or a reliability target (R_T). In general, there are three different integrity reliability approaches that can be utilized to conduct these analyses based on available information:

1. *Historical –based qualitative ranking analysis of slope hazard (*QuRASH*)*: the *PoF* is established based on historical failure rates of similar sites;
2. *Semi-quantitative reliability analysis of slope hazard (*SQuASH*)*: the *PoF* is established through a combination of statistical uncertainty quantification, quantitative explicit limit state-based reliability analysis, and expert based judgements.
3. *Quantitative analysis of slope hazard (*QuASH*)*: the *PoF* is established based on a quantitative implicit limit state-based reliability analysis using the finite element technique.

In regards to the implementation process for reliability analysis, as with any analysis technique these approaches exhibit both advantages and disadvantages. However, it is worth noting that for slope crossings expert-based assessments are currently more utilized than probabilistic modelling due to challenges in collecting and quantifying all of the geological uncertainties typically present at slope sites (e.g. spatial variation in soil strength along a slope). Hence, there is a benefit and need to generate multiple integrity reliability approaches in order to capture different site information scenarios.

QuRASH

Pipelines crossing slopes generally exhibit dramatically varying characteristics, and thus exhibit highly varying site specific *PoF*'s. For example, slopes that have confirmed instabilities are logically more likely to cause failure than those that are stable; slopes prone to toe erosion are more likely to cause failure than those that are not; long slopes are more likely to induce pipeline failure than short, etc. Based on this concept, a qualitative ranking assessment of slope hazards (QuRASH) methodology has been developed. Through literature review and comprehensive discussions with geohazard integrity program managers, geotechnical experts, and reliability experts, the main causes for slope instabilities have been identified and categorized [16]. Additionally main pipeline parameters that affect the likelihood of failure should movements occur have also been identified. Each of these categories has been assigned subcategories containing related slope crossing characteristics, and each of these subcategories has been assigned a score. The specific value for each of these scores has been determined through stakeholder input based on the expected impact to pipeline *PoF* and in consideration of inherent uncertainty in the available information. In order to determine the total score for each slope crossing, each subcategory score is multiplied together. This methodology, as shown in Table 1, results in each slope crossing exhibiting a unique score whereby sites with higher scores exhibit increased *PoF*.

When establishing the slope crossing scores using Table 1, site specific slope characteristics data is required to be provided by the geotechnical engineers who conduct the site monitoring. In order to increase the likelihood of representative results, if no features were identified then '*none of the above*' is selected opposed to assuming the most conservative (e.g. if the subsurface soil strength was unknown, it would be assumed to be competent strength).

Table 1 – QuRASH Characteristic Scores

Historical Slope Movements Identified	
Deep seated movements currently occurring on slope	15
Deep seated movements immediately adjacent to ROW	10
Surficial (> 0.5m scarp height) slope movements identified	9
Slope movements on ROW possible but not confirmed	8
Slope/ROW adjacent to slope exhibited historical movements but no signs of current movements	7
Slope movements in general vicinity (for example at nearby banks for river crossing)	6
Slope movements common at locations with similar geology	5
Surface erosion/degradation/shallow slumping identified	4
None of the above	1
Slope Angle	
Steep (> 15 degrees)	2
Shallow	1
Soil Strength Expected Subsurface	
Low Strength (swamp, weak saturated conditions, melted permafrost, presence of weak seam layer, liquefaction during earthquake, etc.)	2
Competent Strength	1
Triggering Factors	
Significant toe erosion potential / high pore water pressures expected / combination of any 2 active triggers	5
Melting permafrost / Changes due to heavy rainfall previously identified	4
Permafrost at site / poor drainage on slope / minor toe erosion potential	3
Within seismic area / surficial changes identified	2

Copyright © 2019 by ASME

(e.g. Loss of vegetation, surface erosion, ATV degradation, etc) / man-made activities that could affect slope stability	
None of the above	1
Site Mitigations	
Slope stability mitigations recommended / planned but not conducted	1.25
Minor surficial mitigations conducted / slope monitored at daily frequency	0.9
Substantial slope mitigations conducted	0.75
Permanent mitigations conducted	0.5
Pipe is HDD'd (horizontally directionally drilled) / permanent mitigations conducted and negligible movements and strains identified for past 5 years	0.1
None of the above	1
Instrumentation Measurements	
Measured deep seated on ROW movements > 25mm/year / Slope on ROW movements > 150 mm cumulatively	2
Measured deep seated on ROW slope movements < 25mm/year	1.5
Positive pore water pressure / Positive average annual temperature within permafrost	1.25
Slope movement monitoring installed and only negligible measurements identified	0.75
None of the above	1
Slope Ranking	
Very High	3
High/Medium	2
Low/Very Low	1
Pipeline Length along Slope	
> 70m long	3
20 - 70m long	2
<20m long	1
Crossing Angle	
Traverse to slope direction (> 15 degrees from parallel)	2
Parallel to slope direction	1
Pipeline Diameter	
Diameter < NPS 24	2
None of the above	1

An example of QuRASH application and total score evaluation is as follows for a slope crossing that:

- Was at the outside bend of a river, with measured deep seated movements of 30 mm/year, and was thus ranked as very high and planned to be mitigated;
- Exhibited a slope angle of 20 degrees, length of 100 m, and was within a muskeg area;
- The pipeline was NPS 20 and crossed the slope at a skew.

The site score would be:

15 * 2 * 2 * 5 * 1.25 * 2 * 3 * 3 * 2 * 2 = 27000 (which is the maximum possible score)

QuRASH was applied to 520 slopes representing North America's common pipeline slope crossing characteristics. The resulting sorted QuRASH score distribution for these sites is shown in Figure 2. It is demonstrated that the scores exhibited a uniform logarithmic distribution, which would be expected for pipeline slope crossings.

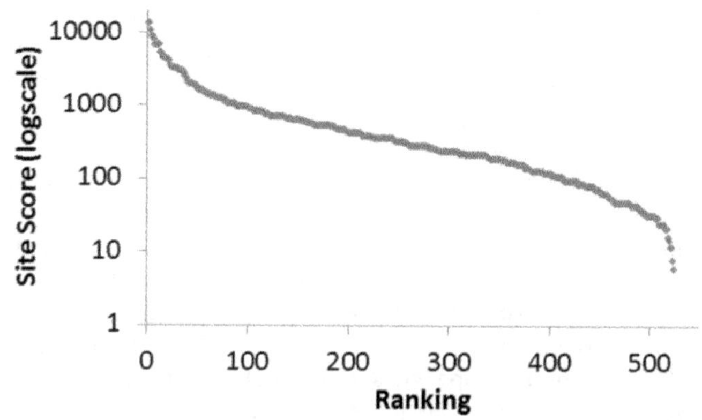

Figure 2 – Distribution of QuRASH Scores for 520 Slope Crossings

A relative *PoF* could be assigned to each QuRASH score by assigning probabilities of failure to 2 scores, and interpolating/extrapolating the *PoF* of the other scores based on these assigned *PoF* values. For example a site with a score approaching the maximum possible score should exhibit a very high *PoF* than a pipeline with a lower score that previously failed, and a site with a score near 1 should exhibit a lower *PoF* than the industry average. This was incorporated herein through the following approach [5]:

- The QuRASH site scores were plotted on a log scale;
- The pipeline system had historically experienced six pipeline buckles at slope crossings, and one of these buckles resulted in a leak. The *PoF*, given a buckle has formed, of 0.9×10^{-2} / site / year was assigned to the median score (8511) of these six buckled sites;
- The median score of all sites (302) was assigned the industry failure rate due to slopes of 2.6×10^{-5} / site / year as reported by BGC Engineering Ltd. [5]
- Based on these two fixed points, the relationship described in Equation (1) was used to define the scores of the remaining sites.

Copyright © 2019 by ASME

$$PoF_{QuRASH} = 10^{(1.75 \times \log(Score) - 8.93)} \quad (1)$$

SQuASH

The *PoF* at slope crossings may also be established using a semi-quantitative assessment of slope hazards (SQuASH) method. This approach combines geotechnical engineer expert based evaluations with an explicit limit state that contains probabilistic distributions for basic random variables (e.g. wall thickness, yield strength, tensile strength, etc.). Correspondingly SQuASH is expected to be highly independent from assessments conducted by the integrity program managers since it utilizes quantitative reliability analysis techniques.

While the vast majority of pipeline slope crossings are stable, every slope exhibits a potential to initiate a movement under certain conditions. Within the SQuASH approach, two types of slope movement were considered: steady state creep movement and sudden mass landslide. Steady state creep movement is what most commonly occurs at pipeline slope crossings whereby slope movements occur at a relatively slow constant rate; and sudden mass landslide is the condition whereby the slope suddenly moves by a large magnitude (herein >300 mm movement within a year). Should slope movements initiate, pipeline failure would occur if the strain demand induced by the moving slope block exceeds the pipelines strain capacity as described in Figure 3. If the pipeline tensile strain capacity is exceeded, a circumferential throughwall crack could form at a girthweld. Similarly, if compressive strain capacity is exceeded, a buckle would form and, with increased slope movement, a circumferential crack may form at the buckle.

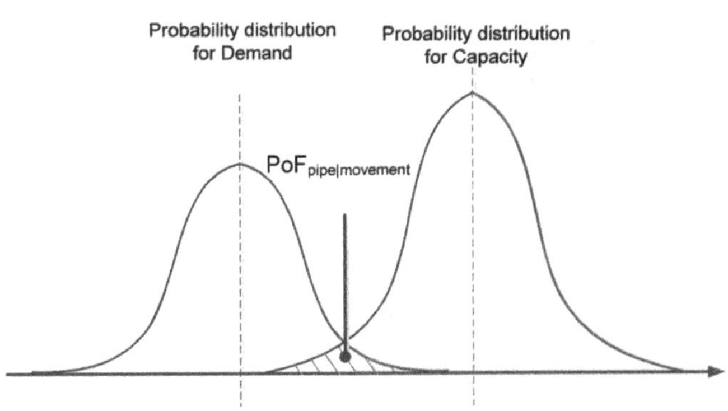

Figure 3 – Limit State for Reliability Analysis

Thus, the *PoF* at a slope crossing can be evaluated as a conditional probability: the probability of slope movement initiating multiplied by the probability of the pipe strain demand exceeding its capacity should movement occur as described in Equation (2).

$$PoF_{SQuASH} = P_{creep\ initiating} \times POF_{creep\ strain\ demand > capacity} + P_{landslide\ initiating} \times P_{landslide\ strain\ demand > capacity} \quad (2)$$

Establishing the probability of slope movement initiating is challenging given the complexity of slope characteristics, limited information that is available from visually based surficial observations, and uncertainty in future environmental conditions. Table 3, modified from a previously published report [2], was utilized to assign the probability of creep movement initiating at each site. The annual probabilities in Table 3 were revised from the table within [2] based on subject matter experts' judgments (resulting in higher probability of movements for each classification). The probability of creep initiating for each slope within SQuASH was assigned by geotechnical engineers, who selected these based on their site observations in comparison to the slope characteristics defined in Table 2.

Table 2 – Relationship between Slope Characteristics and Annual Probability of Creep Movement Initiating

Slope Characteristics	$P_{creep\ initiating}$ (/ site / year)
Deep seated landslide is imminent; landslide form is well defined; material is continuously moving or has evidence of episodic movement within past 5 years; main scarp sharp and unvegetated; landslide expected during the life of the pipeline	$7\,(10)^{-1}$
Deep seated landslide form is well defined; evidence of triggering of movements from events that occur, on average, every 10 years to 50 years; adjacent to ROW main scarp evident and unvegetated; landslide expected during the life of the pipeline	$3\,(10)^{-1}$
Clear evidence of deep seated slope movement within the past 100 years; potential for movement due to mobilizing triggers, shallow slope movements within past 10 years, landslide expected during the life of the pipeline	$3\,(10)^{-2}$
Slopes have no evidence of previous instability; potential from movement based on observations of similar slopes, historical records, or analysis; landslide not expected during the life of the pipeline	$3\,(10)^{-3}$
Slopes have no evidence of previous instability; no evidence of movement based on observations of similar slopes, historical records, or demonstrated by analysis; mobilizing triggers unlikely to initiate movements; extremely remote chance of landslide during the life of the pipeline	$3\,(10)^{-4}$
Slopes have no evidence of previous instability; no evidence of movement based on observations of similar slopes, historical records or demonstrated by analysis; no presence of mobilizing triggers, possibility of landslide during the life of the pipeline can be ignored	$3\,(10)^{-5}$

Copyright © 2019 by ASME

Directly assessing the probability of a sudden landslide is even more complex than the above and the error associated with any such evaluations was expected to be equivalent to or exceed more practicable correlation approaches. It was therefore assumed that the probability of mass landslide was 1/50 of the probability of slope creep initiating. This was determined based on discussion with geotechnical experts.

A previously developed closed-form equation was used to quantify the strain demand induced by slope movements for the SQuASH assessments [13]. The closed-form state is formulated based on a simplified estimation of the relationship between longitudinal ground movements and the resulting pipeline strains. The governing equation is shown in Equation (3) while the details of the analytical model and an example are presented in [13].

$$\frac{dN}{dx} + f = 0 \quad (3)$$

where,

N = axial force; and

f = pipe-soil friction force per unit length (x)

The basic variables for the strain demand model are: pipe diameter, wall thickness, yield strength, internal pressure, temperature differential, pipe bend angle, soil backfill stiffness, soil backfill yield displacement, and slope movement. The pipeline properties should be readily available from operators' records, the operating parameters were based on maximum operation pressure (MOP) and the estimated difference between the installation and pipeline operating temperature, the slope length and bend angle were based on topographical maps, soil properties were estimated by the geotechnical engineers, and the slope movement magnitude was developed herein in consideration of re-inspection interval. Probabilistic distributions were assigned to all of these inputs, except for MOP which was assumed (conservatively) to be deterministic.

The strain demand predicted by Equation (3) is sensitive to the slope movements that affect the pipeline. For creep movements, the re-inspection interval for each slope was reported by the geotechnical engineers. For slopes with no confirmed movements, it was assumed that the geotechnical engineers were unable to identify between 200mm to 400mm of pre-existing movements due to them being subsurface, masked by vegetation overgrowth, etc. Assuming that movements initiated immediately following the last inspection, a slope movement rate with a mean of 12.5 mm / year, a coefficient of variation of 0.5 was multiplied by the inspection frequency based on a joint industry project [7]. Correspondingly, the slope movements modelled in the strain demand equation is shown in Equation (4) and an example of the resulting compressive and tensile strain demand distributions are shown in Figures 4 and 5 respectively.

$$\textit{Creep Movement} = \textit{Unobservable Movement} + \textit{Slope Movement rate x Re-inspection interval} \quad (4)$$

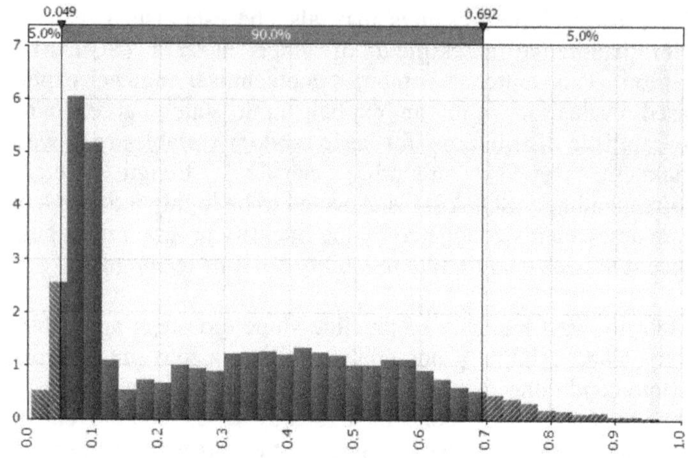

Figure 4- Example Creep Movement Compressive Strain Demand Distribution

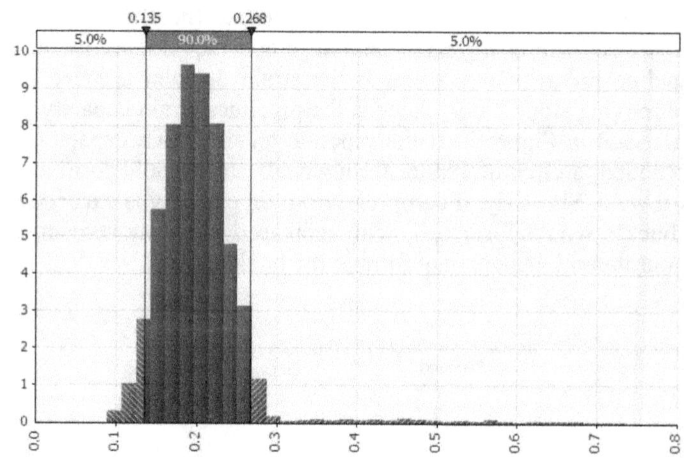

Figure 5- Example Creep Movement Tensile Strain Demand Distribution

Based on discussion with geotechnical experts, it was assumed that if there was a sudden landslide it would range in magnitude from 300 to 4000 mm, with a most likely movement of 800 mm. This movement distribution was applied to the strain demand calculations for landslide movements. As the input movements for the landslide were generally much higher than for creep movements, the strain demand was generally higher as well. However this was counterbalanced by the decreased probability for a landslide event to occur.

The pipeline buckle strain capacity was calculated using the University of Alberta [3, 14] plain pipe rounded stress-strain curve calculation shown in Equation (5). All of the inputs for this equation within the SQuASH approach were probabilistically defined (except for internal pressure which

Copyright © 2019 by ASME

was set at MOP deterministically). As these resulted in distributed results, the strain capacity predictive equations model errors were neglected.

$$\varepsilon^{p}_{crit} = \left(\frac{2.94}{D/t}\right)^{1.59} \cdot \left(\frac{0.01}{1-0.868\left(\frac{P_{int}}{P'_y}\right)}\right) \cdot \left(\frac{E}{F_y}\right)^{0.854} \cdot \left(1.27 - \left(\frac{imp}{100}\right)^{0.150}\right) \quad (5)$$

where,

ε^{p}_{crit}= buckle strain (mm/mm)
D = outside diameter (mm)
t = wall thickness
P_{int} = internal pressure (MPa)

$$P'y = \frac{2 \cdot t \cdot F_y}{(D - 2 \cdot t)}$$

F_y – yield strength (MPa)
E = Modulus of Elasticity (MPa)
Imp= pipe body imperfection such as ovality

The formation of a buckle in a pipeline is considered a serviceability limit state, whereby the pipe section may be repaired upon identification, without a loss of containment. However if a buckle were to form and the slope movements were to continue, a circumferential crack could initiate at the buckle. Correspondingly Equation (6) was generated herein to consider that a buckle could accommodate approximately 2 to 3% additional global strain prior to fracture. An example distribution for the ultimate compressive strain capacity ($\varepsilon_{Ultimate\ Compressive}$) for a pipeline is shown in Figure 6.

$$\varepsilon_{Ultimate\ Compressive} = \varepsilon^{P}_{Crit} + (2\ to\ 3\%) \quad (6)$$

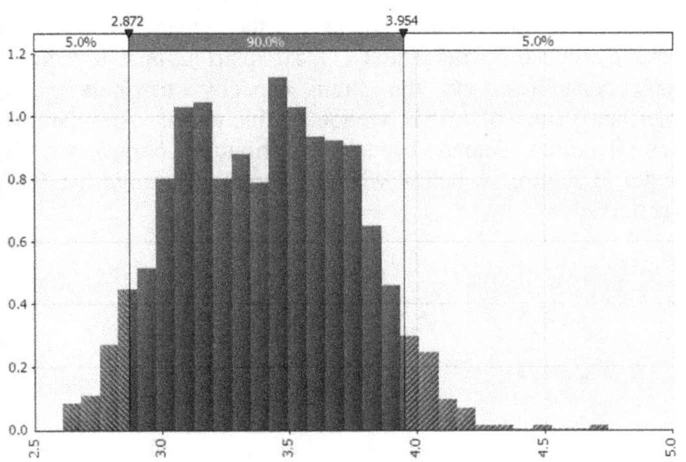

Figure 6 – Example: Ultimate Compressive Strain Capacity Distribution

There is also a potential that slope movements can induce tensile strains that are sufficient to exceed the tensile capacity. This would generally occur near the top of the slope where there are both axial tensile strains and increased bending strains present. More importantly, the tensile strain capacity at girth welds (GW)'s is much lower than that of the pipe body given the potential presence of manufacturing weld flaws, geometric weld imperfections, material property inconsistencies, and other stress concentrators. Correspondingly in order to be conservative, it was assumed that the maximum tensile strain would occur at a GW location. Should a tensile overload at a GW occur, it is expected that a circumferential crack would form. The tensile strain capacity model selected for use in SQuASH was that developed by CRES and sponsored by the US Department of Transportation and PRCI. The equation is overviewed in Equation (7); details are provided in the PRCI report [12].

$$TSC = \min(\varepsilon_u, P(f_p)G(t)TSC_p) \quad (7)$$

where, ε_u represents the pipe uniform strain and the functions $G(t)$ and $P(f_p)$ charactorize the effect of wall thickness and internal pressure; respectively. An example resulting tensile strain capacity distribution is shown in Figure 7.

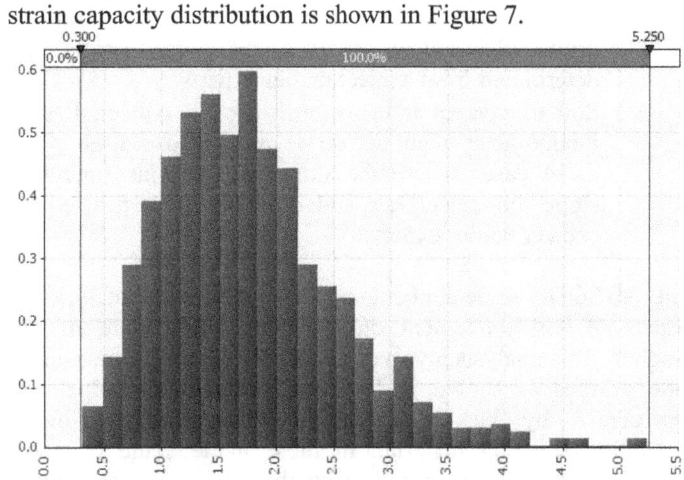

Figure 7 – Example: Tensile Strain Capacity Distribution

The limit state function used in SQuASH is shown in Equation (8).

$$P_{strain\ demand > capacity} = Limit\ State\ (Demand - Capacity) \quad (8)$$

When conducting Monte Carlo simulations to solve this limit state, it was revealed that for all of the slope crossings assessed the probability of the strain demand exceeding the capacity, was always > 0.001 (if the slope movements were to initiate). This is fairly reasonable given that if a pipeline is located on a moving slope, pipeline failure would exhibit increased potential.

Copyright © 2019 by ASME

QuASH

The most representative method for establishing the *PoF* at slope crossings is expected to be through a fully quantitative assessment of slope hazards (QuASH). This methodology is the same as that of SQuASH, except the strain demand is determined using finite element analysis. As this method requires considerable computational resources, it is usually only applied to slopes with measurable deep seated movements in their current condition ($P_{creep\ initiating} = 1$, $P_{landslide\ initiating} = 0$). The strain capacity assessments and limit state evaluation are conducted using the same methodology as described in the SQuASH section above.

The key inputs to the QuASH model include:

- Pipe properties, such as diameter, wall thickness, grade, etc.
- Pipe geometry, which may be obtained from as-built drawings or in-line inspection data
- Soil properties, which are collected from geotechnical consultant reports when available. For most cases the full set of information required (expected values with uncertainties) is not available and the input is estimated based on soil type, following distributions determined from subject matter experts
- Soil movement information, which is collected from inclinometer readings on instrumented slopes. For those cases where the slopes are not instrumented, slope movement rate and direction are estimated by subject matter experts

The ABAQUS finite element software was used to analyze the effects of soil movement on the stress-strain state of the pipeline. This analysis involved generation of three dimensional pipe profiles as shown in Figure 8, whereby the pipe was represented by beam elements comprised of non-linear isotropic hardening material. In these models, the pipe-soil interactions were determined using three dimensional springs [2]. This analysis allowed for the incorporation of soil movement of any magnitude and in any direction relative to the pipeline, making it an ideal tool for scenario analysis. Each finite element model results in a stress and strain distribution across the pipeline for the slope, which was used to identify the peak tensile and compressive strains as well their locations along the slope.

Figure 8 – Example FEA Pipe Profile

Either a simplified or detailed method was used to determine the uncertainties in the FEA strain demand depending on analysis requirements. For the simplified method; a joint industry project [7] reported that FEA strain demand uncertainty follows either a lognormal, gamma or Weibull distribution with COV's ranging between 30% to 100% for tensile strain demand, and 10% to 140% for compressive strain demand. Thus herein a normally distributed 35% COV was applied to both the mean FEA tensile and compressive strain demands during the limit state evaluation. For the detailed method; a response surface was developed by conducting 2n+1 FEA simulation per slope; where n is the number of variables of interest. The response surface was then used to generate a closed-form equation which was used to calculate the strain demand probabilistically as shown graphically in Figure 9. The benefit of the detailed methodology is that each of the critical variables is incorporated directly into the analysis and case-specific uncertainties are defined for each model. The primary limitation of this methodology is that it requires a series of models to be developed which can lead to prolonged computational times.

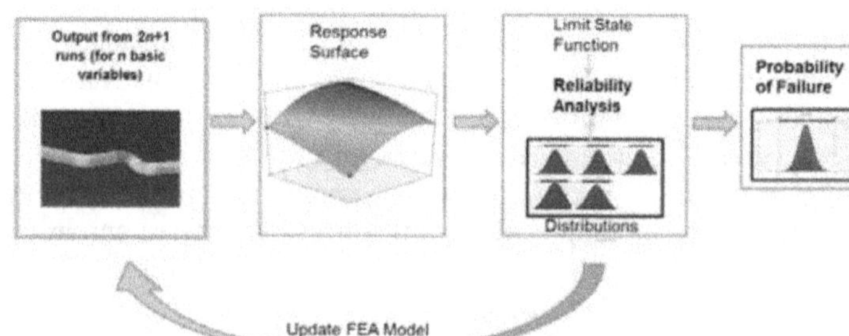

Figure 9 – QuASH Methodology

When evaluating the QuASH limit state, pipe failure occurs if the strain demand exceeds the capacity. Monte Carlo simulations were conducted using the mean FEA strain demand, the strain demand COV's / distributions, the mean strain capacities, and the strain capacity distributions as described in the SQuASH section of this paper. An example QuASH strain demand/capacity distribution comparison is shown in Figure 10 below whereby for this example the *PoF* was 6.6E-01.

Copyright © 2019 by ASME

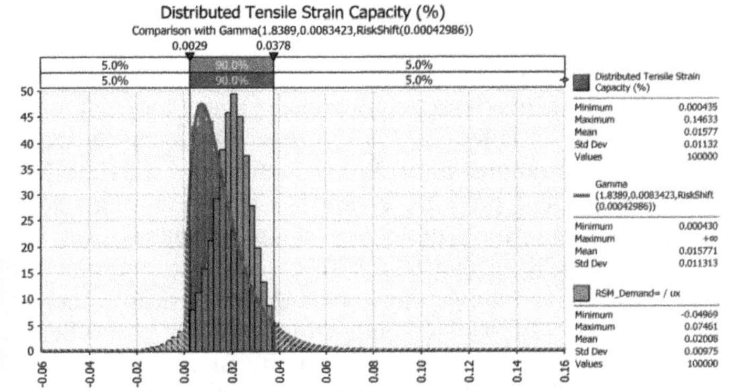

Figure 10 – Example QuASH Comparison of Strain Demand and Capacity

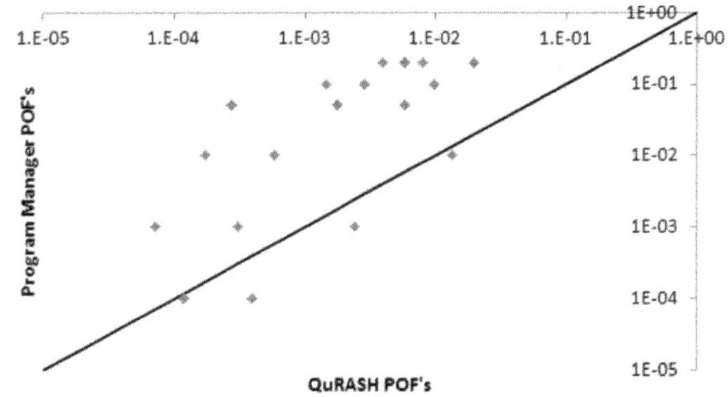

Figure 11 – Comparison between QuRASH and Program Manager POF's

RELIABLITY TARGET AT SLOPE CROSSINGS

Reliability Targets that have been previously used by pipeline operators for mainline leak and rupture are 1E-03 and 1E-05 respectively. The historical failure rate used to convert the QuRASH scores to *POF's* was based on the historic leak failure rate for slopes. Correspondingly the 1E-03 target would be appropriate for the QuRASH reliability target. If rupture rate for QuRASH is required, the *POF's* could be multiplied by 0.1 which is the PHMSA natural force (2002-2012) rupture / leak failure rates [9]. For the SQuASH and QuASH models the limit state is based on the tensile strain demand infinitesimally exceeding strain capacity. This is a leak failure mode and thus the 1E-03 target would be applicable to the SQuASH *POF's* as well. Rupture would occur if the tensile strain capacity was exceeded considerably (2-4% based on discussion with pipe strain experts), and it is evaluated that the probability of this occurring is extremely low and thus the leak Reliability Target would govern.

EVALUATION OF SLOPE RELIABILITY MODELS

The resulting *PoF*'s from the QuRASH evaluations were compared to those for 36 sites that had been previously assigned *PoF*'s based on program manager expertise as shown in Figure 11. It is shown that a reasonable correlation was achieved whereby the QuRASH *PoF*'s tended to increase as the program manager *PoF*'s increased. It is also shown that the program manager *PoF*'s were generally higher than those from QuRASH, which would be expected given program managers general tendencies towards conservative assumptions. Thus based on the correlations presented in Figures 2 and 11, it is demonstrated that the QuRASH methodology provides a reasonable ranking of individual slope crossing threats, and provides a *PoF* for each site within engineering tolerance.

The resulting *PoF*'s from SQuASH were compared to those from the QuRASH analyses, and reasonable correlations were achieved as shown in Figure 12. This is impactful in regards to acceptance because the two approaches vary significantly but agree with both each other and the program managers *PoF* estimations.

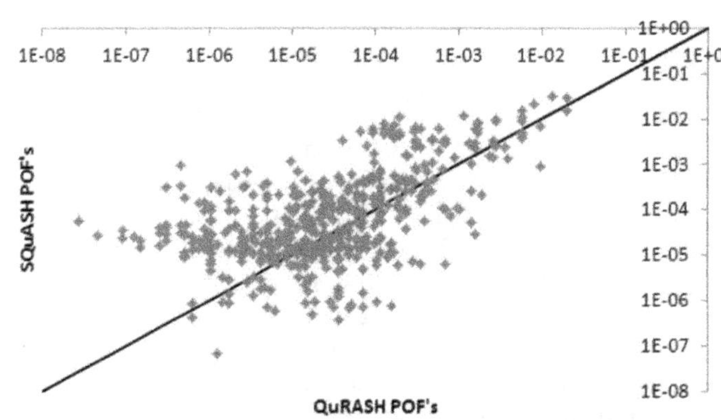

Figure 12 – Comparison of QuRASH and SQuASH *PoF's*

In consideration that the results of both QuRASH and SQuASH compared favorably, the logarthmic average of the two approaches was used for reporting of the QuRASH/SQuASH (semi-quantitative) slope *PoF's*.

The QuASH *PoF's* that were calculated using FEA were compared to those calculated using QuRASH/SQuASH, whereby QuASH was conducted on the 14 pipelines that were crossing the expected 4 highest threat slopes in the simulated system. It is shown in Figure 13 that the QuASH *PoF's* were generally within 3 orders of magnitude of those calculated using QuRASH/SQuASH. For those QuASH *PoF's* that exceeded those calculated using QuRASH/SQuASH, it is expected that this resulted because the measured slope movement rates were much higher than the historical average slope movent rates. For those QuASH *PoF's* that were lower, these are expected to have occurred because several of the pipelines had had undergone stress reliefs which could be

Copyright © 2019 by ASME

considered within the FEA models but not the semi-quantitative models. In any case as FEA models are generally more accurate than closed form solutions when predicting the strain demand at moving slopes, the QuRASH *PoF's* are considered to be more representative.

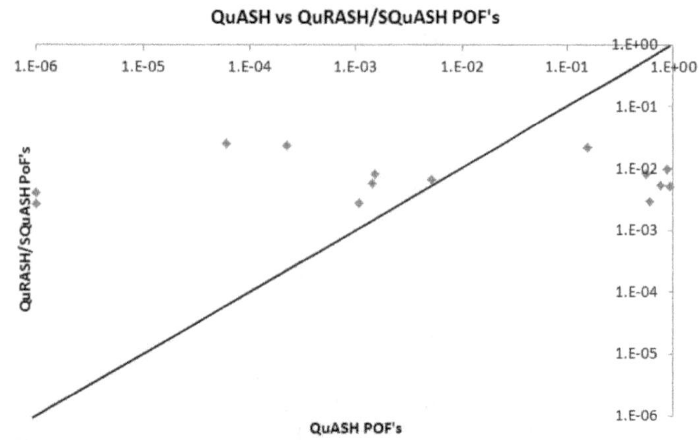

Figure 13 – Comparison between QuASH and QuRASH/SQuASH *POF's*

The QuRASH/SQuASH *PoF*'s were established for the 520 high ranked slope crossings in the simulated system. This resulted in 50 sites exhibiting a *PoF* higher than a proposed permissible *PoF* (PoF_p) of 1.0E-3 per site for the leak failure mode. The resulting top 50 sites were compared to the program manager empirical evaluation of these sites. It was identified that this top 50 list identified every high-threat slope site within 500+ sites evaluated, and that 70% (35 / 50) of the probabilistic top 50 crossings were also evaluated to be high-threat by the progam managers as demonstrated in Figure 14. This agreement is considered to be near optimal in consideration that the probabilistic approaches are intended to act as a check to the program managers expert-based integrity decisions.

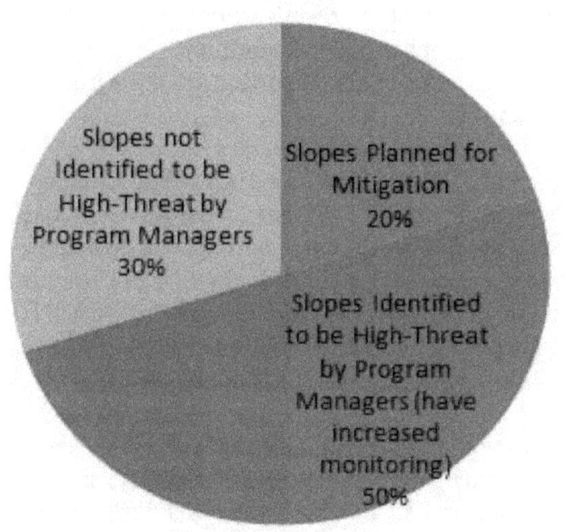

Figure 14 - Comparison of Relability Top 50 sites and Program Manager Evalutation of These Sites

CASE STUDY

The simulated ROW crosses a slope in Northern Alberta that has experienced over 2m of movement since the first of 6 pipelines were installed in 1999. This movement caused a pipeline buckle failure in 2013, and in response a $100MM+ slope management program comprised of monitoring, slope stabilization, and strain mitigation actions is ongoing at the crossing [15]. Given that this ROW is crucial to multiple oil sands shippers, and the pipeline operators low risk tolerance at this location, alternative mitigation options are being considered at this location. Quantification of the post mitigation *PoF's* would provide support to decision makers in evaluating the optimal mitigation option. Some key considerations during the *PoF* analysis was that a failure on any single pipeline would result in shutdown of the entire ROW, and that pipeline operations at this location are expected to continue for at least 25 years. The mitigation options considered and their governing operational threats (failure mode that would cause shut down of the ROW) are described below:

1. Continue existing slope management plan
   - Strain overload on any of the 6 lines on slope
2. Replace all 6 lines on slope with heavy wall
   - Strain overload on any of the 6 heavy wall lines on slope
3. Relocate 2 of the lines, and replace the other 4 in place with heavy wall
   - Strain overload on any of the 4 heavy wall lines remaining on slope
4. Replace all 6 lines on slope above grade
   - 3rd Party Damage on any of the 6 lines at slope, or slope movements sufficient to cause above grade failure
5. Re-route all 6 lines
   - Strain overload caused by other more stable slopes that would be crossed by the pipelines

The resulting *PoF's* and their descriptions are provided in Table 3 below:

Table 3 –Slope Mitigation Option *PoF's*

Option	Operational *PoF*	*PoF* Description
(1)	$10^{-1} – 10^{-2}$	QuRASH/SQuASH/QuASH of all 6 lines
(2)	$10^{-2} – 10^{-3}$	QuRASH/SQuASH/QuASH of all 6 heavy wall lines
(3)	$10^{-2} – 10^{-4}$	QuRASH/SQuASH/QuASH of 4 heavy wall lines remaining on slope

Copyright © 2019 by ASME

(4)	$10^{-2} - 10^{-3}$	3rd Party Damage Risk + Estimated Above Grade threat from Slope Movement on all 6 lines
(5)	$10^{-4} - 10^{-5}$	Historical Failure Rate at Slope Crossings

As shown in Table 4 it was quantified that maintaining the current slope management program resulted in the highest *PoF*, re-routing all 6 pipelines resulted in the lowest *PoF*, either replacing the lines with heavy wall pipe or replacing them above grade would result in similar *PoF's*, and that not all mitigation options resulted in achieving of the 1E-03 Reliability Target. These safety quantifications, in addition to key considerations such as cost, schedule, and operational impacts, provided high value to the risk based decision makers in selecting the optimal mitigation option.

CONCLUSION

Slope crossing management is currently predominantly conducted based on program manager expert judgement that considers on-site monitoring, deterministic assessments, and historical data results. There is benefit to conducting an alternative evaluation that provides a 'check' to expert assessments, and also quantifies the slope crossing threat to support risk based decisions. Three probabilistic approaches for slope assessment have been developed herein.

The QuRASH approach focused on scoring the threats at slope crossing locations based on their site specific characteristics, and assigning them *PoF*'s based on historical failure rates. The SQuASH approach focused on evaluation of the probability of strain demand exceeding capacity should slope movements occur. The QuASH approach compared the FEA calculated strain demand at moving slopes to the pipeline strain capacities. The QuRASH approach provided reasonable agreement with slope *PoF's* that were estimated by program managers. The SQuASH approach showed reasonable correlation with the QuRASH, even though the approaches are very different. Correspondingly the logarithmic average of the QuRASH and SQuASH methods was used to establish the semi-quantitative *PoF*'s. The QuASH *PoF's* were within 3 orders of magnitude of the QuRASH/SQuASH *PoF's*, but is considered to be the most accurate because additional input variables are considered.

The *PoF*'s for 520 slope crossings were calculated, and of the 50 slope sites that exhibited *PoF*'s higher than a proposed PoF_p, 35 of them were evaluated to also be high-threat by the program managers, and there were no program manager evaluated high-threat slopes that were outside of this top 50 list. This level of agreement is near optimal in consideration that the probabilistic methods are intended to encourage a program manager check of alternate sites.

The *PoF* analyses were applied to evaluation of mitigation options at a critical slope crossing that had previously exhibited large movements. QuRASH, SQuASH, QuASH, historical failure rates, risk assessments, and engineering judgement were utilized during the evaluation. The resulting *PoF's* supported decision makers in selecting the optimal mitigation option.

REFERENCES

[1] Abdolrazaghi, M. & Hassanien, S., Geohazard Reliability Analysis and Assessment, Draft Enbridge Internal Report, 2017

[2] C-CORE, D.G. Honegger Consulting and SSD, Inc., Guidelines for Constructing Natural Gas and Liquid Hydrocarbon Pipelines Through Areas Prone to Landslide and Subsidence Hazards., Report of Pipeline Research Council, 2009

[3] Dorey, A., Murray, D., & Cheng, R., An Experiment Evaluation of Critical Buckling Strain Criteria, Proceedings of the International Pipeline Conference, 2000

[4] Ferris, G., Newton, S., & Porter, M., Vulnerability of Buried Pipelines to Landslides, Proceedings of the International Pipeline Conference, 2016

[5] Koduru, S. & Nessim, M., Slope Reliability Review, C-FER Technologies Report for Enbridge, 2017

[6] LeBlanc, L., *et al*, Advanced Integrity Management Framework for Pipelines, Proceedings of the International Pipeline Conference, 2016

[7] Nessim, M. and Koduru, S., 2017. Limit States Design for Pipelines Subject to Geotechnical Loads, CFER Report M193, 2017

[8] Nessim, M. & Stephens, M., PIRAMID™ Technical Reference Manual, 2009

[9] Pipeline and Hazardous Materials Safety Administration (PHMSA), Data and Statistics Overview, www.phmsa.dot.gov, 2017

[10] Porter, M., Ferris, G. Lear, M., Leach, M. and Haderspock, M. 2016. Updated Estimates of Frequencies of Pipeline Failures Caused by Geohazards, Proceedings of the International Pipeline Conference, 2016

[11] Sen., M. & Hassanien, S., Slope Crossing Methodology, Draft Enbridge Internal Report, 2018

[12] Wang, Y.-Y., Liu, M., Long, X., Stephens, M., Petersen, R., and Gordon, R., "Validation & Documentation of Tensile Strain Limit Design Models for Pipelines," PRCI Project ABD-1, 2011.

[13] Yoosef-Ghodsi, N., Zhou, J. and Murray, D., A simplified Model of Evaluation Strain Demand in a Pipeline Subjected to Longitudinal Ground Movement, Proceedings of the International Pipeline Conference, 2008

[14] Yoosef-Ghodsi, N., Ozkan, I., Chen, Q., "Review of Compressive Strain Capacity Assessment Methods," PRCI Project ABD-1-2, 2013

Copyright © 2019 by ASME

[15] Barlow, P. and Richmond, J., "The Cheecham Landslide Event," Proceedings of the International Pipeline Conference, 2016

[16] Guthrie, R. and Reid, E., "Estimating Landslide Induced Probability of Failure to Pipelines using a Structured Reductionist Approach," Proceedings of the International Pipeline Conference, 2018

Copyright © 2019 by ASME

AIM-PIMG2019-1072

Chapter 10

A COMPREHENSIVE STUDY ON THE EFFECT OF SOIL MOVEMENT DUE TO MANMADE OPERATIONS ON SUBSEA PIPELINES

A. Kumar[1]
WS Atkins
Houston, TX, USA

N. O. Akinci
WS Atkins
Houston, TX, USA

N. Simpson
WS Atkins
Houston, TX, USA

X. Dai[2]
WS Atkins
Houston, TX, USA

ABSTRACT

Geohazards for pipelines are typically generated by natural events such as mudslides or earthquakes. Additionally, deformation or energy released due to manmade geohazards near pipelines can inflict significant damage. These types of events may affect a limited pipeline length but can cause substantial local deformation potentially resulting in leakage. Spudding near fixed jackets is a typical example of such risks. Jack-up rigs are often utilized nearby shallow water fixed platforms and pipelines for well intervention operations. The effects of spudcan penetration/extraction operations on nearby platform piles have been studied extensively and generic guidelines are adopted by many regulatory bodies throughout the world. The interaction between such operations or other potential earthwork related activities and nearby subsea pipelines, and potential long-term integrity effects, are lacking in the literature. This study aims to fill the gap by providing findings from numerical assessment of interaction between displaced soil and pipelines.

This paper presents results and observations from recently carried out case studies. Spudcan-soil-pipeline interaction was studied using a commercially available, large scale, non-linear finite element analysis tool using Coupled Eulerian-Lagrangian (CEL) approach. The investigation also covers potential effects of spudcan extraction and subsequent soil collapse potential. In addition, sensitivity studies were carried out by varying soil profile parameters and pipeline end conditions. Buckling of pipeline and associated potential integrity implications due to soil movement are discussed in detail. Based on the findings, a set of generalized best practices were developed.

It was observed that spudcan penetration increases stresses in the pipeline within one spudcan diameter due to seabed soil movement. Soil movement causes localized deformation to the nearby pipeline and can increase the risk of subsequent inelastic buckling under operating conditions. The most notable aspect of these simulations was the effect of spudcan extraction. It was found that spudcan extraction conditions are more critical for soil displacements and degradation.

Results of the case studies and generalized recommendations provided in this study are expected to provide guidelines for screening level assessments and further precautionary requirements. In addition, observations are reported in this study on the effect of accidental drop of heavy equipment or structural components on pipelines during installation/removal procedures. This study would help industry in assessing and mitigating risks to subsea pipelines in a more effective and comprehensive manner. Methodologies discussed in this study can also be utilized for assessment of pipelines subjected to natural geohazards.

Keywords: Pipe-soil-interaction, Subsea Soil Movement, CEL, FEA.

[1] Contact author: abhimanyu.kumar@atkinsglobal.com
[2] Atkins (Member of SNC-Lavalin Group), 17220 Katy Freeway, Ste. 200, Houston, TX 77094, USA

Copyright © 2019 by ASME

NOMENCLATURE

d	pipe outer diameter
D	spudcan diameter

ABBREVIATION

ALE	Arbitrary Lagrangian-Eulerian
CEL	Coupled Eulerian-Lagrangian
CFD	Computational Fluid Dynamics
FE	Finite Element
FEA	Finite Element Analysis
OD	Outer Diameter of Pipe
PE	Maximum Principal Plastic Strain
SCH	Pipe Schedule
w.t.	Wall Thickness

INTRODUCTION

Soil movement due to geohazards can introduce significant threats to pipelines such as large deformation and permanent damage. This can also be induced due to manmade activities, for example earthwork or accidental drop of a heavy object/equipment. A common example in offshore oil & gas industry is jack-ups located near fixed platforms where soil displacements caused by the spudcan penetration and extraction can generate induced action on the nearby piles [1 – 3]. Similarly, the displaced soil can generate a substantial load on the pipelines. While interaction of jack-up spudcans with piles have been studied extensively, and relevant guidelines have been in implemented in serval design codes [4], interaction between displaced soil and nearby pipelines is virtually nonexistent in the available literature.

In order to address this, we present findings from sensitivity studies to assess the effects of displaced soil on the nearby pipelines in this paper. In this study, subsea pipeline response due to a dropped object event and spudcan penetration/extraction were considered. The findings of this study are used to develop generic guidelines to assess damage to the subsea pipelines.

SOIL-STRUCTURE INTERACTION

Effects of soil movement due to penetration of a rigid body into soil and rock have attracted the interest of the research community [5, 6] in recent years. The common approaches utilized in the literature are empirical, theoretical, and computational methods. Empirical formulas are derived from data obtained through tests and, hence, may not be accurate for a wide range of cases. Similarly, due to a limited number of closed form solutions to partial differential equations, theoretical methods are restricted in their applicability to address a vast variety of such problems. However, numerical methods are often utilized in conjunction with the theoretical approach to solve the governing partial differential equations. Therefore, numerical methods offer a more practical technique to study these classes of problems.

For numerical approaches, FE methods have been the most prominent tool among researchers to solve a large variety of engineering problems including geotechnical problems such as soil-structure interaction [5 – 9]. The classical FE method in structural mechanics utilize Lagrangian formulation. In a Lagrangian approach, spatial discretization (mesh) of an object is attached to the material and the mesh follows the material deformation. All material states are calculated with respect to the initial fixed frame of reference. While the Lagrangian approach allows precise tracking of material boundaries and is also easy to implement numerically, a strong dependency on mesh quality constitutes one of its main shortcomings. In addition, large distortion of mesh/elements during an analysis limit their applicability for certain types of engineering problems. As an alternative, Eulerian approach is commonly utilized, where the computational mesh is fixed, and the material moves with respect to the grid. Originally, the Eulerian formulation was predominant in fluid dynamics, where large material flow was anticipated. In this approach, motion of continuum is defined with respect to its instantaneous position. Therefore, elements maintain their initial geometry and material flows through elements. Downsides of an Eulerian approach include difficult numerical implementation due to presence of convective terms in governing kinematic equations, computational expense, and lack of precise representation of material interface.

For soil-structure interaction problems, deformations in soil are much greater than that in the structure (pipe, pile, frame, etc.). In such cases, a purely Lagrangian description, alone, cannot be used to formulate the problem due to excessive mesh distortion in the soil domain. Therefore, CEL (alternatively called ALE) was developed in the last decade to combine the benefits of both Eulerian and Lagrangian approaches, whilst simultaneously minimizing their respective deficiencies [10]. In CEL, the Lagrangian domain occupies void regions inside the Eulerian mesh, and the contact interaction formulation is utilized to couple the response of interface boundary between Eulerian and Lagrangian domains [11]. Thus, compatible mesh requirement between two domains is completely bypassed. For soil-structure interaction, soil is modeled using an Eulerian mesh whereas Lagrangian description is used to define the stiff structures (pipe, pile, spudcan, etc.).

DROPPED OBJECT STUDY

Finite Element Model

The first case study considered the effects of a dropped object on 20 in (~30.5 cm) OD (0.688 in w.t.) subsea pipelines (API 5L Grade X-65) located close to an offshore installation. Impact scenarios were considered both for exposed and buried pipelines. Resulting deformation of the pipelines was compared to typical dent ratio criteria provided within industry codes.

Copyright © 2019 by ASME

The dropped object consequence study was performed using Abaqus [12], a general-purpose FE software package. Lagrangian description was utilized for the pipeline and the dropped object. The pipeline was modeled using shell elements, and the dropped object was modeled as a rigid pipe with a total mass of 400 MT representing typical heavy equipment or structures common to the offshore industry. An internal pressure was 0.7 kips was applied to the pipe.

The seabed was modeled using Eulerian elements in Abaqus. For the soil representation, three layers were modeled (see Table 1), with the corresponding density, shear strength (for clay) and friction angle (for sand). The mesh was biased such that smaller elements were used at the pipe-soil interface regions.

A snapshot of the CEL model for the dropped object study is presented in Fig. 1. The pipe ends were constrained with symmetry boundary conditions, and zero normal velocity boundary conditions were applied to the soil boundaries. An initial downward velocity of 3 m/s was assigned to the dropped object. The drop velocity was obtained from an independent hydrodynamic CFD analysis.

The explicit dynamic analyses were executed in two steps; (1) application of gravity loading in quasi-static step to obtain equilibrium state for the pipeline settlement in soil, followed by (2) dynamic procedure to study the dropped object impact. Two analysis cases were considered for the dropped object study:

1. Pipeline on top surface of subsea soil
2. Pipeline buried in top layer of soil

TABLE 1: SOIL STRATUM DETAILS FOR DROPPED OBJECT ANALYSES

Stratum	Description	Depth (m)
I	Sandy clay	0 – 0.5
II	Soft to firm clay	0.5 – 30
III	Medium dense fine to silty fine sand	30 – 40

For Case 1, a region of 2 in thick concrete (3500 psi strength) coating was applied to the pipe in the impact region, modeled explicitly using solid elements. For regions outside the impact zone, composite shell elements were utilized with steel and concrete properties. Damage initiation and damage evolution along with element deletion features were employed for the concrete coating material model. For Case 2, no concrete coating was considered.

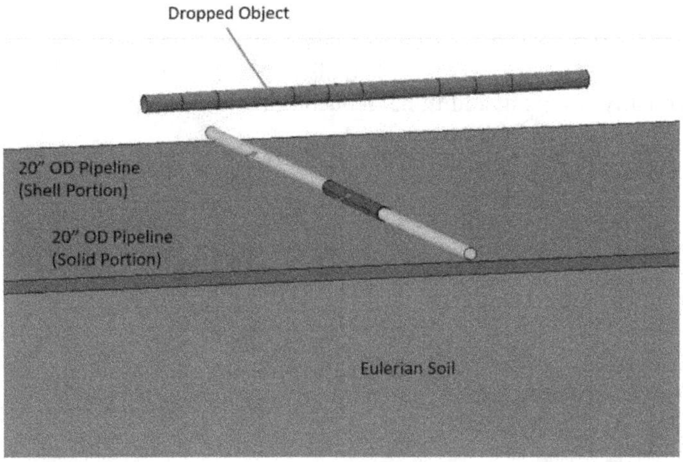

FIGURE 1: CEL MODEL FOR DROPPED OBJECT ANALYSIS FOR CASE 1

Analysis Results

The dropped object analysis cases were conducted in two steps. Firstly, pipeline settlement into soil layer under gravity loading was performed using quasi-static procedure, followed by explicit dynamic analyses to assess the extent of damage due to dropped object. The maximum principal plastic strain contour plots for the two analysis cases are presented in Fig. 3 and Fig. 4.

The maximum dent ratio for each analysis case are summarized in Table 4. The dent ratio, as outlined in DNVGL-RP-F107 [13], is defined as the ratio between the permanent dent in the pipeline and the pipe OD. The dent ratio criteria allow damage to be generalized based on dent depth, and are associated with a set of conditional probabilities that can be used in a subsequent risk assessment. For example, a ratio less than 5% signifies minor damage, whereas 20% dent ratio results in a high probability of rupture of pipe. This criterion, as outlined in DNVGL-RP-F107, is inherently intended to be applied to subsea pipelines, which are generally subjected to a more onerous operating environment.

Based on the dent ratio criteria, pipeline on seabed (Case 1) has a higher probability of rupture leading to operational risk in the event of accidental dropped object. In contrast, pipeline buried in soil (Case 2) is expected to have minor damage. This is also evident from the dent ratios and maximum principal plastic strains (PE) provided in for the two cases. This can be attributed to the fact that a major portion of impact energy was absorbed by the soil covering the pipeline. It is, therefore, critical to account for soil flexibility and associated non-linear effects (soil permanent deformation, plasticity, etc.) appropriately, and any simplification, such as assuming a rigid surface instead of inelastic soil, should be avoided. Note that a sensitivity study was also performed with a rigid seabed surface instead of Eulerian modeling for soil. The dent ratio in this case increased to 87% for 20 in pipeline. This simplification led to overly conservative

Copyright © 2019 by ASME

(and/or incorrect in many instances) results. With a more refined modeling approach, the conservatism in such analyses can be reduced. Accounting for flexibility of the dropped object typically results in additional reduction in dent ratio.

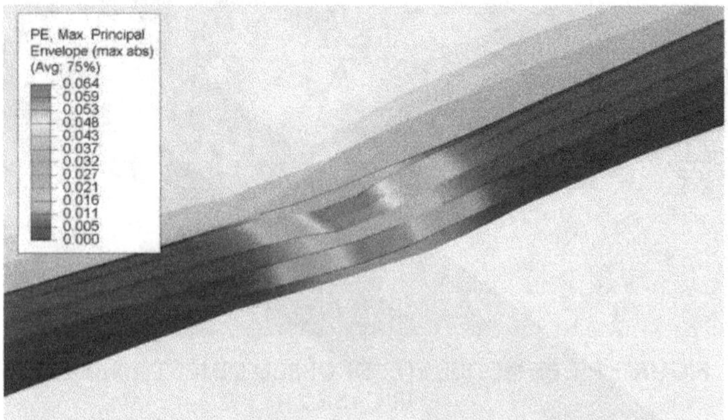

FIGURE 2: MAXIMUM PRINCIPAL PLASTIC STRAIN CONTOURS FOR CASE 1

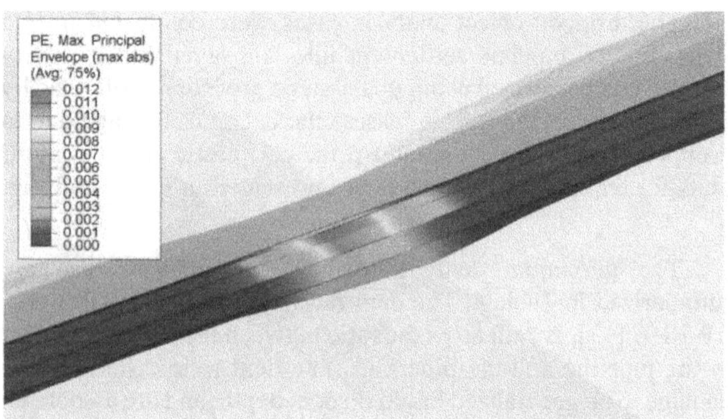

FIGURE 3: MAXIMUM PRINCIPAL PLASTIC STRAINS FOR CASE 2

TABLE 2: DENT RATIO AND MAXIMUM PRINCIPAL PLASTIC STRAIN CONTOURS FOR DROPPED OBJECT ANALYSIS CASES

Case	Dent Ratio	Max. PE
1	25 %	6.4%
2	3 %	1.3%

SPUDCAN PIPE SOIL INTERACTION

Finite Element Model

A detailed FE model of soil, spudcan and subsea pipeline (OD 20 in., SCH 80, API 5L X65) was prepared using Abaqus [12] as shown in Fig. 2. The subsea pipeline was located one spudcan diameter away from spudcan center. The spudcan and pipeline were modeled in Lagrangian domain, whereas soil layers were modeled using Eulerian approach. The spudcan was represented as a rigid object in the FE model owing to its high stiffness compared to the pipeline and soil. In addition, the objective of this study was to understand the interaction between displaced soil and subsea pipeline, therefore the use of rigid spudcan was justified. The actual spudcan geometry was modeled with sufficient level of detail to allow soil flow over the top due to soil collapse effects during penetration and extraction. The subsea pipeline was modeled in detail using shell elements and was assumed to be completely straight; i.e. no imperfection in pipe geometry. The pipe model was assigned elastic-plastic steel properties. It was assumed that the bending stiffness of concrete coating would be much smaller than that of the steel pipe due to potential cracking over time, and therefore the concrete coating was not explicitly represented in the pipe model. However, the density of steel was adjusted to correctly model the pipeline submerged weight accounting for the concrete coating and pipe contents.

Eulerian elements were used to model the soil. In this study, three layered soil strata were considered. Description of different soil layers is provided in Table 2. The Mohr-Coulomb Plasticity model was utilized to describe the inelastic behavior of soil. Shear strength was specified for the clay layer, whereas friction angle was used for sand layers' material properties.

TABLE 3: SOIL STRATUM DETAIL FOR SPUDCAN PIPE INTERACTION ANALYSIS

Stratum	Description	Depth (m)
I	Soft to firm clay	0 – 20
II	Loose to medium dense sandy silt	20 – 30
III	Stiff lean clay	30 – 60

To define the interaction among Eulerian and Lagrangian elements, the general contact capability of Abaqus [12] was utilized. Hard pressure-overclosure was defined for the contact normal direction and frictionless behavior was assumed for the orthogonal shear directions.

The ends of the pipe were restrained axially but released to move vertically. This was done to allow for natural settlement of the pipe due to gravity. For the Eulerian soil domain, zero normal velocity conditions were applied to the boundaries. In addition, a geostatic stress was applied at the mudline. The analysis was conducted in the following three phases:

1. Gravity loading in a quasi-static procedure to obtain embedment of pipe into the soil
2. Spudcan penetration in an explicit dynamic procedure
3. Spudcan extraction phase in an explicit dynamic step

In this study, four analysis cases, as tabulated in Table 3, were considered. The cases investigated the effects of variation in soil properties and, also, examined the sensitivity of results to

Copyright © 2019 by ASME

pipe end boundary conditions. Case 1 represents the lower bound soil stiffness, whereas an upper bound soil estimate is considered in Case 4. The mid-range shear strength from the middle layer of soil is increased by 30% and the friction factor between the pipe surface and the soil is also increased in Case 4.

Analysis Results

Deformation of the pipeline due to soil movement during the spudcan penetration and extraction are presented in Fig. 5 and Fig. 6 for analysis Case 1. The normalized lateral and vertical displacements for each of the analysis cases are presented in Fig. 7 and Fig. 8. For ease in visualization, displacement is normalized with respect to the pipe OD, d, whereas distance from the spudcan is normalized with respect to the spudcan diameter, D. The segment of pipe closest to the spudcan center is located at 0 in the deformation plot. Simulation results showed that the lateral and vertical deformation of the pipe were generally symmetric, with the maximum deformation near the center of spudcan during the spudcan penetration phase. For each case, the pipe moved laterally due to soil movement during spudcan penetration. The maximum lateral deflection in pipe ranged from 0.3d to 0.8d, and that for vertical deformation varied between 0.45d and 1.1d between the ends of modeled pipe segments. In contrast, asymmetric deformation pattern was observed during the extraction phase (see cases 2 – 4 in Fig. 8). Furthermore, the small difference between cases 3 and 4 suggests that pipe end constraint conditions does not significantly affect pipe movement as the length of pipe modeled was sufficiently long.

TABLE 4: SPUDCAN-PIPE-SOIL INTERACTION ANALYSIS CASES

Case	Near Surface Soil Strata	Pipe End Condition
1	Soft Fat Clay	Axial
2	Sandy Silt and Soft Lean Clay	Axial
3	Stiff Clay	Axial
4	Same as Case 3	Axial - Lateral

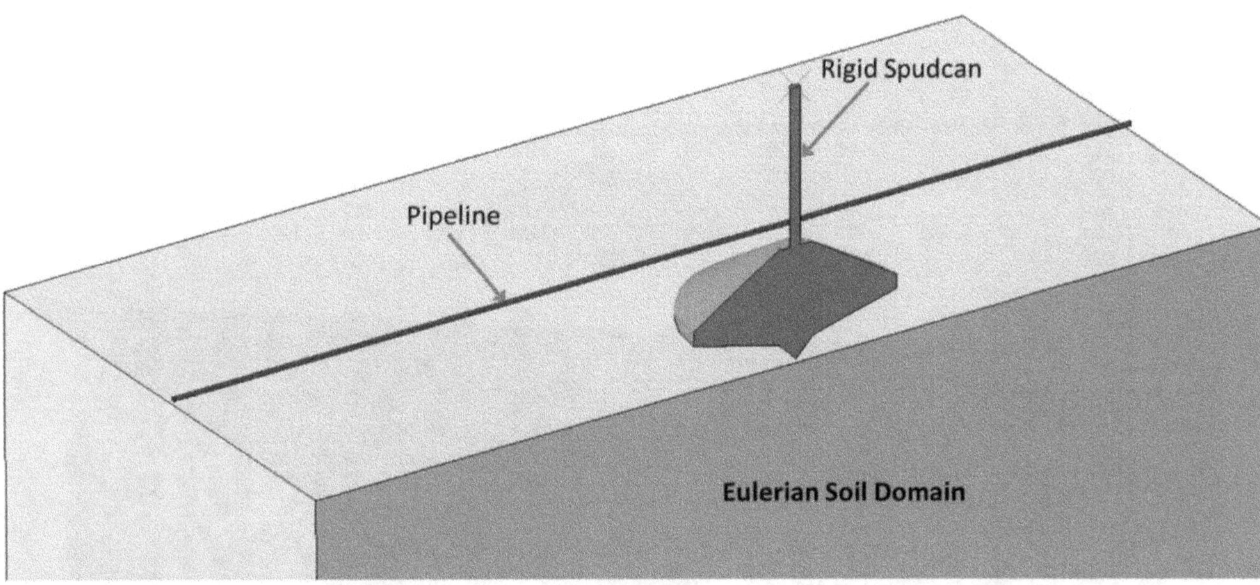

FIGURE 4: CEL MODEL FOR SPUDCAN-SOIL-PIPELINE INTERACTION

Copyright © 2019 by ASME

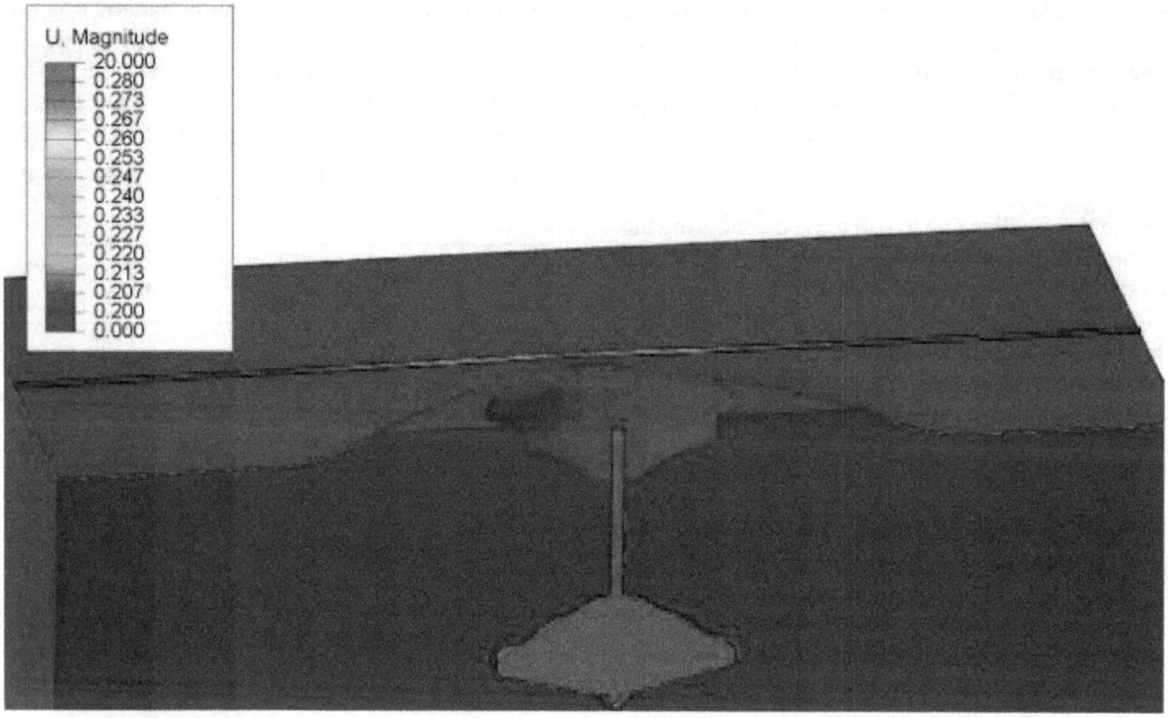

FIGURE 5: DEFORMATION OF PIPELINE AT THE END OF SPUDCAN PENETRATION STEP

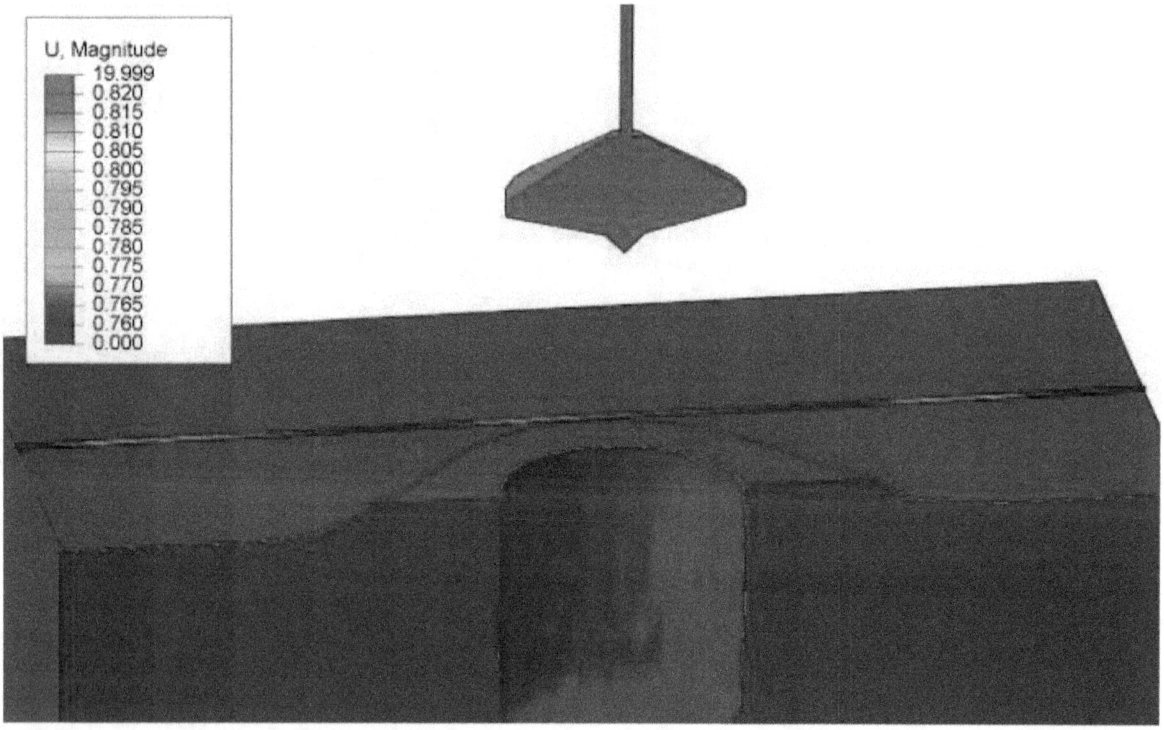

FIGURE 6: DEFORMATION OF PIPELINE AT THE END OF SPUDCAN EXTRACTION STEP (SOIL COLLAPSE NOT SHOWN)

Copyright © 2019 by ASME

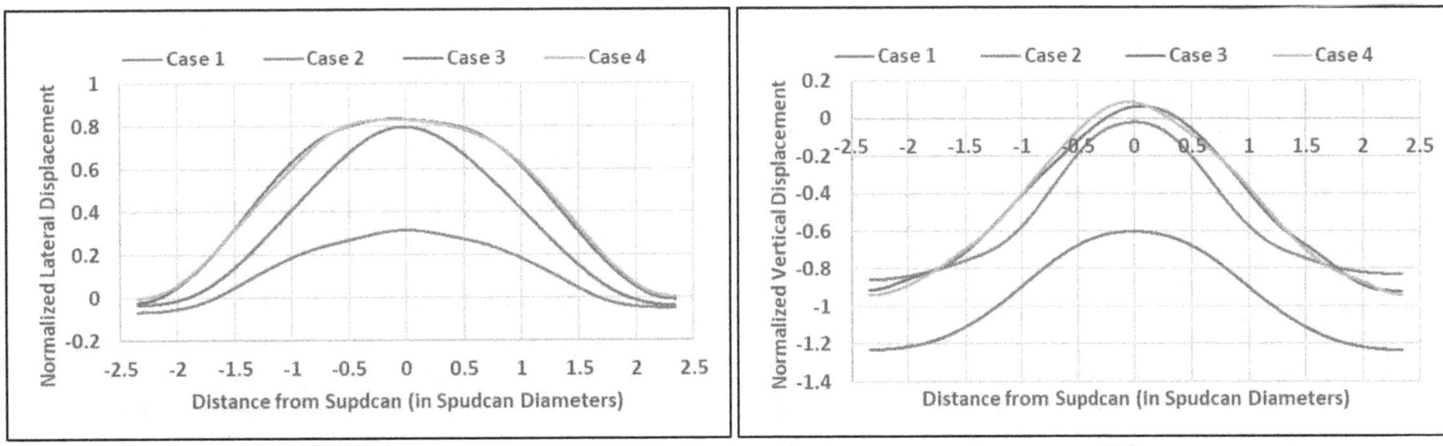

(a) (b)

FIGURE 7: DEFORMED SHAPE OF PIPELINE AFTER SPUDCAN PENETRATION (A) LATERAL DISPLACEMENTS AND (B) VERTICAL DISPLACEMENTS

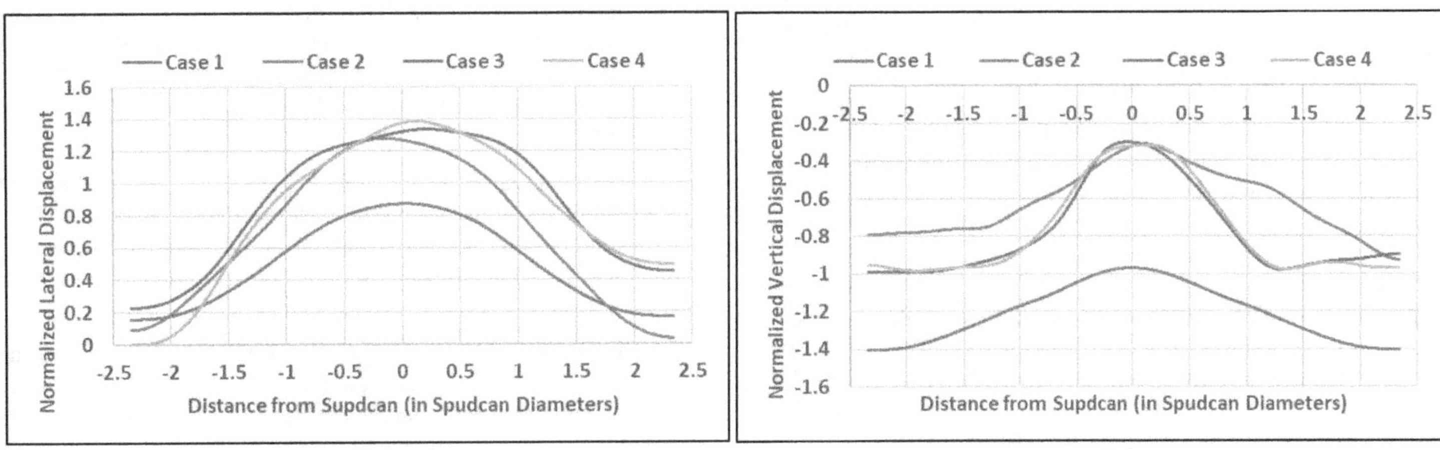

(a) (b)

FIGURE 8: DEFORMED SHAPE OF PIPELINE AFTER SPUDCAN EXTRACTION (A) LATERAL DISPLACEMENTS AND (B) VERTICAL DISPLACEMENTS

CONCLUSION AND DISCUSSION

In this paper, pipeline soil interaction modeling to assess the effect of loading on the pipeline due to manmade activities such as accidental dropped object events, spudcan penetration and extraction nearby subsea pipelines is presented. Detailed FE Models utilizing the CEL capability of Abaqus/Explicit [12] were prepared. The soil was represented using Eulerian elements, whereas dropped object/spudcan and pipelines were modeled using Lagrangian formulation. A series of analysis cases were studied, and the key results obtained for each analysis scenario were presented and discussed.

For the dropped object analyses, dent ratio criteria as outlined in DNVGL-RP-F107 [13] were used to assess the extent of damage. For example, a ratio less than 5% signifies minor damage, whereas 20% dent ratio results in a high probability of rupture of pipe. In contrast, ASME B31.8 [14] allows up to 6% dent ratio to develop without requiring intervention. ASME B31.8 also allows for the severity of a given dent to be assessed by an alternative strain-based criteria. The ASME BPVC [15] allowable plastic strain is based on stress triaxiality and signifies whether local plastic collapse could occur or not at a given plastic strain state. Therefore, the ASME BPVC criteria, in general, concludes, in a deterministic manner, whether the amount of impact energy absorbed by pipe would be expected to cause a failure.

A limiting dent ratio does not directly consider the actual failure mechanism. In general, the evaluation using a strain based approach will have less conservatism when compared to an approach based on dent ratios. A strain based criteria investigates the physical mechanism that would lead to a failure, which is governed by a plastic strain state, and therefore is more accurate than a dent ratio based approach. Regardless of the damage assessment criteria utilized for assessment, high fidelity simulation of such an event should be adopted. Due to inherent large deformation characteristics of subsea soil, a CEL based

Copyright © 2019 by ASME

approach is recommended for these studies that can eliminate the need for simplification of assessment.

Results from the spudcan-pipe-soil interaction study indicated that soil movement due to spudcan extraction is more critical to the pipeline than that due to spudcan penetration. However, the level of stresses and associated deformation in the pipeline were low and did not create a threat to integrity in this case. Furthermore, sudden soil collapse phenomenon was not observed in any of the analysis cases considered. Deformations and curvature in pipeline during spudcan penetration and extraction are directly related to the soil strength. Sensitivity of pipeline deformation is directly related to stiffness of soil; i.e. stiffer soil tends to produce more bending in the pipeline compared to that due to softer soil. Furthermore, the analysis results suggest that the spudcan extraction may affect a longer length of pipeline (in the absence of lateral pipe end restraints the pipeline moved laterally during spudcan extraction unlike penetration), and asymmetric deformation may be observed as the soil stiffness increases. Pipelines within one spudcan diameter can be susceptible to lateral movement and should be assessed prior to any seabed related activities.

In the present study, although the additional loading on pipeline due to soil movement was low, additional scenarios should be considered to assess the integrity of pipeline. The uncertainty in global buckling of pipeline is affected by multiple factors such as pipe-soil resistance and pipeline out-of-straightness, etc. Fundamentally, any activity that imposes additional out-of-straightness in a pipe would increase susceptibility of the pipe to global buckling. The level of lateral pipe movement due to spudding activities is uncertain, as is the potential impact on the available lateral resistance provided by the soil. Due to this uncertainty, it is recommended that the pipeline is shut down during the extraction, and then surveyed prior to operation to understand if the pipeline has moved significantly unless a detailed assessment is performed. The survey data cannot only be used to assess the out of straightness due to penetration and extraction of spudcan, but also be used to calibrate the modeling of the spudcan penetration and extraction that can enable engineers to perform more accurate assessment of pipeline both for the existing operations as well as prior to any future activities.

ACKNOWLEDGEMENTS

The authors thank colleagues from Atkins Oil and Gas for providing support and guidance throughout development of this study.

REFERENCES

[1] ISO 19905-1. "Petroleum and natural gas industries - Site specific assessment of mobile offshore units - Part 1: Jack-ups." British Standards, 2012.

[2] Society of Naval Architects and Marine Engineers (SNAME). "Guideline for Site Specific Assessment of Mobile Jack-up Units, Technical and Research Bulletin 5-5 A." 1994.

[3] MSL Engineering, "Guidelines for jack-up rigs with particular reference to foundation integrity, UK HSE RR 289." 2004.

[4] Engin, Harun K., Khoa, Ho N., Jostad, Hans P. "Finite element analyses of spudcan - subsea template interaction during jack-up rig installation." Frontiers in Offshore Geotechnics III: pp. 1275-1280, 2015.

[5] Sherburn, Jesse A., Hammons, Michael I., Roth, Michael J. "Modeling finite thickness slab perforation using a coupled Eulerian–Lagrangian approach." International Journal of Solids and Structures Vol. 51 no. 25-26 (2014): pp. 4406-4413.

[6] Osborne, Julian, Houlsby, G.T., Teh, K.L., Leung, C.F., Bienen, B., Cassidy, M.J., Randolph, M.F. "Improved Guidelines for the Prediction of Geotechnical Performance of Spudcan Foundations During Installation and Removal of Jack-up Units." Offshore Technology Conference. OTC-20291-MS: pp. C93-1-C93-19. Houston, Texas, May 4-7, 2009.

[7] Dutta, Sujan, Hawlader, Bipul, Phillips, Ryan. "Finite element modeling of partially embedded pipelines in clay seabed using Coupled Eulerian–Lagrangian method." Canadian Geotechnical Journal Vol. 52 No. 1 (2014): pp. 58-72.

[8] Qiu, Gang, Henke, Sascha, Grabe, Jürgen. "Application of a Coupled Eulerian–Lagrangian approach on geomechanical problems involving large deformations." Computers and Geotechnics Vol. 38 No. 1 (2011): pp. 30-39.

[9] Ko, Junyoung, Jeong, Sangseom, Kim, Junghwan. "Application of a Coupled Eulerian-Lagrangian Technique on Constructability Problems of Site on Very Soft Soil." Applied Sciences Vol. 7 No. 10 (2017): pp. 1080.

[10] Donea, J., Huerta, A., Ponthot, J., Rodríguez-Ferran, A. "Arbitrary Lagrangian–Eulerian Methods. Volume 1: Fundamentals." Encyclopedia of Computational Mechanics, John Wiley & Sons (2004).

[11] Tippmann, Jeff D., Prasad, Sharat C., Shah, Parthiv N. "2-D tank sloshing using the coupled Eulerian-LaGrangian (CEL) capability of Abaqus/Explicit." Proceedings of SIMULIA Customer Conference. London, England. 2009.

[12] Dassault Systems, Abaqus Unified FEA software, Version 2016, 2015.

[13] DNVGL. "Risk Assessment of Pipeline Protection.", DNVGL-RP-F107, May 2017.

[14] American Society of Mechanical Engineers. "Gas Transmission and Distribution Piping Systems", ASME B31.8, 2017.

Copyright © 2019 by ASME

[15] American Society of Mechanical Engineers. "Boiler & Pressure Vessel Code" ASME BPVC VIII Division 2", 2017.

Copyright © 2019 by ASME

SECTION 3
GEOHAZARD MONITORING

INTRODUCTION

By Alexander McKenzie-Johnson
(AMckenzieJohnson@Geosyntec.com)
and Don West
(Don_West@Golder.com)

Geohazard monitoring, particularly for pipelines, is the collection of data over time for the following purposes:

- To further understand the monitored geohazard(s) and its relationship to the pipeline,
- To act as a warning system to allow for preemptive intervention to reduce or eliminate potential impacts prior to the occurrence of a geohazard event that could result in a negative outcome to a pipeline,
- To confirm that the implementation of some kind of geohazard mitigation or remediation is functioning as intended (such as confirming that a retaining wall protecting a pipeline has stabilized a landslide), and
- To provide notification that a geohazard event has occurred (such as an earthquake) to facilitate post-event assessment and remediation in order to reduce or minimize post-event consequences on a pipeline or pipeline facility.

Geohazard monitoring encompasses a wide-range of techniques and technologies, from the most basic such as visual observation to cutting-edge technologies such as satellite monitoring by InSAR. The approach and strategy for the selection of geohazard monitoring types should be "fit-for-purpose," that is, the selection should provide the types of data needed at the frequencies needed to address the fundamental purpose of the monitoring. In general, the selection of geohazard monitoring type(s) should be based on answering the following questions:

- What is the overall purpose of the monitoring?
- How will the monitoring da be used?
- What are the characteristics of the hazard being monitored?
- What is the frequency needed (such as seconds, hours, days, months, years) in order to acquire the information needed?
- What is the frequency needed for data analysis and review[1]?
- What is the resolution needed (such as millimeters, centimeters, or meters)?
- What are the limitations to, and uncertainties in the monitoring data?
- What are the resources available to fund and maintain the monitoring?

The focus on pipelines in this volume is identifying the selection and use of geohazard monitoring tools. For instance, a seismic notification system can provide advance warning for high rise buildings to halt elevators prior to the full effects of the earthquake being felt but would not likely provide sufficient warning to halt operations of a pipeline system. In this example, the purpose of a seismic notification system for a pipeline is not to prevent damage to the pipeline from the earthquake itself, but rather to focus post-event assessment and remediation in order to reduce consequences (such as a loss of containment).

A further consideration with pipelines vs. other industries is that in many cases, both the pipeline and the geohazard should be monitored in order to provide a complete picture of the behavior of the geohazard and the effects on the pipeline(s). An example is a large landslide that crosses a pipeline right-of-way; in this case, both the landslide should be monitored to evaluate the rate of movement and enlargement of the landslide, and the pipeline itself should be monitored to evaluate the effects of the landslide on the pipeline. It is plausible that the landslide is affecting the pipeline and it is equally plausible that the landslide is sliding over the top of the pipeline; in this example monitoring of the pipe provides information on the effects on the pipeline that are not determinable by hazard monitoring alone. Conversely, an approach that solely monitors the pipe (such as through strain gauges) could fail to identify expansion or enlargement of the landslide potentially affecting an unmonitored section. Thus, an integrated approach of both pipe and hazard monitoring provides more complete and actionable information than an approach solely focused on one type of monitoring.

Even for monitoring that focuses on the hazard or on the pipe, an approach that utilizes multiple forms of monitoring provides

[1]This is not necessarily the same as the frequency of measurement. For instance, an onsite data logger might record measurements hourly, but only be downloaded quarterly or annually

Copyright © 2019 by ASME

more actionable information than a solitary form of monitoring. Returning to the large landslide example above, an approach to monitor the hazard that relies both on repeat LiDAR and inclinometers will provide more information than one of those monitoring techniques in isolation. In this example, LiDAR will constrain the area and size of the landslide movement, while the inclinometers will provide precise measurements of the amount and depth of landslide movement; together these tools can be used to gain a greater understanding and characterization of the landslide, which can be used to make appropriate decisions about management of the hazard.

Using this same example, pipe monitoring through both strain gauges and repeat in-line-inspection (ILI), Inertial Measurement Unit (IMU) bending strain will provide more information than either technique alone. Strain gauges can provide measurements of changes in strain at very close intervals (up to minutes if there is enough power), but also have a limited length of pipeline that any one set of gauges can measure. Conversely, ILI IMU bending strain provides information about the entire segment of pipeline but is not feasible to run frequently. In this example, strain gauges provide site-specific, detailed and regular monitoring of changes and accumulations in strain induced by the landslide, while ILI IMU bending strain can identify if portions of the pipe not currently monitored by the gauges are also being affected by landslide movement.

The introduction to geohazard monitoring above, was general, and discussed approaches and strategies to geohazard monitoring. The two papers in this chapter provide examples of specific geohazard monitoring techniques for pipelines and for a specific subsidence geohazard at a mine site, respectively.

The first paper, by Clay Johnson, Christopher Markley, Martin Derby, Bailey Theriault, and John deLaChapelle of Golder Associates Inc. "An Overview of Current Methods For Monitoring Landslide Ground Movement To Better Understand Potential Hazards To Buried Pipelines" provides an overview of landslide monitoring techniques and technologies. As summarized in this paper, landslide monitoring can be divided into two classes; those that monitor the landslide and those that monitor the pipe. Landslide monitoring techniques range from remote sensing techniques such as aerial photography, LiDAR, and SAR, to visual monitoring through aerial and ground reconnaissance, to installation of instrumentation in the landslide(s) itself such as piezometers, survey points, and inclinometers. Options for pipe monitoring include strain gauges, in-line inspection techniques, and fiber optic cables.

The second paper, by Danielle Smilovsky and Jeffrey R Keaton of Wood "Detecting and Quantifying Ground Subsidence for Pipeline Geohazard Management Using Satellite Radar Interferometry - Potash Mine Subsidence Example" provides an example of using a cutting-edge monitoring technique (in this case satellite-based InSAR) to evaluate and better understand a geohazard. This paper provides a detailed overview of InSAR and an example of the application to monitor subsidence for pipelines and other linear infrastructure.

Chapter 11 An Overview of Current Methods For Monitoring Landslide Ground Movement To Better Understand Potential Hazards To Buried Pipelines (AIMPIMG2019-1043)
Clay Johnson, Christopher Markley, Martin Derby, Bailey Theriault; and John deLaChapelle, Golder Associates Inc. USA

Chapter 12 Detecting and Quantifying Ground Subsidence for Pipeline Geohazard Management Using Satellite Radar Interferometry - Potash Mine Subsidence Example (AIMPIMG2019-1038)
Danielle Smilovsky and Jeffrey R Keaton, Wood, USA

Copyright © 2019 by ASME

AIM-PIMG2019-1043

Chapter 11

AN OVERVIEW OF CURRENT METHODS FOR MONITORING LANDSLIDE GROUND MOVEMENT TO BETTER UNDERSTAND POTENTIAL HAZARDS TO BURIED PIPELINES

Clay Johnson[1], Christopher Markley, Martin Derby
Golder Associates Inc.
Redmond, WA; Manchester, NH; Buffalo, NY

Bailey Theriault; John deLaChapelle
Golder Associates Inc.
Manchester, NH; Redmond, WA

ABSTRACT

Landslide-related ground movement may occur rapidly and without warning or may occur slowly over time. Both rates of movement, if engaging a buried pipeline, may result in increasing adverse strains to, and possible failure of the pipeline. To effectively manage both the near-term and long-term threat associated with landslide-related movement on pipeline systems, it is important to first identify and characterize the landslide threat, and then develop a mitigation and monitoring program to manage the risk associated with the threat. A landslide threat monitoring program, if tailored correctly for a given pipeline system, can be used to identify newly-developing landslides, as well as monitor movement of known or existing landslides. A monitoring program can also be used to confirm that there are no landslide threats present on a pipeline system that crosses landslide-susceptible terrain. This paper provides a high-level overview of current methods available for monitoring landslide ground movement to better understand potential threat to pipeline systems. The targeted audience includes pipeline owners and operators who are interested in understanding current and common methodologies for monitoring slope movement and pipe response, and how the available methods may assist them to manage the landslide-related risk. Landslide monitoring can generally be carried out in two different forms: (1) monitoring the ground itself, either at or below the surface, and (2) monitoring the pipe. Monitoring ground movement can be accomplished through direct visual observations (e.g., aerial or ground reconnaissance); remote sensing techniques using light detection and ranging (LiDAR) data, interferometric synthetic aperture radar (InSAR) data, or aerial imagery; and measurement of ground conditions through instrumentation such as geodetic survey monuments, slope inclinometers, and groundwater piezometers. Monitoring the pipe is commonly accomplished through use of strain gauges, ILI IMU bending strain data, or fiber optics. Selecting the most suitable form(s) of monitoring for a single landslide or an entire pipeline system depends on many factors, such as the characteristics of each landslide, the local geographic and geologic conditions, the density of landslide sites, site accessibility, the owner's risk tolerance, etc. In addition, it is often beneficial or necessary to combine more than one type of monitoring at a single site, to fully understand and quantify the threat to the pipeline.

INTRODUCTION

Prior to the development of a landslide monitoring program for a single landslide or for an entire pipeline system it is first necessary to produce a landslide inventory. A landslide inventory includes the location, size, and nature of known and potential landslide threats. Generally, preparing a landslide inventory includes the following tasks: 1) desktop review of available remote sensing data, e.g., light detection and ranging data (LiDAR), aerial imagery, public/agency data, 2) visual identification and assessment of potential landslide hazards during an aerial reconnaissance, and 3) ground-based landslide mapping and characterization of identified potential higher threat landslides. Once a landside inventory is prepared, decisions can be made to either mitigate or monitor identified landslide hazards with the ultimate purpose of reducing the potential for these hazards to affect the integrity of the pipeline.

A monitoring program can be used to track ground movement and/or pipeline movement for known higher hazard landslides and can also be used to monitor conditions at landslides with high uncertainty regarding potential future movement. Monitoring ground movement can be completed using a variety of techniques, some of which are the same methods used during the development of a landslide inventory (i.e. remote sensing

[1] Contact author: cjohnson@golder.com

Copyright © 2019 by ASME

review, visual assessments from the air and ground). Monitoring ground movement can also be completed by direct ground measurement with instrumentation. Typical methods to monitor and assess ground movement through instrumentation include the installation and repeated measurement of survey monuments (both geodetic and visual), slope inclinometers, and piezometers. Monitoring the pipe is commonly accomplished through use of strain gauges, Inline Inspection (ILI) Inertial Measurement Unit (IMU) bending strain data, or distributed fiber optics.

If mitigation efforts are implemented for higher threat landslides, it is important to set up a monitoring program to confirm that the mitigation is effectively addressing the hazard(s).

Ongoing monitoring can also be conducted through repeat LiDAR surveys by performing change detection between datasets. Similarly, aerial reconnaissance can be used to monitor pipeline systems on prescribed intervals or after significant natural events (e.g., rain storms, earthquakes) that may affect slope stability.

Figure 1 below shows a typical workflow for initial characterization, mitigation, and monitoring for landslide-related hazards for pipeline systems.

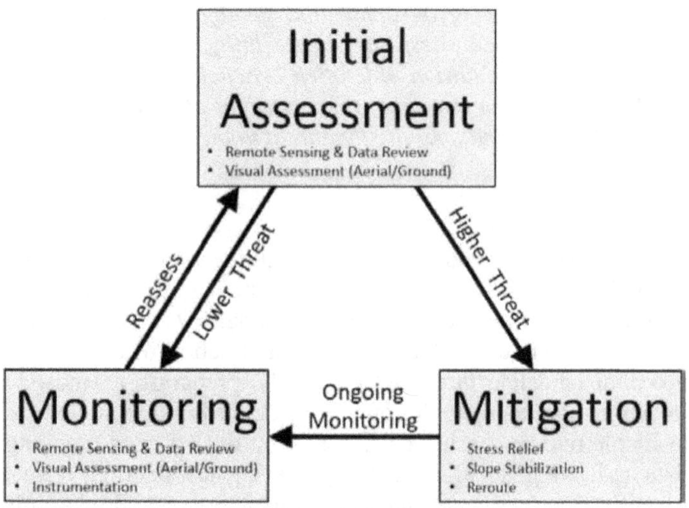

FIGURE 1: PROGRAM FOR ASSESSING AND EVALUATING LANDSLIDE HAZARDS ALONG PIPELINE SYSTEMS

Selecting the appropriate form(s) of monitoring for a given site or pipeline system depends on many factors, such as the characteristics of each landslide, the local geographic and geologic conditions, the number/density of monitoring sites, site accessibility, type and strength of the pipe(s), presence of high consequence areas, the owner's risk tolerance, etc. In addition, it is often beneficial or necessary to combine more than one type of monitoring at a single site, to fully understand and quantify the threat to the pipeline.

This paper covers some of the more common industry-established techniques for monitoring landslide ground movement and related pipeline movement.

1.0 Ground Movement Monitoring Methods

This section describes various methods for monitoring landslide-related ground movement that may present a hazard to buried pipelines. Monitoring ground movement is typically conducted by either conducting visual assessment of the ground surface or by instrumentation. This section also includes a description of some of the most common desktop-based remote sensing techniques used to monitor ground movement, including review of LiDAR data, aerial imagery, and interferometric synthetic aperture radar (InSAR) data. Monitoring ground movement can also be accomplished through field-based assessments such as ground reconnaissance and aerial reconnaissance, both of which can assist with better understanding potential landslide-related hazards. This section also includes an overview of instrumentation methods to monitor ground movement with survey monuments, slope inclinometers, and piezometers.

1.1 Aerial Imagery Review

High resolution aerial imagery is a relatively quick and inexpensive way to obtain information about surface conditions on and adjacent to a pipeline right-of-way (ROW). However, accurate identification of landslides through interpretation of high-resolution aerial imagery can be dependent on many factors. A poorly maintained or overgrown pipeline ROW may limit the visibility of landslide features across, or in the vicinity of the pipeline. Depending on the time of day the imagery is collected, shadows from objects along the ROW (e.g. trees, cliffs, etc.) may cover the area within the pipeline ROW. Data collected in 'leaf off' conditions tend to produce the best results, and snow cover during the winter, or thick vegetation during summer may obscure landslide morphology.

Desktop review of aerial imagery can be conducted in several different ways, each of which has limitations. Review of digital orthophotographs can be conducted digitally, but only in 2-dimensions (2D). It can often be very difficult to visualize terrain (e.g., slope inclinations, etc.) and identify landslides in 2D. Stereoscopic aerial photograph interpretation can be conducted to view aerial imagery in 3-dimensions (3D); however, stereoscopic aerial photograph interpretation has been, and is still generally completed using hardcopy air photos and either a pocket or mirror stereoscope. Thus, mapped features must be transferred from hard copy to digital format, which can be time consuming, and resolution can be lost in the process. In addition, the scale at which stereoscopic aerial photographs can be reviewed is limited to the scale of the photographs. Resources such as Google Earth™ overlay 2D imagery with a digital elevation model, to provide an illusion of 3D; however, the resolution of the imagery is variable and often low resolution and poor quality. Digital photogrammetric workstations and software, such as PurVIEW, can be used to complete air photo interpretation digitally on a 3D computer monitor with the use of 3D glasses. The advantage of using such software is that it allows you to zoom in on the data from traditional imagery scales (e.g., 1:24,000) to scales as large as 1:500. The ability to view imagery at such fine scales allows for better identification of critical

Copyright © 2019 by ASME

geomorphic features that are important to identifying possible landslides.

1.2 LiDAR Analysis and Review

LiDAR is a survey technology that uses pulsed laser light to illuminate a target and then measure the returns of the reflected pulses with a sensor/data recorder. The strength of LiDAR is its ability to "see" through trees and vegetation and produce a bare earth digital elevation model of a given slope. As such, LiDAR is one of the most powerful tools for identifying landslides where tree cover is present. Aerial-based LiDAR surveys are commonly conducted with the sole purpose of monitoring landslide hazards for pipeline systems. High resolution LiDAR datasets are useful in identifying geologic hazards in areas of dense vegetation coverage, as the data can be processed to take the longest LiDAR returns in a determined area (e.g. in each square meter), which can effectively 'remove' the vegetation, resulting in a bare earth model (Figure 2). Recent or active landslides typically have morphology that appears fresh and distinct with exposed soil visible throughout the landslide, and they may be easily identified via aerial patrol or ground reconnaissance. However, over time, these features weather and erode, becoming subtler and more subdued, making identification in the field more difficult. In older/dormant landslides, vegetation may become well-established, making identification during standard aerial patrols unlikely. Where subtle geomorphic indicators of geologic hazards (e.g., landslide features such as scarps, hummocky ground, etc.) may be obscured in other types of data (e.g., aerial imagery, low resolution digital elevation model [DEM] data, etc.), these features can be readily identified at a desktop level using LiDAR data. Bare-earth DEMs produced using LiDAR data present a highly detailed representation of the ground surface, typically at decimeter-level accuracy [1]. It is important to note that geomorphic evidence of landslides can range from easily and readily identifiable, to subtle and difficult to identify, depending on the characteristics of the landslide, such as the age of the landslide, rate of past or current movement, size of the features, the type of landslide, amount of vegetation coverage, etc. Because of the variability in geomorphic expression, the possible landslide locations and limits delineated from remote sensing review can range from well-defined to approximate. As such, follow up ground-based assessments are often conducted to adequately characterize and more accurately map the limits of landslides and also to better understand the potential hazard to buried pipelines.

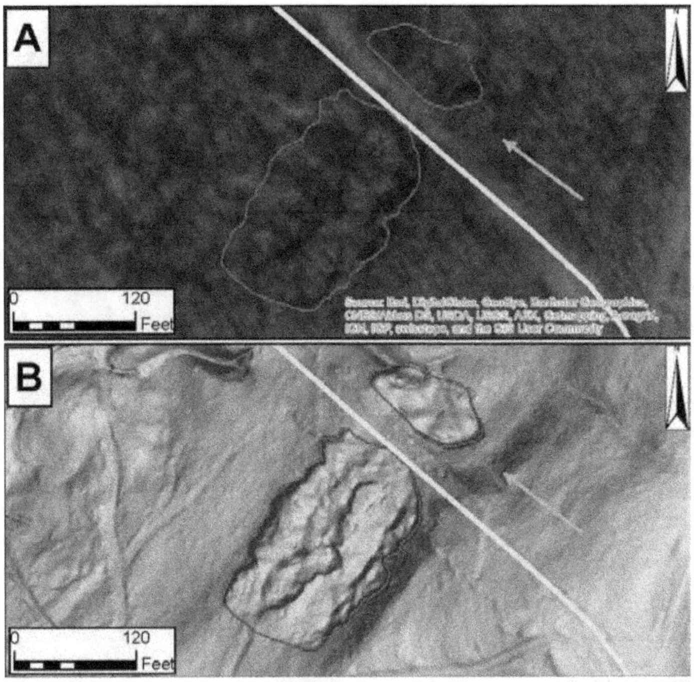

FIGURE 2: A: PUBLICLY AVAILABLE AERIAL IMAGERY OF TWO LANDSLIDES (OUTLINED IN RED). THE SOUTHERN LANDSLIDE ABUTS THE PIPELINE (YELLOW LINE). THE ORANGE ARROW POINTS DOWNSLOPE. NOTE THAT MUCH OF THE LANDSLIDE MORPHOLOGY IS OBSCURED BY VEGETATION. B: PROJECT SPECIFIC LIDAR DATA OF THE SAME TWO LANDSLIDES, WHERE THE LANDSLIDE MORPHOLOGY FLANKING THE PIPELINE IS CLEARLY VISIBLE.

Repeated acquisition of LiDAR data over time and encompassing the same area can be used for the identification of new or developing landslides, as well as for monitoring of known landslides [2]. LiDAR Change Detection Analysis (Change Detection) is a grid-to-grid comparison of successive LiDAR datasets which shows the apparent changes in elevation between the two datasets (Figure 3). By comparing the ground elevations between two datasets, areas of apparent ground movement can be identified, which can assist in the evaluation of the activity level of a landslide (e.g., active vs. apparently stable, at least over the period being assessed). Such information can aid in determining the potential threat of a landslide to a pipeline.

Copyright © 2019 by ASME

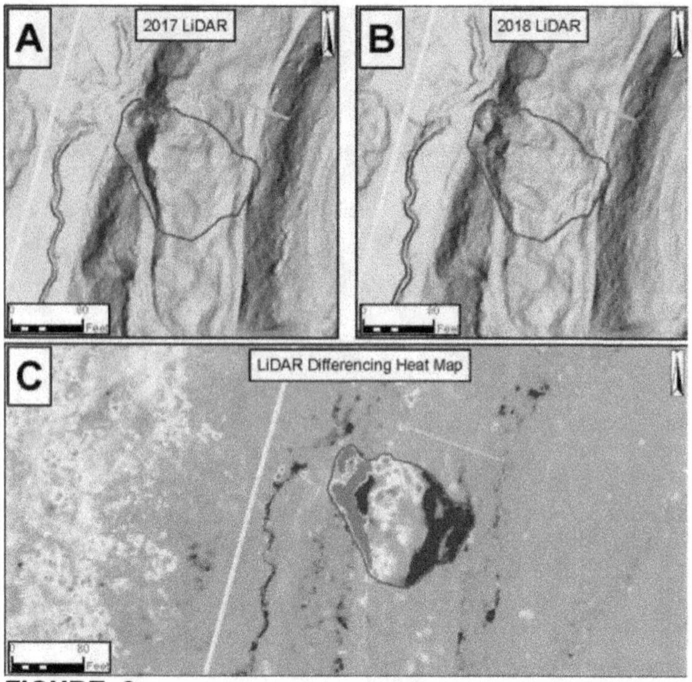

FIGURE 3: A: 2017 LIDAR IMAGERY (PRESENTED AS A PERCENT SLOPE MAP), WITH A LANDSLIDE OUTLINED IN RED. THE YELLOW LINE IS THE LOCATION OF A PIPELINE, AND THE ORANGE ARROW POINTS DOWNSLOPE. B: 2018 LIDAR IMAGERY SHOWING THE LOCATION OF THE SAME LANDSLIDE. C: LIDAR DIFFERENCING HEAT MAP, WHERE COOL COLORS (I.E. BLUES) REPRESENT AREAS OF NET LOSS AND WARM COLORS (I.E. YELLOW TO RED) REPRESENT AREAS OF NET GAIN. GREEN REPRESENTS AREAS OF NO NET LOSS OR GAIN.

1.3 InSAR

Interferometric synthetic aperture radar (InSAR) is a satellite, radar-based remote sensing technique that uses two or more synthetic aperture radar (SAR) images to generate maps of ground surface deformation. This is accomplished by computing the differences in the phase of the radar signal returned to the satellite to produce an interferogram, which is then converted to a ground deformation map. The SAR data are collected by side-viewing, polar-orbiting satellites that pass over the same geographic region on a periodic basis, typically every 4 to 46 days depending on the satellite and orbit characteristics. The SAR data collected during repeated surveys can be used to detect and quantify ground movement over time. InSAR processing techniques can potentially measure+. It has applications for geophysical monitoring of natural hazards, for example pre- and post-earthquake tectonic strains, volcanoes and landslides, and is extensively used in mining and infrastructure monitoring. InSAR data provide ground movement detection and measurements that are line-of-sight to the orbiting side-viewing satellites, however multiple datasets can be combined to provide vertical movement and horizontal east-west movement. Because SAR data are collected over large geographic areas, localized landslide-related ground movement can be distinguished from area-wide ground movement. Because the SAR satellites have a polar orbit (i.e. north-south) and are side viewing (i.e. east west) the horizontal movement detection is constrained to east-west ground movement. Note that the most commonly used frequencies, X band (7.0 to 11.2 GHz) and C band (4 to 8 GHz) cannot penetrate dense vegetation, which limits the density of ground deformation measurements. Lower frequency satellites (L Band, <1 GHz) are becoming available that are capable of penetrating vegetation to measure displacement in forested areas.

There are two general approaches for ground movement detection using InSAR methods, 1) processing of historical datasets and 2) processing of forward-looking datasets. Examples of historical and forward looking InSAR are provided in Figure 4.

Historical SAR

Historical SAR data are available from satellites that have been collecting data from as far back as 1992, and many are free to the public (Figure 4A). One of the strengths of historical SAR is that it provides measurement of past ground movement, which can be extremely valuable during the routing of new pipelines through landslide susceptible terrain. Limitations of low-resolution historical SAR data include 1) low spatial density of the data, especially in rural areas, sometimes insufficient to characterize historical movement on medium to small landslides, and 2) data gaps for some years.

Forward Looking SAR

Forward looking SAR involves tasking satellites to collect higher resolution SAR data in geographic areas of known concern (Figure 4B). This will produce a higher density of measurement points and as such, is a more appropriate method for moderate- and smaller-sized landslides. One drawback to forward-looking high-resolution data is that, unlike the free, publicly available historical datasets, the forward-looking SAR involves tasking a satellite to collect imagery and the resulting datasets require purchase from the satellite companies.

Copyright © 2019 by ASME

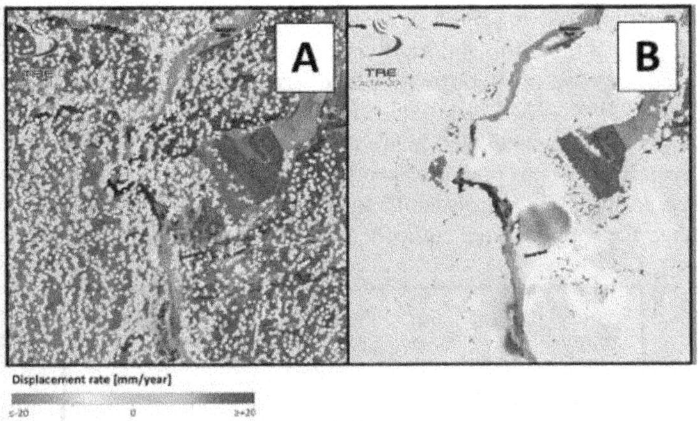

FIGURE 4: TWO INSAR DATA SETS OF THE SAME RIVER VALLEY WITH ACTIVE LANDSLIDES. WARM COLORS INDICATE GROUND DISPLACEMENT RATE IN WESTWARD DIRECTION WHEREAS COOL COLORS INDICATE GROUND DISPLACEMENT RATE IN EASTWARD DIRECTION. GREEN COLOR INDICATES NO APPRECIABLE GROUND MOVEMENT. A: LOWER POINT DENSITY DATASET GENERATED FROM HISTORICAL SAR DATA. B: HIGHER POINT DENSITY DATASET GENERATED FORWARD LOOKING SAR DATA (i.e. TASKED SATELLITE).

1.4 Ground Reconnaissance

Ground reconnaissance monitoring efforts can range in scale from assessment of individual sites to assessment of numerous landslides identified during a desktop assessment. After initial assessment and characterization of landslides, follow up assessments may be recommended as part of a monitoring program. Typically, the purpose of conducting a ground reconnaissance is two-fold and includes: 1) visual assessment by personnel qualified in assessing landslide hazards (e.g., a geologist or engineer) to confirm the presence of landslide morphology relative to the pipeline location, and 2) determination of the potential hazard that a given landslide may pose to a buried pipeline. During a ground reconnaissance, the limits of landslide-related deformation and internal morphology (e.g., cracks, scarps, seeps, etc.) should be geospatially mapped using a handheld GPS unit with sub-meter accuracy so that the data can be later assessed in a GIS, and any future landslide movement can be tracked. Accurate and detailed landslide mapping is also integral to developing mitigation plans (either conceptual or detailed). In order to characterize potential landslide-related pipeline deformation during a ground reconnaissance, accurate pipeline survey information is needed. It is preferred that pipeline field locates be completed by a qualified pipeline technician or surveyor at the same time as the ground reconnaissance. Possible landslide-related pipeline movement can be identified if the field-located pipeline is observed to deviate, typically in the downslope direction, a significant distance (e.g. several feet) within the boundary of a landslide. If possible, the original as-built pipeline survey should be compared in the field with the wand-identified pipeline location and depth in order to identify potential discrepancies. Pipeline pin flags or wooden lathe should be used to mark the pipeline location from both the as-built and wand-identified pipeline location. Any discrepancies observed from the pipeline survey (e.g., a horizontal deviation in the pipeline; Figure 5) should be well documented with field photos and notes, and pin flags should be geospatially located.

FIGURE 5: EXAMPLE OF HOW LOCATING THE BURIED PIPE (YELLOW PIN FLAGS) WAS USED TO IDENTIFY A DEFLECTION OF THE PIPE (WHITE DASHED LINE) AS A RESULT OF LANDSLIDE-RELATED GROUND MOVMENT (RED ARROWS). (2012 PHOTO; D. WEST)

1.5 Aerial Reconnaissance

Repeat aerial reconnaissance surveys can be a very effective slope monitoring tool because it allows for a relatively rapid assessment of possible landslide hazards over regionally extensive pipeline systems in relatively short timeframes. Repeat aerial reconnaissance can be performed on a periodic basis (e.g., annual or monthly) or can be in response to a potentially slope destabilizing event such as a high precipitation storm or an earthquake. Conducting landslide hazard assessments with helicopter or fixed-wing aircraft can also be relatively inexpensive when compared to other monitoring efforts (e.g. multiple ground-based site visits). Visual observations from aerial reconnaissance are not as accurate as instrumental monitoring techniques and have temporal and spatial limitations; i.e., the observations recorded represent a single point in time and only record landslide-related movement near the ground surface. However, they can be used to identify newly developed landslides and to monitor landslides that may not have other types of monitoring instruments already installed. Data collected during aerial reconnaissance typically include geolocation using a GPS device, photographs and notes. More frequent aerial reconnaissance may be warranted if there is uncertainty regarding slope stability, anticipated faster rates of landslide movement, and during periods of high precipitation (e.g. during winter months), particularly for newly constructed pipeline

Copyright © 2019 by ASME

systems located in landslide-susceptible terrain. It is often difficult to fully characterize a landslide, and therefore estimate the potential threat to the pipeline, during an aerial reconnaissance. As such, ground-based field assessments are typically required to better characterize the location, type, level of activity, and amount of displacement of the landslide relative to the pipeline location.

1.6 Survey Monuments

Survey monuments are often used to monitor landslide-related ground movement for targeted sites along pipeline systems. Monitoring with survey monuments is generally accomplished by either using geodetic survey monuments (GSMs), which are typically measured using repeat survey-grade GPS equipment, or with visual survey monuments (VSMs), which generally includes a series of survey markers installed in a line that are visually assessed for movement. For landslide monitoring it is typically necessary to install GSMs and VSMs both within and outside the limits of identified landslides so that 1) potential landslide movement can be tracked, and 2) areas not expected to move can be used as either visual or geodetic survey control.

1.6.1 Geodetic Survey Monuments (GSMs)

GSMs can be installed using a variety of methods such as driving rebar into the subsurface until refusal, or by installing monuments set in shallow concrete pads. Regardless of the type of geodetic monuments used, they should be installed in a manner such that they are reasonably protected from vandalism or accidental displacement if driven over by off road vehicles. After installation, GSMs receive an initial baseline survey which is conducted by a qualified surveyor. The baseline survey represents the initial survey to which future surveys will be compared. If movement is recorded in subsequent surveys, this data can be used to inform follow-up assessments and/or trigger potential mitigation efforts. Repeated measurement of GSM over time can identify ground movement with high precision (i.e. centimeter accuracy) and as such this method of slope monitoring can provide an early warning to pipeline operators that a slope is moving before this movement could be detected with other methods of monitoring such as visual monitoring or VSMs.

1.6.1 Visual Survey Monuments (VSMs)

VSMs are a series of above-ground monuments that can be easily visually assessed from the ground and from the air. VSMs typically consist of a vertical pipe or post that is imbedded into the ground. VSMs can be installed in a straight line across an area of potential ground movement, with the line of VSMs installed perpendicular to the expected direction of landslide movement. VSMs can also be installed along the location of the pipeline so that adverse movement along the path of the pipeline can be assessed. If some of the VSM are displaced over time it may indicate there has been ground movement. Permanent or temporary VSMs can be constructed from a variety of materials, such as wooden stakes, PVC, concrete, steel, etc. VSMs should be installed at a height such that they will not be covered by growing vegetation and high visibility paint or decals should be used for ease of identification and assessment. A diagram of a typical PVC VSM installation is shown in Figure 6 below. The main advantages to VSMs are that they can be rapidly assessed by anybody for potential landslide-related movement during regular pipeline patrols (both aerial and ground) and without the need for sophisticated equipment or trained personnel.

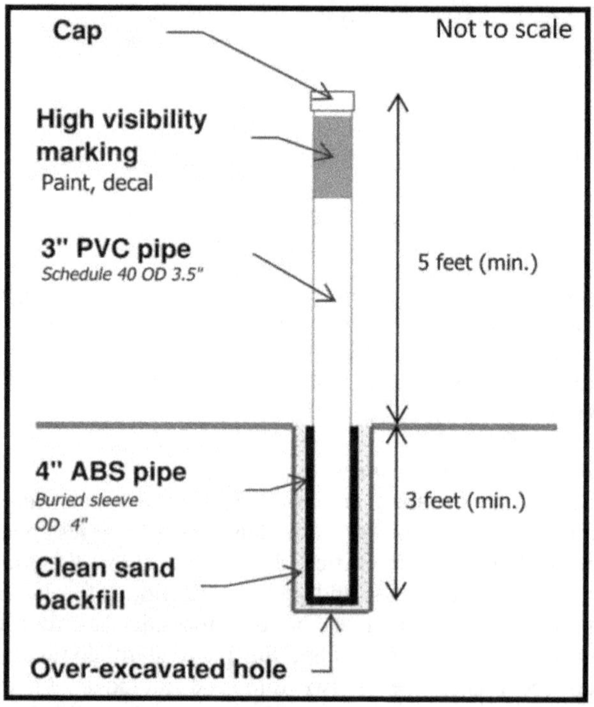

FIGURE 6: TYPICAL DETAIL OF A PERMANENT VISUAL SURVEY MONUMENT.

1.7 Slope Inclinometers (SIs)

SIs are used to detect landslide-related deformation in the subsurface and can be used to measure the magnitude, rate, direction, and total landslide displacement at the slip surface. SIs can also provide an understanding of the landslide geometry and type of movement (e.g., rotational, translational, etc.) if multiple SIs are placed within a landslide. The installation of SIs requires the drilling of a vertical geotechnical borehole followed by installation and grouting of an SI casing (typically made of plastic) within the borehole. Once installed, SI measurements are taken at specified and repeatable depth intervals with either a probe inclinometer (PI), or an in-place inclinometer (IPI).

IPIs are essentially identical to PIs but are left in the SI casing and connected to a data recorder. IPIs can include as few as one IPI or several IPI sensors connected in a string. The main advantage of installing IPI in landslides is that real time detection of landslide movement can be recorded. If warranted, the real time inclinometer data collection can be linked through telemetry to a warning system (e.g., via cellular or satellite network) that can trigger alarms for pipeline shut off if adverse ground movement is measured.

Copyright © 2019 by ASME

Measurements taken with the SI can help identify landslide-related deformation in the subsurface based on the profile generated from successive surveys (Figure 7). SIs are typically quite useful for identifying landslide depth for relatively slow-moving landslides. However, SIs may also be helpful in assessing landslide depth for rapidly moving landslides if the upper portion of SI casing above the landslide slip plane remains intact.

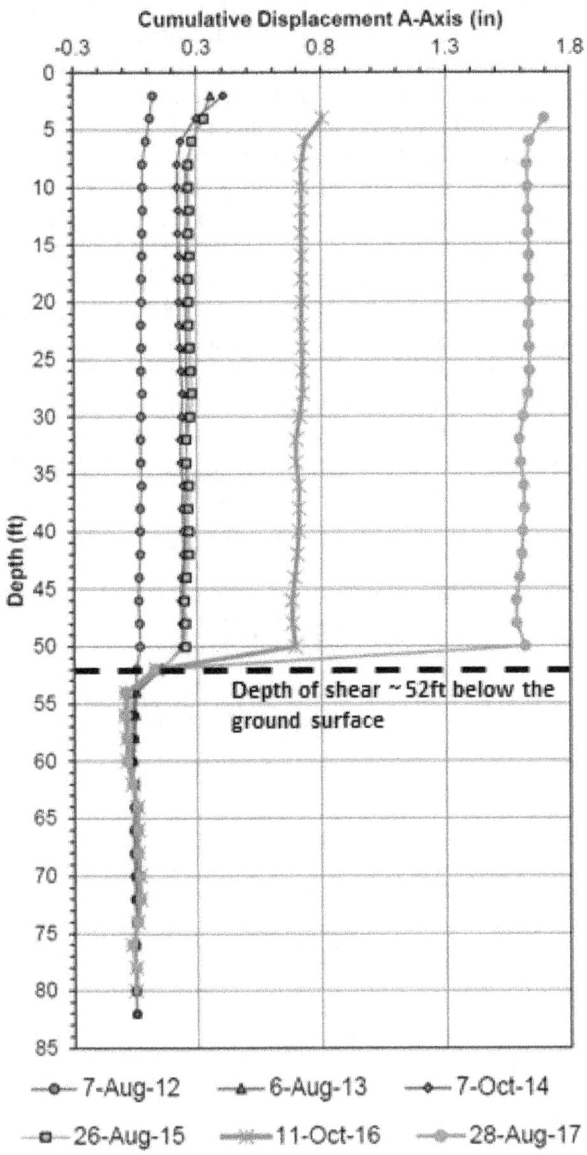

FIGURE 7: SI PLOT SHOWING LANDSLIDE DISPLACEMENT BASED ON SUCCESSIVE ANNUAL READINGS. THE SI PLOT INDICATES THAT THIS LANDSLIDE WITH A FAILURE SURFACE AT ABOUT 52 FEET BELOW GROUND SURFACE HAD MOVED ABOUT 1.7 INCHES OVER A 6 YEAR PERIOD.

1.8 Piezometers

Landslide movement is commonly triggered by elevated groundwater levels/pore water pressures in subsurface materials as a result of surface precipitation and groundwater infiltration into a slope. A piezometer is an instrument that is used to monitor groundwater pressures based on the height that a column of water will rise against gravity, or a devise which measures pressure directly. The purpose of using a piezometer for landslide hazard assessments is to better understand the response of groundwater to individual storm events and seasonal precipitation, and to relate the observed groundwater trends to the triggering of landslide movement. Piezometers are often installed as open standpipe piezometers (also known as an observation well) in geotechnical boreholes, with plastic casing screened at the depth where groundwater pressure is of interest (e.g., near the landslide slip surface, or within a particular geologic formation). Groundwater elevations are typically measured within standpipe piezometers using an electric water level meter, which signals (i.e., 'beeps') when ground water is encountered by an electronic indicator (e.g., by using the water to close a circuit). Piezometers that are installed directly within a geotechnical borehole are generally either 1) vibrating-wire piezometers (VWP), which convert groundwater pressures into an electric signal, or 2) pneumatic piezometers, which use a gas pressure diaphragm to directly measure water pressure. Of these two types, VWPs are the more commonly installed piezometer because they are cost effective, durable, and can be easily read by data loggers at the surface. Typically, piezometers are paired with SIs so that groundwater conditions can be monitored coincident with the inclinometer readings to potentially determine which water elevations/pressures may be triggering landslide movement (Figure 8).

FIGURE 8: PAIRED STANDPIPE PIEZOMETER AND SLOPE INCLINOMETER INSTALLED IN A LANDSLIDE ON A PIPELINE RIGHT-OF-WAY (PHOTO BY: D. WEST, 2005)

Copyright © 2019 by ASME

2.0 Monitoring Pipeline Movement

This section describes methods commonly used for monitoring movement of buried pipelines as a result of landslide-related displacement.

2.1 ILI/IMU monitoring

Inertial measurement unit (IMU) data collected during an in-line inspection (ILI) identify areas of bending strain. The purpose of a bending strain analysis is to identify sections of pipeline bent or deflected by an external force, such as a geohazard (e.g., landslide) or from a 3rd-party (such as anchor drag at water crossings). A baseline ILI run is typically conducted, to which IMU data collected in the future can be compared; however, single-run IMU data can be utilized directly. It is typically assumed that bending strain features imposed by an external force formed or were induced post-construction. Collection of IMU data to map pipe centerlines is a standard service for most ILI runs and the additional costs for performance of bending strain analysis are relatively low, making it an attractive option to evaluate areas where effects from an external force(s) are known or suspected. If IMU bending strain data are put into context (e.g., overlain with known geohazard locations), the results can be used to aid in the prioritization of response, such as monitoring or mitigation of sites. Alternatively, through a close review of the IMU bending strain data and their locations, previously unidentified geohazards may be identified (e.g., the IMU data can be used to prioritize where to look for possible geohazards).

2.2 Strain Gauge Monitoring

Strain gauges (Figure 9) are a common geotechnical instrument for pipeline monitoring and are a valuable component to a successful monitoring plan. The use of geotechnical instrumentation is not merely the selection of instruments, but a comprehensive step-by-step engineering process beginning with definition of the objective and ending with implementation of the data [3]. Strain gauges can be used to monitor pipeline response to both landslide and subsidence ground movements, as well as pipeline wrinkle-bend monitoring.

There are several different types of strain gauges that are available depending on the data that need to be collected. Vibrating wire (VW), resistance, and fiber optic strain gauges are among the more common types of strain gauges used in geotechnical pipeline monitoring.

VW are considered the most common strain gauges used in pipeline monitoring. The units work by plucking a tensioned steel wire and measuring the frequency. Vibrating wire strain gauges are known for their long-term stability and durability. The gauges measure frequency which allows for longer transmitted cable lengths (>2000 m) over other gauges that generate a voltage and can be susceptible to voltage drop. Gauges can be spot welded (recommended) onto steel pipelines or epoxied (not recommended), making for a simple installation (see Figure 10).

Resistance strain gauges are also used on pipeline monitoring projects. They work by measuring strain recorded by the changes in electrical resistance of the gauge. Installation of resistance gauges take greater care in preparing the surface and ensuring a good bond is created between the steel and the gauge. Resistance gauges are not as robust as VW gauges but can provide a greater sampling rate (dynamic) over vibrating wire. Resistance gauges are available for different types of materials and sizes and can be installed longitudinally or circumferentially on a pipeline.

Fiber optic strain gauges work by measuring a deformation distance as a change in reflected wavelength connected to special readout boxes. Fiber optic strain gauges are similar to resistance gauges with the exception that they are using reflected wavelength changes instead of changes in resistance. These units can also be welded on or epoxied. Fiber optic strain gauges do offer two distinct advantages over both vibrating wire and resistance gauges: they are intrinsically safe with no moving parts or electricity, and they are immune to electromagnetic and radio frequency interference. This is often a requirement when working near certain oil and gas pipeline facilities.

Any of the above-mentioned strain gauges can be connected into traditional data logging systems and data can be sent via radio, cellular or satellite communication. The data collected are used to determine the longitudinal strains at that location on the pipe. These data can be displayed on a remote alert website to alert clients via SMS message, or email when targeted alert thresholds are being approached or exceeded.

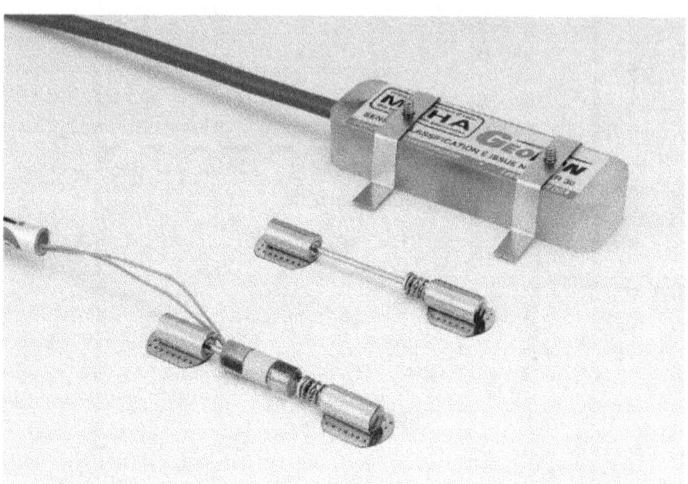

FIGURE 9: VIBRATING WIRE STRAIN GAUGE (GEOKON.COM)

Copyright © 2019 by ASME

FIGURE 10: INSTALLED STRAIN GAUGE WITH PROTECTIVE COVER FOR PIPELINE MONITORING

2.2 Distributed Fiber Optic Sensing

Long-range distributed fiber optic sensing is a pipeline monitoring technique (Figure 11) which uses a fiber optic sensing cable to detect changes in both temperature and strain along the length of the sensing fiber. This is a valuable and cost-effective method for sensing longer segments of pipeline through potential hazardous regions and or terrains. Data can be captured at small intervals (~1m) along the sensing fiber and along the entire length of the fiber (up to ~30km [19 mi]). The fiber optic cables are immune to EM interference and have a long-life cycle, typically the same as the pipeline.

The measuring systems use the principle of optical scattering in optical fibers. Optically-scattered peaks can be used to isolate where either strain or temperature anomalies have occurred over the entire length of the installed fiber optics. This allows for the anomaly to be isolated and its location accurately reported. The temperature anomalies can be caused by potential leaks or even unauthorized digging on the pipeline. Strain sensing is used to measure change in strain of the pipeline from a beginning state and evaluate if the pipeline may be undergoing changes in strain from an initial state which may indicate some movement of the pipe (ground).

Successful integration of distributed fiber optics systems on pipelines involves a careful coordination of planning, preferably during the design and the early construction phase. Understanding the monitoring requirements and expectations of the client and project team is paramount for a successful deployment. River and road crossings will need to be well understood as typically the fiber will need to be spliced at these locations and splice boxes installed. As with any monitoring system, power and security concerns are important considerations and often a separate installation shield is constructed with permanent power and communications. Data interrogators are often housed in existing structures along the pipeline right of way which have access to power and communications.

As with other traditional monitoring methods the data collection system can be set up to monitor and alert the user when predetermined action levels are reached. These alerts can be sent through email or SMS text message. It is also possible to feed this data directly into a supervisory control and data acquisition (SCADA) control room. The alerts will locate the anomaly based on the spatial sampling resolution of the design. The data can then be relayed to teams on the ground to investigate.

Distributed fiber optic sensing is a technology that exploits the fiber optic cable as the sensor and allows for fine sampling over large distances. These benefits now make it possible to monitor long lengths of pipeline for temperature and strain events.

FIGURE 11: INSTALLATION OF FIBER OPTICS FOR PIPELINE MONITORING ON A PIPELINE. . THE FIBER OPTIC CAN BE SEEN AT THE 12 OCLOCK POSITION ON THE PIPELINE.

3.0 DISCUSSION

Once initial landslide assessments have been completed and the relative threat to the pipeline(s) has been determined, management decisions can be made to either 1. conduct monitoring for future adverse landslide movement or 2. mitigate landslides (which may require ongoing monitoring). Implementing a monitoring program can be conducted using a variety of methods and instrumentation techniques to monitor both ground movement and pipeline movement. Table 1 below provides a summary of the commonly used methodologies provided in this paper.

Copyright © 2019 by ASME

Monitoring Method	Movement Monitored (Single Landslide)			Movement Monitored (Pipeline System)		
	Pipe	Ground (Surface)	Ground (Subsurface)	Pipe	Ground (Surface)	Ground (Subsurface)
LiDAR	-	X	-	-	X	-
InSAR	-	X	-	-	X	-
Aerial Imagery	-	X	-	-	X	-
Ground Reconnaissance	-	X	-	-	not typically feasible	-
Aerial Reconnaissance	-	X	-	-	X	-
SI	-	X	X	-	-	-
VSM	-	X	-	-	-	-
GSM	-	X	-	-	-	-
Piezometer	-	proxy	-	-	-	-
ILI/IMU	X	-	-	X	-	-
Strain Gauge	X	-	-	-	-	-
Fiber Optic	X	-	-	X	-	-

TABLE 1: SUMMARY OF MONITORING METHODS AND APPLICATIONS TO PIPELINE, GROUND SURFACE, AND GROUND SUBSURFACE MONITORING

Geologic, topographic, and climatic conditions are important considerations when implementing a monitoring program, e.g. landslide susceptible geologic and topographic conditions coupled with high precipitation generally require more in-depth monitoring efforts. Also, newly constructed pipeline systems, especially if located in landslide susceptible areas, may require considerable efforts to monitor landslide-related hazards during the first few years while vegetation establishes, and soil and near surface hydrologic conditions equilibrate. Ongoing monitoring for pipeline systems may also be conducted at regular intervals, or after potential ground destabilizing events such as high precipitation storms or earthquakes.

Selecting the monitoring frequency is an important consideration when developing and maintaining a monitoring program. Typically, the frequency of measurement or observation should be timed such that the pipeline owner can respond with adequate time to respond to and mitigate the threat. As such, sites or pipeline systems with expected higher rates of change, or with higher uncertainty regarding change, should be monitored more frequently. The monitoring frequency should be adjusted over time based on the behavior of a site or pipeline system. Typically, sites and new pipeline systems are monitored more frequently at first. If there are little or no changes over time, then the monitoring frequency is reduced. Alternatively, if a monitored site or pipeline system show higher rates of change over time then the monitoring frequency is increased to the point where an adequate reaction time is established to maintain management of the hazard.

In summary, the development and management of a landslide monitoring program allows pipeline owners and operators to identify, understand and manage potential threats before they can become pipe integrity issues and as such they can effectively manage landslide-related risks associated with their pipeline systems.

ACKNOWLEDGEMENTS

A special thanks to Don West (Golder) and John Hennessy (Golder) for their contributions to this paper.

REFERENCES

[1] Hodgson, M.E., and Bresnahan, P., 2004, Accuracy of Airborne Lidar-Derived Elevation: Photogrammetric Engineering & Remote Sensing, v. 70, p. 331–339, doi:10.14358/PERS.70.3.331.

[2] Mora, O., Lenzano, M., K. Toth, C., A. Grejner-Brzezinska, D., and Fayne, J., 2018, Landslide change detection based on Multi-Temporal airborne LIDAR-derived DEMs: Geosciences, v. 8, p. 23, doi:10.3390/geosciences8010023Field, E.H., Biasi, G.P.,

[3] Duncliff, John, "Geotechnical Instrumentation for Monitoring Field Performance", John Wiley & Sons, Inc.,

Copyright © 2019 by ASME

AIM-PIMG2019-1038

Chapter 12

DETECTING AND QUANTIFYING GROUND SUBSIDENCE FOR PIPELINE GEOHAZARD MANAGEMENT USING SATELLITE RADAR INTERFEROMETRY - POTASH MINE SUBSIDENCE EXAMPLE

Danielle Smilovsky
Wood
Phoenix, AZ, USA

Jeffrey R Keaton
Wood
Los Angeles, CA, USA

ABSTRACT

The potential value of Synthetic Aperture Radar interferometry (InSAR) to detect and quantify Earth surface changes has been recognized for more than 25 years. The European Space Agency (ESA) relatively recently established the Copernicus Program with a satellite constellation that includes a pair of newer satellites, Sentinel-1A and Sentinel-1B. These satellites are in a near-polar, sun-synchronous orbit. They share the same orbital plane with a 180° orbital phasing difference, and carry a C-band synthetic aperture radar active sensor which is applicable for many scientific purposes, including all-weather, day and night data collection. C-band radar sensors have 6-cm wavelength which is suitable for detecting changes as small as ~3 cm per cycle. The pair of satellites are complimentary with one having an ascending view direction and the other a descending view direction, and acquiring repeated 6-m pixel size coverage of the Earth every 12 days.

Satellite radar interferometry is the difference in distance from the satellite to the ground between two radar scenes acquired from the same position on its orbit. An initial data processing step is calculation of a coherence diagram which compares the distance to all pixels in the pair of radar scenes to each other relative to one wavelength (± 3 cm). Interferograms with good coherence are suitable for small-scale distance change detection, whereas poor coherence may indicate large-scale distance changes or some other type of major change, such as a wild fire. Examples of changes suitable for use in pipeline geohazard management that are related to ground surface subsidence over underground potash mines are in the form of areal changes with color bands of distance change, known as fringes, and profiles of distance change extracted from the 6-m pixel interferograms using geographic information system (GIS) utilities.

INTRODUCTION

The application of repeat-pass synthetic aperture radar (InSAR) technology to characterize the location, rate, and distribution of ground subsidence is significantly helpful in identifying and quantifying geohazards caused by ground subsidence so that risks associated with deformation can be managed. Interferometry has the ability to detect and quantify millimeter-level changes in terrain elevation by comparing radar-signal phase differences from pairs of images acquired from two or more satellite passes on the same orbit separated in time (Figure 1).

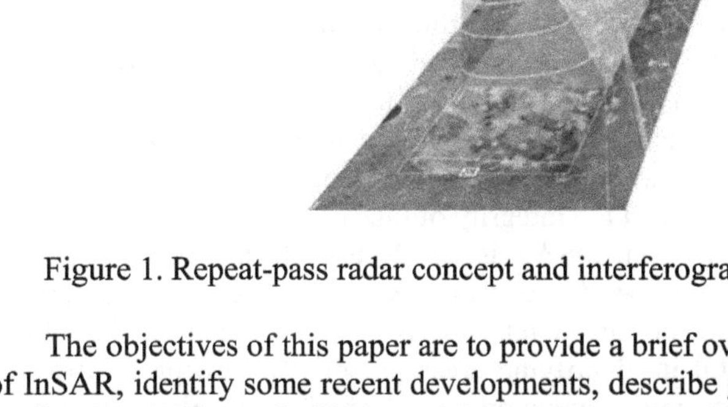

Figure 1. Repeat-pass radar concept and interferogram.

The objectives of this paper are to provide a brief overview of InSAR, identify some recent developments, describe general applications to linear facilities, and focus on ground subsidence in a potash mining area. InSAR is a computationally intensive process of comparing two radar data sets of the Earth's surface acquired from the same location on an orbit. Synthetic aperture

Copyright © 2019 by ASME

radar energy has specific characteristics that allow the phase of the signal returned to a sensor on the satellite to be measured. The difference in return-signal phase between two datasets allows the measurement to be converted into a change in position of the ground relative to the satellite that is a fraction of the wavelength of the radar energy. This change can be described as ground subsidence, deformation, elevation change, or movement depending on the application. The Z values representing this change are in raster form with each pixel holding one change value. These values can then be used to develop maps, visual models, and profiles by extracting values at pixel centroid points along a line. These methods help communicate InSAR data in a compelling way.

Recent developments in synthetic aperture radar include a pair of satellites with C-band active sensors that are on the same orbital plane 180° apart, with ascending and descending look-directions. They each have a 12-day repeat cycle, and have a six-day repeat cycle in some areas where they orbit in tandem. The European Space Agency (ESA) offers data collected by these satellites free of charge. This is revolutionary in the realm of earth observation and change detection because just a few years ago, acquiring satellite data for InSAR change detection purposes was very expensive. The ESA not only offers data free of charge, they also offer InSAR processing software. This software is called the Sentinel Application Platform (SNAP). However, InSAR processing requires a high level of expertise and background knowledge to create correct and applicable data.

GENERAL APPLICATIONS TO LINEAR FACILITIES

Since 1992, colleagues of the authors have been studying land subsidence and deformation for a variety of infrastructure and resource projects. Geohazard management and risk assessment have been the driving purposes for most of these projects. Flood retarding structures, transportation infrastructure, mining facilities, active wells, oil and gas fields, water recharge facilities, earth fissures, sinkholes, and landslides are some of the assets and major geohazards that can be monitored using InSAR technology to inform management decisions.

InSAR has been used to characterize landslides with deformation classified into three-centimeter increments, which highlights the most affected and vulnerable areas. This information was valuable for the Colorado Department of Transportation because these landslides are located above a major interstate highway, making them serious hazards in the effort to protect the integrity of linear infrastructure (Figure 2).

In Arizona, InSAR is widely used to monitor the ground and characterize deformation near long, linear flood retarding structures on an annual basis. This monitoring has been successful in identifying areas potentially susceptible to earth fissuring and inundation by flash floods. Flood retarding structures in Arizona are valuable assets that are managed, in part, with InSAR technology, which has been in use since 2002 for prioritizing dam mitigation and rehabilitation engineering.

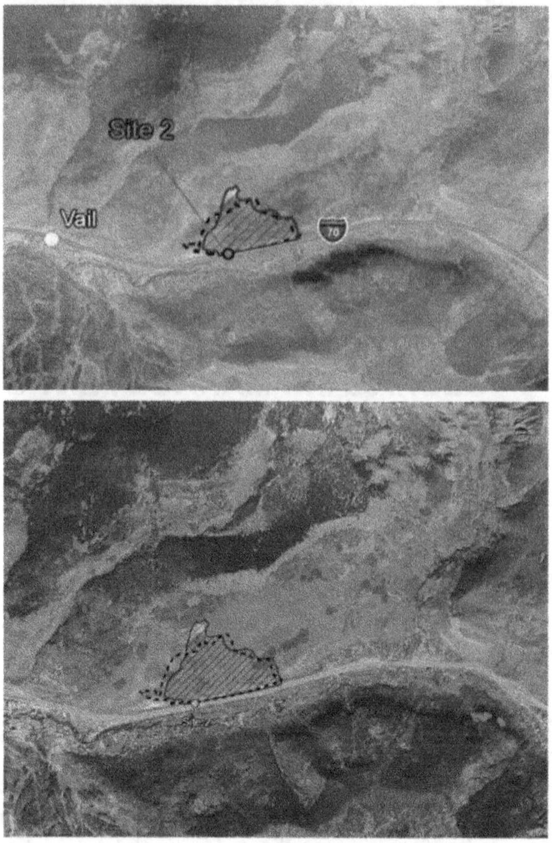

Figure 2. Interferogram showing deformation partially corresponding to mapped landslide (top); classified deformation model over two-year period: 3-6 cm in red, less than 3 cm in orange (bottom).

In 2002 to 2003, a historical InSAR study was performed over McMicken Dam, a long, linear feature in the west valley of the Phoenix metropolitan area. At the time, past ground deformation was not well understood, so the initial step used InSAR to look back six years. During this study, a subsidence feature was discovered at the south end of McMicken Dam using a pair of C-band satellite scenes from December 1996 and December 1999, originally developed by the Center for Space Research at the University of Texas in Austin. This is one of the first interferograms created covering Phoenix (Figure 3). This subsidence was later related to an earth fissure zone [1] after deep resistivity soundings were analyzed. Since the identification of this earth fissure zone, this geohazard, and the ground around it, has been under annual monitoring to protect the integrity of the linear infrastructure.

POTASH MINE SUBSIDENCE EXAMPLES

The Carlsbad Potash District has several active subsidence features between about 25 and 50 km to the east of Carlsbad, NM, in response to underground solution mining. Sinkhole activity has been observed near some mining locations. Potash mining typically extracts mineral-rich brine solutions which are

Copyright © 2019 by ASME

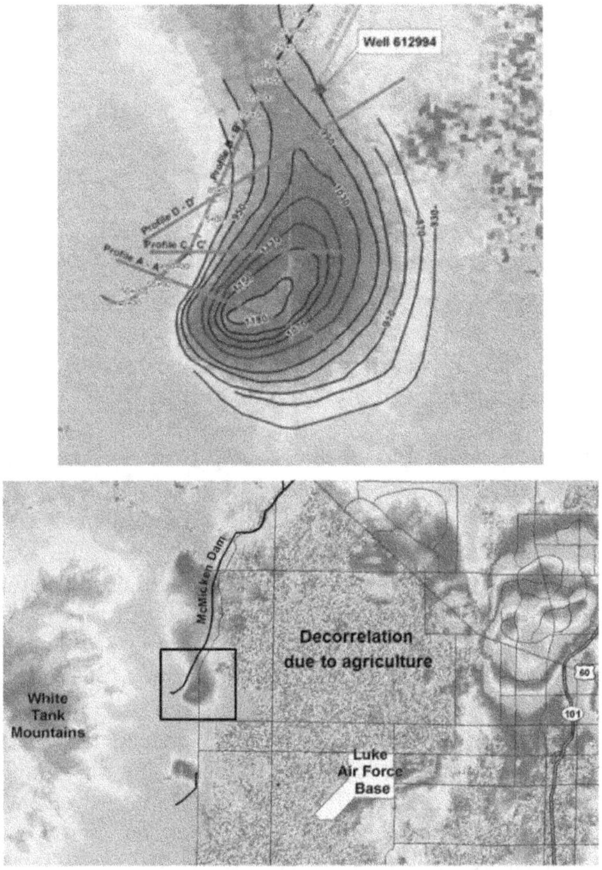

Figure 3. Focused view of subsidence adjacent to flood retarding structure (top); regional view of west Phoenix showing subsidence (bottom).

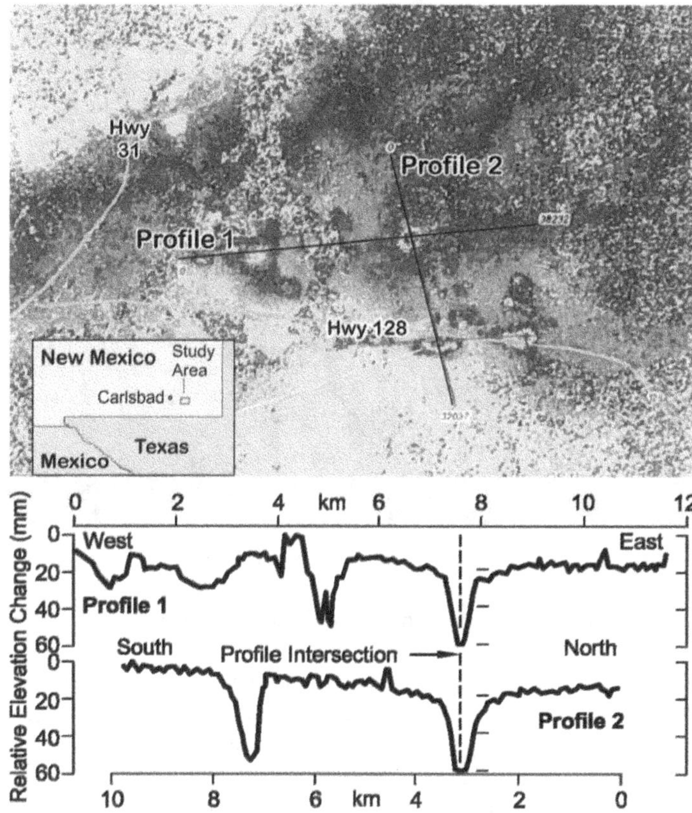

Figure 4. Interferogram (top) and relative elevation-change profiles (bottom) from the Carlsbad Potash District, NM. Based on data from the project described in [2].

transported by pipelines to evaporation ponds. The mining district area has a variety of natural and human-induced ground disturbances, and is crossed by valuable linear infrastructure, notably major highways and transmission pipelines, with alignments near sinkholes and potash mines.

Previous projects to monitor ground deformations near potash mines and identify potential sinkholes locations included an interferogram constructed from two ENVISAT C-band scenes from July 2005 to March 2006, an 8-month time difference [2]. The interferogram (Figure 4), which clearly shows subsidence features at locations determined to be potash mines, was used to extract two elevation-change profiles to help quantify and characterize the subsidence features. C-band sensors allow up to about 3 cm of line-of-sight (LOS) change per color cycle displayed with GIS utilities as values stretched along a color ramp. Elevation-change features that exceed 3 cm repeat the same color ramp again, requiring interpretation of the displayed image. The most prominent feature in Figure 4 is located at the intersection of Profiles 1 and 2, with 40 millimeters of relative subsidence over the 8-month period, corresponding to an average annualized subsidence rate of ~60 mm/yr. The subsidence profile shape at the intersection of Profiles 1 and 2 suggests that the maximum ground elongation over a 200-m distance is about 0.004 mm (0.02 με) in 8 months or an annualized ground elongation of about 0.006 mm (0.03 με) on the west side of the subsidence area. This small rate of subsidence could persist for a number of years without adversely affecting a modern transmission pipeline; however, knowledge of subsidence locations would inform pipeline operators of locations where routine patrols should pay particular attention to evidence of earth fissures or other features of concentrated strain.

Another interferogram covering the Carlsbad Potash District was generated in 2012 by Arizona Department of Water Resources (ADWR). This interferogram used scenes from January 2008 and March 2011, and was generated using an L-band sensor called Alos Palsar (Figure 5). This sensor can capture approximately 11.8 centimeters of change LOS per color cycle. Since L-band is able to capture more change per color cycle, the temporal baseline can be extended using longer periods of time between satellite passes, however, 38 months may be too much time between these two particular passes given the amount and rate of change that is occurring at the potash mines. The major subsidence features in the interferogram have decorrelation in the centers. Decorrelation usually occurs in response to dense vegetation growth or localized water ponding. The decorrelation in Figure 5 appears to have occurred because too much subsidence occurred in the

Copyright © 2019 by ASME

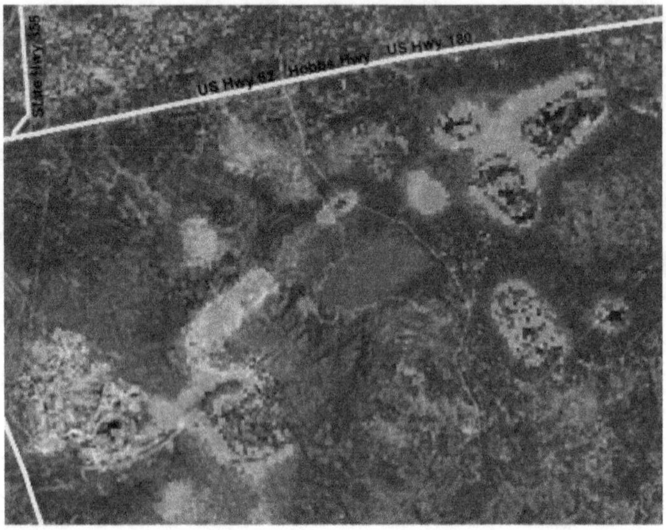

Figure 5. Interferogram of the Carlsbad Potash District showing decorrelation in areas of rapid subsidence. Image courtesy of Arizona Department of Water Resources.

38-month time period between the satellite scenes. Subsidence at these locations might have been displayed more effectively with shorter time periods and then subsidence increments summed. Given the rate of ground movement and the sensor capabilities, a three-month maximum period between satellite passes for C-band sensors, and a twelve month maximum period for L-band sensors may provide better results.

A comprehensive study of ground deformation induced by potash mining in southeast New Mexico used both Alos Palsar (L-band) and the newer Sentinel-1A (C-band) sensor, and correlated subsidence rates with mining production records [3]. This study included Alos Palsar InSAR coverage between December 2007 and January 2011, and Sentinel-1A InSAR coverage from January 2016 to November 2016. Although the InSAR results reflect inconsistent time periods ranging from one to six months, the interferograms showed compelling life-cycle information about subsidence activity in and around the potash mines without exceeding the maximum change amount that affected the interferogram in Figure 5. Potash production showed correlation with subsidence rates in the 2007-2011 Alos Palsar time series analysis, which was anticipated. However, this study also showed continuing subsidence occurring after potash production paused on May 6, 2016 (Figure 6). These results demonstrate the importance of monitoring natural and anthropogenic ground disturbances during periods of known activity, and also after activity has stopped [3].

To examine the subsidence occurring at the potash mines during the end of 2016 and continuing into 2017, a time series analysis was performed using Sentinel-1A interferometric data to compare results using longer temporal baselines to results using shorter temporal baselines. The September 2016 to June 2017 interferogram (Figure 7a; nine months in a single image) shows results in the potash mine area interpreted either as no

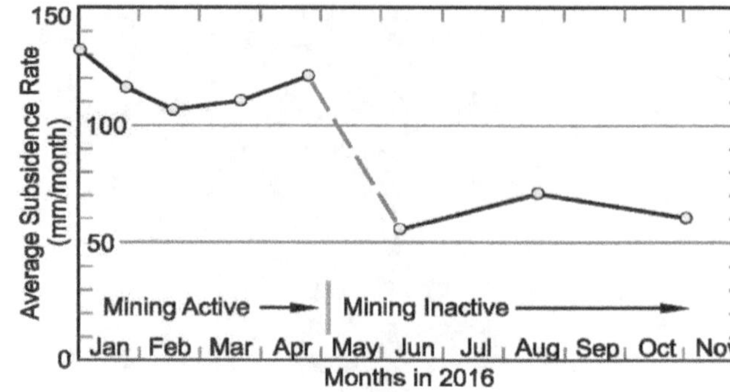

Figure 6. InSAR based average subsidence rate showing correlation to mining activity. Modified from [3].

activity or as unusable decorrelation. However, timeframes shortened to three-month intervals produces interferograms with 85% or better coherence scores and subsidence features can be identified, despite being slightly decorrelated.

This improved coherence score with slight decorrelation in active areas indicated that subsidence is still occurring quickly and the temporal baseline should be shortened further while using this C-band sensor. Although this seemed like quite a bit of processing effort, it actually produced the results that enabled specific months to be identified when more activity was occurring compared to other months. In this case, the December 2016 to March 2017 three-month interval reveals the most prominent subsidence features when compared to the other two three-month intervals (Figure 7b, three images). When this time frame is subdivided into monthly interferograms with a 96% or better coherence score, the behavior of all the subsidence features is captured well in a monthly time series evolution of the area (Figure 7c, four images). Additionally, a time series profile over the subsidence feature closest to US Highway 62 was generated to compare results (Figure 8).

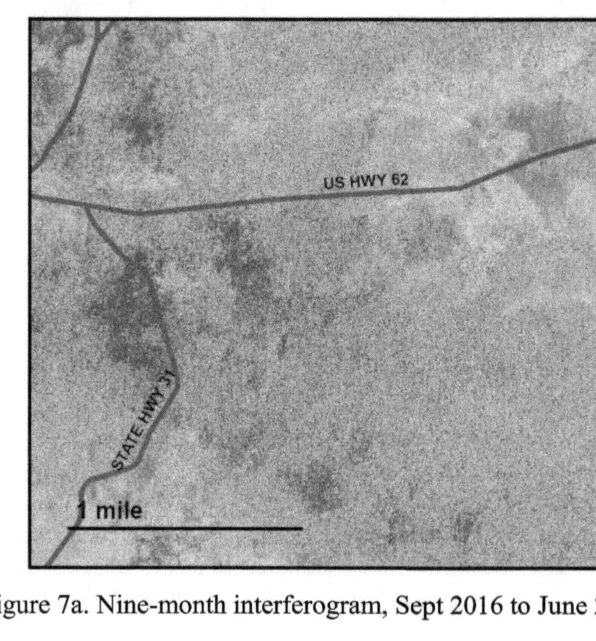

Figure 7a. Nine-month interferogram, Sept 2016 to June 2017.

Copyright © 2019 by ASME

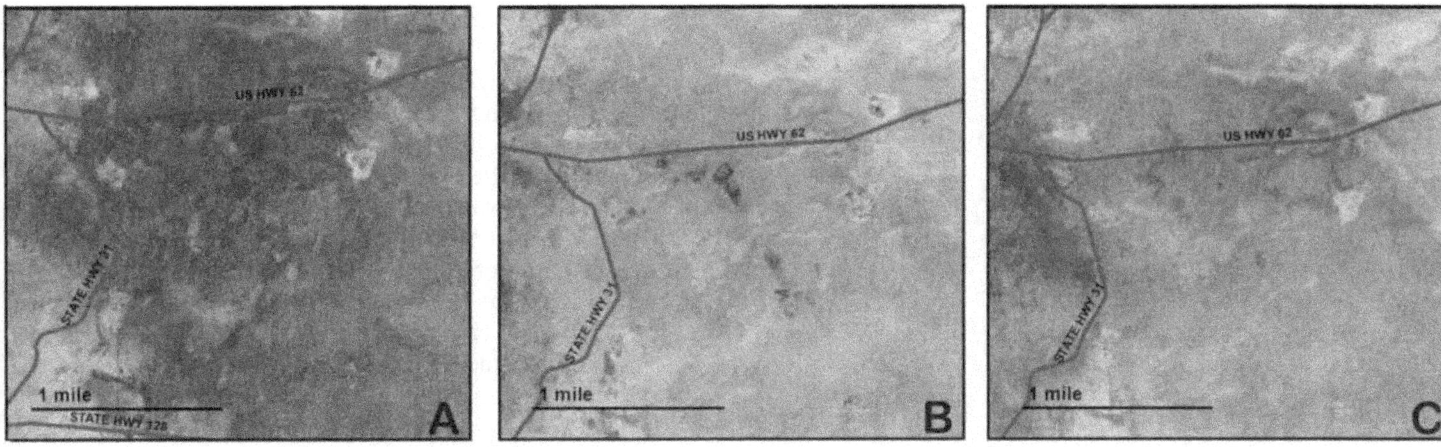

Figure 7b. Three interferograms each spanning three months. A. Sept to Dec 2016. B. Dec 2016 to Mar 2017. C. Mar to Jun 2017.

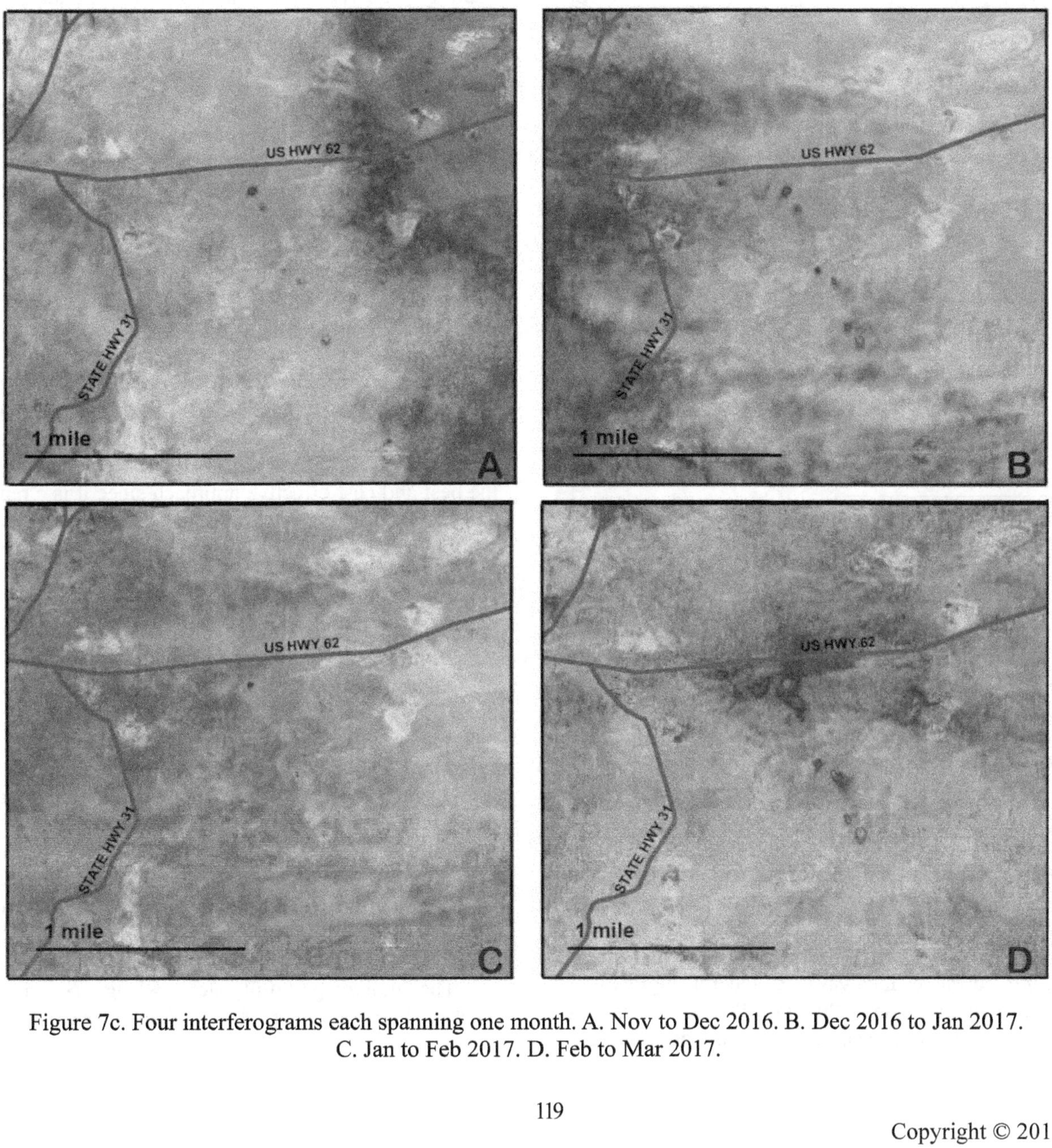

Figure 7c. Four interferograms each spanning one month. A. Nov to Dec 2016. B. Dec 2016 to Jan 2017. C. Jan to Feb 2017. D. Feb to Mar 2017.

Copyright © 2019 by ASME

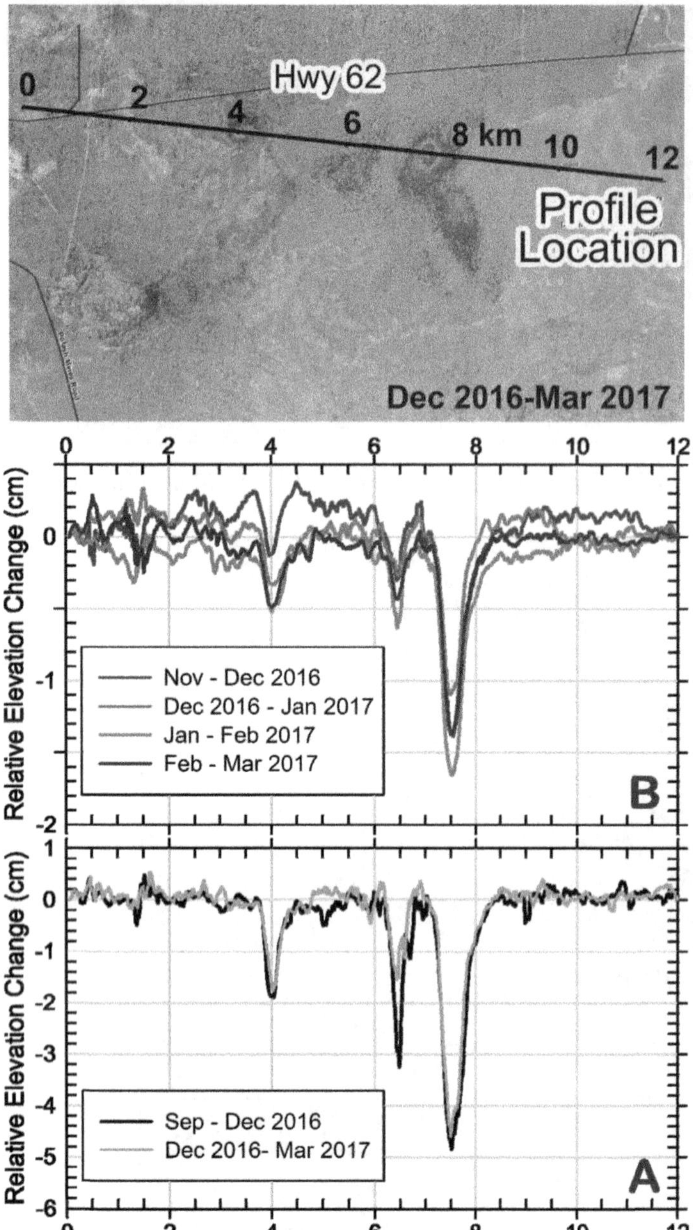

Figure 8. Time series of relative elevation change. Top. Profile location on InSAR image (enlarged portion of Fig. 7b-B). A. Plot of three-month interval relative elevation change. B. Plot of one-month interval elevation change.

CONCLUSIONS

The utility of InSAR results to linear projects in the Carlsbad Potash District would also apply to linear project in other potash mining areas. Linear projects in these examples have been highways, for the most part. A project in New Mexico currently underway involves a relatively small site in close proximity to state highways, a federal highway, a railroad, and an irrigation canal. Pipelines that are nearby are distribution or service pipelines, not transmission pipelines. The nature and magnitude of ground deformations detected and quantified with InSAR need to be interpreted for each application and with knowledge of the geologic setting. Sources of error and uncertainty in the synthetic aperture radar data also need to be recognized. Time of year and weather conditions affect the radar data and the condition of vegetation in the area being evaluated.

Since the launch of the European Space Agency Sentinel-1A mission in April of 2014, synthetic aperture radar data needed to calculate interferometric scenes can be downloaded every 12 days. In tandem with Sentinel-1B, launched in April 2016, the time frame for satellite radar data shortens to 6 days. The descriptions of multiple studies regarding potash mining, and sinkhole activity in the karst environment east and southeast of the Carlsbad, New Mexico area, indicate that InSAR technology is a valuable tool to identify, monitor, and quantify natural and anthropogenic induced subsidence. Furthermore, an evolution of InSAR processing and analysis methodologies seems to be leading to a refined InSAR analysis approach that has been calibrated for geohazards in this area of interest and probably has application to similar environments elsewhere. Twelve-month time-intervals should be a maximum temporal baseline with L-band data, and three-month intervals should be the maximum temporal baseline with C-band data, to yield the most reliable and interpretable results that accurately reveal the maximum subsidence at a given area of study or specific subsidence feature of interest given relatively rapid subsidence rates.

Other environmental pressures and geotechnical processes in geohazard areas may warrant monitoring on a regular basis for asset management purposes, especially considering valuable linear infrastructure nearby, such as highways and pipelines. Several methods of study have been and are being used to help monitor active processes; however, satellite technology maybe the best and cost-effective approach since it is a remote option, and the repeat-pass interval has shortened to 6 to 12 days as satellite technology has advanced.

Development of unconventional oil and gas resources has contributed to an increase in the number of in-service pipelines, with a large number of pipelines in designated corridors and a large number of pipelines crossing other pipelines. One aspect of geohazard management for which interferograms might be particularly useful is ground deformation in the vicinity of pipeline corridor crossings. Detected deformation in such locations, combined with knowledge of some details of the foreign pipelines, such as general age and what is being transmitted, might warrant an earlier response and interaction among operators. Because individual InSAR scenes have substantial footprints on the Earth's surface, collaborative application of the technology might be appropriate to consider.

ACKNOWLEDGMENTS

The authors gratefully acknowledge coworkers Michael Rucker, John Lommler, Bibhuti Panda, and Mark Hartig, all

Copyright © 2019 by ASME

Wood employees in the Phoenix or Las Cruces offices, for discussions, information, and support.

REFERENCES

[1] Rucker, M.L., Greenslade, M.D., Weeks, R.E., Fergason, K.C., and Panda, B., 2008, Geophysical and Remote Sensing Characterization to Mitigate McMicken Dam. GeoCongress 2008, New Orleans, American Society of Engineers. https://doi.org/10.1061/40971(310)26.

[2] Rucker, M.L., Panda, B.B., Meyers, R.A., Lommler, J.C. 2013. Using InSAR to Detect Subsidence at Brine Wells, Sinkhole Sites, and Mines. Carbonates and Evaporites, v. 28, p. 141–147. https://doi.org/10.1007/s13146-013-0134-9.

[3] Kim, J.-W., Zhang, A., Lu, J. 2017. Detecting Mining-Induced Ground deformation and Associated Hazard Using Spaceborne InSAR Techniques. Geomatics, Natural Hazards and Risk. v. 9, no. 1, p. 211-223. https://doi.org/10.1080/19475705.2017.1415229.

Copyright © 2019 by ASME

SECTION 4
GEOHAZARD MANAGEMENT

INTRODUCTION

By Alexander McKenzie-Johnson
(AMckenzieJohnson@Geosyntec.com)

What is geohazard management? This deceptively simple question can have many answers. To some, it might mean managing the after-effects of a geohazard event such as a large landslide, flood, or earthquake. To others, it might mean an integrity-based approach to proactively reduce or eliminate the potential for catastrophic events such as pipeline loss of containment once a pipeline has been constructed. Still, to others, it might mean a complete lifecycle approach where the costs and risks from geohazards are reduced to an "acceptable" level.

At the time of this publication (and writing from a North American perspective), there are no firm regulatory requirements and industry consensus about what constitutes geohazard management. While ample guidance exists in terms of conducting assessments to identify geohazards, methods to mitigate and monitor geohazards, and risk assessment systems (such as the papers in this volume), the fundamental question of what constitutes a "good" or "state of practice" geohazard management program remains largely undefined for pipeline operators.

This presents a tremendous challenge for pipeline operators, because there is no baseline or minimum standard against which to measure themselves. An ISO standard (20074) "Petroleum and natural gas industry -- Pipeline transportation systems -- Geological hazard risk management for onshore pipeline" had been approved in July 2019, shortly before this volume was published. This standard, covering onshore geohazard management by pipeline operators, could serve as a minimum standard, but had not yet been adopted by any regulatory agencies as of the date this introduction was written.

However, the relative lack of regulatory guidance around geohazards also presents an opportunity for pipeline operators. Because the operators have relative freedom to determine the shape and content of their programs, they are not bound by pre-determined structures and regulations that limit what they can do. This gives operators the opportunity to craft programs that fit their own company cultures, local regulatory and political environments, and local geologic conditions.

ISO 20074 requires operators to have complete lifecycle programs to consider and manage geohazards at every stage of the pipeline cycle, from the initial routing to the final closure of a pipeline. However, it leaves the specifics of how to accomplish these tasks largely up to the individual operators.

Returning back to the original question, it is this editor's opinion that geohazard management is the systematic process by which an operator attempts to proactively eliminate the potential for catastrophic events (i.e., loss of containment) from geohazards and to reduce to as low as reasonably possible the overall costs associated with long-term geohazard management. This process involves the complete pipeline lifecycle, in line with the structure espoused in the draft ISO standard.

The papers in this section provide examples of geohazard management approaches from both the perspective of pipeline operators who operate geohazard management programs and the consultants who provide the services needed to execute these programs. The authors describe geohazard management components from the perspective of integrity management programs for operational pipelines and from pipelines in the routing and design phases. Although the papers describe widely different locations ranging from the varied terrain of California, to the plains of Canada, to the tropical mountains of Malaysia; have different regulatory and cultural environments; and have differing approaches, they share common descriptors of the fundamental processes that compose comprehensive geohazard management programs. In short, these elements may include:

- A systematic process by which geohazards are proactively identified, evaluated, and classified.
- A means by which data collected in the identification and evaluation phases are stored and analyzed.
- A systematic process by which the collected information is used to implement some kind of management action to reduce risk from the identified geohazards, such as preemptive mitigation, routine monitoring, or a combination of the two.
- Integration of multiple technologies in the performance of the preceding elements.

Thus, a "good" geohazard management program makes use of almost all of the elements described in this volume. The development and widespread commercialization of several technologies have greatly facilitated the ability to perform the processes that constitute a geohazard management program. From the use of Light Detection and Ranging (LiDAR) digital elevation data to facilitate geohazard identification and characterization, to the use

Copyright © 2019 by ASME

of ILI Inertial Measurement Unit (IMU) data to evaluate bending induced by geohazards, to remotely monitored instrumentation, to evaluation and monitoring by satellites, to data storage and analysis through powerful geographic information systems (GIS) databases, it has never been easier to identify, evaluate, and manage geohazards than ever before. As new ILI, satellite, and other technologies mature and become commercially available, it should only become easier to effectively manage geohazards.

A final point before introducing our authors and papers. When discussing what constitutes a "good" geohazard management program, it is also important to consider what causes geohazard management programs to fail or be inadequate because that helps to inform the considerations in a good program. Leaving aside the obvious answer of operators with no program; geohazard management programs tend to fail for one of the following reasons:

- Limiting the geohazard management program to the vicinity of past failure locations. A common assumption is that only certain areas are prone to geohazards, and that geohazard management should only be confined to those areas. Yet, from a North American perspective, landslide related failures have occurred in areas as diverse as Washington State, Ohio, Kentucky, West Virginia, Ontario, British Columbia, Saskatchewan, Alberta, North Dakota, California, and Arkansas, with a similar, if not greater range for hydrotechnical hazards. A good geohazard management program evaluates and manages the entire pipeline network, and only confines itself to smaller areas if the initial hazard identification process supports this conclusion.
- Use of an inadequate identification process. The initial identification of geohazards is in many ways, the most consequential step in a geohazard program, because it guides all further actions within the program. With widespread commercial availability of high-resolution LiDAR data, the ability to identify geohazards is greater than ever before. If this step is skipped, or if inadequate data (e.g., low resolution, narrow focus) are used, then key hazard locations may not be recognized.
- Limiting the program to solely identification and evaluation. As stated previously, identification of geohazards is the single most important component of a geohazard program, but if there is no plan for monitoring or mitigation, then the information collected does not provide risk reduction. A good geohazard management program has a defined process by which to act on information collected during the identification and evaluation steps.
- Limiting the program to a single portion of the pipeline life cycle, such as only applying it to operations. A complete and comprehensive geohazard management program considers all aspects of a pipeline life cycle, from the routing, through design, construction, and then finally, operation of the pipeline. In particular, ignoring the routing, design, and construction phases misses opportunities for avoiding and reducing costs associated with geohazards, as well as potentially creating new geohazard risks.

In conclusion, the current state of the science makes it possible to proactively manage and reduce the risks associated with geohazards. The specific form or nature of the program, while not completely unimportant, is less important than having a comprehensive program. Our authors describe a diverse set of experiences, locations, and approaches to geohazard management. All of these papers contain concepts, observations, and ideas that can be of value when creating or improving upon a geohazard management program. I found these papers to be interesting, thought-provoking, and I hope that you do as well.

The first paper "Creation and Management of a Landslide and Erosion Geohazards Inventory for Natural Gas Transmission Pipelines in California" by Katherine Herr and Teddy Atkinson of PG&E is an overview of an integrated, phased approach to landslide and erosion characterization and management for almost 7,000 miles of natural gas transmission pipelines crossing a variety of terrains in California. The program uses a suite of integrated detection techniques including geomorphic analysis of Light Detection and Ranging (LiDAR) data, change detection assessment from repeated LiDAR surveys, and correlation of geohazards with in-line inspection (ILI) data for bending strain, along with traditional methods of geohazard detection such as periodic aerial and ground patrols.

Herr and Atkinson make a key point that is worth reproducing here. They note that, "*Well curated geohazard inventories support risk reduction and asset management efforts by providing a tool for systemwide prioritization of geohazard mitigations. Improved early detection and prioritized mitigation of geohazards allows for increased proactive asset management and is expected to reduce the need for reactive response, leading to increased public safety and cost savings.*"

This is a key point. Effective, systematic geohazard management, particularly when conducted proactively, reduces operator risk and by extension, reduces public risk and operator cost.

The second paper "Use of a Geohazard Management Program to Reduce Pipeline Failure Rates" by Sarah Newton, Aron Zahradka, Gerald Ferris, and Michael Porter of BGC Engineering Inc. provides an overview of geohazard programs from the perspective of a prominent pipeline consultant. Similar to Herr and Atkinson, Newton et al., describe an integrated, phased approach to geohazard identification, assessment, and prioritization. Newton et al., reviewed geohazard failure results for the approximately 300,000 kilometers of operating, discontinued, and abandoned North American pipelines contained in BGC's GIS system. Their results suggest that pipeline systems managed through a comprehensive geohazard program have about a four times lower failure rate from geohazards than pipeline systems where geohazard management programs are incomplete.

The third paper, "Dynamic Geohazard Management in Challenging Environment" by Rohani M. Joehan, Wan Ismail Wan M Marzuiki, Kamal Ibrahim, Zazli Bob, and Rahman Azam (all with Petronas), takes us to geohazard management in the mountainous, tropical environment of Malaysia. The paper discusses lessons learned on the recently completed Sabah Sarawak Gas

Copyright © 2019 by ASME

Pipeline (SSGP) following two incidents related to geohazards. These incidents resulted in Petronas reevaluating their pipeline integrity management system and developing a comprehensive geohazard management plan. Similar to the preceding papers, this plan uses an integrated approach to hazard identification, assessment, and monitoring, including geomorphological mapping, LiDAR, aerial surveillance, and ground movement monitoring, all managed and stored using GIS type databases.

Joehan et al., provide detailed checklists and examples for all phases of the Petronas geohazard program. A key component of the program is the requirement for multiple levels of investigation, monitoring, and assessment prior to commissioning a pipeline (i.e., prior to gas-in), and then subsequent continuous monitoring and reassessment once the pipeline is in service.

The fourth paper, "Developing a Geotechnical Hazard Inventory: A Case Study and Best Practice Recommendations" by Joel Van Hove (BGC Engineering), Alex Baumgard (BGC Engineering), Joel Babcock (Pembina Pipeline Corporation), and Jan Bracic (Pembina Pipeline Corporation) discusses the creation of a geotechnical hazard inventory for a North American pipeline system. The authors discuss and provide examples of the use of high-resolution LiDAR to create an effective and accurate landslide inventory as part of a geohazard management approach for a Canadian pipeline system. The authors also point out that relying on excessively simplistic methods, such as slope angle, to identify landslide threat can both over and under represent the actual threat.

The fifth and final paper, "Example of Semi-Quantitative Stream Crossing Hydrotechnical Hazard Assessment for New Pipeline" by Rodolfo Sancio and David Vance of Geosyntec Consultants presents an example of a semi-quantitative system for assessing hydrotechnical threat for a new pipeline, and how such a system can be used for decision making during construction to manage long-term threat from this hazard type. Whereas our previous papers have focused on managing geohazards after pipeline construction has been completed, this paper presents an example of geohazard management during the routing, design, and construction phases of a pipeline project.

Chapter 13 Creation and Management of a Landslide and Erosion Geohazards Inventory for Natural Gas Transmission Pipelines in California (AIMPIMG2019-1010)
Katherine Herr, PG&E Geosciences, and Teddy Atkinson, PG&E Transmission Integrity Management, USA

Chapter 14 Use of a geohazard management program to reduce pipeline failure rates (AIMPIMG2019-1045)
Sarah Newton, Aron Zahradka, Gerald Ferris, and Michael Porter, BGC Engineering Inc., Canada

Chapter 15 Dynamic Geohazard Management In Challenging Environment (AIMPIMG2019-1070)
Wan M Marzuki Wan Ismail, Azam A. Rahman, M, Joehan Rohani, Kamal Azam Ibrahim and Mohd Zazil Mohd Zabidi, Petronas, Malaysia

Chapter 16 Developing a Geotechnical Hazard Inventory: A Case Study and Best Practice Recommendations (AIMPIMG2019-1051)
Joel Van Hove and Alex Baumgard, BGC Engineering Inc.; and Joel Babcock and Jan Bracic, Pembina Pipeline Corporation, Canada

Chapter 17 Example of Semi-Quantitative Stream Crossing Hydrotechnical Hazard Assessment for New Pipeline (AIMPIMG2019-1017)
Rodolfo B. Sancio and David Vance, Geosyntec Consultants, Inc., USA

Copyright © 2019 by ASME

AIM-PIMG2019-1010

Chapter 13

CREATION AND MANAGEMENT OF A LANDSLIDE AND EROSION GEOHAZARDS INVENTORY FOR NATURAL GAS TRANSMISSION PIPELINES IN CALIFORNIA

Katherine Herr
PG&E Geosciences
San Francisco, California, US

Teddy Atkinson
PG&E Transmission Integrity Management
San Ramon, California, USA

ABSTRACT

The Pacific Gas & Electric Company (PG&E) natural gas transmission system consists of approximately 6,800 miles of gas transmission pipeline traversing nine of the eleven unique geomorphic provinces of California. In this dynamic geologic environment, it is inevitable that extensive linear assets such as gas transmission pipelines will be affected by geohazards such as landslides and erosion. Geohazard-related ground movement can result in increased bending strain which in turn can lead to permanent pipe deformation and possibly failure. Asset owners must therefore develop strategies for assessing and managing these geohazards and associated asset risk.

This paper provides an overview of the creation and management of the PG&E Transmission Integrity Management Program (TIMP) Geohazards Program which was developed to assess and quantify exposure of gas transmission assets to landslide and erosion systemwide. The Geohazards Program has, to date, inventoried approximately 14,000 landslide and erosion sites along PG&E gas transmission pipeline corridors. Geohazard sites are identified, evaluated, ranked, and managed via a phased approach consisting of: 1) detection and desktop assessment; 2) initial field assessment and ongoing monitoring; 3) detailed site-specific study; and 4) mitigation, if necessary. This program utilizes a suite of detection techniques including geomorphic analysis of Light Detection and Ranging (LiDAR) data, change detection assessment from repeated LiDAR surveys, and correlation of geohazards with in-line inspection (ILI) data for bending strain; all in addition to traditional methods of geohazard detection such as periodic aerial and ground patrols.

Well curated geohazard inventories support risk reduction and asset management efforts by providing a tool for systemwide prioritization of geohazard mitigations. Improved early detection and prioritized mitigation of geohazards allows for increased proactive asset management and is expected to reduce the need for reactive response, leading to increased public safety and cost savings. The ultimate goal of this program is to develop and maintain a comprehensive geohazard inventory that will support more accurate and reliable characterization of asset risk by providing quantification of geohazard exposure. This, paired with future efforts to quantify associated asset fragility, will allow further development of a quantitative risk-based approach for asset management.

INTRODUCTION

While geohazards are only a subset of hazards that may affect pipeline integrity, they can result in significant damage such as permanent deformation or rupture. Geohazards that can cause permanent ground deformation which may affect pipeline integrity include landslides, erosion, fault creep or rupture, liquefaction or other seismically-induced ground failure, and subsidence. Comprising approximately 6,800 miles of pipeline throughout the state of California, the PG&E natural gas transmission system, like most extensive linear assets, is potentially exposed to a plethora of natural hazards. While avoidance is the optimal geohazard mitigation option (e.g., during the alignment selection of a new pipeline), geohazard exposure must also be continually assessed and managed for existing assets; moreover, 100 percent geohazard avoidance for a state-wide linear asset system is effectively impossible to achieve and changing land use and weather conditions over time can result in emergent geohazard conditions unanticipated by the original asset designers. Therefore, while thoughtful original route selection is important, asset owners must develop tools to

Copyright © 2019 by ASME

assess and manage ongoing or emergent geohazard risks throughout the service life of their assets.

In order to assess the risk of exposure of an asset to geohazards, one must first understand the location and nature of the hazards of concern. To this end, the PG&E TIMP Geohazards Program was initiated in 2014 to characterize and quantify the geohazard threat posed by landslides and erosion, thereby informing risk assessment and management of transmission pipelines. Other geohazards, such as fault rupture, seismically induced ground failure, and subsidence, are addressed in similar parallel programs; the scope of this paper is limited to landslide and erosion hazards. For supplementary information regarding types of geohazards and their wide-ranging potential impacts to pipelines, please refer to more comprehensive works by others such as Sancio et al. [1] and Nyman et al. [2].

The objective of this paper is to provide a snapshot of an owner/operator's method for building and maintaining a geohazard inventory of landslide and erosion hazards effecting pipelines in California in support of a comprehensive risk-based asset management strategy. Risk-based asset management allows for systemwide optimization of risk reduction efforts (e.g., geohazard mitigations), which if diligently and aggressively employed over time can incrementally increase the safety and reliability of the transmission system. Proactive system improvements allow an asset owner to act more consistently in a maintenance capacity rather than an emergency response capacity.

GEOHAZARD MANAGEMENT

The TIMP Geohazards Program utilizes a phased approach to characterize and address each geohazard site that is identified and incorporated in the geohazard inventory. The phase approach consists of: 1) detection and desktop assessment; 2) initial field assessment and ongoing monitoring; 3) detailed site-specific study; and 4) mitigation, if necessary.

A similar phased approach was recently proposed in a guidance document commissioned by the Pipeline Research Council International, Inc. (PRCI) [1] for management of geohazards affecting new pipelines (e.g., geohazards assessment during route selection). While this PRCI guidance document provides a reasonable geohazard assessment program structure and is an excellent reference for new pipeline installation site assessments, our program differs in several ways to better support the risk assessment and risk reduction needs of an extensive ***existing*** collection of assets. Our geohazard management phases are discussed in more detail in the following sections, and a flow chart illustrating the phase classification progression is provided in Fig. 1.

As individual geohazard sites are assessed via the phased management method, they are assigned a numeric geohazard ranking to characterize the site's overall activity and potential to impact a pipeline asset. Geosciences Subject Matter Expert (SME) judgement is required to make these numeric assignments on a site-specific basis. While landslides and erosion typically pose different threats to pipeline assets, parallel threat ranking systems were developed for these geohazards. Table 1 provides a simplified example of the geohazards risk index used for landslide and erosion sites.

TABLE 1 – GEOHAZARD CHARACTERIZATION MATRIX

GRI[1]		Geohazard Description[2]
6	Unknown[3]	• Suspected geohazard within 1,000 feet of pipeline centerline. • Unknown operational significance; potential impacts to operability not characterized.
5	Very High	• Active geohazard within 100 feet of pipeline centerline. • Immediate operational significance; potentially imminent adverse impacts to operability.
4	High	• Geohazard within 100 feet of pipeline centerline. • Near-term operational significance; possible adverse impacts to operability within service life.
3	Moderate	• Geohazard within 100 feet of pipeline centerline. • Long-term operational significance; possible adverse impacts to operability within service life.
2	Low	• Geohazard within 100 feet of pipeline centerline. • Negligible operational significance; unlikely to adversely impact operability within service life.
1	No Hazard	• Active or dormant geohazard that cannot feasibly impact the pipeline. • No anticipated impacts to operability within service life.

[1] Geohazard Risk Index (GRI)

[2] Applicable to both landslide and erosion geohazards.

[3] All emergent geohazard sites detected via remote sensing (e.g. LiDAR change detection), ILI, or direct reporting by local or patrol personnel are assigned this Geohazard Risk Index by default until the site can be assessed and characterized by a geosciences SME.

A Note on Program Support

The management philosophy presented herein is fundamentally built upon early engagement of SMEs throughout all phases, facilitating "cradle-to-grave" (detection-to-mitigation) interdisciplinary support that is custom-fit to the demands of each individual geohazard site; see Fig. 1. This phased approach also provides a programmatic structure to ensure interdisciplinary stakeholder input and consensus at critical hold-points, such as the transition from site specific assessment and study (Phases 2 and 3) to mitigation (Phase 4), and allows a relatively small interdisciplinary support staff of geosciences SMEs and database management personnel to manage thousands of individual geohazard sites affecting thousands of miles of linear assets. This program structure promotes consistent site characterization and treatment while also building and maintaining significant institutional knowledge and asset familiarity year over year. Consistent characterization of geohazard sites is key for consistent assessment of risks across the entire service area.

Copyright © 2019 by ASME

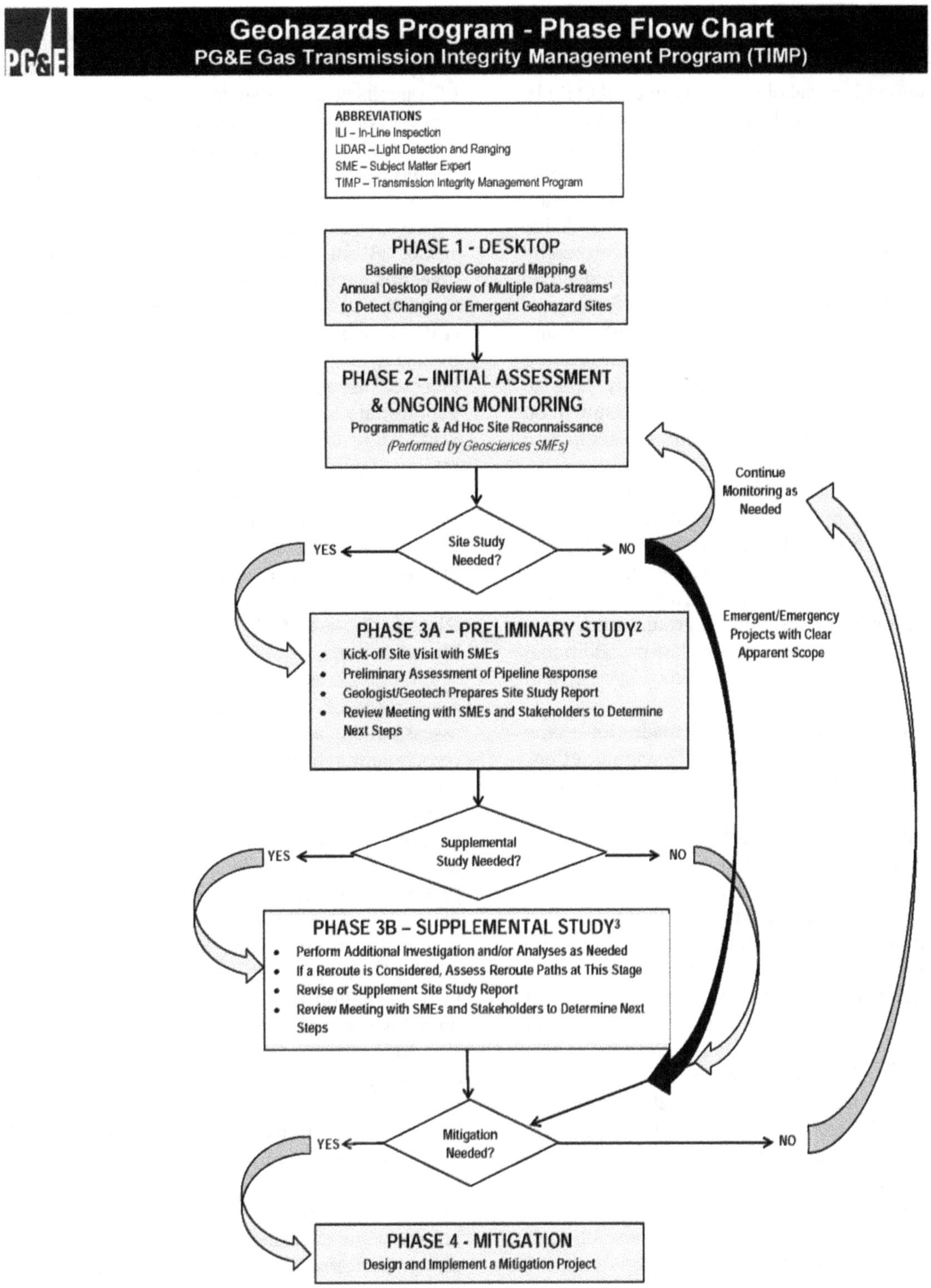

[1] Data reviewed includes (but is not limited to) LiDAR change detection, ILI anomaly data, and direct reports from local personnel.

[2] No ground-disturbing activities at this stage. Preliminary study may consist of, but is not limited to, site-specific survey and detailed site mapping.

[3] Supplemental study may be necessary to better understand the geohazard extent and/or pipeline response. Supplemental study may consist of, but is not limited to, borings, test pits, soil sampling and testing.

FIGURE 1. PG&E TIMP GEOHAZARDS PROGRAM PHASES

Copyright © 2019 by ASME

Program Initiation - Baseline Mapping Effort

The original goal of the TIMP Geohazards Program was to identify and rank landslide hazards that are affecting or could potentially affect transmission pipelines. At the onset of the program in 2014, systemwide aerial photographs and LiDAR (16 points/square meter) were collected in a 2,000-ft wide swath centered on the transmission assets of interest. These photo and topo resources were used to support baseline geomorphic interpretation for landslide hazards throughout the transmission system, supplemented with existing published landslide mapping. The LiDAR-based baseline mapping was informed by the Oregon Department of Geology and Mineral Industries (DOGAMI) Special Publication 42 [3]. Geomorphic features of interest were delineated digitally on an ArcGIS platform and assigned preliminary information and attributes from the desktop assessment effort, including an estimate of activity and an interpretation of proximity to and likelihood of encroachment on the nearest pipeline asset, resulting in a preliminary geohazard risk index (Table 1).

In 2015, sites identified in the baseline desktop mapping effort that had a high or very high hazard ranking were assessed in person by geosciences SMEs. In-person assessments were limited to visual inspection and confirmation of landslide geomorphology, with landslide sites remapped, added, or deleted from the database as needed. The systemwide ground-truthing of the desktop mapping effort also allowed for additional calibration of the geohazard risk index ranking system and revision as appropriate.

Provisional desktop delineations were made for some erosion sites during the landslide baseline mapping effort, however this mapping effort was not comprehensive. During the 2015 ground-truthing efforts, erosion was frequently observed triggering or exacerbating adjacent landslide hazards, or as an independent hazard affecting transmission pipelines by way of cover reduction or exposure. Additional erosion hazards were mapped and ranked as they were encountered during the 2015 landslide assessment work. A comprehensive systemwide baseline desktop mapping effort utilizing PG&E and public LiDAR data, was conducted specifically for erosion hazards in 2018, with ground-truthing and calibration efforts to continue in 2019.

For both landslide and erosion sites, as a geohazard is mapped it receives interpretation or observation notes, a geohazard risk index, and a recommended monitoring interval ranging from annual visual inspection by an SME to desktop-only assessment every five years. Photos from field inspections are also appended in the GIS database for each site to allow visual benchmarking and tracking of site conditions through time. Once the systemwide baseline mapping and assessment effort has been completed, management of sites continues cyclically per the phased approach described in the following sections.

Phase 1 – Detection & Desktop Assessment

Phase 1 provides a mechanism for detection and addition of emergent geohazard sites into the geohazard database post-systemwide-baseline mapping. The TIMP Geohazards Program is intended to provide a living inventory of geohazards as they change or develop through time, rather than just a snapshot of the geohazard risk at one point in time (as is typically the scope of a geohazard assessment for new pipeline siting studies). To this end, ongoing detection and monitoring efforts are necessary. This program utilizes a suite of detection techniques including geomorphic analysis of LiDAR data collected and interpreted at intervals, change detection assessment from repeated LiDAR surveys, and correlation of geohazards with ILI data for bending strain; all in addition to traditional methods of geohazard detection such as periodic aerial and ground patrols.

Once a site has been detected by desktop analysis methods (LiDAR and ILI data interpretation), or identified via reporting by local or patrol personnel, the site location of interest is assessed using desktop resources such as recent and historic geomorphic topographic interpretation (typically derived from LiDAR), historic and recent aerial and/or satellite imagery, and review of published geologic and geohazards maps. A geohazard site is added to the inventory database and populated with any available reporting information or photos. The new site is assigned a geohazard risk index flag of "UNKNOWN" and is prioritized for in-person assessment by a geosciences SME.

Phase 2 – Initial Field Assessment and Ongoing Monitoring

Phase 2 activities typically comprise in-person visual assessment of geohazard sites by geosciences SMEs. These assessments are performed for newly identified sites (precipitating from Phase 1) or for previously identified sites requiring regular visual inspection per their assigned geohazard risk index and monitoring interval. In parallel with the ground-truthing efforts conducted for the baseline mapping efforts, sites are mapped or remapped as needed, observation notes and photos are cataloged, and a geohazard risk index and monitoring interval and method are assigned. The monitoring method assigned may be ongoing in-person visual monitoring, remote sensing/desktop-based monitoring, and/or analysis of instrumentation measurements if a site has been instrumented with devices such as inclinometers or pipeline strain gages.

To support informed and expedient site review, Phase 2 assessments are recorded with GPS-enabled tablets that are outfitted with a GIS mapping interface connected to the geohazards inventory database. This mobile database connection provides access to historic photos, previous site visit notes, and supplementary reporting if available for a given site. The ability to review site history on-demand in the field allows for more informed and consistent assessments since site assessment is not always performed by the same individuals. Additionally, remote access to the database allows for an efficient combination of site history review and assessment update, rather than trying to review hundreds of sites in a desktop setting prior to assessing them *en masse* in the field.

If during a Phase 2 assessment a site is found to be complex enough to warrant additional study (such as but not limited to site specific survey, pipeline loading assessment, subsurface

Copyright © 2019 by ASME

investigation, or instrumentation) then the site is flagged for a Phase 3 study (discussed in the following section).

If an emergent or deteriorating site is determined to be potentially critical during a Phase 2 in-person inspection, the site is flagged as needing an urgent assessment and possibly a mitigation effort (Phase 4).

Phase 3 – Detailed Site Study

Phase 3 is utilized when additional information is necessary to better understand a geohazard site, the pipeline of concern's location and condition, and the combined potential for damage thereof. Since geohazard sites are often remote and difficult to access, and since a systemwide geohazard inventory may consist of thousands of sites, detailed site studies are reserved for sites that have an apparent potential to impact pipeline operations in the near-term (e.g., a landslide that is not yet crossing a pipeline but has the potential for headward retrogression toward a pipeline, or an erosion feature where the pipeline is not yet exposed, but active removal of cover is clearly observed). For our purposes, this phase is further subdivided to support a preliminary non-invasive site-specific study (Phase 3A) and a more intensive supplementary study that may include invasive activities such as subsurface exploration and instrumentation installation (Phase 3B).

Phase 3A may include, but is not limited to, site specific survey and detailed geohazard mapping. This phase also incorporates a review of both the geohazard history (changes through time, etc.) and the pipeline history (installation, retrofit and repair history, etc.). This phase seeks not only to provide a more detailed assessment of the activity and likely progression of the existing geohazard, but how the geohazard may affect the associated pipeline. The Phase 3A effort also develops additional monitoring recommendations (if necessary) and provides a preliminary alternatives analysis for site-appropriate mitigation options for both short-term (if necessary) and long-term mitigation efforts. Mitigation alternatives are ranked with considerations for long-term reliability, ongoing maintenance and monitoring needs, environmental impact, and relative cost.

If after the Phase 3A study it is determined that the site can be appropriately managed with ongoing visual monitoring only, then the site is cycled back into Phase 2 periodic monitoring. For sites of this nature, it is important to develop a monitoring threshold based on the Phase 3A findings to inform future monitoring of this site. For instance, if the geohazard concern was cover loss above a pipeline due to erosion, and after a study of records and non-invasive investigation the existing cover was determined to be adequate, then a monitoring threshold may indicate that the condition of the site is appropriate so long as future erosion results in little or no further cover loss (to a specified elevation).

If after the Phase 3A study a clear and critical potential for the geohazard to impact pipeline operations is identified, then the site advances to Phase 4 for mitigation.

If after the Phase 3A study additional information is necessary to better understand the potential impacts from the geohazard of concern and/or the response of the pipeline of interest, then supplementary studies can be undertaken via Phase 3B. Phase 3B studies may include, but are not limited to, test pits, borings, soil sampling and testing, instrumentation installation, slope stability modeling, modeling of pipeline response to anticipated loading scenarios, hydrologic and hydraulic modeling, and scour analysis. The results of these supplementary analyses will inform whether the geohazard site can be classified for ongoing monitoring (visual or supplemented with instrumentation or other methods, reviewed regularly via Phase 2) or prioritized for mitigation (Phase 4).

Phase 4 – Mitigation

Mitigations may range from minor adjustments in surface water management to pipeline replacements and reroutes. For some sites, it may be necessary to implement a short-term or temporary repair if the long-term repair requires significant design and/or permitting lead time.

Once a geohazard is mitigated, it is either removed from the active geohazard database (archived) or modified with appropriate remapping, geohazard risk index assignment, and monitoring interval if ongoing visual monitoring is necessary.

An additional benefit of tracking geohazard-related mitigations in this program structure is the potential for bundling projects based on proximity, which can have significant cost and project timing efficiencies especially in areas that are remote, difficult to access, or involve complex permitting and environmental review.

Site by site, mitigation of geohazard sites ultimately results in incremental reduction of risk for the pipeline system. The following section further discusses how this phased management approach directly supports risk reduction efforts.

GEOHAZARD RISK MANAGEMENT

The phased management approach as outlined herein fosters improvement in interdisciplinary and interdepartmental coordination and supports more informed (and therefore more efficient) mitigation design by involving SMEs early in the management process. While cold-start mitigation efforts will never be completely eliminated (e.g., for emergency response events without detectable precursors), they can be significantly reduced. Once systemwide geohazard exposure is well characterized, overall asset safety and reliability can be increased over time by way of targeted geohazard repairs, starting with the most urgent sites and progressing to less urgent risk-reduction-investment efforts. Informed risk reduction efforts can be pursued until the overall asset risk is determined by the asset owner to be tolerable. This risk reduction strategy, if diligently and aggressively employed through time, allows an asset owner to act more consistently in a maintenance capacity rather than an emergency response capacity.

To fully characterize the risk posed to pipelines by geohazards, future improvements toward probabilistic assessment and inventorying of the "fragility" or susceptibility to damage of the pipeline assets with respect to the geohazard of concern. Overlaying a probabilistic model of geohazard effects at a given site with an associated pipeline fragility will

Copyright © 2019 by ASME

result in a more informed assessment of pipeline risk of damage or failure and allow for more informed business decisions.

ACKNOWLEDGMENTS

We would like to acknowledge Golder Associates and InfraTerra for their baseline mapping support for the landslide and erosion datasets, respectively.

REFERENCES

[1] Sancio, Rodolfo, Rice, Anthony, Audibert, Jean, Morgan, David, and Rattray, Jordan. "Guidelines for Management of Geohazards Affecting the Engineering and Construction of New Oil and Gas Pipelines." Design Material and Construction Committee of Pipeline Research Council International, Inc., Chantilly, VA. 2018.

[2] Nyman, Douglas, Lee, E. Mark, and Audibert, Jean. "Mitigating Geohazards for International Pipeline Projects: Challenges and Lessons Learned." *Proceedings of the IPC 2008: 7th International Pipeline Conference, AMSE.* IPC2008-64405. Calgary, Alberta, Canada, January, 2008. DOI 10.1115.

[3] Burns, William, and Madin, Ian. "Protocol for Inventory Mapping of Landslide Deposits from Light Detection and Ranging (LiDAR) Imagery." Special Publication 42. Oregon Department of Geology and Mineral Industries, Portland, OR. 2009.

Copyright © 2019 by ASME

AIM-PIMG2019-1045

Chapter 14

USE OF A GEOHAZARD MANAGEMENT PROGRAM TO REDUCE PIPELINE FAILURE RATES

Sarah Newton[1], Aron Zahradka, Gerald Ferris, Michael Porter
BGC Engineering Inc.
Calgary, AB

ABSTRACT

A risk management framework, consisting of geohazard identification, assessment and management, can be used to significantly reduce pipeline failure rates from geohazards compared to industry averages. Since the late 1990s, BGC Engineering Inc. (BGC) has supported pipeline operators with geohazard management programs. In December 2017, geohazard inventories had been completed for over 300,000 km of pipelines, with the data stored in BGC's online geohazard management system, Cambio™. Pipeline failures (defined as leaks or ruptures) resulting from hydrotechnical or geotechnical hazards have been tracked for pipelines at differing levels of risk management, from cases where only inventories were created to where the full risk management life cycle has been undertaken. Inventory creation alone does not appear to have a marked impact on failure rates. However, pipelines that have been inspected by qualified geohazard experts, and have been assessed and prioritized for monitoring and mitigation, show a significant decrease in the rate of geohazard-related failures compared to the average North American and European failure rate of 0.02 per 1000 km per year (for 2005 to 2014). In areas of high geohazard exposure, geohazard failure rates have been shown to decline by a factor of 3 to 4 with proper application of a geohazard management program.

Keywords: Geohazard, Pipeline, Failure, Statistics, Integrity, Risk.

NOMENCLATURE

R	Risk
PoF	Annual probability of pipeline failure
C	Consequences of failure
P(o)	Probability of hazard occurrence
P(s)	Probability of spatial impact
V	Vulnerability

INTRODUCTION

Geohazards can result in pipeline failures in the form of leaks or ruptures where hydrotechnical or geotechnical loading impacts and exceeds the resisting force of a pipeline.

Hydrotechnical hazards occur at pipeline watercourse crossings where river processes such as scour, bank erosion, avulsion and channel degradation expose previously buried pipelines. Hydrotechnical loading occurs in three main forms – hydrodynamic (water pressure) loading, impact loading by the channel substrate and vortex induced vibrations (VIV) [1,2].

VIV results in fatigue failures if high water velocities act on a long enough length of spanning (unsupported) pipeline and if high flows remain in effect for sufficient duration [3]. Pipelines can also be dented by the channel substrate if the water velocities are high enough and if the flow has enough energy to move considerably large bed material (in excess of one meter diameter) [4,5]. Steel welded pipeline is resistant to failure by hydrodynamic loading; however, coupled pipelines have been known to fail following exposure in watercourses under low loading conditions [1].

Geotechnical hazards can occur where pipelines cross sloping ground. Geotechnical hazards consist of earth slides, rock slides, debris slides, earth flows and rockfalls [6]. When moving ground impacts a pipeline, geotechnical hazards can load the pipeline in tension, compression or causing local buckling. With enough strain, failure can occur [7].

Geohazards result in an average of 74 pipeline failures per year globally, with 16 failures per year occurring in the United States. In Canada, the United States, and western Europe, the industry average failure rate is reported to be 0.02 per 1,000 km per year; however, this rate varies by as much as two orders of

[1] Contact author: snewton@bgcengineering.ca

Copyright © 2019 by ASME

magnitude, from 0.003 per 1,000 km per year for Australia to 0.5 per 1,000 km per year for Bolivia [8].

While the report authors are not familiar with regulations that specifically mandate a frequency of failure limit, it is understood that a design frequency of loss of containment was set at 0.00001 per geohazard site on the TransMountain Expansion Project (TMEP), and this design frequency was accepted by the Canadian National Energy Board [9]. The closest code-based mention is likely the Target Reliability Index, which is used for strain- or reliability-based design (CSA Z662 O.1.5.2.1). Reliability and frequency of failure are not the same; but they are similar in cases where the ultimate limit state is considered as rupture. It should be noted that the CSA Z662 Reliability Target Index uses class ratings and pressure and is principally applicable to gas, rather than liquid, systems.

The most significant factors determining the failure rate have been shown to be terrain type and routing/construction practices. However, geohazard risk management practices have also been shown to measurably reduce pipeline failure rates when applied appropriately [8].

GEOHAZARD RISK MANAGEMENT

Geohazard management programs that follow a risk management framework consist of a process of hazard identification, assessment and management.

1.1 Geohazard Identification

Geohazard risk management programs start with the creation of a hazard inventory at the hazard identification stage [10,11]. If high risk geohazard crossing sites are missed during the identification phase, the opportunity for failure prevention and consequence reduction is lost. Methods for building a hazard inventory are variable, based on data and resource availability. The following is a partial list of potential data sources:

- Construction records
- Historical surveys and geohazard inspection records
- Digital elevation models and topographic maps
- Stream network maps
- Landslide susceptibility maps
- Satellite imagery
- Light imaging and ranging (LiDAR) imagery
- Interferometric synthetic aperture radar (inSAR).

Best practice in geohazard identification is to use a combination of the above data sources to create the hazard inventory. As more data sources become available, the inventory expands to include sites not previously identified.

BGC Engineering Inc. (BGC) has been conducting geohazard risk management for the pipeline industry since the late 1990s. At the end of 2017, BGC's pipeline geographic information system (GIS) contained approximately 300,000 km of operating, discontinued and abandoned hydrocarbon pipelines that are owned and operated by BGC's North American pipeline clients (Figure 1). BGC has undertaken some form of GIS and/or desktop study to inventory hydrotechnical and/or geotechnical hazards for most these pipelines. Not included in these statistics are approximately 4,000 km of operating hydrocarbon and mine concentrate pipelines in South America for which BGC has completed similar geohazard inventories.

Previous studies have shown that the average rate of recorded geohazard incidence per kilometer is approximately 0.25 or 0.06 for hydrotechnical and geotechnical hazards, respectively [10,11], suggesting that an operator with 5000 km of pipeline can expect approximately 1,500 geohazard sites, depending on terrain.

These earlier hazard inventories were created primarily using 1:50,000 stream maps and digital elevation maps to identify slopes with angles greater than 10 or 15 degrees. With the inclusion of additional data sources, especially high-resolution LiDAR imagery and stream mapping, the same operator with 5000 km of pipeline would be likely to identify 2,500 geohazard sites. This metric is based on the 154,000 hazards identified on 300,000 km of operating and discontinued pipelines in BGC's geohazard risk management system, Cambio™, and equates to an average rate of 0.5 geohazard sites per kilometer.

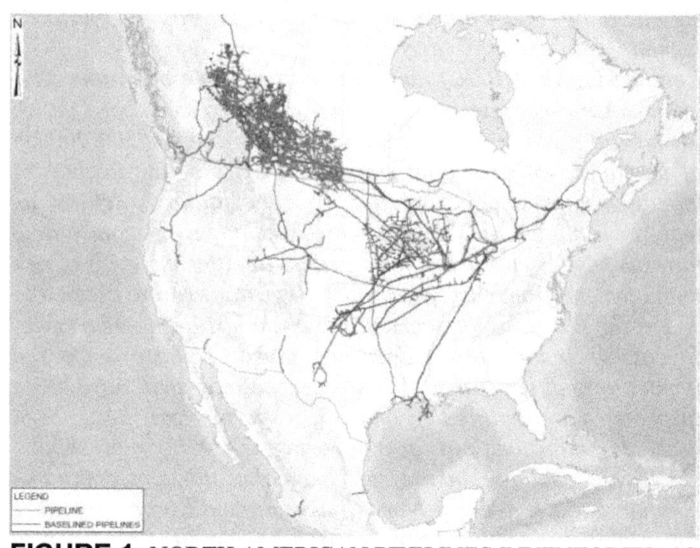

FIGURE 1: NORTH AMERICAN PIPELINES INVENTORIED AS OF DECEMBER 2017. BASELINED[2] PIPELINES ARE SHOWN IN RED.

1.2 Geohazard Assessment and Prioritization

Due to the high frequency of geohazard sites identified per kilometer of pipeline, preliminary assessment and prioritization is necessary before hazard management activities can be effectively conducted, given typical pipeline integrity management program resource constraints.

Geohazard risk assessment consists of estimating the probability and consequences of pipeline failure at each site in the geohazard inventory [Eq. 1].

[2] When field inspections are completed on at least 90% of the geohazards identified on a pipeline section, that section is identified as 'baselined'.

Copyright © 2019 by ASME

$$R = PoF \times C \quad (1)$$

Many methods have been developed to assess geohazard risk [12, 13] and the most useful of these methods provide a quantitative risk estimate that can be used to compare risk between different geohazard types as well as to compare geohazard risk to other integrity threats (such as third-party damage and corrosion).

To reduce pipeline failure frequencies, high probability of failure sites are prioritized for geohazard management. Probability of failure is estimated by taking the product of the probability of hazard occurrence, the spatial probability of the hazard impacting or exposing the pipeline, and the vulnerability [Eq. 2]. Vulnerability is defined as the probability of pipeline failure given pipeline impact or exposure at a geohazard site [3,7]. Details of the methods used to calculate probability of failure have been previously documented [1,3,7].

$$PoF = P(o) \times P(s) \times V \quad (2)$$

As part of developing the hazard inventory, the list of site locations and any characteristics that can be gleaned from the data sources used in the inventory creation are stored in a database and used to complete a preliminary geohazard risk assessment and prioritization for field inspections.

During the initial (baseline) field inspections, the presence of a geohazard is confirmed, pipeline depth of cover (DoC) is measured, observable site and geohazard characteristics are recorded, and the preliminary geohazard assessment is refined based on the collected field data [11]. BGC then stores the inspection results in Cambio™.

As of December 2017, BGC had completed baseline geohazard inspections at approximately 52,500 geohazard sites. Where more than 90% of geohazard sites identified on a pipeline segment have been field inspected, BGC considers the pipeline to have been 'baselined'. At the end of 2017, approximately 125,000 km of pipelines in North America had been baselined (Figure 1).

1.3 Geohazard Management

In addition to the updated risk assessment, the results of the ground inspections provide a recommended action [10], either aimed at reducing remaining uncertainty in the risk assessment or with the objective of reducing the risk (Table 1).

Risk management activities should match the expected risk, with more expensive and time-intensive actions, such as detailed investigation and mitigation undertaken at high risk sites [11]. Some geohazard sites will be assessed as having very low risk which, depending on the operator's risk tolerance, could be considered broadly acceptable and not require additional active geohazard management actions. In these cases, no fixed schedule is associated with the next action for the site. Instead, monitoring of the geohazard trigger events, such as intense rainfall, flooding and seismicity should be conducted and if previously defined thresholds are exceeded then ground inspections should be carried out [1].

In 2017, BGC began tracking actions taken as part of geohazard management to count the number of uncertainty and risk reduction actions. In 2017, BGC completed 8,346 ground inspections, 384 aerial inspections and 4,885 office inspections. Detailed investigations (as defined in Table 1) were completed at 53 and 62 geotechnical and hydrotechnical sites, respectively.

BGC provided geotechnical support for four pipeline stress relief projects in 2017. These were carried out at geotechnical sites where ground movement caused excessive pipe strains that were detected through ground inspection, geotechnical instrumentation data, and/or in-line inspection monitoring.

In 2017, 13 liquids pipelines were temporarily shut in and/or purged at river crossings in response to flood monitoring threshold exceedances. These actions were undertaken to reduce environmental consequences in the event of pipeline failure as a result of the flood event. Mitigation designs were completed for seven geohazard sites in 2017.

TABLE 1. Recommended actions and impacts on uncertainty or risk reduction.

Recommendation	Reduce Uncertainty	Reduce Risk
Inspections – Ground: Site visits to assess geohazard risk and to provide new or revised recommendations.	✓	
Inspections – Aerial: Site visits by air to assess geohazard risk and to provide new or revised recommendations.	✓	
Inspections – Office: An office-based task such as requesting/ obtaining/ reviewing as-built designs or other documents.	✓	
Detailed Investigation: Intensive site-specific investigations to gather detailed information such as: site surveys to determine pipe location and DoC, stream morphology studies, geotechnical drilling to obtain subsurface information or install instruments, or air photograph interpretation.	✓	
Monitoring: Reading and analysis of surface or subsurface instrumentation to quantify the progression of a geohazard and/or its impact on the pipeline (i.e. reading and analysis of inclinometer data or pipeline strain gauge data).	✓	✓
Maintenance: Field actions typically not requiring an engineering design (i.e. cleaning culverts, minor ditch or berm construction, replacement of swamp weights or pipeline markers).		✓
Mitigation: General field actions typically requiring an engineering design such as riprap placement, slope erosion control, slope drainage control, slope stabilization, or relocation of pipeline.		✓

Copyright © 2019 by ASME

RECORDED FAILURES OF PIPELINES IN CAMBIO™

To quantify the impact of BGC's geohazard risk management program on pipeline failure frequencies, BGC has been tracking the failures that occur on our clients' pipeline network since January 2016. BGC's North American pipeline clients experienced six pipeline failures caused by geohazards in 2016 and three failures in 2017. Over the past two years, this equates to an approximate failure frequency of 0.017 failures per 1,000 km per year.

The estimated Canada and US failure rate (0.02 per 1,000 km per year [8]) is 15% higher than BGC's North American client failure rate. Given the short time of observation and low number of failures, there is a high degree of uncertainty in this comparison. However, it is likely that this improvement from the North American average is actually greater than 15%, given the elevated geohazard exposure of BGC's clients.

Of the nine pipeline failures recorded in 2016 and 2017, one failure occurred on a pipeline where preliminary geohazard inventory and baseline surveys had been completed, and the remaining eight occurred on pipelines that either had not had geohazard inventories developed, or where baseline surveys had not yet been completed. The following provides a brief description of the extent to which each failure site had undergone geohazard risk management:

- Two failures occurred on pipeline for which geohazard inventories had not yet been completed.
- Four failures occurred at sites that were identified during recent geohazard inventories, but no baseline inspections had yet to be completed (one of these failures was found during completion of the baseline field assessment).
- Two failures were at sites that were not captured in the geohazard inventory; however, no baseline inspections had yet to be conducted on the pipelines.
- One failure occurred on a site that was not in the geohazard inventory prior to the failure, but where baseline inspections had been completed on the pipeline. In essence, a full geohazard management program was being carried out on this pipeline when the failure occurred.

IMPACT OF GEOHAZARD RISK MANAGEMENT ON MEASURED FAILURE RATE REDUCTION

The corresponding failure frequencies for pipelines that were actively managed in BGC's system, versus those in the early stage of hazard identification are approximately 0.005 per 1,000 km per year and 0.026 per 1,000 km per year, respectively. It should be noted that the exact numbers are difficult to determine because the lengths of inventoried and baselined pipeline changed throughout 2016 and 2017. Despite this, and the inherit limitations of the short timeframe in which the failure case studies have been tracked, the results suggest that pipelines undergoing a full geohazard risk management program have been successful at reducing the rate of pipeline failure by a factor of 4.

A significant impact on geohazard failure rates was also measured at the OSSA-1 pipeline within the Transrede network operated by YPFB in Bolivia. Pipeline failure rates declined by a factor of 3.4 following the application of a geohazard risk management program in 2003. Prior to 2003, 21 geohazard failures occurred in 20 years. In the 13 years following application of the program, only four failures occurred [8].

Similar results were realized along the NorAndino gas pipeline located in northern Argentina and Chile [14]. Prior to 2003, the geohazard failure frequency was approximately 0.67 per 1,000 km per year. Geohazard inventory and risk assessment informed a number of mitigation actions including small re-routes, replacement of vulnerable stream crossings, and enhanced maintenance and monitoring efforts. In 2016, the long-term geohazard failure rate since commencement of operation of the NorAndino pipeline was 0.18 per 1,000 km per year [8], a reduction by a factor of 3.7.

CONCLUSIONS

One of the key conclusions of this work is the importance of the completion of a full geohazard management program in reducing pipeline failure rates. As noted, most of the failures occurred at sites identified in BGC's geohazard management system, but that had not yet undergone field baseline inspections. Baseline inspections were scheduled for completion the following year in most of these cases. Therefore, geohazard identification alone does not contribute to reducing geohazard failure rates.

Notably, three of the failures occurred at sites which were not included in the initial geohazard inventory, either as a result of poor data availability, human error or limitations of the screening algorithms. This emphasizes the need to bring more and higher resolution data sources together at the geohazard identification stage. Geotechnical hazard identification has previously relied almost exclusively on digital elevation models; significantly more data and advanced analysis will be required in the future to reliably detect landslide hazards, particularly in certain geologic settings where deep seated landslides can occur in areas with slopes as low as 3 degrees. There is a need to increase the overall performance of high-risk geohazard site detection, without increasing the proportion of negligible-risk sites which divert resources from risk mitigation efforts.

Finally, assessment and prioritization of geohazard sites does not reduce risk but provides the necessary focus for risk mitigation efforts as part of geohazard management. Failure frequency is only reduced through action – such as the installation of an appropriate instrumentation monitoring program with thresholds that are linked to action plans that will allow an operator to intervene before pipeline failure occurs, construction of a hydrotechnical mitigation design, or replacement of a trenched crossing by suitably investigated horizontal directional drilling.

FUTURE WORK

Additional work is required to continue tracking pipeline failure case studies within and outside of BGC's geohazard

Copyright © 2019 by ASME

database. Learning from case studies will provide additional insight into how geohazard risk management programs can be used to reduce the global rate of pipeline failures caused by geohazards. Additional tracking of cases where affirmative action has been taken to prevent pipeline failures will add greatly to this body of knowledge, such that learnings can be taken from near hits as well as incidents.

ACKNOWLEDGEMENTS

The report authors would like to acknowledge the data contributions from BGC's Cambio™ clients, as well as BGC's engineers, geoscientists and software developers that are contributing to improving geohazard identification, assessment and management.

REFERENCES

[1] Ferris, Gerald, Newton, Sarah, Ho, Minh, Eichhorn, Geoffrey and Bear, Devin. "Flood Monitoring for Buried Pipeline Watercourse Crossings." *Proceedings of Rio Pipeline Conference & Exposition.* IBP1091_15. Rio de Janeiro, Brazil. September 22-24, 2015.

[2] CEPA. "Pipeline watercourse management recommended practices." 1st Edition, Canadian Energy Pipeline Association (2014).

[3] Dooley, Colin, Prestie, Zachary, Ferris, Gerald, Fitch, Murray and Zhang, Hua. "Approaches for evaluating the vulnerability of pipelines at water crossings." *Proceedings of the International Pipeline Conference, Volume 2: Pipeline Integrity Management.* IPC2014-33240: pp. V002T06A084. Calgary, Alberta, Canada, September 29–October 3, 2014. DOI 10.1115. http://proceedings.asmedigitalcollection.asme.org/proceeding.aspx?articleid=2022665.

[4] PRCI. "Enhancements of integrity assessment models/software for exposed and buried pipelines in river channels." Pipeline Research Council International in PR-170-9826 (2000).

[5] Nguyen, Nikki and Ireland, Yvanna. "Managing water crossings from an operator's perspective." *Proceedings of the International Pipeline Conference, Volume 2: Pipeline Integrity Management.* IPC2014-33481: pp. V002T06A085. Calgary, Alberta, Canada, September 29–October 3, 2014. DOI. 10.1115.http://proceedings.asmedigitalcollection.asme.org/proceeding.aspx?articleid=2022666.

[6] Rizkalla, M. Reed, R. "The assessment and management of pipeline geohazards." *Proceedings of Rio Pipeline Conference & Exposition.* Rio de Janeiro, Brazil. October 2-4, 2007.

[7] Ferris, Gerald, Newton, Sarah and Porter, Michael. "Vulnerability of buried pipelines to landslides." *Proceedings of the International Pipeline Conference, Volume 2: Pipeline Safety Management Systems; Project Management, Design, Construction and Environmental Issues; Strain Based Design; Risk and Reliability; Northern Offshore and Production Pipelines.* IPC2016-64071: pp. V002T07A001. Calgary, Alberta, Canada, September 26-30, 2016. DOI 10.1115. http://proceedings.asmedigitalcollection.asme.org/proceeding.aspx?articleid=2583873.

[8] Porter, Michael, Ferris, Gerald, Leir, Mark, Leach, Miguel and Haderspock, Mario. "Updated estimates of frequencies of pipeline failures caused by geohazards." *Proceedings of the International Pipeline Conference, Volume 2: Pipeline Safety Management Systems; Project Management, Design, Construction and Environmental Issues; Strain Based Design; Risk and Reliability; Northern Offshore and Production Pipelines.* IPC2016-64085: pp. V002T07A003. Calgary, Alberta, Canada, September 26-30, 2016. DOI 10.1115. http://proceedings.asmedigitalcollection.asme.org/proceeding.aspx?articleid=2583875.

[9] Baumgard, Alex, Beaupre, Michael and Leir, Mark. "Implementing a quantitative geohazard frequency analysis framework as a component of risk assessment of new pipelines." *Proceedings of the International Pipeline Conference, Volume 2: Pipeline Safety Management Systems; Project Management, Design, Construction and Environmental Issues; Strain Based Design; Risk and Reliability; Northern Offshore and Production Pipelines.* IPC2016-64580: pp. V002T07A020. Calgary, Alberta, Canada, September 26-30, 2016. DOI 10.1115. http://proceedings.asmedigitalcollection.asme.org/proceeding.aspx?articleid=2583892.

[10] Leir, Mark. "Geohazard integrity management program for onshore pipelines." *Proceedings of 9th International Symposium on Environmental Concerns in Rights-of-Way Management.* Portland, Oregon, September 27-30, 2009.

[11] Baumgard, Alex, Coultish, Tara and Ferris, Gerald. "Implementing a geohazard integrity management program – statistics and lessons learned over 15 years." *Proceedings of the International Pipeline Conference, Volume 2: Pipeline Integrity Management.* IPC2014-33559: pp. V002T06A088. Calgary, Alberta, Canada, September 29–October 3, 2014. DOI. 10.1115. http://proceedings.asmedigitalcollection.asme.org/proceeding.aspx?articleid=2022669.

[12] Muhlbauer, K. *Pipeline risk assessment: the definitive approach and its role in risk management.* Expert Publishing, LLC, Austin, Texas (2015).

[13] Porter, Michael, Logue, Clint, Savigny, K.Wayne., Esford, Fiona and Bruce, Iain. "Estimating the influence of natural hazards on pipeline risk and system reliability." *Proceedings of the International Pipeline Conference, Volumes 1, 2, and 3.* IPC2004-0238: pp. 2587-2595. Calgary, Alberta, Canada, October 4-8, 2004. DOI. 10.1115. http://proceedings.asmedigitalcollection.asme.org/proceeding.aspx?articleid=1646760.

[14] Porter Michael, Reale, Rodolfo, Marcuz, Gabriel and Savigny, K. Wayne. "Geohazard Risk Management for the Nor Andino Gas Pipeline." *Proceedings of the International Pipeline Conference, Volume 3: Materials and Joining; Pipeline Automation and Measurement; Risk and Reliability, Parts A and B.* IPC2006-10208: pp. 989-997. Calgary, Alberta, Canada, September 25-29, 2006. DOI. 10.1115. http://proceedings.asmedigitalcollection.asme.org/proceeding.aspx?articleid=1597445

Copyright © 2019 by ASME

AIMPIMG2019-1070

Chapter 15

DYNAMIC GEOHAZARD MANAGEMENT IN CHALLENGING ENVIRONMENT

Rohani M Joehan
PETRONAS
Kuala Lumpur, Malaysia

Wan Ismail Wan M Marzuki
PETRONAS
Kuala Lumpur, Malaysia

Kamal Ibrahim
PETRONAS
Kuala Lumpur, Malaysia

Zazli Bob
PETRONAS
Kuala Lumpur, Malaysia

Rahman Azam
PETRONAS
Kuala Lumpur, Malaysia

ABSTRACT

The Sabah Sarawak Gas Pipeline (SSGP) is a 521 km x 36" O.D. Grade X 70 onshore natural gas pipeline that was completed in 2013 to transport 750 MMscfd of gas from the Sabah Onshore Gas Terminal (SOGT) to the LNG Complex at Bintulu, Sarawak. SSGP traverses through many different types of terrain. Two separate pipe parted incidences in August 2014 and January 2018 resulted in PETRONAS reevaluating its pipeline integrity management system for pipeline constructed in difficult terrain.

The pipe parted incidences triggered a reassessment of the terrain along the whole SSGP pipeline route. In order to manage the geohazards located along the pipeline a comprehensive Geohazard Management Plan was developed to provide immediate actions (prior to gas in), intermediate actions (during commissioning stage) & for more permanent solutions at high risk locations.

This paper presents PETRONAS unique experiences and new solutions developed to assess the difficult terrain and associated stresses impacting the pipeline. The efforts put in place in mitigating and managing geohazards are described not only for purpose of integrity management of the current SSGP pipeline condition but also for longer term aim to further enhance the integrity and reliability of the pipeline.

Summary of geomorphological terrain assessments using Geo Slope Monitoring Assessment and Repair Management (GeoSMART) findings, geomorphological mapping and engineering use of LiDAR, aerial surveillance and ground movement monitoring data incorporated via Intelligent Geohazard Management System (iGeMS) database and details of Inspection Maintenance Monitoring and Repair (IMMR) methods and procedures are presented.

The basic description of SSGP terrain in which the pipeline parted incidences occurred is described herein. Aspects of the site investigation and assessment activities undertaken are shared. Soil information used to assess slope stability analyses, description of geomorphology that causes ground movement or impart additional stresses are described. The method used to qualitatively assess geohazards and terrain impacting pipeline risk are presented.

This paper presents aspects of geohazard assessments, mitigations and management to put in place solutions to enable lower likelihood of failure. Due to experiences, data gathered, analysed and managed via monitoring programmes, costly terrain repairs are able to be rationalized and optimised to ensure geohazard risks are ALARP.

This paper presents PETRONAS updated practices in managing geohazards and their impacts to SSGP pipeline. The comprehensive work done and improvement to the Geohazard Management Plan demonstrates: 1) PETRONAS overall advancement in assessment and monitoring of the SSGP pipeline that can be affected by external stresses due to ground movement, 2) Engineers innovativeness to overcome technical challenges via GeoSMART, iGeMS and IMMR in-house solutions to raise PETRONAS confidence in managing pipeline in challenging terrain and 3) PETRONAS capability to predict pipeline overstressing occurrence and face more complex and demanding pipeline integrity issues in locations with geohazards.

NOMENCLATURE

ALARP	As Low As Reasonable Possibly
GeoSMART	GeoSlope Monitoring and Repair Management
GMP	Geohazard Management Plan

Copyright © 2019 by ASME

iGeMS	intelligent Geohazard Management System
IMMR	Inspection, Maintenance, Monitoring & Repair
KP	Kilometer Point
Lidar	Light Detection and Ranging
MTM	Magnetic Tomography
SPT	Standard Penetration Test
SOGT	Sabah Onshore Gas Terminal

INTRODUCTION

The petrochemical industry is generally aware that loss of mechanical integrity of pipeline can be due to a variety of causes such as corrosion, external impact, defects and natural hazards [3].

The first pipeline parted incident at SSGP, KP 135 triggered a reassessment of the terrain along the whole pipeline route. In order to manage the geohazards located along the pipeline a Geohazard Management Plan was developed to provide immediate actions (prior to gas in), intermediate actions (during commissioning stage) & for more permanent solutions at selected terrain locations.

The geohazard management plans takes into consideration the pipeline sustainability from a strain demand perspective. The strain demand perspective requires that the pipeline be under a specified deformation or movement limits.

Figure 1 below based on [4] shows the limit of scope for purpose of development of a Geohazard Development Plan.

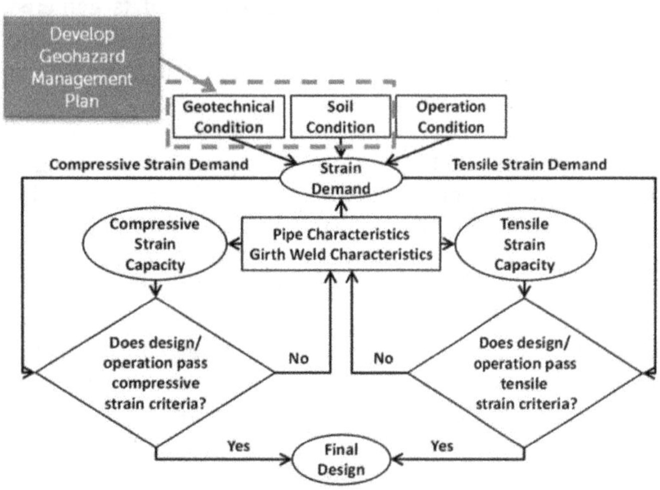

Figure 1: How Geohazard Management Fits Into Overall Pipeline Operation/Design Process

RISK MANAGEMENT FRAMEWORK

A starting point in development of the Geohazard Management Plan (GMP) is the selection of a framework to assist in addressing salient elements are in place to effectively put in place a series of activities to manage aspects of geohazards impacting the pipeline.

The framework shown below is used in the development of the GMP.

Figure 2: Risk Management Framework for basis of GMP

GEOHAZARD MANAGEMENT PLAN

Due to the dynamic and complex nature of the geohazard threats impacting on the SSGP pipeline, a focus approach on geohazard management related to pipeline management strategy was planned in three distinct phases:

1. Immediate (Before Gas-In) Phase 1
2. Intermediate (During Commissioning) Phase 2
3. Long Term (Operation at Designed Pressure) Phase 3

The different phases provide an easier clustering of different work scoping and level of interventions to manage the geohazard issues. The GMP essentially covers the management of the geohazard as well as pipeline integrity due to geohazard threats that can potentially lead to a pipe overstressing condition.

Essentially in totality the GMP looks into overall planning, implementation, monitoring to establish a response strategy that factors also the following major considerations facing SSGP:

1. Mountainous terrain creating logistic constraints
2. Local and native land practices and preferences
3. Pipeline inherent technical characteristic

The summarized GMP activities over various phases leading to pipeline operational gas in status are shown following Figure 3 below.

Copyright © 2019 by ASME

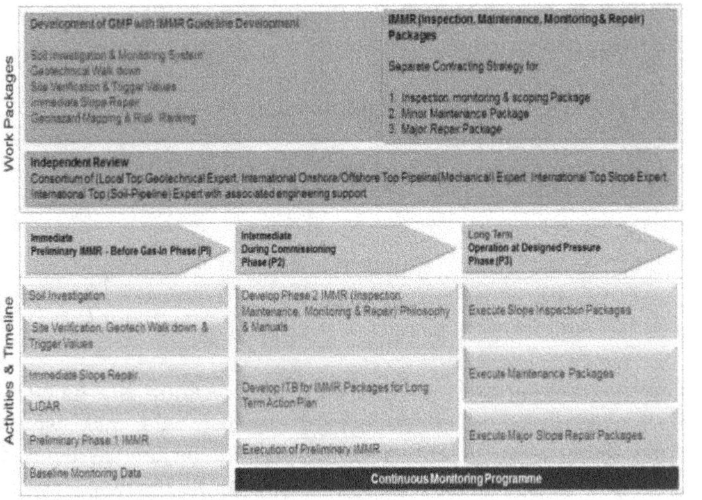

Figure 3: General Programme Schedule to Meet SSGP Gas In Date.

The GMP requires various activities to be implemented immediately prior to gas-in. These activities are shown in Figure 4 below includes:

1. Geotechnical site assessment with MTM walkdown
2. Slope stability analysis
3. Root cause analysis
4. Develop trigger values and preliminary ALARP
5. Slope instability assessment and terrain mapping using LiDAR
6. Soil investigation
7. Installation of instrumentation
8. Slope immediate rectification works
9. Develop Inspection, monitoring, maintenance and repair (IMMR) manual and guideline.

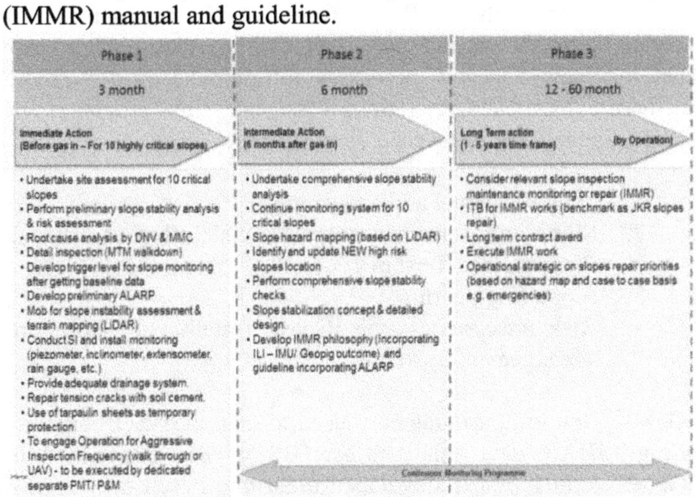

Figure 4: Activity List in Phases for GMP

One of the conditional requirements for purpose of Gas-In is the development of a preliminary IMMR manual with trigger values as shown in Figure 5 below. This IMMR manual provides the following key guidance:

1. The trigger action plan for terrain affecting pipeline integrity.
2. Inspection and maintenance frequencies for different risk levels.
3. Basic early reporting and rectification recommendation recommendations
4. Roles and responsibilities matrix for various phases and different geohazard mitigation activities.

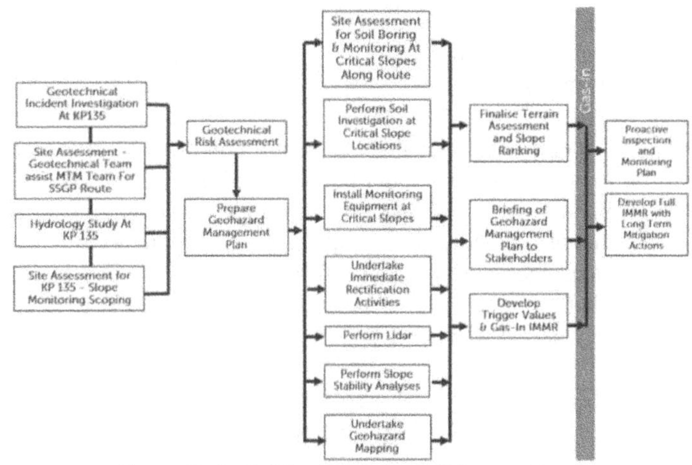

Figure 5: Activities Flowchart Prior to Gas-In

The IMMR can be enhanced during the commissioning stage (after gas in) once more data from the following activities are undertaken:

1. Reanalyzes of survey data
2. Pipeline stress analyses.
3. Data from soil pipeline interaction study is necessary.

The flowchart in Figure 6 below shows the various activities required to be undertaken in selection of the final mitigation actions for any particular site that has geohazard related threats.

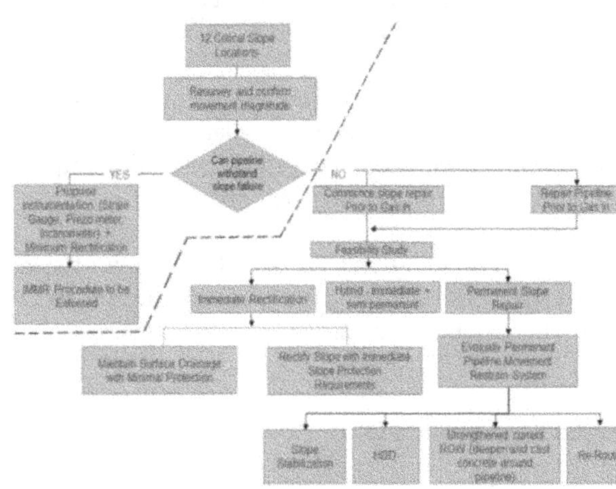

Figure 6: Activities Flowchart for Commissioning Stage and Long Term Basis

Copyright © 2019 by ASME

TERRAIN ASSESSMENT

For purpose of SSGP, major geohazard conditions affecting pipeline integrity are identified as:

1. Slope Instability
2. Gullying and soil erosion
3. Compressible Ground
4. Flooding
5. Stream Channel Migration
6. Hydrodynamic Forces

A basic general description of the terrain in which the two pipe parting incidents occurred can be described as mainly covered by soil conditions which can be categorized as colluvium (i.e heterogeneous transported soil with variable particle grain size) along slopes inclination ranging from 5° to 30°. It cannot be negated that slope failures or potential hazards can even occur at relatively gentle slope angles.

To expedite assessment of risk associated to locations with geohazards, the following features were considered to expeditiously assess the severity and extend of geohazard threat at any particular location poses:

1. Geomorphological features
2. Geological history
3. Slopes related to pipeline (perpendicular or longitudinal) greater than 40°
4. Slope vulnerability and stability
5. Slope failure history during construction
6. Major soil or slope distress indication

Typically, Standard Penetration Test (SPT) investigation boreholes are performed for purpose of identifying soil characteristics at any particular site. Water tables are measures where SPT boreholes are performed.

Where closer investigation are required seismic cross hole logging has been performed to identifying lateral soil variation. Where fast and basic soil investigations are required Mackintosh Probes are assigned and performed.

Hydrology studies providing volume, speed and flow of steam and river water are also derived from available historical data whilst considering catchment area.

RISK RANKING ASSESSMENT

As of when this document is prepared, there are 957 numbers of geohazard location along the 512km SSGP pipeline. A risk assessment was undertaken using expert judgement and critical criteria to categorize the slopes into Very High, High Risk, Medium Risk and Low Risk. Risk ranking shall consider the following aspects:

1. Vegetation surface and coverage
2. Slope gradient
3. Drainage condition
4. Serviceability
5. Scouring
6. Slope Distress (Signs of instability and seepage)
7. Geological discontinuity
8. And Level of Impact (Consequence)

The risk ranking ascribed to a particular location is a function of the likelihood of occurrence of any particular risk aspect from item 1 to 7 above and factored by an associated consequence of that particular risk aspect to the pipeline.

As an example, for risk aspect of higher slope gradient, larger volume of soil mass can potentially dislodge and cause bigger impact loading on the pipeline. Thus a higher risk category can be associated to the site. When there are more than one geohazard risk aspects, the combine risks are factored to provide an even higher level of risk.

Where mitigation can be demonstrated a risk level can be lowered accordingly.

For SSGP purpose, a method was developed to qualitatively assessed the terrain type and damage potential of the pipeline and integrated with associated risk stabilization and reduction solutions. The system is known as GeoSMART (Geo Slope Monitoring and Repair Management).

Together with the IMMR, GeoSMART represented by schematic as shown in Figure 7 below, provides a framework that includes tools and processes for purpose of performing onshore slope and terrain assessment and provision of mitigation actions. In essence it constitutes:

1. IMMR Manual and Procedures
2. Slope and Terrain Assessment via Rating of Damage Potential and Terrain Type
3. Slope Engineering
4. Risk Assessments via iGeMS (intelligent Geohazard Management System).

iGeMS is an in-house database used to map the SSGP pipeline using LiDAR data using an ArcGIS system. Embedded in iGeMS is an in-house developed algorithm based on pairwise comparison that is able to provide categorization of terrain vulnerability to impose potential threats to the pipeline.

Copyright © 2019 by ASME

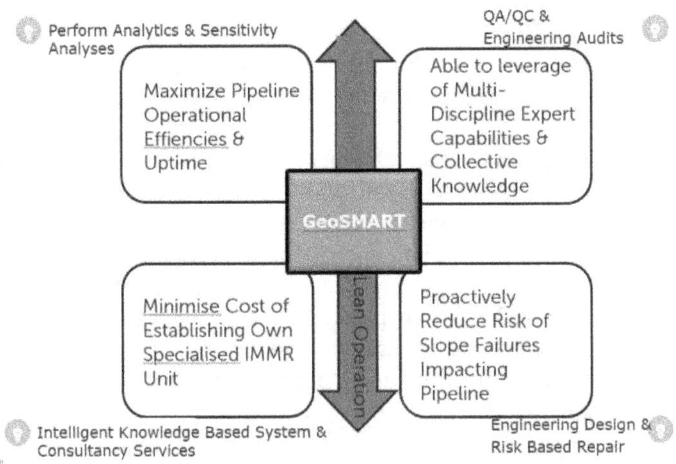

Figure 7: Schematic Representing GeoSMART

Based on the GeoSMART system, slopes which have been qualitatively risk categorized can be clustered where prescribed solutions are developed to deliver appropriate response and mitigation plan to any particular location.

TRIGGER ACTION PLAN

In general, a Trigger Action Plan as shown in Figure 8 below is required for purpose of managing the pipeline based on the conservative assumption that soil movement results in direct pipeline movement. These criteria have been prepared to trigger the general mitigation plan for operation team response to manage the pipeline in the event that the soil moves excessively causing pipeline displacement.

Where strain gauges are not available, the soil movement is monitored at the most critical slopes via instrumentation (i.e. inclinometer, piezometer, rain gauge, settlement marker, tilt meter and extensometer) as a basis of pipeline movement.

No	Soil Movement by Inclinometer for every 50m span	Pipeline Strain Based on Strain Gauge	Action
1	Displacement <250mm	Less than 0.15%	Classified as Green condition and follow routine inspection schedule.
2	250 ≤ Displacement <400mm	0.15%≤ S <0.19%	Classified as Yellow-TA and plan & perform remedial action.
3	400 ≤ Displacement <500mm	0.19%≤ S <0.23%	Classified as Orange-NA and execute remedial action.
4	Displacement ≥ 500mm	Greater than 0.23%	Classify as Red and standby for shutdown procedure.

*Note:

TA- Tolerable Anomaly

NA – Near Anomaly (reaching max strain limit)

Figure 8: Summary of Trigger Values & Action Plan

INSPECTION, MONITORING, MAINTENANCE, REPAIR (IMMR)

The IMMR shown in Figure 9 [2] below is used to provide the implementation strategy to establish a program to continuously react to terrain related issues and record geohazards that can be monitored to continuously registered and provide appropriate levels of repairs to pipeline threatened to movement. The trigger values shall be developed to provide appropriate actions to the movement detected on the pipeline.

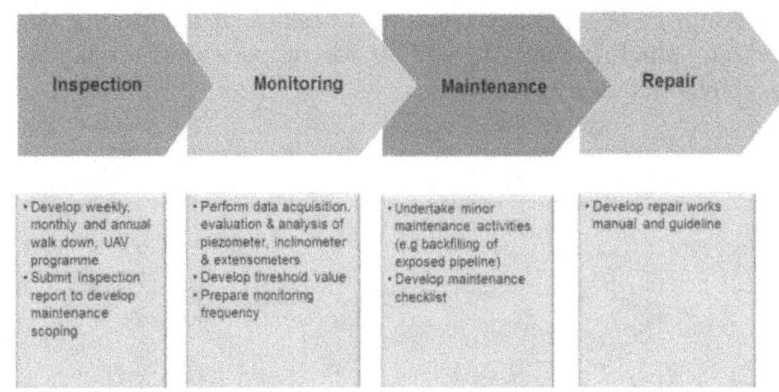

Figure 9: IMMR

ADVANCE RESPONSE PLAN, CONTINOUS MONITORING & IMPROVEMENT

Since it is noted that the nature of geohazard occurrences within the SSGP route is dynamic in nature, the number of geohazard incidences affecting pipeline integrity location may change subject to findings from IMMR proforma reports, latest geohazard advanced mapping (LiDAR) data, inline inspection, (geopig) and observations from instrumentation data.

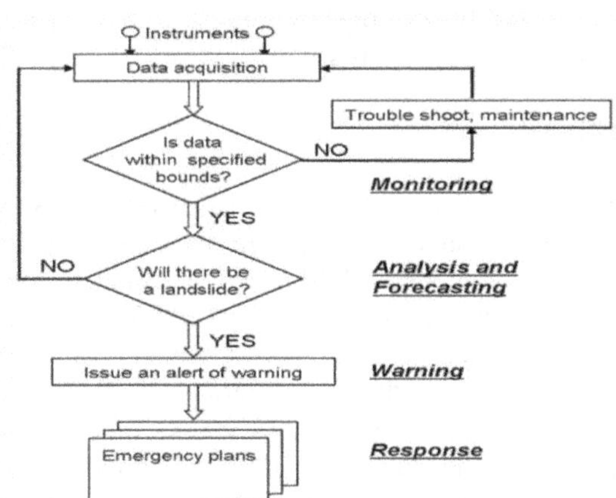

Figure 10: Warning Systems for Response Plan (DiBiagio & Kjekstad, 2007 [1])

As and when the new data is obtained the geohazard locations can be reassessed and the geohazard register can be amended accordingly. When new data is obtained appropriate ALARP

Open

Copyright © 2019 by ASME

mitigation action to manage the occurrence of pipeline overstressing due to soil movement or impact of geohazard needs to be activated based on trigger values being observed.

A response plan incorporating understanding of geohazard occurrence rates at all relevant locations along the pipeline needs to be trigged as per the flowchart shown in Figure10 above.

Figure 11 below demonstrates continuous improvement to the GMP over the years developed based on need to understand trending of occurrences rates, relationships of monitoring data sets to pipeline integrity and fit for purpose operational and technical requirements.

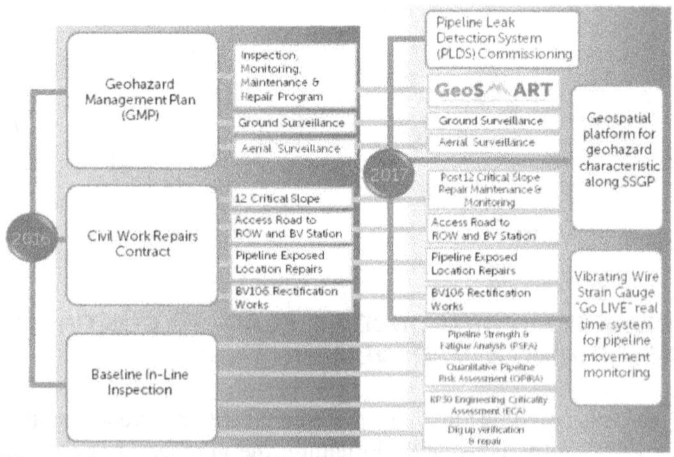

Figure 11: Continuous Improvement Program

Much more work has been done since 2018 such as the development of a new Pipeline Inspection and Maintenance Plan to replace the IMMR Manual and Guideline providing new trigger values and fit for purpose response plan to various class of pipeline located at various sectors.

DISCUSSION

This paper presents aspects of geohazard assessments, mitigations and management to put in place solutions to enable the SSGP to be operated whilst meeting stringent operational limits and requirements. Due to the SSGP experiences, understanding of the rate of geohazard related occurrences (e.g. slope failures perpendicular and longitudinally to the pipeline), instrumentation data (geopig, strain gauges and inclinometers analyzed and managed via monitoring programs (IMMR reports, LiDAR data) costly terrain repairs are able to be rationalized and optimized to meet ALARP.

This paper presents PETRONAS practices in managing geohazards and its impact to the SSGP pipeline. The comprehensive work done and improvements to the Geohazard Management Plan demonstrates:

1) PETRONAS overall advancement in assessment and monitoring of the SSGP pipeline that can be affected by external stresses due to ground movement,

2) Engineers innovativeness to overcome technical challenges via GeoSMART, IGEMS and IMMR in-house solutions to raise PETRONAS confidence in managing pipeline in challenging terrain and

3) PETRONAS capability to manage pipeline overstressing occurrence and face more complex and demanding pipeline integrity issues in location with geohazards.

As a way forward, it needs to be mentioned that continuous training and retraining of the processes above is critical for the safe operation of the SSGP pipeline.

Continuous assessment of suitability and applicability of new technology such as usage of drone for continuous surveillance and advance satellite technology once made fit for purpose will ensure that geohazard management practices be further enhanced to ascertain that the pipeline is safely operated under the dynamic nature of the geohazard challenges faced.

ACKNOWLEDGMENTS

The authors would like to express our utmost appreciation to Dr Salama, Mamdouh, M. for his patience, technical advice and wise guidance provided to the team. Our thanks also goes to these colleagues (M Nazmi Ali and M Jeffri Yunos, numerous colleagues from the Pipeline and Geotechnical Unit) and organizations (DnV, MAA & GnP) who has assisted us in delivering our work to ensure that this pipeline remains operational.

REFERENCES

1) DiBiagio, E., Kjekstad, O., 2007, Early Warning, Instrumentation & Monitoring Lanslides, Reclaim II.
2) Petronas, 2015, Geohazard Inspection, Maintenance, Monitoring & Repair Manual, MY ALL 00735, Malaysia.
3) Varia De Stefani et al, 2009, A Model to Evaluate Pipeline-Failure Frequencies Based on Design & Operation Condition, 5th Global Congress on Process Safety.
4) Yong-Yi Wang & Ming Lu et al, 2014, Overall Framework of Strain Based Design & Assessment of Pipeline, 10th International Pipeline Conference.

Copyright © 2019 by ASME

AIM-PIMG2019-1051

Chapter 16

DEVELOPING A GEOTECHNICAL HAZARD INVENTORY: A CASE STUDY AND BEST PRACTICE RECOMMENDATIONS

Joel Van Hove[1], Dr. Alex Baumgard
BGC Engineering Inc.
Vancouver, Canada

Joel Babcock, Jan Bracic
Pembina Pipeline Corporation
Calgary, Canada

ABSTRACT

Geotechnical hazards are a subset of geohazards concerning mass wasting mechanisms (slope movement), creep, subsidence, and other forms of moving ground. Moving ground may present a hazard to pipelines by applying unintended loads or removing support, potentially causing pipeline damage or failure [4]. As with all integrity management programs, a foundational component of managing risk is to develop an inventory of potential hazards. For geotechnical hazards, evaluating the potential for ground to move is the basis for geotechnical hazard identification and development of a hazard inventory. Prior to the availability of new tools, geotechnical hazard inventories were typically developed with an emphasis on slope angle, identifying the steepest slopes crossed by the pipeline. However, slope angle and geotechnical hazard occurrence are poorly correlated [3]. Where weak geology is present, even gently inclined slopes can be prone to landslides [1]. These shallower landslides tend to be longer, which is important as the length of pipe affected by ground movement is one of the key factors which elevates pipeline vulnerability [3]. The presented case study demonstrates that overlooking low angle geotechnical hazards may result in a significant proportion of high-rated hazards missing from the inventory. An improved methodology for developing an inventory of geotechnical hazards is proposed which addresses limitations of earlier methodologies by leveraging the increased coverage and reduced cost of LiDAR to create an initial inventory and encourages the use of alternate data sources, detailed studies, and new technology to further enhance geotechnical hazard inventories. This paper provides best practice recommendations for developing a geotechnical hazard inventory and provides a case history for a pipeline system where the hazard inventory was initially developed using traditional methodologies but was enhanced by completing a LiDAR based inventory, and lastly by a variety of data sources, tools and methodologies to create a more rigorous inventory.

Keywords: Geotechnical, Geohazard, Inventory, Slope

INTRODUCTION

Over the past two decades an increasing proportion of pipeline operators in Western Canada have established formal geohazard management programs which aim to reduce geohazard risk through identifying, assessing and managing geohazards. Developing an inventory of geohazards is a foundational component of any geohazard management program. Hazards which exist outside of an inventory cannot be proactively managed since management actions (e.g. mitigation or formalized inspection) are driven by risk assessments completed within the geohazard management program. Geohazard management actions are less effective when inventories are incomplete as management effort may not be prioritized on the highest risk hazards since only a subset of highest risk hazards are known to the operator.

Geohazards are generally divided into three categories, geotechnical (i.e. ground movement) hazards, hydrotechnical (i.e. watercourse) hazards, and seismic hazards (i.e. ground motion and seismically trigger permanent ground displacement). This paper proposes best practices for developing and enhancing a geotechnical hazard inventory and does not address hydrotechnical or seismic hazard inventories.

Past standard of practice

Developing a hydrotechnical hazard inventory is a simpler task than developing a geotechnical hazard inventory because the hazard, flowing surface water, is typically observable in

[1] Contact author: jvanhove@bgcengineering.ca

Copyright © 2019 by ASME

aerial imagery and watercourse maps are often readily available. Improvements in hydrotechnical hazard inventories in the past two decades have included increased resolution of mapping small watercourses and developing digital watercourse networks. Generally, if watercourses are missing from a hydrotechnical hazard inventory after an inventory study has been completed it is because they are too small to be identified in publicly available aerial imagery, and of little consequence to a geohazard management program because the pipeline vulnerability of small watercourses is very low [2].

The production of geotechnical hazard inventories has not developed as quickly as hydrotechnical hazard inventories because the data used to identify geotechnical hazards has not been readily available or economical to obtain and evaluate. Traditionally geotechnical hazard inventories were developed using air photo interpretation and geological maps. The major challenges with air photo interpretation include having the specialist skill required to accurately and efficiently complete a terrain assessment, that trees and vegetation can obscure slope morphology, and that completing a hazard inventory is highly time consuming due to the need to obtain and view hundreds to often thousands of aerial images in stereo, depending on the length of the pipeline system being assessed. Through a lack of appreciation for geotechnical hazards and the cost of undertaking these studies, they were rarely completed. Hazard inventories were typically incomplete or non-existent, and often only included sites where geohazard activity has been observed during the life of the pipeline (e.g. previously broken or damaged pipe).

Around two decades ago, the standard methodology was modified to be more time and cost effective by taking advantage of publicly available aerial imagery and digital elevation models (DEMs). The modified approach uses publicly available (typically low resolution 1:50,000 scale) DEMs to identify slopes which exceed slope angle and slope height criteria, and reviews publicly available aerial imagery (typically from satellite) for indications of instability. This methodology is more time efficient than air photo interpretation and requires less specialized skill, but also has limitations. Landslide features are less apparent in aerial imagery compared to air photos, trees and vegetation still obscure slope morphology, and slopes are added to the inventory based on geometry without knowledge of whether a landslide is present.

Slope angle is poorly correlated with geotechnical hazards because variability in site conditions and material strength generates landslides on a wide range of slopes, as low as only a few degrees [3]. Using a high slope angle threshold results in screening out many potentially relevant hazards and using a low slope threshold results in the inventory being overwhelmed with geotechnically benign slopes. This is shown in Figure 1 by the grossly disproportional number of low angle slopes relative to the number of low angle landslides, particularly for slopes below about 10° where the number of slopes with landslides present dramatically decreases. The data provided in Figure 1 is based on the pipelines and study area in the presented case study.

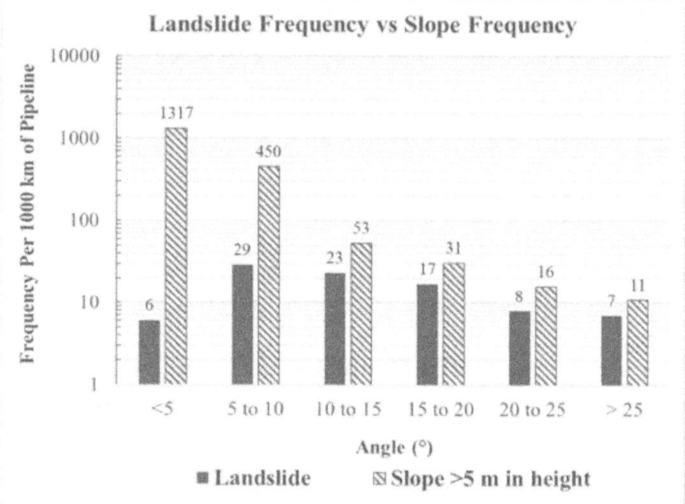

FIGURE 1: LANDSLIDE FREQUENCY VS SLOPE FREQUENCY NORMALIZED FOR 1000 KM OF PIPELINE USING DATA FROM THE CASE STUDY PRESENTED IN THIS PAPER.

Within the last five years new technology has become increasingly available which overcomes many of the limitations of the previous standard methodologies. Light detection and ranging (LiDAR) datasets generate high resolution DEMs of bare earth, permitting observation and interpretation of the slope morphology without obstruction from vegetation. LiDAR has become widely and economically available and is the principle tool currently used for developing geotechnical hazard inventories. Programs at both the federal and state/provincial levels are increasingly making this data freely available. Additionally, technologies such as inline inspection (ILI) tools and interferometric synthetic aperture radar (InSAR) also offer valuable data that can be combined with the LiDAR assessments to further enhance the ability to correctly identify significant ground movement sites.

Case Study

Pembina Pipeline Corporation (Pembina) is a pipeline operator with midstream assets located primarily in Western Canada. Pembina established a formal geohazard management database in 2007. Geotechnical hazards were added to the inventory initially by identifying hazards from existing documents, studies and inventories. A second stage of geotechnical hazard identification was completed by overlaying pipeline center lines with a DEM and using geographic information systems (GIS) scripts to identify sloping ground with average angles and heights above set thresholds. The thresholds were variable based on topography and locally observed landslide conditions, but generally were between lower bound angles of 10° to 20°, and lower bound slope heights of 5 m to 15 m. The DEM varied between a 1:20,000 and 1:250,000 scale dataset from the Canadian Digital Elevation Model (CDEM) from Natural Resources Canada.

Copyright © 2019 by ASME

By 2015, Pembina's geohazard management program covered 6543 km of pipeline and included 462 geotechnical hazard sites. At the end of 2015 and with the availability of many of the new tools described above, Pembina decided to undertake an enhanced geotechnical inventory (EGI) study using LiDAR bare earth images to identify geotechnical hazards which would not have been identified by the methodology used to date. The study area is shown in Figure 2.

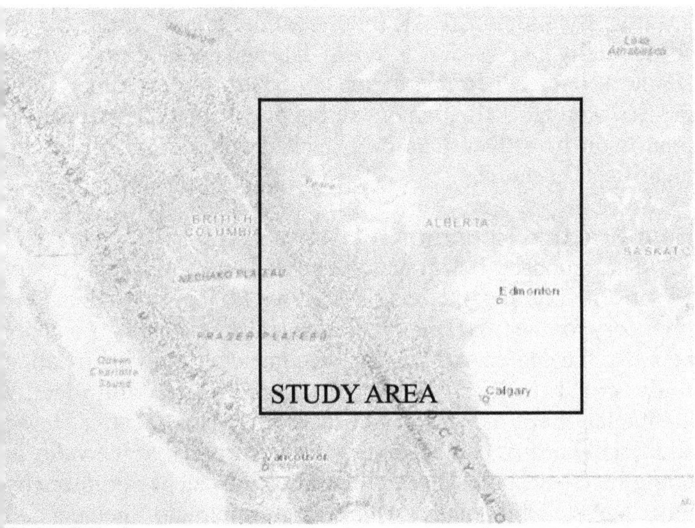

FIGURE 2: THE WESTERN CANADIAN STUDY AREA FOR THE PIPELINES WITHIN THE CASE STUDY SCOPE.

Pembina recognized the importance of the study because of a growing awareness of the risk of excluding low angle slopes in the Western Canadian sedimentary basin. Deep-seated bedrock-controlled landslides tend to form long shallow slopes, which are difficult to identify and often have high pipeline vulnerability because of their length [3]. Slopes as shallow as 4° have caused pipeline failures in areas where Pembina operates [1]. Reducing the lower bound angle threshold to capture these landslides with GIS would produce an inventory with a impractically large number of mainly geotechnically benign slopes and no simple way to differentiate slopes affected by landslides (Figure 1). The study utilized "off the shelf" LiDAR, which by the time of the study was available in most of Pembina's areas of operation.

The EGI study added 443 hazard sites based on hazard morphology observed in LiDAR bare earth images, approximately doubling the geotechnical hazard inventory. Of the 443 new hazard sites 169 were not identified by the previous methodology because the slope angle was below 10°, 135 were not identified because errors/coarseness in the DEM reported the slope as less than 10° when the slope was slightly greater, and 139 were not identified because the landslide was not intersecting the right of way (i.e. upslope hazards, encroaching hazards). Table 1 summarizes the change to the inventory and highlights the importance of the newly identified hazards as fewer than half of the 100 most important hazards currently in the inventory (i.e. highest probability of failure (PoF) [3]) were identified before completing the EGI.

TABLE 1: A SUMMARY OF HAZARD SITES IN THE INVENTORY PRIOR TO AND AFTER THE EGI STUDY.

Metric	Inventory Pre-EGI	Inventory after EGI
Number of hazard sites	462	905
Number of confirmed landslides	269	539
Number of low angle sites (<10°)	60	229
% of sites in the top 100 ranked by PoF	45	100

Figure 3 below illustrates the advantages of utilizing LiDAR imagery by comparing a site from the case study in a publicly available DEM, aerial imagery, and LiDAR bare earth imagery. The site in Figure 3 was not identified by GIS screening criteria because the slope measures less than 10° and does not intersect the pipeline at the low resolution of the DEM (Figure 3 – A). In the aerial imagery (Figure 3 – B), landslide morphology is not clear, however, in the LiDAR bare earth imagery it is apparent that the pipeline crosses a landslide (Figure 3 – C). The landslide, measuring 130 m long, 450 m wide, and 9.5°, is traversed by the pipeline for 250 m. This hazard was identified concurrently by the EGI study and inertial measurement unit (IMU) data from an ILI tool run which measured bending strains and pipe movement caused by active ground movement.

Copyright © 2019 by ASME

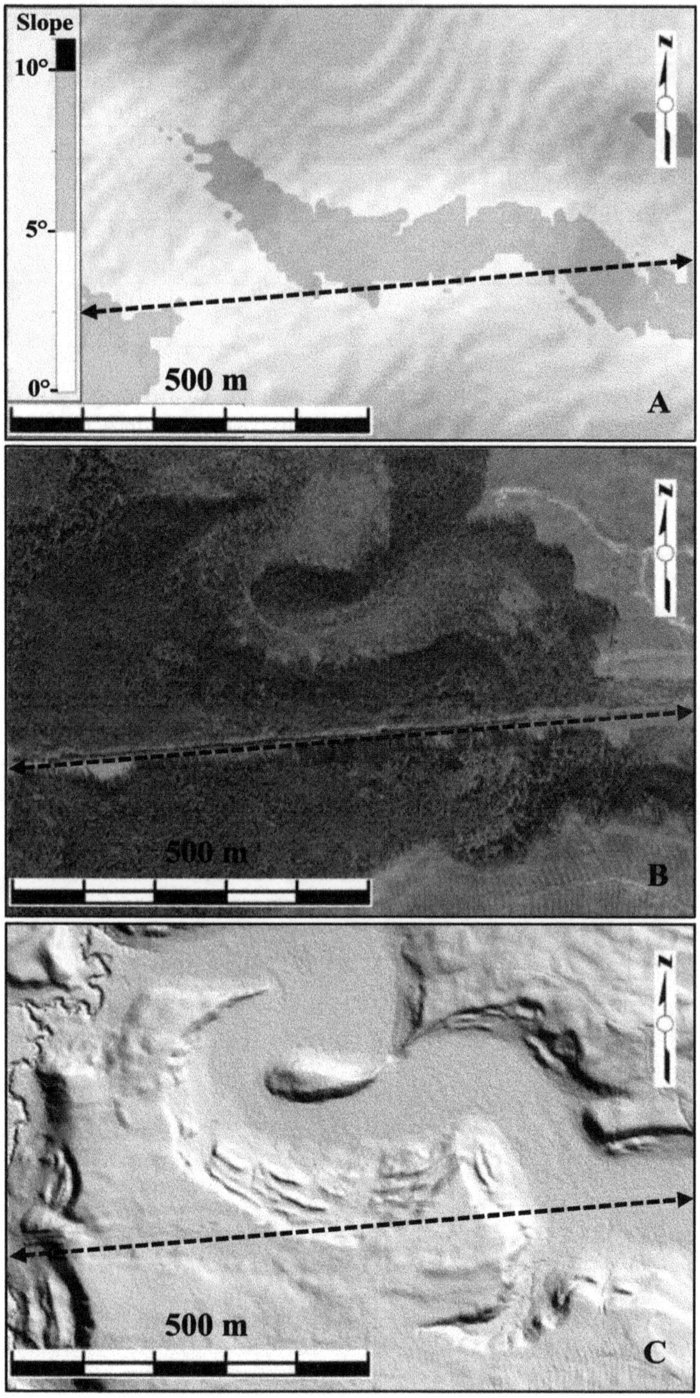

FIGURE 3: A COMPARAISON OF IMAGERY FROM A) A DEM FROM THE CANADIAN DIGITAL ELEVATION MODEL BASED ON THE NATIONAL TOPOGRAPHIC SYSTEM AT THE 1:50,000 SCALE. B) AERIAL IMAGERY FROM GOOGLE EARTH DATED OCTOBER 9, 2003., AND C) A BARE EARTH IMAGE CREATED FROM LIDAR DATA COLLECTED OCTOBER 2015.

Since completion of the EGI study, Pembina has continued to add hazards to the inventory through detailed studies, and review of other data such as ILI, field inspections, and aerial patrols. In total 44 sites have been added bringing the total number of geotechnical hazards associated with the original 6543 km of pipeline to 949. Around 95% of the hazards in the inventory were identified by the initial GIS based hazard inventory and the subsequent EGI study.

This case study demonstrates that using aerial imagery interpretation with DEMs and GIS based criteria is likely to omit a significant portion of hazards from the inventory, both in terms of the total number of hazards and the number of highest PoF hazards. By adding LiDAR interpretation, the inventory can be substantially increased. Additional data and detailed studies with similar scopes added approximately 5% to the inventory since the EGI study in 2015, demonstrating that in general diminishing returns on investment can be expected with additional hazard identification studies.

Best Practice Recommendations

The proposed best practice establishes an initial inventory by prioritizing the methods which yield the most complete inventory for the least amount of time and cost. There is no upper bound to the degree of detail a geotechnical hazard identification study could include. Highly detailed studies will identify additional hazards, however the effort to identify those additional hazards can be grossly disproportional to the value in doing so. Geotechnical hazard identification can be optimized by utilizing a methodology which at a minimum includes A) compilation of all existing hazard inventories and hazard data from existing reports and studies, B) identification of topography above a slope and slope height threshold established by a qualified and experienced geohazard specialist, and C) identification of hazard morphology with the potential to impact the pipeline through review of LiDAR imagery by a qualified and experienced geohazard specialist. These three components are illustrated in Figure 4.

Copyright © 2019 by ASME

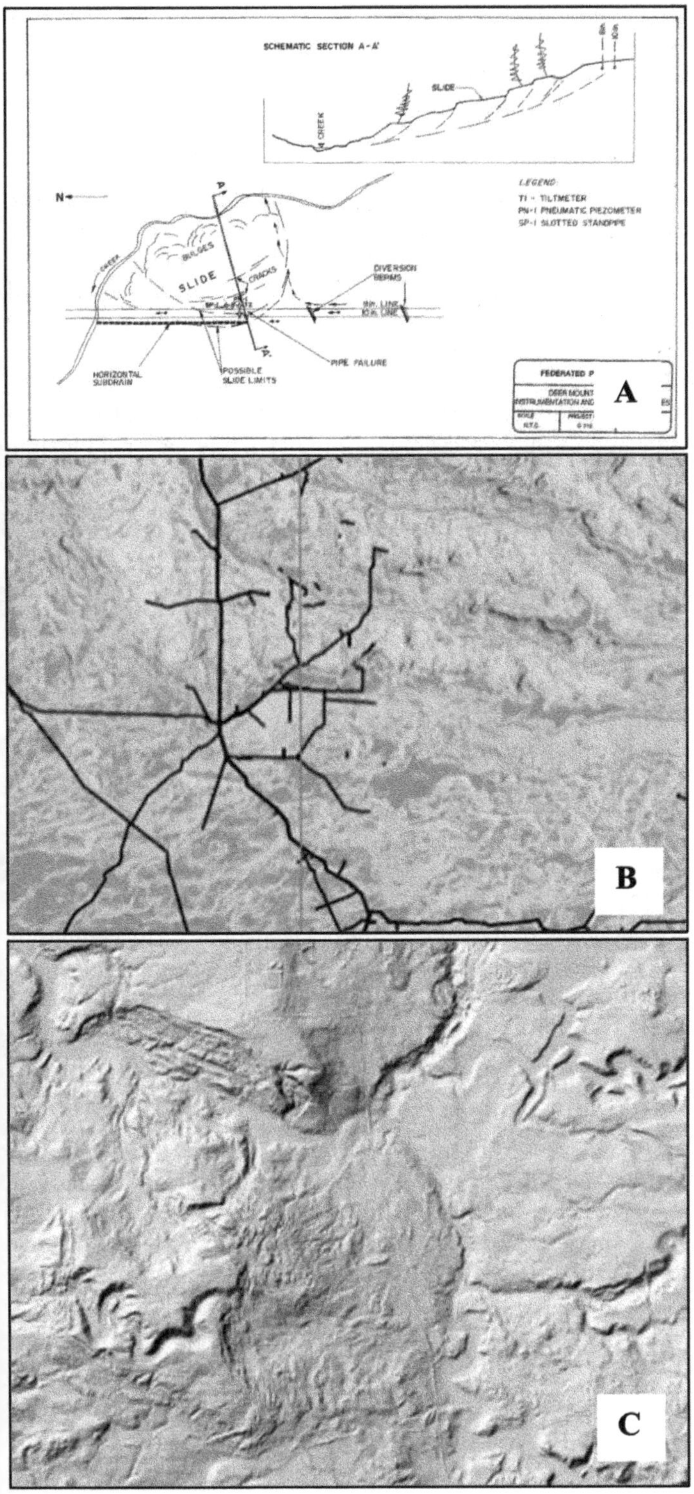

FIGURE 4: THE RECOMMENDED SCOPE FOR INITIALLY DEVELOPING A GEOTECHNICAL HAZARD INVENTORY, INCLUDING A) EXISTING GEOHAZARD DATA, B) IDENTIFICATION OF TOPOGRAPHY MEETING GIS CRITERIA USING A DEM, AND C) BARE EARTH LIDAR INTERPRETATION.

After a LiDAR based hazard identification study, some hazards will still be missing from the inventory. A review of the data and methodology by a geohazard specialist familiar with the regional geology and stability conditions may help to identify possible sources of omission from the inventory, and the level of effort and value of completing additional studies. Operators will have to determine whether conducting additional hazard identification studies is necessary to achieve the hazard management objectives of the program. Additional studies should be completed in a carefully considered manner which focusses on addressing the source of the omissions rather than completing studies with broad scopes.

Ongoing inventory enhancement should be completed by utilizing ILI data, and integrating right-of-way observations from routine right-of-way patrols including aerial reconnaissance and ground inspection. The incremental cost of incorporating these data into the geohazard management program is typically low since the data is often already collected, and hazards which are not observable by a LiDAR based study may be identified using these data (e.g. recent landslides, landslides obscured by surface grading, and ground movement from subsidence and frost processes).

This proposed best practice should continually be evaluated and updated by reviewing the performance of the methodology and the benefits provided by utilizing emerging tools such as machine learning based landslide prediction models, ILI axial strain tools, UAV remote monitoring, and satellite monitoring. Regardless of how the future standard practice changes, it is critical to recognize the importance of a substantially complete geotechnical hazard inventory and the tendency for significantly reduced marginal value to the geohazard management program with more detailed study which is not carefully considered.

Limitations

The proposed best practice relies on readily available data and qualified geohazard specialists experienced in terrain interpretation. If LiDAR has not been flown, it may be necessary to evaluate the cost and benefit of acquiring the data. To estimate the benefit, some estimate of the number of unidentified hazards will need to be developed.

The primary cause of unidentified hazards for the GIS based method is low angle landslides which comprise a small proportion of the total number of slopes with similar angles (Figure 1). The case study covers many kilometers of pipeline and multiple physiographic regions (prairie, foothills, mountains, etc.) and can be considered as a representative average for pipeline systems within the western Canadian study area, however isolating physiographic regions within the dataset will likely yield variable levels of inventory enhancement. For example, a GIS screening methodology may work well in physiographic regions where slope angle and landslide frequency have a better correlation, such as the Prairies, where the topography is either flat or valley slopes. In this terrain the slopes have a bimodal distribution and a slope angle threshold may be selected which identifies all the valley slopes while excluding flat topography. Conversely, in terrain dominated by

Copyright © 2019 by ASME

low angle landslides a GIS screening methodology is expected to miss many hazards since a low angle threshold cannot be used because it captures an impractical number of benign hazard sites.

In mountainous terrain, a GIS screening methodology will identify many slopes since the topography is steeper, and slope angle is expected to have a poor correlation with landslide frequency. Additionally, landslides are expected to be omitted from the inventory because they are upslope or encroaching hazards (e.g. debris slides), which represents the second largest group of landslides added to the hazard inventory in the presented case study.

A future study could utilize the data of the presented case study to identify trends in physiographic regions and hazard types which could help operators determine where LiDAR acquisition and interpretation is most critical for developing geotechnical hazard inventories.

CONCLUSIONS

Establishing and continuing to enhance a geohazard inventory is foundational to executing a geohazard management program. A reliable inventory enables proactive management by identifying hazards which have not yet caused damage or failure to pipelines or right-of-ways. Program effectiveness is improved by allowing all sites to be ranked relative to one another, so hazard management activities can prioritize the most important hazards.

Prior to the widespread availability of LiDAR data, hazard inventories were often based on coarse resolution DEMs and a review of publicly available aerial imagery. Inventories were supplemented by existing geohazard knowledge from engineering reports and operational records. Inventories created with this methodology may be missing a significant number of hazards. The case study presented was missing more than half of the highest PoF rated geotechnical hazards and the total number of hazard sites was approximately doubled, highlighting the significant improvement in hazard inventory by utilizing LiDAR to identify existing hazard morphology.

The proposed best practice is aimed at initially producing a substantially complete hazard inventory for a minimum amount of time and cost. The proposed methodology includes: 1) compilation of all existing hazard inventories and hazard data from existing reports and studies, 2) identification of the topography above a slope and slope height threshold established by a qualified and experienced geohazard specialist, and 3) identification of hazard morphology with the potential to impact the pipeline through review of LiDAR imagery by a qualified and experienced geohazard specialist.

Geotechnical hazard inventories should continually be enhanced by utilizing ILI data, integrating right-of-way observations from routine right-of-way patrols including aerial reconnaissance and ground inspection. Consideration should be given to completing detailed studies where limitations or uncertainties with the inventory are unacceptable to the operator. Emerging technologies may provide valuable additional data and the industry should continue to evaluate their effectiveness. The value of any subsequent hazard identification study should be evaluated based on the expected enhancement to the inventory and the cost of the study, recognizing that cost can easily become disproportionate to the value if the study scope and methodology is not carefully considered.

ACKNOWLEDGEMENTS

The authors would like to acknowledge Pembina Pipeline Corporation for sharing the results of this study and the pipeline operators who host data on BGC Engineering Inc.'s Cambio™ platform and have undertaking similar enhanced geotechnical hazard inventory studies.

REFERENCES

[1] Barlow, J. Peter and Richmond, John A. "The Cheecham Landslide Event." Proceedings of the 11th International Pipeline Conference. Calgary, Alberta. September 29 – October 3, 2016. Paper No. IPC2016-64515, pp. V003T04A022. doi:10.1115/IPC2016-64515. https://proceedings.asmedigitalcollection.asme.org/proceeding.aspx?articleID=2583927

[2] Dooley, Colin, Prestie, Zachary, Ferris, Gerry, Fitch, Murray and Zhang, Hua. Proceedings of the 10th International Pipeline Conference. Paper No. IPC2014-33240, pp. V002T06A084. Calgary, Alberta. September 29 – October 3, 2014. doi:10.1115/IPC2014-33240. http://proceedings.asmedigitalcollection.asme.org/proceeding.aspx?articleid=2022665&resultClick=3

[3] Ferris, Gerald, Newton, Sarah, and Porter, Michael. "Vulnerability of buried pipelines to landslides." Proceedings of the 11th International Pipeline Conference. Paper No. IPC2016-64071. pp. V002T07A001. Calgary, Alberta. September 29 – October 3, 2016. doi:10.1115/IPC2016-64071. http://proceedings.asmedigitalcollection.asme.org/proceeding.aspx?articleid=2583873

[4] Porter, Michael, Ferris, Gerald, Leir, Mark, Leach, Miguel, and Haderspock, Mario. 2016. "Updated estimates of frequencies of pipeline failures caused by geohazards." Proceedings of the 11th International Pipeline Conference. Paper No. IPC2016-64085. pp. V002T07A003. Calgary, Alberta. September 29 – October 3, 2016. doi: 10.1115/IPC2016-64085. http://proceedings.asmedigitalcollection.asme.org/proceeding.aspx?articleid=2583875

Copyright © 2019 by ASME

AIMPIMG2019-1017

Chapter 17

EXAMPLE OF A SEMI-QUANTITATIVE STREAM CROSSING HYDROTECHNICAL HAZARD ASSESSMENT FOR A NEW PIPELINE

Rodolfo B. Sancio
Geosyntec Consultants, Inc.
Houston, Texas, USA

David Vance
Geosyntec Consultants, Inc.
Atlanta, Georgia, USA

ABSTRACT

Industry guidance and recommended practice documents, such as CEPA's 2014 "Pipeline Watercourse Management - Recommended Practices", and the more recent version of API RP 1133 "Managing Hydrotechnical Hazards for Pipelines Located Onshore or Within Coastal Zone Areas", provide helpful information related to conducting hydrotechnical hazard and/or risk assessments for stream crossings. However, little direction is provided in these documents, or in other readily available published literature, on specific steps or processes that may be followed to semi-quantitatively estimate the hydrotechnical hazard level of a stream crossing.

This paper describes the steps that were followed in a Phase 1 hydrotechnical hazard desktop study to estimate the hazard level for a proposed new gas pipeline that traverses various physiographic provinces in mountainous terrain in the eastern US. The Phase 1 work comprised the compilation and analysis of stream channel and watershed characteristics and geomorphic attributes at each stream crossing along the pipeline, together with susceptibility to flood flows and other geologic hazards (e.g., debris flows) that affect the likelihood for vertical and horizontal movement of the stream bed, as well as potential pipeline exposure. Threshold values and scores were assigned to each attribute, the data was analyzed in a GIS platform, and then streams were sorted into three semi-quantitative hydrotechnical hazard levels according to their scores: 1) low hazard stream crossings, which could be addressed through commonly used industry rules of thumb such as 5 feet of cover below thalweg and 15 feet of sag-bend set-back beyond the ordinary high water mark, 2) moderate hazard stream crossings, which would require additional engineering assessment, design and mitigation such, as additional cover, greater sag-bend set-backs and/or bank protection, and 3) high hazard stream crossings, which would require specialized engineering assessment, design, and mitigation, or specialized construction methods such as horizontal directional drilling (HDD).

The results of the Phase 1 study were used to support selection of sites requiring later field verification (Phase 2). Data from the field verification was then used to support the design of hydrotechnical hazard mitigation including specific recommendations for burial depth, sag-bend setbacks, bank or bed protection, and other measures.

INTRODUCTION

Segment AP-1 of the Atlantic Coast Pipeline (ACP) will comprise a 42-inch diameter pipe that will start in the southwest corner of Harrison County approximately 9 miles southwest of West Milford, WV, (MP 0.0) and will traverse through West Virginia and Virginia in a southeasterly direction for approximately 329 miles to a point near the Virginia/North Carolina state line (Figure 1). Along this route, the pipeline will cross over 660 streams that include rivers, perennial, intermittent, and ephemeral creeks, runs, drafts, etc., as well as drainage ditches.

At each one of these streams, the pipeline needs to be buried in a manner that mitigates against the threat of pipeline exposure, because when the pipeline becomes exposed within the stream, it is subject to potential loss of integrity that may be caused by hydrodynamic loading, debris impact, fatigue caused by vortex-induced vibration, loss of support caused by stream bed erosion, and loss of cathodic protection or increased corrosion [1]. The likelihood of exposure and subsequent damage of the pipeline in the stream is reflected in the hydrotechnical hazard level (low, medium, high).

A previously buried pipeline may become exposed due to changes in the channel morphology that include vertical movement (e.g., scour, degradation, and aggradation), horizontal movement (e.g., bank erosion and encroachment), and complete channel relocation (e.g., avulsion and meander cutoffs). These stream changes can occur over long timeframes due to natural channel evolution or in response to episodic events such as floods and mass wasting events (i.e., debris flows).

Copyright © 2019 by ASME

Categorization of the hazard at each stream is the first step required for mitigation. This paper describes the steps that were followed during a desktop study that was conducted to categorize the hazard at the streams crossed by the pipeline.

GEOLOGICAL AND HYDROLOGICAL SETTING

ACP will traverse five physiographic provinces (Figure 2), or large areas with characteristic landforms and similar geology, and hydrology including: the Appalachian Plateau Province in West Virginia; the Valley and Ridge Province in West Virginia and north-central Virginia; the Blue Ridge Province in central Virginia; the Piedmont Province in eastern Virginia and North Carolina; and, the Coastal Plain Province in south-central Virginia and North Carolina.

The Appalachian Plateau Province forms the northwestern flank of the Appalachian Mountains from western New York to northern Alabama and is characterized by elevated, planar sedimentary rocks with differing levels of stream dissection. The Valley and Ridge Province consists of folded sedimentary bedrock that forms a long and narrow belt of parallel mountain ridges and valleys trending in a northeast to southwest direction. Differential weathering of these folds and faults has produced the distinctive repeating landscape of ridges and valleys that contain moderately large rivers. The Blue Ridge Province is a narrow zone of mountain ridges trending in a northeast to southwest direction, inclusive of the Blue Ridge Mountains. The Blue Ridge Province is comprised of Precambrian granite and gneiss, late Precambrian to Cambrian age meta-basalt, and Cambrian age limestone, conglomerate, and shale. The Blue Ridge Province contains the highest peaks in the Appalachian Highlands and steep gradient mountainside streams. The Piedmont Province is characterized by gently rolling topography, underlain by saprolite or crystalline bedrock, with a scarcity of bedrock outcrops and large, low gradient, sometimes poorly confined rivers. The Coastal Plain Province occupies relatively lower elevations of the southeastern United States and is characterized as having relatively low relief with rivers and streams discharging through large estuaries into the Atlantic Ocean.

PHASED APPROACH FOR HAZARD ASSESSMENTS

Geohazard assessments may be conducted in three phases, namely: phase 1, that comprises a desktop study using publicly and privately obtained remotely sensed data, phase 2, which involves a reconnaissance level field study in which site specific data is collected by subject matter experts (SMEs) through non-intrusive methods, and phase 3, which involves detailed field characterization through various more intrusive techniques [2]. Uncertainty in the characterization of the hazard is progressively reduced as the SMEs evaluating the hazard advance from one phase to the next.

The approach described herein was primarily desktop based, but a limited number of phase 2 visits to streams was used to calibrate the desktop evaluation.

STANDARD OF PRACTICE FOR HYDROTECHNICAL HAZARD EVALUATION

There are several documents that describe the hazards at stream crossings and provide general guidance and design steps for hazard mitigation, e.g., [3]. However, there is no standard of practice or guidance document that describes the process for identification and characterization of hydrotechnical hazard at pipeline stream crossings through specific stream attributes. Guidance such as [4] focuses on recommendations for mitigation of hazard such as pipeline burial rather than the initial process to identify and filter streams more or less likely to be subjected to hazardous conditions. [3] describes classification into major, intermediate, and minor or typical river crossings based on hydrotechnical complexity of the crossing and its required design components. Other guidance documents such as the "Pipeline Watercourse Management - Recommended Practices" published by the Canadian Energy Pipeline Association (CEPA) [1], and the more recent version of API RP 1133 "Managing Hydrotechnical Hazards for Pipelines Located Onshore or Within Coastal Zone Areas" [5], provide helpful information related to conducting hydrotechnical hazard and/or risk assessments for stream crossings. However, little direction is provided in these documents, or in other readily available published literature, on specific steps or processes that may be followed to qualitatively or quantitatively estimate the hydrotechnical hazard level of a stream crossing or hundreds of stream crossings along a pipeline alignment.

CHARACTERISTICS OF A STREAM THAT AFFECT HAZARD

The hydrotechnical hazard at a specified crossing is a function of the watershed hydrology, mass wasting characteristics within the watershed that is tributary to that crossing, the steepness of the watershed, the watershed size, land use in the watershed, and the characteristics of both the channel and floodplain which can include channel width, active floodplain width, longitudinal slope, depth to bedrock, flow characteristics, and erodibility of bed and bank materials. Channel characteristics are therefore key inputs into the hydrotechnical analysis process. In addition to the current physical properties of the crossing and its watershed characteristics, the historical migration behavior of the channel, a channel's susceptibility to conveying potential debris flows, and a channel's proximity to potential slope instability features is further utilized in assessing potential hazards.

QUALITATIVE HAZARD LEVELS

For the purposes of this study three qualitative hydrotechnical hazard levels were defined as follows:

- Low Hazard: the stream has a low likelihood of horizontal or vertical mobility over the design life of the project (i.e., water crossings that can be addressed through standard design rules such as 5 feet of cover, 15 feet of set-back, and no bank protection).
- Moderate Hazard: the stream has a moderate likelihood of horizontal or vertical mobility over the design life of the

Copyright © 2019 by ASME

project (i.e., water crossings that will require additional engineering assessment, design and mitigation such as additional cover and/or bank protection).

- High Hazard: the stream has a high likelihood of horizontal or vertical mobility over the design life of the project (i.e., water crossings that will require specialized engineering assessment, design, and mitigation, or specialized construction methods such as directional drilling).

HAZARD ANALYSIS APPROACH

A five-step process was developed and used in the desktop-based hydrotechnical hazard assessment of the crossings. A GIS-based screening tool was used to conduct this analysis and to develop a semi-quantitative hazard potential ranking for each stream crossing.

The first step involved identifying stream crossings followed by selection of parameters most relevant to hazard assessment and compilation of a database with attributes. Next, a weighted hazard classification scheme was used to assign preliminary high / moderate / low hazard designations to each crossing. A geomorphological evaluation was then used to assign manual hazard classifications to a subset of crossings in each of the initial high, moderate, and low hazard categories. The weighted hazard classification scheme was adjusted iteratively to improve agreement between automated hazard assignments and the expert manual evaluation assignments. Field evaluation of a selected subset of accessible stream crossings was conducted via aerial surveys and ground reconnaissance; findings were then used to further refine the automated weighted classification methodology. These steps are described in more detail in the following sections.

Step 1: Stream Crossing Identification

Stream crossings along the route of interest were identified on a GIS platform using the National Hydrography Dataset (NHD) high resolution flowline data.

Step 2: Selection of Parameters and their Estimation

Seven watershed and channel attributes were selected for hazard assessment: drainage area, slope at crossing, depth to bedrock, percent sand (i.e., in soils), debris flow, and mountainous area. These attributes were selected as they take into consideration sufficient geomorphic, geologic, hydrologic, and hydraulic parameters that influence the potential for vertical and horizontal stream migration that can affect pipeline integrity throughout all the physiographic provinces along the route. Moreover, factors that exacerbate hazard such as the presence of cohesionless soils in stream banks and debris flow susceptibility as well as mitigating factors such as the presence of shallow bedrock were included. Below is a discussion on how each parameter affects stream hazard and the source for the data.

Drainage Area

Drainage area is the land surface area enclosed by the watershed upstream of the pipeline stream crossing. The drainage area is the single parameter with the greatest effect on the hydrology of streams in the study area. For example, studies conducted for the West Virginia Appalachian Plateaus physiographic province [6], for the Valley and Ridge physiographic province in Virginia [7], and for the Piedmont physiographic province in Virginia [8] to develop regional curves for non-urban streams have found a strong correlation between stream parameters (i.e., bankfull discharge, width, depth, and cross sectional area) and drainage area. A direct relationship between drainage area and stream peak discharge, which has a strong influence on scour potential, and thus hazard, was demonstrated for Virginia [9].

The drainage area at stream crossings was calculated using StreamStats [10] and manual delineations. StreamStats is a GIS-based application with a web-browser interface that combines ArcGIS Server technology, ArcHydro Tools, as well as Python scripts to provide stream parameters and statistics. StreamStats uses the elevation data from the USGS 3D Elevation Program, streams from the National Hydrography Dataset (NHD), and previously determined watershed boundaries from the Watershed Boundary Dataset to automatically delineate watersheds at the points of interest.

Table 1 summarizes the distribution of stream crossings according to drainage area and physiographic province. This analysis shows that 490 streams out of 660, about 74%, have a drainage area of less than 0.5 square mile (320 acres) and 81% are less than 1 square mile, sizes that are relatively small and typical of headwater streams (first and second order streams) that lack the power and potential to reach cumulative scour depths (e.g., local scour, long-term degradation, etc.) greater than 5 ft below existing grade (i.e., thalweg) within the design life of the project. As described later in Step 5, the hazard scores were adjusted according to the size of the drainage area.

Although modifications of land use in the future can affect the magnitude of discharge in streams, estimation of hazard for this attribute was solely based on the drainage area and not on land use. The mitigation of the hazard, which is conducted in the design stage should consider, as appropriate, the potential for modifications of land use over the project design life to establish appropriate burial depth and set back distances from the current stream banks.

A low hazard was assigned to watersheds with a drainage area smaller than 4 square miles, a high hazard to watersheds with a drainage area greater than 10 square miles, and a moderate hazard to watersheds with a drainage area between 4 and 10 square miles. These thresholds were chosen because, regardless of physiographic province along the pipeline alignment, drainage areas smaller than 4 square miles are headwater streams (e.g., first, second, third order, etc.) that are less likely to develop the ability to migrate vertically beyond a minimum burial depth of 5 ft below the current thalweg that was established for low hazard streams. Streams with drainage area between 4 and 10 square miles have an increased likelihood to scour vertically more than 5 ft below the existing thalweg and to migrate horizontally, and thus represent moderate hazard. The likelihood for stream scour deeper than 5 ft increases further when the drainage area is greater than 10 square miles. Additionally,

Copyright © 2019 by ASME

streams greater than 10 square miles are wider and will require evaluation of the sag bend set-backs on a case-by-case basis to address lateral migration potential of the channel.

Slope at Crossing

The slope of the stream at the location of interest affects hydraulic factors such as flow velocity and therefore scour potential. Moreover, the product of drainage area and slope provides an indication of stream power. Therefore, the average slope across a 200-ft long reach at the stream crossing (100 ft upstream and 100 ft downstream) was measured from project acquired topography from lidar data.

Streams in the project area with an average slope less than 2% were considered to exhibit low hazard, whereas streams with an average slope greater than 4% were considered to exhibit high hazard. Streams with an average slope between 2% and 4% were considered to exhibit moderate hazard. These thresholds were selected based on geomorphic relationships of channel slope to channel form developed by [11].

Depth to Bedrock

Although dependent on orientation, condition, and type, as well as the design life of a given project, competent bedrock typically represents structural grade control for streams; in most cases negating the need to derive alluvial equilibrium slopes (i.e., to evaluate long-term degradation) and the related potential scour depths. Assuming good construction practice (i.e., excavation of a relatively narrow trench), burial of the pipeline within bedrock would therefore provide long-term protection from erosion hazards and significantly reduce any potential for channel migration to detrimentally affect the pipeline at a stream crossing given that bedrock will remain upstream and downstream of the trench thus protecting the narrow trench from the hazard of vertical or horizontal migration.

The depth to lithic bedrock for the project area was compiled by from the Soil Survey Geographic Database (SSURGO), which contains information collected by the National Cooperative Soil Survey, which is part of the Natural Resources Conservation Service (NRCS) of the United States Department of Agriculture (USDA).

Streams in the project area with a depth to bedrock of less than 5 ft (150 cm) were considered to exhibit low hazard, whereas streams with bedrock depth exceeding 6.6 ft (200 cm) were considered to exhibit high hazard. Streams with bedrock depth between 5 ft and 6.6 ft are considered to exhibit moderate hazard.

The choice of 5 ft as a threshold for establishing low hazard assumes that the minimum pipeline burial depth across streams will be 5 ft below the existing thalweg. Therefore, bedrock at a depth of 5 ft or less has a greater likelihood of protecting the pipeline from vertical scour and may mitigate potential lateral migration. Bedrock at a depth between 5 ft and 6.6 ft, which approximately coincides with the springline of the pipeline, provides moderate protection.

Percent Sand

In general terms, sand-sized particles (sizes between 0.075 mm and 4.75 mm), requires less hydraulic force to initiate motion and erode the stream banks than silt and clay. Therefore, horizontal migration hazard increases with an increase in the sand content of the stream banks. As with the depth to bedrock, sand content in the vicinity of the stream crossing was obtained from county-level soils data at stream crossings from SSURGO.

Soils at stream crossings in the project area with percent sand less than 40% were considered to exhibit low hazard, whereas soils with percent sand exceeding 60% were considered to exhibit high hazard. Soils with percent sand between 40% and 60% were considered to exhibit moderate hazard. These thresholds were selected because soils with sand content of 40% or less may also contain fine-grained soils (clay and silt) or coarser grained soils such as gravel and cobbles that can mitigate the vertical scour potential. The range between 40% and 60% is transitional (i.e., moderate hazard), whereas sand contents exceeding 60% describe conditions of a highly erodible stream channel (e.g., higher likelihood of vertical scour exceeding 5 ft).

Debris Flows

As defined by the USGS [12], a debris flow is a form of rapid mass movement in which a combination of loose soil, rock, organic matter, air, and water mobilize as a slurry that flows downslope. Debris flows carry significant erosive power along their path and have the potential to significantly change post-event channel characteristic including bed elevations, planform morphology, floodplain morphology, and the hydraulics at given points where, for example, pre-event conveyance geometries may become constricted or re-shaped due to structural obstructions (e.g., mass movements, large trees, debris jams and boulders). Moreover, streams within recent debris flow deposits may be actively scouring vertically (i.e., undergoing degradation) to establish an equilibrium slope.

Steep gullies, similar to those within the Appalachian Plateaus, Valley and Ridge, and Blue Ridge physiographic provinces, are prone to debris-flow, and morphologies associated with the occurrence of past debris flow deposits are noticeable. As documented in, for example, [13], [14], [15], [16], and [17] debris flows are common in the project area having been caused by intense surface-water flow, due to heavy precipitation which erodes and mobilizes loose soil or rock on steep slopes.

Debris flow events similar to those in the past are likely to occur within the project area over the design life of the project. This attribute was therefore incorporated in the analysis assuming their occurrence within susceptible areas of the pipeline alignment. Upon analysis of the topographic, geologic, and hydrologic conditions along the pipeline alignment as well as mapped areas of past occurrence of debris flows (e.g., [17]), the portions of the pipeline segment listed in Table 2 were considered susceptible to debris flows. Although debris flow susceptibility was mapped through approximately MP 207.0 [17], the low-lying topography starting from MP 179.0 renders a lower likelihood for development of debris flows and therefore

Copyright © 2019 by ASME

the range of debris flow susceptible areas was terminated at MP 179.0.

Mountainous Areas

As described above, the drainage area of the watershed upstream of the pipeline stream crossing has a significant influence on peak flow and therefore vertical migration potential (i.e., scour) and lateral migration (e.g., meander migration, avulsion, etc.). However, the shape of the watershed also has a significant influence on stream hydrology. For all else equal (e.g., similar soil conditions, vegetation, and precipitation) a watershed in a mountainous area with steeper slopes will exhibit a hydrograph with a shorter duration and higher peak flow (i.e., flashy) than a watershed of the same size located in a low lying coastal area with gentler slopes. Stream flashiness has a significant detrimental effect on the migration and scour potential of a stream and therefore increases the hazard.

To incorporate the effect of mountainous areas in the hazard assessment, the application of a hazard factor to streams located in the Appalachian Plateaus, Valley and Ridge, and Blue Ridge physiographic provinces was considered appropriate. The first 74 miles of the pipeline are within the Appalachian Plateaus. The Valley and Ridge physiographic province encompasses between MP 74 and MP 147, and the Blue Ridge physiographic region extends from MP 147 to MP 168 within AP-1.

Step 3: Automated Scoring Classification

Following the development of a database populated with attributes for each of the parameters listed in Step 2, initial scores were assigned for the low, moderate, and high hazard categories of drainage area, slope at crossing, percent sand, and depth to bedrock discussed above and a score for streams within mountainous areas and areas susceptible to debris flows. The individual scores were added, and the total score was then evaluated and classified as exhibiting high, moderate, or low hazard. The initial scores were selected by expert opinion and later adjusted based on comparisons with the results of the manual assessment which is described next.

Step 4: Manual Expert Classification

Following the initial automated weighted hazard rating assignments, a manual screening was conducted to provide hazard ratings for a subset of crossings (approximately 6%). At each manually evaluated crossing, the lidar-based digital earth model and historical aerial photos from Google Earth™ were studied by an expert fluvial geomorphologist to assess stream migration, other visible indicators of instability, riparian land use characteristics at the crossing, and channel and floodplain geomorphology. Stream width, bank material, watershed relief, stream slope, physiographic province, local geology, and other factors were also considered in making the manual evaluation.

Step 5: Field Reconnaissance and Adjusted Automated Scoring Classification

Following preliminary identification of sites as described above, a field program to assess hydrotechnical hazards was conducted along a small number of stream crossings. Efforts included both aerial surveys via helicopter and ground-based visits. Terrestrial reconnaissance included qualitative assessments of floodplain extents, debris flow potential (through identification of evidence of past occurrences), geomorphic characteristics of stream system, bed/bank materials and overall observations of recent/historical channel migration that all together indicate the need for further understanding to confidently characterize potential hazards for which engineering mitigation may be recommended. Aerial reconnaissance, where it was possible to achieve, also provided qualitative assessment at a larger scale and helped support and cross-reference terrestrial field assessments. Following field reconnaissance, additional adjustment of the automated classification process was carried out by adjusting the scores assigned.

The manual assessment resulted in a set of relatively high confidence hazard assignments for several crossings in all hazard categories (low, moderate, high). With automated and manually computed hazard ratings side-by-side, the screening criteria and scores established in Step 3 and used for the initial automated classification were then iteratively adjusted to reduce the hazard rating assignment differences between the automated and manual processes.

The final scores used in the evaluation are listed on Table 3. Additionally, watershed areas < 0.5 square mile were assigned a weight of -10 and areas > 100 square miles +10 such that sufficient bias is introduced for small and large watershed sizes. Using the criteria in Table 3, the highest total score at a crossing is 21, and the lowest value is -15. Crossings with a total score of < 2 were considered to exhibit Low Hazard potential. Crossings with a score between 2 and 6 were considered to exhibit Moderate Hazard potential, and crossings with score > 6 were considered to exhibit High Hazard potential.

Table 4 provides hypothetical parameters for three streams to describe the development of scores and categorization of the hazard.

- Stream 1: The relatively small drainage area of 1.5 square miles of Stream 1 located in the Valley and Ridge physiographic province and within a debris zone prone area exhibits low hazard mostly because of the small size of its drainage area.
- Stream 2: This stream is also located in the Valley and Ridge physiographic province but exhibits a drainage area about an order of magnitude larger than Stream 1. Although the slope of the channel is shallower than Stream 1, the hazard designation for this stream is high.
- Stream 3: This stream with a 22.5 square mile drainage area within the Piedmont physiographic province exhibits moderate hazard because of the size of its drainage area and sand bed, which combined increase the likelihood of vertical scour exceeding 5 ft as well as its potential for lateral migration.

Copyright © 2019 by ASME

RESULTS

After applying the previously described approach, of the 660 streams along the pipeline alignment, the vast majority, 566 or about 86%, were classified as exhibiting low hazard, 69 streams (about 10%) were classified as exhibiting moderate hazard, and 4 streams as exhibiting high hazard (4%). Table 5 provides a summary of the streams hazard classification by physiographic province and Tables 6, 7, and 8 provide the frequency distribution by drainage area.

Results provided in these tables indicate that:

- Relative to the other provinces, more streams in the Piedmont exhibit low hazard (92%) than in the other mountainous provinces, where between 75% and 85% of the streams exhibit low hazard.
- With the exception of one stream in the Valley and Ridge physiographic province, all low hazard streams in the Mountainous Area exhibit watershed drainage areas less than 4 square miles. Regarding high hazard streams, 19 out of 25 exhibit drainage areas exceeding 4 square miles.
- Streams that exhibit moderate hazard have a wide distribution of drainage area and physiographic province.
- The Valley and Ridge physiographic province exhibits the highest ratio of high hazard streams, likely because of the mountainous character of this region and its susceptibility to debris flows as well as the larger drainage areas crossed compared to other physiographic regions in the Mountainous Area.

SUMMARY

This paper describes the steps that were followed in a Phase 1 hydrotechnical hazard desktop study to estimate the hazard level for a proposed new gas pipeline that traverses various physiographic provinces in mountainous terrain in the eastern US.

The results of the Phase 1 study were successfully used as a tool to support selection of sites requiring later field verification (Phase 2). Data from the field verification was then used to support the design of hydrotechnical hazard mitigation including specific recommendations for burial depth, sag-bend setbacks, bank or bed protection, and other measures.

ACKNOWLEDGEMENTS

The authors appreciate the support of the owners and operator of this pipeline for allowing the publication of this work. Mr. Tony Rice of Geosyntec Consultants provided significant input early in the development of the approach and throughout its execution. The authors are grateful for his support.

The authors also wish to thank James Fisher for preparing the figures for this paper and two anonymous reviewers that provided thoughtful feedback and edits that improved the quality of the paper.

REFERENCES

[1] Canadian Energy Pipeline Association, CEPA. 2014. Pipeline Watercourse Management, Recommended Practices, 1st Edition, December 2014. 70 pages.

[2] Pipeline Research Council International, Inc., Guidelines for Management of Geohazards Affecting the Engineering and Construction of New Oil and Natural Gas Pipelines. Catalog No. PR-616-164506-R01, 2 May 2018.

[3] Veldman, W. 2008. Open Cut and Elevated Pipeline Crossings, in Pipeline Geo-Environmental Design and Geohazard Management edited by Moness Rizkalla. ASME, New York, NY. 353 pages.

[4] Fogg, J. and Hadley, H. 2007. Hydraulic Considerations for Pipelines Crossing Stream Channels. Technical Note 423. BLM/ST/ST-07/007+2880. U.S. Department of the Interior, Bureau of Land Management, National Science and Technology Center, Denver, CO. 20 pp. http://www.blm.gov/nstc/library/techno2.htm

[5] API, 2017. Recommended Practice for Managing Hydrotechnical Hazards for Pipelines Located Onshore or Within Coastal Zone Areas.

[6] Messinger, T. 2009. Regional Curves for Bankfull Channel Characteristics in the Appalachian Plateaus, West Virginia: US Geological Survey Scientific Investigations Report 2009-5242, pp. 43.

[7] Keaton, J.N., Messinger, T., and Doheny, E.J. 2005. Development and Analysis of Regional Curves for Streams in the Non-Urban Valley and Ridge Physiographic Province, Maryland, Virginia, and West Virginia: U.S. Geological Survey Scientific Investigations Report 2005-5076, 116 pages.

[8] Lotspeich, R.R., [2009]. Regional Curves of Bankfull Channel Geometry for Non-urban Streams in the Piedmont Physiographic Province, Virginia: U.S. Geological Survey Scientific Investigations Report 2009– 5206, pp. 51.

[9] Austin, S.H., Krstolic, J.L., and Wiegand, U. 2011. Peak-Flow Characteristics of Virginia Streams: U.S. Geological Survey Scientific Investigations Report 2011-5144, 106 p.

[10] Ries, K.G.III, Guthrie, J.D., Rea, A.H., Steeves, P.A., and Stewart, D.W. 2008. StreamStats: A Water Resources Web Application. USGS Fact Sheet FS 2008-3067. 6 p.

[11] Rosgen, D.L. 1994. A Classification of Natural Rivers, Catena, v. 24, pp. 169-199.

[12] USGS. 2004. Landslide Types and Processes. Fact Sheet 2004-3072. July 2004. 4 p.

[13] Jacobson, R.B., Cron, E.D., and McGeehin, J.P. 1989. Slope Movements Triggered by Heavy Rainfall, November 3-5, 1985, in Virginia and West Virginia, U.S.A. Geological Society of America, Special Paper 236. pp. 11-13.

[14] Gryta, J.J., and Bartholomew, M.J. 1989. Factors Influencing the Distribution of Debris Avalanches Associated with the 1969 Hurricane Camille in Nelson County, Virginia. Geological Society of America, Special Paper 236. pp. 15-27.

[15] Morgan, B.A., Wieczorek, G.F., Campbell, R.H., and Gori P.L. 1997. Debris Flow Hazards in Areas Affected by the June 27, 1995, Storm in Madison County, Virginia: U.S. Geological Survey, Open File Report 97-438, 15 p.

[16] Morgan, B.A., Iovine, G., Chirico, P., and Wieczorek, G.F. 1999. Inventory of Debris Flows and Floods in the Lovingston and Horseshoe Mountain, VA, 7.5' Quadrangles,

Copyright © 2019 by ASME

from the August 19/20, 1969, Storm in Nelson County, Virginia. USGS Open File Report 99-518.

[17] Wieczorek, G.F., Eaton, L.S., Morgan, B.A., Wooten, R.M., and Morrissey, M. 2009. An examination of selected historical rainfall-induced debris-flow events within the central and southern Appalachian Mountains of the Eastern United States: U.S. Geological Survey Open-File Report 2009–1155, 25 p.

Copyright © 2019 by ASME

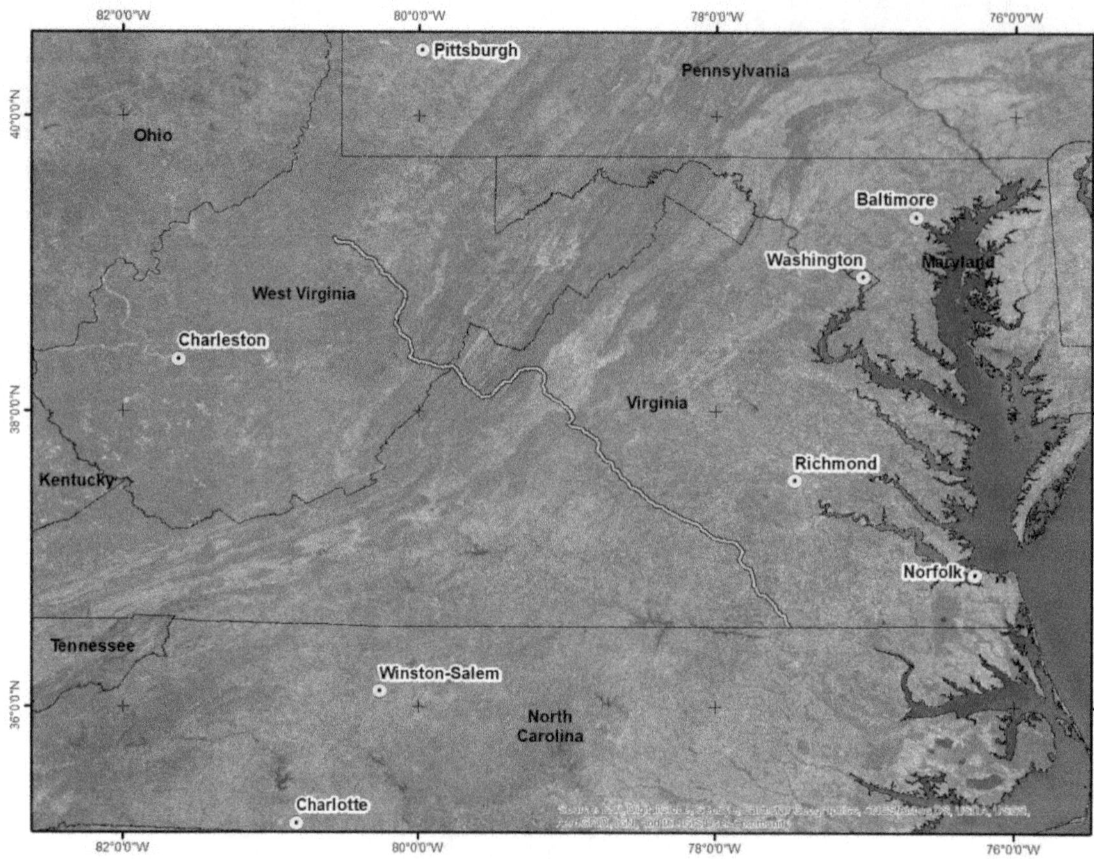

Figure 1 – Location map of pipeline

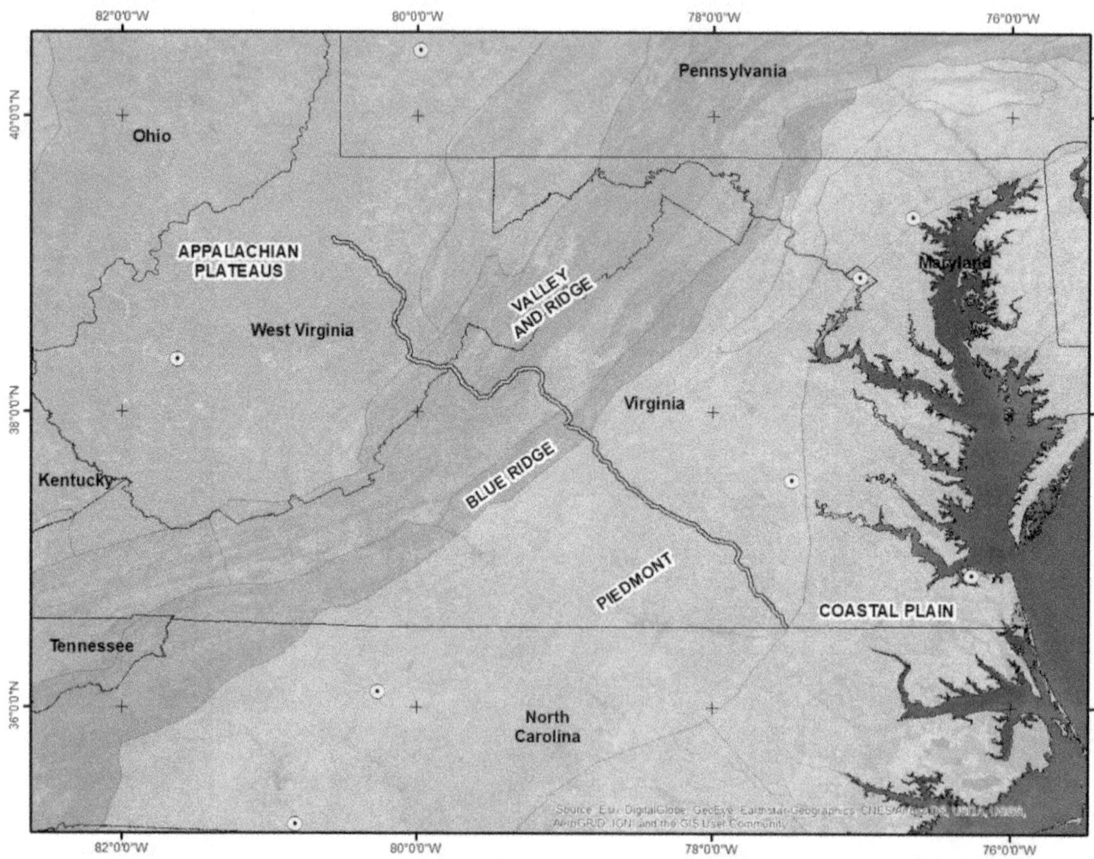

Figure 2 – Map of physiographic provinces traversed by the pipeline

Copyright © 2019 by ASME

Table 1 – Distribution of drainage area of each stream in each physiographic province

Drainage Area Bins	Appalachian Plateaus	Valley and Ridge	Blue Ridge	Piedmont	Total
	Frequency				
< 0.5	54	142	46	248	490
0.5 - 1	10	20	3	13	46
1 - 2	7	10	3	12	32
2 - 4	9	10	3	10	32
4 - 8	3	5	1	9	18
8 - 10	0	1	0	7	8
10 - 20	3	3	2	1	9
20 - 30	0	2	0	2	4
30 - 40	0	1	0	2	3
40 - 50	1	1	0	1	3
> 50	2	7	1	5	15
Total	**89**	**202**	**59**	**310**	**660**

Table 2 – Sections of the pipeline considered susceptible to debris flow hazards

State	County	Pipeline Milepost	
		From	To
West Virginia	Harrison County	0.0	1.1
	Lewis County	1.1	21.4
	Upshur County	21.4	43.9
	Randolph County	43.9	63.0
	Pocahontas County	81.0	83.9
Virginia	Highland County,	83.9	91.6
	Bath County	91.6	106.8
	Augusta County	106.8	123.0
	Augusta County	152.5	158.2
	Nelson County	158.2	179.0

Table 3 – Automated weighted hazard rating applied to stream crossings

Attribute	Effect on Hazard	Hazard	Attribute Value	Score
Watershed Drainage Area	Areas with larger watershed drainage areas are expected to cause higher hazard since larger volumes and flow rates are expected from rainfall	Low	< 0.5 mi^2 < 4 mi^2	-10 -5
		Moderate	4 to 10 mi^2	0
		High	> 10 mi^2 > 100 mi^2	+5 +10
Slope at Crossing over 200-ft long Reach	Steep crossings are more susceptible to erosion due to increased velocities	Low	< 2%	-2
		Moderate	2% to 4%	0
		High	> 4%	+2
Depth to Bedrock	Bedrock limits channel migration, therefore shallow depth to bedrock represents lower hazard of channel migration	Low	< 150 cm	-1
		Moderate	150 to 200 cm	0
		High	> 200 cm	+1
Percent Sand	Sand is highly erodible, therefore higher percent sand at the crossing increases hazard of channel migration	Low	< 40%	-2
		Moderate	40% to 60%	0
		High	> 60%	+2
		High	Inside	+3

Copyright © 2019 by ASME

Attribute	Effect on Hazard	Hazard	Attribute Value	Score
Mountainous Regions	Mountainous regions have increased hazard since mountainous watersheds are flashier than watersheds in flatter areas	None	Outside	0
Debris Flow Potential	Debris flows that occur after pipeline installation can degrade and remove soil cover. Installation of the pipeline in a recent debris flow deposit will require sufficient burial depth to mitigate the progressive vertical scour hazard	High	High	+3
		None	Low	0

Table 4 – Three examples of hazard scores

Stream Characteristics	Drainage Area	Slope	Depth to Bedrock	Percent Sand	Mountainous Area	Debris Flow	Total Score
Stream 1 DA = 1.5 mi² Valley and Ridge 2.1% Slope Bedrock at 6 ft Gravel armor, hence 10% Sand MP81	-5	0	0	-2	+3	+3	-1 (Low)
Stream 2 DA = 10.5 mi² Valley and Ridge 1.6% Slope Bedrock at 10 ft Gravel armor, hence 10% Sand MP100	+5	-2	+1	-2	+3	+3	+8 (High)
Stream 3 DA = 22.5 mi² Piedmont 0.3% Slope Bedrock at 6 ft 45% Sand MP194.1	+5	-2	+1	0	0	0	+4 (Mod)

Table 5 – Summary of results of hazard analysis by physiographic region

Appalachian Plateaus		
Hazard	Count	Amount
Low	76	85%
Mod	9	10%
High	4	5%
Total	89	100%

Valley and Ridge		
Hazard	Count	Amount
Low	161	80%
Mod	28	14%
High	13	6%
Total	202	100%

Blue Ridge		
Hazard	Count	Amount
Low	44	75%
Mod	12	20%
High	3	5%
Total	59	100%

Piedmont		
Hazard	Count	Amount
Low	285	92%
Mod	20	6%
High	5	2%
Total	310	100%

Copyright © 2019 by ASME

Table 6 – Frequency of low hazard streams by physiographic province

Drainage Area Bins	Appalachian Plateaus	Valley and Ridge	Blue Ridge	Piedmont	Coastal Plains	Total
	Frequency					
< 0.5	54	137	43	248	0	482
0.5 - 1	8	14	0	13	0	35
1 - 2	7	5	1	12	0	25
2 - 4	7	4	0	6	0	17
4 - 8	0	0	0	3	0	3
8 - 10	0	1	0	3	0	4
10 - 20	0	0	0	0	0	0
20 - 30	0	0	0	0	0	0
30 - 40	0	0	0	0	0	0
40 - 50	0	0	0	0	0	0
> 50	0	0	0	0	0	0
Total	**76**	**161**	**44**	**285**	**0**	**566**

Table 7 – Frequency of moderate hazard streams by physiographic province

Drainage Area Bins	Appalachian Plateaus	Valley and Ridge	Blue Ridge	Piedmont	Coastal Plains	Total
	Frequency					
< 0.5	0	5	3	0	0	8
0.5 - 1	2	6	2	0	0	10
1 - 2	0	3	0	0	0	3
2 - 4	2	5	3	4	0	14
4 - 8	3	3	1	5	0	12
8 - 10	0	0	0	3	0	3
10 - 20	1	3	2	1	0	7
20 - 30	0	1	0	2	0	3
30 - 40	0	1	0	2	0	3
40 - 50	1	0	0	1	0	2
> 50	0	1	1	2	0	4
Total	**9**	**28**	**12**	**20**	**0**	**69**

Table 8 – Frequency of high hazard streams by physiographic province

Drainage Area Bins	Appalachian Plateaus	Valley and Ridge	Blue Ridge	Piedmont	Coastal Plains	Total
	Frequency					
< 0.5	0	0	0	0	0	0
0.5 - 1	0	0	1	0	0	1
1 - 2	0	2	2	0	0	4
2 - 4	0	1	0	0	0	1
4 - 8	0	2	0	1	0	3
8 - 10	0	0	0	1	0	1
10 - 20	2	0	0	0	0	2
20 - 30	0	1	0	0	0	1
30 - 40	0	0	0	0	0	0
40 - 50	0	1	0	0	0	1
> 50	2	6	0	3	0	11
Total	**4**	**13**	**3**	**5**	**0**	**25**

Copyright © 2019 by ASME

SECTION 5
GEOHAZARD MITIGATION

INTRODUCTION

By Azam B A Rahman
(azamrahman@petronas.com)

Whenever a pipeline system is laid across natural land, hydro-geo-mechanical issues may be encountered. In such a case, issues of concern include accelerated erosions, disturbance of surface and subsurface flows, ground movement, settlement, slope failures, colluvium, foundations problems, and so on. This suggests that geohazard mitigation is not confined to geotechnical aspects. Therefore, technical factors such as geotechnical, hydrology, geomorphology, geography, meteorology, botanical science, civil and environmental engineering need to be considered as part of geohazard mitigation process and thus require multidisciplinary approach. In addition, the mitigation approach requires consideration of the "right of way " (ROW) corridor since in many cases mitigations may be required within as well as outside the ROW.

The ROW issue is relevant when the integrity against geohazard is being established an already installed pipeline. In such a case, mitigation work may require extended resources and time to execute. Thus, it is also beneficial to consider a "term based" mitigation plan so as to ensure that while a "final state" mitigation is in place, the geohazard conditions do not deteriorate further due to the uncontrollable dynamic factors that influence it, such as precipitation, surface and subsurface flows, erosion, etc. Practices of utilizing tarpaulin sheets, geobags, concrete canvas, temporary diversions of surface runoff etc. should be available as options and part of the "meanwhile" mitigation. Figure 1 presents some "meanwhile" mitigation approaches.

This chapter includes three papers covering the key aspects of geohazard mitigation strategies. This paper by Morris (AIMPIMG2019-1014) describes a case study of a slope failure with multiple constraints which requires the mitigation to be confined within ROW. A system utilizing soil nailing was used to overcome the constraints of the site in question. Some highlights that would affect the effectiveness of the mitigation chosen are as follows:

- Identifying up front the constraint within the ROW
- Mobility considerations for materials and machines

Figure 1. "Meanwhile" mitigation options

Copyright © 2019 by ASME

- Robustness of the method applied against the elements of nature
- Maintenance flexibility of the system as well for the pipeline
- Stabilization systems that utilize vegetation, reducing erosion and able to promote the surroundings to its original state.

This paper by Brant and Sancio (AIMPIMG2019-1046) underlines the technical challenges facing new pipelines on slope terrains and provides some insights into the performance-based design criterion as mitigation to instability of natural steep slopes on pipeline ROWs. The paper provides an alternative to the purely standard practice of evaluating the factor of safety or likelihood of landslide triggering factors which is rarely being evaluated due to the variance of the soil characterization (and other elements) associated with the nature of the pipelines being laid linearly.

Generally, pipeline integrity can be affected to a variable degree whenever a permanent ground displacement (PGD) occurs and how far it extends to the depth of the pipeline. This PGD imposes additional external loads to the buried pipeline, e.g. from landslides, i.e. with greater impact from deep-seated slides or minor impact from shallow landslides. Factors that contributes to slope instability are discussed, among others:

- Soil shear strength
- Impact of water
- Slope inclination.

The authors provided insights on Performance-Based Design Criterion, its practicability approach and considerations when dealing on the variability of the soils along the slope, its dynamic changes during construction and operations and also the fact that it is seasonably influenced. Other factors that adds into Performance-Based Design Criterion includes:

- Steep Slopes on Most Civil Engineered Infrastructure
- Steep Slopes on Pipeline ROW – citing the constraint within ROW as oppose to other types of civil engineered slopes
- Steep Slope Geohazard Avoidance
- Mitigation of Permanent Ground Displacement- emphasizing the importance of minimizing ground disturbance, surface water controls and subsurface drainage
- Mitigation for Pipeline Integrity – steps in improving resistance of the pipeline from PGD
- Geohazard Monitoring and Maintenance – emphasizing that monitoring geohazard alone is not a substitute for mitigation

The authors concluded that performance-based criterion approach is more appropriate for design and construction of pipelines along natural steep slopes and its objective is to manage the geohazard risks through the restoration of slopes on pipeline ROW to a level of performance comparable to the preconstruction condition.

This paper by Keaton (AIMPIMG2019-1036) describes ground subsidence and surface fault rupture as examples of geohazard that cannot be avoided (or entirely rectified) but provide geohazard management alternative that can be applied during pipeline operations. The element of geohazard management alternative includes:

- Modifying the hazard
- Modifying what is at risk of damage
- Modifying operation or procedures
- Compensate for potential losses arising from inability to operate and cost of repair damage.
- Avoid the hazard thru new infrastructure or relocation.

The paper included two studies. Case 1 addressed new pipeline that has documented ground subsidence due to groundwater pumping. Finite element model and desktop research was applied to estimates the strain limits of the pipeline. Operation alignment patrols and satellite radar interferometry was used to monitor and record the ground subsidence and rate. From the operation information taken, it was demonstrated that the strain registered was within its allowable range and thus allowing it to continue to be in operation.

Case 2 addressed new pipeline that crosses numerous active and potentially active faults. The normal design approach of pipelines at faults crossings ranges from prescriptive (i.e. no analytical verification) to analytical approach. Operation approach consists of the availability of an emergency action plan and triggered by occurrence of an earthquake that meets the magnitude and distance criteria that would impact the pipeline alignment. The paper concluded that even with effective design, some geohazard mitigation would require operational treatments to monitor conditions and enable detection for timely implementation of mitigation measures.

- Operational alternative requires:
- Knowledge of geological details
- Engineering geological model requires constant updating and validation.
- Both items are essential in providing information that would make operational alternative touching base with reality.

Chapter 18 Stabilization of a Natural Gas Pipeline Using Soil Nails and High Tensile Steel Wire Mesh (AIMPIMG2019-1014)
Matthew B. Morris, Gannett Fleming, Inc., USA

Chapter 19 Performance-based Approach for Stabilizing Steep Slopes on Pipeline Right of Way (AIMPIMG2019-1046)
Logan C. Brant and Rodolfo B. Sancio, Geosyntec Consultants, Inc., USA

Chapter 20 Operational Alternative to Design Treatment for Some Ground Subsidence and Surface Faulting Geohazards (AIMPIMG2019-1036)
Jeffrey R Keaton, Wood, USA

Copyright © 2019 by ASME

AIM-PIMG2019-1014

Chapter 18

STABILIZATION OF A NATURAL GAS PIPELINE UTILIZING SOIL NAILS AND HIGH TENSILE STEEL WIRE MESH

Matthew B. Morris[1]
Gannett Fleming, Inc.
Pittsburgh, PA

ABSTRACT

A landslide investigation and mitigation was conducted for a pipeline in southeastern Ohio. The landslide, with a lateral extent of 185 feet and a length more than 300 feet had encroached on the pipeline right-of-way and impacted an existing 26-inch natural gas pipeline. The subsurface conditions were observed in test pits advanced parallel to the pipeline. The design parameters used in the stability evaluations were based on limit equilibrium and a back-analysis approach. A range of soil parameters were evaluated along with published literature and our prior experience working with similar landslides in the region to establish the likely failure conditions. Based on the field observations, the water level in the slope was modeled to ensure the assumed slide plane is below the water table.

It was determined that the most appropriate stabilization system for this application was a combination of soil nails and high tensile strength steel mesh on the ground surface. The soil nail stabilization elements were designed with their spacing and lengths adjusted to ensure a minimum global Factor of Safety of 1.5 was achieved. The wire mesh facing was designed to retain smaller wedges of soil between the nails and was installed with turf reinforcement mat to promote the growth of vegetation on the slope face. The wire mesh was connected to the soil nails by use of plates and nuts and tensioned to provide confining pressure on the landslide mass. The benefits of this system included:

- *The soil nails and wire mesh provide long term stabilization of the slide with no maintenance.*
- *The soil nails were installed using a crawler mounted drill rig, creating minimal disturbance to the slope following the installation of the new pipeline.*
- *Should future maintenance of the pipeline be required, the mesh can be detached from the soil nails and rolled back allowing for excavation/backfill prior to replacement.*
- *The mesh allows for vegetation to grow on the surface, reducing the likelihood of soil erosion, and is aesthetically pleasing.*

Both verification and proof testing of the soil nails was completed to ensure the nail capacities assumed during the design process were achieved in the construction of the soil nails. Design and construction of the slope stabilization system was completed over a three-month period in 2015. The design and installation of this system allowed the owner to reopen the pipeline on schedule and will allow for future maintenance and operation of the line.

Keywords: Landslide, Soil Nail, Slope Stabilization.

NOMENCLATURE

ROW	Right-Of-Way
FS	Factor of Safety
γ	Soil Unit Weight (pounds per cubic foot)
c	Soil Cohesion (pounds per square foot)
ϕ	Soil Friction Angle (degrees)

INTRODUCTION

The upper Ohio River Basin is characterized physiographically by relatively flat uplands separated by steep hillsides and dendritic drainages containing tributaries that lead to major rivers that ultimately feed into the Ohio River [1]. The bedrock stratigraphy in the region is comprised of gently dipping sedimentary bedrock, with moderate to low-amplitude open folds. The typical sedimentary sequences contain sandstone, shale, claystone, and coal. Bedrock exposures on the steep hillsides weather over time, forming sandy and clayey residual soils. As these residual soils continue to weather, both physically and chemically, their strength degrades, and they are subject to

[1] Contact author: mmorris@gfnet.com

Copyright © 2019 by ASME

downslope movement as they are saturated from groundwater and heavy precipitation, effectively becoming colluvial soils. Ongoing strength degradation and downslope movement continues in these colluvial soils. Movement can be nearly imperceptible over many years (a process called creep) or can occur rapidly in a matter of days or hours, moving significant distance downslope on slide surfaces that range from shallow to deep. This continuum of movement is called landslide development and can pose a significant threat to infrastructure located above, on, or below colluvial slopes in the region. Linear features such as pipelines and roadways are particularly adversely impacted by landslides. This paper discusses a project where a relatively new approach to stabilizing shallow landslides, utilizing soil nails and high tensile strength wire mesh, was used to protect a high-pressure natural gas pipeline being impacted by a landslide in shallow colluvial soils at a site in southeastern Ohio.

1.0 PROJECT BACKGROUND

The natural gas transmission owner selected an engineer/contractor team to provide a turnkey mitigation solution that would protect the pipeline Right-Of-Way (ROW) from a large landslide that was impacting a natural gas pipeline in the Ohio River Basin. The landslide, thought to have originated between 2006 and 2014, was first noted on the pipeline ROW in 2014. The ROW contains two high pressure parallel lines, the western pipeline is a 26-inch high pressure line and the eastern is a 36-inch high pressure line. Figure 1 shows the limits of the landslide in relation to the ROW and gas pipelines. The right flank of the slide encroached on the pipeline ROW and induced stress on 260 lineal feet of the existing 26-inch gas line as it translated downslope. It appears that the right flank of the slide was likely bounded by the original pipeline trench excavation, as evidenced by the linear nature of the boundary and the parallel gas line located 25 feet to the east was not impacted by the slide. In whole, the landslide measured 185 feet wide and 300 feet long, with a 90-foot elevation change between the headscarp to the toe of slide, which terminates in the ravine bottom at the edge of an unnamed creek. Along with the colluvial soils present on the slope, erosion from the toe of the slope along the outside bank of the meander likely contributed to the activation of the slide.

2.0 GEOTECHNICAL INVESTIGATION SUMMARY

A geotechnical report was prepared that summarized the field conditions observed shortly after the landslide occurred on the pipeline ROW. Part of that investigation included the advancement of six test pits along the pipeline spanning the landslide area. The test pits were advanced to a depth of 15 feet, generally 10 feet below the pipeline invert elevation. The location of the test pits and other landslide features observed are shown on Figure 1 and Figure 2 shows the extent of the 6-foot high headscarp on the ROW prior to excavation of the pipeline.

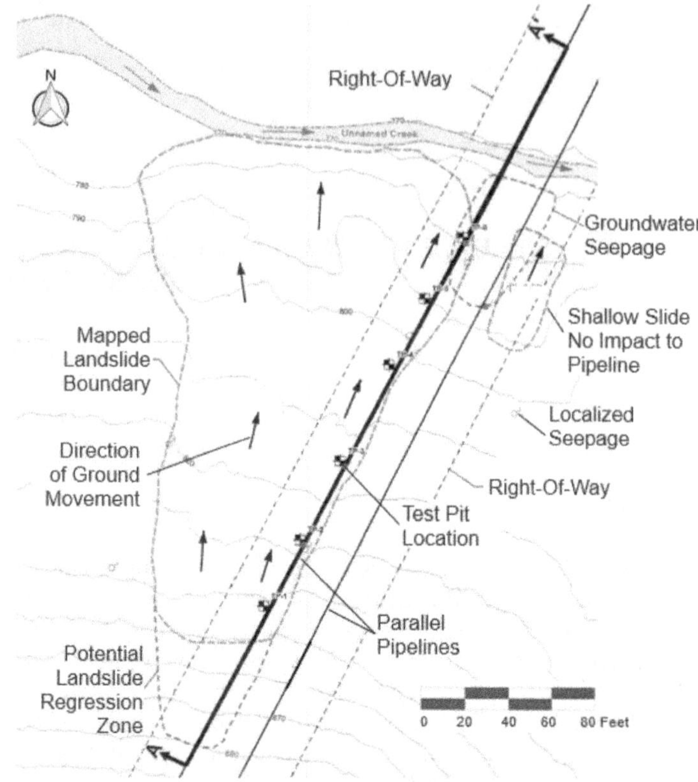

FIGURE 1: MAP OF LANDSLIDE AREA FEATURES AND RELATIONSHIP OF LANDSLIDE TO THE PIPELINE RIGHT-OF-WAY.

The contractor and their engineer performed additional field reconnaissance to observe the various features at the site and to compare these findings with those included in the geotechnical report. In general, the features observed during this investigation were consistent with the previously prepared report. These features included the landslide boundaries, surface drainage pathways and groundwater seep locations.

FIGURE 2: HEADSCARP OF THE LANDSLIDE ON THE PIPELINE ROW PRIOR TO EXCAVATION OF THE PIPELINE.

Copyright © 2019 by ASME

Based on the geologic conditions observed and test pit data the landslide impacting the pipeline is characterized by soft, clayey colluvial soils sliding along a 5 to 10-foot-deep slide plane, running nearly parallel to the ground surface. The soil beneath the slide plane was stiff to hard silty clay, residual in nature, and extended to depths more than 10 to 15 feet below the existing ground surface. These residual soils did not appear to have experienced prior movement based on the test pit observations. Weathered bedrock, comprised of siltstone, shale and coal, was observed at a minimum depth of six feet near the headscarp of the active slide area to a maximum of 15 feet near the toe of the slide. Figure 3 is an illustration of the general subsurface conditions along profile A-A' from Figure 1.

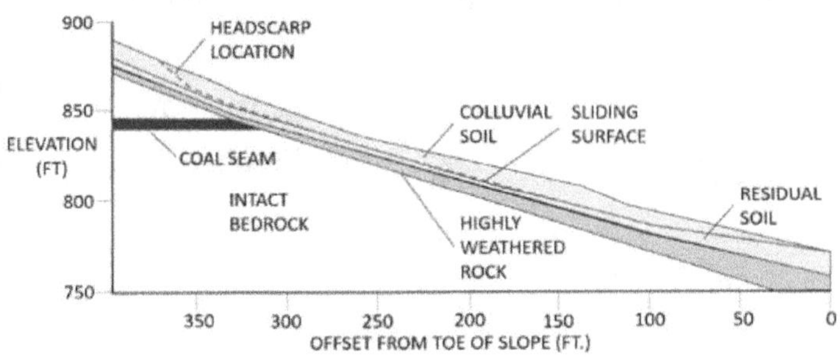

FIGURE 3: CROSS SECTION A-A' SHOWING THE INTERPRETED SUBSURFACE CONDITIONS.

3.0 STABILIZATION CONSIDERATIONS

Stabilization of shallow landslides can be accomplished by several methods including; excavation and replacement of the slide materials with more competent soil or aggregate, installation of a retaining wall(s), installation of buried concrete inclusions such as drilled caissons or auger cast piles and installing passive ground inclusions such as soil nails. The structural methods essentially improve the stability of the slope by intercepting the slide plane and increasing the resisting forces (shear strength) on the slope using steel and/or concrete elements. Typically, the standard of practice includes constructing or mitigating such that a FS = 1.3 to 1.5 is achieved. Using limit equilibrium methodology, this equates to forces resisting sliding being 30 to 50 percent greater than the forces driving the slide. For this project the owner established that a minimum FS = 1.5 would be required for the stabilization system selected by the engineer/owner contracting team. In addition, all elements of the landslide stabilization had to be constructed from and entirely within the pipeline ROW and outside of the riparian zone near the toe of the slope.

Given much of the landslide is located outside of the ROW, neither excavation and replacement of the landslide mass, nor the construction of retaining walls within the right of way were considered feasible options to prevent future movement of the slide outside the ROW from continued impacts to the pipeline. The stabilization solution used in this case would need to stabilize the slide on the ROW and also prevent any future movement of the larger slide from encroaching on the pipeline.

It was determined that the most efficient, least invasive, stabilization solution that would prevent the landslide from impacting the pipeline would be to install soil nails, that intercept the slide plane and stabilize the surficial soils using high tensile strength steel wire mesh. The relatively tight spacing of the soil nails along with the vertical confining force induced from the steel mesh creates a soil arching between the nails, preventing the loss material should adjacent movement occur in the slide mass to the west of the ROW. The steel mesh facing serves to retain smaller wedges of soil between the nails and is typically installed with turf reinforcement mat to promote the growth of vegetation. The steel mesh is connected to the soil nails by use of spike plates and nuts that are tensioned to provide confining pressure on the landslide mass. The limits of the soil nail and steel mesh stabilization installed on this project are shown in Figure 4.

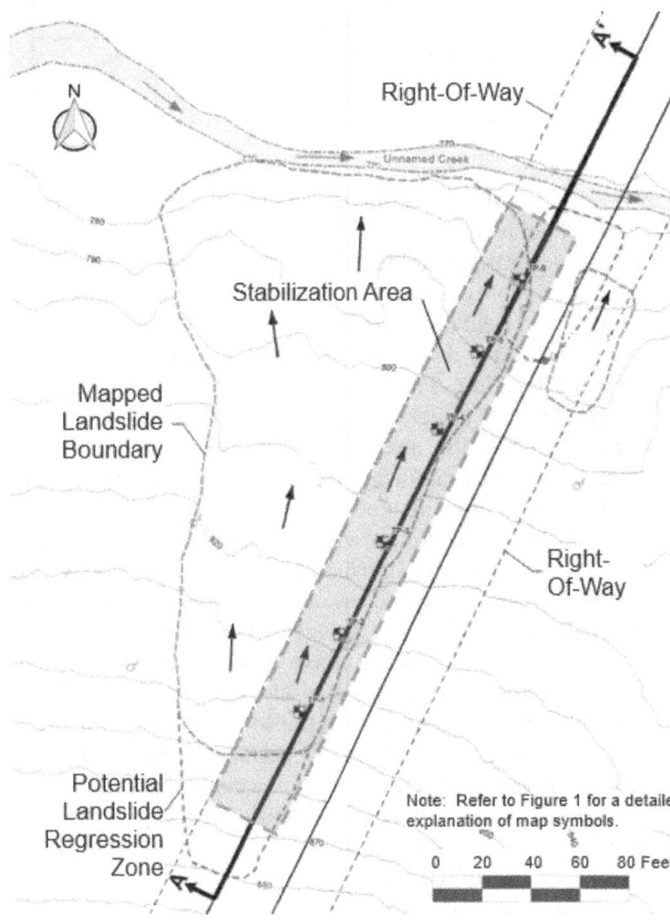

FIGURE 4: MAP OF LANDSLIDE AREA SHOWING LIMITS OF SLOPE STABILIZATION.

Copyright © 2019 by ASME

4.0 STABILIZATION DESIGN

To establish the appropriate design parameters for the stability evaluations, the existing slide was evaluated using limit equilibrium and a back-analysis approach with the GSTABL software program [2]. A range of soil parameters were evaluated and referenced to published literature [3, 4] and our prior experience working with similar landslides in the region to establish the representative soil properties at the time of failure. This methodology facilitated the use of conservative soil parameters for design (Table 1), particularly for the colluvium, which was considered to have only residual shear strength given its history of previous movement.

TABLE 1: BACK CALCULATED SOIL PARAMETERS USED IN DESIGN OF STABILIZATION SYSTEM.

Soil Type	ϒ (pcf)	c (psf)	φ (degrees)
Colluvium (Clay)	110	0	18
Residual (Silty Clay)	115	0	24
Pipe Trench Backfill	120	0	24
Highly Weathered Rock	120	150	34

Based on the geotechnical report and field observations, the water level in the slope was modeled such that the sliding plane was founded below the water table. Additionally, the water was assumed to generate from the observed coal seam in Test Pit 1 as coal is commonly a source of groundwater seepage. The geotechnical and field observations noted seepage near the toe of the slope; therefore, the water was modeled near the ground surface in this location. The back-analysis included adjusting the soil parameters to generate a FS of just under 1.0, near equilibrium. Once the analysis of the failed condition was complete, the soil nail stabilization elements were included in the model with their spacing and lengths adjusted to ensure a minimum FS of 1.5. Ruvolum, a proprietary software developed by the steel wire mesh manufacturer [5], was used to verify that the steel wire mesh facing and associated connections to the soil nails are compatible with the soil nail spacing to contain the soil between the soil nails should it move downslope. The design analysis determined that a minimum global FS of 2.2 could be achieved through the installation of the soil nails in combination with the tensioning of the steel mesh surface facing. While the calculated FS was higher than the minimum required for the project, the design and construction team felt a higher FS was warranted given the risk that additional movement could occur in the slope outside of the pipeline ROW.

5.0 STABILIZATION SUMMARY

Construction of the stabilization system began in September 2015, immediately following the replacement of the 26-inch gas line by the owner. At the onset of construction, five sacrificial test nails were installed to verify the design grout-to-ground bond capacity was achievable using the planned installation methods. In total, 230 nails were installed in the 280-foot-long by 30-foot-wide stabilization zone. A total of 52 rows of nails were installed, spaced 8 feet apart extending down the slope. One nail was installed on the east side of the pipeline and 4 nails on the west side of the pipeline, effectively cutting off any failures from continued movement of the slide mass west of the right-of-way. The nails had a minimum length of 25 feet, each grouted into a 5-inch diameter drill hole that extended into the residual soil/bedrock beneath the failure plane. The nails were comprised of 1.25-inch diameter epoxy coated Grade 75 steel bars and galvanized hardware with full grout encapsulation to provide long-term protection against corrosion. Refer to Figure 5 for an illustration of the typical soil nail arrangement.

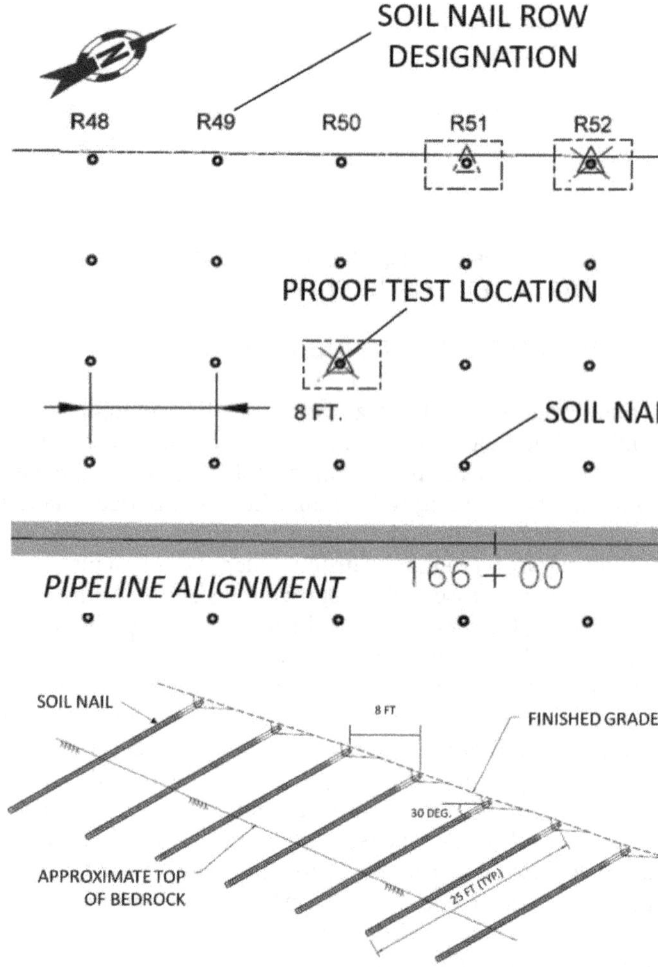

FIGURE 5: SLOPE PLAN VIEW AND CROSS SECTION SHOWING SOIL NAIL SPACING AND ARRANGEMENT.

The soil nails were installed using an excavator mounted drilling rig positioned between the newly replaced western line, which was out of service during construction, and the eastern line, which remained in service throughout the construction.

Copyright © 2019 by ASME

This equipment allowed the entire stabilization area to be reached from between the pipelines, while inducing little additional load to the gas lines. The nails were spaced horizontally on either side of the westernmost pipeline to ensure a minimum 5-foot minimum offset from the centerline. Construction was started at the top of the slope and the nails were installed downslope in rows to the bottom of the treatment area. This method ensured that the slope would be stable above the work area.

The general soil nail installation process included:

- Drilling the 5-inch diameter hole.
- Installing of the soil nail assembly.
- Grouting of the soil nail to the ground surface using neat cement grout and tremie techniques.
- Proof Testing of the soil nails

FIGURE 6: SOIL NAIL INSTALLATION USING A TRACK MOUNTED EXCAVATOR.

Upon completion of the soil nail installation, 5 percent of the total nails installed were load tested to 150 percent of the design load, ensuring adequate capacity was achieved. After testing, an erosion control blanket was placed on the soil slope and the steel mesh was installed and attached to the soil nails using a spike plate and spherical nut. The nut was torqued to impart a load of 11 kips on the mesh and soil nail system at each nail head. Refer to Figures 7 and 8 for the steel wire mesh layout and a photograph of the installed nail, mesh, spike plate and nut.

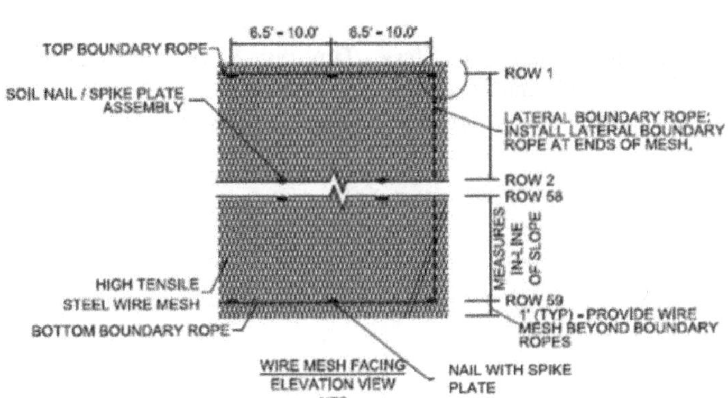

FIGURE 7: STEEL WIRE MESH DETAIL.

FIGURE 8: SOIL NAIL HEAD ASSEMBLY SHOWING THE NAIL HEAD, STEEL MESH, SPIKE PLATE, AND NUT.

6.0 PROJECT COMPLETION

The design of the project was completed in one month during the replacement of the western pipeline through the slide area. Construction of the soil nail and steel mesh system was completed in the two months immediately following the design and replacement of the western pipeline. The slope was seeded, and site completely restored so that vegetation would grow through the erosion control blanket and steel mesh prior to winter. The photographs included as Figures 9 and 10 show the stabilization system in place and the vegetative growth on the ROW following the completion of construction.

FIGURE 9: STABILIZED SLOPE SHOWING VEGETATION GROWING THROUGH THE WIRE MESH.

Copyright © 2019 by ASME

FIGURE 10: PIPELINE ALIGNMENT FOLLOWING RESTORATION.

7.0 DISCUSSION AND CONCLUSIONS

Shallow landslides within colluvial soils are a common occurrence on utility rights-of-way throughout the Ohio River Basin. This phenomenon often adversely impacts shallow pipelines that traverse these slopes. In many instances of shallow slides, the limits of the slide are well defined, and the depth of the sliding plane can be ascertained through simple investigation methods such as test pits. Once these factors are understood, the geometry of the slide can be established. Back-analysis of the slope failure is a proven method for establishing representative mobilized shear strength parameters, which in most shallow slides is equivalent to the drained residual strength [6]. These parameters can then be used to design a stabilization solution that best fits the site conditions.

Soil nail and high tensile steel wire mesh systems can be used to effectively stabilize shallow colluvial slides. In this case the stabilization system was advantageous for the following reasons:

- The system could be constructed entirely within the ROW, incorporating relatively small drilling and grouting equipment capable of traversing the steep terrain.
- Materials for the soil nails and steel mesh could be procured quickly and moved around the site with limited support equipment.
- The soil nails, wire mesh and associated hardware include corrosion resistant coatings and will provide long term stabilization of the slide mass on the ROW with no future maintenance required.
- The installation of the soil nail system minimally disturbed the recently regraded slope surface and did not impact other drainage and erosion control features following the installation of the new pipeline.
- Should future maintenance of the pipeline be required, the mesh can be detached from the soil nails and rolled back allowing for maintenance and then replaced and tensioned.
- The mesh allowed for vegetation to grow on the surface, reducing the likelihood of soil erosion, and does not adversely impede wildlife from traversing the slope.

REFERENCES

[1] Ohio Division of Geological Survey, 1998, Physiographic regions of Ohio: Ohio Department of Natural Resources, Division of Geological Survey, page-size map with text, 2 p., scale 1:2,100,00.

[2] GSTABL V2.004. Gregory Geotechnical, Stillwater, Oklahoma, USA, 2003; software available at http://www.gregeo.com.

[3] Naval Facilities Engineering Command (NAVFAC), 1986, DM7.02 Foundations and Earth Structures, pp.7.2-39.

[4] Bowles, Joseph E., 1988, Foundation Analysis and Design, p. 84 and 141.

[5] Ruvolum, GeoBrugg AG, Romanshorn, Switzerland, 2015; software available at www.geobrugg.com.

[6] Hussain, Manzoor, Stark, Timothy & Akhtar, Kamran.; 2010; Back-analysis Procedure for Landslides, International Conference on Geotechnical Engineering. pp. 159-166.

Copyright © 2019 by ASME

AIM-PIMG2019-1046

Chapter 19

PERFORMANCE-BASED APPROACH FOR STABILIZING STEEP SLOPES ON PIPELINE RIGHT OF WAY

Logan C. Brant
Geosyntec Consultants, Inc.
Houston, Texas, USA

Rodolfo B. Sancio
Geosyntec Consultants, Inc.
Houston, Texas, USA

ABSTRACT

Pipelines around the world are built through rough terrain that includes natural slopes that can be marginally stable. The standard of practice is that the stability of the slopes along the pipeline route is rarely formally evaluated to calculate the factor of safety or likelihood of landslide triggering. An evaluation that would render reliable values of factors of safety (FOS) would require a degree of characterization that cannot be achieved for these long linear structures given that subsurface conditions along the slope prior to pipeline construction are spatially and seasonably variable and can be affected by weathering over the design life of the project.

Best practice is to conduct geomorphological evaluations to identify areas that are prone to landslides, such as steep slopes or gentler slopes with relatively weak soil or rock formations. Where it is practical, areas that are subject to landslides before pipeline installation are avoided. When steep areas cannot be avoided, the pipeline is typically installed with best management practices that are intended to mitigate the landslide hazard. These include minimizing ground disturbance during pipeline construction and application of surface water diversions and subsurface drainage, along with other stabilization measures, such as re-establishing vegetation and soil compaction. When enough of these mitigation measures are applied to steep slopes disturbed by pipeline construction the result can generally offset soil strength losses attributed to ground disturbance.

In steep slope environments, such as in Appalachia, some natural slopes become unstable following periods of intense precipitation and ground saturation. When slope instability does occur, shallow landslides and soil creep are among the most common mechanisms of permanent ground displacement (PGD). Mitigating these natural processes from occurring on the pipeline right of way (ROW) is not always practical, particularly where they may have occurred naturally regardless of pipeline construction. Emphasis should be placed on the overall risk to the integrity of the pipeline and addressing those with appropriate mitigation measures, such a burying the pipe deeper than the minimum requirement, if needed, to reach stable soil or rock formations and maintaining the stability of the trench backfill.

Since the computation of a FOS is highly uncertain and may not reflect the likelihood of initiation of instability, a performance-based criterion is therefore considered to be a more appropriate approach for the design and construction of pipelines along natural steep slopes. Using this approach, the risk of landslides occurring and their possible effects on pipeline integrity, although unknown, is tolerable to the pipeline operators because the slope is expected to perform in a way that is similar to its performance prior to the installation of the pipeline.

This paper discusses the technical challenges facing construction of new pipelines on steep soil-covered slopes and provides recommendations for implementation of a performance-based design criterion to mitigate instability of steep slopes on pipeline ROWs.

INTRODUCTION

The design and construction of steep slopes on pipeline ROWs presents unique challenges which generally require a different approach compared to most other types of civil engineered infrastructure. This paper provides recommendations for the implementation of a performance-based design criterion which targets a minimum level of performance comparable to the pre-construction condition for steep slopes on pipeline ROWs. This level of performance can generally be achieved by minimizing ground disturbance during pipeline construction and by applying surface water diversions and subsurface drainage, along with other stabilization measures, such as re-establishing vegetation, and soil compaction.

STEEP SLOPE INSTABILTY

Natural steep slopes are highly variable, and as a result, there are many forms of slope instabilities [1, 2]. The risk to pipeline integrity from landslides on steep slopes varies widely, but it generally is from deep-seated landslides, i.e., landslides with sliding surfaces below the pipeline, since they transfer PGD

Copyright © 2019 by ASME

to the pipeline. Shallow landslides and surficial soil creep in the soil veneer are among the most common geologic processes contributing to PGD, but they often occur above the crown of the buried pipe where the risk to the integrity of the pipeline is negligible. However, shallow slips can still make the pipeline ROW impassable for maintenance operations, be problematic to the pipeline's cathodic protection system, are not aesthetic, convey to the community the perception that the pipeline is not safe and can have negative environmental consequences if sediment is transported and released into adjacent streams and wetlands.

Figure 1 shows a schematic illustration of shallow and deep sliding surfaces and shows the shallow sliding surfaces may remain above the pipe while the deep sliding surface typically extends below and surrounds the buried pipe, resulting in greater potential for pipeline rupture. Thus, the consequence of landsides with deep sliding surfaces is generally more severe than that of landslides with shallow sliding surfaces.

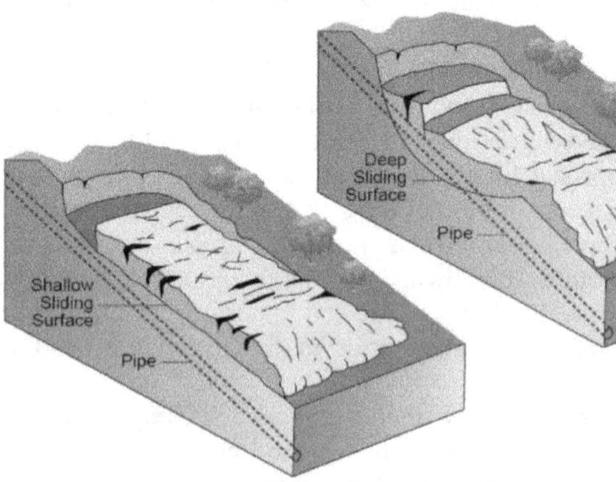

FIGURE 1: SHALLOW AND DEEP SLIDING SURFACES

Other types of slope instability, including debris flows and rock falls, are generally a lower risk to buried pipelines as well, unless the ensuing PGD extends to the depth of the pipe and overstresses or exposes the pipe.

LIKELIHOOD OF SLOPE INSTABILITY

The likelihood slope instability is affected by slope inclination, soil shear strength, soil unit weight, groundwater conditions, and the depth of the potentially unstable soil layer. These parameters are spatially and seasonably variable and can be affected by weathering over the design life of the project.

Soil Shear Strength

Partially saturated fine-grained soils disturbed by grading and excavation during pipeline construction may exhibit shear strengths consistent with the fully softened condition. The estimation of fully softened effective angle of internal friction for slope stability analyses represents the peak shear strength of normally consolidated remolded (i.e., disturbed) fine-grained soil [3]. The native soils encountered on a pipeline ROW are generally residual in nature and partially saturated. They have been subjected to numerous cycles of desiccation throughout the years, resulting in an overconsolidated condition. Depending on the degree of disturbance due to pipeline construction, the actual shear strengths may be higher than the fully softened condition assumed when the fully softened peak strength is used in stability analyses.

For slopes containing pre-existing sliding surfaces, where past shear deformations have caused the re-alignment of clay minerals and the development of preferential sliding surfaces, residual effective angles on internal friction are appropriate for stability analyses [3]. The residual value has a lower strength than the fully softened condition.

The effective internal friction angle for slope stability calculations can be estimated using empirical relationships [3, 4, 5] which approximate the friction angle of fine-grained soils as a function of the clay-size fraction, effective confining pressure, and liquid limit. These relationships are available for both the fully softened peak and residual shear strength conditions.

The parameters needed as inputs for these relationships can be obtained from laboratory index tests of samples collected during conventional geotechnical investigations (e.g., boreholes or test pits) and, for preliminary analysis, can be estimated from information available in soil survey databases, such as the United States Department of Agriculture (USDA) National Resource Conservation Service (NRCS) web soil survey [6]. When analyzing the slope stability of a long pipeline that traverses many steep slopes throughout its route, representative values and simplifying approximations must be considered with an understanding of the uncertainty in the calculations from not using site specific data.

Soil veneers on natural steep slopes are often the product of in situ weathering of the parent rock (i.e., residual soil). These soils may therefore retain some apparent cohesion, provided by cementation for example. Additionally, they are partially saturated, thus exhibiting apparent cohesion caused by interstitial pore water tension. Moreover, at shallow depths (e.g., < 2 ft), they also exhibit apparent cohesion caused by the root mat of trees, shrubs, and grasses. Upon saturation, however, the apparent cohesion caused by interstitial tension is likely to decrease or disappear. Also, the removal of vegetation during grading of the pipeline ROW will decrease the effect of the root mat, at least until vegetation re-establishes following restoration. The selection of parameters for slope stability evaluation should thus consider these effects.

For temporary slopes a modest value of apparent cohesion may be attributed to matrix suction of partially saturated fine-grained soils, however for the permanent condition, zero apparent cohesion is typically assumed. This offers a conservative approach, since over time as vegetation re-establishes apparent strength gain may be attributed to root-induced cohesion, but since it is difficult to quantify the amount and rate of strength gain from root cohesion, this stabilizing force is usually neglected when analyzing slope stability for permanent slopes.

Copyright © 2019 by ASME

Water

In partially saturated fine-grained soils, the shear strength is affected by the degree of saturation within the soil, particularly near the ground surface. Changes in the degree of saturation result in changes to the soil's pore water pressures and thus its shear strength and directly affect the slope's likelihood of instability [7]. The shear strength of partially saturated fine-grained soil decreases as the degree of saturation increases and as a result, slopes are more likely to move following periods of intense precipitation and ground saturation.

On the pipeline ROW, water can infiltrate soil from the top down, when there is precipitation or ponded surface water, or it may rise from the bottom up, with seepage and changes to the groundwater level, both of which can fluctuate seasonally. Whether originating from the surface or the subsurface, water must be controlled on the pipeline ROW through a combination of surface water diversions and subsurface drains. When these mitigation measures are applied, seepage can be managed, and internal seepage pressures can generally be assumed to be negligible. These mitigation measures are needed to maintain unsaturated conditions and maximize the shear strength of fine-grained soils to mitigate slope instability.

Slope Inclination

Slope inclination can be obtained from field measurements, topographical maps, or from digital elevation models (DEM) developed from high resolution LiDAR surveys.

PIPELINE INTEGRITY RISK FROM LANDSLIDES

The risks to pipeline integrity from slides that develop on steep slopes are greatest when soil in contact with the pipeline deforms causing either additional load transfer to the pipe or the removal of bedding material supporting the pipe. Load is transferred to the pipe as soil in contact with the pipe deforms. Support for the pipe can be lost if the surrounding soil moves away, exposing the pipe and causing the pipe to span across an unstable zone. When the stress on the pipe resulting from landslides on steep slopes is combined with other types of pipe stresses, such as internal pressures, external loads, and thermal loads, the combined resultant may overstress the pipe segments or the pipe weld connections and, in severe cases, may cause the pipe to rupture.

The risk to pipeline integrity from landslides on steep slopes is influenced by many factors, including the configuration of the pipeline with respect to soil movement, character and depth of PGD, the type of soil surrounding the pipe, and the pipe's physical properties and condition. The character and depth of the PGD is important, since shallow landslides occurring above the crown of the buried pipe present negligible risk to the integrity of the pipeline, while deeper landslides with PGD extending to the soil surrounding the pipe present a greater integrity risk (Figure 1). The type of soil in contact with the pipe is also important, since clayey soils tend to adhere to the pipe and transfer more load onto it, while loose granular soils, such as sand, have lower shear strength at shallow depths and transfer less load onto the pipe across the soil-pipe interface.

The pipe's resistance to loading depends on factors including the orientation of soil displacements with respect to the pipe, pipe diameter, wall thickness, surface texture, grade of steel, internal operating pressures, degree of corrosion and quality of the girth welds. For example, a newly installed thick-walled large diameter high strength steel pipe will have a significantly greater capacity to resist landslides on steep slopes compared to an older and partially corroded smaller diameter conventional steel pipe.

Pipeline Integrity Failures from Natural Forces

Historical records for the United States over the past several decades show that natural forces, a category which includes landslides, on average make up only about 4 to 6 percent of all pipeline integrity failures [8]. The estimated annual frequency of pipeline failures in the United States caused by geohazards is 0.2 per 1,000 km and corresponds to about 16 failures per year in the United States [9]. While these numbers as a percentage of total pipeline integrity failures are relatively low, the consequences of these pipeline failures can be significant in terms of safety, environment, and lost revenue. It is therefore important to mitigate landslide hazard.

Risk Tolerance

Decisions regarding risk tolerance should be made by the pipeline operator, and in many cases must also account for government-mandated performance requirements. The decision requires assessment of costs and benefits. Models have been proposed to support these decisions and to optimize the process [10, 11]. Costs include financial considerations, but may also include human, environmental, property, legal, and, reputation considerations, as well. Ultimately, risk tolerance is a business decision which should be made by the pipeline operators and clearly communicated to their managers, consultants, and contractors, who are to implement an efficient and consistent program to manage risk to pipelines from landslides on steep slopes.

PERFORMANCE-BASED DESIGN CRITERION

FOS-based design of natural steep slopes traversed by pipelines, which are long linear structures, is not practical. An evaluation that would render reliable values of FOS would require a degree of characterization that cannot be achieved given that subsurface conditions along the slope prior to pipeline construction are spatially and seasonably variable and can be affected by weathering over the design life of the project. Therefore, the computation of a FOS is highly uncertain and may not reflect the likelihood of initiation of instability.

A performance-based criterion is therefore considered to be a more appropriate approach for the design and construction of pipelines along natural steep slopes. Steep slopes on pipeline ROWs are designed to exhibit a level of performance similar to the slope's pre-construction condition. When enough of the appropriate types of non-structural mitigation measures are applied to steep slopes disturbed by pipeline construction, the result can offset soil strength losses caused during pipeline

Copyright © 2019 by ASME

construction. This level of performance can generally be achieved by minimizing ground disturbance during pipeline construction and by the application of surface water controls and subsurface drainage measures, along with other stabilization measures, such soil compaction and re-establishing vegetation.

An expectation that steep slopes on pipeline ROWs should be more stable than prior to construction or in nearby undisturbed slopes is not reasonable. When shallow landslides occur in the soil veneer above the pipe crown, those situations require grading repairs. However, instead of seeking to prevent all shallow landslides, which is neither a reasonable nor practical standard for design, emphasis should be on managing the overall risk to the integrity of the pipeline, which can be achieved through the use of appropriate mitigation measures, such as burying the pipe deeper than the minimum requirement in order to reach stable soil or rock formations and maintaining the stability of the trench backfill.

The remainder of this section on the performance-based design criterion is subdivided into six parts. The first part describes how steep slopes for most types of civil engineered infrastructure are designed and constructed. The second part explains why the design and construction of steep slopes on pipeline ROW should be addressed differently. The third part describes why it is best to identify and avoid steep slope geohazards through route selection, whenever it is practical. The fourth part describes common techniques used to mitigate the occurrence of PGD on the pipeline ROW. The fifth part describes some of the mitigation measures which can be taken to benefit pipeline integrity on steep slopes. And finally, the sixth part describes the importance of monitoring geohazards and using that data to inform decisions regarding the types of mitigation measures needed and effectively allocate resources for geohazard management and to prioritize the implementation of maintenance and repairs when corrective actions are necessary.

Steep Slopes on Most Civil Engineered Infrastructure

Engineered steep slopes designed and constructed for many other types of civil engineered infrastructure, including cuts, embankments, and dams, are typically constructed from homogenous soils with well understood engineering properties. Designers are generally able to choose the slope inclination or source materials with strength properties needed to achieve a prescribed minimum FOS against slope instability. For example, for the design of earthen dams, a minimum FOS against slope instability of 1.5 is typically prescribed for long-term conditions and 1.3 typically for temporary conditions during construction [12]. These prescribed minimum design factors of safety are intended to account for variability in loading and material properties and the consequence of failure, among other considerations.

Steep Slopes on Pipeline ROW

Unlike many other types of civil engineered infrastructure, the pipeline industry geohazard mitigation references [13] do not explicitly recommend factors of safety for slope stability evaluation. The design criteria for steep slopes is thus performance-based and not based on the magnitude of the factor of safety. Steep slopes on the pipeline ROW are instead designed to achieve a performance similar to the pre-construction condition and may exhibit factors of safety that are lower than what is traditionally used for road cuts, embankments, and dams. This means that the FOS for steep slopes on pipeline ROW are generally not formally computed [14].

While there are numerous mitigation measures frequently used to stabilize steep slopes on pipeline ROWs, there also exist some unique constraints which can make stabilizing steep slopes on pipeline ROWs particularly challenging. For example, one cost-effective approach to slope stabilization which is used extensively in many other types of civil engineered infrastructure is to cut back the slope inclination with grading, however, regulatory requirements, such as those in the FERC's "Upland Erosion Control, Revegetation, and Maintenance Plan", call for the restoration of pre-construction contours following pipeline construction [15]. This constraint also makes it difficult to buttress the base of a slope for improved stability. In addition, the practice of replacing native soil materials on steep slopes on pipeline ROWs on a wholesale scale to improve engineering properties is generally discouraged by regulators and contractors alike.

These constraints complicate efforts to stabilize steep slopes on pipeline ROWs and contribute to the need for a performance-based design criterion. As a result, the standard of practice within the pipeline industry for slope stability is not the same as for engineered slopes used in many other types of civil engineered infrastructure, such as road cuts, embankments, and dams.

Steep Slope Geohazard Avoidance

For steep slope geohazard avoidance, geomorphological evaluations are conducted to identify areas that are prone to landslides, such as steep slopes or gentler slopes with relatively weak soil or rock formations. This work is commonly achieved through multi-phased geohazard management programs which are conducted to identify and evaluate potential geohazards, including steep slope geohazards, along new pipeline alignments. These multi-phased geohazard assessment programs are becoming the state of practice within the pipeline industry [16, 17].

Even when areas prone to landslides are identified by multi-phased geohazard assessment programs, hazard avoidance is not practical in all situations. When steep slopes cannot be avoided, the pipeline is typically installed with best management practices that are intended to mitigate the landslide hazard.

Mitigation of Permanent Ground Displacement

In steep and rugged terrain, mitigation of landslide hazards on pipeline ROW can be achieved by minimizing ground disturbance during pipeline construction and through the application of surface water controls and subsurface drainage measures, along with other stabilization techniques, such as soil compaction and re-establishing vegetation. When enough of these appropriate mitigation measures are applied to steep slopes disturbed by pipeline construction, the result can generally offset

Copyright © 2019 by ASME

soil strength losses attributed to ground disturbance and removal of vegetation.

Since groundwater and surface water are common triggers for landslides, soil saturation should be avoided on the surface and within the ground. Slope breakers are a type of surface water diversion which captures water on the ground surface and conveys it to the edge of the pipeline ROW, away from the slopes disturbed by pipeline construction. Trench drains are a type of subsurface drain which collects and removes water from inside the pipe trench. Seep drains are another type of subsurface drain installed at targeted locations to intercept sources of groundwater seepage within the pipeline ROW. Trench breakers resist groundwater piping through the backfilled pipe trench, but on steep slopes they should include drains, which are needed to prevent the build-up of hydrostatic pressures. The use of sand bag trench breakers or sack-crete structural breakers can help to maintain the stability of the trench backfill, particularly when they are keyed at the base and sides into surrounding stable ground for improved stability.

In certain situations, increasing the strength of soil on slopes is needed to improve resistance to slope instability. This can be achieved by improving the existing materials through compaction, the addition of lime or cement, or by replacing unsuitable materials with engineered fills, flowable fills, crushed stone, or sack-crete. Scarifying, roughening, and/or benching the ground surface prior to backfilling can improve the bond between the existing material and the overlying backfill. Re-establishing vegetation can improve the stability of shallow soils through the addition of root cohesion. Erosion and sediment control best management practices can help with the management of surface water, but not all contribute to improving the geotechnical stability of steep slopes. Figure 2 shows a schematic illustration of some commonly used techniques for mitigating permanent ground displacement during construction and restoration on pipeline ROW.

When conventional mitigation techniques are not enough to reduce the likelihood of landslide-induced PGD, then geo-structural stabilization, such soil nails with wire mesh or engineered retaining structures, are viable options, however, their high cost and complexity limit their application to all but the most challenging steep slope sites.

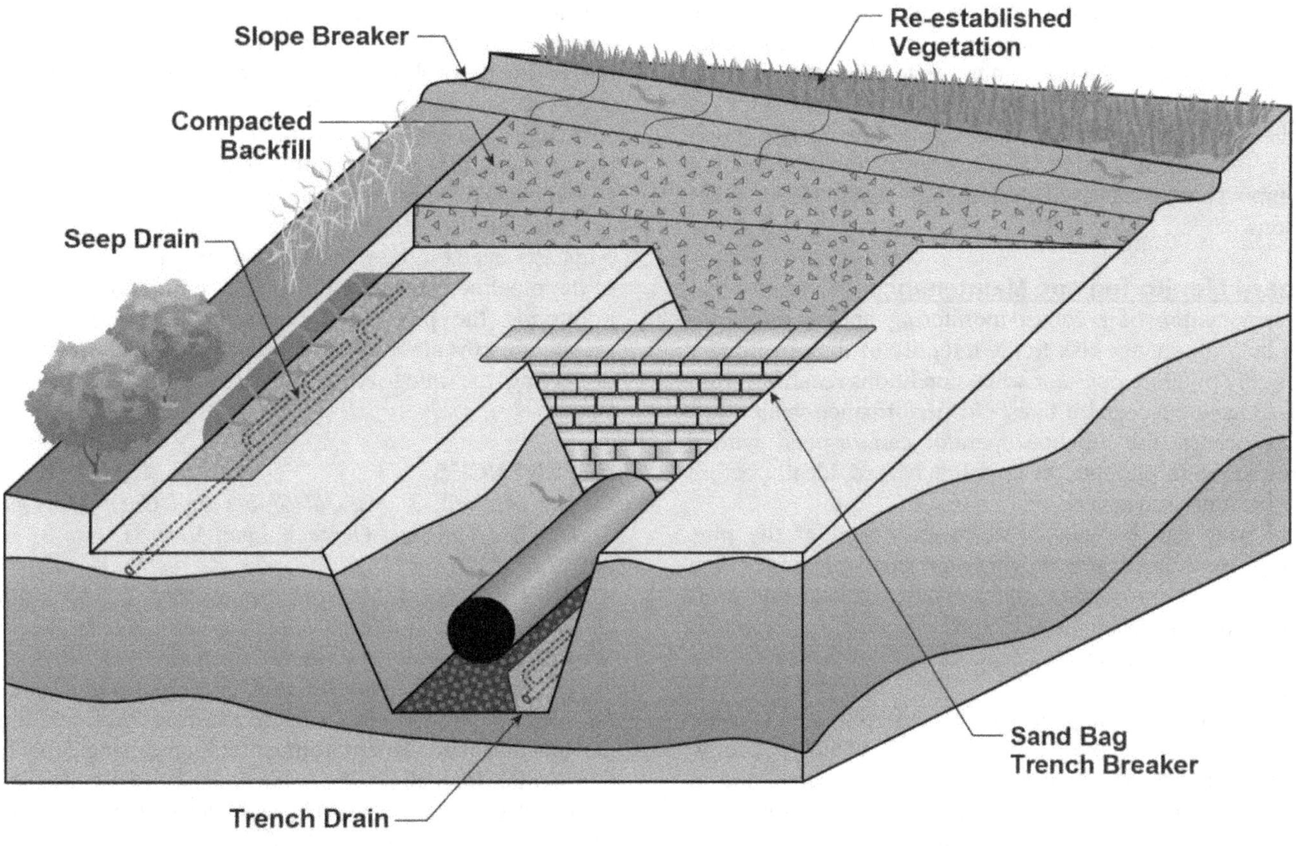

FIGURE 2: ILLUSTRATION SHOWING COMMONLY USED TECHNIQUES FOR MITIGATING PERMANENT GROUND DISPLACEMENT ON PIPELINE ROW

Copyright © 2019 by ASME

Mitigation for Pipeline Integrity

Pipelines are robust structural elements, but their ability to resist steep slope induced PGD depends on numerous factors. Increasing the pipe wall thickness and the grade of the steel, or decreasing internal operating pressures are just a few of the options available to increase the pipe's resiliency [18]. Monitoring for corrosion and inspecting the quality of girth welds is also important. To limit risk to the integrity of the pipeline, it is important to bury the pipe within stable soil or rock formations and to prevent the trench backfill from becoming unstable through the use of trench drains, trench breakers or sack-crete structural breakers and utilizing only appropriate materials for trench backfill.

If the pipeline alignment must extend through an area of pre-existing instability or an area with a high potential to become unstable after ground disturbance from pipeline construction, it is best to orient the pipeline parallel to the direction of anticipated PGD, so that loading on the pipeline will occur axially, because structurally, pipes are more resistant to axial loading compared to shear loading. Also, installation of pipe fittings or tees in the vicinity of anticipated PGD should be avoided, as they may become a source of stress concentrations [18]. Since pipeline integrity risks resulting from steep slope geohazards occur when accumulated load from soil surrounding the pipeline is transferred onto the pipe, techniques are available to reduce that load transfer, such as wrapping the pipe in geotextile with a low interface shear strength or encapsulating the pipe in loose clean sand [18]. If necessary, external loading on the pipe can be released periodically using stress relief excavations.

Geohazard Monitoring and Maintenance

Implementation of a robust monitoring and maintenance plan can help to manage risk to the integrity of the pipeline by informing the pipeline operator when conditions resulting from a landslide have exceeded a threshold performance value [17]. Once this occurs, the pipeline operator can respond with a corrective action to mitigate the landslide hazard, ideally before the PGD becomes too severe.

Monitoring can include direct measurement of the pipe using strain gages, optical surveying, and smart pigging with a three-dimensional gyroscope. Alternatively, the landslide itself can be monitored for changes in inclination, position (terrestrial or satellite methods), groundwater level or fluid pressures, and even ground shaking.

Reasons for conducting geohazard monitoring are to ensure that the geohazard potential has been properly characterized, to confirm the efficacy of long-term mitigation measures, and to identify the need for additional mitigation when the performance of existing mitigation measures is not satisfactory [17].

Monitoring the geohazard is not a substitute for mitigation, but it does help pipeline operators understand the performance of existing mitigation techniques and to make informed decisions regarding the types of mitigation measures needed and the urgency of their installation.

CONCLUSIONS

FOS-based design of steep slopes traversed by pipelines, which are long linear structures, is not practical. An evaluation that would render reliable values of FOS would require a degree of characterization that cannot be achieved given that subsurface conditions along the slope prior to pipeline construction are spatially and seasonably variable and can be affected by weathering over the design life of the project. Therefore, the computation of a FOS is highly uncertain and may not reflect the likelihood of initiation of instability.

A performance-based criterion is therefore considered to be a more appropriate approach for the design and construction of pipelines along natural steep slopes. Management of geohazard-driven risk on new pipeline construction can generally be achieved through the restoration of slopes on pipeline ROW to achieve a level of performance comparable to the pre-construction condition. Through the application of mitigation measures including surface water diversions and subsurface drains, along with other stabilization measures, such as soil compaction and re-establishing vegetation, the soil strength losses attributed to ground disturbance can generally be offset.

Because even natural steep slopes sometimes become unstable following intense precipitation and ground saturation, there should not be an expectation that the stability of relatively thin soil veneers above the crown of the pipeline on steep slopes should be made more stable following pipeline construction, particularly given the constraints imposed on most pipeline construction that pre-existing contours of steep slopes on the pipeline ROW be maintained.

Instead of treating all slope instability in the same manner, emphasis needs to be placed on the overall risk to the integrity of the pipeline, and the utilization of mitigation measures, such a burying the pipe in stable soil or rock formations and maintaining the stability of the trench backfill, which are keys to protecting the integrity of pipelines from landslide on steep slopes.

REFERENCES

[1] Leroueil, S. *Natural Slopes and Cuts: Movement and Failure Mechanisms,* Geotechnique, Vol. 51, No. 3, 197-243, 2001.

[2] Highland, L.M., and Bobrowsky, P., *The Landslide Handbook — A Guide to Understanding Landslides*, Reston, Virginia, U.S. Geological Survey Circular 1325, 2008.

[3] Stark, T.D., and Hussain, M. *Empirical Correlations: Drained Shear Strength for Slope Stability Analyses,* Journal of Geotechnical and Geoenvironmental Engineering, Vol. 139, No. 6, 853-862, June 2013.

[4] Stark, T.D., Choi, H., and McCone, S. *Drained Shear Strength Parameters for Analysis of Landslides,* Journal of Geotechnical and Geoenvironmental Engineering, Vol. 131, No. 5, 575-588, May 2005.

[5] Gamez, J.A. & Stark, T.D., *Fully Softened Shear Strength at Low Stresses for Levee and Embankment Design,* Journal of Geotechnical and Environmental Engineering, Vol. 140, No. 9, 2014.

Copyright © 2019 by ASME

[6] USDA NRCS, *Web Soil Survey,* https://websoilsurvey.sc.egov.usda.gov, 2019.

[7] Anderson, S.A., and Sitar, N., *Analysis of Rainfall-Induced Debris Flows,* Journal of Geotechnical Engineering, Vol. 121, No. 7, 544-552, 1995.

[8] Groeger, L., *Pipeline Safety Tracker,* ProPublica, https://projects.propublica.org/pipelines/, 2012.

[9] Porter, M., Ferris, G., Leir, M., Leach, M., and Haderspock, M., *Updated Estimates of Frequencies of Pipeline Failures Caused by Geohazards,* 2016 11th International Pipeline Conference, Paper No. IPC2016-64085, 2016.

[10] Wenping W., Xia F., *Risk Acceptance Criteria of Oil and Gas Pipelines Projects in Construction Period*, In: Zhang Y. (eds) Future Communication, Computing, Control and Management, Lecture Notes in Electrical Engineering, Vol 141. Springer, Berlin, Heidelberg, 2012.

[11] Dawotola, A., van Gelder, P.H.A.J.M., Vrijling, J.K., *Design for Acceptable Risk in Transportation Pipelines,* International Journal of Risk Assessment and Management, Vol. 16, Nos. 1/2/3, 2012.

[12] U.S. Army Corps of Engineers, *Slope Stability,* EM 1110-2-1902, 31 October 2003.

[13] INGAA Foundation. *Mitigation of Land Movement in Steep and Rugged Terrain for Pipeline Projects,* Final Report No. 2015-03, April 2016.

[14] Duncan, J. M. and Wright, S. G. *Soil Strength and Slope Stability,* John Wiley & Sons, 1st Ed., 2005.

[15] Federal Energy Regulatory Commission, *Upland Erosion Control, Revegetation, and Maintenance Plan,* May 2013.

[16] Leir, M., Baumgard, A., and Johnson, K. *A Staged Approach for Managing Terrain and Geohazards on New Pipeline Projects*, 2016 11th International Pipeline Conference, Paper No. IPC2016-64175, 2016.

[17] Pipeline Research Council International, Inc., *Guidelines for Management of Geohazards Affecting the Engineering and Construction of New Oil and Natural Gas Pipelines.* Catalog No. PR-616-164506-R01, 2 May 2018.

[18] Pipeline Research Council International, Inc., *Guidelines for Constructing Natural Gas and Liquid Hydrocarbon Pipelines Through Areas Prone to Landslides and Subsidence Hazards,* January 2009.

Copyright © 2019 by ASME

AIM-PIMG2019-1036

Chapter 20

OPERATIONAL ALTERNATIVE TO DESIGN TREATMENT FOR SOME GROUND SUBSIDENCE AND SURFACE FAULTING GEOHAZARDS

Jeffrey R Keaton
Wood
Los Angeles, CA, USA

ABSTRACT

Some geohazards pose risk to buried pipelines that can be managed with pipeline operations instead of requiring design treatments. The range of geohazard management alternatives consists of 1) modifying the hazard, 2) modifying what is at risk, 3) modifying operations or procedures, 4) compensating for potential losses, and 5) avoiding the hazard. Ground subsidence and surface fault rupture are two geohazards that may have the potential to be managed with operational instead of constructed approaches. Pipelines extending across areas susceptible to ground subsidence caused by groundwater pumping or underground mining could experience strain at levels that could adversely affect pressure integrity or the maximum allowable operating pressure (MAOP). Geologic settings with a potential for ground subsidence can be identified, and methods are available for forecasting the rate of subsidence, the potential for earth fissures to form, and the strain capacity of the pipeline. Monitoring with aerial and ground-based surveillance is routine for pipelines, and also includes occasional inline inspections. An element in a pipeline emergency action plan triggered by a threshold amount of subsidence being reached could be to excavate the affected reach of a pipeline to relieve accumulated strain, check the corrosion protection coating, and rebury. For new projects, engineering economics could be applied to weigh the cost of an initial design treatment to the cost of future mitigation of an affected reach and the likelihood that it would be needed. If initial design treatment is selected, the emergency action plan would still include an element for monitoring and mitigation, if needed.

Treatment of surface faulting geohazards may be regulated, particularly in proximity to population or environmentally sensitive resources. In areas remote from population or high consequence areas, certain categories of faults may be suitable for an operational approach to geohazard management. Potentially active faults may pose little risk of loss of pressure integrity. A possible element in a pipeline emergency action plan triggered by the occurrence of earthquakes of magnitude 5.5 and larger would be to use online resources to determine the epicenter location and proximity to the pipeline, assuming that operations detected no loss of pressure. Certain magnitude and distance combinations would result in a ground inspection to check for ground cracks. Detection of cracks would result in an appropriate response, such as excavation of the pipe for relief of strain, inspection, repair, and reburial.

INTRODUCTION

Transmission pipelines traverse landscapes that were formed by natural processes, including some that may be hazardous to infrastructure (i.e., geohazards). Some geohazards can be relatively discrete, such as a landslide-prone slope or a cut bank at a meander bend on a stream channel, whereas other geohazards are areal or linear, such as broad alluvial valleys with potential for groundwater subsidence or an active earthquake fault. Alignments for new pipelines may be adjusted to avoid discrete geohazards, but most areal and linear geohazards cannot be avoided by linear projects. This paper uses ground subsidence and surface faulting as examples of geohazards that cannot be avoided and describes management alternatives that can be applied as part of pipeline operations.

GEOHAZARD MANAGEMENT ALTERNATIVES

Five alternatives tend to encompass the range available for managing infrastructure exposed to geohazards. This assumes that the geohazards have been recognized and characterized, to at least some extent. The management alternatives are 1) modify the hazard, 2) modify what is at risk of damage, 3) modify operations or procedures, 4) compensate for potential losses arising from inability to operate and cost to repair damage, and 5) avoid the hazard by initial location of new infrastructure or relocation of existing infrastructure.

Copyright © 2019 by ASME

In most cases, modifying hazardous processes is impossible or impractical. Landslide-prone slopes can be stabilized, but active tectonic faults cannot. Stabilizing a landslide can be costly and efforts may not truly stabilize the slope, even if stabilization measures could be implemented outside the limits of a right-of-way. Depending on the nature of the unstable slope and anticipated deformations, pipe with extra wall thickness or geofabric wrap might be used to make the pipeline better able to withstand some ground movement.

Even with heavy-wall pipe, operation of the pipeline on the unstable slope would likely involve monitoring the amount and rate of slope movement and having an emergency action plan to excavate the pipe to relieve strain, and rebury it if a triggering strain or deformation threshold were reached. Some sort of insurance might be acquired to compensate for lost revenue or possibly environmental damage, but that would be in addition to other management alternatives. An alternative alignment could be selected for a new pipeline or a horizontal directional drill path could be considered to avoid the unstable slope.

Geohazard management alternatives for slow processes or processes that have exceptionally low probability of occurring, depending on the severity of potential consequences to the facility and regulatory requirements, can be operation of the system with a specific element in the facility's emergency action plan. The emergency action plan element must be developed with knowledge of the nature of the geohazard process – basically, an engineering geologic model. Routine patrols of pipeline alignments can be augmented with specific observations at managed locations that are documented and reviewed for change. Some inspections might be periodic or episodic with documentation to permit change detection. Elements in the emergency action plan would be implemented if and when a threshold condition was reached. Potential subsidence induced by groundwater pumping and potential deformation along faults are geohazards discussed in this paper.

POTENTIAL GROUND SUBSIDENCE

A new pipeline was planned in an arid region that had documented ground subsidence caused by groundwater pumping. The alignment did not cross areas of known subsidence, but it did cross alluvial basins with groundwater resources. A regulatory agency, in anticipation of groundwater pumping in proximity to the pipeline to support population growth and possibly for power plant cooling water, wanted the geohazard considered specifically in pipeline design. Standard line pipe parameters were NPS36, API 5L Grade X70 steel, 0.429 in. (10.9 mm) wall thickness, and 1,200 psig (8.27 MPa).

Finite element model analyses were used to determine strain in standard line pipe across a hypothetical symmetrical subsidence zone that was approximately 15, 60, and 150 m wide with the subsidence occurring abruptly, as a graben between a pair of vertical normal faults (Fig. 1), with displacements incremented up to 3 m. This analysis revealed that calculated strain was nearly identical for subsidence-zone widths of 60 and 150 m (Fig. 2). Strain equal to the threshold of pressure integrity (1.5%) was reached at a modeled vertical subsidence amount greater than 1.5 m.

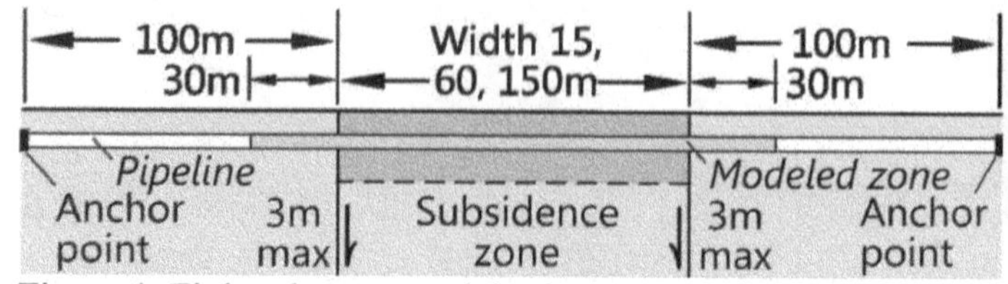

Figure 1. Finite element model schematic of a profile view of a pipeline crossing a hypothetical subsidence zone.

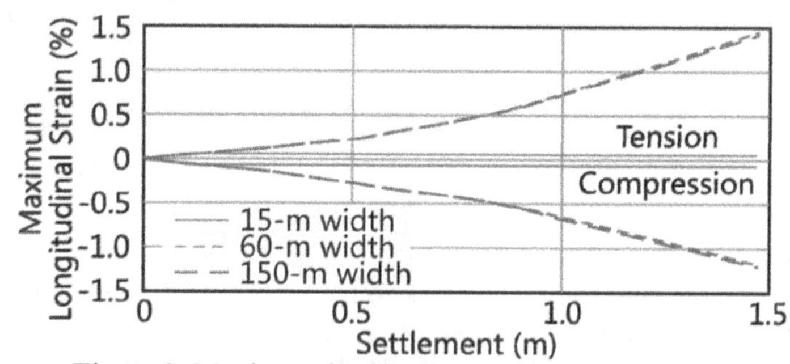

Figure 2. Maximum longitudinal strain in tension and compression for the schematic profile in Figure 1.

The region was subjected to desktop research to determine the largest amount of documented subsidence. Published reports indicated the maximum documented subsidence in the region was 2.0 m, which occurred between 1930 and 1992 at a location more than 30 km from the alignment (Fig. 3; [1]). The average annual subsidence rate was 0.032 m/yr. However, the subsidence history included the period 1962 to 1965 during which the maximum annual subsidence rate was 0.1 m/yr (Table 9 in [2]). The maximum subsidence annual rate applied to the allowable strain level indicated that subsidence would not affect the pipeline for at least 15 years after it went into operation.

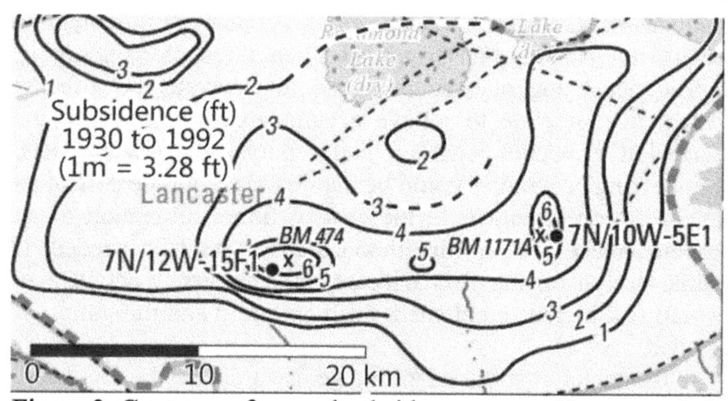

Figure 3. Contours of ground subsidence over a 62-year period. Extracted from [1].

The regulatory agency was satisfied with the results of engineering geologic model regarding potential subsidence.

Copyright © 2019 by ASME

The pipeline operations requirement included a subsidence analysis of the alignment across the alluvial basin 15 years after it went into service. The routine alignment patrols would look for signs of ground cracks along the alignment across the potential subsidence area, with particular attention being paid to the margins of the alluvial basins.

The 15-year subsidence analysis was performed in 2018. Satellite radar interferometry, developed in the 1990s, was applied to ground subsidence in this area [3], bringing documented subsidence in the region up to 2009, and identifying locations of subsidence on dry lake beds that were affected by groundwater pumping. The maximum rate of subsidence determined on dry lake beds no closer than 10 km from the pipeline alignment increased from 0.004 m/yr between 1993 and 1999 to 0.011 m/yr between 2004 and 2009 [3]. Satellite radar interferometry showed that small amounts of subsidence occurred along parts of the pipeline alignment. The alluvial basins still have the potential for subsidence in response to groundwater pumping. Pipeline operations continue to include inspections; the inspectors continue to look for signs of ground cracks. However, no additional subsidence analyses were recommended to demonstrate that the pipeline remained within allowable strain unless ground cracks or other signs of subsidence were noticed along or in close proximity to the pipeline alignment.

POTENTIAL DEFORMATION ALONG FAULTS

The alignment of a new pipeline proposed across the Basin and Range province in the western United States crossed numerous active and potentially active faults. Almost all of the alignment was in Department of Transportation Population Class 1 (49 CFR 192.5) with fewer than 10 buildings intended for human occupancy within 200 m of the centerline along any 1600-m long section; most of the alignment was remote from all buildings. A philosophy for treatment of natural gas pipelines at fault crossings is based on recency and recurrence of surface rupture on active faults and proximity to population [4] and pipeline design guidelines and details are available [5, 6]. Mitigation options consist of design treatment or operational treatment, depending on proximity to population and high-consequence areas (HCAs) and degree of fault activity as an index of fault hazard. Faults that have low to moderate hazard and are remote from population and HCAs may need no treatment to mitigate an already low risk of damage to the pipeline. The decision to mitigate low to moderate hazard faults in remote locations might be made by the pipeline owner based on economic and other considerations, such as reputation.

Design treatment of pipelines at fault crossings ranges from prescriptive to analytical. Prescriptive options are those that do not have analytical verification, and include avoiding transitions in pipeline wall thickness within one hundred meters of fault traces. Analytical options are those that have been verified by engineering calculations, including finite-element analyses [5, 6]. The finite-element method tools that are available allow calculation of stresses and strains in the pipe material as a result of discrete displacement across a fault plane or zone. The conditions of the pipeline are modeled specifically, including steel grade, wall thickness, internal pressure, pipeline geometry (bends and tees), weight coating, ditch geometry, and soil parameters for natural and backfill materials. Iterative analyses can be performed with adjustments in selected parameters to develop design details that allow fault rupture to occur without overstressing the pipe. The capacity of the pipeline to withstand fault rupture (displacement capacity) is a function of the pipeline diameter and wall-thickness, and taken to be strain at the threshold of pressure integrity, which is 4% strain in tension and strain associated with buckling in compression.

Operational treatment of pipelines at fault crossings consists of an emergency action plan that includes specific information about the faults, namely location, orientation with respect to the pipeline, and the direction of anticipated displacement. The emergency action plan is triggered by the occurrence of an earthquake that meets certain criteria in terms of magnitude and distance from the pipeline alignment (Tab. 1), or loss of pressure in the pipeline system. A maximum magnitude for crustal earthquakes in the seismically active Basin and Range Province of the United States that were not accompanied by surface rupture on fault traces was on the order of moment magnitude (Mw) 6.5 [7]; therefore, the minimum earthquake magnitude that triggers immediate inspection is a full magnitude interval lower (Tab. 1).

Table 1. Schedule of earthquake magnitude and distance that triggers an immediate inspection of the alignment (after [4]).

Earthquake Magnitude (Mw)	Distance of Epicenter from Pipeline Alignment (km)
≥ 5.5	≤ 10
≥ 6.0	≤ 20
≥ 6.5	≤ 30
≥ 7.0	≤ 40

Occurrence of earthquakes without loss of pressure triggers field inspection of the pipeline alignment in the vicinity of the earthquake epicenter; loss of pressure results in immediate emergency response. Sufficient automated scientific information is generated immediately following earthquakes in the United States that it may be relied upon for the operational treatment in the emergency action plan. The U.S. Geological Survey maintains an online Earthquake Notification Service (https://earthquake.usgs.gov/ens/) that will send automated information to subscribers at no cost. The information includes preliminary determination of earthquake magnitude and geographic epicenter location, and can be set for a rectangular area of interest. Commonly available resources, such as Google Earth, can be used for plotting earthquake epicenters with reference to pipeline alignments. Immediate inspection of the pipeline alignment in proximity to earthquakes of magnitude 5.5 and greater that occur within 40 km of the pipeline alignment could be done according to the schedule in Tab. 1.

Copyright © 2019 by ASME

The earthquake magnitudes in Tab. 1 could be produced by shallow crustal faults or deeper subduction zone or interpolate faults. Shallow crustal seismic environments are where consideration of operational treatment of minor fault rupture hazards is reasonable; major active faults pose high enough hazard that pipelines will be designed specifically for anticipated displacements.

Engineering analyses can be performed in a parametric manner to assess stresses and strains in pipelines crossing potentially active fault traces to inform those making decisions about mitigation options. Many tectonic faults in the Basin and Range Province have geologic features that suggest low degree of activity (Fig. 4). These faults may have geomorphic expression suggesting that the most recent displacement occurred more than several hundred thousand years ago, or even more than one million years ago. Geologic evaluations to determine the degree of fault activity typically are time consuming and expensive. Pipeline alignments typically cross fault traces at a single location on any one fault, which may not have suitable geologic details to permit assessment of degree of activity. Geologic estimates can be made of fault displacement details that can be used as input into finite-element analyses of several generic fault crossing geometries. Twelve such generic fault crossing geometries were developed for a pipeline in the Basin and Range Province. Standard line pipe material properties and operating pressure were used in finite-element analyses that demonstrated that strain would not exceed the threshold for pressure integrity by the unlikely occurrence of displacement on the modeled faults. These results supported the conclusion that the operational treatment was sensible.

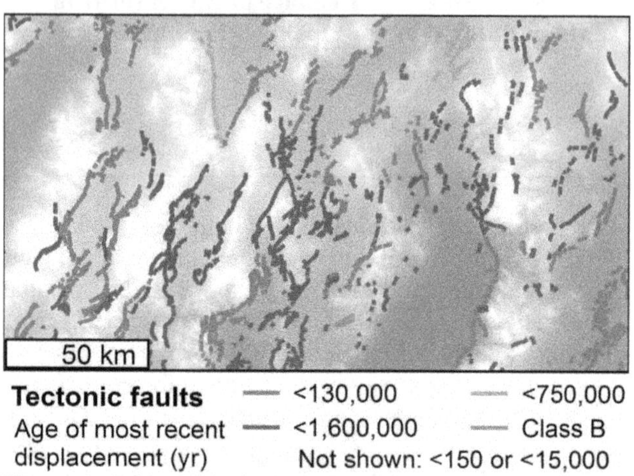

Figure 4. Selected faults in the Basin and Range Province. Source: [8] fault database; hillshade base map. Class B applies to faults that have geologic evidence too strong to confidently assign to non-tectonic or pre-Quaternary, but not strong enough to assign to a Quaternary age of most recent displacement.

Projected traces of growth faults on the Coastal Plain of the Gulf of Mexico (Fig. 5) would be crossed by the alignment of a proposed new pipeline. Displacement along growth faults has caused damage in Houston [9]. These are geologically active faults, but considered incapable of generating earthquakes of engineering significance. The displacement occurs gradually and progressively, cracking concrete pavement and straining buried pipelines. The proposed pipeline crossed the projected growth faults in agricultural areas where no topographic expression of faulting was detectable in GIS analysis of 10-m digital elevation data (DEM), and no dense vegetation was present that might mask small-scale topographic features.

No field evaluations of the twelve projected traces crossed by the alignment were undertaken. It would be reasonable for geologic inspection of the construction trench to be performed where the projected traces would be crossed. The operational treatment of these projected fault crossings is similar to the treatment for ground subsidence. If geologic observation were performed and fault traces identified, then the locations of the crossings could be refined for the emergency action plan. Annual patrols would know where the crossings were located and make specific notes for comparison on subsequent patrols. Notes about the nature of observed ground cracks, if any, and about the soil moisture conditions at the time of the patrol would be important for the Gulf Coastal Plain area because most soils are clayey and have high shrink-swell potential.

In areas where operational alternatives are reasonable, but the ground is not readily visible, other monitoring options for detecting incipient deformation could be considered. These options include geodetic survey monuments at accessible locations on the ground, strain gauges fixed to the buried pipe with leads to the ground surface, and quantitative remote sensing, such as airborne laser swath mapping or satellite radar interferometry. At some point, the total cost of the more elaborate operational alternatives may exceed the initial cost of a design treatment. An engineering economics evaluation may be warranted for conditions with decisions that are not clear.

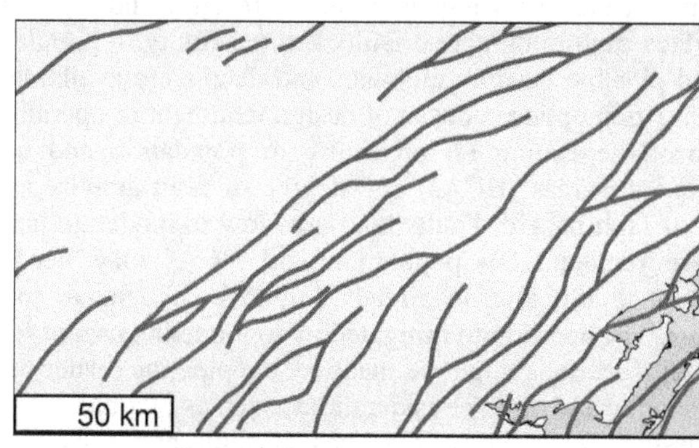

Figure 5. Normal faults in the Gulf Coastal Plain. Gulf of Mexico in lower right corner. Source: [10].

CONCLUSIONS

Some geohazards, particularly those that develop progressively and gradually or those with exceptionally low

Copyright © 2019 by ASME

probability of occurring, can be managed with pipeline operations instead of design treatments if they are present along alignments in areas that are remote from population and high consequence areas. Even with effective design treatments, some geohazards also deserve operational treatments to monitor conditions and enable change detection for timely implementation of mitigation measures. Operational alternatives require knowledge of the geologic details so that operations activities are focused and remain relevant. The operational treatment element of the emergency action plan must be developed with knowledge of the nature of the geohazard process – in essence, an engineering geologic model is necessary for the pipe-geohazard system. As with all models, the engineering geologic model of the hazardous processes must be reviewed and updated as new information is acquired or existing information is revalidated, or as new concepts about hazardous geologic processes emerge. The refined or revalidated engineering geologic model is needed to keep decisions about pipeline operational alternatives current.

REFERENCES

[1] Siade, A.J., Nishikawa, T., Rewis, D.L., Martin, P., and Phillips, S.P., 2014, Groundwater-flow and land-subsidence model of Antelope Valley, California: U.S. Geological Survey, Scientific Investigations Report 2014–5166, 136 p., http://dx.doi.org/10.3133/sir20145166.

[2] Ikehara, M.E., and Phillips, S.P., 1994, Determination of Land Subsidence Related to Ground Water-Level Declines Using Global Positioning System and Leveling Surveys in Antelope Valley, Los Angeles and Kern Counties, California, 1992: U.S. Geological Survey Water-Resources Investigations Report 94-4184, 101 p. https://pubs.usgs.gov/wri/1994/4184/report.pdf

[3] Brandt, J. and Sneed, M., 2017, Land Subsidence in the Southwestern Mojave Desert, California, 1992–2009: U.S. Geological Survey Fact Sheet 2017-3053, unpaginated, https://doi.org/10.3133/fs20173053.

[4] Keaton, J.R., Robison, R.M., Beckwith, G.H., and Slemmons, D.B., 1991, Philosophy for treatment of high-pressure natural gas pipelines at active fault crossings, in Cassaro, M.A., ed., Lifeline Earthquake Engineering: Technical Council on Lifeline Earthquake Engineering Monograph No. 4, New York, American Society of Civil Engineers, p. 898-906.

[5] Honegger, D.G., and Nyman, D.J., 2004, Guidelines for the seismic design and assessment of natural gas and liquid hydrocarbon pipelines: Pipeline Research Council International PRCI PR-268-9823.

[6] O'Rourke, M.J., and Liu, X., 2012, Seismic design of buried and offshore pipelines: MCEER, University of Buffalo, Buffalo, NY, MCEER-12-MN04, 380 p.

[7] dePolo, C.M., 1984, The maximum background earthquake for the Basin and Range Province, western United States: Bulletin of the Seismological Society of America, v. 84, n. 2., p. 466-472.

[8] USGS, 2016, Quaternary Fault and Fold Database of the United States: US Geological Survey https://earthquake.usgs.gov/hazards/qfaults/.

[9] Engelkemeir, R.M., and Khan, S.D., 2008, Lidar mapping of faults in Houston, Texas, USA: Geosphere, February 2008; v. 4; no. 1; p. 170–182; doi: 10.1130/GES00096.1.

[10] US NRC, 2015, Final Safety Evaluations for South Texas Project, Units 3 and 4 Application, Chapter 2, Site Characteristics, Figure 2.5S.1-2, page 2-278. https://www.nrc.gov/reactors/new-reactors/col/south-texas-project/documents/ser-final.html.

Copyright © 2019 by ASME

SECTION 6
STRAIN CAPACITY PREDICTIONS

INTRODUCTION

By Guiyi Wu
(guiyi.wu@twi.co.uk)

Onshore pipelines installed and operated in challenging environments, where geohazards such as ground movements, landslides and earthquake are likely to occur, could be subjected to large plastic strains. Although a number of fracture mechanics-based procedures, e.g. BS 7910 and API 579, are available for engineering critical assessment (ECA) of pipeline girth welds, most of these methods are often limited to primarily elastic response and stress-based assessments and are therefore not directly applicable to cases where the loading generates large amounts of plastic deformation. Hence, strain-based design and assessment (SBDA) technology is needed to ensure the integrity for pipelines against the geohazards. The papers presented in this Chapter bring the recent development of SBDA technology.

SBDA requires appropriate definition and determination of limit state, crack driving force and material properties (e.g. yield and tensile strength and fracture toughness). Treating the above three factors differently will lead to distinct SBDA methods. Finite element analysis (FEA) is normally carried out for industrial projects to determine crack driving force which is dependent on the pipe dimensions, weld geometry information, flaw size and position, tensile properties, applied strain level and internal pressure. The idea of developing SBDA technology is to simplify the design and assessment procedures by implementing analytical solutions without the need of performing FEA. It is appreciated that several strain-based methods have been developed in the last decade, most of which are intended for assessing pipeline girth welds containing flaws which are a subset (but an important one) of strain-based design and assessment in general. It is also acknowledged that various code and standardization bodies form strain-based design subcommittees to work on the incorporation of suitable methodologies into the respective documents. The papers presented here will contribute to the standardization of SBDA technology.

To assess fracture of pipelines subjected to plastic straining, proposals have been made to use BS 7910:2005 Level 2 failure assessment diagram (FAD) with adjustment to the cut-off line, L_{rmax}. In January 2006, DNV published a recommended practice DNV-RP-F108 "Fracture control for pipeline installation methods introducing cyclic plastic strain". This recommended practice followed BS 7910:2005 Level 2 assessment method but with some modifications. In Sections 3, 4 and Appendix A of DNVGL-RP-F108, advice on assessment of pipeline girth welds subjected to large strains is provided. For strains beyond 0.4%, an ECA is advised based on the BS 7910:2013 Option 2 failure assessment diagram (FAD), taking account of tearing and other minor adjustments.

Modifications to the present stress-based methods such as BS 7910 Option 2 FAD has been also proposed by other researchers by proposing correction factors, In addition, extension to the EPRI scheme and reference strain methods have been developed to determine the crack driving force.

Two major efforts have been undertaken in North America by Center for Reliable Energy Systems (CRES) and ExxonMobil to develop SBDA methods which enable the tensile strain capacity (TSC) to be evaluated. These methods are being presented by their respective papers in this chapter. In Summary, the existing SBDA methods can be classified into two major categories, i.e.

- C1: Crack driving force methods
- C2: Strain capacity methods

For a specific flaw size, C1 determines the crack driving force in terms of *J*-integral or CTOD, for example, DNVGL-RP-F108. The tangency method is normally used to determine the unstable point at which strain capacity is determined assuming that plastic collapse does not occur. The flaw size corresponding to the strain capacity is considered as the maximum allowable flaw size. C2 directly predicts the strain capacity without the need to determine crack driving force, for example, CRES and ExxonMobil. C2 methods are principally for design but they can be used for assessment to determine the maximum allowable flaw sizes. The paper by Wu, Wang, London and Pisarski (AIMPIMG2019-1013) presented compares crack driving force and strain capacity estimated using the existing SBDA methods with those obtained from full-scale pipe tests.

The paper by Wu, Wang, London and Pisarski (AIMPIMG2019-1013) provides detailed information on two large-scale tests containing girth welds with notches at the weld centerline subjected to large tensile strains with and without internal pressure, respectively. The strain capacity and the crack driving force obtained from the tests were compared with the existing strain-based methods which covers those described above. The test results show that lower strain capacity was obtained for the pipe sample under bi-axial loading. Most existing SBDA methods are conservative or over–conservative with regard to the test

Copyright © 2019 by ASME

results in terms of strain capacity. However, some methods give non-conservative crack driving force at specific strain levels.

The paper by Wang, Liu and Wang (AIMPIMG2019-1066) covers strain-based design of pipelines via explaining three key elements: (1) linepipe specifications/qualification, (2) girth welding procedure qualification and field girth welding practice, and (3) tensile strain models. Brief discussions were provided to the linepipe specifications and girth welding procedure qualification before presenting PRCI-CRES tensile strain models in details. The fundamental principles, structure, applicable range and limitations of PRCI-CRES tensile strain models were discussed. Finally, examples showing the application of PRCI-CRES tensile strain models on designing new pipelines and assessing vintage pipeline girth welds were provided.

The paper by Østby, Törnqvist and Bjerke (AIMPIMG2019-1015) presents a perspective on the developments in strain-based fracture assessment procedures mainly based on knowledge and results developed in the "Fracture Control Offshore Pipelines" JIPs. The basic principles for strain-based fracture assessment were described. Discussions were provided with regard to the key factors in strain-based fracture assessment, i.e. crack driving force, crack growth calculations, definition of critical conditions, role of stress-strain curve, mismatch, misalignment and bi-axial loading. This is followed by presenting the application of numerical modelling in strain-based fracture assessment. Finally, the authors proposed a reliability-based scheme which allows the users to consider the uncertainties in the strain-based fracture assessment.

The paper by Fairchild, Crapps, Panico, Cheng, Cook and Macia (AIMPIMG2019-1039) presents ExxonMobil's tensile strain capacity prediction model with a safety factor derived from more than 100 full-scale pipe tests. The safety factor was generated from a 95% upper confidence limit on the ratio of predicted-to-measured tensile strain capacity. The ExxonMobil's model was used to predict strain capacity for ten hypothetical pipeline design cases which covers a range of pipe geometries, pipe and weld physical properties and hi-lo weld misalignment. The effect of mismatch and misalignment on maximum allowable flaw sizes was clearly presented. Brief discussions on the use of the tensile strain capacity prediction technology were provided. A key information delivered from the paper is that the tensile strain capacity prediction model can be used to assist the design phase of a strain-based design pipeline project and to carry out strain-based fracture assessment during pipeline installation and operation where large strain is a concern.

The paper by Wang (AIMPIMG2019-1067) provides an overview of conceptual understanding and practical implementations of SBDA. The difference between stress- and strain-based design was explained. Key steps for strain-based design of new pipelines were introduced. It is important to evaluate strain demand and target strain capacity for designing new pipelines which are likely to be operated in areas with potential geohazards. Brief discussions were also provided on compressive strain design. Then, the author introduced key steps for carrying out strain-based assessment of in-service pipelines. It is suggested that careful consideration should be given to the nature of the loads over the entire service life of a pipeline. At the end of the paper, the author highlighted the gaps in the following areas for SBDA: (1) sufficiency of standards in ensuring good girth weld strain capacity, (2) consistent measure of strains, (3) interaction of strain capacity and strain demand, (4) flaw interaction rule, and (5) roles of large-scale testing.

Copyright © 2019 by ASME

AIM-PIMG2019-1013

Chapter 21

PIPE GIRTH WELDS UNDER PLASTIC STRAINING: FULL-SCALE TESTING AND STRAIN-BASED APPROACHES

Guiyi Wu*
TWI Ltd.
Granta Park
Great Abington, Cambridge, UK, CB21 6AL
Email. Guiyi.wu@twi.co.uk

Longjie Wang
TWI Ltd.
Granta Park
Great Abington, Cambridge, UK, CB21 6AL
Email: Longjie.wang@twi.co.uk

Tyler London
TWI Ltd.
Granta Park
Great Abington, Cambridge, UK, CB21 6AL
Email. Tyler.london@twi.co.uk

Henryk Pisarski
TWI Ltd.
Structural Integrity Consultant
Cambridge, UK
Email: H.pisarski@btinternet.com

ABSTRACT

Strain-based design and fracture assessment of pipelines is becoming increasingly important because large pipeline networks tend to be placed in challenging environments where they might undergo large strains due to ground movement, for example. The integrity of pipeline girth welds in such conditions should be assessed using strain-based design and fracture assessment approaches instead of traditional stress-based engineering critical assessment. This paper describes two large-scale pipe tests containing girth welds with notches at the weld centreline subject to large strains with and without internal pressure, respectively. The experimental results are used to investigate existing strain-based design and assessment methods. The results show that most existing methods are conservative or over-conservative with regard to the test results in terms of strain capacity. However, some methods give non-conservative crack driving force at specific strain levels. Recommendations for future work in order to reach a satisfactory approach are proposed.

INTRODUCTION

In the last decade, extensive efforts have been made by industry and researchers to develop strain-based design and assessment methods. This is due to the need for installing pipeline networks in challenging environments where the pipelines are likely to undergo plastic straining. The fracture mechanics based approaches in the existing structural integrity codes/standards, eg, BS 7910 [1], API 579 [2] and R6 [3], are primarily stress-based assessments and are therefore not directly applicable to the cases where the applied strain introduced by a displacement-controlled or strain-controlled loading is greater than the yield strain. In these circumstances, strain-based design and fracture assessment methods should be considered.

Although there are various strain-based design/assessment methods available in the public domain [4-18], there is not a unified strain-based design/assessment approach that is universally recognised by the oil and gas industries. This may be attributed to the different applications for which strain-based methods were developed. For example, some methods address the design stage whilst others may focus on assessing installation and others on operation. The outcomes of the existing methods can be separated into two categories: tensile strain capacity methods and crack driving force methods. In general, the strain-based design methods involve characterisation of strain demand and strain capacity but the strain-based assessment methods normally give the results of crack driving force at a specific strain level. Nevertheless, some methods can be used for pipeline design and installation as well as operation.

Pisarski [19] carried out a critical review on stress- and strain-based methods for assessing flaws in pipeline girth welds. Later, Wu and Wang [20] carried out a detailed review on strain-based methods on assessing flaws in pipelines and compared them with the experimental results from a full-scale parent pipe test subjected to large plastic straining. It was found that there were errors in the results using CRES and

*Corresponding author: Guiyi.wu@twi.co.uk

Copyright © 2019 by ASME

ExxonMobil methods in Figure 8 of paper [20] due to typos in the Python codes. The difference in tensile strain capacity (TSC) using CRES method is -60% instead of 28% in that figure and the difference in TSC using ExxonMobil method is -26% instead of -61%. These codes were upgraded and corrected in this paper which extends Wu and Wang's work [20] by comparing the existing strain-based solutions with the experimental results of two full-scale pipes containing flaws in the girth welds. First, the tests on full-scale girth welded pipes are described. This is followed by classification of the strain-based methods available in the public domain. Next, the large-scale test results are presented together with the predictions from the strain-based methods. Finally, some conclusions are drawn, and recommendations are proposed for future work.

NOMENCLATURE

$R_{P0.5}$	Yield strength (σ_y)
R_m	Tensile strength (σ_u)
a	Crack length
B	Single edge notched tension (SENT) specimen thickness, perpendicular to the width
W	SENT specimen width
δ	Crack tip opening displacement
a_o	Initial notch height
$2c$	Notch length
V_1	Crack mouth opening measured at a height of z_1, above the notch mouth
V_2	Crack mouth opening measured at a height of z_2, above the notch mouth
z_1, z_2	Heights of clip gauges
$\Delta\varepsilon\%$	Difference in strain capacity
$\varepsilon_{predicted}$	Predicted strain capacity
$\varepsilon_{experiment}$	Measured strain capacity

LARGE-SCALE EXPERIMENTS

Overview

A pair of tests was conducted on pipes containing girth welds. Surface notches were introduced along the weld centreline by electro-discharge machining (EDM) in the circumferential direction. One of the pipes, which contained a single notch (5.7mm deep and 47.6mm long), was plastically loaded in tension until failure. The second pipe contained two notches on opposite sides of the pipe (each was approximately 4.8mm deep and 49mm long). This was first internally pressurised to 620barg (62MPa) with water to produce a hoop stress of approximately 460MPa representing 90% of parent pipe yield strength ($R_{P0.5}$ = 512MPa). Subsequently, the pipe was axially loaded in tension until either through pipe wall tearing or a leak occurred from one of the notches. Each pipe was instrumented to enable recording of pressure, applied force, local strain, overall strain, and crack mouth opening at each of the notches. The results have been analysed to provide the relation between stress and strain, local strain and remote strain, crack tip opening displacement (CTOD), J-integral and strain capacity.

It may be noted that these full-scale pipe test form part of a larger test programme which included tests and analyses on plain pipe. The latter are described by Pisarski et al [21] and Wu and Wang [20].

Material properties: pipe material and girth welds

The pipe is seamless to API 5L PSL2 Grade X65 with an average outside diameter of 273.4mm and average wall thickness of 18.4mm. The girth weld was made into a J edge preparation from one side by pulsed Gas Metal Arc Welding (PGMAW), except for the weld cap, which was deposited by Flux-cored arc welding (FCAW). The welds were deposited at typical heat inputs of 1kJ/mm. The weld widths were approximate 8.7mm at the OD and 6.2mm at the ID. The two welded pipes selected for these full-scale tests were designated TWI 9 for which axial straining only was applied, and TWI 10 for which internal pressure plus axial straining was applied.

Tensile testing has shown that the actual average yield strength ($R_{P0.5}$) of parent material is 512MPa and tensile strength (R_m) 597MPa in the longitudinal direction. All weld metal tensile tests showed that the average tensile and yield strengths were 625 and 690MPa, respectively. Figure 1 shows the average stress-strain curves of parent and weld metals. It can be seen that the weld metal strength over-matching, relative to the parent pipe, was 22% and 16% with respect to the yield and tensile strengths. A Lüders plateau is present in both parent and weld metal curves. The stress-strain information was used later in the strain-based analysis methods.

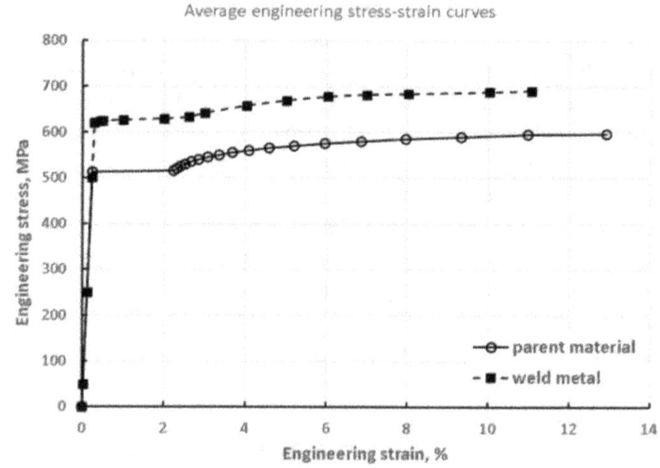

Figure 1 ENGINEERING STRESS VERSUS ENGINEERING STRAIN CURVES FOR WELD METAL AND PARENT PIPE

Fracture toughness was determined using SENT specimens using a multiple specimen method to derive CTOD and J resistance curves (R-curves) at room temperature. The tests were compliant with BS 8571. The specimen design employed

Copyright © 2019 by ASME

an over-square cross-section (2BxB, where B is 16mm), with an initial crack depth to width ratio of approximately 0.3 (a/W) and was notched from the pipe OD. The specimens were notched by electro discharge machining (EDM) and then fatigue pre cracked to achieve the final crack depth. Tests were conducted on the parent pipe and on specimens notched at the weld metal centreline. In addition, a set of tests was conducted using EDM notches only into the weld metal, ie without fatigue pre-cracking. The radius of the notch tip was 0.12mm and similar to the radius employed to make the notches in the full-scale pipe tests. The best fit J and CTOD R-curves obtained from these tests of weld metal are given below (units in mm, N/mm):

- With fatigue pre-cracking

$$\delta = 1.248\ \Delta a^{0.60} \quad (1)$$

$$J = 833 \Delta a^{0.62} \quad (2)$$

- EDM notch without fatigue pre-cracking (0.12mm tip radius)

$$\delta = 1.303\ \Delta a^{0.549} \quad (3)$$

$$J = 857 \Delta a^{0.565} \quad (4)$$

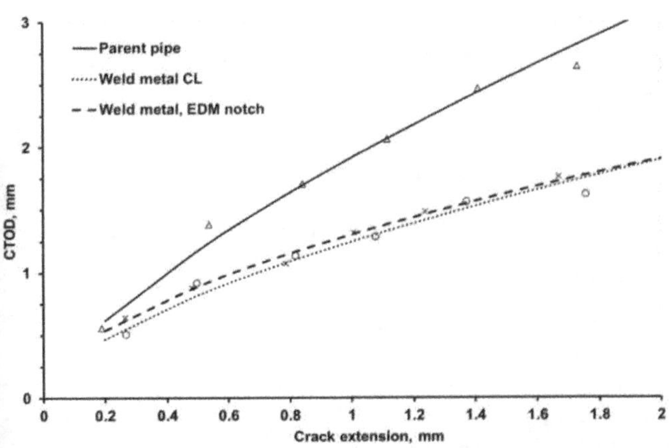

Figure 2 CTOD RESISTANCE CURVES FOR PARENT PIPE AND WELD METAL (FATIGUE PRECRACKED AND EDM NOTCH) OBTAINED FROM *2BxB* SENT SPECIMENS WITH INITIAL *a/W*=0.3

As illustrated in Figure 2, the two weld metal R-curves are almost identical, indicating that, in this case, the SENT R curve is insensitive to crack tip radius. Furthermore, this means that it would be acceptable to compare the SENT R-curve obtained from fatigue precracked specimens with the R-curve obtained from the pipe tests themselves containing EDM notches. The SENT R-curve obtained from fatigue precracked specimens was used later in the strain-based analysis methods to take into account the effect of tearing and to determine the strain capacity at the unstable point using the tangency approach.

Manufacture and preparation of pipe test specimens

For the purposes of testing, the pipes were cut to a length of 2000mm. The girth weld was located in the middle of the pipe length and the weld cap at the designated notch location was ground flush with the original pipe surface and etched to reveal the weld. This procedure enabled the notch to be located along the weld centreline and also facilitated fitting of the knife edges close to the notch mouth, for subsequent instrumentation with clip gauges, as illustrated in Figure 3. The first cardinal point (12 o'clock) was positioned at the weld start-stop position and a single EDM notch located at 4 o'clock in pipe test TWI 9. In the second test, on TWI 10, two notches were introduced at 4:30 and 10:30 o'clock. In both tests the notches avoided the start-stop position at 12 o'clock.

EDM was conducted in two stages, using a thicker blade to start. The final notch size was achieved with a nominally 0.2mm thick blade to provide a notch tip radius similar to that achieved in the SENT specimens. The notch was of a canoe shape and was essentially of constant depth with semi-circular ends. Table 1 provides details of the actual crack measured after completing all the tests. The initial notch depths were chosen to avoid testing the FCAW capping weld (at the deepest part of the crack), see Figure 3. This was to ensure that the region sampled by the notch tips in the pipe and SENT tests was the same.

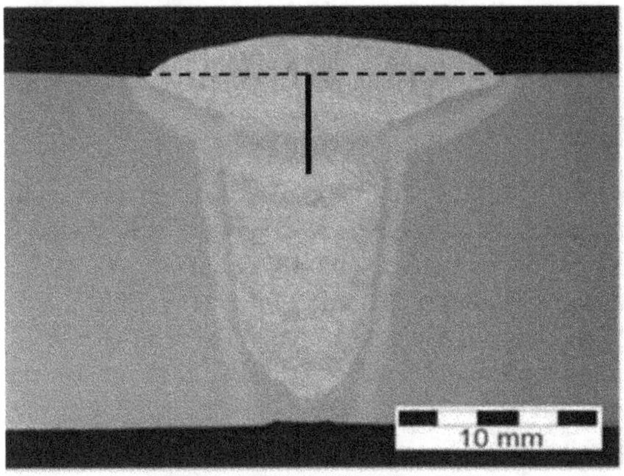

Figure 3 NOTCH POSITION EMPLOYED IN PIPE TESTS TWI 09 AND 10 (WELD OVERFILL MACHINED-OFF PRIOR TO EDM NOTCHING)

Table 1 NOTCH SIZES IN FULL-SCALE PIPE TESTS

Pipe test	Notch position	Clock position	Notch size* a x 2c, mm	Final size ** a x 2c, mm	Δa, mm
TWI 9 (Axial loading)	Weld centre line	4:00	5.74 x 47.6	18.4 x 52	12.7
TWI 10 (Internal pressure and axial loading)	Weld centre line	10:30 (notch 1) 4:30 (notch 2)	4.84 x 49.63 4.81 x 49.79	7.75 x 54.42 12.69 x 63.93	2.91 7.88

Notes:
*Notch size measured after test. The depth measurement (a) represents the average of the inner five of nine measurements made along the notch length (2c).
**At end of test.

Copyright © 2019 by ASME

After inserting the notches, flanges were welded to the ends of the pipes to enable them to be bolted to the face plates of the tension test rig. The pipe designated for internal pressurisation had thin steel sheets welded to the ends to retain pressure. Internal pressurisation was achieved by drilling a hole through the flange weld and into the pipe.

Steel shims were spot welded to the pipe surface on either side of the notch. A pair of knife edges was then bolted to each shim. This arrangement enabled mounting of a pair of clip gauges so that near crack mouth opening could be measured and CTOD could be inferred by extrapolating the pair of opening readings (see Equation 5). The internal pressure and axial force applied to the pipe were also recorded using appropriate transducers.

$$CTOD = V_1 - \left(\frac{a_0 + z_1}{z_2 - z_1}\right)(V_2 - V_1) \qquad (5)$$

Where

a_0 = initial notch height (average of inner five of nine measurements). This term is set at zero when estimating CMOD;

V_1, V_2 = crack mouth opening measured at a height of z_1, z_2, respectively, above the notch mouth;

z_1, z_2 = heights of knife edges (plus shim).

Testing of pipe specimens

The pipes were instrumented with strain gauges (5mm gauge length) located between the notches and on the same plane as the notches. Additional strain gauges were located remote from any influence that the notch may have on the strain field at a distance of 300mm either side of the notch. Strain was also measured over a gauge length of 700mm using a pair of displacement transducers which did not straddle the notches, on opposite sides of the pipe. The gauge plans for each test are shown in Figures 4 and 5.

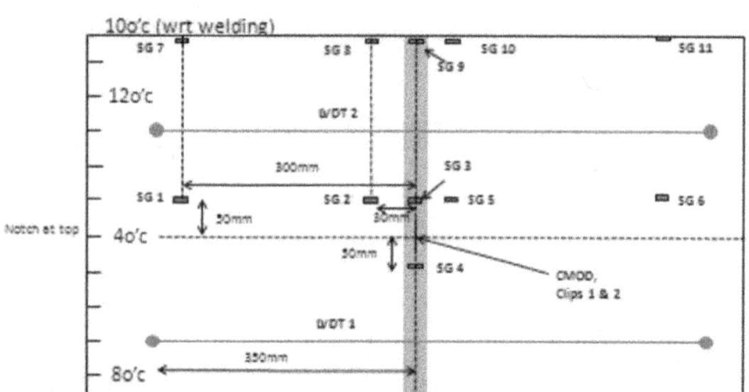

Figure 4 GAUGE PLAN FOR PIPE TEST TWI 9 (OVERALL STRAIN IS MEASURED OVER A GAUGE LENGTH OF 700mm USING LVDT 1 AND 2)

The specimens were tested in a machine employing a pair of pistons, each capable of exerting 1000 tonne axial force, which reacted against the face plates to which the test pipes were bolted via the flange arrangement. The set-up for TWI 9 and TWI 10 pipe specimens can be seen in Figures 6 and 7, respectively. The set-up for the second test was located in TWI's dedicated 8m x 3m pressure test facility, which is designed to safely contain any release of content or fragmentation products (should that arise).

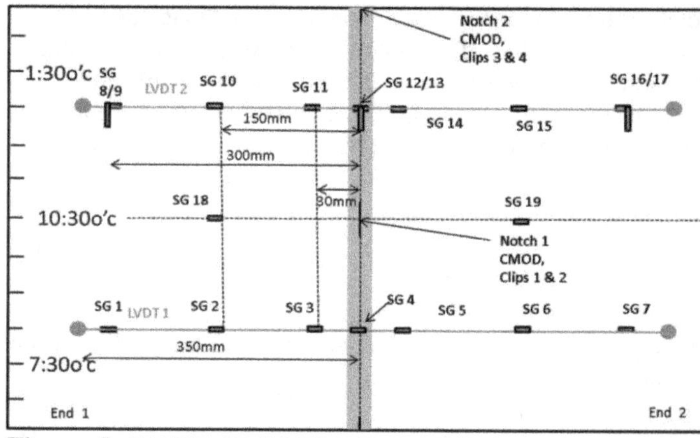

Figure 5 GAUGE PLAN FOR PIPE TEST TWI 10 (OVERALL STRAIN IS MEASURED OVER A GAUGE LENGTH OF 700mm USING LVDT 1 AND 2)

Figure 6 PIPE TEST TWI 9 INSTALLED IN TENSION MACHINE FOR TESTING WITH NOTCH IN GIRTH WELD AT 4 O'CLOCK

The two pipe tests were conducted at ambient temperature which was 20°C. When internal pressure was employed, the pipe was internally pressurised to 620barg before axial straining commenced. During straining, pipe volume changes required manual adjustments to be made to maintain constant pressure; typical variations were ±10barg of the nominal value. Axial straining was continued until "failure" which was by stable, ductile tearing through wall in TWI 9 and tearing followed by a leak resulting in depressurisation in TWI 10. However, for TWI 10, tearing was almost through wall and the maximum load was marginally affected by a small flaw near the root which caused the leak. The strain capacity was defined as the

Copyright © 2019 by ASME

average overall strain (measured over the 700mm gauge length) at the first maximum load.

Figure 7 PIPE TEST TWI 10 INSTALLED IN TENSION MACHINE FOR PRESSURISATION AND AXIAL STRAINING WITH NOTCHES IN GIRTH WELD AT 10.30 AND 4.30 O'CLOCK

STRAIN-BASED APPROACHES

The existing strain-based approaches available in the public domain [4-18] have been reviewed by Wu and Wang [20]. These strain-based approaches were divided into five groups shown below:

- Group 1: Modification to existing stress-based approaches;
- Group 2: Extension to the EPRI scheme for fully plastic J estimate;
- Group 3: Reference strain method;
- Group 4: Strain capacity method;
- Group 5: Others.

Table 2 shows the strain-based methods considered in each group in the paper. These methods were coded using Python and validated using the case examples in the publication. The geometric information of the full-scale pipe specimens, flaw size, tensile properties and R-curve presented above were used to calculate the crack driving force and strain capacity. For TWI 10, the actual size of the notch 2 (in girth weld at 4.30 o'clock) resulting in the failure of the test was used to calculate crack driving force and strain capacity.

Table 2 STRAIN-BASED ANALYSIS METHODS CONSIDERED IN EACH GROUP

Group	Methods
Group 1	DNV GL [4], Smith [5]
Group 2	Paredes and Ruggieri [6]
Group 3	Nourpanah and Taheri [7], Budden and Ainsworth [8], Jia et al. [9]
Group 4	CRES [13-14], ExxonMobil [15-17]
Group 5	SINTEF [18]

The welding residual stresses were not considered in the calculation of crack driving force and strain capacity as it was assumed that the welding residual stresses would be relaxed and become insignificant when the plastic strain was large. Since the direct outputs of the methods in groups 1-3 and 5 were the crack driving force, the tangency approach was adopted to calculate the strain at the unstable point which defined the strain capacity. The crack driving force was expressed in terms of J-integral for convenience. For full-scale test results, the measured CTOD was converted into J using the equation below [21]:

$$J = m\delta\sigma_y \tag{6}$$

where m was 1.3 for TWI 9 and was 0.9 for TWI 10. These values were derived from finite element simulation. σ_y is the yield strength ($R_{P0.5}$) of the weld metal.

RESULTS AND DISCUSSION

TWI 9

This test was conducted on the pipe under axial loading only. A maximum load was achieved at an overall strain of 5.07%. However, prior to reaching this strain the stroke limit on the pistons was reached, so the pipe was unloaded and packing added to the ends of the piston push rods and the specimen re-loaded. The test was continued until through wall crack extension by ductile tearing was clearly visible from the notch mouth at an overall strain of 5.46%. The J, which was converted from the CTOD, versus global strain plot is shown in Figure 8. The effects of the Lüders plateau are evident here so that after yield strain was reached, J remained constant until the end of the plateau and work hardening began at a strain of approximately 2.2% coupled with the initiation and growth of tearing (stable ductile crack extension) at higher strains. Thereafter, J increased almost exponentially as applied strain was increased. The J at the attainment of the maximum load plateau was 3279N/mm (corresponds to the CTOD of 4.03mm).

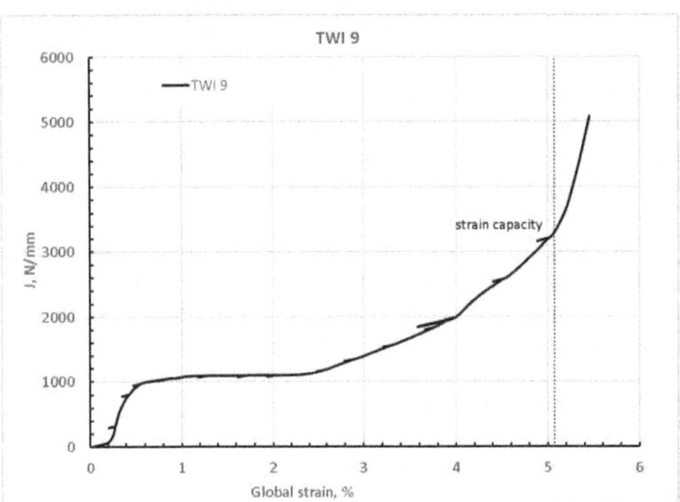

Figure 8 J VERSUS OVERALL STRAIN FOR PIPE TEST TWI 9

Copyright © 2019 by ASME

The fracture surface appearance, revealed after cooling the notched region and breaking open to cause cleavage, is shown in Figure 9. The grey region below the EDM notch is ductile crack extension (tearing) during axial loading. It is evident that tearing preferentially occurred through the wall thickness rather than along the surface. This meant that during the test the notch aspect ratio (height/surface length) changed from 0.12 at the start of the test to almost 0.35 at the end. The final test results in terms of strain and CTOD are summarized in Table 3.

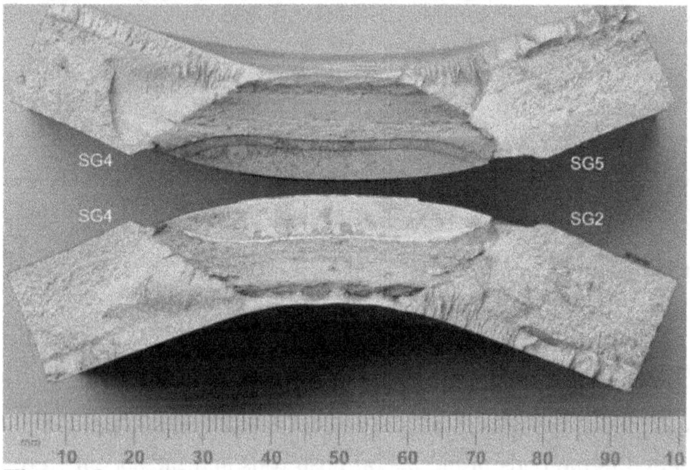

Figure 9 APPEARANCE OF EDM NOTCH AND TEARING IN PIPE TEST TWI 9

Table 3 SUMMARY OF RESULTS FROM FULL-SCALE PIPE TEST TWI 9 (STRAIN AND CTOD AT MAXIMUM "LOAD")

Pipe test	Number of notches	Notch location	Internal pressure	Strain at maximum load %	CTOD, mm	Initial notch size, mm		Type of failure
						a	2c	
TWI 9	1	Weld metal	N	5.07	4.03	5.74	47.6	Through wall tearing

The predicted relation between J and global strain using the strain-based methods presented in Table 2 is compared with the experimental data in Figure 10. ExxonMobil's method [15-17] is not compared in Figure 10 because it does not offer an output of crack driving force. Without consideration of tearing in the calculation, all the strain-based methods presented in Figure 10 predicted lower crack driving force than that calculated from the test results. One of the reasons is that the strain-based methods considered a static flaw but tearing occurred during the testing. Only the methods from DNV GL [4], Smith [5] and Budden and Ainsworth [8] were able to capture the effect of Lüders plateau on crack driving force, but the magnitude of crack driving force predicted using these three methods was far less than the experimental results when the global strain was greater than 0.5%. The other methods did not predict the effect of Lüders plateau on crack driving force. Figure 10 indicates that the strain-based methods under-predict the crack driving force of TWI 9 pipe test when the calculations are based on the initial flaw size (ie, ductile crack extension by tearing is not considered).

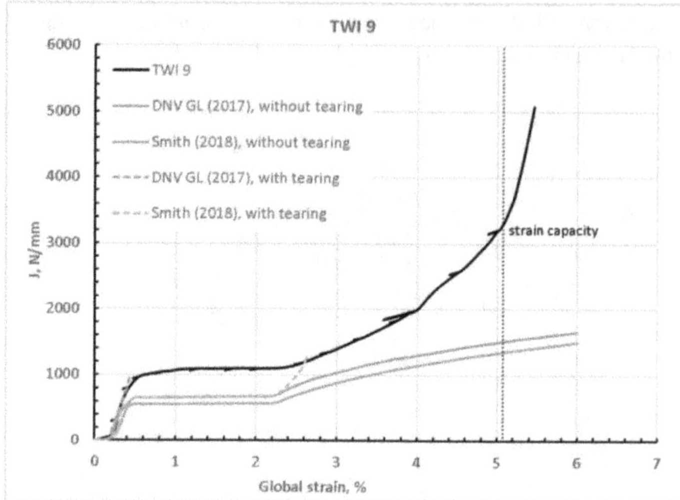

(a)

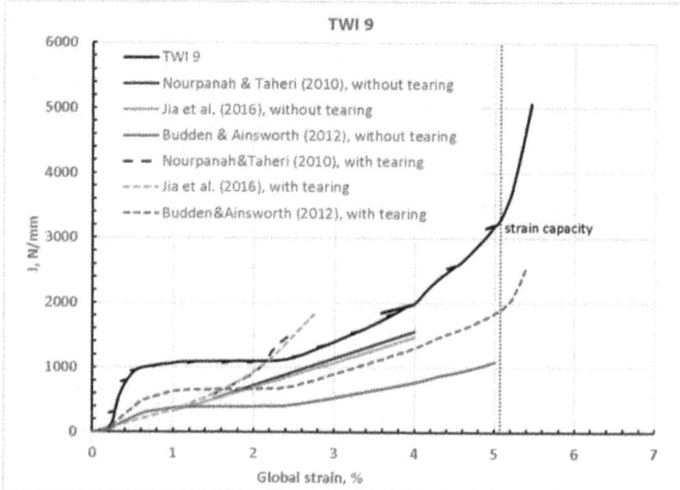

(b)

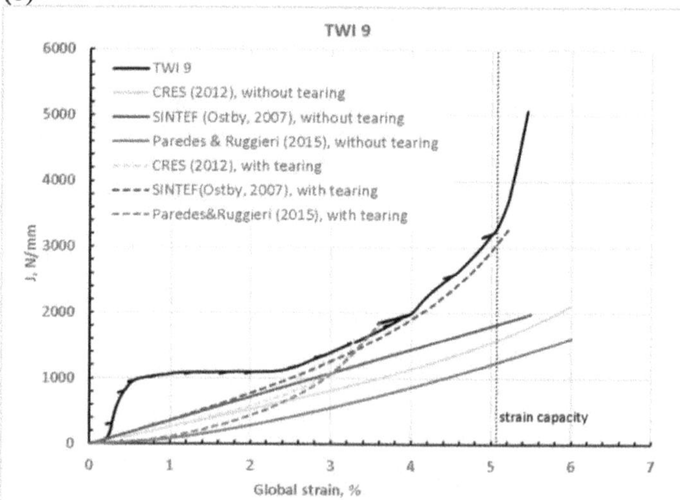

(c)

Figure 10 COMPARISON OF CRACK DRIVING FORCE OBTAINED FROM TEST AND STRAIN-BASED METHODS

Copyright © 2019 by ASME

In order to consider tearing, J for flaw depths, $a_0+\Delta a$, was calculated at a fixed global strain. Then, the J value at the intersection point with the R-curve (Equation 2) was used to update the J at the specific global strain. It may be noted that the SINTEF method [18] has its own method to account for tearing. Figure 10 also compares the J values considering tearing effect with experimental results and J values from a stationary flaw. The crack driving force from DNV GL [4] generally matched with test results before 0.3% applied strain but ductile instability was reached at a strain level of 0.45%. The J values from Smith [5] increased at the end of plateau when work hardening began. The other methods gave J values greater than the test results in the work hardening regime except for the Budden and Ainsworth [8] and SINTEF [18] methods. This suggests that conservatism is achieved by most of the strain-based methods provided that the unstable fracture occurs in the work hardening regime. This is in agreement with the strain capacity prediction shown in Figure 11. Except for the methods from ExxonMobil and SINTEF, the strain capacity was defined when the unstable point was reached using the tangency approach. The strain at the maximum load was defined as the strain capacity in ExxonMobil's methods. The strain capacity predicted using SINTEF approach was based on a criterion for CTOD at maximum load carrying capacity of the crack ligament.

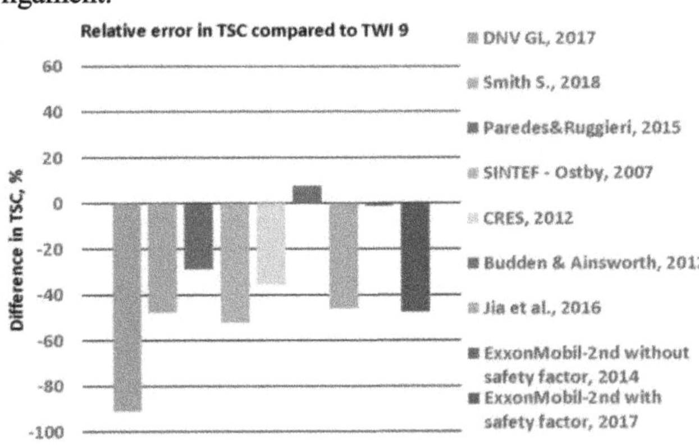

(a)

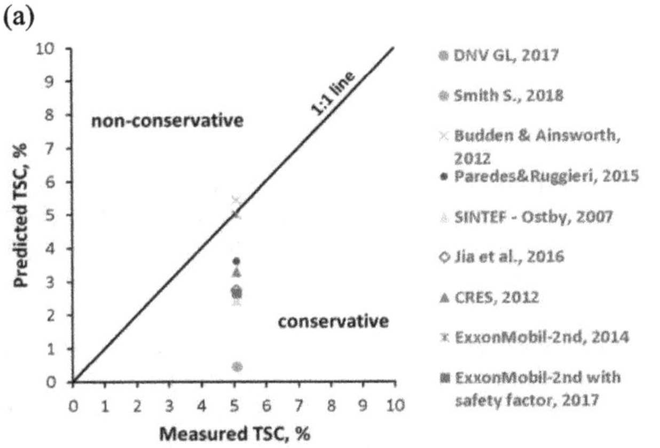

(b)

Figure 11 COMPARISONS OF STRAIN CAPACITY OBTAINED FROM DIFFERENT STRAIN-BASED ASSESSMENT METHODS WITH TWI'S PIPE TEST RESULTS (TWI 9 AXIAL STRAINING)

The relative error in tensile strain capacity (TSC) is calculated using:

$$\Delta\varepsilon\% = \left(\varepsilon_{predicted} - \varepsilon_{experiment}\right)\times 100\% / \varepsilon_{experiment} \quad (7)$$

where $\Delta\varepsilon\%$ is the difference in TSC, $\varepsilon_{predicted}$ is strain capacity estimated using the existing strain-based methods and $\varepsilon_{experiment}$ is strain capacity measured from the full-scale pipe test.

Figure 11 shows that most of the methods are conservative with degree of conservatism less than 50%. However, DNV GL's method [4] is over-conservative compared to the other methods since the strain capacity predicted is about 90% less than the test result. On the other hand, Budden and Ainsworth's method [8] predicted higher strain capacity than the measured data. ExxonMobil's method [15-17] is intended to predict the failure if a safety factor is not included. It can be seen that the prediction from ExxonMobil's method matched with the experimental measurement very well. A safety factor of 1.88 which corresponds to 95% confidence from ExxonMobil's full-scale test results [22] was also considered. This increases the conservatism of ExxonMobil's prediction to a similar level as that obtained using the methods from Smith [5], Jia et al. [9] and SINTEF [18]. Because a safety factor can be set by the users for different applications, ExxonMobil's method has the merit of providing a good prediction and allowing for flexibility. Although the DNV GL and Smith methods do not have such flexibility, they have their advantage of simplicity and are user-friendly since they can be used within the framework of BS 7910 procedures.

TWI 10

This pipe test was conducted by first internally pressurising the pipe, holding the pressure constant and then applying axial strain. Figure 12 shows the development of J from each of the two notches N1 and N2 against the global strain. Both notches were of very similar size (see Table 1) and consistent results were obtained up to approximately 3% strain. Thereafter, a higher J was generated at notch 2 located at 04:30 o'clock, indicating that tearing was occurring preferentially compared to notch 1. As in the previous test, the effects of the Lüders plateau are seen on J versus strain graphs. At a strain of 4.44% a leak was detected from notch 2 and the test was discontinued. At the time it was thought that the leak occurred because ductile tearing had extended through the pipe wall from notch 2. However, as will be shown later, this was not the case. The test results are summarised in Table 4.

By looking at the appearance of the notches after breaking the pipe sample, it was evident that there was significantly more tearing from notch 2 compared to notch 1, as indicated by the J versus strain graph in Figure 12. Examination of the fracture surface at notch 2 (Figure 13) reveals the reason why a leak was detected from this notch. Ductile tearing had not extended fully

Copyright © 2019 by ASME

through the pipe wall as originally thought. Instead, tearing occurred towards one end of the notch linked with a welding flaw located near the weld root which then resulted in the leak. Furthermore, the length of the flaw on the ID of the pipe was very small, around 1mm, which explains why the leak observed was small. Overall, the impact of leak on the test strain capacity was likely marginal and this can be seen from the steep J versus strain curve.

Table 4 SUMMARY OF RESULTS FROM FULL-SCALE PIPE TEST TWI 10 (STRAIN AND CTOD AT MAXIMUM "LOAD")

Pipe test	Number of notches	Notch location	Internal pressure	Strain at maximum load %	CTOD, mm	Initial notch size, mm		Type of failure
						a	2c	
TWI 10	2	Weld metal	Y	4.44	7.0	4.81	49.79	Leak

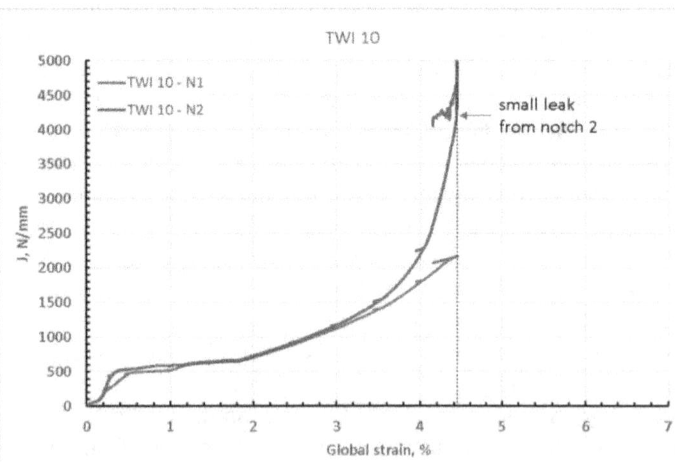

Figure 12 J VERSUS GLOBAL STRAIN FOR THE TWO NOTCHES (N1 AT 10.30 O'CLOCK AND N2 AT 4.30 O'CLOCK) IN PIPE TEST TWI 10

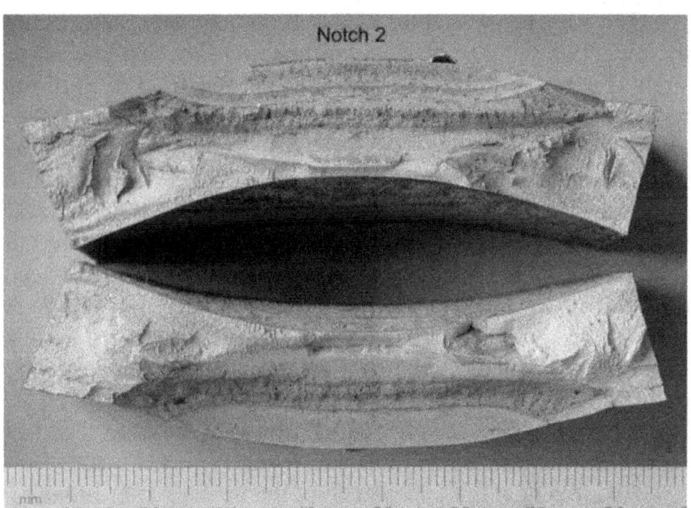

Figure 13 APPEARANCE OF NOTCH 2 AT 4.30 O'CLOCK IN TWI 10

Only a few of the existing methods take into account the effect of biaxial loading on crack driving force and strain capacity. Figure 14 compares the crack driving force in terms of J predicted from the strain-based methods developed by CRES, DNV GL and SINTEF with TWI's test results of TWI 10 notch 2. Without considering tearing, CRES' method under-predicted the crack driving force while DNV GL's method over-predicted the crack driving force in the strain range of 0.5% - 4% and SINTEF's method predicted higher crack driving force than obtained from test in the strain range of 1% - 4.15%.

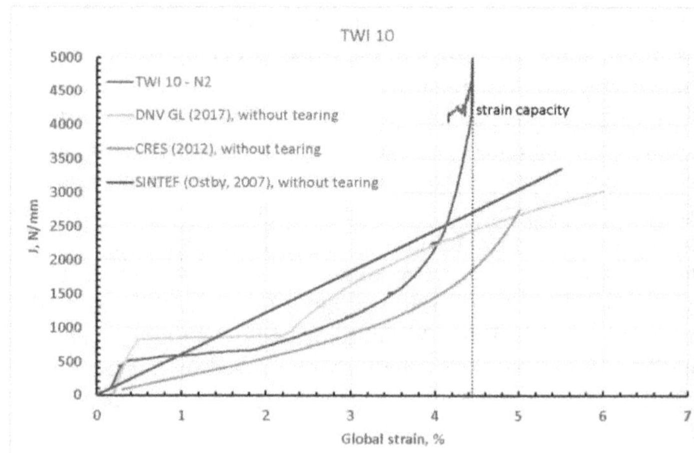

(a)

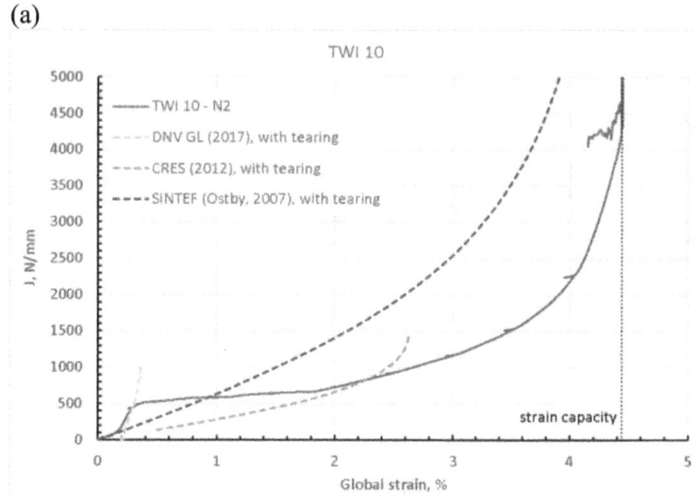

(b)

Figure 14 COMPARISONS OF DIFFERENT STRAIN-BASED ASSESSMENT METHODS (WITH AND WITHOUT CONSIDERING TEARING) WITH TWI'S TESTING RESULTS OF TWI 10 NOTCH 2

If tearing is considered, Figure 14(b) shows that DNV GL's method over predicts J when yield strain is reached but CRES predicts lower J at strains up to 2.2%, before work hardening but a higher J at larger strain when work hardening has commenced. SINTEF's method estimated higher J than test data when global strain exceeds 1%. This suggests that conservative strain capacity may be predicted if the unstable fracture occurs

Copyright © 2019 by ASME

in the regime of work hardening. This agrees with strain capacity prediction shown in Figure 15. Nevertheless, compared with a girth welded pipe under axial loading only, more effort is needed to develop a robust strain-based method to predict the crack driving force for a girth welded pipe subjected to biaxial loading.

Figure 15 shows that the DNV GL, SINTEF, CRES and ExxonMobil methods under-estimate the strain capacity from the experiment but ExxonMobil's method without a safety factor predicted a closer result to the experimental data than the other methods. A safety factor of 1.88 was also applied and this increased the degree of conservatism from 23% to 59%. The strain capacity predicted using DNV GL's method was over-conservative. This may in turn reduce the maximum allowable flaw size and increase the repair rate unnecessarily.

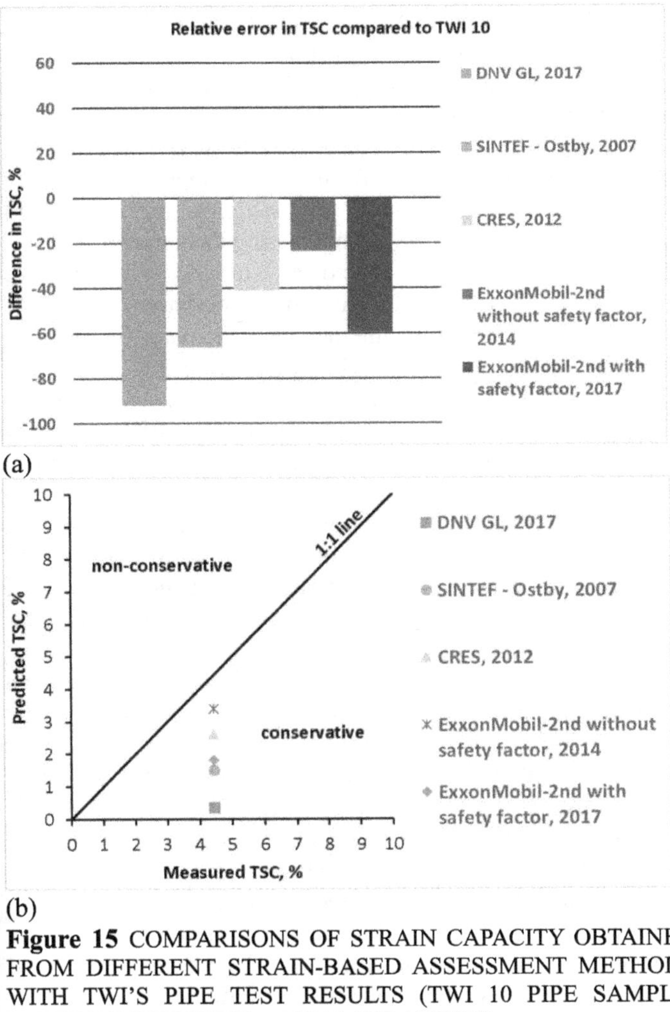

Figure 15 COMPARISONS OF STRAIN CAPACITY OBTAINED FROM DIFFERENT STRAIN-BASED ASSESSMENT METHODS WITH TWI'S PIPE TEST RESULTS (TWI 10 PIPE SAMPLE: INTERNAL PRESSURE + AXIAL STRAINING)

Comparison of the strain-based methods with test results of TWI 9 and 10 pipe specimens shows that the existing methods cannot predict the crack driving force properly if tearing is ignored. In terms of strain capacity, DNV GL's method is over-conservative for pipes under axial loading and biaxial loading. ExxonMobil's method predicted the failure very well when no safety factor is included. However, the conservatism of ExxonMobil's method is flexible by considering a user-defined safety factor. CRES model predicted the strain capacity with reasonable conservatism for TWI 9 and TWI 10. Please note that the resistance curve was considered to estimate the ductile instability in the CRES model (ie, level 3b method in [14]) instead of a single apparent toughness. The strain capacity predicted using SINTEF method is 1.5 times less than that predicted using CRES's method for both cases. The conservatism in the approaches may come from (i) built-in conservatism and (ii) conservatism arising from the use of the SENT R curve. Although the Smith [5], Paredes and Ruggieri [6] and Jia et al. [9] methods predicted conservative strain capacity for the pipe specimen under axial loading, these methods do not consider biaxial loading condition and therefore further development is suggested.

CONCLUSIONS AND FUTURE RECOMMENDATIONS

The following conclusions can be drawn:

- The effect of the Lüders plateau in the parent pipe and weld metal stress-strain curve was apparent in the curve of J versus remotely applied strain, where J remained fairly constant after yield up to approximately 2.5% strain in the axially loaded test and 1.8% strain in the test which included internal pressure. At higher strains, J increased in an exponential manner with increases in applied strain until through wall tearing/leakage occurred.
- Pipe test TWI 09, which was axially loaded only, attained a maximum load plateau at a strain of 5.07% (this was defined as its tensile strain capacity) and CTOD of 4.03mm.
- Pipe test TWI 10, which was internally pressurised to give a hoop stress of 460MPa and then axially loaded, reached a strain of 4.44% (this was considered to be very close to the tensile strain capacity of the pipe and treated as the strain capacity in the comparison work) when tearing caused a leak from one of the notches at a CTOD of approximately 7mm; the CTOD from the other, nominally identical, notch was 3.51mm (the extent of tearing through wall was 2.91mm).
- The strain-based methods (Groups 1-3, 5 and CRES) under-predict the crack driving force for pipes subjected to tension alone and combined internal pressure and tension when tearing was not accounted for in the analyses.
- Most of the strain-based methods conservatively predicted the strain capacity for pipes under tension only.
- For internal pressure plus axial tension, the crack driving force predicted using different strain-based methods did not match with the experimental data and

Copyright © 2019 by ASME

the degree of conservatism of the tensile strain capacity predicted was not consistent.

The following recommendations are proposed:

- Strain-based methods providing an output of crack driving force need to be re-examined in order to improve their accuracy, especially when pipes are under combined internal pressure and tension loading.
- Additional full-scale pipe tests are needed and these should include a repeat of TWI 10 but at a lower internal pressure. A lower pressure would be more representative of typical operation and would also maximise the effects of biaxiality.
- Additional full-scale pipe test data will enable statistical analysis to be carried out and will help in evaluating strain-based fracture methods and recommending them for inclusion in codes/standards.

ACKNOWLEDGMENTS

The experimental work presented in this paper was undertaken in a JIP led by Henryk Pisarski at TWI Ltd. The support provided by the members of TWI's strain-based assessment team and Justin Crapps from ExxonMobil is gratefully acknowledged.

REFERENCES

[1] BS 7910:2013+A1:2015, "Guide to methods for assessing the acceptability of flaws in metallic structures", Incorporating Corrigenda Nos. 1 and 2, 2015.

[2] API 579-1/ASME FFS-1, "Fitness-for-service", 2016.

[3] EDF Energy, R6 Revision 4, "Assessment of the integrity of structures containing defects", with amendments to 2014, 2014.

[4] DNVGL-RP-F108, "Assessment of flaws in pipeline and riser girth welds", 2017.

[5] Smith S., 2018, "Consideration of a proposed SBAD method for BS7910", International Journal of Pressure Vessel and Piping, Vol. 168, pp. 142-147.

[6] Paredes M. and Ruggieri C., 2015, "Engineering approach for circumferential flaws in girth weld pipes subjected to bending load". International Journal of Pressure Vessel and Piping, Vol. 125, pp. 49-65.

[7] Nourpanah N. and Taheri F., 2010, "Development of a reference strain approach for assessment of fracture response of reeled pipelines". Engineering Fracture Mechanics, Vol. 77, pp. 2337-2353.

[8] Budden P. J. and Ainsworth R.A., 2012. "The shape of a strain-based failure assessment diagram". International Journal of Pressure Vessels and Piping, Vol. 89, pp. 59-66.

[9] Jia P., Jing H., Xu L., Han Y. and Zhao L., 2016. "A modified reference strain method for engineering critical assessment of reeled pipelines". International Journal of Mechanical Sciences, Vol. 105, pp. 23-31.

[10] SwRI (Southwest Research Institute), 2006, "Theoretical background to the engineering critical assessment methodology for reeled pipe incorporated in the computer program FlawPRO version 3".

[11] Bezensek B., Chell G., Popelar C. and Luo M., 2018, "Recent development and validation of the latest FlawPRO ECA methodology". Proceedings of the ASME 2018 37th International Conference on Ocean, Offshore and Arctic Engineering, OMAE2018-78068, Madrid, Spain.

[12] Wang Y., Kibey S. and Tang H., 2011, "Strain-based design-advances in prediction methods of tensile strain capacity". International Journal of Offshore and Polar Engineering, Vol. 21.

[13] Wang Y., Liu M., Zhang F., Horsley D. and Nanney S., 2012, "Multi-tier tensile strain models for strain-based design: Part 1 – Fundamental basis". Proceedings of the 9th International Pipeline Conference, pp.

[14] Liu M., Wang Y., Song Y., Horsley D. and Nanney S., 2012, "Multi-tier tensile strain models for strain-based design: Part 2 – development and formulation of tensile strain capacity models". Proceedings of the 9th International Pipeline Conference, pp. 415-425.

[15] Fairchild D., Tang H., Shafrov S., Cheng W. and Crapps J., 2014, "Updates to ExxonMobil's modelling approach for tensile strain capacity prediction". Proceedings of the 24th International Ocean and Polar Engineering Conference.

[16] Tang H., Fairchild D., Panico M., Crapps J. and Cheng W., 2014. "Strain capacity prediction of strain-based pipelines". Proceedings of the 10th International Pipeline Conference, American Society of Mechanical Engineers.

[17] Fairchild D., Panico M., Crapps J., Cheng W. and Cook M., 2017, "Benchmark examples of tensile strain capacity prediction and strain-based engineering critical assessment calculations". Proceedings of the 27th International Ocean and Polar Engineering Conference, San Francisco, CA, USA.

[18] Ostby E., 2007. "Fracture control – offshore pipelines JIP proposal for strain-based fracture assessment procedure". Proceedings of the 17th International Offshore and Polar Engineering Conference.

[19] Pisarski H., 2013, "Assessment of flaws in pipeline girth welds – a critical review". Welding in the World, Vol. 57, Issue 6, pp. 933-945.

[20] Wu G. and Wang L., 2018, "An overview of strain-based fracture assessment of pipelines". Proceedings of the ASME 37th International Conference on Ocean, Offshore and Arctic Engineering, OMAE2018-78685, Madrid, Spain.

[21] Pisarski, H., Smith S. and London T., 2014. "Flaw tolerance of pipelines containing circumferential flaws subjected to axial straining and internal pressure – tests and analyses". International Journal of Offshore and Polar Engineering, Vol. 24, pp. 199-205.

[22] Fairchild D., Crapps J., Cheng W., Tang H. and Shafrova S., 2016, "Full-scale pipe strain test quality and safety factor determination for strain-based engineering critical assessment". Proceedings of the 2016 11th International Pipeline Conference, Calgary, Alberta, Canada.

Copyright © 2019 by ASME

AIMPIMG-1066

Chapter 22

TENSILE STRAIN MODELS AND THEIR APPLICATIONS

Yong-Yi Wang
Center for Reliable Energy Systems
Dublin, OH, USA
ywang@cres-americas.com

Banglin Liu
Center for Reliable Energy Systems
Dublin, OH, USA
bliu@cres-americas.com

Bo Wang
Center for Reliable Energy Systems
Dublin, OH, USA
bwang@cres-americas.com

ABSTRACT

This paper covers tensile strain design, which has at least three key elements, including (1) linepipe specifications / qualification, (2) girth welding procedure qualification and field girth welding practice, and (3) tensile strain models. A brief summary of linepipe specifications, girth welding procedure qualification, and field girth welding is provided, with features specifically important to SBD highlighted. The focus of the paper is on tensile strain models and their applications. The fundamental principles and structure of the PRCI-CRES tensile strain models are presented. Processes and procedures for obtaining non-conventional input parameters such as apparent CTOD toughness are introduced for users to apply the models. Various ways of applying the models and the limits of the models are given. Finally, examples are shown for the strain-based design of new pipelines and strain-based assessment of existing pipelines.

KEYWORDS

Pipeline, strain-based design and assessment (SBDA), tensile strain models, applications of SBDA

INTRODUCTION

The general concept of strain-based design and its relation to the traditional stress-based design is given in a companion paper [1]. This paper focuses on tensile strain models and their applications for large and small construction projects.

KEY ELEMENTS IN TENSILE STRAIN DESIGN

The tensile strain capacity of a pipeline is usually controlled by the tensile strain capacity of its girth welds. The girth welds here refer to the entire weld region, including the weld metal, the fusion boundary, and the heat-affected zone (HAZ). Girth welds tend to be the weakest link due to the possible existence of weld defects and often deteriorative metallurgical and / or mechanical property changes from welding thermal cycles. Consequently, tensile strain capacity (TSC) is intrinsically related to the metallurgical and mechanical properties of linepipes and girth welding procedures. Therefore, tensile strain design broadly consists of three key elements: (1) linepipe specifications / qualification, (2) girth weld specifications / qualification, and (3) tensile strain models.

Linepipe Specifications

In addition to normal linepipe property requirements, such as strength, ductility, Charpy, DWTT, etc., linepipes manufactured for SBD must also meet the following requirements.

1. Select a pipe grade that is conducive to girth weld strength overmatching. This is achieved by choosing a grade such that the strength of the weld can overmatch the upper limits of the yield and ultimate tensile strengths (UTS) of the grade.
2. Test and specify longitudinal tensile properties. Large diameter pipes are typically qualified by tensile testing

Copyright © 2019 by ASME

in the hoop direction. The longitudinal tensile properties are more relevant to the strain tolerance in the longitudinal direction. Typical specifications may include a minimum and maximum yield strength and UTS to facilitate weld strength overmatching. Weld strength overmatching is critical for guarding against strain concentration in the weld, minimizing the negative impact of HAZ softening, and having workable weld flaw acceptance criteria [2].

3. Have a sufficient strain hardening capacity, often by defining an upper limit of the yield-to-tensile-strength (Y/T) ratio.
4. Ensure sufficient ductility and strain capacity in the absence of weld flaws. This is done by specifying a minimum uniform elongation and total elongation,
5. Have certain stress-strain curve characteristics to increase the compressive strain capacity, such as defining stress ratios that are meant to have so-called "round-house" stress-strain relations [3].
6. Reduce the joint-to-joint pipe strength variations to reduce the possibility of strain concentration in lower-strength joints. This can be achieved by limiting the permissible range of strength variations for a grade and having high strain hardening. For a large pipeline project with small segments needing SBD, pipes meeting stringent SBD requirements may be selected after production from the available pools of pipes.
7. Minimize the level of girth weld HAZ softening. This may be achieved by using a relatively richer chemistry and less reliance on rolling and accelerated cooling to achieve the desired strength [4].

Steel making processes aimed at achieving the SBD requirements might negatively impact other conventional requirements, such as DWTT. Consultations between pipe / steel mills and project procurement are often necessary to achieve the optimal outcome.

It should be recognized that the current test procedures are often inadequate both in scope and rigor for meeting SBD qualification requirements [5].

Girth Welding Qualification and Field Practice

Requirements for welding procedure qualification in current industry standards, such as API 1104, are inadequate for SBD. A minimum level of weld strength overmatching against the higher end of the pipe strength distribution is necessary unless special weld profile designs are taken (see Section *Examples of Applications*). HAZ softening must be minimized such that there should not be failure in the HAZ in cross-weld tensile tests. The toughness should be sufficient to ensure upper-shelf behavior at the minimum design temperature.

A quality assurance and verification program are necessary for field welding, particularly for manual welding. The selection of welding processes and consumables must consider the difficulties in such programs and build sufficient margin to account for field variations.

Many modern microalloyed steels are susceptible to HAZ softening. HAZ softening tends to be aggravated with multiple thermal cycles, including those from repair welds. Tight tolerances for inspection indications can lead to excessive repairs, which may not be beneficial for the overall weld quality.

Tensile Strain Models

Tensile strain models correlate key parameters affecting the tensile strain capacity. Considerable efforts have been devoted to the development of tensile strain design models worldwide in recent years [6,7,8]. The prevailing tensile strain models include those from CRES/PRCI [9,10,11,12,13], SINTEF [7,14,15], ExxonMobil [8,16,17,18,19], JFE [20,21,22], University of Ghent [23], and TWI [24].

The PRCI-CRES models are fully documented in two comprehensive reports to PHMSA [25,26] and in conference publications [9,10,11,12,13]. The models have been shown to provide consistent trends with experimental test data for pipe grades of X56, X65, X80, and X100. The models have been used in the design and construction of X70 and X80 large-diameter pipelines in areas with mine subsidence [27,28].

PRCI-CRES TENSILE STRAIN MODELS

Building Blocks of PRCI-CRES Models

The three essential elements of the PRCI-CRES models are: (1) limit states, (2) crack-driving force, and (3) toughness.

Limit States

Two limit states are recognized in the PRCI-CRES models.

The initiation limit state is achieved when the crack driving force ($CTOD_F$ or δ_F) reaches the material's apparent toughness ($CTOD_A$ or δ_A). The apparent toughness is the toughness corresponding to the onset of stable tearing. Both $CTOD_F$ and $CTOD_A$ are represented by the crack tip opening displacement (CTOD). The strain corresponding to $CTOD_A$ is the tensile strain capacity (TSC or ε_t^{crit}). This limit state is illustrated in Figure 1.

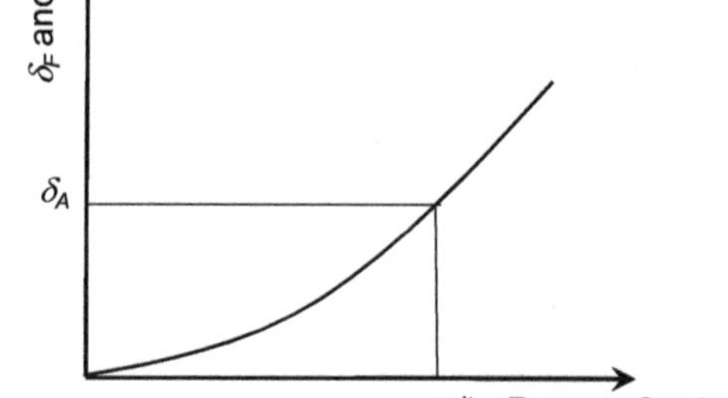

Figure 1 Illustration of the initiation limit state

Copyright © 2019 by ASME

The ductile instability limit state is achieved when CTOD$_F$ is about to exceed the resistance to crack extension (CTOD$_R$ or δ_R) as the flaw grows. This limit state is illustrated in Figure 2. A convenient way of implementing the ductile instability using driving force relations is shown in [29].

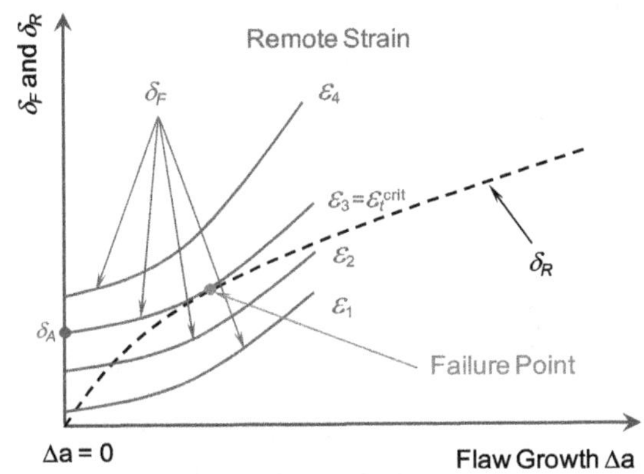

Figure 2 Illustration of the ductile instability limit state

Crack-Driving Force

CTOD$_F$ is given as a function of remotely applied nominal strain, ε_F, as shown in Figure 1 and Figure 2. The relationship between CTOD$_F$ and ε_F, is obtained through extensive finite element analysis (FEA). The FEA incorporated the following parameters:

(1) Geometric parameters:

t	pipe wall thickness, mm,
a	flaw height, mm,
$2c$	flaw length, mm, and
h	girth weld high-low misalignment, mm,

(2) Material parameters:

σ_y	pipe yield strength, MPa,
σ_U	pipe ultimate tensile strength, MPa,
σ_U^W	weld metal tensile strength, MPa, and

(3) Loading parameter:

f_p	pressure factor, ratio of applied hoop stress to pipe yield strength.

Weld profiles representing narrow-groove mechanized GMAW and standard groove SMAW and FCAW processes were incorporated into finite element models. A database of crack-driving forces was generated from the FEA. Two sets of parametric equations, one for GMAW and one for FCAW/SMAW, were developed from the database. These equations use the follow normalized parameters:

$\eta = a/t$	normalized flaw depth,
$\beta = 2c/t$	normalized flaw length,
$\psi = h/t$	normalized weld high-low misalignment,
$\xi = \sigma_y/\sigma_U$	pipe Y/T ratio,
$\phi = \sigma_U^W/\sigma_U$	weld metal strength mismatch ratio,
f_p	pressure factor, ratio of applied hoop stress to pipe yield strength.

The crack driving force relationship takes the following generic form,

$$\varepsilon_F = \min\left(\varepsilon_u, P(f_p)G(t)\varepsilon_P\right) \tag{1}$$

where ε_u is the pipe uniform strain (uEL) and the functions $G(t)$ and $P(f_p)$ characterize the effect of wall thickness and internal pressure, respectively.

The function $G(t)$ is given as,

$$G(t) = \left(\frac{15.9}{t}\right)^{0.8096(1+1.503\psi^{1.229})} \tag{2}$$

and the function $P(f_p)$ is given as,

$$P(f_p) = \begin{cases} P_{max} - \frac{5}{3}f_p(P_{max} - 1), & if\ 0 \le f_p < 0.6 \\ 1, & if\ 0.6 \le f_p \le 0.8 \end{cases} \tag{3}$$

A conservative and fixed value of 1.5 was initially used for P_{max} in [26]. A more refined relation was developed later, given as,

$$P_{max} = 2.25 - 2\frac{a}{t} \tag{4}$$

The function ε_P in Eq. (1) is given as,

$$\varepsilon_P = A\frac{F(\delta_F)}{1+F(\delta_F)} \tag{5}$$

where the function $F(\delta_F)$ is

$$F(\delta_F) = (C\delta_F)^{B\delta_F^D} \tag{6}$$

A, B, C, and D in Eqs. (5) and (6) are fitted functions of normalized geometry and material properties [10].

Toughness

Two forms of toughness, (1) initiation toughness as represented by the apparent toughness, and (2) resistance curves, may be used in the PRCI-CRES tensile strain models as shown in Figure 1 and Figure 2. The toughness values may be obtained from multiple types of tests, including traditional Charpy impact, standard three-point bend CTOD, low-constraint test of shallow-notched three-point bend CTOD, SENT (single edge notched tension), and CWP (curved wide plate) tests. Images of some toughness test options are given in Figure 3.

Copyright © 2019 by ASME

Figure 3 Images of various toughness test options that may provide input to the PRCI-CRES tensile strain models. Left: standard three-point bend CTOD; Middle: SENT; and Right: CWP

As the apparent toughness represents the initiation toughness of girth welds under predominantly tensile loading, the relevant toughness is that from low-constraint specimens. A conversion process is provided for toughness obtained from traditional specimens.

Conversion for Mechanized GMAW Welds

$CTOD_A$ may be estimated from the Charpy upper-shelf impact energy, U_{CVN}.

For welds of X52 to X65 linepipes, the conversion relation is given as,

$$\delta_A = \left(0.00590\frac{\sigma_y}{\sigma_U} - 0.00103\right) U_{CVN} \quad (7)$$

For welds of X70 and X80 linepipes, the conversion relation is given as,

$$\delta_A = \left(0.00634\frac{\sigma_y}{\sigma_U} - 0.00155\right) U_{CVN} \quad (8)$$

In the above equations, σ_y is the yield stress at 0.5% total strain and σ_U is the ultimate tensile strength in the pipe longitudinal direction.

These imperial equations are dimensionally inconsistent. The units of the Charpy energy and the $CTOD_A$ are Joule and mm, respectively. The maximum $CTOD_A$ that may be used after the conversion is 1.2 mm.

When upper-shelf CTOD toughness is available from standard CTOD test specimens, a multiplication factor of 1.5 to 2.0 may be applied to obtain the $CTOD_A$. The recommended default conversion factor is 1.75. The conversion factor is applicable to standard SENB specimens without side grooves. It should be emphasized that the relevant CTOD toughness for conversion must be from the upper-shelf CTOD. A single CTOD toughness value from the maximum load is not sufficient to indicate upper-shelf behavior [30].

The $CTOD_A$ may be obtained from resistance curves of low-constraint specimens, such as SENT, at a flaw growth Δa = 0.5 – 1.0 mm as illustrated in Figure 4. The selection of the amount of flaw growth depends on pipe wall thickness and possibly other parameters. Generally, the recommended Δa is 0.5 mm for pipe wall thickness of 12.7 mm or 1/2 inch and 1.0 mm for pipe wall thickness of 25.4 mm or 1.0 inch. When the pipe wall thickness is between above two extreme values, a linear interpolation may be used to determine the appropriate Δa value.

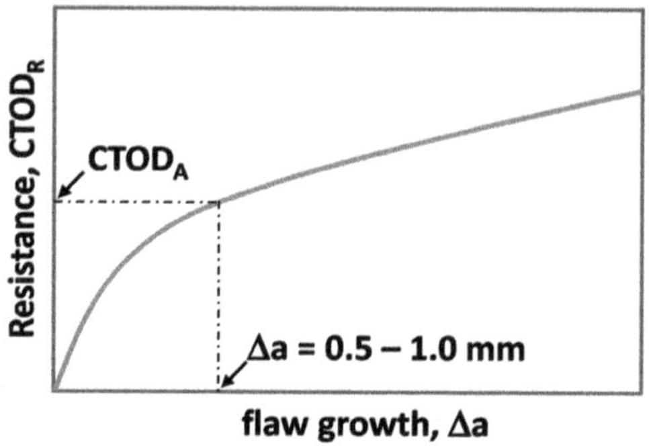

Figure 4 Schematic illustration of obtaining $CTOD_A$ from a resistance curve

Conversion for Manual Welds of Cellulosic Electrodes

The Charpy to $CTOD_A$ conversion for manual welds made with cellulosic electrodes has been developed within a PRCI program. This conversion is being validated.

Obtaining $CTOD_A$ Directly from Tests

$CTOD_A$ may be obtained directly from crack-tip deformation profile. Examples of such an approach for mechanized GMAW welds are given in [9]. Such an approach has been applied to vintage girth welds as well [31].

Structure of PRCI-CRES Models

The PRCI-CRES models are organized in four levels as shown in Figure 5.

The Level 1 models are intended for quick estimations of the likely tensile strain capacity. The TSC is tabulated for selected pipe dimensions, material properties, and flaw size [26]. The apparent toughness is estimated from the upper-shelf Charpy impact energy.

The Level 2 models are given in a library of parametric equations. The apparent toughness can be estimated from either the upper-shelf Charpy impact energy or the upper-shelf standard CTOD toughness.

The Level 3 models have two options. Level 3a uses an initiation limit state. Level 3b uses a ductile instability limit state. In Level 3a, the TSC can be obtained from the same library of parametric equations as in Level 2. The apparent toughness can be obtained from a number of low constraint test options, including shallow-notched SE(B), SENT, or CWP. In Level 3b, the TSC is solved by matching the crack driving force and resistance curves, as shown in Figure 2.

Copyright © 2019 by ASME

The Level 4 models are structured in the same format as Level 3 models, except that the crack-driving force relations are directly obtained from FEA as opposed to from parametric equations. The toughness options are the same as in Level 3. This level allows for special cases when the specific weld geometry and material properties do not allow the use of the first three options. The Level 4 models should only be exercised by seasoned experts.

Level	Limit State	Target Application	Format of Models	Toughness Option
1	Initiation	Initial screening, feasibility study	Tabular	Upper shelf Charpy energy
2	Initiation	Nominal analysis	Equations	Upper shelf Charpy energy, CTOD or J from SENB
3a	Initiation	Advanced analysis	Equations	Single value CTOD or J from shallow-notched SENB, SENT, or CWP
				CTOD or J resistance curves from shallow-notched SENB, SENT, or CWP
3b	Ductile instability	Advanced analysis	Equations	CTOD or J resistance curves from shallow-notched SENB, SENT, or CWP
4a	Initiation	Case specific-analysis	Expert finite element analysis	Single value CTOD or J from shallow-notched SENB, SENT, or CWP
				CTOD or J resistance curves from shallow-notched SENB, SENT, or CWP
4b	Ductile instability	Advanced analysis	Expert finite element analysis	CTOD or J resistance curves from shallow-notched SENB, SENT, or CWP

Figure 5 Overall structure of the PRCI-CRES tensile strain models

Applications of Tensile Strain Design Models

Role of Tensile Strain Models

Tensile strain models may be used in the design phase of a pipeline or in the integrity management of an existing pipeline. In the design phase, tensile strain models may be used for:

- Setting linepipe specifications,
- Setting requirements for girth weld qualification,
- Determining flaw acceptance criteria, and
- Helping to develop field inspection procedures, such as treatment of high-low misalignment.

For the integrity management of existing pipelines, tensile strain models may be used for

- Assessing the risk of tensile rupture,
- Facilitating decision-making on mitigation options, and
- Determining the intervention threshold in allowable strain demand.

Applicable Range of PRCI-CRES Tensile Strain Models

The applicable ranges of the input parameters are:

η	0.05 – 0.5,
β	1.0 – 20.0,
ψ	0.0 – 0.2,
ξ	0.75 – 0.94,
ϕ	1.0 – 1.3,
δ_A	0.2 – 2.5 mm, and
f_p	0.0 – 0.8.

There is no inherent limit on the pipe grade, provided that the pipe and welds meet the material property requirements.

Summary of PRCI-CRES Models

The essential features of the PRCI-CRES tensile strain models are

(1) Two limits states are recognized: (a) initiation and (b) ductile instability.

(2) Two weld bevel geometries are recognized: (a) narrow-groove, typical of mechanized GMAW welds and (b) standard groove, typical of FCAW and SMAW welds.

(3) Toughness may be obtained from multiple types of specimens. The use of initiation toughness, particularly the ability to obtain such toughness from Charpy impact tests is advantageous for the assessment of in-service pipelines.

(4) The target optimum strain range of the models is from 1.0% to one half of the pipe's uniform elongation.

(5) The fundamental basis of the tensile strain models does not assume or rely on material property data of any particular grade. The parametric representations of the tensile properties of the pipes and welds were developed using a material database which covered grades from X65 to X100. The tensile strain design models are, in principle, applicable to all GMAW and FCAW/SMAW processes, provided that appropriate mechanical property data are within the applicable range of the models.

(6) Buried flaws can be conservatively treated as surface-breaking flaws of the same dimensions. The flaw re-characterization rules for near-surface flaws, such as those in API 1104 Annex A, may be used. Treating buried flaws as surface-breaking flaws introduces additional level of conservatism [32].

Limitations

(1) Welds should not have gross strength undermatching that allows for strain concentration across the girth welds when there are no flaws in the welds.

(2) The models are applicable to one single flaw in a girth weld. If there are multiple flaws in a single girth weld, appropriate interaction rules should be applied and the application of the models in such an environment should be revised and updated [1].

(3) No embedded safety factor is applied or implied in the tensile strain models. Establishing an appropriate safety factor requires the understanding of the

Copyright © 2019 by ASME

uncertainties in the input parameters to the tensile strain models, uncertainties in the estimated strain demand, and the risk tolerance in the location of interest [1].

(4) The models should not be used for flaw acceptance after repair welding without further evaluation.

(5) The potential impact of material anisotropy on the tensile strain capacity is not considered in the models. The potential impact of material anisotropy on the tensile strain capacity is discussed elsewhere [33].

CONDITIONS NOT COVERED BY MOST TENSILE STRAIN MODELS

The TSC of a pipeline is often controlled by the TSC of its girth welds as the welds may contain welding imperfections and have deteriorative material properties in comparison to the base pipe material. Most tensile strain models focus on the interaction of weld flaws, material properties, and dimensional parameters during development. However, when the material property variations can be well defined, an effective strategy can ensure that the girth welds are not the weakest link. For instance, a consistent and high degree of weld strength overmatching, a limited extent and degree of heat-affected zone (HAZ) softening, a low limit on high-low misalignment, sufficient toughness and ductility, and a tight flaw acceptance level can be exercised so the resistance of the tensile deformation of the girth welds would always be higher than that of the base pipe [34].

On the other side of the spectrum is the fact that tensile strain capacity can be low in the absence of weld flaws. If the weld strength undermatches the strength of the pipe by more than 10-15% (both yield and UTS), the strain can accumulate predominantly in the weld area when the pipeline is subjected to longitudinal tensile strain beyond about 0.25-0.35%. The TSC is low even in the absence of weld flaws and materials having good toughness [35,36]. This condition should be prevented in strain-based design. However, there are no effective test methods in welding procedure qualification to preclude the potential existence of high levels of HAZ softening. Traditional cross-weld hardness surveys do not provide enough resolution and accuracy to assess the overall level of HAZ softening. The zones with the maximum HAZ softening vary from weld to weld, depending on the characteristics of linepipe steels and the welding process. These zones can be much further from the fusion boundary than many would expect.

EXAMPLES OF APPLICATIONS

Selecting Appropriate Parameters to Achieve Target Strain Capacity

The Levels 2 and 3 procedures of the PRCI-CRES models can be used to examine "what-if" scenarios to meet a target level of TSC, as shown in Figure 6. For instance, increasing the level of weld strength overmatching would increase the upper limit of the Y/T ratio of the pipe. Similarly, reducing the high-low misalignment could increase the limit of acceptable flaw dimensions.

A software implementation of the Levels 2 and 3 procedures of the PRCI-CRES models are being developed under the PRCI project API-3-1.

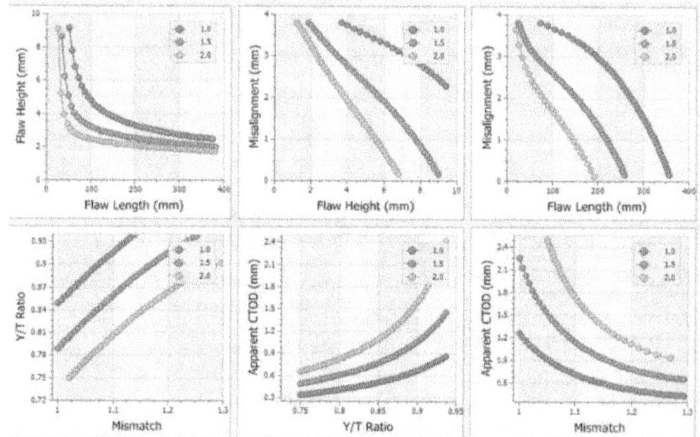

Figure 6 A sample application showing the impact of various parameters on various levels of target tensile strain capacity (1.0%, 1.5%, and 2.0%)

Estimating TSC of Vintage Girth Welds

Using the Level 4 procedures of the PRCI-CRES models, it has been shown that a weld cap reinforcement with sufficient width can prevent the formation of a plastic shear band along the softened HAZ, leading to an increase in strain capacity [37]. The same procedure can be used to examine the effects of weld cap reinforcement on manual welds with a standard bevel geometry. The Level 4a approach of the PRCI-CRES models was used to estimate the TSC of a vintage girth weld, whose macro is shown in Figure 7. The weld cap is relatively wide, spreading beyond the weld groove by approximately 2.2 mm. The crack-driving force relations are obtained using a finite element model similar to that shown in Figure 8. Typical driving force relations are shown for two different weld cap widths (Figure 9) and two weld strength mismatch levels (Figure 10). As expected, the driving force is reduced with the increase of weld cap width and weld strength mismatch level. The TSC of the welds can be estimated using the apparent CTOD toughness obtained from testing of similar welds [31].

Copyright © 2019 by ASME

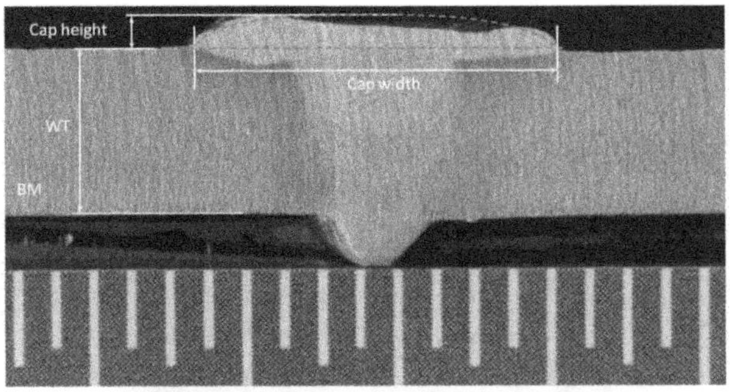

Figure 7 Macro of a vintage girth weld

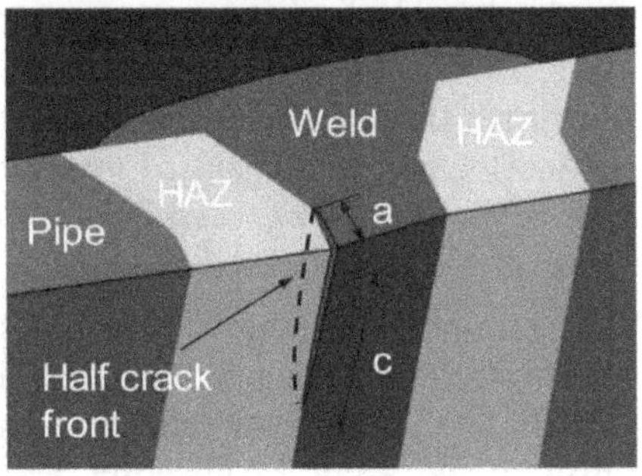

Figure 8 An example case-specific FEA model for Level 4a analysis

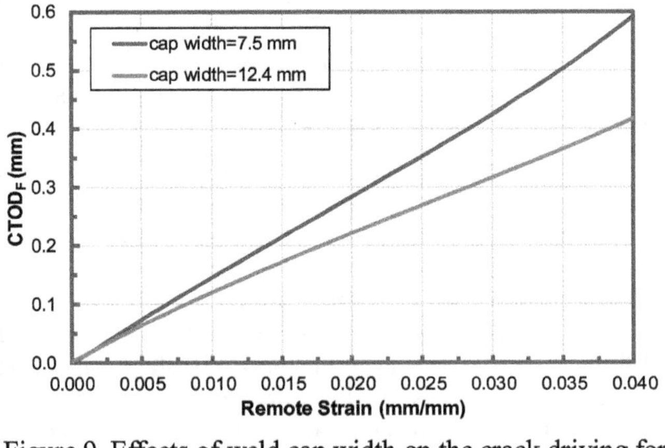

Figure 9 Effects of weld cap width on the crack driving force for a weld on a pipe of 5.2 mm wall thickness

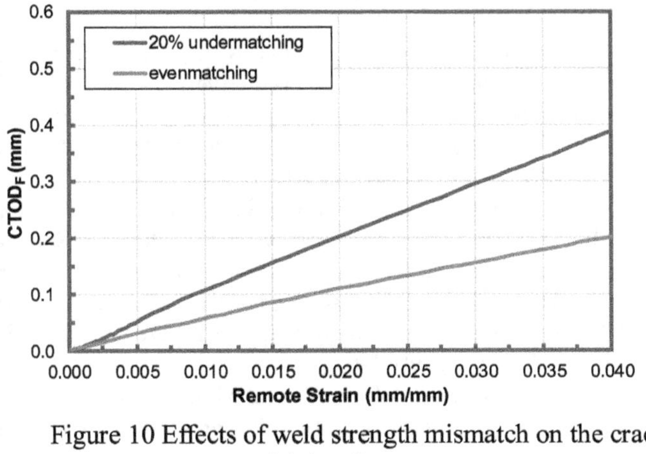

Figure 10 Effects of weld strength mismatch on the crack driving force

CONCLUDING REMARKS

Strain-based design and assessment (SBDA) is a multi-faceted task. The focus of this paper is on tensile strain models; in particular, the PRCI-CRES models.

The significant advantage of the PRCI-CRES models is their conceptual structure and flexibility. The separation of crack-driving force and toughness and the availability of two limit states offer a clear and robust framework for quick design analysis and case-specific assessment for special circumstances. Furthermore, this framework allows for continued development and refinement as linepipe manufacturing and welding processes continue to evolve.

In the current state of the PRCI-CRES models, the crack driving force relations can be obtained from available equations for narrow-grooved mechanized GMAW and standard-bevel welds with well-defined ranges of parameters. The models can be exercised to examine "what-if" scenarios, so the trade-offs and practicality of various requirements are understood and planned for in the early stages of a new project. For conditions beyond the applicable range of the equations, the crack driving force relations can be developed through material and geometry-specific finite element analysis. Having the two limit states in the models allows the use of multiple toughness options. This approach provides opportunities to make full use of existing data and new data to be generated in the future. The models can be applied to the design of new pipelines and the assessment of existing pipelines.

Conditions covered by most tensile strain models, including those of the PRCI-CRES models, are idealized so the influence of the various parameters can be rationalized and quantified. Real world conditions can be more complex than the assumed conditions in the models [38]. For instance, variations of tensile properties within a joint and joint-to-joint can affect both strain demand and strain capacity. The distribution of strain demand in the vicinity of bends and fittings can be quite different from that in a long and straight pipe segment. Field girth welds, particularly manual welds, may have greater variability than that seen in the welding procedure qualification. Field inspection and flaw acceptance criteria, including the interpretation and application of flaw interaction rules, must strike a balance between having small acceptable flaw dimensions and not requiring unnecessary repairs. Small acceptable flaw dimensions are often necessary for a high tensile strain capacity. However, repairs could lead to further deterioration of weld and HAZ properties, such as a greater level of HAZ softening, in comparison to unrepaired welds. The interactions among steel metallurgy, welding, NDT, fracture mechanics, and field conditions must be understood to successfully apply SBDA.

Copyright © 2019 by ASME

ACKNOWLEDGMENT

The work reported here represents long-term efforts supported by US DOT PHMSA (US DOT Contract No. DTPH56-06-T000014 and DTPH56-10-T-000016), PRCI, and CRES. The authors are grateful for the assistance of Dr. Dan Jia, Dr. Dunji Yu, and Mr. Alex Wang of CRES in preparing this paper.

REFERENCES

1 Wang, Y.-Y., "Strain-Based Design and Assessment – Concepts and Gaps," *Proceedings of the 1st AIM-PIMG Conference*, AIMPIMG-1067, Houston, TX, March 25-28, 2019.

2 Horsley, D. and Wang, Y.-Y., "Weld mismatch effects on the strain limits of X100 girth welds," in Pipeline Technology, *Proceedings of the 4th International Conference on Pipeline Technology*, Edited by Rudi Denys, Oostende, Belgium, May 9-12, 2004.

3 Wang, Y.-Y., Liu, M., Gianetto, J., and Tyson, B., "Considerations of Linepipe and Girth Weld Tensile Properties for Strain-Based Design of Pipelines," *Proceedings of the 8th International Pipeline Conference*, Paper No. IPC2010-31376, Calgary, Alberta, Canada, September 27 – October 1, 2010.

4 Rapp, S. and Wang, Y.-Y., "Low Strain Capacity of X70 Pipeline Girth Welds and its Implications on Linepipe Steels," *Proceedings of the 2nd International Symposium* on the Recent Developments in Plate Steels, June 3-6, 2018, Orlando, Florida, USA.

5 Wang, Y.-Y. and Liu, M., "Status and Applications of Tensile Strain Capacity Models, *Proceedings of the 6th Pipeline Technology Conference*, Ostend, Belgium, October 7-9, 2013.

6 Wang, Y.-Y., Liu, M., Stephens, M., Petersen, R., and Horsley, D., 2009, "Recent development in strain-based design in North America," *Proceedings of the 19th International Offshore and Polar Engineering Conference*, Osaka, Japan, July 21-26.

7 Sandivk, A., Østby, E, Naess A., Sigurdsson G., and Thaulow, C., 2005, "Fracture control – offshore pipelines: probabilistic fracture assessment of surface cracked ductile pipelines using analytical equations", *Proceedings of the 24th International Conference on Offshore Mechanics and Arctic Engineering*, Halkidiki, Greece, June 12-17.

8 Kibey, S.A., Minnaar, K., Cheng, W., and Wang, X., 2009, "Development of a physics-based approach for the predictions of strain capacity of welded pipelines", *Proceedings of the 19th International Offshore and Polar Engineering Conference*, p132, Osaka, Japan, June 21-26.

9 Wang, Y.-Y., Liu, M., Zhang, F., Horsley, D., and Nanney, S., "Multi-Tier Tensile Strain Design Models for Strain-Based Design Part I - Fundamental Basis," Proceedings of the 9th International Pipeline Conference, Calgary, Alberta, Canada, September 24-28, 2012.

10 Liu, M., Wang, Y.-Y., Song, Y., Horsley, D., and Nanney, S., "Multi-Tier Tensile Strain Models for Strain-Based Design Part II - Development and Formulation of Tensile Strain Capacity Models," *Proceedings of the 9th International Pipeline Conference*, Calgary, Alberta, Canada, September 24-28, 2012.

11 Liu, M., Wang, Y.-Y., Horsley, D., and Nanney, S., "Multi-Tier Tensile Strain Design Models for Strain-Based Design Part III - Model Evaluation against Experimental Data," *Proceedings of the 9th International Pipeline Conference*, Calgary, Alberta, Canada, September 24-28, 2012.

12 Wang, Y.-Y., Liu, M., and Song, Y., 2012, "Tensile Strain Models for Strain-Based Design of Pipelines," Proceedings of the ASME 2012 31st International Conference on Ocean, Offshore and Arctic Engineering, Rio de Janeiro, Brazil, July 1-6.

13 Wang, Y.-Y., Zhang, F., Liu, M., Cho, W., and Seo, D., "Tensile Strain Capacity of X80 and X100 Welds," *Proceedings of the ASME 2012 31st International Conference on Ocean*, Offshore and Arctic Engineering, Rio de Janeiro, Brazil, July 1-6, 2012.

14 Østby, E, "New strain-based fracture mechanics equations including the effects of biaxial loading, mismatch and misalignment", *Proceeding of the 24th International Conference on Offshore Mechanics and Arctic Engineering*, Halkidiki, Greece, June 12-17, 2005.

15 Nyhus, B., Østby, E, Knagenhjelm, H.O., Black, S., and RØstadsand, P.A., "Fracture control – offshore pipelines: experimental studies on the effect of crack depth and asymmetric geometries on the ductile tearing resistance", *Proceedings of the 24th International Conference on Offshore Mechanics and Arctic Engineering*, Halkidiki, Greece, June 12-17, 2005.

16 Minnaar, K. Gioielli, P.C., Macia, M.L., Bardi, F., Biery, N.E., and Kan, W.C., 2007, "Predictive FEA modeling of pressurized full-scale tests", *Proceedings of the 17th International Offshore and Polar Engineering Conference*, p3114, Lisbon, Portugal, Jul 1-6.

17 Fairchild, D.P., Cheng, W., Ford, S.J., Minnaar, K., Biery, N.E., Kumar, A., and Nissley, N.E., 2007, "Recent advances in curved wide plate testing and implications for strain-based design," *Proceedings of the 17th International Offshore and Polar Engineering Conference*, p3013, Lisbon, Portugal, July 1-6.

18 Kibey, S., Issa, J.A., Wang, X., and Minnaar, K., 2009, "A simplified, parametric equation for prediction of tensile

Copyright © 2019 by ASME

strain capacity of welded pipelines", *Pipeline Technology Conference*, Ostend, Belgium, October 12-14.

19 Fairchild, D., Kibey, S., Tang, H., Krishnan, V., Macia, M., Chen, W., and Wang, X., 2012, "Continued advancements regarding capacity prediction of strain-based pipelines," *Proceedings of the 9th International Pipeline Conference*, Calgary, Alberta, Canada, September 24-28.

20 Igi, S. and Suzuki, N., 2007, "Tensile strain limits of X80 high-strain pipelines," *Proceedings of the 17th International Offshore and Polar Engineering Conference*, Lisbon, Portugal, July 1-6.

21 Igi, S., Sakimoto, T., Suzuki, N., Muraoka, R., and Arakawa, T., 2010, "Tensile strain capacity of X80 pipeline under tensile loading with internal pressure," *Proceedings of the 8th International Pipeline Conference*, Calgary, Alberta, Canada, September 27 – October 1.

22 Tajika, H., Igi, S., Sakimoto, T., Muraoka, R., Endo, S., Tsuyama, S., Arakawa, T. and Suzuki, N., 2012, "Compressive and tensile strain capacities of 48" X80 line pipes," *Proceedings of the 9th International Pipeline Conference*, Calgary, Alberta, Canada, September 24-28.

23 Verstraete, M., Waele, W., Denys, R., Hertele, S., 2012, "Pressure correction factor for strain capacity predictions based on curved wide plate testing," *Proceedings of the 9th International Pipeline Conference*, Calgary, Alberta, Canada, September 24-28.

24 Smith, S., 2012, "Development of the BS 7910 failure assessment diagram for strain based design with application to pipelines," *Proceedings of the ASME 2012 31st International Conference on Ocean, Offshore, and Arctic Engineering*, OMAE2012-83527, Rio de Janeiro, Brazil, July 1–6.

25 Wang, Y.-Y., Liu, M., Long, X., Stephens, M., Petersen, R., and Gordon, R., 2011, "Validation & Documentation of Tensile Strain Limit Design Models for Pipelines," US DOT Contract No. DTPH56-06-T000014, final report, http://primis.phmsa.dot.gov/matrix/PrjHome.rdm?prj=200

26 Wang, Y.-Y., Liu, M., and Song, Y., 2011, "Second Generation Models for Strain-Based Design," US DOT Contract No. DTPH56-06-T000014, final report, http://primis.phmsa.dot.gov/matrix/PrjHome.rdm?prj=201

27 Zhang, F., Liu, M., Wang, Y.-Y., Yu, Z., and Tong, L., "Strain Demand in the Areas of Mine Subsidence," *Proceedings of the 9th International Pipeline Conference*, Paper No. IPC2012-90641, September 24-28, 2012, Calgary, Alberta, Canada.

28 Zhang, F., Liu, M., Wang, Y.-Y., Yu, Z., and Tong, L., "Integrity of Pipeline in Area of Mine Subsidence," *Proceedings of the 9th International Pipeline Conference*, Paper No. IPC2012-90642, September 24-28, 2012, Calgary, Alberta, Canada.

29 Liu, M., Wang, Y.-Y., and Long, X., "Enhanced Apparent Toughness Approach to Tensile Strain Design," *Proceedings of the 8th International Pipeline Conference*, Paper No. IPC2010-31386, September 27 – October 1, 2010, Calgary, Alberta, Canada.

30 Wang, Y.-Y., Liu, M., Tyson, B., Gianetto, J., and Horsley, D., "Toughness Considerations for Strain-Based Design," *Proceedings of the 8th International Pipeline Conference*, Paper No. IPC2010-31385, September 27 – October 1, 2010, Calgary, Alberta, Canada.

31 Kotian, K. and Wang, Y.-Y., "Mechanical Properties of Vintage Girth Welds," *Proceedings of the 11th International Pipeline Conference*, Paper No. IPC2016-64420, September 26-September 30, 2016, Calgary, Alberta, Canada.

32 Wang, Y.-Y., Cheng, W., and Horsley, D., "Tensile Strain Limits of Buried Defects in Pipeline Girth Welds," Proceedings of the International Pipeline Conference 2004, Calgary, Alberta, Canada, October 4-8, 2004.

33 Wang, Y.-Y. and Liu, M., "The Role of Anisotropy, Toughness Transferability, and Weld Misalignment in the Strain Based Design of Pipelines," Proceedings of the 17th International Offshore and Polar Engineering Conference (ISOPE 2007), Lisbon, Portugal, July 1-6, 2007.

34 Wang, Y.-Y., Horsley, D., Mamdouh, S., and Sen, M., "General framework for Strain-Based Design and Assessment of Pipelines," *Proceedings of the 10th International Pipeline Conference*, Paper No. IPC2014-33745, September 9-October 3, 2014, Calgary, Alberta, Canada.

35 Wang, Y.-Y., Rapp, S., Horsley, D., Warman, D., and Gianetto, J. "Attributes of Modern Linepipes and their Implications on Girth Weld Strain Capacity," *Proceedings of the 12th International Pipeline Conference*, Paper No. IPC2018-78809, Calgary, Alberta, Canada, September 24-September 30, 2018.

36 Wang, Y.-Y., Jia, D., Rapp, S., Johnson, D., "Low Strain Capacity Girth Welds of Newly Constructed Pipelines and Mitigative Approaches," *Proceedings of the 1st AIM-PIMG Conference*, AIMPIMG-1064, Houston, TX, March 25-28, 2019.

37 Wang, Y.-Y., Chen, Y., and Salama, M., "Enhancing Tensile Strain Capacity through the Optimization of Weld Profiles," *Proceedings of the 10th International Pipeline Conference*, Paper No. IPC2014-33589, September 9-October 3, 2014, Calgary, Alberta, Canada

38 Wang, Y.-Y. and Liu, M., "Real World Considerations for Strain-Based Design and Assessment," *Proceedings of the 6th Pipeline Technology Conference*, Ostend, Belgium, October 7-9, 2013.

Copyright © 2019 by ASME

AIM-PIMG2019-1015

Chapter 23

DEVELOPMENTS IN STRAIN-BASED FRACTURE ASSESSMENTS – A PERSPECTIVE

Erling Østby, Rikard Törnqvist, Steinar L. Bjerke
DNV GL
Høvik, Norway

ABSTRACT

When assessing geohazards for pipelines, scenarios where the pipeline may be subjected to strains well into the plastic regime is likely to be encountered. It is of importance to assess the integrity of girth welds in this respect. Most existing fracture assessment procedures were primarily developed for assessing elastic conditions, and there is a need to develop assessment models to provide more accurate estimates of the performance of girth weld with flaws subjected to large strains.

This paper presents a perspective on the development of strain-based fracture assessment procedures. There has been a significant activity in the field resulting in valuable contributions from several parties in the last 10-15 years. The perspective presented in this paper is, however, mainly based on knowledge and results developed in the Fracture Control Offshore Pipelines JIPs.

Assessment of cracks under globally applied large strains is characterized by plasticity governing most of the opening of crack. This poses some challenges related to the applicability of general assumptions in fracture mechanics. Special features related to crack driving force, crack growth calculations and critical conditions are discussed in the paper. Further, the roles of stress-strain curve, mismatch, misalignment, and bi-axial loading in strain-based assessments, and possible ways to account for them, are highlighted.

The paper also contains discussions associated with the use of numerical modelling, strain capacity prediction models, and probabilistic assessments. The paper is concluded with a general discussion on the state of fracture assessments for scenarios with global loading in the plastic regime. It is concluded that considerable progress has been made, however, a more active coupling of recent models or numerical modelling schemes with probabilistic approaches will probably increase the benefit of the developments.

Keywords: Pipelines, fracture, strain-based

INTRODUCTION

Design scenarios associated with geohazards typically involves possible strain states in the pipeline well within the plastic region. Thus, assessment schemes must be able to provide applicable predictions of pipeline integrity in this regime. Traditional assessment procedures have often been limited to primarily elastic response and stress-based approaches. The challenge outlined above point to strain-based approaches being of particular interest for evaluation of geohazard scenarios. Especially the limits for utilization of girth welds with potential flaws is of importance to predict in a reliable way. There has been a large focus towards development of strain-based fracture assessment schemes the last 10-15 years, and several important contributions have been made in this respect. It is beyond the scope of this paper to provide a detailed overview of all this work, but some examples are referenced below. Strain-based assessment approaches are by no means a new invention, and the design curve in the PD6493 document had the applied load expressed as a strain value. The recent years have, however, seen development of more detailed approaches targeting applications to pipelines.

Adaptation of FAD-based ECA procedures for large strain scenarios has been discussed and proposed in e.g. [1-4]. Proposals for dedicated models to estimate strain capacity for pipelines have been presented in [5-8]. The development has resulted in an improved understanding of the effects of features like bi-axial loading, mismatch, and misalignment. There have been several large-scale testing campaigns reported [6,9,10], and results from these serve as a valuable basis for validation of strain-based fracture assessment model for pipelines. An increased use of advanced numerical modelling taking into account features like crack growth has also been the case [11,12], and it can be noted that the strain capacity model presented in [8] was established directly from on a large parametric study using

Copyright © 2019 by ASME

such an approach. There has also been considerable effort towards development of experimental techniques to establish relevant material data, with the development of the SENT specimen as a means to estimate ductile tearing resistance as a prominent example [13].

While the broad range of contribution from different parties is fully acknowledge, the perspective discussed in this paper is mainly taken from developments and results from the Fracture Control – Offshore Pipelines JIP and associated projects carried out in Norway the last 15 years. The outline of the paper is as follows. The primary part of the paper is dedicated to discussions concerning basic understanding of key aspects related to strain-based fracture assessments. After this the potential role of numerical modeling is illustrated with some examples, followed by a brief summary of proposed strain capacity predicting approaches. Key ingredients in a probabilistic approach to fracture assessments for pipelines are the discussed, together with possible benefits of such an approach. At the end a general discussion and concluding remarks are presented.

1 GENERAL ISSUES AND BASIC UNDERSTANDING OF STRAIN-BASED FRACTURE ASSESSMENT APPROACHES

This primary part of the paper addresses general issues and basic understanding related to strain-based fracture assessments of pipelines. The purpose is to discuss the basis for the knowledge related to key aspects of the problem. This will consider crack driving force predictions, significance of fracture mechanics parameters, crack growth aspects, failure criteria, plasticity, mismatch, misalignment, bi-axial loading and some additional aspects.

1.1 Crack Driving Force

Although failure scenarios where pipeline leakage could occur without pre-existing cracks or defects, the main focus in strain-based fracture assessment of pipeline is considering scenarios where cracks are the triggering feature of the failure. Thus, application of fracture mechanics is a natural starting point. In order to carry out fracture mechanics assessments the crack driving force, i.e. the loading of the crack as a function of the global loading, must be evaluated. Engineering fracture mechanics was initially developed around use of the linear elastic stress intensity factor, K_I, as the crack driving force parameter. For relevant defect sizes (typically less than 25-30% of the wall thickness) and pipe geometries, the magnitude of the elastic stress intensity factor is usually limited to a crack tip opening displacement (CTOD) in the order of 0.02-0.03 mm. At the same time, typical critical CTOD values observed experimentally for more ductile materials is in the range 0.5-1.5 mm (see e.g. results reported in [22]), see Figure 1. Thus, for typical strain-based fracture assessment scenarios, the contribution from the linear elastic stress intensity factor is of secondary importance. The main contribution to the crack opening rather stems from the localization of plastic deformation in the crack ligament as illustrated in Figure 2. Estimation of crack driving force for the strain-based scenarios is then mainly relying on how well the contribution to the crack opening from the localized plastic deformation is captured. This is reflected in some proposals having been made for the use of slip-line theory to develop fracture mechanics for scenarios with fully yielding in the crack ligament [14]. One key geometrical parameter in this respect is the depth of the crack, a, or relative depth, a/t, of the crack. As illustrated in Figure 3, the opening or the crack, or the crack driving force, as a function of the applied global strain increases as the relative depth of the crack increase. However, the crack driving force will also be influenced by the width of the crack in the hoop direction, 2c, see Figure 4. For cases where the crack is much wider than the depth, the influence of the width of the crack on the crack driving force in the middle of the crack, is limited. As the width become smaller, the influence from the edges of the crack is felt along the whole crack front, and the crack driving force starts to display a significant dependence on the width of the crack in addition to the relative crack depth.

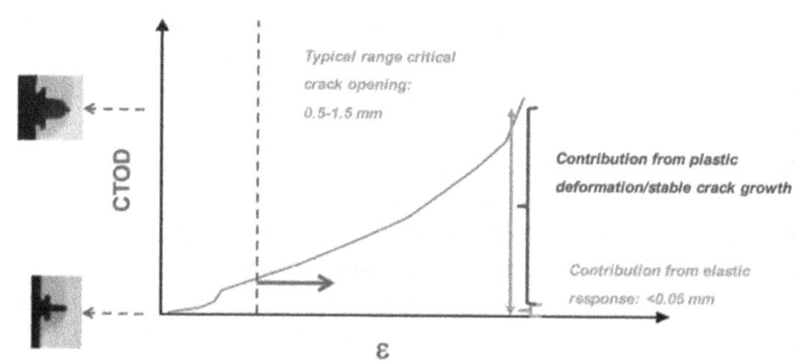

Figure 1. Contributions from elastic and plastic crack opening.

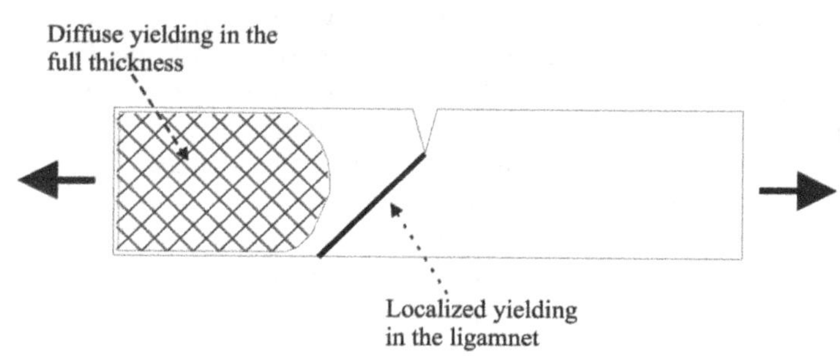

Figure 2. Schematic of localization of plastic deformation in the crack ligament.

Copyright © 2019 by ASME

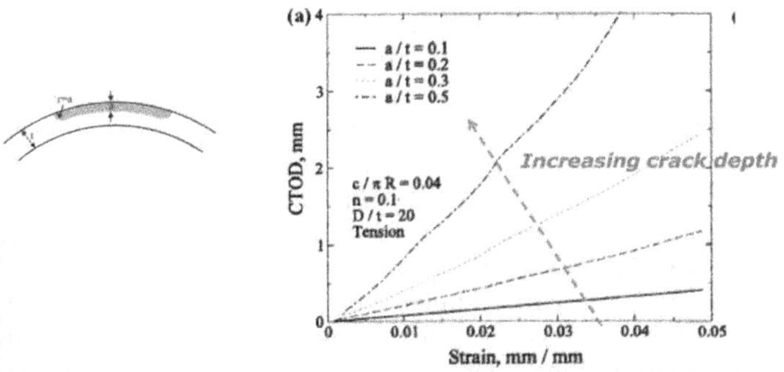

Figure 3. Crack driving force, expressed as CTOD, as a function of applied strains for different relative crack depths, a/t.

From the above it is evident that plasticity is a key factor for the crack driving force in strain-based fracture assessments. Plasticity will also have some additional effects that could be of importance to consider in strain-based assessments. As plastic deformations associated with the localization of strains in the ligament can be large, they may lead to geometrical changes of significance for the crack driving force as e.g. shown in Figure 5. Whether such effects are accounted for or not may lead to rather different crack driving force behavior as illustrated in Figure 6. Neglecting large deformation effects results in a linear relation between the crack opening and the applied strain, whereas including the effect results in non-linear and faster evolution of the driving force due to the necking formation on the opposite side of the pipe wall.

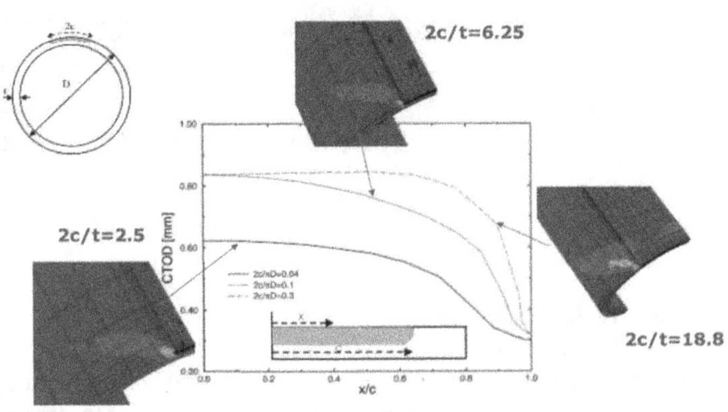

Figure 4. Influence of crack width, 2c, on the crack driving force, expressed as CTOD, along the crack front for different crack widths evaluated at the same globally applied strain.

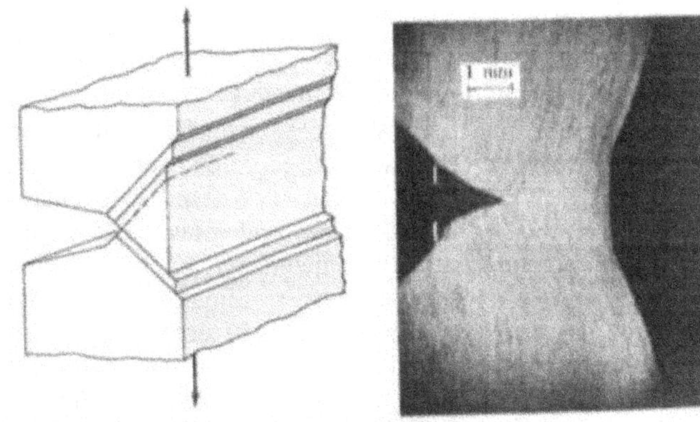

Figure 5. Large geometry changes associated with crack opening and growth of cracks. [15]

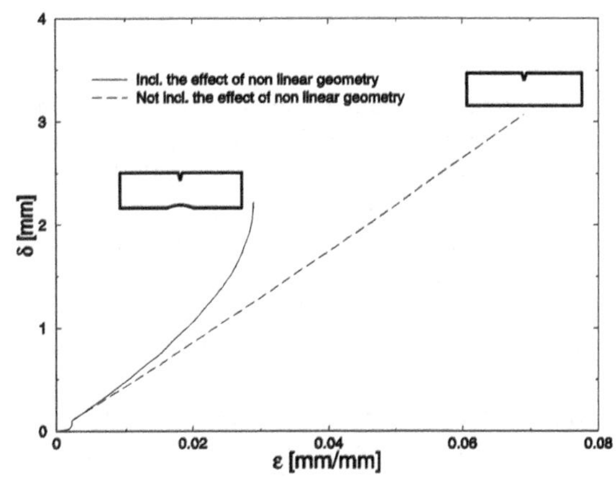

Figure 6. Effect of including large deformation or non-linear geometry considerations in crack driving force calculations. Crack driving force expressed as CTOD denoted δ.

1.2 Significance of Fracture Mechanics Parameters

A basic concept in fracture mechanics is that a crack remains stable until the applied crack driving force becomes equal to the materials resistance to crack propagation. The critical conditions for crack propagation has been associated with a critical energy being released from the material during crack propagation, i.e. a critical energy release rate has been achieved. However, energy release only takes place from elastic unloading of the material. As discussed above, the elastic deformation only contributes to a small part of the driving force, thus, the concept of a critical energy release rate becomes ambiguous for scenarios where large-scale plastic deformation takes place. In elastic-plastic fracture mechanics parameters like the J-integral [16] and CTOD [17] have been introduced to consider the contribution from plastic deformation to the crack driving force. For small scale yielding it has been shown that the J-integral, J, and the CTOD are related as follows [18]:

$$J = m * \sigma_y * CTOD \quad (1)$$

Copyright © 2019 by ASME

where m is a constant primarily depending on the work hardening of the material and σ_y is the yield stress. By applying concepts from slip-line theory it is possible to show that the same relation apparently also holds for large deformation scenarios. This is outlined further in Appendix A. Two main aspects may be extracted from this. First, under large scale plastic deformation scenarios, fracture mechanics parameters may not be interpreted as energy release estimates, but rather as estimates of the opening of the crack. Second, relations between different fracture elastic-plastic mechanics parameters are also to a first order approximation valid under large-scale yielding conditions, however, their physical meaning may have to be interpreted differently from the basic concepts in fracture mechanics. These are important aspects to keep in mind when discussing strain-based fracture assessment schemes.

1.3 Crack Growth

For strain-based scenarios the assessment of crack growth plays a vital role. Traditionally a distinction has been made between brittle unstable crack propagation and potentially stable crack extension. Brittle crack extension can be associated with a cleavage fracture propagation mode. Cleavage fracture is typically not acceptable for strain-based fracture scenarios, and should be avoided. Some limited ductile crack extension may however be acceptable as long as it can be controlled to be small. The mechanism of ductile crack extension is believed to be due to void nucleation/growth and eventually coalescence of voids. The mechanism can be well described using models like the Gurson-Tvergaard-Needleman model [19] or similar concepts. Such models predict that the resistance to crack propagation is sensitive to the local stress triaxiality in addition to the materials parameters. Although, as seen below, it is possible to explicitly modelling crack growth in FEA, the resistance to crack extension is usually measured experimentally. The use of SENT specimens to obtained crack growth resistance for materials has become widely adopted in the pipeline industry, and the method has quite recently been standardized [20]. Crack extension due to influence of hydrogen may also be of relevance to assess, however, this is currently subject to active research and there are still several uncertainties related to the best way to perform such assessments.

In parallel to determination of resistance to crack growth, procedures to calculate crack growth must also be available. Whereas the basis for static fracture mechanics is rather well established analytically, the situation regarding crack growth is less so. There are some semi-analytical results for simple cases assuming perfectly plastic material, but in general no rigorous analytical approach exists for crack growth calculations. Some conceptual challenges exist in that the crack tip moves and the applicability of the concepts from static fracture mechanics is associated with some limitations. Some researchers have argued that the CTOA (crack-tip-opening-angle) may be a better measure of the resistance to crack propagation for growing cracks. The CTOA tends to increase for materials with increasing crack growth resistance as illustrated in Figure 7. The CTOA concept is however not developed in detail for assessment of crack growth in pipelines. It has been argued that as long as crack extension remains limited compared to other length scales in the problem (crack depth, ligament height, etc.) use of fracture mechanics based on concepts for stationary cracks should be sufficiently accurate. Thus, from an analytical point of view, crack growth calculations today are primarily carried out based on assumed or measured crack growth resistance curves and general crack driving force concepts. An alternative approach is to use numerical simulations with micromechanical models for materials failure embedded in the analysis. As discussed in Section 2 such approaches are potentially powerful alternatives, and there are now examples on large-scale applications to pipeline fracture assessment. Application of numerical analyses explicitly including crack growth calculations circumvent some of the conceptual challenges identified above when applying basic fracture mechanics concepts to crack growth analysis (but are associated with challenges both related to computational resources and calibration of micromechanical models).

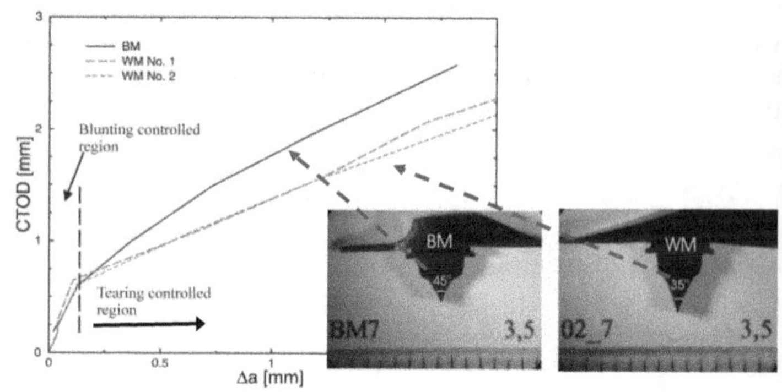

Figure 7. Correlation between CTOA and CTOD-R curves.

1.4 Failure Conditions

The definition of failure or critical condition is also of crucial importance for the problem. Traditional FAD approaches have operated with two failure conditions: 1) unstable crack propagation and 2) plastic collapse of the crack ligament. Initiation of crack growth has also been used in some cases as the critical condition. For materials displaying a rising R-curve behavior, i.e. a situation where the toughness increases with crack extension, the tangency criterion has been applied to determine the onset of unstable fracture. Although this provides an operational definition of practical importance, it has some conceptual challenges associated with its application. As discussed above, the fracture mechanics parameters calculated under large plastic deformation do not directly represent genuine energy release measures. Also, the history effects embedded in the problem add some complicating factors. In [21] a more theoretical discussion regarding assessment of critical conditions

Copyright © 2019 by ASME

for crack growth under highly non-linear material response, is presented. It is beyond the scope of this paper to go into details around this, however, one special result is mentioned. The critical point regarding crack growth coincides with the maximum load bearing capacity in the specimen. In the Fracture Control Offshore Pipelines projects this result formed the background for the formulation of alternative failure criteria [22]. The basis for the failure criteria was reaching the CTOD corresponding to the maximum load in the crack ligament. Two different formulations were considered: 1) stationary crack without crack extension and 2) growing cracks. The CTOD at maximum load, δ_c, for stationary crack was estimated to be:

$$\delta_c = \sqrt{2}L\varepsilon_{TS} \tag{2}$$

where L is the crack ligament height and ε_{TS} is the strain corresponding to the tensile strength of the material. The motivation was that CTOD could be estimated as the general strain in the crack ligament multiplied by $\sqrt{2}L$. Invoking the Consideré condition, the maximum load should coincide with the strain reaching ε_{TS}. In **Figure 8** the predictions from Eq. (2) (dashed line) are compared with axi-symmetrical FEA CTOD vs applied strain curves for materials with different Y/T, i.e different ε_{TS}. It can be noted that the model quite well predicts the onset of rapid growth of CTOD, thus it provides a first order estimate of the CTOD at maximum load for stationary cracks.

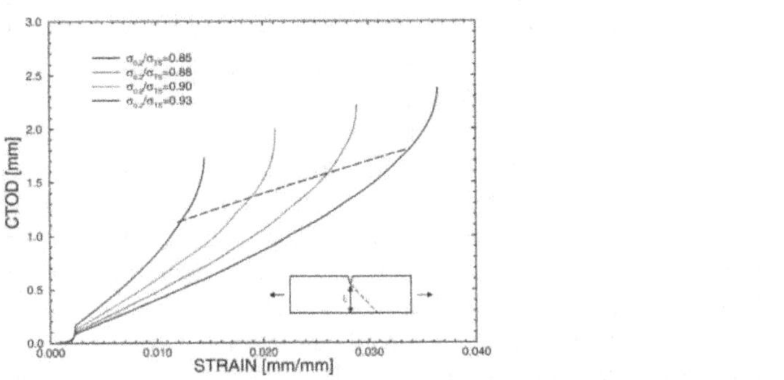

Figure 8. Comparisons of predicted CTOD at maximum load from Eq. (2) (dashed line) and FEA CTOD vs. strain curves for stationary cracks with different Y/T-ratios.

The situation is more complicated when accounting for possible crack growth. No semi-analytical model was developed for this case, however, the following equation for CTOD at maximum load in the crack ligament was calibrated from numerical analyses including crack growth modelling:

$$\delta_{max} = (0.03L + 2.84\delta_t - 0.61)(-12.1(\sigma_{0.2}/\sigma_{TS})^2 + 18.9(\sigma_{0.2}/\sigma_{TS}) - 6.28) \tag{3}$$

where L is the current crack ligament and δ_t was the CTOD at 1 mm of ductile tearing. The new failure criterion, together with a new approach to calculate ductile tearing, was compared against numerical results and use of the tangency criterion, see **Figure 9**. In general, it was observed that the new model results in strain capacity predictions that are in better agreement with numerical simulation results compared to application of the tangency criterion, especially in case of larger strain capacities. This was mainly taken as an indication of limitations of the tangency approach for large strain scenarios when using crack driving force relations based on stationary cracks.

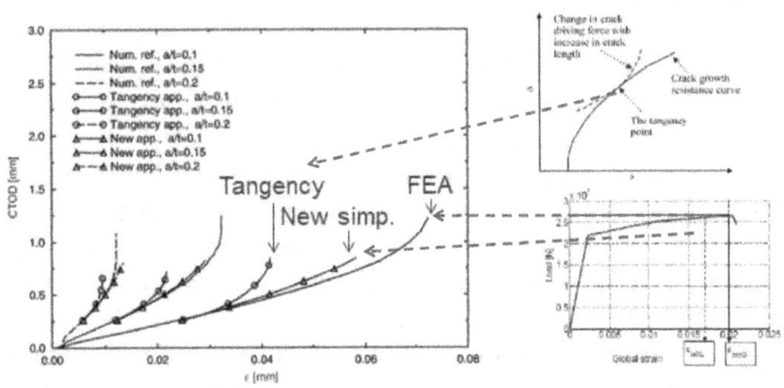

Figure 9. Comparison of strain capacity predictions using different models.

Despite the progress made there are still some uncertainties and complicating aspects related to determination of the critical conditions in strain-based assessments. While the basis for fracture mechanics is the calculation of a local parameter and determining when this reaches a critical value, the failure in strain-based scenarios (assuming the material not to be brittle) is more a combination of critical crack growth and the load carrying capacity of the crack ligament. In this respect the whole crack geometry plays a more important role. It can be mentioned that the failure criteria discussed above yield rather good predictions when compared to cracks where the width of the crack is much longer than the depth, but conservative predictions for cracks with smaller widths where the edges of the crack have a significant influence on the load carrying capacity. Again, these challenges may to some extent be circumvented if numerical simulations where the crack growth process is explicitly included, are applied.

1.5 The Role of Plasticity Properties

As the primary part of the crack opening is due to plastic deformation, the plasticity properties of the material are important in strain-based scenarios. Plasticity properties comprise yield stress, work hardening and yield surface representation. The absolute yield stress is of secondary importance when considering the evolution of CTOD vs applied global strain. The J-integral scales with the yield stress, and thus, the absolute yield stress will affect this parameter (see Eq. (1)). The work hardening is, however, of primary importance. As the work hardening decreases, the crack driving force as function of applied strain increases, as strains more easily localize in the crack ligament for lower work hardening materials. The stress-strain response of materials may also display features like Lüders plateau. Both experimental observations and numerical simulations have shown that the presence of Lüders plateau can

Copyright © 2019 by ASME

affect the crack driving force (and eventually strain capacity) significantly. It is seen that a strong increase in crack driving force happens just before general yielding in the wall thickness is reached. The additional increase in CTOD, $\Delta\delta_L$, for this case can to first order be estimated as:

$$\Delta\delta_L \approx \sqrt{2}\, L\varepsilon_L \qquad (4)$$

where L is the crack ligament height and ε_L is the length of the Lüders plateau. The presence of a Lüders plateau can have a stongly non-linear effect on the strain capacity. For shallow cracks and/or moderate Lüders strain the effect may be beneficial, whereas for deeper cracks and/or long Lüders plateau the strain capacity may be significantly reduced (see e.g. sensitivity to Lüders plateau for strain capacity in SENT specimens reported in [34]). The influence of yields surface representation has to little extent been investigated with regards to the impact on strain-based fracture assessments, and an assumption that the von Mises model applies is dominating in reported results in the literature.

1.6 Mismatch

Mismatch in properties between base material and weld metal can potentially have a large impact on the strain capacity of pipeline girth welds with defects. Mismatch has primarily been considered concerning its effect on the crack driving force but could potential also affect crack growth resistance due to constraint effects or modification of crack growth paths. Regarding the crack driving force, it has been recognized for some time that the effect of mismatch may be estimated by providing modified expression for the plastic limit load. The plastic limit load for welds are governed by the geometry of the weld and the level of mismatch. In the Fracture Control Offshore Pipelines project [23] it was shown that the modified limit load, see Figure 10, could be expressed through a rule of mixture where the limit load could be calculated as a weighted average of the limit load for the two materials (base and weld). This result was further used to modify crack driving force relations by introducing an effective crack depth that was dependent on the mismatch/weld geometry. Modified limit load may also be used in stress-based FAD formulations, e.g. as basis for effective stress-strain curve definition. One challenging aspect in this respect is however, that if a J-integral approach is applied, the evaluation of the J-integral will depend on where the contours for the evaluation are defined. Also, the mismatch phenomenon is more complicated due to features like mismatch in work hardening and asymmetrical crack opening.

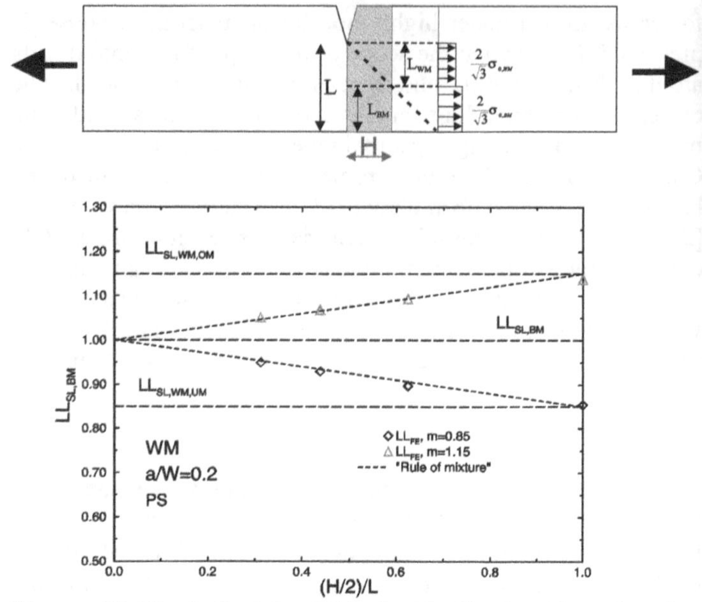

Figure 10. Plastic limit load, LL, for idealized welds as function of weld geometry, (H/2)/L, following a rule of mixture relation. Both overmatch, m=1.15, and undermatch, m=0.85, cases shown.

1.7 Misalignment

Due to tolerances, out-of-roundness features and challenges in aligning subsequent linepipe segments perfectly during welding, misalignment has to be considered in fracture assessment of pipelines. Under globally elastic conditions the primary effect of the misalignment is to introduce a secondary bending moment over the wall thickness. For scenarios where the global loading results in large scale plastic deformation the picture is complicated by the fact that the secondary bending moment becomes dependent on the applied loading. The Neuber rule has been proposed as a potential way of accounting for the effect of misalignment in fracture assessment of large scale yielding scenarios (see e.g. [1]). In the Fracture Control Offshore Pipelines project some success was obtained by considering how the misalignment affected the limit load of the crack ligament, assuming this to be the primary influence [23]. In as similar fashion as for mismatch, the modified limit load was used to estimate an effective crack depth to be used in crack driving force estimations. Apart from affecting the crack driving force, misalignment can introduce a significant asymmetry in the crack opening, as illustrated in Figure 11. The asymmetrical crack opening can again affect crack propagation leading to crack growth deviation. An attempt to address this was made in the Fracture Control Offshore Pipeline projects by considering an asymmetrical CTOD at maximum load carrying capacity in the crack ligament. This approach was, however, never validated in more detail. While there are several experimental studies including misalignment, there are no analytical formulations considering the possibility of crack growth deviations.

Copyright © 2019 by ASME

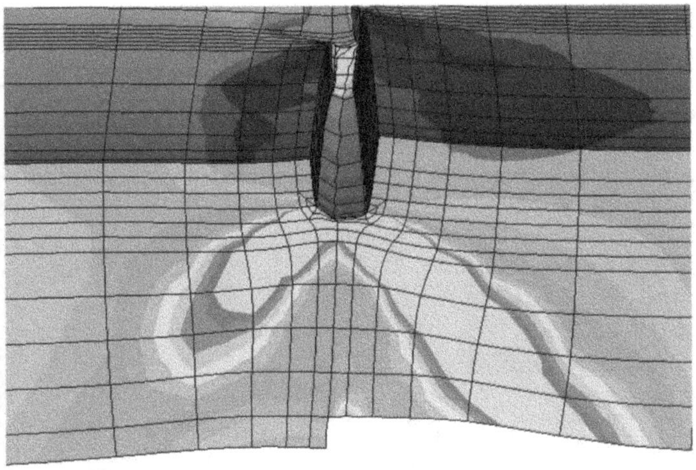

Figure 11. Asymmetrical crack opening from FE simulation of the effect of misalignment.

1.8 Bi-axial Loading

Assessment of pressurized pipes has received much attention in development of strain-based fracture procedure. In traditional fracture assessment procedures, the effect of bi-axial loading, due to an out-of-plane stress, has only had a marginal effect on the outcome of the analysis. However, both large-scale tests [9,10] and numerical simulations [24,25] have pointed to a significantly different strain capacity for pressurized pipes compared to pipes without internal pressure. This difference has been assessed both from a fracture resistance and crack driving force perspective. Some earlier results opened for the possibility that bi-axial load could reduce the resistance to initiation of brittle fracture [26], but later studies have not supported this proposal. For ductile materials both numerical simulations [27] and experimental observations [28] have shown an insignificant effect of bi-axial loading on the tearing resistance. Thus, the fracture toughness is apparently not strongly influenced by the bi-axial loading. A qualitative explanation for this can be provided by considering the stress and strain fields in front of the crack-tip. The out-of-plane loading only affects the stress field far ahead of the crack tip for this loading condition (bi-axial loading in the crack plane, however, has a much stronger effect). In the large deformation zone, where the primary part of the void growth takes place, there is no significant effect of the bi-axial loading, i.e. little effect on the resistance to ductile tearing could be expected. Based on this, the primary effect of bi-axial loading is thus believed to be due to a difference in crack driving force. In the Fracture Control Offshore Pipelines project this difference was explained by the different effect the bi-axial loading had on the resistance to plastic deformation in the crack ligament and full wall thickness, respectively. In the crack ligament, where the strains are highly localized, this deformation always takes place under predominantly plane strain conditions. Due to this, the load carrying capacity of the ligament is not sensitive to the out-of-plane stress level. The effect of bi-axial loading in the full pipe wall thickness is different. Here the axial stress at yield is given by the shape of the yield surface. Assuming a von Mises yield surface the effective axial stress at yield will be higher for a positive bi-axial loading ratio. In strain-based assessment procedures the crack driving force is evaluated against the globally applied strain. This results in that when the crack driving force is evaluated against the global strain, the crack ligament will have to carry a higher load under bi-axial loading compared to the case of uniaxial loading (i.e. an unpressurized pipe). The effect of this is illustrated in Figure *12*, where it can be observed that for the same applied global strain, the strain localization in the crack ligament, and thus the crack driving force, is strongly dependent on the level of bi-axial loading (or hoop stress).

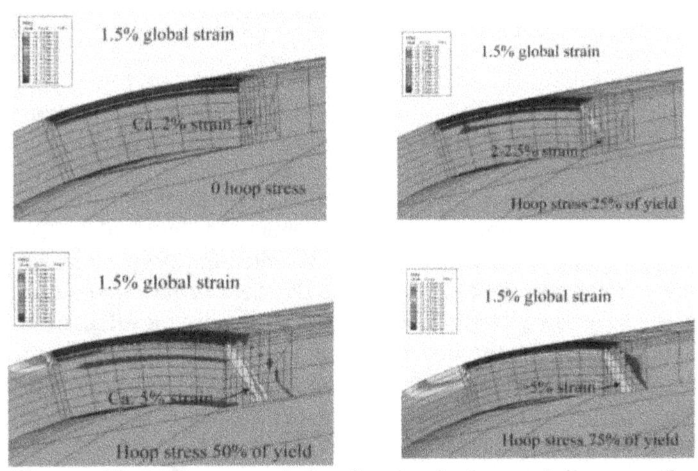

Figure 12. Different strain localization in the crack ligament for the same applied global strain under different bi-axial loading ratios/different applied hoop stress values.

The effect of bi-axial loading on the crack driving force as predicted from FEA is shown in Figure 13. It is observed that the effect of the ratio between the hoop stress, σ_h, and the yield stress, $\sigma_{0.2}$, strongly affects the crack driving force as the ratio increases from 0 to ca. 0.5, after which is more or less saturates. In the Fracture Control Offshore Pipelines projects it was proposed that the effect of bi-axial loading on the crack driving force could be accounted for by introducing an effective wall thickness that depended on the bi-axial loading ratio. This would eventually affect the effective relative crack depth, and thus the crack driving force [23]. The proposal was shown to be a reasonably good approximation by comparisons with large-scale test results [5].

Copyright © 2019 by ASME

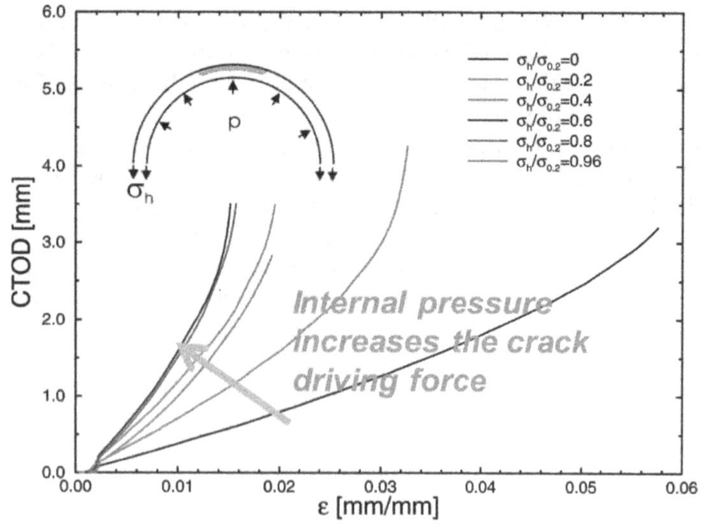

Figure 13. The effect of bi-axial loading, expressed as hoop over yield stress, on the CTOD vs. applied global strain.

Based on the discussion above the effect of bi-axial loading on the crack driving force should depend on the yield surface shape. Numerical analysis results presented in [29] have shown this to indeed be the case. Changing the yield surface shape from a von Mises to a Tresca model virtually removes the effect of bi-axial loading on the crack driving force as a function of the applied global strain. It should, however, be mentioned that most comparisons with large-scale test results have pointed towards an effect of bi-axal loading in the range predicted by assuming a von Mises yield surface.

1.9 Other Aspects

The discussion above has addressed most of the key aspects related to strain-based fracture assessment of pipelines. There are, however, more aspects of potential importance, and some of these are very briefly mentioned in this section. Most of the results obtained have been for surface breaking defects. There are some uncertainties related to how to treat embedded defects, and there are indications that existing procedures to re-categorize such defects to surface braking defect may be non-conservative for strain-based assessment [30]. Procedures for how to carry out assessment of cyclic crack growth also have some uncertainties attached to them. In cases where slant crack growth is governing the capacity, how to predict and assess this is currently not very well established. As mentioned above, the yield surface or flow potential could possibly affect the behavior, and some indications of a significant effect in this respect have been reported [30]. Also, the quantitative influence of residual stresses in association with strain-based fracture assessment is not very well established.

2 NUMERICAL MODELLING

The non-linear effect of many of the parameters, and their interactions, makes the use of numerical tools like FE modelling an interesting approach in strain-based fracture assessments, see **Figure *14***. There is now a large body of papers reporting on the general use of FEA for assessing crack driving force calculations, and this problem will not be given a general discussion in this paper (see. e.g. [31] for a broader discussion). Instead, focus is being put on the capabilities of FEA to capture some more advanced features illustrated by a couple of examples.

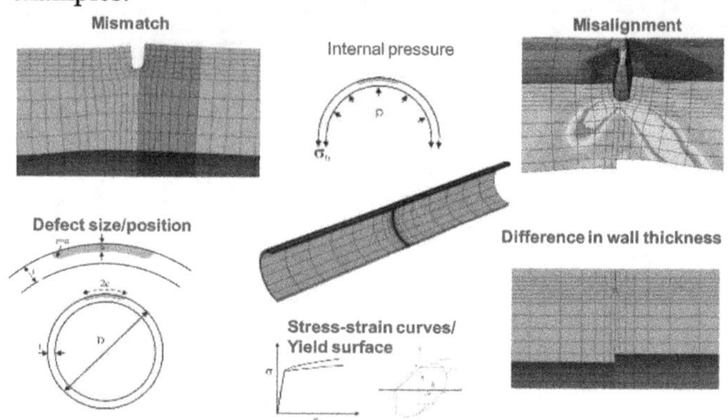

Figure 14. Application of FEA in crack driving force calculations in strain-based fracture assessment.

The first example considers modelling of ductile tearing in a SENT test of an X65 seamless pipe [32]. The crack growth was modelled using the complete Gurson model proposed in [33]. A silicone replica technique was used to monitor the CTOD and crack growth during the test. **Figure *15*** shows comparisons between Load vs CMOD (directly measured parameters) and CTOD vs. Δa (parameters inferred from silicone replicas). It can be noted that very good correspondence between experimental measurements and numerical simulation results are obtained in both case. Especially the capability of the simulations to capture the transition from crack blunting to ductile tearing is noted.

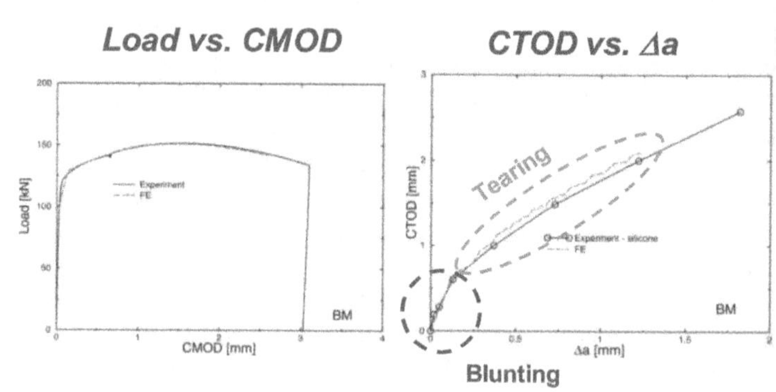

a) b)

Figure 15. Comparison of experimental and FEA results for a SENT specimen. a) Load vs. CMOD. b) CTOD vs crack extension, Δa.

The second example considers mismatch, Lüders plateau, and scatter in material properties [34]. Numerical modelling was carried out for a SENT specimen with the crack in the center of a highly overmatched weld metal, see Figure 16. Tensile testing was performed to establish the scatter in the properties for both

Copyright © 2019 by ASME

the weld metal and the base material (which displayed a Lüders plateau in the stress-strain curve). When applying the scatter in measured tensile curves in the FEA, it can be observed that the scatter in experimental CMOD vs applied strain results are well captured. Further, the highly non-linear response in the CMOD vs applied strain curves due to the Lüders plateau in the base material is also well described by the FEA. Thus, the example demonstrates the capability of the FEA to handle highly non-liner material stress strain behavior and non-homogeneous material properties.

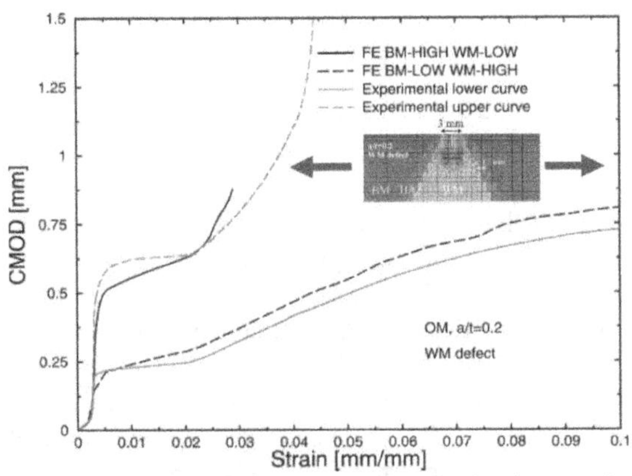

Figure 16. Comparison of experimental observations and numerical modelling of CMOD vs strain in SENT specimen with crack in overmatched weld metal.

The last example concerns comparison with large-scale tests of pressurized pipes. The pipe in the examples was an X65 seamless pipe with an internal pressure corresponding to a hoop stress of 60% of yield. Detailed discussions regarding FEA can be found in [12,35]. As shown in Figure 17 the numerical simulations are capable of capturing both the CTOD vs. global strain and CTOD vs Δa response (including the transition from blunting to ductile tearing.)

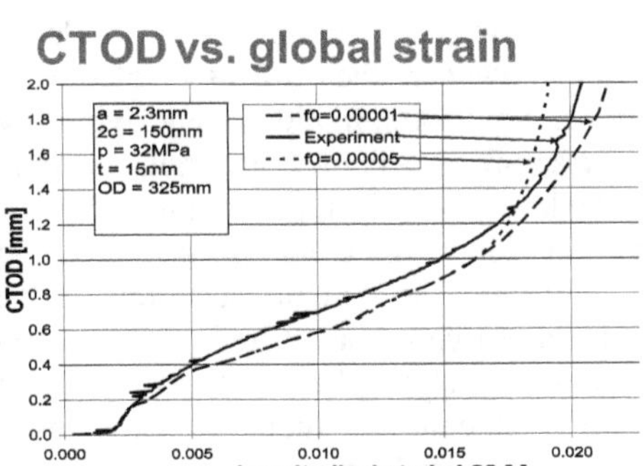

a)

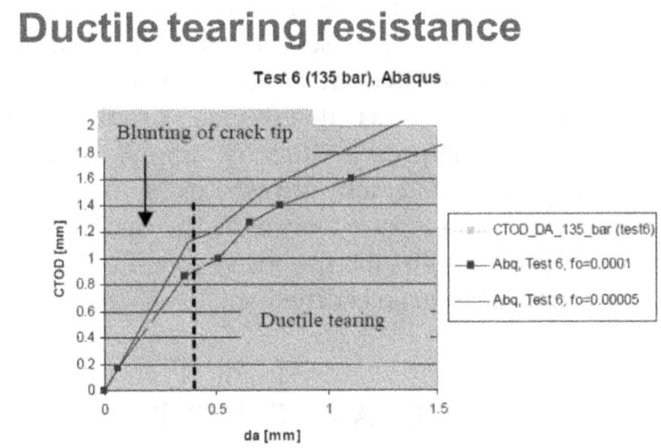

b)

Figure 17. Comparison between large-scale test result and FEA of pressurized pipe. a) CTOD vs. global strain. b). CTOD vs. Δa.

The examples above serve to illustrate the potential capability of numerical simulation to capture most features observed in association with strain-based fracture assessments with very good precision. It is, however, underlined that the topic is not trivial and still poses challenges both in terms of computational resources, calibration and definition of scatter in input parameters. The second-generation strain capacity model presented in [8] represents the largest effort as of today in applying a modelling approach along these lines for practical purpose. Less computationally expensive approaches based on estimation of crack driving force from simulations of stationary cracks can be of significant value too. However, use of such results are associated with general uncertainties in the applicability of traditional fracture mechanics as discussed in the sections above. Also, alternative approaches combining the use of shell elements and line-spring elements have been shown to hold potential for assessment of pipelines [36].

3 APPROACHES FOR STRAIN CAPACITY PREDICTIONS

It is not the intention of this paper to provide an in-depth review of different approaches for strain capacity predictions, however, some general remarks are made. The majority of fracture assessment for strain-based assessment scenarios are still carried out with basis in the FAD approach outlined in BS7910. The main procedures for evaluation of the fracture and fatigue limit state for pipelines presented in DNVGL-RP-F108 is in general founded on the use of a FAD approach, however, specific guidance towards the use for situations with large strain and internal pressures are included in the procedures. The use of numerical modelling in association with FAD approach is also seen to an ever-increasing extent. Proposals for dedicated strain-based fracture assessment models have also been presented in the literature [5-8]. The model presented in [5] considered the use of FEA to derive general crack driving force relations, which was modified to account for mismatch, misalignment, and bi-axial loading through the concept of effective crack depths.

Copyright © 2019 by ASME

Critical conditions were assessed against the criteria discussed in Section 1.4. The proposal in [6] used FEA directly to capture the effects of different parameters in the crack driving force. The ultimate strain capacity was determined by the tangency criterion. The only approach directly applying explicitly numerical modelling of crack growth in association with development of a strain-capacity model is the one presented in [8]. It is noted that none of the proposed specific strain-based assessment models have been codified yet.

4 RELIABILITY BASED ASSESSMENT SCHEMES

General pipeline code requirements for different limit states are often formulated as acceptable failure probabilities, with levels depending on the consequence of failure. Whereas probabilistic fracture mechanics is by any means not a new concept, estimating the reliability of fracture assessments are in general challenging. Thus, fracture assessments are in most cases carried out as semi-deterministic analyses. This has partly been driven by a general understanding of the approaches being applied to be conservative, with experience mainly supporting this assumption. However, with the current trend of challenging this and aiming for more accurate assessment models, this could require some further need for better reliability assessment as earlier "built-in", but not quantified, safety factors are being reduced or removed.

The problem of carrying out probabilistic fracture assessments under strain-based conditions has been discussed in more detail in [37], and a possible approach demonstrated. The approach used has been to apply the simplified strain-based fracture assessment model presented in [5] in a probabilistic framework. In addition to the prediction of probability of pipeline failure, the approach is also used to calibrate features like partial safety factors.

The general considerations needed for such an approach are shown in Figure 18. Basically, two main aspects should be considered: the uncertainty in the model applied and the uncertainty in input parameters. The first aspect, often termed model uncertainty, is addressing the accuracy to which the physics of the problem is captured by the models applied. Although historically the model uncertainty in fracture assessment approaches should be considered as high (although probably biased towards the conservative side), the recent development and validations of strain-based models now opens for much more accurate model uncertainty estimations. Definition of scatter in input parameters is not trivial, however, often reasonable estimates should be possible to provide. One main challenge is related to uncertainty in defect distributions. Results presented in [37] illustrates how such estimates may be derived from AUT qualifications and estimates of PoD (probability of detection) for the methods applied. There are additional effects, e.g. system effects, that should be incorporated in the probabilistic models.

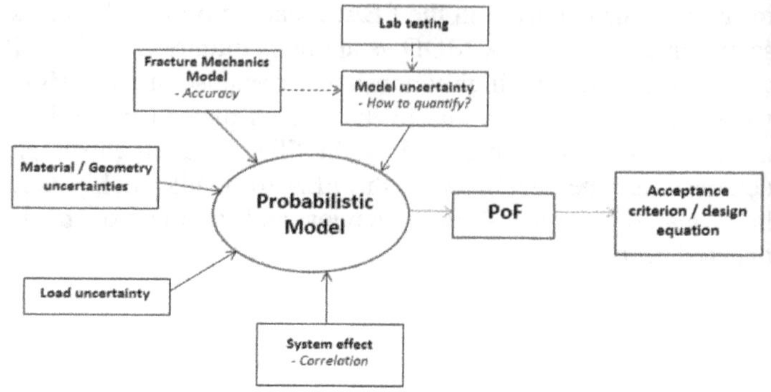

Figure 18. General considerations for development of probabilistic fracture assessment schemes

It is argued that a more wide-spread application and further development of probabilistic approaches would be beneficial for a more optimized use of the recent developments. Application of new approaches/models can be easier if reliable estimates that failure probabilities are sufficiently low can be presented. Further, the approaches can form the basis for determination of partial safety factors for use in design. Linked up with more accurate fracture assessment models, the possibilities for performing better sensitivity studies may also help in targeting the primary parameters governing most of the failure probability. Finally, often there is a need to consider multiple loading steps, including cyclic loading, and a probabilistic approach could help in a more rational consideration of such scenarios.

5 DISCUSSION

Above a perspective regarding the general understanding of aspects of primary importance for strain-based fracture assessments, together with some general remarks concerning numerical modelling, strain capacity prediction models, and probabilistic analyses, has been presented. Existing fracture assessment procedures are in general believed to be applicable, at least if necessary modifications to account for large plastic strain scenarios are made, however, they are associated with some potential limitations. Some assumptions underlying the general application of fracture mechanics are not, or only approximately, valid under large globally applied strains (e.g. the energy release rate interpretation of the crack driving force and the tangency criterion). Further, the general accuracy of the assessment schemes is to little degree quantified. This renders rational use of e.g. probabilistic approaches challenging. Some of the shortcomings of conventional methods have been removed, or reduced, with the development of special strain-based fracture assessment approaches. Especially the focus on the key physics governing and rather extensive validation of such models allows a better quantification of the model uncertainty. This again opens for more rational use of reliability-based approaches, and possibly better ways to assess multiple load steps scenarios. It is the view of the authors that further benefits

Copyright © 2019 by ASME

of the recent developments may be harvested if probabilistic fracture approaches are further developed. Numerical analysis is also seen to be increasingly used in relation to fracture assessment for large strain scenarios. Potentially such modeling may embed the actual crack growth process, and critical conditions may be extracted directly from the analysis results. If failure mechanisms are correctly modelled such procedures offer the possibility for highly accurate predictions. However, the approach is somewhat hampered by being computationally expensive and depending of sufficient variability in properties being considered to provide significant results. Less advanced use of FEA like estimation of crack driving force relations is also of value. Although several companies have developed internal procedures for how to carry out such analyses, there is still, however, a lack of general guidance or code recommendations in this respect. When it comes to the detailed physical understanding and available models to describe the phenomena, there are still room for improvements. The critical conditions or estimation of limits for failure modes primarily governed by slant fracture is less well understood. The quantitative impact of residual stresses for scenarios with large applied plastic strains is subject to some uncertainty. Also, models for predicting cyclic crack growth could possibly be revised. Difference between e.g. offshore and onshore pipeline design, e.g. in inspection qualification requirements, must also be recognized. Working with physically sound fracture models and formulating the assessment within a probabilistic framework, would, however, help in handling such differences in rational ways, and optimizing the general utilization of the advances in knowledge.

6 CONCLUDING REMARKS

This paper has presented a perspective on the development of strain-based fracture assessment procedures for pipelines. Significant progress in understanding and proposal for how to carry out such analyses have been made in the recent years. The authors are of the view that further benefits of these developments can be harvested if the models or numerical modelling approaches are linked with probabilistic assessments. Certain aspects of relevance would also benefit from further development of the physical understanding of the phenomena.

ACKNOWLEDGMENTS

The support from participants in The Fracture Control Offshore Pipelines JIP, Phase I and II, is gratefully acknowledged. The additional support for Equinor (former Statoil) for development of reliability-based fracture assessment schemes is specially acknowledged.

References

[1] DNVGL-RP-F108. Assessment of flaws in pipeline and riser girth welds. October 2017.

[2] Tkaczyk, T (2010). Development of Fracture Mechanics Criteria for failure of Reeled Pipes. Ph.D. thesis

[3] Sriskandarajah, T., Zhou, D.(2013). Engineering critical assessment of pipeline under axial straining with internal pressure. Proceedings of the International Offshore and Polar Engineering Conference, pp. 40-44.

[4] Bruschi, R., Torselletti, E., Vitali, L., Hauge, M., Levold, E.(2005b).Fracture control - Offshore pipeline current status of fracture assessment for pipelines limitations and the need for development. Proceedings of the International Conference on Offshore Mechanics and Arctic Engineering - OMAE, 3, art. no. OMAE2005-67519, pp. 659-668.

[5] Østby, E. (2007). Fracture Control - Offshore Pipelines JIP Proposal for strain-based fracture assessment procedure. Proceedings of the International Offshore and Polar Engineering Conference, pp. 3238-3245.

[6] Liu, M., Wang, Y.-Y., Song, Y., Horsley, D., Nanney, S. Multi-tier tensile strain models for strain-based design part 2 - Development and formulation of tensile strain capacity models (2012) Proceedings of the Biennial International Pipeline Conference, IPC, 4, pp. 415-425.

[7] Fairchild, D.P., Tang, H., Shafrova, S.Y., Cheng, W., Crapps, J.M. (2014). Updates to Exxonmobil's modelling approach for tensile strain capacity prediction. Proceedings of the International Offshore and Polar Engineering Conference, pp. 487-496.

[8] Tang, H., Fairchild, D., Panico, M., Crapps, J., Cheng, W. (2014a). Strain capacity prediction of strain-based pipelines. Proceedings of the Biennial International Pipeline Conference, IPC, 4,

[9] Østby, E., Hellesvik, A.O. (2007). Fracture control - Offshore pipelines JIP results from large scale testing of the effect of biaxial loading on the strain capacity of pipes with defects. Proceedings of the International Offshore and Polar Engineering Conference, pp. 3231-3237.

[10] Gioielli, P.C., Minnaar, K., Macia, M.L., Kan, W.C. (2007). Large-scale testing methodology to measure the influence of pressure on tensile strain capacity of a pipeline. Proceedings of the International Offshore and Polar Engineering Conference, pp. 3023-3027.

[11] Sandvik, A., Østby, E., Thaulow, C. (2008b). Fracture control - Offshore pipelines JIP use of Abaqus/Explicit to simulate ductile tearing in pipes with defects loaded beyond yielding. Proceedings of the International Offshore and Polar Engineering Conference, pp. 72-78.

[12] Dybwad, J., Törnqvist, R., Østby, E., Thaulow, C. Simulations of ductile tearing at large strains of biaxially loaded pipes (2009) Proceedings of the International Conference on Offshore Mechanics and Arctic Engineering - OMAE, 6, pp. 115-122.

[13] Nyhus, B., Loria, M. P. and Ørjasæter, O. L., "SENT specimens an alternative to SENB specimens for fracture mechanics testing of pipelines", Proceedings of OMAE03, 22ND International Conference on Offshore Mechanics and Arctic Engineering, June 8–13, 2003, Cancun, Mexico

Copyright © 2019 by ASME

[14] McClintock, F.A. (2002). Slip line fracture mechanics: A new regime of fracture mechanics. ASTM Special Technical Publication, (1417), pp. 3-54.

[15] McClintock FA. Plasticity aspects of fracture. In: Liebowitz H, editor. Fracture, vol. 3. New York: Academic Press; 1971. p. 47–225.

[16] Rice, J. R. (1968). A path independent integral and the approximate analysis of strain concentration by notches and cracks. Journal of Applied Mechanics, V. 35, pp. 379-386.

[17] Wells, A. A. (1961). Unstable crack propagation in metals: Cleavage and fast fracture. Proceedings of the crack propagation symposium, Vol. 1, paper 84.

[18] Shih CF. Relationship between the J-integral and crack opening displacement for stationary and growing cracks. J. Mech. Phys. Solids, 29, 1981, pp 305-326.

[19] Tvergaard, V., Needleman, A. Analysis of the cup-cone fracture in a round tensile bar (1984) Acta Metallurgica, 32 (1), pp. 157-169.

[20] BS 8571:2014. Method of test for determination of fracture toughness in metallic materials using single edge notched tension (SENT) specimens.

[21] Bilby BA, Howard IC, Li ZH. Failure assessment diagrams I: The tangency condition for ductile tearing instability. Proc. Roy. Soc. London A444, 1994, pp. 461-482.

[22] Østby, E., Thaulow, C., Nyhus, B. (2007) A new approach to ductile tearing assessment of pipelines under large-scale yielding. International Journal of Pressure Vessels and Piping, 84 (6), pp. 337-348

[23] Østby, E. (2005). Fracture control - Offshore pipelines: New strain-based fracture mechanics equations including the effects of biaxial loading, mismatch and misalignment. Proceedings of the International Conference on Offshore Mechanics and Arctic Engineering - OMAE, 3, art. no. OMAE2005-67518, pp. 649-658.

[24] Jayadevan, K.R., Østby, E., Thaulow, C. (2004). Fracture response of pipelines subjected to large plastic deformation under tension. International Journal of Pressure Vessels and Piping, 81 (9), pp. 771-783.

[25] Østby, E., Jayadevan, K.R., Thaulow, C. (2005). Fracture response of pipelines subject to large plastic deformation under bending. International Journal of Pressure Vessels and Piping, 82 (3), pp. 201-215.

[26] Bass BR, McAfee WJ, Williams PT, Pennell WE. Fracture assessment of shallow flaw cruciform beams tested under uniaxial and biaxial loading conditions. Nucl Eng Des 1999;188:259–88.

[27] Nyhus, B., Østby, E., Knagenhjelm, H.O., Black, S., Røstadsand, P.A. (2005). Fracture control - Offshore pipelines experimental studies on the effect of crack depth and asymmetric geometries on the ductile tearing resistance. Proceedings of the International Conference on Offshore Mechanics and Arctic Engineering - OMAE, 3, art. no. OMAE2005-67532, pp. 731-740.

[28] Østby, E., Hellesvik, A.O. Large-scale experimental investigation of the effect of biaxial loading on the deformation capacity of pipes with defects (2008) International Journal of Pressure Vessels and Piping, 85 (11), pp. 814-824.

[29] Kane, A., Østby, E., Lademo, O.-G., Berstad, T., Hopperstad, O.S. (2010). Effects of the yield criterion on predicted fracture responses of pipelines subjected to large plastic deformation with and without internal pressure. Proceedings of the International Conference on Offshore Mechanics and Arctic Engineering - OMAE, 6, pp. 251-259.

[30] Østby, E., Nyhus, B., Thaulow, C.(2011). Small-scale testing of the effect of embedded defect position in the thickness direction of an X65 UOE pipe. Proceedings of the International Offshore and Polar Engineering Conference, pp. 593-599.

[31] Østby, E.Strain-based design of pipelines - The important role of numerical modelling in fracture assessments (2015) Proceedings of the International Offshore and Polar Engineering Conference, 2015-January, pp. 543-550.

[32] Østby, E., Hauge, M., Levold, E., Sandvik, A., Nyhus, B., Thaulow, C. Strain capacity of SENT specimens - influence of weld metal mismatch and ductile tearing resistance (2008) Proceedings of the International Offshore and Polar Engineering Conference, pp. 64-70.

[33] Zhang, Z. L., Thaulow, C., Ødegård, J (2000). "A complete Gurson model approach for ductile fracture". *Eng. Frac. Mech* 67, pp 155-168

[34] Østby, E., Nyhus, B., Kane, P.-A., Thaulow, C. (2010). The effect of weld metal mismatch level on the failure mode in small-scale sent testing of an X80 material. Proceedings of the International Offshore and Polar Engineering Conference, 4, pp. 460-467.

[35] Sandvik, A., Østby, E., Thaulow, C.Fracture control - Offshore pipelines JIP use of Abaqus/Explicit to simulate ductile tearing in pipes with defects loaded beyond yielding (2008) Proceedings of the International Offshore and Polar Engineering Conference, pp. 72-78.

[36] Berg, E., Østby, E., Thaulow, C., Skallerud, B. (2008).Ultimate fracture capacity of pressurised pipes with defects - Comparisons of large scale testing and numerical simulations. Engineering Fracture Mechanics, 75 (8), pp. 2352-2366.

[37] Agrell, C., Østby, E., Levold, E., Hauge, M., Bjerke, S.L. The use of probabilistic fracture assessment procedures in design of pipelines subjected to large strains (2016). Proceedings of the International Offshore and Polar Engineering Conference, 2016-January, pp. 429-435.

Copyright © 2019 by ASME

Appendix A – Relationship Between CTOD and the J-integral for a Perfectly Plastic Material

This appendix contains a simplified outline the relationship between CTOD and the J-integral for a perfectly plastic material and a crack subject to tensile loading. The basic scenario is shown at the top of Figure 19. It considers a specimen with a crack loaded in tension. The following main assumptions are made:

- The material is perfectly plastic
- Upon reaching the yield load, all plastic deformation takes place along a shear band of height h along a 45° line to the crack ligament.
- The displacement between the two crack flanks, u, is only due to sliding along the shear bands
- From kinematic considerations, the CTOD, δ, is equal to u.
- The secondary bending action is neglected

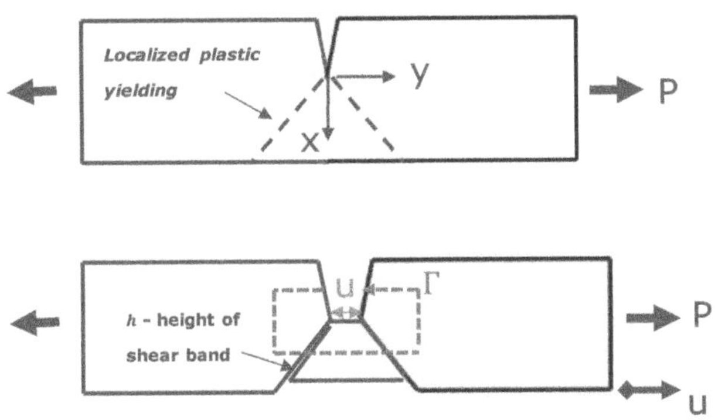

- *Neglecting elastic deformations*
- *Only consider contribution form localized yielding*

Figure 19. Outline of the basis to evaluate relationship between CTOD and J for cases governed e by full crack ligament yielding.

The shear yield stress, τ_y, is:

$$\tau_y = \frac{\sigma_y}{\sqrt{3}} \tag{5}$$

where σ_y is the yield stress. The shear strain, γ, in the shear band is:

$$\gamma = \frac{\sqrt{2}}{2}\frac{u}{h} \tag{6}$$

The plastic work, w, is:

$$w = \tau_y \gamma \tag{7}$$

The J-integral, J, is expressed as:

$$J = \oint_\Gamma \left(w * dy - T_i \frac{\partial u_i}{\partial x} ds \right) \tag{8}$$

where T_i is the traction normal to the contour Γ (see bottom Figure 19), u_i is the displacement in the i direction, and ds is a line segment along the contour Γ. By making the observations that

$$T_i \frac{\partial u_i}{\partial x} = 0 \tag{9}$$

everywhere around the contour Γ. T_i=0 along the horizontal parts of the contour and $\frac{\partial u}{\partial x} = 0$ along the vertical parts (all deformation assumed to take place in the shear band. Then the J-integral may be evaluated as:

$$J = \oint_\Gamma wdy = \frac{\sigma_y}{\sqrt{3}} u \frac{\sqrt{2}}{2h} \oint_\Gamma dy = \frac{2}{\sqrt{3}} \sigma_y \delta \tag{10}$$

In calculating the integral it has been used that w is zero everywhere except within the shear band, leaving the las integral equal to $2\sqrt{2}h$. Finally, the condition $u = \delta$ has been invoked. It can then be noted that Eq, (10) has the same form regarding the relationship between CTOD and J as was established under small-scale yielding in Eq. (1). However, it should be noted that the contribution to the J-integral only stems from magnitude of the localized deformation in the crack ligament. Thus, the J-integral in this case is a measure of the accumulated deformation in the localized deformation band. This energy associated with this deformation is dissipated as plastic work and cannot be recovered, i.e. the magnitude of the J-integral does not represent a true energy release rate.

Copyright © 2019 by ASME

AIM-PIMG2019-1039

Chapter 24

TENSILE STRAIN CAPACITY MODEL, FULL-SCALE TESTING, SAFETY FACTOR DERIVATION, AND BENCHMARK EXAMPLE CALCULATIONS

D.P. Fairchild[1], J.M. Crapps[2], M. Panico[1], W. Cheng[3], M.F. Cook[4], M.L. Macia[1]

ExxonMobil Production Company[1], ExxonMobil Upstream Research Co.[2], ExxonMobil Pipeline Co.[3], ExxonMobil Research and Engineering Co.[4], Spring, Texas, USA

ABSTRACT

This paper describes a tensile strain capacity (TSC) prediction model and how ~100 full-scale pipe strain capacity tests were used to derive a safety factor for use in either pipeline design or strain-based engineering critical assessment (SBECA). The safety factor accounts for the difference between model predictions and test results thus providing a quantifiable degree of reliability. The safety factor is generated from a 95% upper confidence limit on the ratio of predicted-to-measured TSC. The practical meaning of this is that if either TSC or tolerable defect size is calculated using the model, then the probability of being non-conservative is estimated to be 5%.

The model is used to estimate TSC for a variety of hypothetical pipeline design cases. These scenarios cover a range of pipe geometries, pipe and weld physical properties, and high-low weld misalignment. Using the model as a design tool is discussed and the model is used to create flaw assessment diagrams showing trends in defect tolerance as a function of variables like weld strength overmatch and high-low misalignment. It is the intent of this paper to (1) show the utility of the model in assessing the influence of variables that affect girth weld performance in SBD scenarios, and (2) provide these calculations as benchmark cases to ensure that other users have accurately programmed their spreadsheets.

Keywords: pipelines, strain-based design, tensile strain capacity, engineering critical assessment, girth weld defects.

INTRODUCTION

Worldwide energy demand brings the need for pipelines that cross challenging environments and potentially encounter geohazards such as seismic faults, landslides, offshore ice features, and the freeze/thaw cycles of discontinuous permafrost. Ground movements can create significant longitudinal strains in pipelines and strain-based design (SBD) technology is needed to design against these threats. From the late 1990's to the present, ExxonMobil has designed and/or installed about 16 SBD pipelines. Tensile strain demands of 3% and beyond have been addressed [1]. To prepare for and assist these SBD projects, research has been conducted on many topics including tensile strain capacity (TSC) prediction, engineering critical assessment (ECA) procedures, and full-scale testing (FST) to measure pipe strain capacity.

ExxonMobil's TSC prediction technology has evolved over a period of about 15 years. In the late 1990s and early 2000's the engineering approach was often empirical with dependence on the curved wide plate test [2-5]. Emphasis soon shifted to a modelling approach based on stationary crack finite element analysis (FEA) [6-8]. This led to a first TSC prediction model and a strain-based ECA approach [9]. During the early modelling period, the biaxial loading effect was realized [10, 11]. This and the influence of other important performance variables (weld strength overmatch, high-low misalignment, etc.) motivated increased use of FSTs to measure TSC [12-14]. FSTs were also increasingly used by SBD projects [1]. By comparison to FST results, it was realized that stationary crack FEA and the first model [9] has limitations [15] and can, in some circumstances,

[1] Contact author: doug.p.fairchild@exxonmobil.com

Copyright © 2019 by ASME

result in over prediction of fracture driving force and under prediction of TSC (i.e., over conservative). To overcome the limitations, a new TSC prediction model based on damage mechanics FEA was developed [16] and the old model [9] was retired. The new model and ~6500 FEA runs provided the basis for a robust TSC prediction equation. The model FEA explored the following parameter ranges.

- Pipe Diameter: 8" - 30"
- Wall thickness: 13 - 22 mm
- Pipe Y/T Ratio: 0.76 - 0.87
- Pipe Uniform Elongation: 6.7% - 13%
- Flaw height: 3 - 4.5 mm
- Flaw length: 25 - 200 mm
- Weld strength overmatch (at UTS): 0 - 40%
- Weld single edge notch tension (SENT) δ value: 0.9 - 2.3 mm
- Misalignment: 0 - 3 mm
- Pipe Pressure: 0 - 80% specified minimum yield stress (SMYS)

The new model equation was validated by comparison to 101 FSTs and a safety factor (SF) for capacity prediction or strain-based ECA (SBECA) was derived from a statistical analysis of the data [17]. The TSC prediction equation with the SF embedded is provided in the Appendix.

The key data on which the SF is based is shown in Figure 1. This 45° plot compares the TSC predicted using the new model [16] to the TSC measured in FSTs [17]. A best fit statistical distribution to the ratio of predicted-to-measured TSC results in the Gamma distribution shown in Figure 2. A 95% upper confidence limit (CL) on the Gamma distribution (red dashed line in Fig. 2) serves as a quantified safety factor. The value of this CL is 1.88 and it appears as the reciprocal value 0.53 in Equation (Eq.) A1 shown in the Appendix. The practical meaning of the safety factor is that if TSC is predicted using Eq. A1, then the probability of being non-conservative (over predicting the capacity) is estimated to be 5%. Likewise, if this equation is used for SBECA to predict acceptable flaw size, then the probability of being non-conservative (over predicting flaw size) is estimated to be 5%.

Using the new TSC model and SF, a protocol for SBECA was developed and currently this protocol is being prepared for an API RP by an API 1104 Task Group [18]. The protocol includes procedures for pipe and girth weld testing, outlines a step-by-step approach for girth weld defect assessment, and also provides worked examples. Recognizing that others may be interested in this technology and that the equation is complex, a number of hypothetical SBD pipeline scenarios are provided below along with calculations of TSC. These "benchmark" cases provide checks for spreadsheet programming.

To demonstrate the utility of the TSC prediction technology, selected benchmark cases are used for study. First, an example of TSC vs. flaw height is given to show how the technology might be used in an early project design scenario. Subsequent examples are given to explore the effect of critical performance variables (weld strength overmatch and high-low misalignment) on tolerable flaw size. These examples could be used for early design or they could be used during a pipeline project to guide planning for FST proof testing. The TSC prediction technology can also be used post-construction for integrity management objectives and potentially for fitness-for-service assessments. These topics are addressed elsewhere [19].

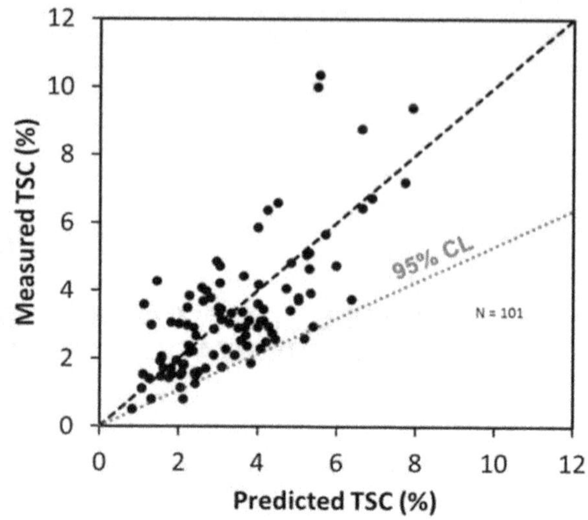

Figure 1: 45° plot showing the 101 FST data points to be used for safety factor calculation.

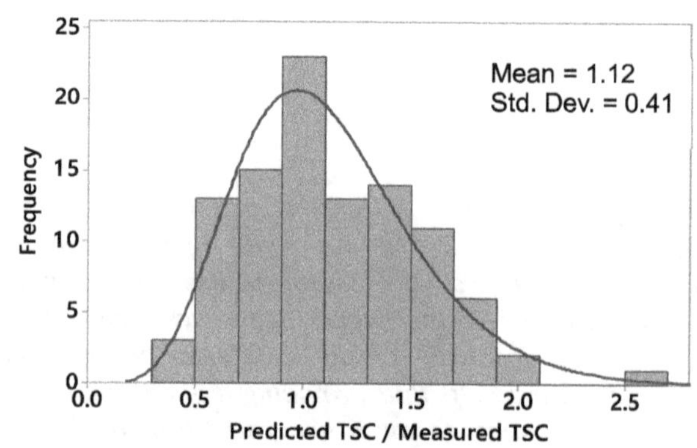

Figure 2: Gamma distribution comparison to the ratio of predicted-to-measured TSC for the 101 FSTs of Fig. 1.

OVERVIEW OF FULL-SCALE TESTING

Full-scale pipe strain tests consist of loading a piece of pipe to failure in tension and/or bending. The primary measurement is strain capacity; i.e., how much longitudinal strain can be absorbed by the specimen prior to failure, which is often defined as maximum load. The specimen might be parent pipe or it may contain one or more girth welds. Girth welds are typically populated with manmade defects (notches) and it can be useful to include varying degrees of weld joint MA. The specimens can be pressurized to simulate service conditions. When girth weld defects are included, the failure scenario often involves ductile tearing from the defect until the remaining ligament collapses.

Copyright © 2019 by ASME

Figure 3 shows a schematic of a typical FST specimen involving girth welds. Adding together the cost of specimen design and fabrication, then testing and analysis, a per-test cost can be tens of thousands of dollars. Months can be consumed in conducting a single test.

The most important aspect of test specimen design and fabrication is preparing a specimen with known properties. Because the specimen is damaged during testing, the actual properties cannot be known. It is necessary to conduct small-scale tests on additional material to generate data representative of the specimen. These tests are called companion tests and they provide key inputs for TSC model predictions. Successful strategies for companion material preparation are provided elsewhere [13, 14]. The layout in Figure 4 produces the design in Figure 3. Rings are first cut from the test pups to provide tensile data for overmatch calculations. The FST girth welds join the three pups at the locations where the rings were removed. Companion welds are made at the ends of the FST specimen or they can both be placed on one end of the specimen.

The most challenging phase in the preparation of companion material is to perform girth welding so that it generates representative data. It has proven beneficial to conduct the companion girth welds using additional pups welded directly onto the ends of the FST specimen (subsequently cut off). It is recommended that the companion girth welds are made as soon as possible (say, within 24 hours) after the FST welds are made and that they are produced using the same welders, welding equipment, and welding procedures.

Specimen instrumentation is another important part of full-scale testing [14]. Instrumentation is used for monitoring longitudinal strain and notch opening. Strain is typically monitored using linear variable displacement transducers

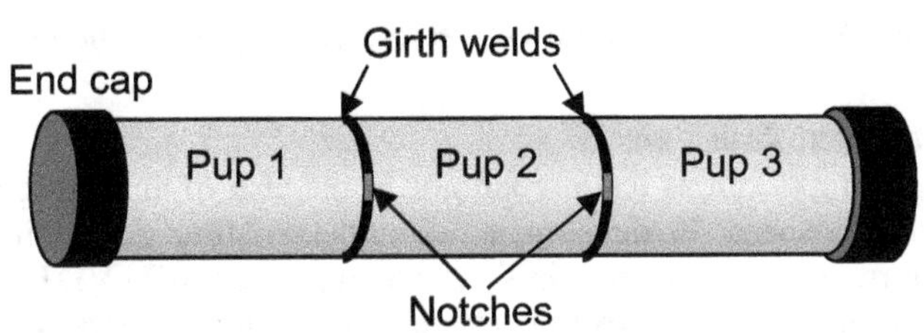

Figure 3: Schematic design of a FST specimen.

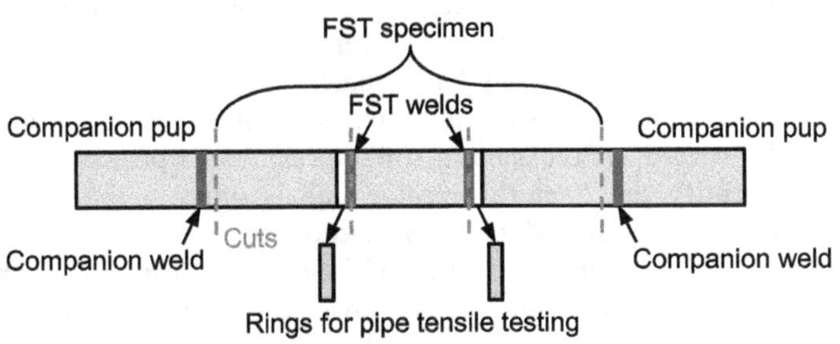

Figure 4: Strategy for FST specimen and companion material preparation.

(LVDTs) and clip gages are used for crack mouth opening displacement (CMOD). Figure 5 shows an instrumentation layout that has proved successful. The schematic involves a hypothetical unrolling of the pipe specimen. Three large LVDTs are used to measure global strain across the specimen length. The anchor points are as close as practically possible to the end cap juncture with the pipe specimen. Strain capacity of the FST is defined as the average of these LVDTs at the point of maximum load.

Shorter LVDTs are positioned on each pup and these LVDTs can monitor potential non-uniform straining in the pups. The least sophisticated (but very useful) of all strain monitoring methods is the use of scribe lines [1] (Fig. 5). Scribe lines with regular (100-150 mm) cross hairs can be positioned along the same linear paths as the LVDTs. The distance between, and wall thickness at, the markings are recorded before and after the test to provide strain measurement. Scribe line data can be used to double check the LVDTs and to monitor non-uniform straining.

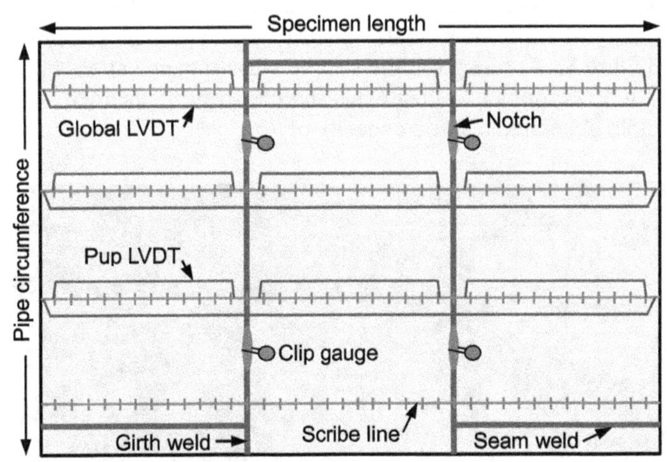

Figure 5: FST instrumentation (unrolled pipe view).

Full-Scale Test Examples

As part of a pipeline project in Papua New Guinea that involved numerous sections where SBD engineering was necessary [1], FSTs were conducted on 8" and 32" diameter pipe. The strain demand for this project varied amongst the different SBD sections. Suffice it to say that the estimated strain demand was approximately 3%. A design margin was applied which lead to a target TSC for the FSTs of 4.8%.

A load trace output from one of the 8" tests is shown in Figure 6. Axial stress (y-axis) is a result of pipe geometry and load cell output. Axial strain (x-axis) is taken from the global LVDTs (Fig. 5). Periodic unloadings were incorporated into the testing. This, in combination with the clip gauge data (Fig. 5), enabled compliance measurements, estimates of crack growth, and verification that tearing behavior was similar between FSTs and small-scale companion tests (SENT). The results in Figure 6 show a strain capacity of about 11% which is quite high, easily greater than the 4.8% target, but not uncharacteristic of high quality seamless pipe with overmatched welds.

Copyright © 2019 by ASME

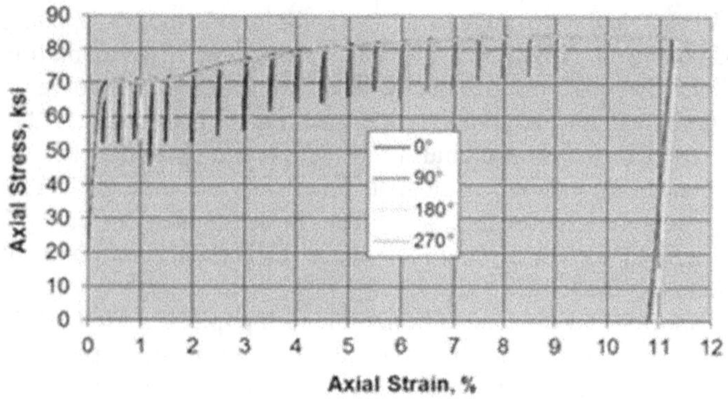

Figure 6: Axial stress vs. strain plot for 8" diameter FST.

One of the 32" diameter FST samples positioned in the load frame is shown in Figure 7. Due to the large size of this pipe, the load frame tensile capacity (applied through welded end caps) was augmented with external hydraulic rams positioned regularly around the test sample circumference. This enabled a capacity of 8 million pounds. The stress vs. strain plot is shown in Figure 8. At one point the sample was unloaded at about 2% strain for equipment adjustments and then testing resumed. This sample generated a strain capacity of 7%.

Figure 7: 32" diameter FST sample mounted in load frame.

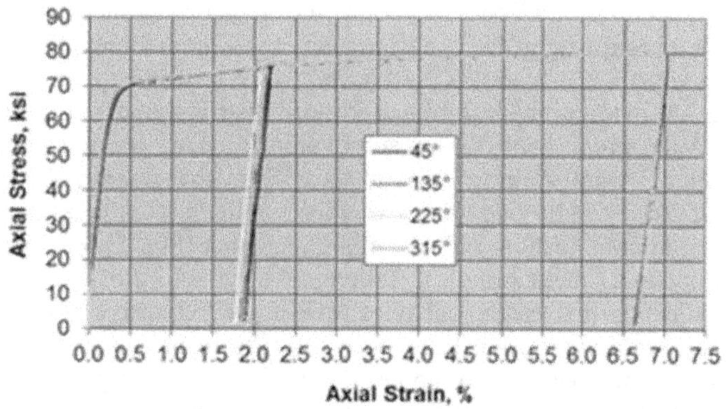

Figure 8: Axial stress vs. strain plot for 32" diameter FST.

FSTs are generally conducted for one of two purposes, (1) to verify the accuracy of a TSC prediction model, or (2) to test materials being considered for use as a SBD pipeline; i.e., for project work. The second purpose includes screening candidate materials (qualification) and/or proof testing final materials and welding procedures as shown in the examples above. Because project FSTs are often designed to verify TSC in light of chosen defect sizes (i.e., a vetting of the ECA approach), it is often the case that little-to-no tearing from inserted defects occurs in these tests and the failure mode is pipe wall collapse (necking). On the other hand, it is often the case for FSTs designed to explore tearing phenomena (useful for model validation), that one or more FST notches exhibits active tearing during the test. Figure 9 shows crack mouth opening displacement (CMOD) data from the clip gauges of a FST. When a FST ends as a result of a notch tearing through wall, then the CMOD data of the critical notch typically shows a rapid increase near the end of the test. The red curve in Figure 9 shows that Notch C was rapidly tearing at the end of the test – the CMOD trace is nearly vertical at the strain capacity of 4.5%.

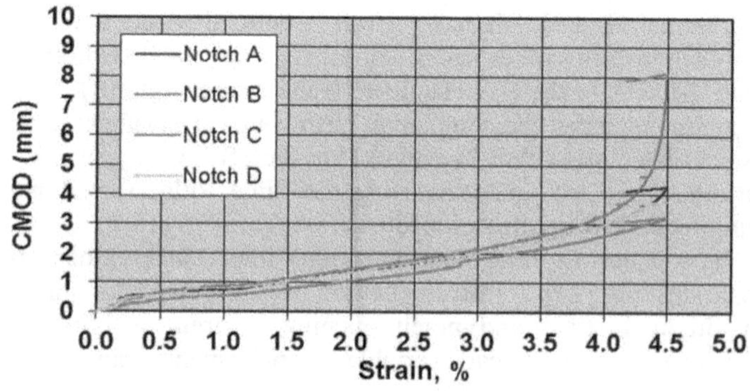

Figure 9: CMOD data showing active opening at one notch.

Post Test Analysis

As important as test preparation is the task of analyzing the results post-test to develop a full understanding of specimen performance. This requires inter-comparison of LVDT and scribe line data. The exercise should search for the presence of non-uniform straining. If such is suspected, then additional wall thickness measurements by ultrasonics can prove useful. The CMOD data should be reviewed in tandem with the other notch examination techniques (fractography, sectioning). A primary goal is to determine if any notches were "active" during the test, meaning that ductile tearing had initiated.

If the specimen does not fracture, the CMOD data is usually straightforward. If the specimen fractures, then a primary goal is to identify the fracture origin which is often at the notch showing the largest CMOD at the end of the test. It should be noted that cases of brittle fracture initiation away from the notch with largest CMOD have been observed. Specimens that fracture can be challenging to interpret. Fractography is a valuable exercise that creates an understanding of the failure event often determining the significance of the FST. The

Copyright © 2019 by ASME

objective of fractography is to determine (1) the initiation site of fast fracture, (2) whether propagation was ductile, brittle, or mixed, and (3) the direction of crack propagation.

There are about 180 FSTs available to ExxonMobil. About 110 were conducted by ExxonMobil and about 70 were conducted by Joint Industry Projects. Not all tests are suitable for validating the model described above and given in the Appendix [17]. Whereas the model simulates a single, surface breaking weld flaw, some of the FSTs contained interacting or embedded flaws. Other tests included welded sleeves to examine repair scenarios. Also, some FSTs contained unintentional weld defects, other tests failed by brittle fracture, and still others experienced welding problems rendering them unsuitable for model validation. After the tests were screened for suitability in model validation [17], a total of 101 tests were used (Figures 1, 2). The following variable ranges were explored in the FSTs:

- Pipe grades: X60, X65, X70, X80
- Pipe manufacturing methods: seamless, ERW, spiral, UOE
- Diameter: 8", 10", 12", 16", 20", 24", 30", 32", 42", 48"
- Wall thickness: 10.3 - 36.3 mm
- Flaw height: 1.3 - 6.7 mm
- Flaw length: 15 - 200 mm
- Weld strength overmatch (at UTS): 0 - 59%
- Misalignment: 0 - 3.5 mm
- Pressure: 0-119% of specified minimum yield stress (SMYS)
- Pressure: 0-95% of actual measured yield stress
- Welding methods: manual shielded metal arc welding (SMAW), flux-cored arc welding (FCAW), gas tungsten arc welding (GTAW), and mechanized gas metal arc welding (GMAW) with and without pulsing.
- Notch positions: base metal, weld metal centerline, heat affected zone, outside and inside diameter

The reader is cautioned that these ranges do not imply acceptable selections for SBD engineering. Some of the variables tested would be poor choices for SBD and were only tested to investigate the effects of certain variables on TSC. One example is that of 0% weld strength overmatch. While this test was useful to examine a regime where the TSC model is technically valid, even matched welds (and to an even greater degree, undermatched welds) are not suitable for SBD.

PIPELINE DESIGN AND TSC PREDICTION

A (very) simplified flowchart of the SBD process is shown in Figure 10. For a real pipeline project, the design phase will involve many more steps than shown in this Figure; however, the objective of Figure 10 is to highlight the basic use of the TSC prediction model. The initial steps in the flowchart are the same as for conventional pipelines, often referred to as the stress-based design process (aka allowable stress design), the primary purpose of which is pressure containment. Initial selections are made of pipe size, pipe grade, design pressure, wall thickness, and pressure-related design safety factor as a percent of the specified minimum yield stress (SMYS).

The SBD process often begins with geohazard characterization of the pipeline route via field surveys and then pipeline routing adjustments can be made to reduce the potential for, and the magnitude of, ground movements along the pipeline. This is followed by detailed pipe-soil analysis to estimate how much load will be transferred from the ground into the pipe during a geohazard event(s). This results in an estimate of the strain demand. Both compressive and tensile strain demand are dealt with during design. In the limit state, the compressive mode typically results in wrinkling of the steel (which is serviceable), but tensile loading can result in tearing, rupture, and loss of containment; i.e., an ultimate limit state. This paper will concentrate on tensile loading and the reader is directed to [19] for additional discussion about design.

An estimate is made of the TSC and this can be done using Eq. A1. A key step in the SBD process is comparing tensile strain demand to TSC and in Figure 10 a double carrot (<<) is used to indicate that the criticality of this high load design scenario justifies that the TSC is suitably and reliably larger than the demand. Many engineering assumptions and decisions are made during design and certain constraints will be realized. For example the right-of-way terrain (mountainous, offshore, subarctic, etc.) may drive selections of welding, inspection, and other construction technologies which, in turn, affect quality and reliability. Other factors may include the return period of the strain event(s), the ease of mitigating the strain event(s), the ease of monitoring the pipeline, and the consequences of failure. Each project will be unique and the designers and project team must take all aspects into consideration to engineer the comparison of demand versus capacity.

It is extremely useful to the designers if the tools they use are quantified. It brings a higher degree of certainty to design decisions. Because an estimate of TSC plays a central role in the design process, it is imperative that the accuracy of the TSC prediction model be known. As explained above, the model given by Eq. A1 has been validated by comparison to FST data and the derived SF provides quantification of the model's

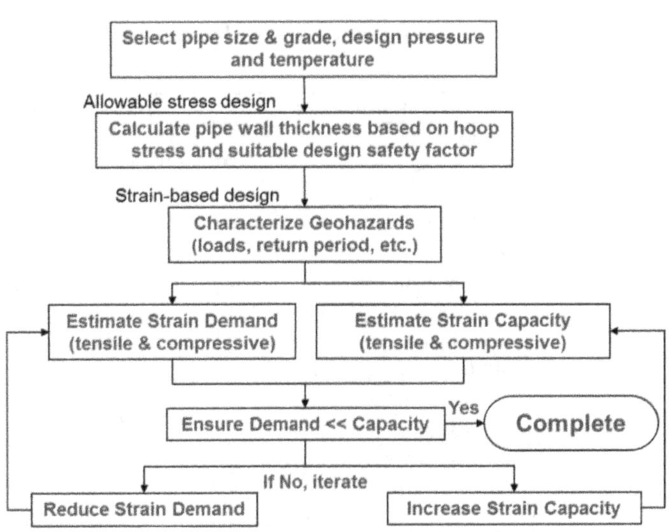

Figure 10: Flowchart of a simplified SBD process.

Copyright © 2019 by ASME

accuracy. The SF, as it is embedded in Eq. A1, essentially reduces the predicted TSC. The portion of Eq. A1 after (to the right of) the SF 0.53 provides an unfactored prediction of TSC.

Continuing with the flowchart, there are many ways to manage the design margin (the double carrot in Fig. 10) between strain demand and capacity. The TSC prediction SF is just one component. Once all factors are accounted for, if a suitable margin is achieved, then the process is complete. However, if not, then either the strain demand must be reduced, or TSC must be increased. The most reliable way to reduce strain demand is to route the pipeline away from geohazards. Other means are also possible, for example, through trench design [1]. For most SBD projects, adjustments to the demand-to-capacity challenge will involve increasing TSC. The most common ways to increase TSC include increasing wall thickness, increasing weld metal strength and/or toughness, limiting high-low misalignment, reducing tolerable flaw size and setting a minimum value for longitudinal uniform elongation in the pipe specification. It is in contemplation of all these factors that the TSC model of Eq. A1 has utility. Once a spreadsheet is programmed, many iterations of TSC calculation can be conducted to explore tradeoffs in steel versus construction costs with the goal being to determine an optimum set of pipeline design parameters.

It is worth mentioning a scenario where the design process is completed and the project is approaching execution when it is realized that the targeted demand vs. TSC cannot be met. One example is that a project decides to conduct a full-scale proof test(s) using project pipe/welds and the results generate a lower than desired TSC. When such a challenge comes late in the project, often the only viable recourse is to reduce the acceptable girth weld flaw size. The TSC model (Eq. A1) can help quantify the choice of reducing flaw size.

THE BENCHMARK CASES

To aid the accurate programming of Eq. A1 into spreadsheets and to create hypothetical design cases for further study, ten benchmark cases are provided in Table 1. These cases are listed in ascending order of pipe diameter and the diameters range broadly from 219 mm (~8 inch) to 1219 mm (48 inch). Variations of all inputs required in Eq. A1 are included. As explained elsewhere [16-18], tensile failure can occur by either ductile tearing (DT) from a girth weld defect or by plastic collapse (PC) in the pipe. These modes compete to define minimum TSC. Both failure modes are accounted for in the TSC model through Eq. A2 and the predicted failure modes for the benchmark cases are listed in Table 1. The **Predicted TSC** in Table 1 is achieved by using Eq. A1 without the 0.53 safety factor term while the **Factored TSC** is achieved by using Eq. A1 exactly as it appears in the Appendix.

A few of the variables, individual cases, and trends in Table 1 deserve further discussion. The wall thicknesses vary from 16 mm to 25 mm and no cases are listed that might be considered "thin wall" pipe. Although the TSC model is validated by full-scale testing down to 10 mm wall thickness, the SBECA protocol [18] requires that for design scenarios involving wall thicknesses below 12 mm, validation by full-scale testing must be conducted. In the experience of the authors, when wall thickness decreases below ~12 mm, the relationship between wall thickness and the size of a single weld bead creates increased sensitivity for minor upsets in welding to affect TSC. Unintentional variations in variables like heat input, preheat, interpass temperature, or even the accidental use of the wrong welding consumable can result in local occurrences of low strength material and a reduction in TSC. The wall thicknesses in the benchmark cases demonstrate a preference for ample wall thickness because this variable is the most robust and reliable way to achieve the desired TSC.

Flaw heights in Table 1 range from 1.5 mm to 4.5 mm. This range is related to the relationship between flaw height and the size of weld beads. Typical pipeline girth weld beads (not including double joint welds) vary in height from about 3-5 mm; therefore, natural constraints on flaw height exist because common flaws are typically associated with a single bead. In the situation when weld defects are stacked, flaw heights often occur in discrete steps governed again by bead height. Stacking two

Table 1: Benchmark design cases for ten hypothetical SBD pipelines. Abbreviations given for each input parameter are consistent with Appendix nomenclature.

Input Parameters		Case 1	Case 2	Case 3	Case 4	Case 5	Case 6	Case 7	Case 8	Case 9	Case 10
Diameter: mm (inches)	D	219 (8.625)	324 (12.75)	406 (16)	610 (24)	762 (30)	762 (30)	762 (30)	1067 (42)	1067 (42)	1219 (48)
Wall Thickness (mm)	t	16	16	16	22	16	16	16	22	22	25
Flaw Height (mm)	a	2.5	4	4	4	4.5	3	4.5	4	4	1.5
Flaw Length (mm)	$2C$	50	50	50	50	100	50	100	150	50	75
Hi-Lo Misalignment (mm)	e	2	1	0	0.75	0	2.8	0	1	3.5	1.5
Pipe yield/tensile Ratio	Y/T	0.85	0.85	0.84	0.88	0.88	0.88	0.88	0.75	0.9	0.8
Uniform Elongation (%)	UEL	10	12	11.5	8	10	8	10	10	5.5	5.5
SENT R-curve Delta (mm)	δ	1.3	1.3	1.4	1	1	1.4	1.2	1	1.4	1.4
Weld Strength Overmatch (%)	λ	5	10	25	12	25	6	16	55	35	35
Pipeline Pressure (% SMYS)	R_b	80	30	80	80	80	80	72	72	72	60
Predicted Failure Mode		DT	DT	DT	DT	DT	DT	DT	DT	DT	PC
Predicted TSC (%)		2.3	3.0	5.8	3.2	1.6	1.1	1.5	3.0	2.2	3.7
Factored TSC (%)		1.2	1.6	3.0	1.7	0.8	0.6	0.8	1.6	1.2	--

DT = ductile tearing, PC = plastic collapse

Copyright © 2019 by ASME

defects means encountering flaw heights in the range of about 6-10 mm which is large and deleterious to TSC. Despite the fact that the TSC prediction model can work with flaw heights characteristic of stacked defects, it is suggested that such occurrences be avoided because stacked defects are potentially large in flaw height and deleterious for SBD scenarios.

Case 3 produces the highest TSC of the benchmark cases. The primary reasons are: small misalignment (0 mm), good overmatch (25%), good toughness (R-curve δ = 1.4 mm), and a low Y/T ratio (0.84). The lowest TSC's are displayed by Cases 5, 6 and 7. Case 6 is the lowest due to the controlling factors of low weld strength overmatch (6%) and 2.8 mm of misalignment. The 6% overmatch in this case is well below the recommended minimum value of 15% in the SBECA protocol [18]. The TSCs of Cases 5 and 7 are mostly controlled by the large flaw size (4.5 mm x 100 mm).

Case 10 is the only scenario that has an estimated capacity controlled by PC in the pipe. The reason is that the capacity associated with DT is higher than the capacity estimated for PC. The DT capacity is controlled by a small flaw height, (1.5 mm), small misalignment (1.5 mm), high overmatch (35%), and low Y/T ratio (0.8). As noted by Eq. A2, the PC capacity is estimated by 2/3 of the pipe's uniform elongation (UEL) and because the UEL for Case 10 is somewhat low for strain-based pipe (only 5.5%), the predicted TSC is 3.7%. There is no factored entry for DT listed in Table 1 because the capacity is governed by PC. It is interesting to note that Case 9 has the same pipe UEL as Case 10; however, for Case 9 the DT failure mode is predicted to be more critical than PC. The DT capacity for Case 9 is lower due to the combination of notable misalignment (3.5 mm) and relatively large flaw height (4 mm).

The primary controlling variable for Case 1 is the low overmatch (5%), the lowest of all cases; however, this low strength weld is counterbalanced by a relatively small flaw (2.5 mm x 50 mm) and thus the TSC is not as low as the other low overmatch scenario, Case 6. For Case 4 the important variables are moderate overmatch (12%) and significant flaw height (4 mm). For Case 8 the performance is controlled by a large flaw size (4 mm x 150 mm).

When considering the controlling factors for the benchmark cases certain trends of estimated TSC performance (high, medium, or low) can be reasoned by examining the values of dominant variables like misalignment, overmatch, flaw size and wall thickness; however, this reasoning extends from the benefit of pre-calculation of TSC. In situations without pre-calculation of TSC, it may not be easy to recognize the dominant factors. TSC can be difficult to understand through intuition and review of the variables. One reason for this is because there are a large number of controlling variables (nine) for TSC according to the model (Eqs. A1, A2). It is difficult to guess the performance of a solid mechanics outcome in nine-dimensional space.

Design Example: TSC vs. Flaw Height

As an example of using the TSC prediction model as a design tool, benchmark Case 4 was used as a basis to explore the relationship between predicted TSC (factored) and flaw height. Numerous iterations of flaw height and TSC were calculated using Eqs. A1 and A2 (see Figure 11). Four sub-cases of weld strength overmatch are included. The x-axis begins at 1 mm because relevant flaw heights are not likely below this value once inspection sizing accuracy is accounted for (on the order of 1 mm, additional discussion below). The plateau that appears at a TSC of 5.3% in the case of 35% overmatch is caused by the competition between DT and PC. When the DT capacity exceeds the PC capacity, then PC becomes the limiting failure mode and the predicted TSC plateaus at 2/3 of the pipe UEL.

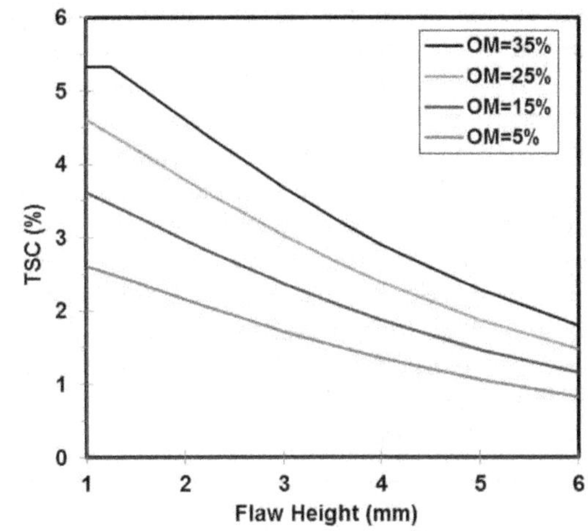

Figure 11: The relationship between factored TSC and flaw height for four values of weld strength overmatch. The basis for this graph is benchmark Case 4.

Two trends in Figure 11 are worth noting. First, there is an obvious decreasing trend in TSC as flaw height increases because flaw height effectively reduces the pipe wall cross section which is the primary load carrying structure in situations of longitudinal pipeline strain. Just a few millimeters increase in flaw height for this 16 mm wall thickness example results in nearly halving the estimated TSC. Second, the trend of increasing TSC with increasing overmatch is significant. This example shows that weld strength overmatch is useful in mitigating low TSC (say, TSC < ~1%). Each increase of overmatch by 10% results in an increase of ~1% TSC at small values of flaw height and produces an increase of ~0.5% TSC at larger values of flaw height. Good weld strength can compensate for the presence of weld flaws. Furthermore, because overmatch significantly affects TSC, it places importance on the weld procedure qualification process and the monitoring of welding during pipeline fabrication.

SBECA Example: Flaw Tolerance and Overmatch

Much of the conversation above was from a design perspective where predicted TSC was calculated using Eq. A1 with flaw size as inputs. For SBECA the opposite occurs. Estimates of tolerable flaw size are calculated using Eq. A1 with a target TSC as a primary input. Target TSC for ground movement scenarios is determined from an estimated strain

Copyright © 2019 by ASME

demand that is established using geotechnical information (ground movements, soil properties, etc.). The goal of the geotechnical work is to estimate the magnitude of load transfer from soil-to-pipe. Other design scenarios might include determining strain demand associated with in-service thermal transients. Regardless of the load origin, uncertainties in demand should be accounted for and then a target TSC established.

Target TSC is input on the left side of Eq. A1. The influence of the safety factor (SF = 0.53 in Eq. A1) is to increase the value of the target capacity because when 0.53 transfers to the left side of the equation, it becomes a multiplier of value 1.88 (i.e., reciprocal of 0.53). Increasing the left side of Eq. A1 produces smaller flaw sizes as output from the right side of the equation. Through iterative calculations, many height-length data pairs can be generated to create flaw tolerance curves.

As an example of using the TSC prediction model as an ECA tool, benchmark Case 4 was used as the basis to calculate the flaw tolerance curves shown in Figure 12. To maintain consistency with the benchmark values in Table 1, the factored capacity for Case 4 (TSC_f = 1.7%) was used as the target capacity in this example. The truncations that appear in the curves of Figure 12 are a result of the SBECA protocol limits on maximum acceptable flaw size [18]. The following quote is from the pertinent section of the protocol.

10.1.1 MAXIMUM ACCEPTABLE IMPERFECTION SIZE
For all four Options of this RP, the maximum acceptable length of a single imperfection shall not be greater than the lesser of 125 mm or 5% of the pipe circumference. The maximum acceptable height of a single imperfection shall not be greater than the lesser of 5 mm or 25% of the pipe wall thickness.

This clause indicates that regardless of the protocol assessment calculations, there are limits placed on maximum flaw size, both height and length. These values are based on workmanship "rules-of-thumb" from other standards and from the collective experiences of the authors in working with SBD pipelines. In the opinion of the authors, the SBECA limits on maximum flaw size represent a level of workmanship quality that is prudent considering the critical nature of the loading in SBD pipelines.

Figure 12 can be thought of in the same way as traditional flaw tolerance diagrams where flaw sizes above (outside) the curves are deemed too large and sizes below (or inside) the curves are acceptable. The SF mentioned above provides a basis for risk quantification regarding flaw size calculations. In analogy to the previously stated meaning of the SF, if flaw dimensions fall inside the curves in Figure 12, then the risk that the flaw is a non-conservative prediction (i.e., it is actually a flaw that should be assessed as too large) is less than 5%.

Figure 12 shows trends with weld strength overmatch that are similar to those shown in Figure 11. As overmatch increases, flaw tolerance increases. Within the region of Figure 12 where flaw size varies with overmatch, each increase of 10% overmatch results in roughly a millimeter increase in flaw height tolerance. Similar to the previous example, this indicates that good weld strength is beneficial for flaw tolerance. And again it emphasizes the importance of quality controls for the welding operations associated with SBD pipeline fabrication.

SBECA Example: Flaw Tolerance and Misalignment

As a next SBECA example, the same case that is displayed in Figure 12 is used with a fixed overmatch of 12% (Table 1, Case 4) and high-low misalignment is varied as the sensitivity sub-cases. The results are shown in Figure 13. This graph shows clearly the significant effect of misalignment on flaw tolerance (and TSC). Case 4 involves a pipe geometry of 610 mm (22 in.) diameter and 22 mm wall thickness, which is not a small pipe by any means. Yet, just 3 mm of misalignment creates the bottom curve in Figure 13. One selection from this curve shows a maximum tolerable flaw size of about 2 mm x 25 mm and this is restrictive for a pipe of this size. Obvious alternatives to this

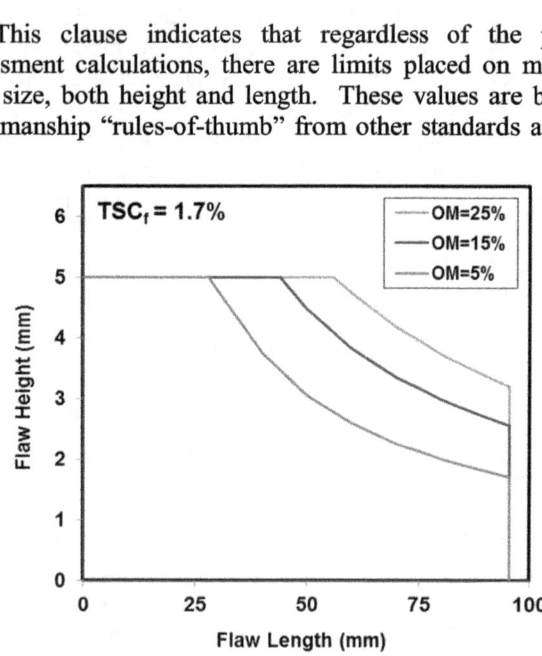

Figure 12: The relationship between flaw height and length for three cases of weld strength overmatch (OM). The basis for this graph is benchmark Case 4.

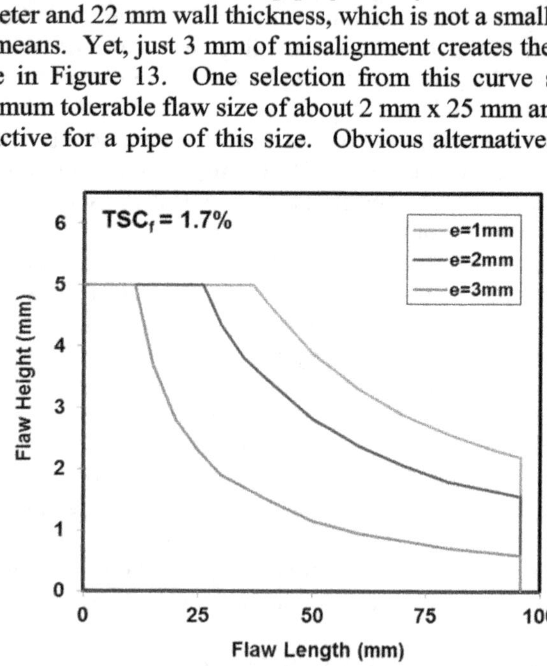

Figure 13: The relationship between flaw height and length for three cases of high-low misalignment (e). The basis for this graph is benchmark Case 4.

Copyright © 2019 by ASME

specific example are to restrict misalignment to smaller values and/or to increase weld strength overmatch. Considering that further restrictions on misalignment means 2 mm or 1 mm which would undoubtedly bring difficulties for production weld fit-up, the better choice would likely be to increase overmatch.

As additional demonstration of the influence of misalignment, consider the scenario from Figure 13 of flaw length in the range of 25 - 30 mm. At this length, the tolerable flaw height changes from about 2 mm to 4 mm with only 1 mm change in misalignment (from 3 mm to 2 mm of high-low). In other words, flaw height tolerance doubles with only 1 mm decrease in misalignment. Likewise, if a flaw height of about 2 mm is considered, then tolerable flaw length changes from about 25 mm to nearly 75 mm with only 1 mm decrease in misalignment. This represents an increase in flaw length tolerance of almost triple for 1 mm decrease in misalignment.

The primary reason that misalignment is detrimental to TSC and flaw tolerance is related to the interaction between the misaligned geometry and weld flaws during the extreme loading of a high strain scenario. A schematic showing a misaligned weld and several possibilities of flaw location are shown in Figure 14a. Misalignment causes (1) a stress/strain concentration, and (2) effectively reduces the cross section as shown by the red arrows in Figure 14a. Furthermore, at the levels of loading that can happen in SBD scenarios (cross section is fully plastic), the 45° shearing patterns that are characteristic of ductile failure are well oriented to "connect" an inner diameter weld toe with weld flaws, particularly if the flaws are positioned on a fusion line. These geometric considerations also explain why softened HAZs can be deleterious; because they also align well with the angled shearing pattern that accompanies failure. The stress/strain concentration that occurs due to misalignment during fully plastic loading can be extreme to the extent that the weld attempts to rotate which creates bending stresses that exacerbate the situation.

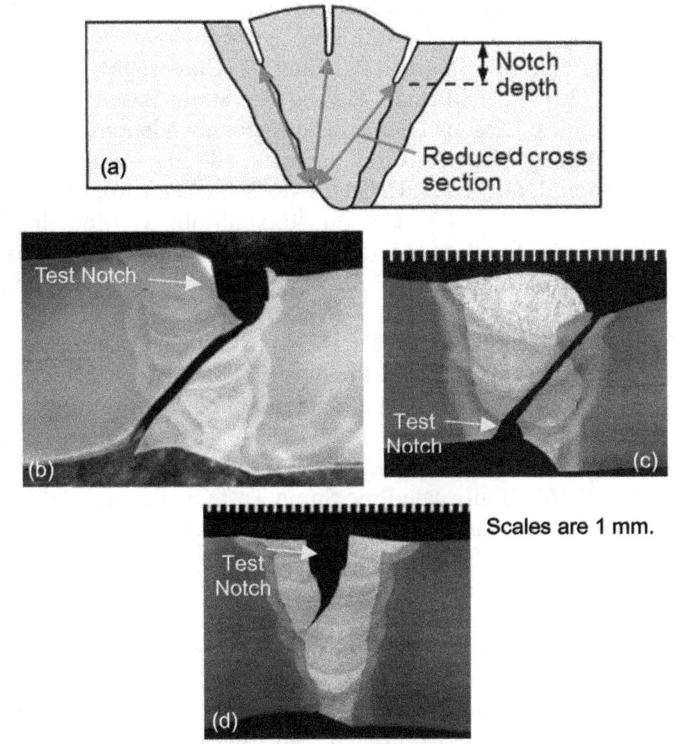

Figure 14: A schematic showing how high-low misalignment can reduce effective cross section and thus influence failure during SBD loading scenarios. Also included are three sections from full-scale tests.

Figures 14b – c show three cross section images from FSTs that contained high-low misalignment. These sections demonstrate the tendency for the failure path to coincide with both the smallest cross section, which is significantly influenced by the high-low, and a (roughly) 45° shearing angle. In Figure 14c, the ductile tearing had not yet consumed the full wall thickness at the time the FST was stopped. The weld metal overmatch in this case was about 30%. It has been noticed that in cases of high overmatch (say, > 25%), there is a tendency for the tearing path to deviate notably from the weld metal - at a sharper angle than 45°. This tendency is shown in the last photo where it is apparent that the tearing path is traveling generally in the direction of the high-low (and the necking); however, the path is not traveling directly towards the weld toe.

USE OF THE TSC PREDICTION TECHNOLOGY

While the TSC prediction and SBECA technology discussed in this paper is supported by many FSTs and is factored using a 95% confidence limit, this alone is not sufficient for good SBD engineering. Many quality challenges exist during the various phases of linepipe procurement and pipeline construction. There are ample opportunities for unintended quality lapses. It is imperative to maintain rigorous controls or other conservative selections/decisions throughout all phases. This is particularly the case for girth welding. While the safety factor provides some safeguard against welding upsets, there is no substitute for a well-qualified, monitored, and inspected approach to pipeline field welding.

One other aspect concerning the use of the TSC prediction technology is worth mentioning. When flaw tolerance calculations are conducted for SBECA, it is prudent to account for non-destructive examination (NDE) sizing errors before the ECA criteria are used. This may involve conducting NDE qualification trials and quantifying the accuracy of the chosen method in measuring flaw size. Seeded defect samples, immersion scans, and serial sectioning are often used in these qualifications. Current day ultrasonic technology can result in sizing accuracies on the order of 1 mm. The sizing accuracy is typically subtracted from the calculated tolerable flaw size which means, for example, that the flaw assessment curves in Figure 12 and 13 would be reduced by, say, 1 mm to account for the NDE system being used.

SUMMARY

This paper provides an overview of approximately 20 years of ExxonMobil's strain-based design technology development. Over the years, two tensile strain capacity (TSC) prediction

Copyright © 2019 by ASME

models have been development. One has been retired [9], while a second model (provided in the Appendix of this paper) has been validated using the results of 101 full-scale tests (FSTs). An overview of FST specimen preparation and example results is given. A safety factor (SF) for use with the TSC model has been derived from statistical analysis of FSTs results.

The second model has been used to create 10 hypothetical "benchmark" pipeline design cases. For each design case, the variables controlling TSC are given (pipe geometry, pipe and weld properties, high-low misalignment, pipe pressure), and the predicted TSC is calculated using the model. The benchmark cases can be used by others to check spreadsheet programming. The use of the SF is explained for the benchmark cases. Also, explained is how the TSC model can be used to assist the design phase of a strain-based design (SBD) pipeline project.

The TSC model can be used for strain-based engineering critical assessment (SBECA); i.e. to calculate tolerable weld flaw sizes. Several examples of use as a SBECA tool are given in the paper and flaw assessment diagrams are created. From these examples, it is concluded that weld strength overmatch and high-low misalignment are significant to TSC and flaw tolerance. In one example tolerable flaw height doubles and flaw length nearly triples due to a decrease in misalignment from 3 mm to 2 mm.

REFERENCES

[1] Noecker, FF, Fairchild, DP, Cook, MF, Macia, ML, and Kan, WC (2014). "A Case Study in High Strain Capacity Pipeline Qualification: PNG LNG Project", Proc. of 10th Int'l Pipeline Conf., Calgary, Alberta, Canada.

[2] Hukle, MW, Hoyt, DS, Horn, AM, LeBleu, JB, (2005). "Girth Weld Qualification for High Strain Pipeline Applications," Offshore Mechanics and Arctic Eng., Paper 67573, Halkidiki, Greece.

[3] Hukle, MW, Hoyt, DS, LeBleu, JB, Dwyer, JP, Horn, AM, (2006). "Qualification of Welding Procedures for ExxonMobil High Strain Pipelines," Offshore Mechanics and Arctic Eng., Paper 92503, Hamburg, Germany.

[4] Lillig, DB, Hoyt, DS, Hukle, MW, Dwyer, JP, Horn, AM, Manton, K, (2006). "Materials and Welding Engineering for ExxonMobil High Strain Pipelines", Proc. of 16th Int'l Offshore and Polar Eng. Conf., San Francisco, California, pp. 8-17.

[5] Fairchild, DP, Cheng, W, Ford, SJ, Minnaar, K, Biery, NE, Kumar, A, Nissley, (2007). "Recent Advances in Curved Wide Plate Testing and Implications for Strain-Based Design", Proc. of 17th Int'l Offshore and Polar Eng. Conf.., Lisbon, Portugal, pp. 3013-3022.

[6] Minnaar, K, Gioielli, P, Macia, M, Bardi, F, Biery, N, Kan, W, (2007). "Predictive FEA Modeling of Pressurized Full-scale Tests," Proc. of 17th Int'l Soc. Offshore and Polar Eng. Conf., Lisbon, Portugal, pp. 3114-3120.

[7] Kibey, S, Minnaar, K, Cheng, W, and Wang, X (2009). "Development of a Physics-Based Approach for the Prediction of Strain Capacity of Welded Pipelines", Proc. of 19th Int'l Offshore and Polar Eng. Conf., Osaka, Japan.

[8] Kibey, S, Wang, X, Minnaar, K, Macia, ML, Kan, WC, Ford, SJ, and Newbury, B, (2010). "Tensile Strain Capacity Equations for Strain-based Design of Welded Pipelines", Paper No. IPC10-31661, Proc. 8th Int'l Pipeline Conf., Calgary, Canada.

[9] Fairchild, DP, Kibey, S, Tang, H, Krishnan, VR, Wang, X, Macia, ML, Cheng, W (2012). "Advancements Regarding Capacity Prediction of Strain-Based Pipelines", Proc. of 22nd Int'l Offshore and Polar Eng. Conf., Rhodes, Greece, pp. 591-598.

[10] Mohr, WC, (2007). "Pressure Effects on Strain Concentration and Constraint for Strain-Based Design" Proc. of 17th Int'l Offshore and Polar Eng. Conf., Lisbon, Portugal.

[11] Gordon, JR, Zettlemoyer, N, Mohr, WC, (2007). "Crack Driving Force in Pipelines Subjected to Large Strain and Biaxial Stress Conditions", Proc. of 17th Int'l Offshore and Polar Eng. Conf., Lisbon, Portugal, pp. 3129-3140.

[12] Gioielli, P, Minnaar, K, Macia, M, Kan, W, (2007). "Large-Scale Testing Methodology to Measure the Influence of Pressure on Tensile Strain Capacity of a Pipeline," Proc. of 17th Int'l Soc. Offshore and Polar Eng. Conf., Lisbon, Portugal, pp. 3023-3027.

[13] Kibey, SA, Lele, SP, Tang, H, Macia, ML, Fairchild, DP, Cheng, W, Noecker, FF, Wojtulewicz, P, Newbury, BD, Kan, WC, Cook, MF, Hukle, M, (2011). "Full-scale Test Observations for Measurement of Tensile Strain Capacity of Welded Pipelines", Proc. of 21st Int'l Offshore and Polar Eng. Conf., Maui, Hawaii, USA, pp. 660-667.

[14] Fairchild, DP, Shafrova, S, Tang, H, Crapps, JM, Cheng, W, (2014). "Full-scale Testing for Strain-Based Design Pipelines: Lessons Learned and Recommendations," Proc. of 10th Int'l Pipeline Conf., Calgary, Alberta, Canada.

[15] Fairchild, DP, Tang, H, Shafrova, SY, Cheng, W, Crapps, JM, (2014). "Updates to ExxonMobil's Modelling Approach for Tensile Strain Capacity Prediction", Proc. of 24th Int'l Offshore and Polar Eng. Conf., Busan, Korea, pp. 487-496.

[16] Tang, H, Fairchild, DP, Panico, M, Crapps, JM, Cheng, W, (2014). "Strain Capacity Prediction of Strain-Based Design Pipelines," Proc. of 10th Int'l Pipeline Conf., Calgary, Alberta, Canada.

[17] Fairchild, DP, Crapps, JM, Cheng, W, Tang, H, Shafrova, S, (2016). "Full-scale Pipe Strain Test Quality and Safety Factor Determination for Strain-based Engineering Critical Assessment," Proc. of 11th Int'l Pipeline Conf., Calgary, Alberta, Canada.

[18] Strain-Based Engineering Critical Assessment of Girth Welds, Draft API RP. As of the writing of this paper, this draft is being progressed by an API 1104 Task Group.

[19] Panico, M, Macia, ML, Fairchild, DP, Cheng, W, (2017). "A Case Study of Design and Integrity Management Framework for Strain-Based Design Pipelines", Proc. of 27th Int'l Offshore and Polar Eng. Conf., San Francisco, CA., USA.

Copyright © 2019 by ASME

APPENDIX

This appendix provides a TSC prediction equation (A1) that is a combination of the work in [16] with the addition of a safety factor (0.53). This technology is in use by ExxonMobil for SBECA. The equation is used in conjunction with other pipe and weld testing requirements and strategies for quality monitoring [18]. The TSC equation takes the form:

$$\varepsilon_c = 0.53 \cdot PF \cdot \frac{\left[\phi(N,\kappa,\lambda,e/t)\frac{\delta}{t} + \beta_7\right]\left(\beta_1 \frac{D}{t} + \beta_2\right)}{\beta_6 + \beta_8\left(\frac{a/t}{1-a/t}\right)^{\beta_3} tanh\left(\beta_4\left(\frac{2C}{t}\right)^{\beta_5}\right)} \quad \text{(A1)}$$

And the competition between failure by tearing vs. plastic collapse in the pipe is handled as:

$$\varepsilon_c = \begin{cases} \varepsilon_c, & \text{if } \varepsilon_c < 2/3 \cdot UEL \\ 2/3 \cdot \mathrm{UEL}, & \text{if } \varepsilon_c \geq 2/3 \cdot \mathrm{UEL} \end{cases} \quad \text{(A2)}$$

The pressure factor, *PF*, accounts for the maximum anticipated internal pipe pressure and is determined as:

$$PF = \begin{cases} 3 \times 10^{-4} R_h^2 - 3.3 \times 10^{-2} R_h + 1.85, & R_h < 40 \\ 1, & R_h \geq 40 \end{cases} \quad \text{(A3)}$$

$$R_h = \sigma_h / SMYS \quad \text{(A4)}$$

The function ϕ takes the form:

$$\phi(N,\kappa,\lambda,e/t) = [W_1(N,\kappa)\lambda + W_2(N,\kappa)]\,(e/t) + W_3(N,\kappa)\lambda + W_4(N,\kappa) \quad \text{(A5)}$$

where the function $W_n(N,\kappa)$ with $n = 1,2,3,4$ takes the form:

$$W_n(N,\kappa) = h_{3n-2}(\kappa)N^2 + h_{3n-1}(\kappa)N + h_{3n}(\kappa) \quad \text{(A6)}$$

N is the strain hardening coefficient and can be determined for given tensile properties with the form:

$$N = \frac{(1+UEL)ln\,\frac{1+UEL}{1.005}}{(1+UEL) - 1.005 \cdot Y/T} \quad \text{(A7)}$$

The function $h_m(\kappa)$ with $m = 3n\text{-}2,\ 3n\text{-}1$ and $3n$ takes the form:

$$h_m(\kappa) = \beta_{6+3m}\kappa^2 + \beta_{7+3m}\kappa + \beta_{8+3m} \quad \text{(A8)}$$

Where κ is the relative strength coefficient and can be determined for given tensile properties with the form:

$$\kappa = \frac{\frac{1+UEL}{1.005 \cdot Y/T} - 1}{\left(ln\frac{1+UEL}{1.005}\right)^N} \quad \text{(A9)}$$

$$\lambda = UTS_{weld}/UTS_{pipe} - 1 \quad \text{(A10)}$$

The equation nomenclature is as follows:

- ε_c = tensile strain capacity (calculated as the decimal value; e.g., 1% is output as 0.01)
- *PF* = pressure factor
- *N* = strain hardening coefficient
- κ = relative strength coefficient
- λ = weld strength overmatch (entered as the decimal value; e.g., 10% is input as 0.1)
- δ = fitting parameter of CTOD R-curve (mm)
- *e* = high-low misalignment (mm)
- *t* = pipe wall thickness (mm)
- *D* = pipe outside diameter (mm)
- *a* = imperfection height (mm)
- 2*C* = imperfection length (mm)
- *UEL* = pipe uniform elongation (entered as the decimal value; e.g., 10% is input as 0.1)
- R_h = ratio of hoop stress over SMYS (entered as a percentage of SMYS; e.g., 10% is input as 10)
- σ_h = hoop stress created by the pipe pressure of interest
- *Y/T* = pipe yield-to-tensile ratio
- β_i = fitting coefficients (see Table A-1)
- UTS_{weld} = ultimate tensile strength measured from all-weld metal tensile test
- UTS_{pipe} = ultimate tensile strength measured from pipe longitudinal tensile test

Table A-1 – TSC Equation Coefficients

β_1	0.01972675	β_{12}	20.614713	β_{23}	-0.95868731	β_{34}	-0.76243717
β_2	4.5057389	β_{13}	-28.576332	β_{24}	0.12268418	β_{35}	0.22541011
β_3	1.4507932	β_{14}	7.3007237	β_{25}	-0.26825317	β_{36}	-0.41409428
β_4	0.04783402	β_{15}	-2.6518808	β_{26}	0.08219822	β_{37}	0.60967636
β_5	1.5786387	β_{16}	3.2891875	β_{27}	5.9561721	β_{38}	-0.26655138
β_6	0.10629258	β_{17}	-0.98321262	β_{28}	-8.4261999	β_{39}	0.24183907
β_7	0.000181	β_{18}	2.1397272	β_{29}	2.1570579	β_{40}	-0.33417047
β_8	6.3886415	β_{19}	-3.6478384	β_{30}	-4.2723879	β_{41}	0.16285831
β_9	-29.989814	β_{20}	1.6154649	β_{31}	6.235062	β_{42}	-0.03185441
β_{10}	41.766677	β_{21}	-1.0490341	β_{32}	-1.566328	β_{43}	0.05716533
β_{11}	-11.183184	β_{22}	1.8431406	β_{33}	0.57519391	β_{44}	-0.0150489

Copyright © 2019 by ASME

AIMPIMG2019-1063

Chapter 25

ANALYSIS OF EARTHQUAKE-RESISTANT DUCTILE IRON PIPE JOINT AT FAULT LOCATION USING ABAQUS – A CASE STUDY FOR SIMULATION BASED PRODUCT QUALIFICATION

Jeff Mason[1],
U.S. Pipe
Portland, OR

Srikanth Srigiriraju[2]
VIAS
Houston, TX

Arindam Chakraborty[3]
VIAS
Houston, TX

Burak Ozturk[4]
VIAS
Houston, TX

ABSTRACT

A seismic event can cause severe damage, including ruptures and leaks to pipelines and distribution systems, leading to environmental hazard, loss of life, and impairment of the infrastructure. An Earthquake-Resistant Ductile Iron Pipe (ERDIP) is capable of absorbing large ground displacements that occur during severe earthquakes by movement of its joints through expansion, contraction, and deflection. This engineering solution brings great potential to reduce environmental hazard and disruption of critical public services for cities with expansive infrastructure that are on a seismically active zone. However, to ensure safe operation and feasibility of design such pipeline products are to be subjected to rigorous verification and validation process. Although physical test based design validation is assuring it is often time not feasible to perform full scale tests of large diameter pipe joints, especially considering the effect of a fault movement within the test framework. Alternative is to use physics based realistic simulation to model the pipe joint behavior and use numerical results to determine if the design meets allowable criteria. A recent study involved design qualification of ERDIP large diameter joint against seismic event using simulation based approach. Utilizing the advanced non-linear modeling capabilities within Abaqus (a Dassault Systèmes FEA simulation software), first the pipe joint stiffnesses were calculated using a 3D model. Validation of the FEA model was performed using physical test data for a small diameter pipe joint. Further analysis was carried out wherein the pipe-soil interactions and the pipe-joint behaviors are represented by sets of non-linear springs. A stretch of the pipeline consisting of many pipe joints was undertaken to ascertain integrity under seismic fault movement conditions. The pipeline was acted upon by a fault rupture at specified azimuth angles and fault length. The simulation predicted that the pipe joints would meet the design criteria if the fault azimuth angle was within a certain threshold.

1. INTRODUCTION

Seismic events like earthquakes cause considerable damage to life and property. The differential movement of the ground during an earthquake leads to significant movement of pipelines leading to their rupture affecting supplies to a vast population besides incurring heavy repair expenses. Moreover, earthquake prone areas like California in USA, Japan etc. experience such disruptions in supplies often making their repair expenses exorbitant raising the need for a better technological solution of pipelines that remain unaffected during earthquakes.

The concept of Earthquake Resistant Ductile Iron Pipe (ERDIP) is described by K. Oda et al of Kubota Corporation[2]. The fundamental basis of this ERDIP technology is to have some flexibility in pipeline to absorb the deformation of the ground around the pipe. ERDIP is a misnomer as the pipe itself is not significantly deformable to be earthquake resistant, but the joints between the pipes move axially and laterally to give the requisite flexibility for the pipeline to absorb the energy from the shocks of tremors. Various ERDIP pipelines have been installed in Japan and they have been very successful to withstand earthquakes without damage or leakage over many years. Moreover, this ERDIP system works well for non-seismic events also that involve temporary or permanent ground deformations.

[1] Contact Jeff Mason: jmason@uspipe.com
[2] Contact Srikanth Srigiriraju: ssrigiriraju@viascorp.com
[3] Contact Arindam Chakraborty: achakraborty@viascorp.com
[4] Contact Burak Ozturk: bozturk@viascorp.com

Copyright © 2019 by ASME

The ERDIP system is designed to be modular and utilize standard nominal pipe lengths (i.e. 20' length) with joints so that it meets top tier criteria set forth in ISO 6134 (standard for ERDIP). The joint assembly allows the system to have 1% axial strain capacity which has been shown by empirical data to accommodate the great majority of seismically induced permanent ground deformations.

The installation process of ERDIP is also straightforward. The gasket and the bell section of the pipe end (where gasket sits) are cleaned free of any debris or foreign matter. Lubricant is added after the gasket is placed in the bell section of the pipe. The spigot end of the other pipe is then inserted and locked in position. After a series of steps involving bolt and nut, the joint between the two pipe segments will be fully assembled.

2. OBJECTIVE

The cost of physical testing of a large diameter ERDIP to verify if it can withstand seismic loads is complex and expensive hindering its development and installation process. The objective of this study was to establish a validated FEA methodology as an alternative to physical test that a regulator can have enough confidence in accepting results from such an FEA simulation in lieu of a physical test.

As an illustration, in this study, a new design of 48" ERDIP with two types of joints (called Type-1 and Type-2) was evaluated for its ability to maintain the structural integrity in satisfying specific design criteria in resisting the earthquake tremors. This ERDIP has Type-1 and Type-2 joint alternately connecting each pipe segment of the pipeline. The schematic diagram of these joints (proprietary) are shown in Figure 1.

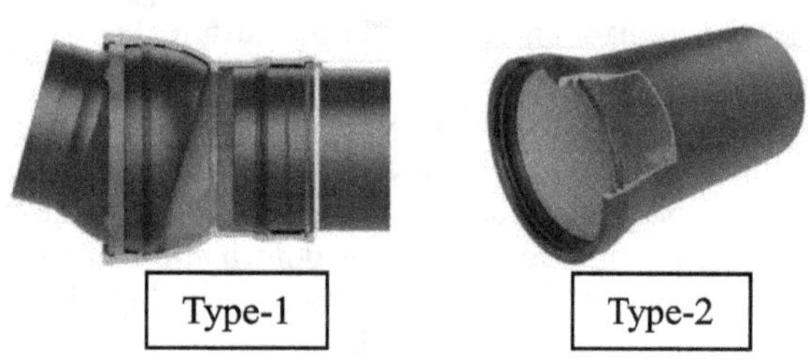

Figure 1 Two types of joints in 48" ERDIP

The behavior of this ERDIP in a seismic fault rupture was simulated using Finite Element Analysis (FEA) commercial software Abaqus (of Dassault Systemes). The methodology in developing the model with all its assumptions and simplifications was first validated using experimental results of a smaller 6" ERDIP with a specific branded joint that also involved a fault rupture simulation. The experimental tests were performed at Cornell University [1]. This validated methodology was then employed for a large diameter ERDIP with Type-1 & Type-2 joints to check if it passed the specific design criteria.

3. METHODOLOGY

Developing and running a simulation using a full 3D model of ERDIP pipeline and its interaction with surrounding soil in FEA is computationally resource intensive. So, the methodology involved simplifying the model geometry to a 2D model but retaining the functional characteristics of a full 3D model for the fault rupture simulation. The pipes were modeled using pipe elements in Abaqus which include the effects of full cross-section, but not in the finite element mesh.

3.1 Soil-Pipe Interaction

The soil-pipe interaction was included as bilinear springs for axial (along the pipe) and transverse (perpendicular to the pipe) directions as shown in Figure 2.

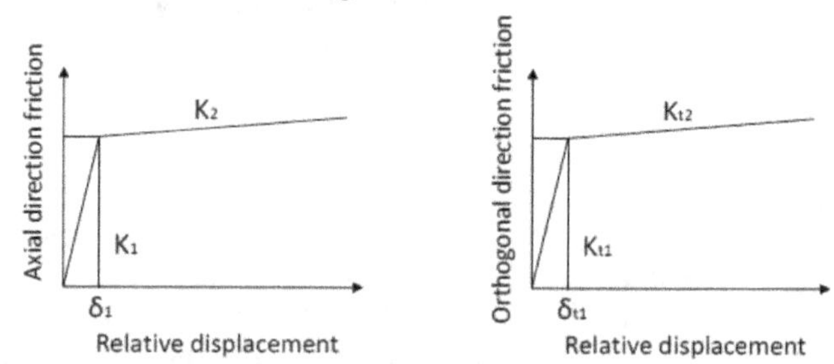

Figure 2 Axial and Transverse joint springs

The axial soil spring stiffness, K_1, before the change of slope:

$$K_1 = \frac{\pi D\gamma(h + 0.5D)(1 + K_0)\tan(\Delta)}{2\delta_1},$$

where

D: Diameter of the pipe
γ: Soil unit weight
h: Burial depth
K_0: Coefficient of lateral soil pressure at rest
Δ: Internal friction angle
δ_1: Turning point of axial spring stiffness

$K_0 = (1\text{-}\sin(f\Delta))$,

where f is a factor that relates the soil internal friction angle to friction angle at the soil-pipe interface. For this study, $f = 1.0$.

The turning point of axial spring stiffness was chosen as 0.019685" (0.005 m) for this study[2].

The spring stiffness after the turning point, K_2, in the axial spring stiffness profile was chosen such that it was a small arbitrary factor of the initial axial spring stiffness of the soil K_1:

$$K_2 = 7x10^{-7} K_1$$

The transverse spring stiffness was chosen to be a factor of pipe diameter, D:

$$K_{t1} = K\,D,$$

Copyright © 2019 by ASME

where K is subgrade reaction modulus (215338 pcf [2]).

The spring stiffness after the turning point, K_{t2}, in the axial spring stiffness profile was chosen such that it was a small arbitrary factor of the initial axial spring stiffness of the soil K_{t1}:

$$K_{t2} = 1.8x10^{-6}\, K_{t1}$$

3.2 Pipe Joint Behavior

The joints connecting the pipes were also modeled as nonlinear springs for their behavior in axial tension-compression, shear and bending/deflection deformation modes. This nonlinear spring behavior was calibrated from the results of simulation of full 3D model of these pipe segments with the joints under appropriate loads independently for each deformation mode.

3.3 Fault Rupture Model

The fault rupture simulation in FEA involved pipe elements connected with nonlinear joint and soil springs running along the pipe as shown in Figure 3.

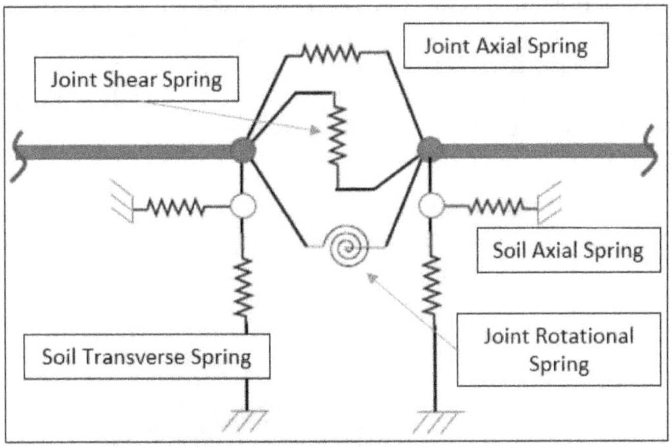

Figure 3 Schematic representation of joint and soil springs along the pipeline

The fault rupture was applied as appropriate displacement in the plane for a given fault orientation and the behavior of pipeline and its joints was studied.

4. DESIGN CRITERIA

The design criteria for large diameter ERDIP Type-1 & Type2 joints are as follows:

Maximum Axial Displacement – 6” (Type-1), ~0” (Type-2)
Maximum Deflection Angle – 15° (Type-1), 0.5° (Type-2)
Axial Load without Leakage – 822 kips
Allowable Stress – 40.5 ksi for both joints and the pipes (based on proportional limit)

The design criteria need to be satisfied for specific conditions of the fault rupture test. The displacement of the fault was considered 1.5m with the pipe laying 5 ft below the ground. Two orientations of fault were considered for the test: 90°; pipe azimuth 90° with fault orientation azimuth 156° as shown in Figure 4.

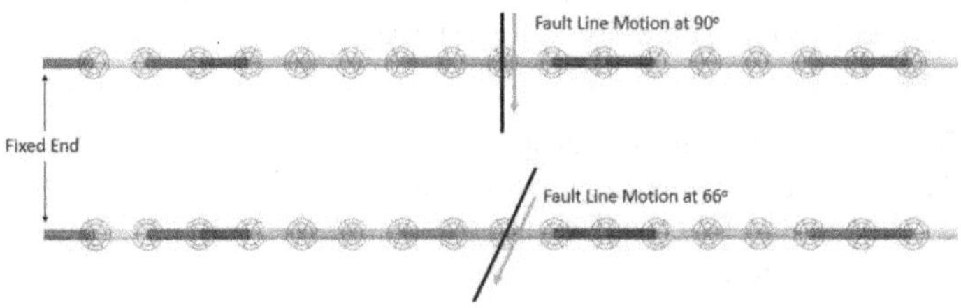

Figure 4 Two scenarios of fault orientations considered along the 48" ERDIP. Pink and Red circles denote each type of joint.

The backfill material for trench was taken as cement slurry, concrete class 100-E-100 (considered same as sand in simulation). The soil beyond the trench was considered as natural sandy soil. A safety factor of 1.1 was applied to model the fault displacement.

5. VALIDATION USING 6” ERDIP PIPE TEST DATA

The validation procedure involved the following:

1. Develop the joint springs through FEA simulations of 3D model pipeline for following loads:
   a. Bending
   b. Axial Tension-Compression
   c. Shear
2. Develop soil springs
3. Run 2D FEA simulation with pipeline, joint and soil springs in a fault rupture load scenario.

A representation of a half-section of geometry of 6” ERDIP joint in FEA is shown in Figure 5.

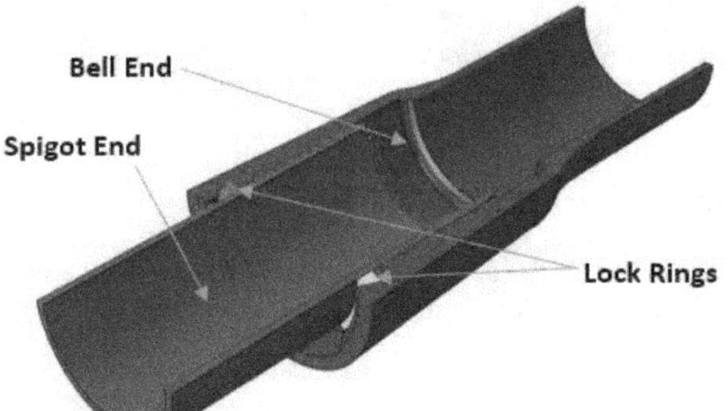

Figure 5 Geometry of the half-section of the pipe

Due to symmetry conditions, the axial tension-compression tests were conducted with half model (shown in Figure 5) while the bending test was done with a full 3D model. The shear model was done with half-model too as there was no physical test data to compare it with.

Copyright © 2019 by ASME

5.1 FEA Model & Results

The material properties of ductile iron were used for the pipeline which was meshed with C3D8R and C3D10 elements in Abaqus. The half-model had a few C3D6 elements also.

5.1.1. Four-point Bending

The resistance of the joint towards an external bending moment was evaluated using a four-point bending test. The set-up is as shown in Figure 6.

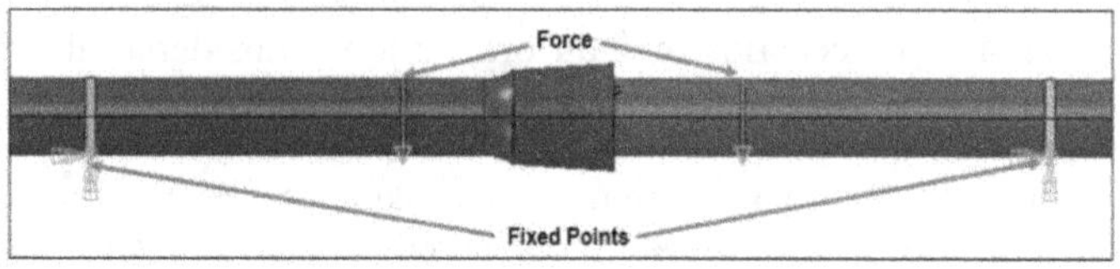

Figure 6 Four-point bending test

Rigid clamps were held at both ends with a contact interaction defined between the pipe and the clamps. An external force was applied at two locations as shown in Figure 6. The simulation was run for 2 models with locking rings at 2 different positions as shown in Figure 7. Test A had the locking ring at 3 and 9 o'clock positions with the loads applied at 6 and 12 o'clock positions while Test B has the locking positions in 6 and 12 o'clock positions with the applied load also in the same locations as shown in Figure 7.

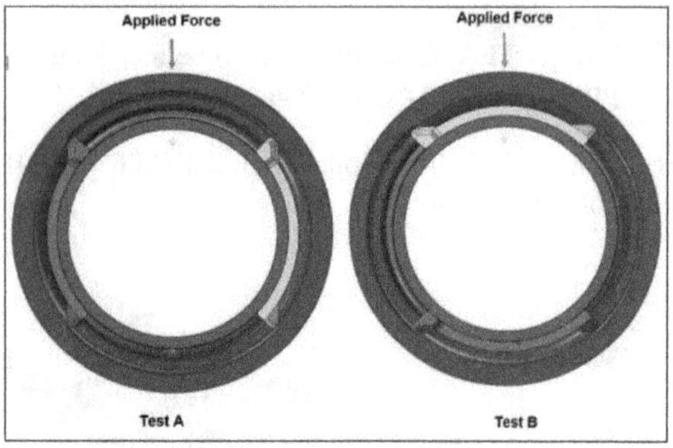

Figure 7 Locking rings and applied positions

The deflection in the beam was measured as the change in anlge of the pipeline axis at the joint as shown in Figure 8.

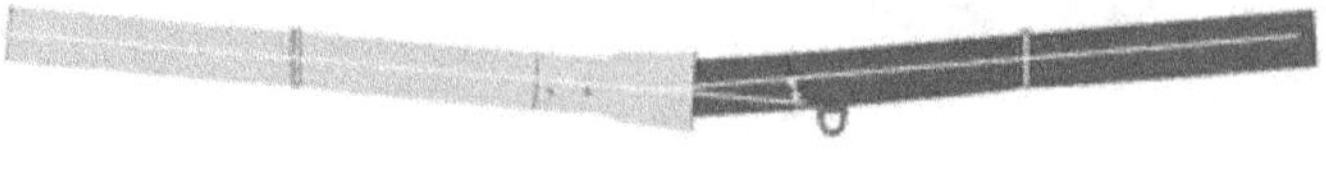

Figure 8 Deflection angle (theta) in the four-point bending test

The simulation result for Test A and Test B were compared with the physical test results as shown in Figure 9.

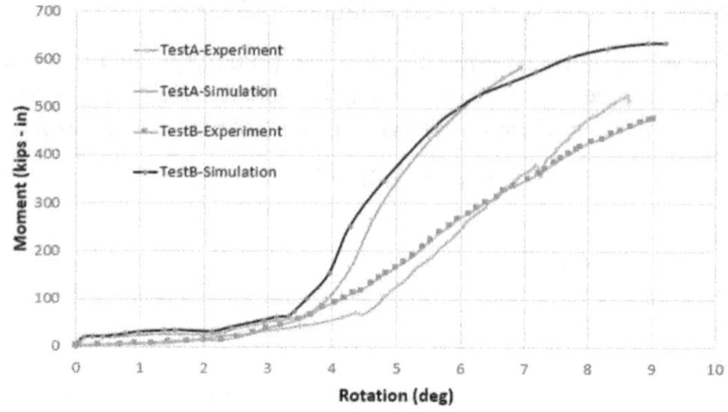

Figure 9 Moment vs deflection curve comparison between simulation and physical tests

The simulation results and the physical test data showed similar trends. However, the moment values for deflection angles greater than 3.5° were higher from simulation than from tests. During both the tests, it was observed that a fracture failure occurred as the deflection angle reached about 9°. It is likely that a crack may possibly have initiated early in the test and had propagated to failure as the stress progressed (a feature of ductile iron) and therefore, showed a lower bending resistance from the tests.

5.1.2. Axial Compression-Tension

As mentioned before, a half-symmetry model was considered for this simulation as shown in Figure 10.

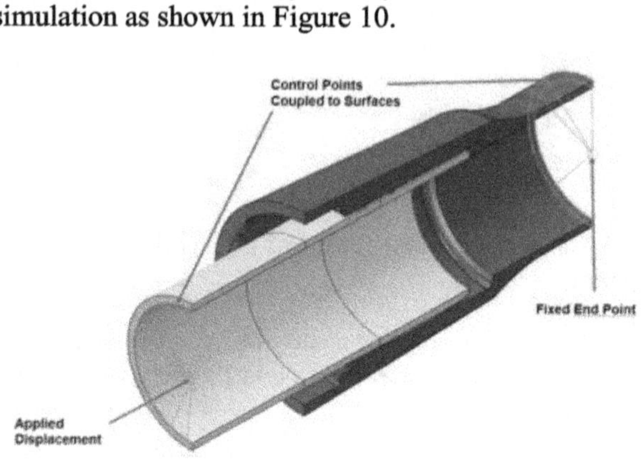

Figure 10 FEA model for measuring response of the joint in axial compression and tension.

The comparison between the physical test and FEA simulation for axial compression is and tension are shown in Figure 12 and Figure 12 Comparison between simulation with two different coefficients of friction and physical test for axial tension, respectively.

Copyright © 2019 by ASME

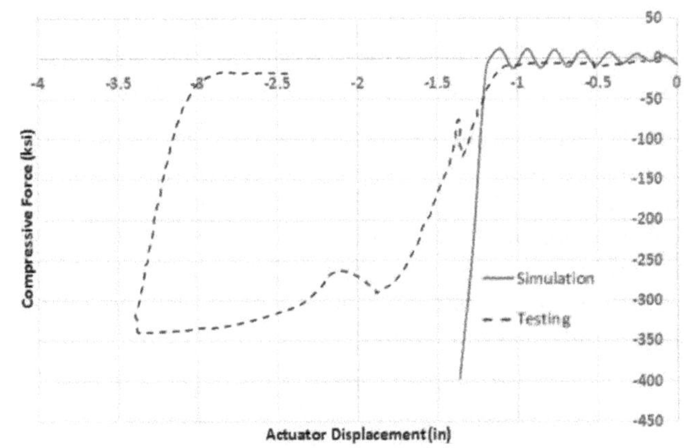

Figure 11 Comparison between simulation and physical test for axial compression

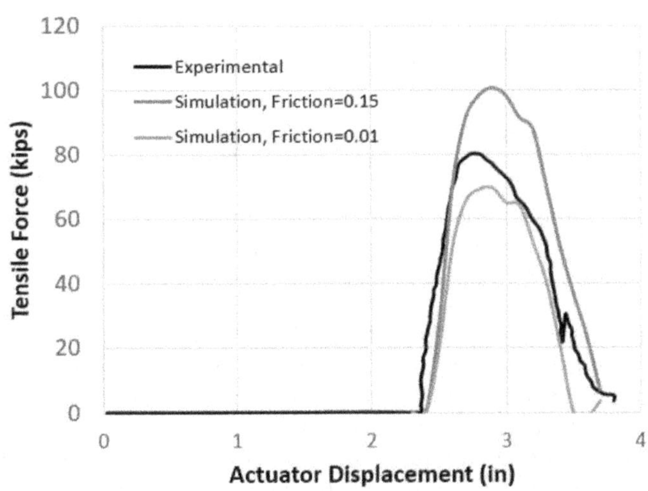

Figure 12 Comparison between simulation with two different coefficients of friction and physical test for axial tension

The simulation results and the physical test data showed similar trends for axial compression and tension. The simulation was run only for the loading part in the case of compression simulation. The simulation result for compression loading scenario showed a stiffer response than the test data due to possible higher local yielding in physical test. In the tensile loading case, the friction coefficient played an important role in determining the peak tensile force. The physical test data was within the bounds of reasonable coefficients of friction used in simulation.

5.1.3. Shear Loading

A physical test on shear loading of 6" ERDIP was not done. The model and results from FEA simulation of shear test is shown in Figure 13.

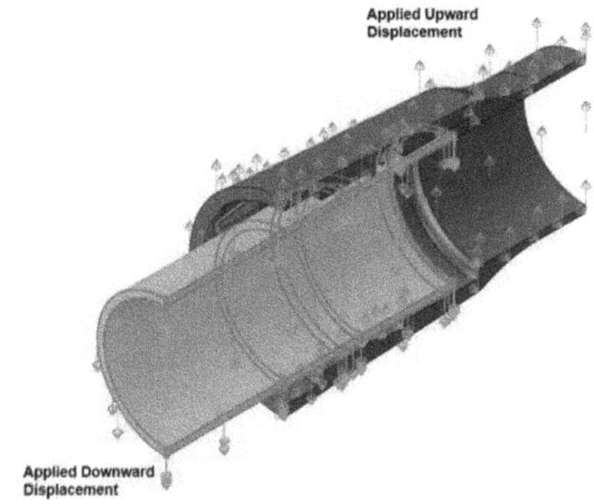

Figure 13 FEA model for shear response of the joint with applied displacements

The shear simulation has no corresponding physical test to compare with. The result from simulation is shown in Figure 14.

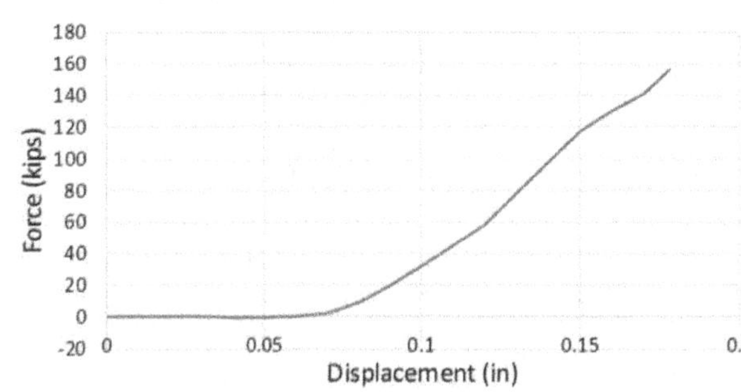

Figure 14 Force vs displacement plot from shear test simulation

5.1.4. Fault Rupture

The fault rupture simulation was done as a 2D model with pipe elements PIPE31 representing four pipe segments connected with joint springs. The joint springs stiffness for all deformation modes: bending, axial compression-tension and shear were evaluated from FEA simulation results. The internal friction angle for soil was 42° with its unit density 106 lb/ft^3. The pipeline was held at a depth of 33" in the soil bed in the test. The soil-pipe interaction was included in the FEA model as bilinear springs using the methodology mentioned in Sec. 3.1.

The fault rupture was at 50° with the azimuth. The internal pressure in the pipe was maintained at 80 psi. The comparison between the physical test and simulation for axial force variation is shown in

Copyright © 2019 by ASME

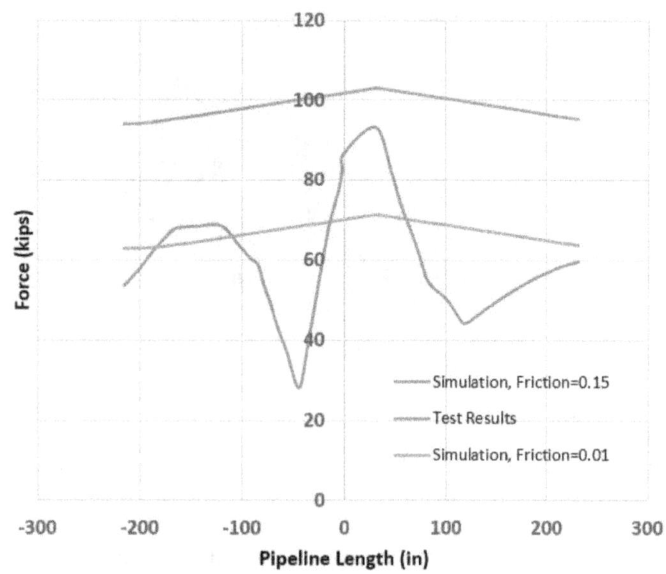

Figure 15 Comparison between simulation with different friction coefficients used for axial tension and physical test for axial force along the pipeline for fault rupture scenario.

The plot in Figure 15 shows that the variation of the axial force along the pipeline length in simulation follows a similar trend as seen in the experimental result. The physical test had sudden showed that the force dropped sharply between 18 in. (457 mm) and 19 in. (483 mm) of fault displacement because the circumferential deformation of the pipe caused the corruption of the axial strain gages [1]. The maximum axial force across the pipeline in the physical test is within the bounds of the simulation results for two different coefficients of friction.

The close match between the experimental results for bending, axial-tension, axial-compression and fault rupture gives enough confidence to use the same methodology of FEA to test 48" ERDIP in a fault rupture scenario.

6. 48" ERDIP PIPE FEA

The workflow of 6" ERDIP simulation as mentioned in Sec. 5 was followed exactly for 48" ERDIP Type-1 & Type-2 joints to evaluate the joint and soil spring stiffness.

6.1 Joint springs for various deformation modes

Without presenting the details of simulation of 48" ERDIP for bending, axial compression-tension and shear loads, it would suffice to mention that following the same methodology as done for 6" ERDIP mentioned in Sec. 5, the joint spring stiffness was evaluated for the bending, axial tension-compression and shear loads independently using FEA for both Type-1 and Type-2 joints.

6.2 Fault rupture

The FEA model for fault rupture simulation was built in the same way as done for 6" ERDIP with the exception that Type-1 and Type-2 joint were alternately placed along the pipeline. The pipe was placed at the depth of 5ft in the simulation. For soil spring stiffness, the internal friction angle of soil used was 34° with the unit density as 120 lb/ft^3 and the stiffness of the soil for axial and transverse directions were evaluated as mentioned in Sec. 3.1.

Two fault rupture scenarios were considered for the simulation as shown in Figure 4. The fault was placed at specific locations in the pipeline and the simulation results from each scenario of fault orientation and fault location were compared with the design criteria mentioned in Sec.4. Table 1 summarizes the scenarios and joints that pass or fail in the simulation when compared with the design criteria.

Fault Location	Joint Type	Fault 90°		Fault 66°	
Design Criteria →		A	R	A	R
At Type-1 joint	1	P	P	F	P
	2	P	F	F	F
At Type-2 joint	1	P	P	F	P
	2	P	P	F	F
Between 2 joints	1	P	P	F	P
	2	P	F	F	F

Table 1 Comparison of results of joint springs from fault rupture simulations for various scenarios of fault locations and orientations. A: Axial, R: Rotational, P:Pass, F:Fail.

The results show that Type-2 joint fails the design criteria for all the fault locations when the fault orientation is 66°. Type-1 joint also fails in the axial deformation design criterion for all fault locations when the fault orientation is 66°. However, as seen in the 6" ERDIP fault rupture test, the coefficient of friction (0.15) chosen for 48" ERDIP is probably higher than what the actual case would be leading to stiffer joint springs. Also, the soil parameters chosen were somewhat arbitrary that could potentially transfer higher forces to the pipeline leading to failure of the design criteria of its joints. Lastly, the Type-1 joint has very strict design criteria which needs to be rerated through modifying its design or tuning it to closer towards leakage criteria which usually is less stringent than the design criteria.

7. CONCLUSION

Testing a pipeline with large diameter in a laboratory is cumbersome and expensive. Moreover, the variation of any joints during the design iterative process can lead to multiple physical tests that make it financially unviable and physically unappealing. This study has successfully demonstrated that FEA could be used as a valid tool for testing large pipelines in seismic events obviating the heavy costs of labor and machinery besides accelerating the design process.

REFERENCES

[1] H. E. Stewart, C. Pariya-Ekkasut, B. P. Wham, T. D. O'Rourke, C. Argyrou, T.K. Bond, 2015. "Hazard Resilience Testing of US Pipe Ductile Iron TR-

Copyright © 2019 by ASME

XTREME™ Pipe Joints", School of Civil Engineering, Cornell University.

[2] K. Oda, T. Ishihara, M. Miyajima, "Research of Earthquake Resistant Ductile Iron Pipe (ERDIP) for fault crossing," Technical Report, Kubota Corporation.

Copyright © 2019 by ASME

SECTION 7
STRAIN CAPACITY MANAGEMENT

INTRODUCTION

By Jens Petter Tronskar
(Jens.P.Tronskar@dnvgl.com)

Decisions made during the pipeline routing and design stages of a pipeline project are important for the management of the pipeline strain capacity to resist soil movement. In this context we are talking about the pipe and girth weld capacity to handle axial or bending strain that may far exceed the yield strain of the pipe i.e. 0.5% strain. Ideally the pipe should buckle before a full-bore rupture occurs in the girth weld. Once the pipeline is constructed the strain capacity of the pipeline and girth welds cannot be changed without sectional replacements or local mitigations that may involve various options for strengthening the pipe and girth weld strain capacity. Hence, we may separate the management of strain capacity into two phases;

1) the route selection stage, design stage, pipe qualification and procurement, welding procedure qualification and the construction
2) after commissioning during the operations phase

Conventional pipeline design based on codes such as ASME B31.8 for gas pipelines, sets the minimum wall thickness to withstand pipeline internal pressure. To design a pipeline to accommodate longitudinal and bending strain above 0.5%, associated with soil movement induced by slopes, arctic permafrost or tropical rain, seismic fault crossings, etc. requires use of strain-based design (SBD) approach. In conventional design the pipe wall thickness is determined based on hoop stresses for pressure containment. When dealing with geohazards causing soil movement, additional design considerations are needed to ensure that longitudinal strains can be safely accommodated. The design margin for longitudinal loads is defined in terms of allowable strain. In addition, for improving the strain capacity of the pipe use of steels with lower yield to tensile (Y/T) ratio, higher strain hardenability or higher uniform elongation are beneficial. This topic i.e. the development of pipeline steels with increased strain capacity is discussed in paper (AIMPIMG2019-1020) of this section.

Thicker pipe with smaller diameter is beneficial for the strain capacity and use of long seam welded (SAWL) is preferred for high strain applications. SAWL pipe is cold expanded and end adjusted to improve weld fit up. Reduction of hi-lo is very important to ensure high strain capacity of the girth welds as there is solid evidence to support that increasing hi-lo leads to deterioration of the girth weld strain capacity. Use of spiral welded pipe for pipeline construction may introduce large hi-lo adversely affecting the strain capacity of the girth welds. This is because such pipe is not cold expanded, end adjusted, and have larger dimensional tolerances i.e. out of roundness and local out of roundness than long seam welded pipe.

Whether using SAWL or SAWH pipe manufactured from thermo-mechanically control processed plate, HAZ softening may become an issue. HAZ softening is linked to the chemistry of the plate and occur especially in pipe with low carbon 0.04-0.07 %weight and low carbon equivalents, i.e. Pcm 0.14 - 0.16 values. It may in particular become an issue for thinner pipe walls welded with high heat inputs.

It is important for the strain capacity to select welding processes and procedures that give upper shelf toughness at the operating temperature and high tearing resistance. Automatic and semi-automatic GMAW are preferred processes for pipeline construction due to the easy control of the welding parameters and consistency of the weld properties. Manual welding using cellulosic electrodes generally gives lower fracture resistance, more weld defects, uses a wide weld preparation which may promote shear fracture at high axial or bending strain. A better alternative with respect to strain capacity is to use basic coated SMAW electrodes that are specifically developed for pipeline welding such as AWS E9045. The first paper of this section explains how such electrodes were used to develop a high strain capacity repair procedure for a ruptured API 5L grade X70 gas transmission pipeline.

Another factor that guides procurement of high strain linepipe is the hoop strength relative to the longitudinal strength of the pipe. Generally, pipe mills follow the industry leading linepipe codes (APL 5L and DNV OS-F101) which only specify pipe properties in the hoop direction. When measuring the tensile properties in pipe manufactured from thermo- mechanically control processed plate it is found that there is mechanical anisotropy and the longitudinal tensile strength may be higher than in the hoop direction.

Weld overmatch should be ensured i.e. the weld tensile strength should overmatch the longitudinal strength of the pipe. This is very important to avoid strain concentration in the weld and potentially full-bore rupture if the pipeline is exposed to

Copyright © 2019 by ASME

displacement due to soil movement. The paper by Wan Hamat, Wan Ismail, Ibrahim, Zin, and Aris, (AIMPIMG2019-1068) presents a pipeline project that experienced two full bore ruptures during the first 4 years of operation, essentially caused by weld strength undermatching. Also, the paper by Wang, Dan Jia, Rapp, and Johnson, (AIMPIMG2019-1064) in this chapter discusses both the effect of weld strength under matching and the effect of HAZ softening.

Planar flaws in the girth welds with undermatching or HAZ softening, accentuates the adverse effect on the strain capacity of the weld. It is therefore important during the project feed and detailed design to plan for use of inspection methods that can consistently detect and size planar flaws with high accuracy. Radiography has traditionally been used for many pipeline projects but has major limitations for sizing and detection of planar flaws. Hence, for pipelines where high strain capacity is required, it is advocated to select AUT in combination with auto or semi-automatic welding processes; a combination which has very good track records for pipeline installation of offshore pipelines that may be subject to large strain from the reeling installation or from lateral buckling in service.

Strain capacity management process is divided into two phases; the first phase addresses how the strain capacity can be managed through the process of pipeline design, qualification and procurement of line pipe and through the welding process and construction QA/NDT. The second phase is in the operations where an operator may select several options to strengthen the pipeline strain capacity.

If during the operations phase regions of active geohazards are discovered before the pipeline has been damaged or ruptured, sectional replacement by shut down and cutting, followed by sectional replacement of pipe and girth welds with higher strain capacity is an option. Another alternative is re-routing or using HDD. Sectional replacement may also be carried out using hot-tap and by-pass, without shutting down the pipeline.

In cases where only a few joints are affected by the soil movement, the operator may select various local strengthening methods such as weld overbuild, use of Type-B welded split sleeve or various grouted clamps and composite wrapping techniques. Weld over build using AWS E9045 electrodes has proven to be effective but would not be advisable for large hi-lo and for welds with presence of deep planar flaws. In such cases welded sleeve repair with specially developed buckle restrainers is a method that can achieve strain capacity only limited by the pipe buckling capacity.

The four papers in this chapter highlight the lessons learnt, in terms of the various approaches for how to improve the strain capacity of pipelines for avoiding pipeline failures caused by geohazards and limitations in the pipe and girth weld strain capacity in future pipeline projects.

The paper by Wan Hamat, Wan Ismail, Ibrahim, Zin, and Aris (AIMPIMG2019-1068), presents PETRONAS' experiences, challenges and new solutions developed to mitigate low strain capacity mainly associated with girth weld undermatching issues. The paper discusses the Sabah Sarawak Gas Pipeline (SSGP), which is a 512 km x 36" O.D. Grade API 5L X 70 mainly spiral welded onshore natural gas pipeline. It was completed in 2013. The SSGP traverses through many different types of terrain which includes mountainous tropical environment with steep slopes, seasonal heavy rainfall, numerous river and stream crossings, swamps and road crossings. The pipeline experienced two full bore rupture incidents; one in June 2014 and one in January 2018. These two incidents caused the operator to re-evaluate its pipeline repair and maintenance practices for pipelines constructed in geohazard active environments.

The SSGP pipeline was designed in accordance with ASME B31.8:2007 edition for non-sour service throughout its 25 years design life and was internally protected against corrosion via fusion bonded epoxy coating. During the construction stages, API Std 1104; Welding of Pipelines and Related Facilities was employed to establish all relevant WPSs, which utilized cellulosic SMAW welding consumables (E6010 root and E8010 hot, fill and capping passes), meeting the requirements of the code and the company specifications. However, API Std 1104 allows fracture of transverse tensile specimens in the weld metal, as long as the tensile test fails at a tensile stress which is equal or higher than the pipe tensile strength. For spiral welded pipe manufactured from coil strip TMCP steel, the actual strength in the pipe longitudinal direction, may be higher than the tensile strength in the hoop direction. The result is a girth weld that meets the code requirements but may grossly undermatch the pipe tensile strength. To manage and reduce the risk of geohazard's effect on the girth welds weld, a new welding procedure has been designed, qualified and validated by full scale bend testing. The aim was to introduce a girth weld with improved strain capacity, that exceeded the buckling strain capacity of the linepipe to avoid full bore girth weld ruptures. The new girth welding procedure and also other "live" repair methods such as weld overbuild on existing girth welds and use of a specially designed Type B welded sleeve, are intended to use for future SSGP weld repair and maintenance purposes

The paper by Shimamura, Yasuda, Igi, Muraoka and Kondo (AIMPIMG2019-1020) discusses the need for pipelines with high strain capacity against buckling and high weld fracture capacity for installation in areas with seismic and permafrost regions. To improve strain capacity of linepipe the authors propose to use steels with lower yield to tensile (Y/T) ratio, higher strain hardenability or higher uniform elongation. The paper describes the development of high strain linepipe steels by the dual phase microstructure controls such as ferrite-bainite and bainite- MA (Martensite-Austenite constituents). Tensile properties can be controlled by controlling the volume fraction of second phase and strength difference between soft and hard phases. Grades X80-X100 high strength linepipe for high- strain application, have been developed through the thermo- mechanical controlled processing (TMCP) with accelerated cooling and on-line heat treatment process. Trial production test of heavy wall Grade X70 high strain linepipe was conducted by applying

Copyright © 2019 by ASME

the advanced TMCP with Super-OLACTM-A and HOPTM in the actual plate and pipe mill. The trial production Grade X70 linepipe, showed that line pipe with sufficient strength, superior toughness and strain capacity meeting the target level could be successfully rolled using the developed TMCP process.

The paper by Sakimoto, Handa, Tajika, Igi and Kondo (AIMPIMG2019-1026) describes the need for high-strength linepipe such as Grades X70 and X80 for application in landslide-prone mountain areas, reclaimed land where liquefaction is an issue and polar regions where frost heave occurs. Buried pipelines in these regions are expected to be subjected to large deformation due to the large ground movement caused by these types of geohazards. The paper recognizes that it is not possible to apply conventional stress-based design when the longitudinal stress of pipe material may greatly exceed the yield stress and the pipeline material is subjected to large plastic strain. The paper presents the residual tensile strain capacity after local buckling has occurred in a full-scale pipe bending tests on X80 girth welded UOE high strain pipe with an outside diameter (OD) of 48 inches and X70 girth welded UOE high strain pipe with OD of 36 inches. To study the loading condition on the tension side after local buckling near a girth weld, the longitudinal tensile strain distributions were investigated. After a large wrinkle had formed on the compression side of bending, the wrinkle worked precisely like a plastic hinge, and longitudinal tensile strain developed only in the exactly opposite area to the large wrinkle on the compression side of bending. The tensile strain capacity in full-scale pipe bending is discussed from the viewpoint of the effect of the tensile strain distribution enhanced by wrinkle occurrence on the compressive side on final fracture in comparison with the pure tension loading condition.

The paper by Wang, Jia, Rapp, and Johnson (AIMPIMG2019-1064) highlights that there has been several unexpected girth weld failures in newly constructed pipelines. The paper very much tells the same story as the first paper in this section. Similar to the first paper, it is found that most of the failures occurred in X70 pipelines with manual welds, although failures have been observed in linepipe grades ranging from X52 to X80. Similar to for the paper (AIMPIMG2019-1068), post-incident investigations indicated that the linepipe, girth weld qualification, and field girth welding and inspection met the requirements of relevant codes and standards. It was found that characteristics of the incidents and main contributing factors were (1) weld strength undermatching, (2) heat- affected zone (HAZ) softening, (3) weld bevel geometries of manual stick welding processes that favor plastic straining along the softened HAZ, and (4) elevated stresses/strains from normal settlement and other loads. The causes of weld strength undermatching and HAZ softening are further examined from the viewpoint of linepipe steel manufacturing, pipe testing and qualification requirements, and a response to a regulatory advisory is proposed.

Chapter 26 High Strain Weld Solutions for Geohazard Active Environment (AIMPIMG2019-1068)
Wan Abdullah Wan Hamat, Wan M Marzuki B Wan Ismail, Kamal Azam B Ibrahim, Ghazali B M Zin, and Noraina Suyanti Bt Md Aris, Petronas, Malaysia

Chapter 27 Development of High Strength High Strain Linepipe Steels (AIMPIMG2019-1020)
Junji Shimamura, Kyono Yasuda, Satoshi Igi, Ryuji Muraoka and Joe Kondo, JFE Steel Corporation, Japan

Chapter 28 Tensile strain Capacity after Local Buckling of High Strain Pipeline in Full-Scale Bending Tests (AIMPIMG2019-1026)
Takahiro Sakimoto, Tsunehisa Handa, Hisakazu Tajika, Satoshi Igi and Joe Kondo, JFE Steel Corporation, Japan

Chapter 29 Low Strain Capacity Girth Welds of Newly Constructed Pipelines and Mitigative Approaches (AIMPIMG2019-1064)
Yong-Yi Wang and Dan Jia, Center for Reliable Energy Systems; Steve Rapp, Enbridge; and David Johnson, Energy Transfer, USA

Copyright © 2019 by ASME

AIM-PIMG2019 1068

Chapter 26

HIGH STRAIN WELD SOLUTIONS FOR GEOHAZARD ACTIVE ENVIRONMENT

Wan Abdullah WAN HAMAT
PETRONAS
Kuala Lumpur, Malaysia

Wan M Marzuki WAN ISMAIL
PETRONAS
Kuala Lumpur, Malaysia

Kamal Azam IBRAHIM
PETRONAS
Kuala Lumpur, Malaysia

Ghazali M ZIN
PETRONAS
Kuala Lumpur, Malaysia

Noraina Suyanti MD ARIS
PETRONAS
Kuala Lumpur, Malaysia

ABSTRACT

Objectives/Scope:

The Sabah Sarawak Gas Pipeline (SSGP) is a 512 km x 36" O.D. Grade X 70 onshore natural gas pipeline that was completed in 2013 to transport 750 MMscfd of gas from the Sabah Offshore Gas Terminal (SOGT) to the Malaysia Liquified Natural Gas (MLNG) Complex at Bintulu, Sarawak, Malaysia. SSGP traverses through many different types of terrains. Two pipe parted incidents in June 2014 and January 2018 resulted in PETRONAS re-evaluating its pipeline repair and maintenance practices for pipeline constructed in geohazard active environment.

The pipe parted incidents triggered a reassessment of the weldment robustness against the geohazard active environment along the whole SSGP pipeline route. One of the issues needed to address in SSGP failure incidents was on the weld strength under-matching phenomena. Managing the active geohazard impact to the pipeline to minimum effect as practically possible was a daunting task and will be a never ending maintenance activity at SSGP. On top of that, a number of actions were also undertaken for strengthening the girth welds. Thus, in order to manage, reduce the risk and dampen the geohazard's effect on the weld, a new welding procedure has been designed, tested at full scale and qualified. The aim is to have a weld with improved & higher strain capacity that can 'match' the pipe base metal strain capacity. The new girth welding procedure is also intended to be used for any SSGP weld repair and maintenance purposes.

This paper presents PETRONAS experiences, challenges and new solutions developed to mitigate the inferiority of the strain capacity of the weld (mainly due to weldment strength under-matching) compared to the higher pipe strain capacity.

Methods, Procedures, Process:

Summary of the immediate cause of the parted incident at SSGP at the geohazard active areas, the strain capacity and the weld strength under-matching situation of the original weld from pipeline construction, the new improved design of the weld, the new improved welding procedure, qualification via full scale bend test, the challenges faced during welding of the repaired section and the precautions of welding done under temperature-fluctuation-affected pipeline which impacting the expansion and contraction of the pipeline.

Results, Observations, Conclusions:

This paper presents aspects of weld strain capacity improvement – designed, tested and implemented for geohazard active environment. The new improved weld design and procedure have been deployed ever since it was qualified, for any weld repair and maintenance for this SSGP Pipeline.

Open

Copyright © 2019 by ASME

Novel/Additive Information:

This paper presents PETRONAS updated practice in improvement weld strain capacity to increase the robustness of the weld and the pipeline in against the geohazard active environment of SSGP pipeline. The additional work done and improvement performed demonstrates: 1) Options for improving weld strain capacity 2) Full scale test for validating and qualifying the new weld design and welding procedure.

Introduction

Sabah-Sawarak Gas Pipeline (SSGP) is a 512 kilometers natural gas pipeline that linked Kimanis in Sabah to Bintulu in Sarawak, Malaysia. The construction phase started in 2009 and was completed in 2013. The 36 inches OD spiral welded pipeline was constructed using API 5L X70 steel grade pipe, with thickness ranging between 14 and 20mm. The pipeline right of way is located with anapproximate length of 90km in Sabah and 422km in Sarawak, with multiple crossings, mountainous slopes with a higher altitude, dense forests, rivers, swamps and palm oil plantations.

The pipeline was designed to transport 750 MMscfd of gas from the Sabah Offshore Gas Terminal (SOGT) to the Malaysia Liquefied Natural Gas (MLNG) Complex at Bintulu, Sarawak, Malaysia.

PIPELINE DESIGN CRITERIA

The SSGP pipeline was designed in accordance with ASME B31.8:2007 edition for non-sour service throughout its 25 years design life and was internally protected against corrosion via fusion bonded epoxy coating.

During the construction stages, API Std 1104; Welding of Pipelines and Related Facilities was employed to establish all relevant WPS which utilized cellulosic electrode and other parameters that meet the specifications. Based on API Std 1104, presence of stresses that provide axial load to pipeline during construction, operations and environment condition shall be considered.

This pipeline was designed in accordance with API Std 1104 which is capable of withstanding up to 0.5% strain. Geohazard active areas that exert strain exceeding 0.5% will therefore represent serious threats to this pipeline.

GEOHAZARD ACTIVE AREA

SSGP pipeline route was having unpredicted geohazard threats especially during heavy rains. Multiple locations were identified by geohazard study which require closed monitoring and/or slope strengthening to minimize undesirable strain effect to this pipeline.

PIPELINE INCIDENT RECORDS

Two (2) incidents and loss of containment were recorded in June 2014 and January 2018, respectively. The first incident happened in Lawas, Sarawak and pipeline operation was halted for almost 2 years to allow for a complete failure investigation, technical studies and site repair works. Among the activities conducted during this period were as follows:

1 Geohazard study for the whole pipeline route.
2 Re-review of all of the radiographic films (taken during the construction stages).
3 Laboratory analysis of the fractured pipe girth welds.

Hundreds of weld joints were successfully repaired prior to recommissioning of the pipeline after the first incident. The pipeline was successfully reinstated and gas export was resumed in 2016.

The second pipeline parted incident occurred in Long Luping, Lawas, Sarawak in January 2018. The pipeline operation once again was halted temporarily to allow geotechnical trench/soil studies and weld repair work at site. Among the activities conducted during this period include the followings:

1 Failure investigation by DNV GL Singapore
2 Development and qualification of high strain capacity weld (WPS)
3 Qualification of welders for high strain capacity weld

Apart from that, all the identified geohazard risk areas were re-inspected by non-contact magnetometry method (MTM) and inline inspection (ILI) to identify areas with pipe distortion and strain deformed state. The identified high risk weld joints were then re-inspected by Ultrasonic Testing (PAUT, ToFD), Magnetic Particle Testing (MT), eddy current Test (EC) and Radiography Testing (RT) for confirmation of presence of defects.

Figure 1: First incident with estimated lateral deflection of 1.3 meters

Figure 2: Second incident with estimated lateral deflection of 3 meters

Copyright © 2019 by ASME

FULL SCALE SIMULATION TESTING OF ORIGINAL WELD QUALIFIED TO API 1104 (EXISTING WPS)

Surplus linepipe from SSGP project were used to simulate the weld strength under external load with presence of internal pressure. The weld was produced with a presence of weld misalignment and root defect simulating the condition of the failed pipeline girth weld.

Based on mechanical test result of the produced samples, two (2) parameters were subject to discussion in this journal i.e. charpy impact test and tensile properties. These two (2) parameters will contribute to the weld under matching and thus the girth welds ability to withstand axial load from environment.

A number of welds were reproduced in accordance with original WPS for full scale bend test. From the mechanical testing report, tensile strength was recorded in between 83,541-92,098 psi against API acceptance of 82,671 – 110,228 psi, hence, it was accepted. Similarly, for impact properties, the recorded energy values were 62.7 Joules,

Joules and 262.7 Joules for weld metal, heat effected zone (HAZ) and base metal respectively, meeting the minimum acceptance standard of impact properties of 40 Joules.

Figure 3: Actual test setup for the four-point full scale bend test

Results from the full scale bend test showed that the pipe started to fail by full bore rupture at an average tensile strain of below 0.5% with the presence of a semi-elliptical surface breaking root defect of dimensions 3.0mm in height x 190mm in length in the presence of 2mm hi-lo. The average tensile strain was calculated as the average measurement recorded from the strain gauges located at the tension side of the test pipe. The water pressure and the maximum deflection at the girth weld recorded at the fracture instant were 92.04 bars and 200.94mm, respectively. Full bore unstable fracture took place as show in Figure 4. After fracture initiation, the girth weld had no residual fracture capacity and a full bore rupture occurred.

FULL SCALE SIMULATION TEST WITH HIGH STRAIN CAPACITY WELD

High strain capacity girth welding procedure was developed using basic coated SMAW electrodes, i.e. E7016-H4 for root pass and E9045-P2 H4R for hot, fill and cap pass and with an additional minimum of 3mm height and 60mm width weld overbuild on the external surface following cap pass. The weld was prepared with a root defect of 190mm in length, 2.9mm in height and 0.7mm in width together with 2mm of misalignment (hi-lo at the root).

Figure 4: Full bore unstable fracture of the weld at an average of 0.48% strain level

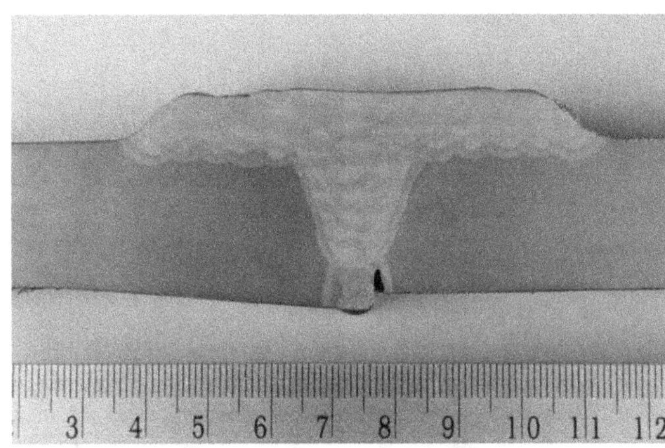

Figure 5: weld configuration with root defect and misalignment

Tensile strength was recorded in between 88,908- 97,175 psi. The recorded energy values for impact properties were 160 Joules, 281 Joules and 298 Joules for weld metal, heat effected zone (HAZ) and base metal respectively.

From the full scale test result, the girth weld did not fracture when the pipe was bend to about 3% strain level with 96 bar internal pressure. The improved strain capacity was mainly attributed to the weld metal overmatching compared to parent pipe strength and also from the improvement imparted by the weld overbuild.

DISCUSSION

Weld strength under matching needs to be addressed besides minimizing risks of geohazard. Weld under matching is defined

Copyright © 2019 by ASME

where the weld metal having lower tensile value compare to base metal as shown in figure 8 below.

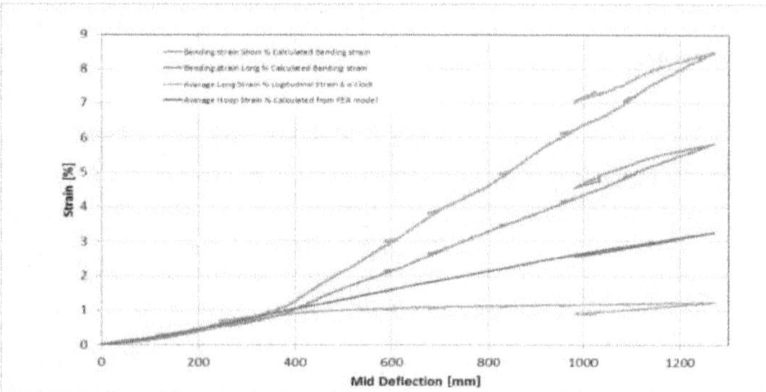

Figure 7: Applied strain assuming no buckling and ovalisation for the full scale bend test

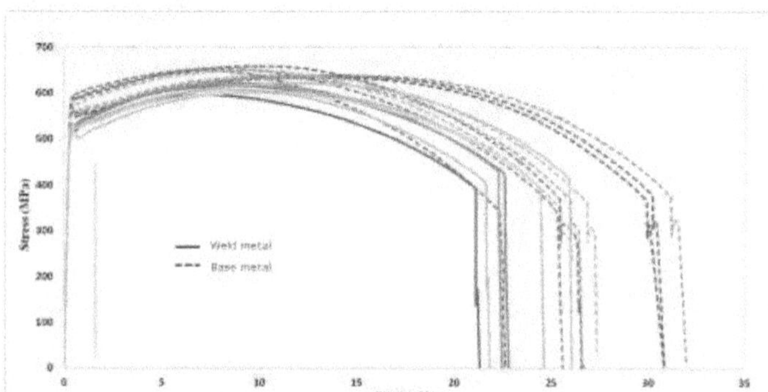

Figure 8: base metal and weld metal tensile SS curve

In the presence of higher strain at a specific time, the weld metal yielded beyond its elastic limits which later causes the weld to shear tearing.

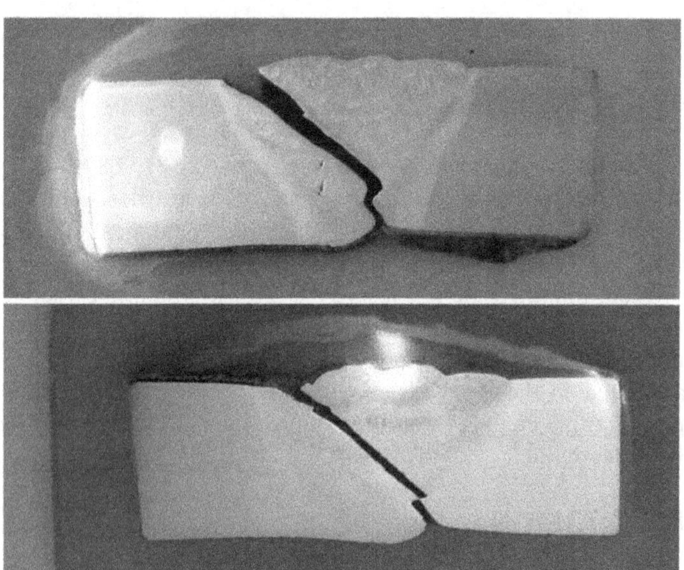

Figure 9 and 10: shear tearing weld failure

Among the strategies that have been considered for strain improvement are:

- Use of better and proper handling of electrode such as storage, baking and holding. This will minimize the risk of hydrogen embrittlement during welding activities and delayed cracking.
- Use of basic covered electrodes instead of cellulosic electrodes. The tensile strength of basic electrode is increased compared to the cellulosic electrode used for the original fill and capping passes from 80,000 psi (E-8010) to 90,000 psi (E-9045) to match the tensile strength of the base metal.
- Combination of E-6010 root and E-8010 hot, fill and cap using E-8010 cellulosic electrodes provides a maximum strain capacity of 0.48%.
- Used of weld overbuild together with the right selection of electrode can make the whole girth weld capable to withstand higher strain up to more than 3%. This was proven by 3D non-linear FEA analysis and full scale bend testing with internal pressure.
- For locations along the pipeline right of way with high strain demand use strain based design as an alternative to traditional stress based design. The strain based design utilizes CTOD fracture toughness and strain level to develop the joint design with high strain capacity girth weld welding specification.
- Live monitoring of geohazard effects. This will allow quick action by identification of soil movement and for mobilization to execute mitigation works whenever required.
- Improve geohazard management via e.g. slope strengthening and introduction of activities to restrain soil movement restraining.

CONCLUSION

High strain capacity girth welds can be an alternative option for providing higher pipeline strain capacity at geohazard active environments, where the strain is expected to exceed 0.5% due to various types of geohazards.

RECOMMENDATIONS

With the presence of geohazard sensors or vibrating wire strain gauge sensors at the identified strategic locations, an effective geohazard management can be achieved. If the risk of geohazard is beyond the permissible limit, "live repair" can be an alternative solution. However, qualification, testing and finite element (FE) simulation should be carried out when the girth weld is subjected to strain beyond the elastic limit.

Other options for weld repair solutions are as listed below.

- Welded split sleeve:
- Grouted connections:
- Composite wrapping

Copyright © 2019 by ASME

NOMENCLATURE

Weld strength under matching, shear tearing, strain capacity, full scale testing, welding procedure specification, geohazard active environment.

ACKNOWLEDGMENTS

The authors would like to express our utmost appreciation to Dr Salama, Mamdouh, M. for his patience, technical advice and wise guidance provided to the team. Ourthanks also goes to Mr. Abd Khalid b Jaafar and Mr. Hambali B Chik from PETRONAS GTS for their continuous supports. Special thanks is expressed to Dr. Jens P. Tronskar, Senior Vice president and CTO, DNV GL Singapore and his team for their technical support and assistance during WPS development, qualification and full scale bend testing.

REFERENCES

1 SSGP-KP182-ENG-TNP-018 PETRONAS technical note
2 Full scale bend test for SSGP pipeline project from DNV GL
3 Geotechnical analysis report from DnV GL
4 Geotechnical desk study report from DNV GL
5 Root cause analysis of SSGP pipeline failure from DNV GL
6 Review of engineering documentation report from DNV GL
7 Material Testing report from DNV GL
8 Metallurgical analysis of cracks in repair weld of KP182 HR 14T report
9 API Std 1104 – Welding of pipelines and related facilities
10 ASME B31.8 – Gas transmission and distribution piping systems.
11 API 5L – Specification for line pipe
12 SSGP existing WPS (19 nos) from PunjLloyd- Dialog-Petrosab Consortium.
13 High strain weld (WPS) from PETRONAS-TMM.

Copyright © 2019 by ASME

AIMPIMG2019-1020

Chapter 27

DEVELOPMENT OF HIGH STRENGTH HIGH STRAIN LINEPIPE STEELS

Junji SHIMAMURA
Steel Research Lab., JFE Steel Corporation
Fukuyama, Japan

Kyono YASUDA
Steel Research Lab., JFE Steel Corporation
Fukuyama, Japan

Satoshi IGI
Steel Research Lab.,
JFE Steel Corporation
Kurashiki, Japan

Ryuji MURAOKA
West Japan Works,
JFE Steel Corporation
Fukuyama, Japan

Joe KONDO
Plate Business Planning
Dept., JFE Steel Corporation
Tokyo, Japan

ABSTRACT

In order to achieve safety and reliability of long-distance gas transmission pipeline installed in seismic and permafrost region, it is quite important to apply high strength linepipes with sufficient strain capacity against buckling and weld fracture by large ground movement. Grades X70-X80 linepipe steels with bainite-MA (Martensite-Austenite constituent) microstructure have been developed by applying advanced plate manufacturing technologies with the combination of accelerated cooling (AcC) and heat treatment on-line process (HOP™). This microstructure shows higher strain capacity which has low yield to tensile (Y/T) ratio and high elongation in the longitudinal direction even after pipe coating heat treatment. In this paper, the material design concept of Grades X70-X80 high strength high strain linepipe steels are introduced. In Grade X70 high strain linepipe steels, adequate MA volume fraction was required even in lower carbon equivalent (CE) level compared with Grade X80 high strain linepipe steels. Mn and Ni addition was useful to obtain the adequate MA volume fraction efficiently, resulting in superior strain capacity which has low Y/T ratio even after the external coating. High strength with Grade X70 level and high elongation have been achieved by control of homogeneous material property such as hardness distribution. Advanced accelerated cooling device (S-OLAC™-A) enables it to obtain the homogeneous microstructure because of homogeneous temperature control after accelerated cooling. Higher elongation has been achieved by lower surface hardness and homogeneous hardness control. In developed Grade X80 high strain linepipe steels, MA volume fraction and its size distribution were controlled simultaneously to achieve higher strength, low Y/T ratio and higher Charpy absorbed energy in full ductile fracture region. It is very important to control MA morphologies to achieve these mechanical properties. HOP™ type bainite-MA dual phase linepipe steels show the highest strain capacity even after pipe coating heat treatment. On the other hand, AcC type bainite-MA dual phase linepipe steels have the highest Charpy absorbed energy because of lower MA volume fraction and finer size of MA by applying appropriate accelerated cooling stopping temperature. Detailed laboratory test results and actual mill test data will be introduced in the presentation.

INTRODUCTION

In order to achieve safety and reliability of the long-distance gas transmission pipeline installed in seismic or permafrost region, it is quite important to apply the linepipes with sufficient strain capacity against buckling and weld fracture caused by large deformation of pipe. Using the steels with lower yield to tensile (Y/T) ratio, higher strain hardenability or higher uniform elongation, is effective to improve strain capacity of linepipes and to prevent buckling or weld fracture [1-6].

High strain linepipe steels are obtained by the dual phase microstructure controls such as ferrite-bainite [7-8] and bainite-MA (Martensite-Austenite constituent) [9-10]. Tensile properties can be controlled by controlling the volume fraction of second phase and strength difference between soft and hard phases [11]. Grades X80-X100 high strength linepipes for high-strain application have been developed through the thermo-mechanical controlled processing (TMCP) with accelerated cooling and heat treatment on-line process [12-13].

Copyright © 2019 by ASME

Accelerated cooling (AcC) is the essential process for producing high strength and high toughness linepipe steels. JFE Steel has long time experience in TMCP technology. Application of the accelerated cooling process to heavy gauge steel plate production was put to practical use in Fukuyama plate mill for the first time in the world in 1980 [14]. "Super-OLAC™" (on-line accelerated cooling device) was installed in Fukuyama plate mill in 1998, which gives extremely higher cooling rate and uniform temperature distribution [15]. More than 380,000 metric ton of Grades X80 and X100 linepipe steels were produced by applying the Super-OLAC™. Plate for the high strain linepipe steel with ferrite-bainite microstructure is available by precise temperature control in AcC stage.

In order for producing the high-performance linepipe steels for recent stringent requirements, the heat treatment on-line process (HOP™), was developed and installed in Fukuyama plate mill in 2003 [16]. The combination of Super-OLAC™ and HOP™ has enabled novel metallurgical controlling that cannot be achieved by the conventional TMCP. Bainite-MA microstructure control is one of the most significant features of the HOP™. Resistance to strain aging was improved by reducing dislocation density and solute carbon. Mass production of Grades X80 and X100 high strain linepipes was successfully conducted by applying Super-OLAC™ and HOP™ and those linepipes were installed in the seismic and permafrost regions.

Continuing development in plate manufacturing process has been conducted. Accelerated cooling system has recently been improved for achieving further higher cooling rate and smaller temperature scattering and for expanding the production of heavier gauge linepipes. The renewed system "Super-OLAC™-A (advance)" started operation in 2011 in Fukuyama plate mill. By applying the Super-OLAC™-A and the HOP™ in plate production, the availability of high strain linepipe products has been expanded in the thickness of pipe and in the strength grade from X60 to X120.

Heavy wall Grade X70 linepipes for high strain application have been required in the seismic region, recently. In designing Grade X70 linepipe, leaner chemistry than Grades X80-X100 linepipe should be applied to achieve optimum strength level. Leaner chemistry, however, may make it tough to obtain adequate amount of hard MA since the transformation temperature need to be low to get martensite by preventing bainite transformation. In order to balance strain capacity and strength in Grade X70 linepipe, the effect of alloy elements on MA volume fraction was studied in lean chemistry condition.

Additionally, toughness requirements for linepipes have become stringent, as well. Low temperature toughness is the essential property to prevent brittle fracture initiation and propagation for the pipeline installed in cold region. The requirement for Charpy absorbed energy is getting higher, recently, in order to prevent running ductile fracture under the increased operation pressure for economical gas supply. As for the bainite-MA dual-phase steel, one may think the MA deteriorates toughness same as the way HAZ (heat affected zone) toughness is deteriorated by MA [17-18]. However, it was demonstrated that the MA formed in the HOP™ does not affect toughness since the MA are finely dispersed in the fine-grained bainitic matrix formed by controlled rolling and accelerated cooling (Shinmiya et al., 2007). In order to ensure the reliability of the high strain linepipe with the bainite-MA dual-phase microstructure, extensive study on the effect of MA on the Charpy impact properties was carried out, and close investigation on the fracture behavior was conducted from the micro-mechanical point of view, in this paper.

HIGH STRAIN AND HIGH STRENGTH LINEPIPE STEELS WITH BAINITE-MA MICROSTRUCTURE

It is well known that the dual-phase microstructure shows higher stain limit. Ferrite-bainite type dual phase linepipe steel was first developed for the application of the gas pipeline installed in seismic area [19]. Optimization of the microstructure of ferrite-bainite steels has been conducted to enhance strain hardenability based on the microscopic deformation mechanisms [20-21]. Recently, dual-phase steels consist of bainite and dispersed MA have been developed by applying HOP™ to the plate manufacturing process [10, 12] and applied to the pipelines installed in the seismic and permafrost regions [22]. Figure 1 shows the schematic illustration of temperature profiles in plate manufacturing process together with the conventional TMCP. In the conventional TMCP, the steel plate is controlled rolled and accelerated cooled and then cooled in the air. On the other hand, in HOP™, the plate is rapidly heated by the induction coils immediately after accelerated cooling. The MA is formed by applying HOP™, since carbon enrichment into the untransformed austenite is enhanced by the heat treatment after the accelerated cooling as shown in Figure 2. Figure 3 shows the example of SEM microphotographs of the plates produced by the conventional TMCP and HOP™. Chemical compositions and plate thickness were the same for either condition. The

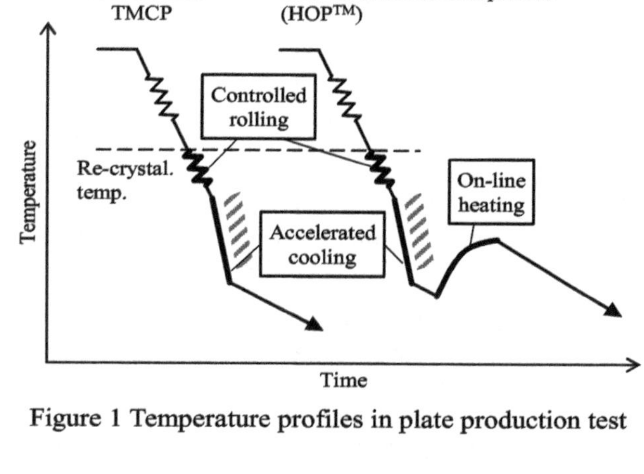

Figure 1 Temperature profiles in plate production test

Copyright © 2019 by ASME

microstructure of the conventional TMCP steel shows bainite, and small amount of MA was observed. On the other hand, large number of MA, which is seen as white region, was observed in the HOP™ applied steel. Figure 4 shows longitudinal stress strain curves of Grade X80 linepipe with the size of 1214 mm OD x 22.0 mm WT produced by HOP™ and the conventional TMCP. It is obvious from the shape of stress-strain curves that Grade X80 linepipe produced by HOP™ shows lower Y/T ratio and longer uniform elongation than that of the linepipe by the conventional TMCP. Additionally, in order to investigate the strain aging behavior of Grade X80 linepipe produced by HOP™, coating simulation tests were conducted on the trial Grade X80 linepipes by using salt bath furnace. Figure 5 shows the stress-strain curves for Grade X80 linepipe by HOP™ and the conventional TMCP. Slight increase in flow stress was observed, but the stress-strain curves remained round-house shape up to 230 °C. Stress-strain curve showed plateau (Luder's elongation) when heated up to 250 °C. On the other hand, in the case of conventional Grade X80 linepipe, Luder's elongation is seen when heated at 230 °C and higher temperature. Upper yield point is clearly seen when heated at 250 °C. Appearing upper yield point can be an evidence of strain aging which dislocation movement was obstructed by carbon atoms. Therefore, bainite-MA microstructure produced by HOP™ shows superior strain capacity and resistance against strain aging.

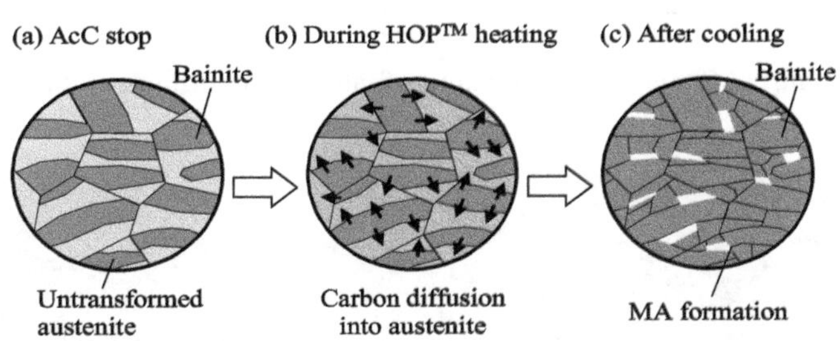

Figure 2 Schematic illustration of microstructure change in the HOP™ for dual phase microstructural control

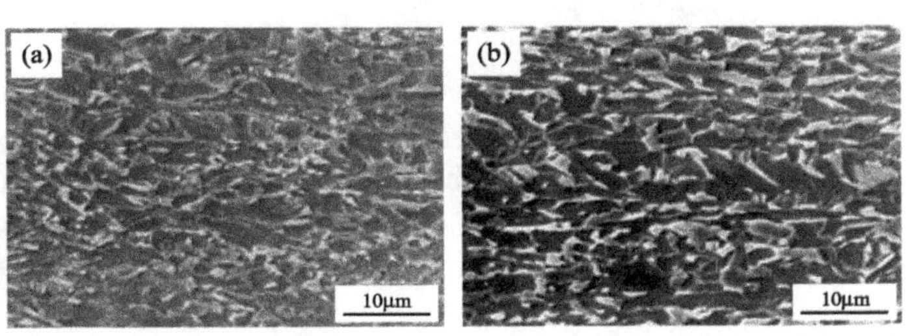

Figure 3 SEM microphotographs of the plates produced by (a) the conventional TMCP and (b) HOP™

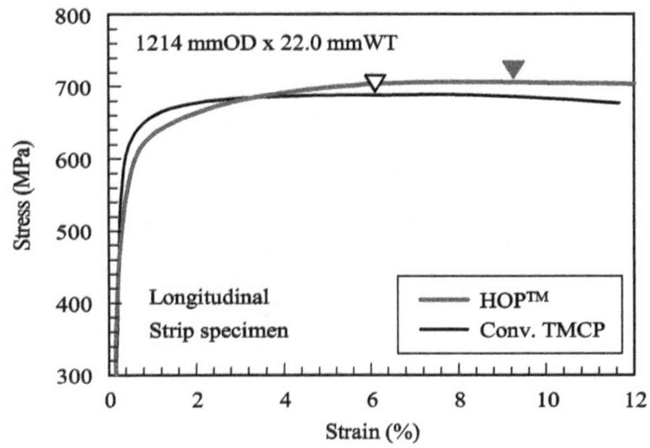

Figure 4 Longitudinal stress strain curves of Grade X80 linepipes produced by HOP™ and the conventional TMCP

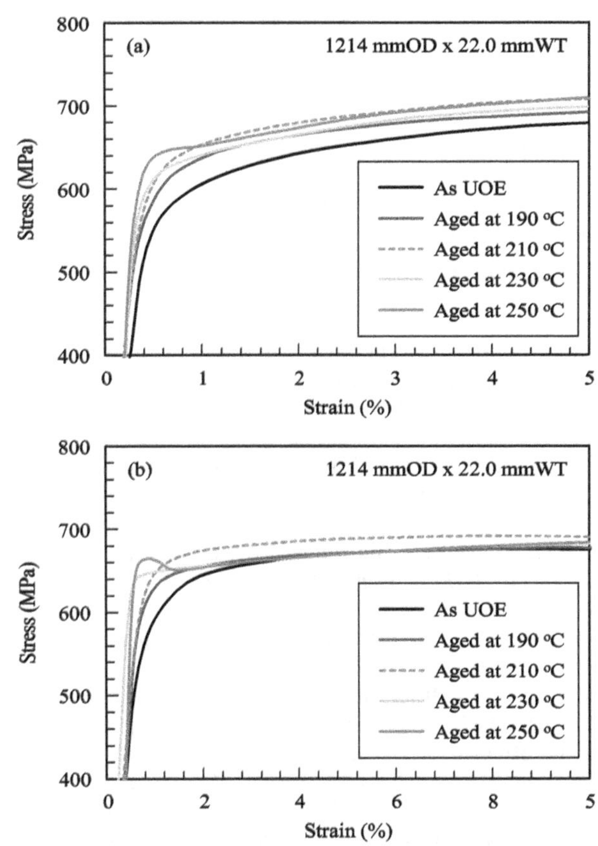

Figure 5 Effect of coating heating on stress-strain curves of Grade X80 linepipes produced by (a) HOP™ and (b) the conventional TMCP

Copyright © 2019 by ASME

MATERIAL DESIGN FOR GRADE X70 HIGH STRAIN LINPIPES

Figure 6 shows the effect of MA volume fraction on longitudinal Y/T ratio of Grade X80 linepipes with the chemistry of CE_{IIW} of .44%. MA volume fraction is required more than 3% to achieve Y/T ratio lower than 85%. However, in the case of Grade X70, it is challenging to obtain the sufficient MA volume fraction because of its lower alloy contents compared with Grade X80 or higher. Therefore, the effect of alloy element on MA formation was investigated.

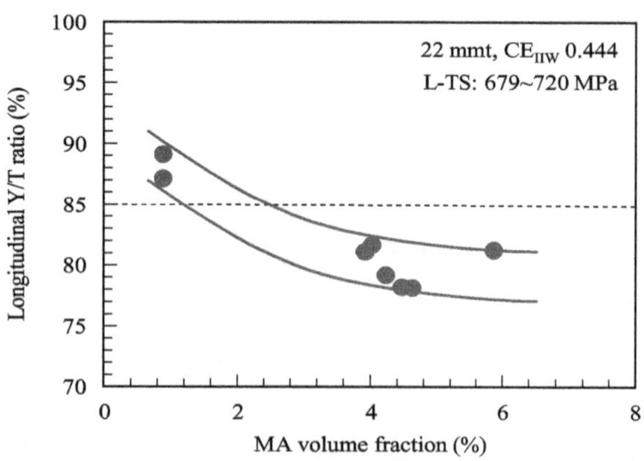

Figure 6 MA volume fraction and longitudinal Y/T ratio in Grade X80 high strain linepipes

Table 1 shows the chemical compositions of the steels used in this study. Carbon equivalent, Ceq, was lower than conventional Grade X80 linepipe. Steel ingots were prepared and hot rolling was conducted using laboratory equipment. Rolling conditions were shown in Figure 7. The steels were controlled rolled to the plate with 20 mm thickness, and then, accelerated cooling and subsequent heating were applied in order to simulate the actual plate production process by Super-OLAC™ and HOP™. Figure 8 shows the SEM images of the steels differing in alloying elements. The samples were etched by 3% Nital acid, then electro-chemical etching was applied for dissolving cementite and remaining only MA as the second phase. MA volume fraction was measured by the image analysis from several SEM microphotographs for each steel. Figure 9 shows the effect of alloy element on MA volume fraction. It is obvious that Mn and Ni addition is effective to increase MA volume fraction even in low CE_{IIW} condition. While Si and C are thought to be harmful for HAZ toughness by increasing MA volume fraction [23-24], both elements showed almost no effect on MA formation in the HOP™. Therefore, Mn and Ni addition was adopted for Grade X70 high strain linepipes. In order to investigate the appropriate chemistry for Grade X70 linepipe, another laboratory test result was conducted with the chemistry of both Mn and Ni addition. Figure 10 shows the relationship between tensile strength and CE_{IIW}. Considering the specified minimum tensile strength, CE_{IIW} for Grade X70 high strain linepipe steel should be 0.39% or higher. And, these show low Y/T ratio.

Table 1 Chemical compositions of the steels used (mass%)

Steel	C	Si	Mn	Ni	Others	CE_{IIW}
1	0.05	0.12	1.58		Mo, Nb, Ti, Ca	0.34
2	0.05	0.14	1.49	**0.29**	Mo, Nb, Ti, Ca	0.34
3	0.05	0.10	**1.79**		Mo, Nb, Ti, Ca	0.37
4	**0.08**	0.10	1.50		Mo, Nb, Ti, Ca	0.35
5	0.05	**0.50**	1.50		Mo, Nb, Ti, Ca	0.32

CE_{IIW}= C+Mn/6+(Cu+Ni)/15+(Cr+Mo+V)/5

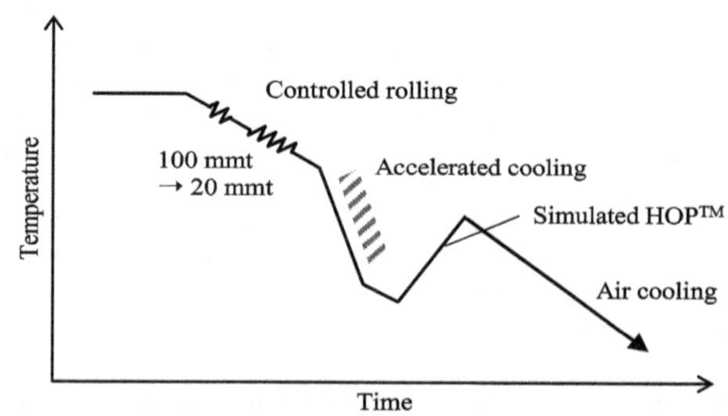

Figure 7 Schematic illustration of laboratory rolling condition

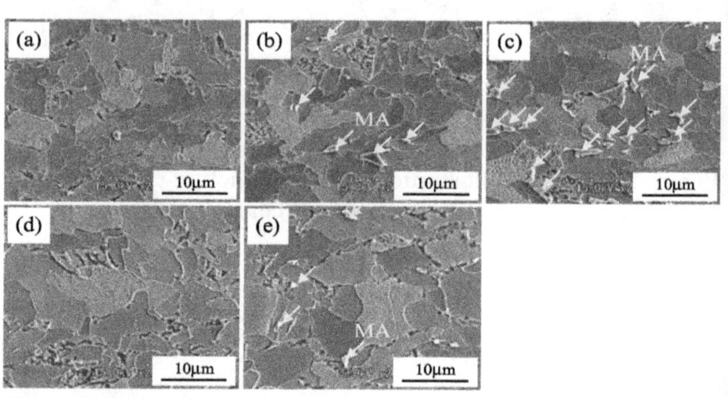

Figure 8 SEM microphotographs of (a) ~ (e) Steels 1 to 5 with different chemistry

Copyright © 2019 by ASME

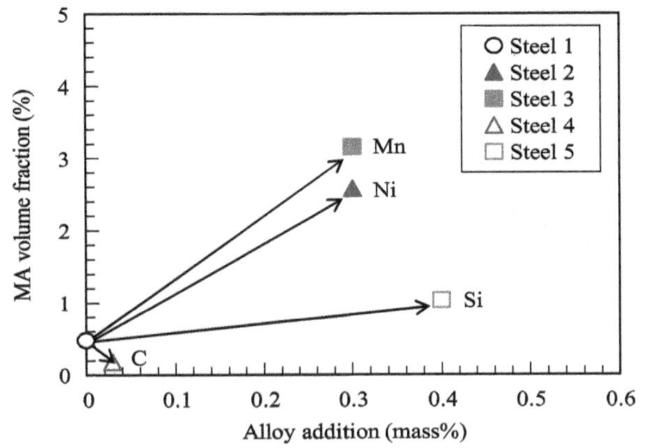

Figure 9 Effect of alloy element on MA volume fraction

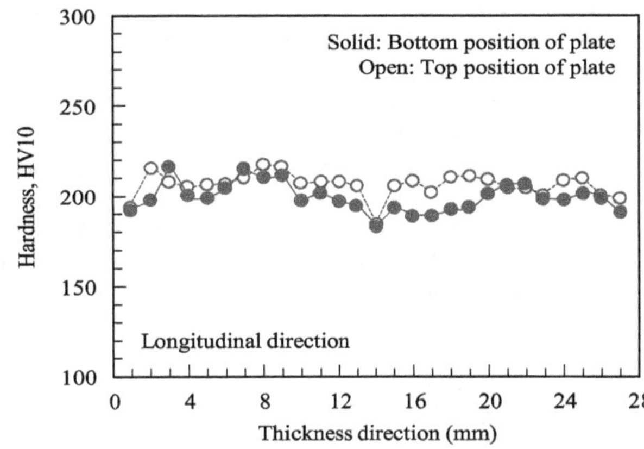

Figure 11 Hardness distribution of thickness direction

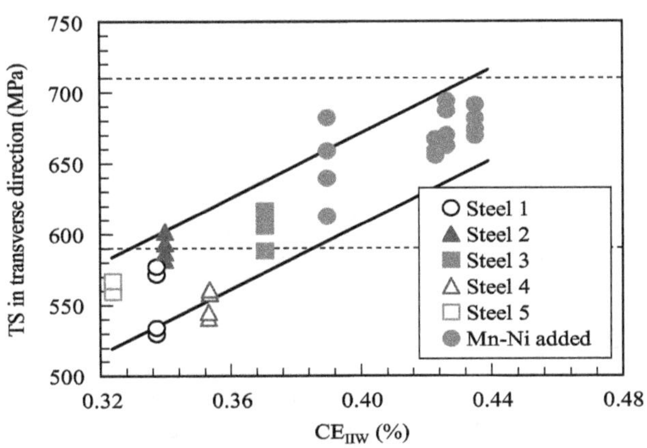

Figure 10 Tensile strength in transverse direction of sample plate steels and CE$_{IIW}$ level

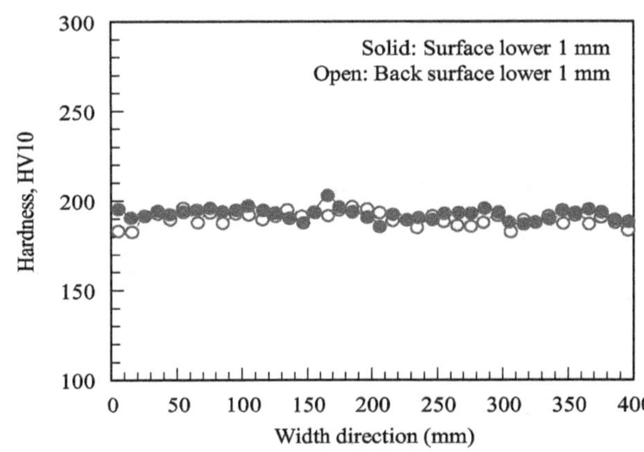

Figure 12 Hardness distribution of width direction

HOMOGENEITY OF MATERIAL PROPERTIES WITH SUPER-OLAC^{TM}-A

Homogeneous material property distribution such as hardness is effective for improving elongation. S-OLAC^{TM}-A is the advanced accelerated cooling device which can uniformly control temperature during accelerated cooling.

Figure 11 shows hardness distribution in thick direction of Grade X70 linepipe manufactured by S-OLAC^{TM}-A. Figure 12 shows surface hardness distribution in wide direction. From these results, the plates manufactured by S-OLAC^{TM}-A have high homogeneity of hardness and other material properties such as tensile strength.

TRIAL PRODUCTION OF GRADE X70 HIGH STRAIN LINEPIPE

Trial production test of heavy wall Grade X70 high strain linepipe was conducted by applying the advanced TMCP with Super-OLAC^{TM}-A and HOP^{TM} in the actual plate and pipe mill. Table 2 shows the chemical compositions for trial production. Plates with 28.0 and 32.0 mm wall thickness were manufactured in the plate production mill. Plate manufacturing condition in the controlled rolling, the accelerated cooling and HOP^{TM} were precisely controlled so that sufficient strength and high strain capacity were achieved in pipes. Table 3 shows the mechanical properties of base metal of each thickness pipes. The trial production Grade X70 linepipes show sufficient strength, superior toughness and strain capacity which have target level. Figure 13 shows the microstructure of base metal in quarter thickness position of pipes. There are many finer MA in quarter

Copyright © 2019 by ASME

thickness position, and have good strength-toughness balance. Figure 14 shows the nominal stress-strain curve of the trial pipes. The trial pipe had round-house type curve and showed low Y/T ratio in the longitudinal direction. Figure 15 shows the total elongation of the trial pipes. The pipes manufactured with Super-OLAC^{TM}-A showed higher uniform elongation and total elongation compared with the conventional pipes manufactured with Super-OLAC^{TM}. From this result, Super-OLAC^{TM}-A, which has much cooling uniformity, is much effective cooling device for obtaining high strain pipes with higher uniform elongation and total elongation.

Table 2 Chemical compositions of the trial Grade X70 high strain linepipes (mass%)

Grade	C	Si	Mn	Ni	Others	CE_{IIW}
X70	0.06	0.14	1.69	0.18	Cu, Mo, Nb, Ti, Ca	0.41

Table 3 Mechanical properties of the trial Grade X70 high strain linepipes

Pipe No.	Pipe size		Pipe body						
			Transverse			Longitudinal			
	OD (mm)	WT (mm)	YS (MPa)	TS (MPa)	El (%)	YR (%)	uEl (%)	El (%)	σ_2/σ_1 (%)
1	1220	28.0	529	644	22	83	10.9	27	1.05
2	1420	32.0	511	649	22	79	11.0	24	1.05

Pipe No.	Pipe body		Welded joint	
	Charpy vE-40 (ave.) (J)	DWTT SA-20 (ave.) (%)	TS (MPa)	HAZ Charpy vE-40 (ave.) (J)
1	266	90	643	86
2	207	93	679	86

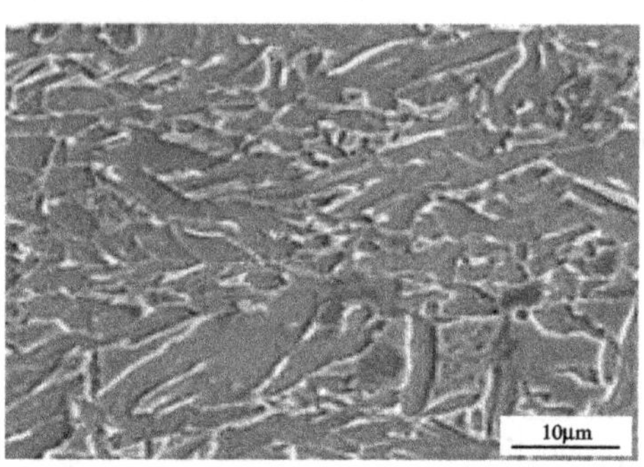

Figure 13 SEM microphotographs of the trial Grade X70 high strain linepipes at quarter thickness position

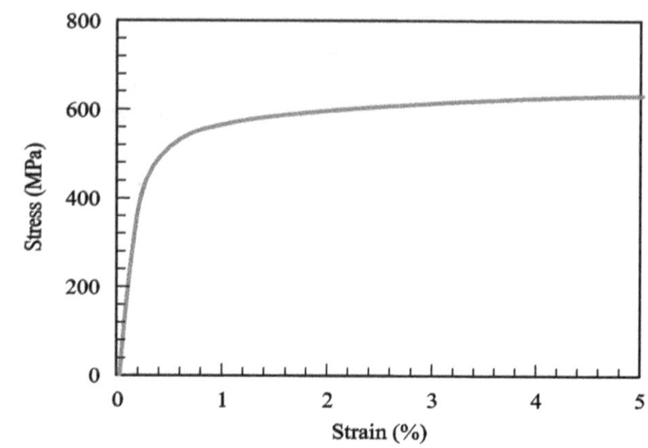

Figure 14 Stress-strain curve of the trial Grade X70 high strain linepipes as UOE with 28.0 mm WT

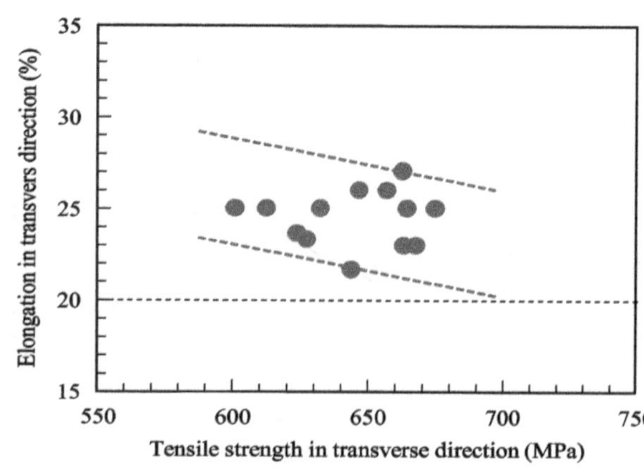

Figure 15 Elongation in transvers direction of the trial Grade X70 high strain linepipes with long gauge length type specimen

IMPROVING CHARPY IMPACT PROPERTIES WITH HIGH STRENGTH AND HIGH STRAIN LINEPIPES

MA is usually thought to deteriorate HAZ toughness of steels, because MA is very hard and brittle phase and it may become an initiator of brittle fracture [23]. Figure 16 shows the effect of MA volume fraction on the ductile to brittle transition temperature (vTrs) and Charpy absorbed energy at –20 °C (vE_{-20}). It was seen that the effect of MA formed by HOP^{TM} does not affect the ductile to brittle transition temperature in the Charpy impact test [10]. It is considered that bainite matrix is very fine because of the controlled rolling and subsequent accelerated cooling. Another important reason for a good toughness of the bainite-MA steel is its crystallographic property, as described in Figure 17. In the case of HOP^{TM}, MA is surrounded with many short length K-S orientation boundaries compared with the long

Copyright © 2019 by ASME

and random boundaries produced by the internal-critical heat treatment.

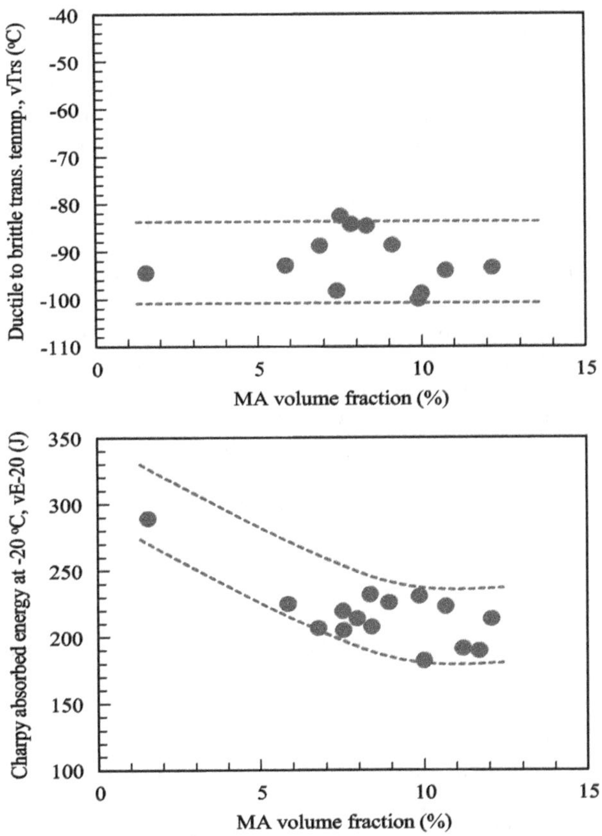

Figure 16 Effect of MA volume fraction on ductile to brittle transition temperature (vTrs) and Charpy absorbed energy at -20 °C (vE-20)

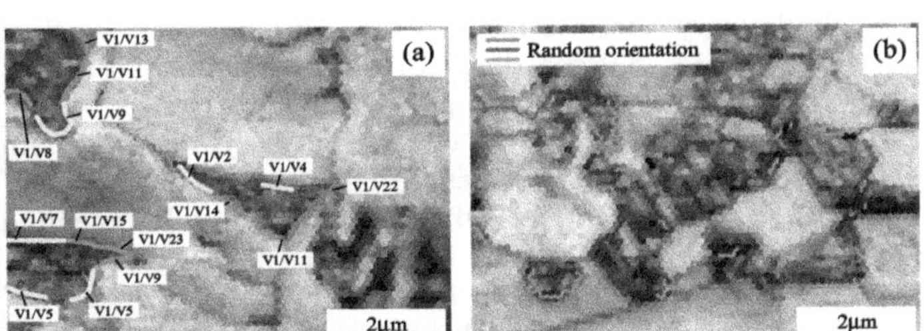

Figure 17 Crystal orientation analyses between MA and matrix in steels produced by (a) HOP™ and (b) the inter-critical heat treatment

On the other hand, Charpy absorbed energy decreased with increasing MA volume fraction as shown in Figure 16. All the fractured surfaces were 100% ductile even at –20 °C, and the absorbed energy thought to be controlled by void growth and coalescence mechanism. Therefore, in order to improv Charpy absorbed energy of bainite-MA steels, the effect of MA morphology on void formation was investigated.

Two different steel plates with different MA size, which were produced by the plate mill, were selected for the investigation on the effect of MA morphology. Both steel plates were produced by applying HOP™, and have strength level of 650 MPa. Figure 18 shows the microphotographs of the steels with differing in MA size. Sample A contains small MA with the size of approximately 0.7 μm, while Sample B has large MA of approximately 1.7 μm; both sizes are the average diameter of equivalent circle. The instrumented Charpy impact tests were conducted on these steels. Crack initiation and propagation energies were evaluated from the load-displacement curves, and results were shown in Figure 19. From these results, MA size affects both the crack initiation energy and the crack propagation energy. In order to investigate the void formation behavior, the interrupted Charpy test was conducted. Smaller angle of the impact hammer was applied to hit the specimen with small impact energy as illustrated in Figure 20. Same impact energy of 25 J at -100 °C, below crack initiation energy, were provided to Steels A and B. Figure 21 shows SEM microphotographs in section of the notch root. It was obviously that Steel A with small MA shows fewer voids, and many of the voids were nucleated close to coarse MA. As above mentioned results, it can be stated that Charpy energy of the steel with bainite-MA microstructure can be controlled by properly controlling the morphology of the MA and also base metal toughness by the controlled rolling.

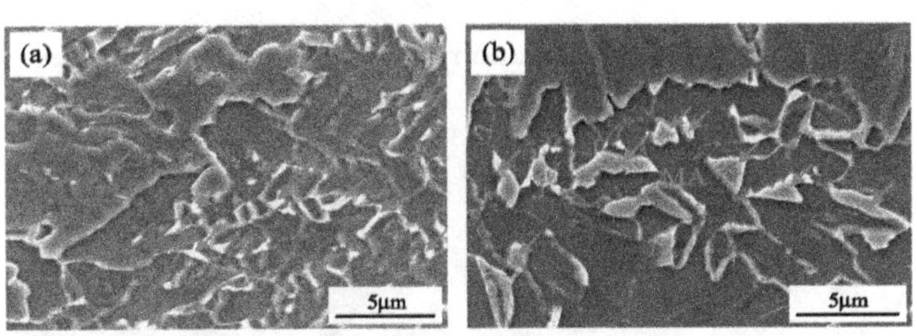

Figure 18 SEM microphotographs of the steels with different MA size (Samples (a) A with small MA and (b) B with large MA)

Copyright © 2019 by ASME

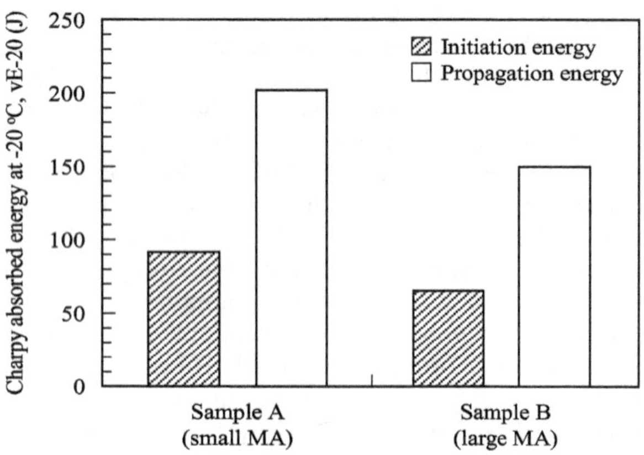

Figure 19 Crack initiation and propagation energies tested at -20 °C of Samples A and B

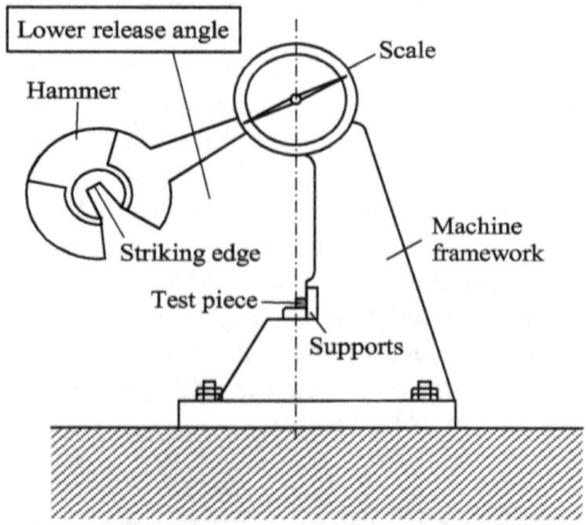

Figure 20 Schematic illustration on the interrupted Charpy test

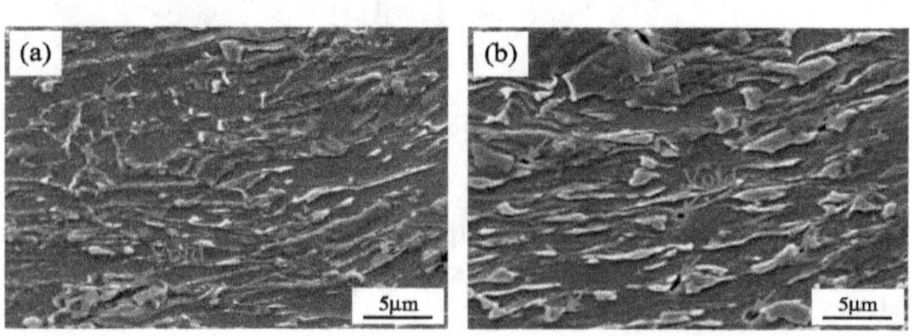

Figure 21 SEM microphotographs of the interrupted Charpy specimen in the section of the notch root region of Samples (a) A and (b) B

CONCLUSIONS

High strain linepipes have been required in the seismic regions around the world for strain based design. In this paper, the material design concept of Grade X70 high strain linepipes with bainite-MA microstructure was proposed. Especially, the effects of alloy elements and manufacturing condition on base metal properties were investigated. The following results were obtained:

(1) High strain linepipe with resistance to strain aged hardening was developed. The developed pipe has bainite microstructure containing dispersed fine MA, and this microstructure can be obtained by applying HOP™ immediately after accelerated cooling.
(2) Mn and Ni addition was effective to obtain the adequate MA volume fraction over 3% in low CE_{IIW} condition to achieve high deformability in Grade X70 grade linepipes.
(3) Super-OLAC™-A, which has much cooling uniformity, is much effective cooling device for obtaining high strain pipes with higher uniform elongation and total elongation.
(4) Developed Grade X70 high strain linepipes in actual pipe mill, which had adequate MA volume fraction in low CE_{IIW} and smaller size MA, show superior strain capacity and low temperature toughness.
(5) Steel plate samples with lower rolling finishing temperature showed smaller size of MA, and showed lower brittle to ductile transition temperature in the Charpy impact test.

REFERENCES

[1] R. M. Denys, et al., "Failure Characteristics and Defect Tolerance Levels of Girth Welds in Large Diameter X65/X70 Steel Linepipes: Experimental Verification through Wide Plate Testing and Comparison with ECA Prediction Models," Proceedings of Pipeline Technology Conference, Ostend, Belgium, vol. 1, (2000), pp. 151-167
[2] Y-Y. Wang, D. Rudland, R. M. Denys and D. Horsley, "A Preliminary Strain-Based Design Criterion for Pipeline Girth Welds," Proceedings of 4th International Pipeline Conference, (2002), Paper No. IPC2002-27169
[3] Y-Y. Wang, M. Liu, D. Rudland and D. Horsley, "Strain Based Design of High Strength Pipelines," Proceedings of the 17th International Offshore and Polar Engineering Conference, (2007), Paper No. ISOPE-2007-SDB-07
[4] N. Ishikawa, S. Endo, S. Igi, A. Glover, D. Horsley, M. Ohata and M. Toyoda, "Ductile fracture behavior of girth welded joints and strain based design for high strength linepipes," Proceedings of the international Pipeline Technology Conference, vol. 1, (2004), pp. 81-98
[5] N. Suzuki and M. Toyoda, "Critical Compressive Strain of Linepipes Related to Workhardening Parameters," Proceedings of 21st International Conference on Offshore Mechanics and Arctic Engineering, (2002), Paper No. OMAE2002-28253

Copyright © 2019 by ASME

[6] N. Suzuki and M. Toyoda, "Bending Capacity of High Strength Line Pipe," Proceedings of the International Pipeline Technology Conference, (2004), pp.1360-1374
[7] S. Endo, N. Ishikawa, J. Kondo, N. Suzuki and K. Omata, "Advance in High Performance Linepipes with Respect to Strength and Deformability," Proceedings of Pipe Dreamer's Conference, (2002), pp.273-288
[8] N. Ishikawa, N. Endo and J. Kondo, "High Performance UOE Linepipes," JFE Technical Report, No. 7, (2006), pp. 20-26
[9] M. Okatsu, T. Shinmiya, N. Ishikawa, S. Endo and J. Kondo, "Development of High Strength Linepipe with Excellent Deformability," Proceedings of 24th International Conference on Offshore Mechanics and Arctic Engineering, (2005), Paper No. OMAE2005-67149
[10] T. Shinmiya, N. Ishikawa, M. Okatsu, S. Endo, N. Shikanai and J. Kondo, "Development of High Deformability Linepipe with Resistance to Strain Hardening by Heat Treatment On-line Process," Proceedings of the 17th International Offshore and Polar Engineering Conference, (2007), Paper No. ISOPE-2007-SDB-60
[11] T. Huper, S. Endo, N. Ishikawa and K. Osawa, "Effect of Volume Fraction of Constituent Phases on the Stress-Strain Relationship of Dual Phase Steels" ISIJ International, Vol.39, No.3, (1999), pp. 288-294
[12] N. Ishikawa, M. Okatsu, J. Shimamura, S. Endo, N. Shikanai, N. Suzuki, R. Muraoka and J. Kondo, "Material Development and Strain Capacity of Grade X100 High Strain Linepipe produced by Heat Treatment Online Process," Proceedings of 7th International Pipeline Conference, (2008), Paper No. IPC-64507
[13] N. Ishikawa, H. Sueyoshi and S. Igi, "Application of Damage Mechanics Modeling to Strain Based Design with Respect to Ductile Crack Initiation," Proceedings of 8th International Pipeline Conference, (2010), Paper No. IPC-31506
[14] K. Tsukada et al., "Application of on-line accelerated cooling process for plate mill," NKK Technical Report, No.89, (1981), pp.121–132
[15] K. Omata, H. Yoshimura and S. Yamamoto, "Leading High Performance Steel Plates with Advanced Manufacturing Technologies," NKK Technical Review, No.88, (2003), pp.73–80
[16] A. Fujibayashi and K. Omata, "JFE Steel's Advanced Manufacturing Technologies for High Performance Steel Plates," JFE Technical Report, No.5, (2004), pp.10–15
[17] A. Liessem and M. Erdelen-Peppler, "A Critical View on the Significance of HAZ Toughness Testing," Proceedings of International Pipeline Conference, (2004), Paper No. IPC-0315
[18] N. Ishikawa, M. Okatsu, S. Endo and J. Kondo, "Design Concept and Production of High Deformability Linepipe," Proceedings of 6th International Pipeline Conference, (2006), Paper No. IPC-10240
[19] S. Endo, M. Kurihara, N. Suzuki, A. Kato and M. Yoshikawa, "Development of the Linepipe with Earthquake Resistance," Materia Japan, Vol.39, (2000), pp.166-168 (in Japanese)
[20] N. Ishikawa, H. Sueyoshi, S. Endo and N. Shikanai, "Effect of Microstructure on Deformation Behavior of Ferrite-Bainite Steels," Tetsu-to-Hagane, Vol.91, (2005), pp.809-815 (in Japanese)
[21] N. Ishikawa, J. Shimamura, H. Sueyoshi, K. Yasuda, S. Endo, T. Morikawa and K. Higashida, "Micromechanical Aspects of Deformation and Fracture Behavior of Dual-Phase Steels," Proc. 6th International Pipeline Technology Conference, (2013), Paper No. S22-03
[22] R. Muraoka, J. Kondo, N. Ishikawa, M. Okatsu, L. Ji, H. Chen and Y. Feng, "Production of Grade X80 High Strain Linepipes for Seismic Region Application," Proc. 8th International Pipeline Conference, (2010), Paper No. IPC2010-31556
[23] Y. Ohno, R. Habu and S. Sekino, "Thermal Cycle Test of High Strength Steels and Steel Plates for Low Temperature Service," Tetsu-to-Hagane, Vol. 58, (1973), pp.306-316 (in Japanese)
[24] Y. Kasamatsu, S. Takashima and T. Hosoya, "Influence of Martensite-Austenite Constituent on Toughness of Heat-Affected Zone of High-Strength Weldable Structural Steels," Tetsu-to-Hagane, Vol. 65, (1979), pp.1222-1231 (in Japanese)

Copyright © 2019 by ASME

AIMPIMG2019-1026

Chapter 28

TENSILE STRAIN CAPACITY AFTER LOCAL BUCKLING OF HIGH STRAIN PIPELINE IN FULL-SCALE BENDING TESTS

Takahiro Sakimoto
JFE Steel Corporation
Chiba, Chiba, Japan

Hisakazu Tajika
JFE Steel Corporation
Chiba, Chiba, Japan

Tsunehisa Handa
JFE Steel Corporation
Chiba, Chiba, Japan

Satoshi Igi
JFE Steel Corporation
Okayama, Kurashiki, Japan

Joe Kondo
JFE Steel Corporation
Chiyoda, Tokyo, Japan

ABSTRACT

Demand for higher grade linepipe that can help reduce the total cost of long-distance gas pipelines has increased recently. As a result, pipelines using high-strength linepipes such as Grades X70 and X80 have been planned in hostile environments such as landslide-prone mountain areas, reclaimed land where liquefaction is an issue and polar regions where frost heave occurs. Buried pipelines in these regions are expected to be subjected to large deformation due to the large ground movement caused by these geohazards. It is not possible to apply conventional stress-based design in cases where the longitudinal stress of pipe material greatly exceeds the yield stress and the pipeline material is subjected to large plastic strain. The Strain Based Design (SBD) approach is required to complement the conventional allowable stress methods in order to ensure a reliable limit state design approach that uses the longitudinal strain capacity of pipelines. In SBD, the tensile strain capacity of the pipeline in the longitudinal direction is very important. The tensile strain capacity under a pure tension loading condition in the pipe axial direction has been studied, taking into account the effects of internal pressure, pipe geometry, flaw size, misalignment of girth welds, weld overmatch, the yield to tensile strength (Y/T) ratio and elongation in the base material. Since loading predominantly takes the form of bending displacements, two different strain capacity limit states must be addressed: compressive strain capacity, which corresponds to the onset of local buckling characterized by the formation of a wrinkle on the compressive side, and tensile strain capacity, which corresponds to final fracture characterized by the ultimate limit state of necking from the base metal or ductile tearing from a girth weld defect on the tensile side. However, the tensile strain capacity after local buckling is not fully understood because few data are available from full-scale bending tests with final fracture after local buckling.

This paper presents the tensile strain capacity after local buckling in full-scale pipe bending tests of X80 girth welded UOE high strain pipe with an outside diameter (OD) of 48 inches and X70 girth welded UOE high strain pipe with OD of 36 inches. In order to discuss the loading condition on the tension side after local buckling near a girth weld, the longitudinal tensile strain distributions were investigated. After a large wrinkle formed on the compression side of bending, the wrinkle worked precisely like a plastic hinge, and longitudinal tensile strain developed only in the exactly opposite area to the large wrinkle on the compression side of bending. The tensile strain capacity in full-scale pipe bending is discussed from the viewpoint of the effect of the tensile strain distribution enhanced by wrinkle occurrence on the compressive side on final fracture in comparison with the pure tension loading condition.

INTRODUCTION

Demand for higher grade linepipe that can help reduce the total cost of long-distance gas pipelines has increased recently. As a result, use of high-strength linepipes such as Grades X70 and X80 has also increased in recent years [1,2]. Grades X100 and X120 linepipes have been the object of continuing development [3,4], and many research and development projects have been carried out around the world. On the other hand, pipeline projects have expanded into hostile

Copyright © 2019 by ASME

environments such as permafrost and seismic regions. Pipelines constructed in those regions must have sufficient resistance against failure caused by large plastic strain, such as local buckling or failure in tension. Because the conventional stress-based design method cannot be applied when strain greatly exceeds the yield stress of the material, application of strain-based design (SBD) is under active study. A number of studies have been carried out on strain-based design and integrity assessment methods. The tensile strain limit of a pipeline is usually characterized by using uniaxial tests such as the curved wide plate (CWP) test [5]. Accordingly, the most critical fracture mode in pipelines is considered to be ductile fracture originating from defects in girth welds, and a technique for prediction of ductile crack initiation in the CWP test by using the critical equivalent plastic strain at the notch tip obtained from SENT tests has been proposed [6]. The tensile strain capacity of ductile crack initiation decreases as the Y/T ratio of the base metal increases [7]. There is also increasing demand for high pressure operation to improve gas transportation efficiency. However, CWP tests cannot consider the effect of internal pressure in strain capacity evaluations. Therefore, it is important to investigate the effect of internal pressure on ductile fracture behavior. Full-scale pressurized and unpressurized tensile tests have been conducted to investigate the influence of internal pressure on the tensile strain capacity of pipelines [8,9]. Dozen strain-based designed pipelines has been designed and/or installed. To prepare for and assist these SBD projects, research has been conducted on many topics, including tensile strain capacity (TSC) prediction and engineering critical assessment (ECA) procedures. Verious kind of equations for tensile strain capacity prediction have been published along with strain-based ECA procedures based on numerous full-scale tension tests [10, 11, 12, 13,14,15].

In large-diameter shore gas transmission pipelines, ground induced displacements generally cause bending loading, which triggers local buckling in a given pipeline section. In this case, different phases can be recognized: start of plasticization, full plasticization, maximum bending moment due to local bucking, post buckling behavior and final failure in tension. For the compressive strain limit under bending, full-scale bending tests have been conducted to investigate the compressive strain limit due to local bucking [16, 17, 18, 19, 20]. However, the tensile strain capacity after local buckling is still not fully understood because few data are available from full-scale bending tests with final fracture after local buckling.

Against the above background, this paper presents the tensile strain capacity after local buckling in full-scale pipe bending tests of X80 girth welded UOE high strain pipe with an outside diameter (OD) of 48" and X70 girth welded UOE high strain pipe with an OD of 36".

EXPERIMENTAL PROCEDURE

Full-scale bending test rig

The authors developed a large-scale bending test rig in order to verify the bending capacity and study the post buckling behavior of high-strength linepipes with OD up to 48" (1219 mm). The test rig consists of a main frame with a built-in hydraulic jack, two bending arms and hydraulic pumps. Figure 1 shows schematic plane views of the full-scale bending test rig, where the upper and lower figures represent the conditions before and during the bending test, respectively. As illustrated in the figures, the test pipe is placed between and welded to the two moment arms. The moment arms are pin-connected to the main frame, where the right side pin is connected to the head of the hydraulic jack and is movable along the main frame, and the left side pin is fixed to the main frame to withstand the force of the hydraulic jack. After the specimen is set, the pipe is fully filled with water, which is pressurized with a hydraulic pump. The jack then pushes the pivot end to generate bending moment (Max. 35 MNm). This moment can be calculated as follows:

$$M_{pipe} = F_{jack} \times (L_{arm} + \delta_y) \qquad (1)$$

where, M_{pipe} is the moment work on the pipe, F_{jack} is the force powered by the jack, L_{arm} is the distance of the pipe from the center of the pivot in the Y-direction and δ_y is the displacement of the pipe to the Y-direction.

To measure force, strain, deformation and the water pressure inside the pipe, pressure gauges, deformation gauges and strain gauges were set on the test rig and specimen. Table 1 shows the specifications of the pipe bending test rig.

Table 1 Specification of bending test rig

Maximum Pipe Diameter	1219 mm (48 inch)
Maximum Pipe Length	8 m
Load Capacity	6000 kN
Moment Capacity	35 MNm
Water Pressure	30 MPa
Test temperature	Ambient temperature

Copyright © 2019 by ASME

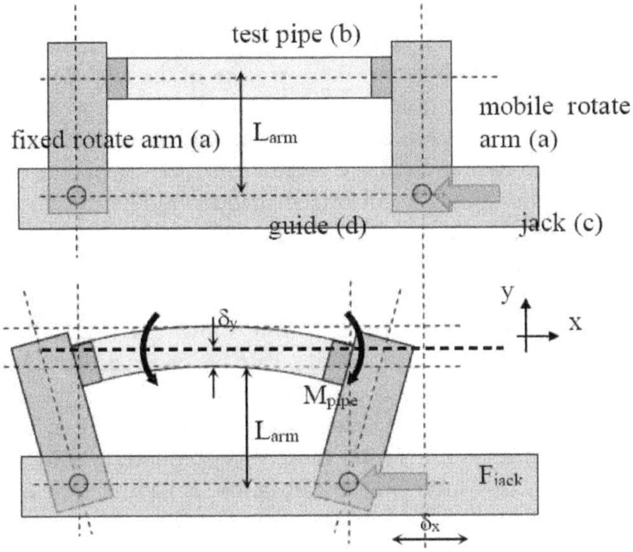

Fig. 1 Schematic view of bending test rig

Test procedure

Full-scale bending tests were conducted using Grade X80 high strain pipes with an OD of 48" and wall thickness of 22 mm, and Grade X70 high strain pipes with OD of 36" and wall thickness of 23 mm [21, 22, 23]. The tested pipes were applied to a simulated anti-corrosion coating heating cycle which the peak heating temperatures was 210 °C. Table 2 shows the typical tensile properties of the base material. All stress-strain (S-S) curves of the base material showed roundhouse-type work hardening characteristics, in which yield elongation is not observed. Two strength levels (overmatched and undermatched) of weld materials were used to evaluate the effect of strength mismatch on fracture behavior. In test No. 1, the same pipe was welded with a gas metal arc welding (GMAW) machine and the welding procedure used for girth welding in actual pipeline construction in order to achieve more than 10 % overmatching in the girth weld (GW). In test No. 2, two different pipes were used, and these specimens were welded with 10 % overmatch in the GW. In test No. 3, the same pipe was welded with 10 % undermatch in the GW.

The test pipes were cut to a length of approximately 8 m for the bending tests and fixed between the arms of the test machine with their ends welded to the arms. In test No. 1, a relatively large artificial surface notch (length 2c: 80 mm, depth a: 4.5 mm) was introduced on the outer surface at HAZ of the GW by electrical discharge machining (EDM). The weld reinforcement was left in the as-welded condition. In both tests No. 2 and No. 3, an EDM notch with a=5 mm and 2c=50 mm was intentionally made at HAZ of the GW on the outer surface on the tension side under bending.

During bending test No. 1, the test pipe was constantly pressurized to 12 MPa with a water pump, and a hoop stress of 60 % SMYS of Grade X80 was introduced in the pipe. In tests No. 2 and No. 3, the test pipes were pressurized with the water pump up to 17.5 MPa, which is equivalent to a hoop stress of 72 % SMYS of Grade X70, and this pressure was kept during the bending test.

During the full-scale bending tests, the load and displacement of the jack, the rotation angle of the arm, the internal pressure in the pipe and the local strain in the longitudinal direction (50 mm intervals) on the compression and tension sides of the pipe were measured.

Table 2 Longitudinal tensile properties of bending test specimens

No.	Grade	YS* [MPa]	TS [MPa]	Y/T [%]	Strength mismatch in GW
1	X80	565	708	80	Overmatch
2	X70	487	623	78	Overmatch
		501	634	79	
3	X70	487	624	78	Undermatch

*Measured as 0.5 % onset yield stress

EXPERIMENTAL RESULTS

Moment and average compressive strain

Figure 2 shows the relationship between the bending moment M and 2D average compressive strain, where 2D is a gauge length of two times the pipe diameter. In all cases, 2D average compressive strain increases linearly as the 2D average compressive strain increases in the elastic region, and after yield, 2D average compressive strain tends to show large increases relative to small increments of bending moment. After passing the peak bending moment, the bending moment decreases steeply as 2D average compressive strain increases. The peak bending moment, M_{max}, measured in bending tests No. 1, 2 and 3 was 18.0 MN-m, 10.0 MN-m and 10.2 MN-m, respectively. The corresponding critical 2D average compressive strains of buckling were 1.2 %, 3.1 % and 3.0 %, respectively.

Regardless of the strength mismatch of the GW, i.e., overmatch or undermatch, one of the wrinkles began to develop at a certain distance from the girth welded area due to local buckling, as shown in Fig. 3. The positions of local buckling were 400 mm, 20 mm and 65 mm in test No. 1, 2 and 3, respectively. It was found that the wavelength of the wrinkle coincided with cyclic changes in longitudinal flatness, which may be one of the triggers for the development of shell wrinkle. Furthermore, it is supposed that strength mismatch and/or a geometric imperfection due to the weld reinforcement or shrinkage in the pipe diameter influenced the generation of the buckling wrinkle at this position.

Copyright © 2019 by ASME

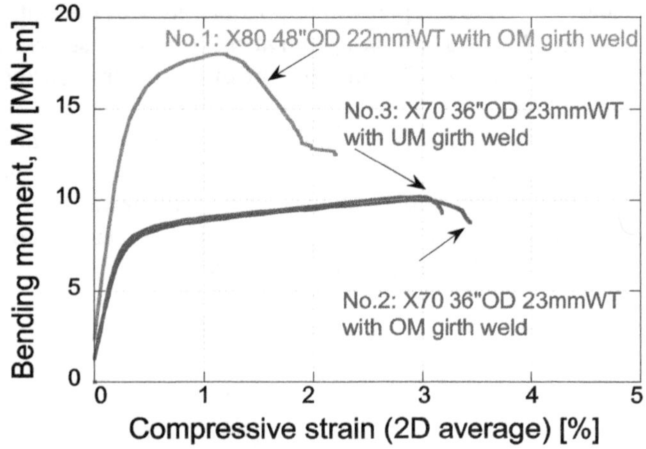

Fig. 2 Relationships between bending moment and 2D average compressive strain

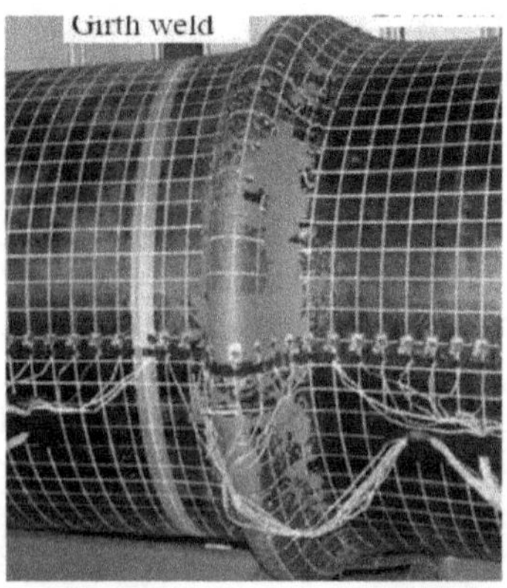

(a) Test No. 1 (X80, overmatched joint)

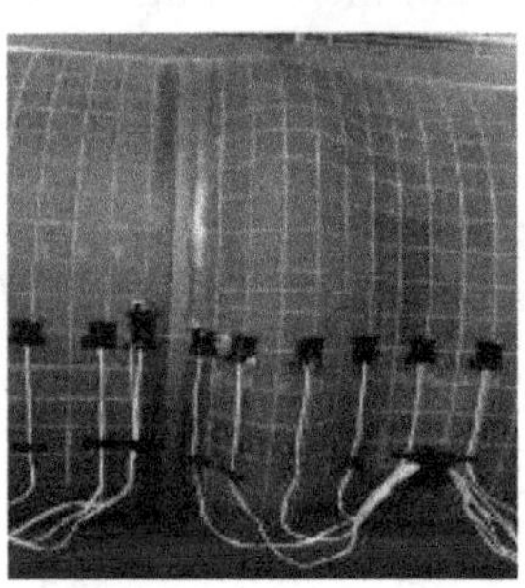

(b) Test No. 2 (X70, overmatched joint)

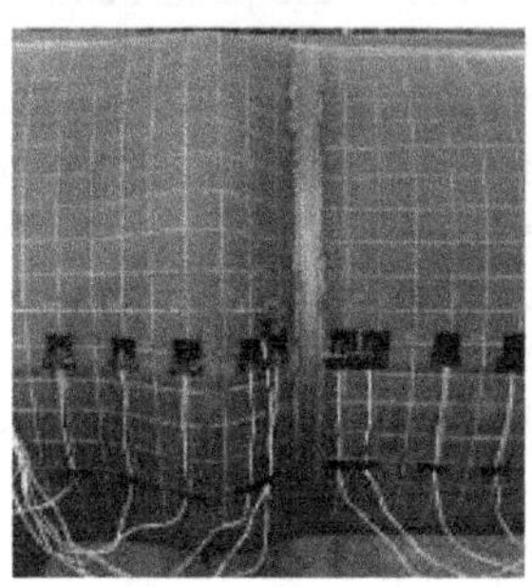

(c) Test No. 3 (X70, undermatched joint)

Fig. 3 Positions of buckling wrinkle

Strain distribution in longitudinal direction

Figure 4 shows the distribution of compressive strain in the longitudinal direction at the peak bending moment, M_{max}. The longitudinal compressive strain was concentrated where the wrinkles can be observed in Fig. 3.

Figure 5 shows the distribution of tensile strain in the longitudinal direction at M_{max} and just before final failure. The longitudinal tensile strain plotted in Fig. 5(a) displays a uniform distribution at M_{max}. Even in test No. 2, there is a small distribution between the two sides of the GW, as this is a joint between two different pipes, but a uniform distribution can be observed within each pipe. Although compressive strain tends to increase around the large developed wrinkle at M_{max}, tensile strain shows a uniform distribution. The development of small wrinkles cannot be recognized from the distribution of tensile strain, which is different from that of the compressive strain shown in Fig. 4.

The longitudinal tensile strain shown in Fig. 5(b) was concentrated where the rupture can be observed just before final failure in the overmatched joints in tests No. 1 and No. 2, however this was not observed in test No. 3 with the undermatched joint. From these longitudinal tensile strain tendencies, the following deformation behavior was clearly observed: After a large wrinkle formed on the compression side of bending, the wrinkle worked precisely like a plastic hinge, and longitudinal tensile strain developed only in the exactly opposite area to the large wrinkle on the compression side of bending.

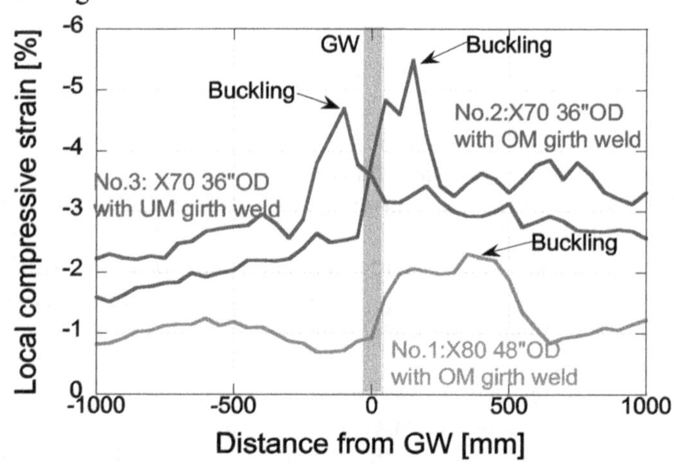

Fig. 4 Distribution of longitudinal compressive strain at M_{max}

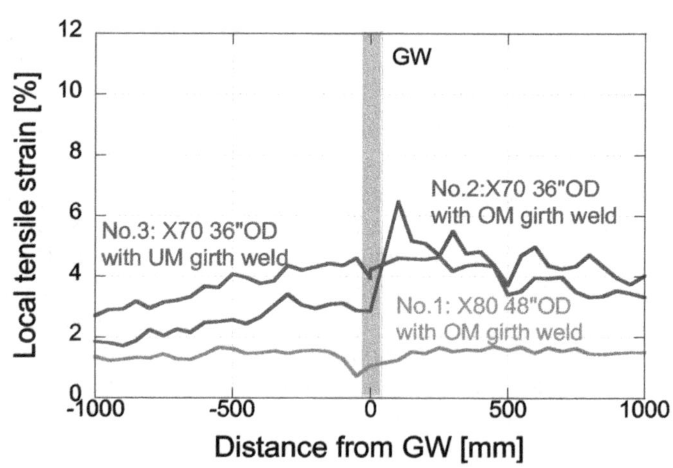

(a) At peak bending moment, M_{max}

Copyright © 2019 by ASME

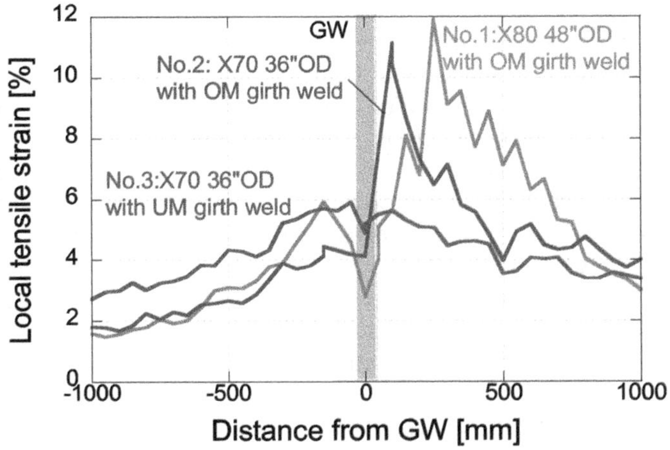

(b) Just before failure

Fig. 5 Distribution of longitudinal tensile strain

Final failure of girth welded pipe under pressurized bending condition

The final failure on the tension side was different in tests No. 1 and No. 2 with the overmatched joints and test No. 3 with the undermatched joint, as shown in Fig. 6. In the overmatched joints, final fracture occurred on the tension side at a certain distance from the GW, and the position of fracture corresponded to the exactly opposite position to the large wrinkle on the compression side of bending. An arc-shaped through-the-wall crevice was observed on the fracture surface in tests No. 1 and No. 2. From detailed observation of the fracture surface, it can be classified into two fracture surface morphologies. In one, the fracture surface is normal to the direction of bending stress, and a decrease in thickness is observed due to necking. In the other, the fracture surface is a shear fracture surface which is formed on both sides of the fibrous fracture surface, and is recognized as a propagating fracture surface that occurs in the subsequent stage of failure.

In the undermatched joint in test No. 3, the initial EDM notch opened as deformation increased, and its opening finally became larger. This led to final fracture due to ductile tearing through the wall thickness.

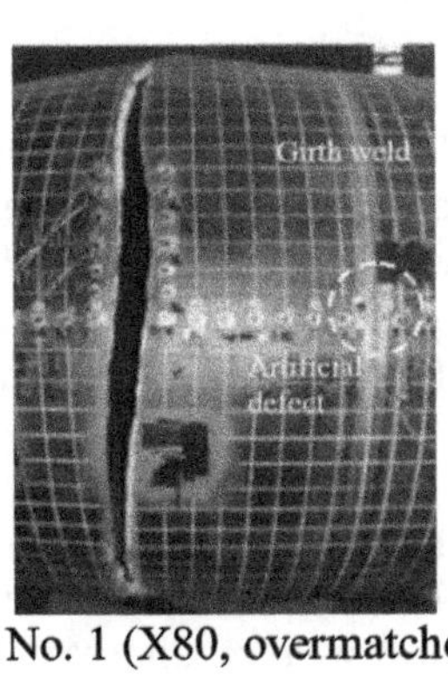

(a) Test No. 1 (X80, overmatched joint)

(b) Test No. 2
(X70, overmatched joint)

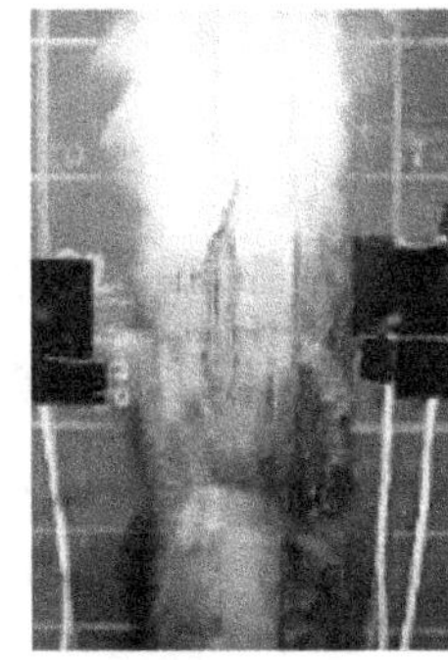

(c) Test No. 3
(X70, undermatched joint)

Fig. 6 Final failure in tension under bending

DISCUSSION

Post-buckling behavior of girth welded joint of high strain linepipe

The difference in the final failure on the tension side in test No. 1 and No. 2 with the overmatched joints and test No. 3 with the undermatched joint were investigated. Figure 7 shows the relationship between 2D average tensile strain and the CMOD (Crack Mouth Opening Displacement) of the initial EMD notch. In both the overmatched and undermatched joints, CMOD increases linearly as 2D average tensile strain increases. However, after buckling, CMOD increases drastically as 2D average tensile strain increases only in the undermatched joint. This could be due to the effect of ductile tearing from the initial notch. In the undermatched joint, tensile strain is concentrated locally in the lower strength region under the uniaxial tension mode after local buckling, as shown in Fig. 5, and this tensile strain concentration in the lower strength weld metal accelerates opening of the notch located in the HAZ. Finally, the strain concentration decreases due to notch opening in the lower strength region of the GW.

In the overmatched joints, tensile strain is not concentrated in the higher strength region of the GW under the uniaxial tension mode, but is concentrated in the opposite position to the large wrinkle on the compression side of bending after the large wrinkle forms on the compression side of bending due to the effect of the plastic hinge on wrinkle formation. Figure 8 shows the relationship between the 2D average tensile and compressive strains. After buckling, 2D average tensile strain increases steeply only in the overmatched joints. It is clear that this remarkable concentration and development of tensile strain causes necking and final fracture in the base material.

Copyright © 2019 by ASME

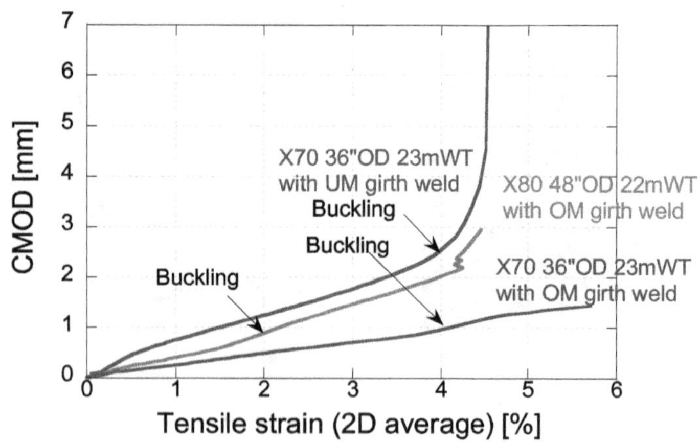

Fig. 7 Relationship between average tensile strain and CMOD

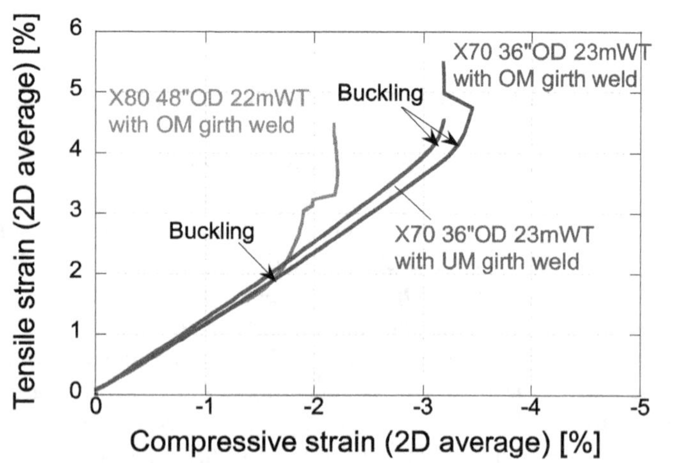

Fig. 8 Relationship between average tensile and compressive strains

Strain capacity of girth welded joint of high strain linepipe under bending deformation

Figure 9 summarizes the final failure modes of the girth welded pipes obtained in the full-scale bending test. Final rupture to leakage occurred in the base material at a certain distance from the girth weld on almost the opposite side to the developed wrinkle on the compression side. In this case, the rupture limit of the base material in the bending test can be predicted from the mechanical property of elongation of the base material. Another failure mode is tensile fracture at a girth weld. When a large weld defect exists in a girth welded joint or large deformation is induced in the tension mode, final rupture may occur by ductile crack propagation from an initial flaw. In this case, the tensile strain limit can be evaluated by using the ductile tearing resistance curve.

In this study using a full-scale bending test with a relatively large artificial surface notch introduced on the outer surface at HAZ of the GW, the difference in the failure mode is considered to be due to the difference in the strength properties of the weld metal. It is thought that buckling accelerates tensile deformation due to effect of the plastic hinge on wrinkle formation. An increase in tensile strain near a girth weld affects the increase in the initial notch mouth opening, because a strength mismatch and/or a geometric imperfection due to the weld reinforcement or shrinkage in the pipe diameter influences the generation of the buckling wrinkle at this position. Because of the weakness of the weld material in undermatched joint, the opening of the initial notch located in HAZ or weld metal is larger than that in an overmatched joint.

On the other hand, in overmatched joints, an initial notch in HAZ or weld metal is not likely to open because of the sealing effect of strength mismatch. Therefore, remarkable concentration and development of tensile strain leads to necking and final fracture in the base material of pipe body.

The tensile strain capacity of the joints with overmatched girth weld is influenced by the formation of the buckle. However, for joints with under-matched girth weld, the formation of the buckle has less effect of the tensile strain capacity. It is clear that the tensile strain limit in overmatched joints is much higher than that in undermatched joints because the final fracture phenomenon is different in tensile fracture in the base metal and in ductile tearing from an initial notch in HAZ or weld metal. The elongation property of the base metal or ductile tearing resistance of HAZ is important for preventing these two fracture modes on the tensile side under bending. Use of high strain linepipe also helps prevent initial local bucking, and this can ultimately lead to a higher tensile strain limit.

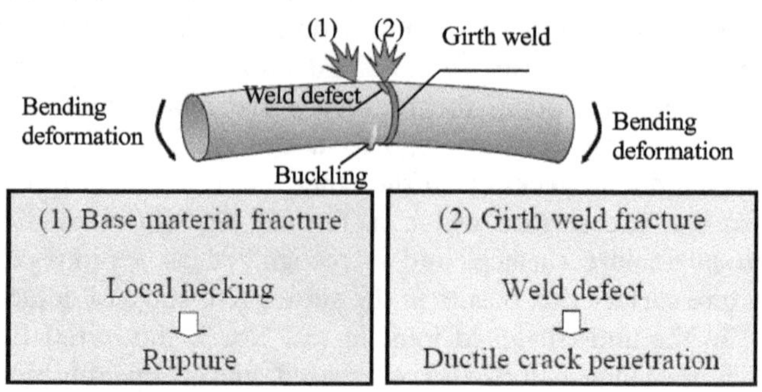

Fig. 9 Final failure mode of girth welded pipe under bending

CONCLUSIONS

Full-scale pressurized bending tests of Grade X80 42" OD and Grade X70 36" OD high strain linepipe with girth welds were conducted in order to investigate the tensile strain limit after buckling. The effect of the strength mismatch between the base metal and the girth weld on the tensile strain limit after buckling was also discussed. The following conclusions were obtained:

- In the overmatched joints, final rupture to leakage occurred in the base material at a certain distance from the girth weld on almost the opposite side from the developed wrinkle on the compression side. This

Copyright © 2019 by ASME

occurred because the wrinkle worked precisely like a plastic hinge, and longitudinal tensile strain developed only in the exactly opposite position to the large wrinkle on the compression side of bending after the large wrinkle formed on the compression side. In this case, remarkable concentration and development of tensile strain led to the necking and final fracture in the base material of the pipe body.

- With the undermatched joint, ductile crack initiation and propagation from an artificial defect in HAZ was obtained in the full-scale bending test. Due to the weakness of the weld material in the undermatched joint, opening of the initial notch in HAZ or weld metal was larger than that in the overmatched joints.
- The tensile strain capacity of the joints with overmatched girth weld is influenced by the formation of the buckle. But, for joints with under-matched girth weld, the formation of the buckle has less effect of the tensile strain capacity.

REFERENCES

[1] T. Masuda, T. Watanabe, N. Hagiwara, H. Yatabe, S. Kawaguchi, H. Motohashi, N. Fukuda, "Application of X80 in Japan," *Proceedings of the International Pipeline Technology Conference, Vol. 1, pp31-40,* 2004.

[2] Y. Wang, J. Pan, "The situation of advanced grade pipe steel in China," *Proceedings of the International Pipeline Technology Conference, Vol. 1, pp. 41-54,* 2004.

[3] A. Glover, J. Zhou, D. Horsley, N. Suzuki, S. Endo, J. Takehara, "Design, Application and Installation of an X100 Pipeline," *Proc. of 22nd International Conference on Offshore Mechanics and Arctic Engineering, OMAE2003-37429,* 2003.

[4] C. W. Petersen, K. T. Corbett, D. P. Fairchild, S. Papka, M. L. Macia, "Improving long distance gas transmission economics," *Proceedings of the International Pipeline Technology Conference, Vol. 1, pp. 3-30,* 2004.

[5] R. Denys, W. D. Waele, A. Lefevre, P. D. Baets, "An Engineering Approach to the Prediction of the Tolerable Defect Size for Strain-Based Design," *Proc. of 4th International Conference on Pipeline Tech,* 2004.

[6] T. Sadasue, S. Igi, T. Kubo, "Ductile Cracking Evaluation of X80 High Strength Linepipes," *Proc. of 5th International Pipeline Conference, IPC04-0249,* 2004.

[7] S. Igi, N. Suzuki, "Tensile Strain Limits of X80 High-strain Pipelines," *Proceedings of the Sixteenth Proceedings of the International Offshore and Polar Engineering Conference,* 2007.

[8] T. Sakimoto, S. Igi, T. Kubo, M. Ohata, F. Minami, "The Influence of Internal Pressure on Ductile Fracture Behavior from a Surface Defect on a Pipe," *Proc. of 19th International Offshore and Polar Engineering Conference,* 2009.

[9] S. Igi, T. Sakimoto, N. Suzuki, R. Muraoka, T. Arakawa, "Tensile Strain Capacity of X80 Pipeline Under Tensile Loading With Internal Pressure," *8th International Pipeline Conference, IPC2010-31281,* 2010.

[10] S. A. Kibey, K. Minnaar, W. Cheng, X. Wang, "Development of a Physics-based Approach for the Prediction of Strain Capacity of Welded Pipelines," *Proceedings of the 19th International Offshore and Polar Engineering Conference,* 2009.

[11] D. P. Fairchild, S. Kibey, H. Tang, V. R. Krishnan, X. Wang, M. L. Macia, W. Cheng, "Advancements Regarding Capacity Prediction of Strain-Based Pipelines," *The Twenty-second International Offshore and Polar Engineering Conference,* 2012.

[12] D. P. Fairchild, H. Tang, S. Y. Shafrova, W. Cheng, J. M. Crapps, "Updates to ExxonMobil's Modelling Approach for Tensile Strain Capacity Prediction," *Proceedings of the Twenty-fourth International Ocean and Polar Engineering Conference,* 2014.

[13] Wang Y., Cheng W., Horsley D., Glover A., Zoe D. and McLamb M., "Tensile strain limits of girth welds with surface-breaking defects – Part I: an analytical framework and part II: experimental correlation and validation". *Proceedings of the 4th International Pipeline Technology Conference, Vol. 1,* 2004.

[14] Wang Y., Liu M., Song Y., Stephens M., Petersen R. and Gordon R., "Second generation models for strain-based design". *US DOT Contract No. DTPH56-06-T000014, final report to US DOT and PRCI,* 2011.

[15] Wang Y., Kibey S. and Tang H., "Strain-based design-advances in prediction methods of tensile strain capacity". *International Journal of Offshore and Polar Engineering, Vol. 21,* 2011.

[16] C. Timms, D. Swanek, D. DeGeer, A. Meijer, P. Liu, E. Jurdik, J. Chaudhuri, "TurkStream Collapse Test Program" *Proceedings of the 37th International Conference on Ocean, Offshore and Arctic Engineering,* 2018.

[17] N. Suzuki, H. Tajika, S. Igi, M. Okatsu, J. Kondo, T. Arakawa, "Local Buckling Behavior of 48", X80 High-Strain Line Pipes," *8th International Pipeline Conference, IPC2010-31637,* 2010.

Copyright © 2019 by ASME

[18] H. Tajika, S. Igi, T. Sakimoto, S. Tsuyama, "Full-Scale Bending Test of 48" X80 Linepipes," *Proceedings of the Twenty-first International Offshore and Polar Engineering Conference,* 2011.

[19] S. Höhler, H. Karbasian, A. Gering, C. Kalwa , B. Ouaissa, "Strain Capacity of Large Diameter Pipes: Full Scale Investigation With Influence of Girth Weld, Strip End Weld and Ageing Effects," *11th International Pipeline Conference, IPC2016-64151,* 2016.

[20] G. Mannucci, A. Lucci, C. M. Spinelli, A. Baldi, G. Mascia, "Full Scale Bend Testing of Strain Based Designed High Grade Buried Gas Pipeline," *Proceedings of the Twenty-first International Offshore and Polar Engineering Conference,* 2011.

[21] S. Igi, T. Sakimoto, H. Tajika, S. Endo, R. Muraoka, N. Suzuki, "Strain Capacity of 48" OD X80 Pipeline In Pressurized Full-scale Bending Test," *The Twenty-first International Offshore and Polar Engineering Conference, ISOPE-I-11-606,* 2011.

[22] H. Tajika, T. Sakimoto, T. Handa, S. Igi, R. Ikeda, J. Kondo, "Strain Capacity Investigation on Grade X70 High Strain Line Pipe with Girth Weld," *Proceedings of the ASME Pressure Vessels & Piping Conference, PVP-2018,* 2018.

[23] H. Tajika, T. Sakimoto, T. Handa, R. Ikeda, J. Kondo, "Girth Weld Strength Matching Effect on Tensile Strain Capacity of Grade X70 High Strain Line Pipe," *Proceedings of the 2018 12th International Pipeline Conference, IPC2018-78778,* 2018.

Copyright © 2019 by ASME

AIMPIMG-1064

Chapter 29

LOW STRAIN CAPACITY GIRTH WELDS OF NEWLY CONSTRUCTED PIPELINES AND MITIGATIVE APPROACHES

Yong-Yi Wang
Center for Reliable Energy Systems
Dublin, OH, USA
ywang@cres-americas.com

Dan Jia
Center for Reliable Energy Systems
Dublin, OH, USA
djia@cres-americas.com

Steve Rapp
Enbridge
Houston, TX, USA
Stephen.Rapp@enbridge.com

David Johnson
Energy Transfer
Houston, TX, USA
David.Johnson@energytransfer.com

ABSTRACT

There have been a number of unexpected girth weld failures in newly constructed pipelines. Girth weld failures have also been observed in pre-service hydrostatic testing. Most of the failures occurred in X70 pipelines with manual welds, although failures have been observed in linepipe grades ranging from X52 to X80. Post-incident investigations indicated that the linepipes, girth weld qualification, and field girth welding and inspection met the requirements in relevant codes and standards. Traditional causes of weld failures such as weld flaws and high-low misalignment played no or insignificant roles. In other words, these failures are beyond expectations based on historical experience. More significantly, the causes of these failures have not been the focus of the relevant codes and standards that the industry has been relying on for quality assurance and long-term integrity

The features of the recent girth weld incidents are first reviewed. The key characteristics of those incidents is summarized. Overall, the main contributing factors to the incidents are: (1) weld strength undermatching, (2) heat-affected zone (HAZ) softening, (3) weld bevel geometries of manual stick welding processes that favor plastic straining along the softened HAZ, and (4) elevated stresses/strains from normal settlement and other loads. The causes of weld strength undermatching and HAZ are further examined from the viewpoint of linepipe steel manufacturing, pipe testing and qualification requirements, and a response to a regulatory advisory. The last part of the paper focuses on mitigative measures the industry can take to reduce the risk of similar failures in future pipeline construction. These measures include responses to the causes of weld strength undermatching and HAZ softening. The systematic gaps in current standards and industry practice are examined and mitigative proposals are provided. It is critically important that the responses to the incidents cover both technical improvements and, more significantly, systematic improvements that minimize the disconnect among different functional departments responsible for various stages of a pipeline's life.

KEYWORDS

Linepipe, modern microalloyed steels, girth weld incident girth welding, HAZ softening, weld strength undermatching, strain capacity, standards revision, systematic approach

INTRODUCTION

There have been a number of unexpected girth weld failures in newly constructed pipelines. Girth weld failures have also been observed in pre-service hydrostatic testing. Post-incident investigations indicated that the linepipes met the requirements of industry standards, such as API 5L. The welds were qualified per accepted industry standards, such as API 1104. The field girth welding was performed, inspected, and accepted per industry standards, such as API 1104. These failures are beyond historical experience. More significantly, the causes of these failures have not been the focus of the

Copyright © 2019 by ASME

relevant codes and standards that the industry has been relying on for quality assurance and long-term integrity.

GIRTH WELD FAILURE INCIDENTS

Typical Failure Path of Girth Weld Incidents

An in-service failure, occurring less than two years after the pipeline was put in service, was covered in prior publications [1, 2]. Another example of a failed girth weld is shown in Figure 1 [3].

The typical failure paths are schematically shown in Figure 2. The paths connect the weld toe on the pipe OD and the weld toe on the pipe ID. In some cases, the path stays on one side of the deposited weld metal. In other cases, the path crosses the deposited weld metal near the ID. On the OD side, the path usually is on the side of the weld with shorter extent of the weld cap (in comparison to the extent of the weld cap on the other side of the weld).

A typical microhardness map of a failed girth weld is shown in Figure 3 [1]. The hardness of the HAZ is lower than the hardness of deposited weld metal in the fill and cap passes and the pipe. The hardness of the root region is the lowest.

The failure paths shown in Figure 2 are associated with HAZ softening and weld strength undermatching. In the absence of weld strength undermatching, a failure can occur in the softened HAZ away from the fusion boundary, as shown in Figure 4.

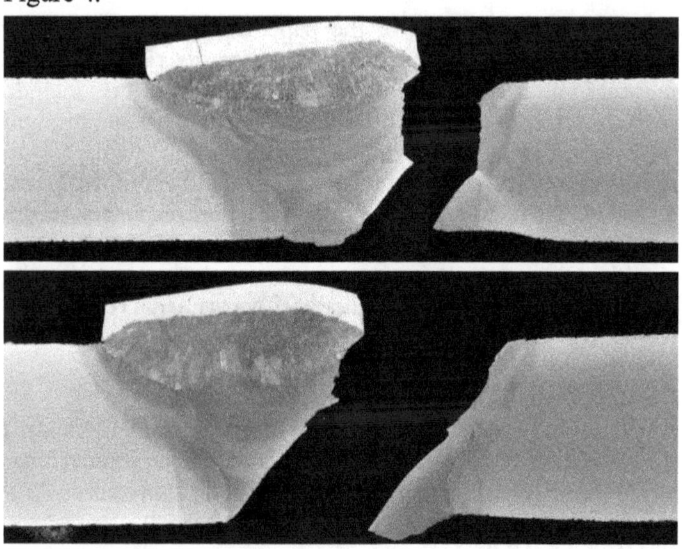

Figure 1 Examples of failure paths of a girth weld [3]. The light-colored material on top of the weld is the field coating.

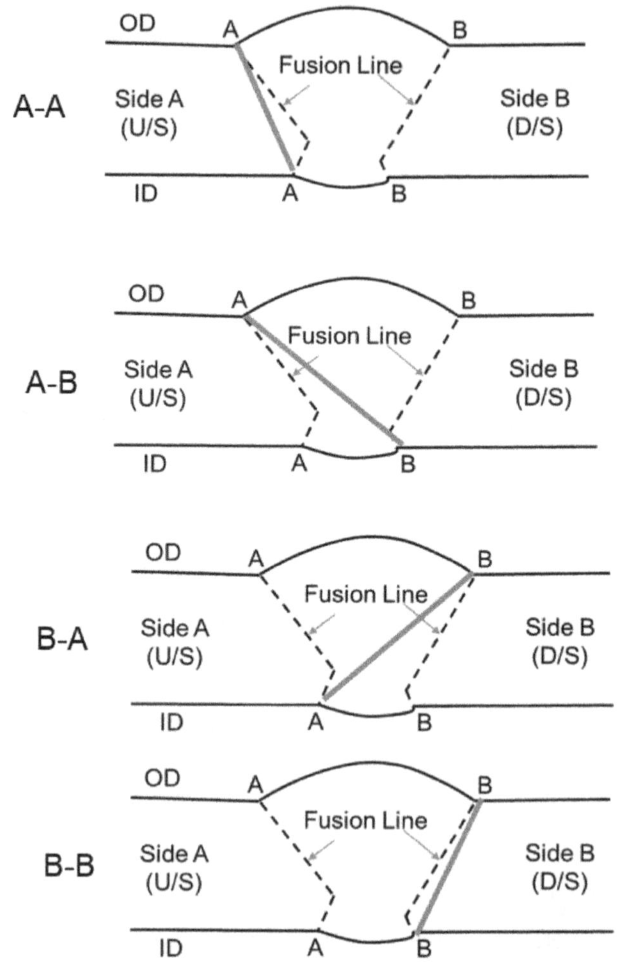

Figure 2 Schematic illustration of typical failure paths [3]

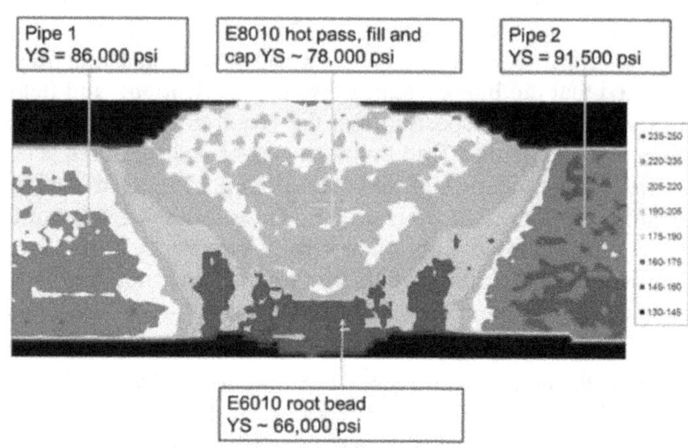

Figure 3 Microhardness map of a failed weld [1]

Copyright © 2019 by ASME

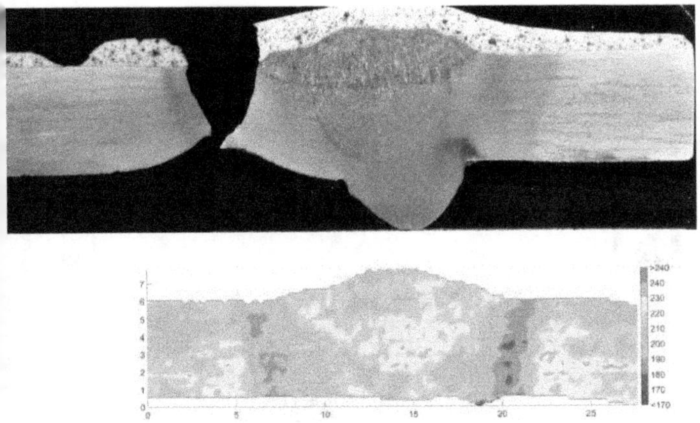

Figure 4 A girth weld failed in the softened HAZ and the corresponding microhardness map

Characteristics of Girth Weld Incidents

The key characteristics of the girth weld incidents occurred in the past few years are:

1. All failures were from either ERW or spiral pipes, with the steels coming from hot strip mills[1].

2. Most of the incidents were from X70 pipes. A few involved grades as low as X52 and as high as X80 pipes.

3. The carbon content and Pcm of most of the X70 pipes were in the range of 0.040-0.053% and 0.13-0.17, respectively. The small diameter ERW pipes had carbon of 0.05-0.07 and Pcm of 0.12-0.14.

4. All welds were made with manual processes, mostly with SMAW processes using E6010 electrodes for the root pass and E8010 electrodes for hot, fill, and cap passes.

5. For the X70 pipes, the actual UTS levels of the pipes were ~10-20 ksi above the specified minimum UTS. For one X52 pipe involved in the incidents, the actual UTS of the pipe was 3-13 ksi above the specified minimum UTS. The actual UTS of the other X52 pipe was 25-37 ksi above the specified minimum UTS.

6. HAZ softening was present in all welds. The level of the softening and the shape, width, and location of the softened zones varied from weld to weld.

Contributors to the Girth Weld Incidents

Girth weld failures occur when the strain demand on the girth welds exceeds the strain capacity of these welds. The major contributors to the low strain capacity are (1) weld strength undermatching and (2) heat-affected zone (HAZ) softening. In addition, plastic shearing along the softened HAZ and undermatched weld metal can occur due to the bevel geometry and the low root pass strength of manual stick welding processes. The softened HAZ and the soft root pass align well with the 45° plastic shear plane.

On the strain demand side, the magnitude of tensile loading and its contribution to the incidents varied. In some cases, there were elevated bending loads from normal ground settlement and from the pipe profile not conforming to the profile of the trench. In a minority of cases, an appreciable amount of ground movement occurred. Post-incident investigations revealed that some of the welds involved in the ground movement events had low strain capacity. Although the ground movement was a triggering event, it can be argued that the low strain capacity was as much a contributor as the ground movement event. In two cases, the strain capacity of the welds was considered adequate, i.e., the welds would not have failed in the absence of large-scale ground movement. The ground movement in those two cases were judged to be more of a contributor than the low strain capacity of the girth welds.

Other than one hydrogen crack that contributed to a hydrostatic test leak, weld flaws were not a contributor to those incidents. Almost all welds had a minimal level of high-low misalignment. The mechanical properties and chemical composition of the linepipes met the requirements of API 5L. The welding procedure qualification, field welding, and NDT met industry standards, such as API 1104. The toughness of the welds was good. The failures were dominated by the strength mismatch, not related to toughness or weld flaws.

CAUSES OF WELD STRENGTH UNDERMATCHING

Changes in Pipe Tensile Properties and Impact on Weld Strength Mismatch

Manual SMAW processes with E6010/E8010 electrodes have been used on X70 pipes for decades. The basic welding processes and consumables have not changed. One of the logical questions is why these failures are occurring in recent years.

Impact of Changes in Overall Pipe Strength

There are some indications that the strength of the pipe of the same grade has change over time. The impact of this change on the weld strength mismatch is shown in Figure 5 with four sets of stress-strain curves. The "All Weld Metal – Native Strength" was obtained from an all weld metal tensile test of a girth weld. The "All Weld Metal – Apparent Strength" is the estimated strength after considering the "strengthening" effects of the weld bevel and the weld cap reinforcement. This strength represents the composite strength exhibited by the weld when the weld is loaded in the cross-weld direction, taking into account the native strength and the strengthening effects. Two

[1] Incidents involving straight-seam UOE pipes have occurred. However, the authors don't have direct knowledge of those incidents. These incidents are not included in this table.

Copyright © 2019 by ASME

other curves are X70 pipes tested in the longitudinal direction. One is a UOE pipe produced around 2002 and the other one is an ERW pipe produced in 2013. Relative to the 2002 UOE pipe, the native weld strength is very close to the strength of the pipe. The apparent strength of the weld is higher than the strength of the pipe, indicating the weld behaves like an overmatching weld. Relative to the 2013 ERW pipe, both the native and apparent weld strength levels are below the strength of the pipe. The undermatching level is greater at the yield strength than at the UTS. The weld behaves like an undermatching weld.

While the trends shown in Figure 5 do not necessarily represent all conditions involved in the girth weld incidents, they demonstrate the impact of the pipe strength change on the weld strength mismatch level even when the same welding procedures are applied to the same grade of pipes.

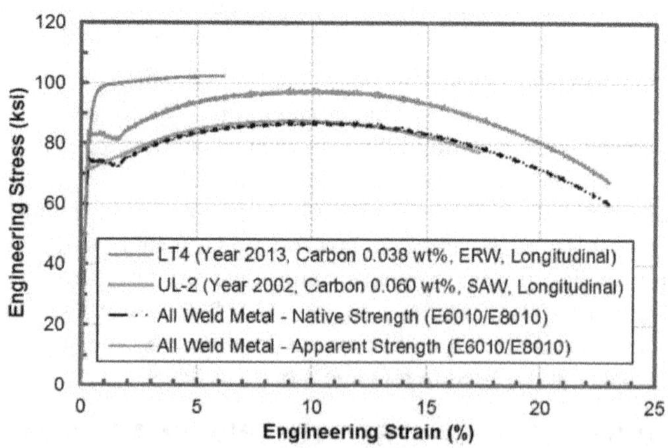

Figure 5 Impact of the X70 pipe strength change on the weld strength mismatch

Impact of Changes in Yield Strength and Strain Hardening

The increased reliance on rolling and accelerated cooling have led to steels with much higher Y/T ratios (thus low strain hardening) than the steels produced before TMCP processes. The hoop and longitudinal stress-strain curves of a vintage X52 and modern X52 are compared in Figure 6 and Figure 7, respectively. In both directions, the UTS levels of the pipes are close. However, the vintage X52 pipe has a much lower yield strength and much greater strain hardening.

The welding procedure qualification requirements in the current API 1104 do not differentiate between those two X52 pipes. A welding process and consumables that can overmatch the yield strength of the vintage X52 pipe could undermatch the modern X52 pipe due to the increase in the yield strength of the pipe.

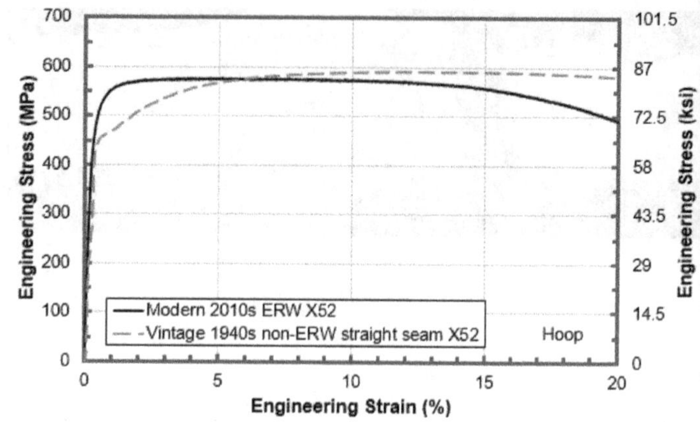

Figure 6 Comparison of hoop stress-strain curves of a vintage late 1940s X52 straight-seam non-ERW pipe and a modern 2010s X52 ERW pipe

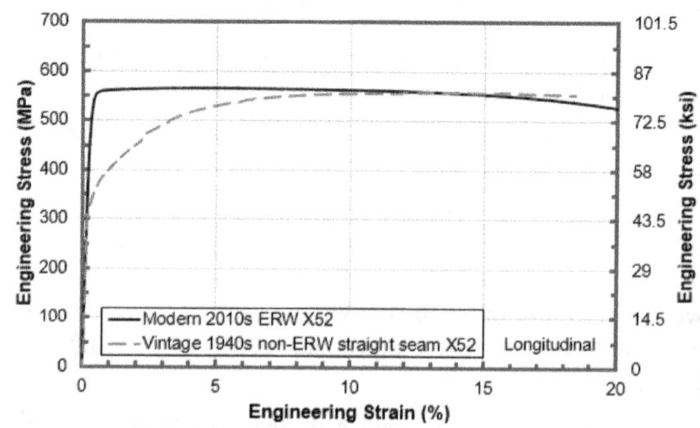

Figure 7 Comparison of longitudinal stress-strain curves of a vintage late 1940s X52 straight-seam non-ERW pipe and a modern 2010s X52 ERW pipe

Contributors to Increased Pipe Strength

A number of factors have contributed to the increased pipe strength within the same grade, including but not limited to,

1. *Large ranges of strength permitted in standards like API 5L.* For PSL2 pipes grades X52 and above, the yield strength can be 20-25 ksi above the SMYS and the UTS can be 40+ ksi above the specified minimum. Given the Bauschinger effects of flattened test specimens, the actual yield strength could be 30+ ksi above the SMYS [1].
2. *Test method systematically under-represents actual yield strength, see below.*
3. *Changes in steel making, test method, and definition of yield strength lead to large variations in reported yield strength for the same joint of pipe.* Mills have to increase the average strength to meet the minimum strength requirements [2].
4. *Industry response to expanded pipes due to "understrength" pipes after PHMSA ABD, see below.*

Copyright © 2019 by ASME

5. *The negative consequence of higher strength was not well recognized.* Having actual strength much higher than the specified minimums has not been viewed as a negative factor on pipeline integrity.

Contribution of Test Methods

It is known in the industry that the flattening process of the hoop specimens tends to produce yield strength values that are lower than the actual YS of the pipe materials. EPRG reviewed a substantial body of test results and compared tensile test results from flattened rectangular specimens with those from round bar specimens as shown in Figure 8 [4]. The strength values from round bar specimens were generally higher than those from flattened rectangular specimens. This difference became more pronounced as the strength increased. The round bar specimens generally represented the yielding behavior in full-scale pipe tests more closely than the flattened rectangular specimens. It should be noted that the difference in yield strength produced by the three types of specimens varies among different pipe manufacturing process (UOE, ERW, spiral, expanded vs. non-expanded) and *D*/*t* ratio. The results shown here represent general trends.

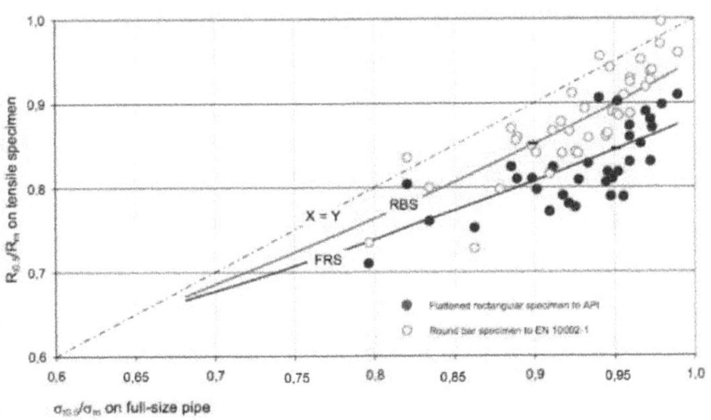

Figure 8 Comparison of yield strength measured using flattened rectangular and non-flattened round bar specimens (*y*-axis) with those measured in pipe samples (*x*-axis) [4]. The graph is edited from original publication. The numbers on the *x*-axis of the original publication are not shown correctly.

The possible impact of under-representation on the actual yield strength is shown in Table 1. On the average, the yield strength from flattened straps was 12% lower than the yield strength from "full-size" pipe specimens. In some upper bound cases, the yield strength from flattened straps could be 18% lower than the yield strength from "full-size" pipe specimen. Table 1 shows the possible "actual" yield strength if the "reported" yield strength from the strap tests is to stay within the specified ranges for PSL 2 pipes.

Table 1 Possible impact of yield strength representation on the actual strength of the pipe

Grade	API 5L PSL 2 YS (ksi)		Possible Actual YS Range after Considering 12% Under-Representation (ksi)		Actual Minimum YS Pipe Mills Would/Could Target Considering 18% Under-Representation (ksi)
	Min	Max	Min	Max	
X52	52	77	59	87	63
X70	70	90	79	102	85

Conversations with industry experts indicated that the flattened strap could reduce the reported yield strength for X70 pipe by 5-12 ksi. Extensive testing from the pipes involved in those incidents and other X70 pipes indicated that the high-end of the observed actual yield strength was in the range of mid-80s to low 90s ksi. The actual ultimate tensile strength was the range of low 90s to low 100s ksi.

For X52 pipes, the observed high-end of actual yield strength was in the range of low 80s to low 90s ksi. This high level of yield strength is often a surprise to most people in the industry. The upper limit of yield strength of X52 pipe is 77 ksi by API 5L. A steel with a yield strength of 93 ksi after taking a 18% reduction (due to the yield strength under-representation) becomes 77 ksi. Similarly, a steel with a yield strength of 87 ksi yield strength becomes 77 ksi after taking a 12% reduction. Thus, the actual yield strength of the pipe can be 30+ ksi higher than the specified minimum yield strength. One of the X52 pipes involved in the incident has a yield strength in the lower 90s ksi. It was not clear if this pipe was qualified as a PSL 2 pipe. PSL 1 pipes have no upper limits on strength.

Impact of PHMSA ABD

In 2009, US DOT PHMSA issued an advisory bulletin advising "pipeline system owners and operators of the potential for high grade line pipe installed on projects to exhibit inconsistent chemical and mechanical properties" [5]. The advisory also stated: "This advisory bulletin pertains to microalloyed high strength line pipe grades, generally Grade X-70 and above" [5]. It was noted that pipeline hydrotest failures and excessive bulging deformation were encountered in newly constructed pipelines as a consequence of the pre-service commissioning hydrostatic test. Some linepipe purchasers added a few ksi to their acceptable minimum yield strength to guard against "under-strength" pipes. Some pipe mills and their suppliers added a few ksi more to give them an extra margin of safety. The actual yield strength of the X70 pipe is likely 10+ ksi above the SMYS.

Copyright © 2019 by ASME

CAUSES OF HAZ SOFTENING

Up until the late 1960s, linepipes of up to API Grade X60[2] were manufactured from hot rolled coils and normalized plates. The evolution of the chemical composition and processing routes as a function of pipe grade and time is shown in Figure 9 [6]. The development of high-strength pipeline steels has followed the route of progressively reduced hardenability through lower carbon and alloying element contents [7].

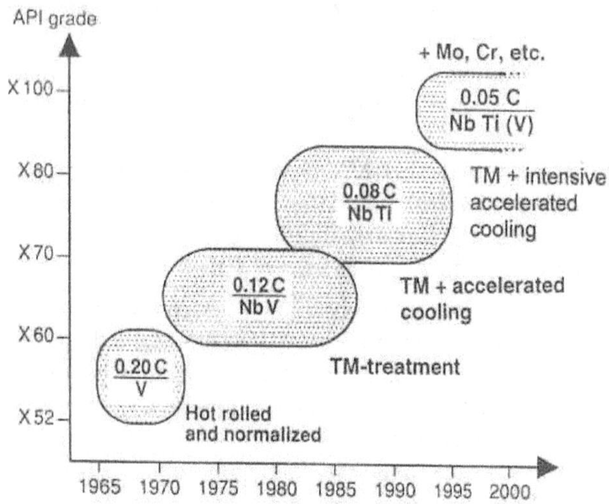

Figure 9 Evolution of chemical composition and rolling practice over time for various grades of linepipes as seen in late 1990's [6]

The evolution of linepipe manufacturing technology has been largely driven by the demand from pipeline operators to increase pipe strength, improve fracture resistance characteristics, and enhance weldability (i.e., reduce the likelihood of hydrogen-assisted cracking). The coiled skelp used for ERW and SAWH linepipes is produced with increasingly lean compositions (i.e. low Pcm) to achieve optimal weldability and a lower transition temperature (and reduced alloy cost). Carbon content is generally low in X70 skelp from a strip mill to enhance toughness and provide good Drop Weight Tear Test (DWTT) performance. Because the rolling forces available in a strip mill are small relative to those in a four-high plate mill, a strip mill relies more heavily on severe cooling and lean alloy composition to achieve yield strength and DWTT properties. As a consequence of lean alloy composition, linepipe steels produced in hot strip mill steels have reduced hardenability and therefore experience a higher degree of HAZ softening.

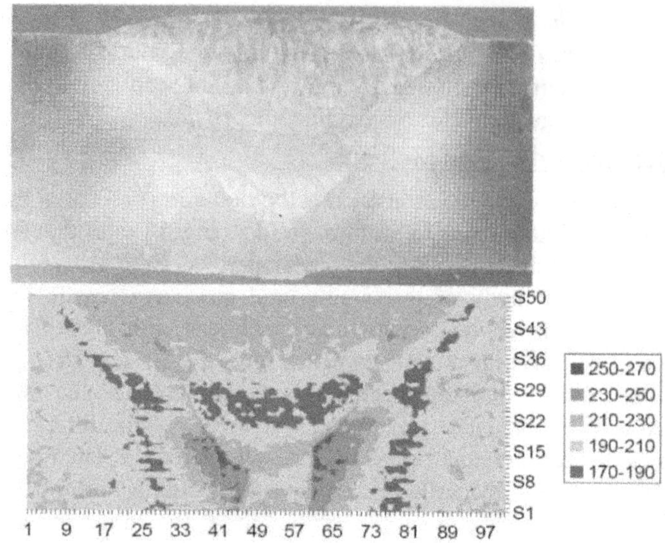

Figure 10 Weld macro and corresponding microhardness map of a girth weld of an X70 UOE pipe produced circa 2000

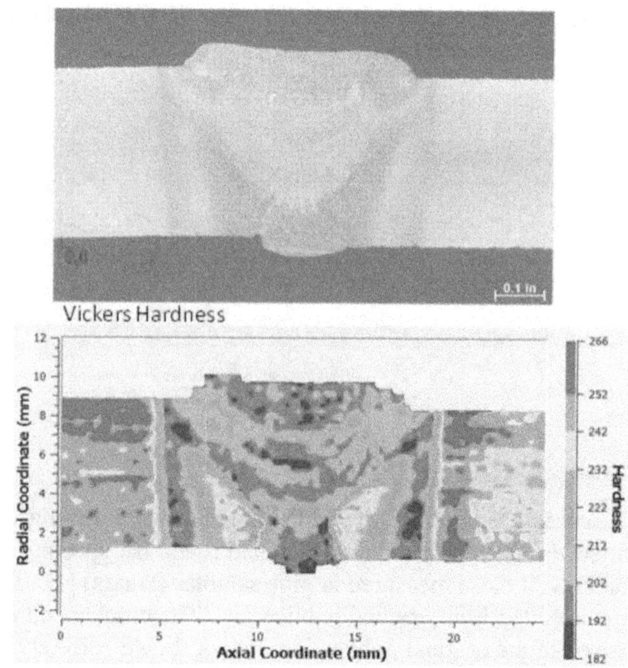

Figure 11 Weld macro and corresponding microhardness map of a girth weld of an X70 ERW pipe produced circa 2013

A comparison of the weld macro and microhardness maps of two girth welds of X70 pipes is given in Figure 10 and Figure 11. The pipe in Figure 10 is an UOE pipe produced around 2000. The pipe in Figure 11 is an ERW pipe produced in 2013. The hardness of the deposited weld metal in fill passes is about the same, around 190-200 Hv. The hardness of the UOE pipe is around 200 Hv. The hardness of the ERW is around 230-250

[2] References are made to API standards. However, comments are generally applicable to other equivalent standards such as those by CSA.

Copyright © 2019 by ASME

Hv. While weld strength approximately even-matches the strength of the 2000-vintage UOE pipe, the same weld strength undermatches the 2013-vintage ERW pipe by approximately 20%. Interestingly, the hardness of the HAZ is approximately the same, at approximately 190 Hv. The softened HAZ in the ERW pipe forms continuous zones, while the softened HAZ in the UOE pipes is isolated islands. On a relative scale, the level of HAZ softening in the ERW pipe is much greater than that in the UOE pipe. Consequently, the softened HAZ in the ERW pipe is expected to have a greater negative impact on the strain capacity of the girth weld.

MITIGATATION – TECHNICAL SOLUTIONS

Linepipe Specifications

Linepipe Strength

One obvious approach to remedy weld strength undermatching is reduce the upper bound strength of linepipes for a particular grade. This approach can be difficult to achieve unless the current test procedures to qualify pipe strength are changed.

There is considerable evidence that the tensile test methods currently used in the industry can produce significant variations in reported yield strength [8, 9, 10]. For a standard flattened strap tensile test to consistently meet the minimum strength requirements of a particular grade, the overall strength has to move up to compensate for the large variability.

The other possibility is exploring the differences in the hoop and longitudinal tensile properties and somehow specifying lower tensile strength in the longitudinal direction. This approach is based on the observations that the yield strength in the hoop direction is lower than the yield strength in the longitudinal direction. If the yield strength in the longitudinal direction can be lowered, this would reduce the level of weld strength undermatching. Tests conducted at CRES indicated that the observed lower yield strength in the hoop direction is largely the result of testing flattened straps (Bauschinger effects) and the imprecision in testing such specimens. There is little evidence that the tensile strength in the hoop direction is actually lower than the tensile strength in the longitudinal direction in SAWH pipes when non-flattened specimens are used. Indeed, the hoop and longitudinal specimens are exactly symmetrical with respect to the strip rolling direction if the spiral angle is 45°. For ERW and SAWL pipes, the longitudinal tensile strength is often lower than the hoop tensile strength if non-flattened specimens are used for hoop strength testing.

To effectively reduce the tensile strength in the longitudinal directions, the following steps are necessary.

1. Have practical test methods that produce yield strengths that are representative of the yield strength of the material in the pipe configuration, i.e., without systematic underrepresentation of the yield strength.
2. The test methods must be robust and minimize the inaccuracy caused by the test methods. This can include a more precise definition of specimen dimensions, instrumentation, post-test data qualification and analysis, etc.
3. Any possible negative impact of having lower strength pipes than the pipes of recent production within the same grade must be understood for the overall integrity of pipelines. For instance, the possibility of "under-strength" pipes within the same grade must be avoided. It is expected that having more consistent test methods would allow the pipe manufacturer to produce pipes with a narrower band of strength variation within the same grade.

Linepipe Chemical Composition

The mechanical properties of the HAZ are principally driven by the chemical composition of the steels and the welding thermal cycles experienced by these steels. The extent of HAZ softening can be reduced by fast cooling from low heat input processes. At the same time, enhancing steel's hardenability should also reduce the propensity for HAZ softening for the same welding cooling rates. The chemical composition analysis of the pipes involved in the incidents indicated that some of the ERW and spiral linepipes delivered in recent years could have low carbon and low alloying contents, leading to low hardenability. These steels have contributed to the increased level and extent of HAZ softening. Maintaining minimum levels of carbon and hardenability should lead to a reduction of the potentially harmful effects of HAZ softening.

Welding Processes and Qualification Requirements

Welding Process - Weld Strength

One obvious way to reduce weld strength undermatching is to increase the weld strength. For instance, using E8010 electrodes can increase the root strength, which can be highly beneficial for enhanced strain capacity.

The strength of the fill and cap passes can be increased by using high strength low-hydrogen electrodes. The strength increase is also possible by using gas-shielded FCAW processes.

Welding Process - Weld Profile

Prior work has shown that the weld cap reinforcement with sufficient width can prevent the formation of a plastic shear band along the softened HAZ, leading to an increased strain capacity [11]. Such a wider cap can be achieved by having a split cap, as shown in Figure 12.

Copyright © 2019 by ASME

Welding Process – Welding Parameters and Process Control

Limiting the heat input could theoretically reduce the level of HAZ softening. Controlling the heat input in manual field welds can be a challenge. Indirect methods of reducing heat input include using smaller-diameter electrodes and specifying more passes for a given pipe wall thickness. This would lead to increased completion time for a weld.

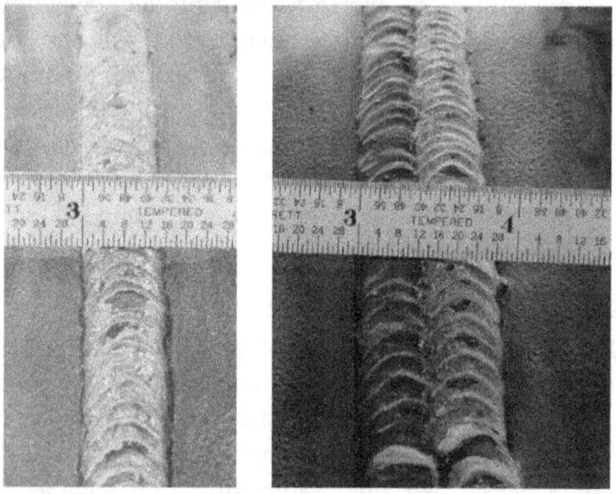

Figure 12 Comparison of girth welds with two cap widths. Left: regular cap width, Right: enhanced wider split cap

MITIGATION – SYSTEMATIC SOLUTIONS

Girth Welding Procedure Qualification Requirements

Factors affecting girth weld performance are summarized in Table 2. The inclusion of these factors in the qualification requirements of API 1104 is also shown. Much of the standard focuses on historical causes of girth weld failures, i.e., weld flaws. Many factors that are contributors to poor girth weld performance, such as weld strength undermaching and HAZ softening, have not been the focus of the standard. In procedure qualification tests in accordance with API 1104, undermatching welds are permitted, provided that the failure stress of cross-weld tensile tests is greater than or equal to the specified minimum tensile strength (SMTS) of the grade of pipe being qualified. Welds can pass qualification tests but have lower strength than the pipe, leading to potentially low strain capacity.

Practical welding procedure qualification requirements that guard against excessive weld strength undermatching need to be implemented. One option is requiring that the failures in cross-weld tensile tests must be in the base pipe, not in the weld metal or HAZ. The other option is requiring no strain concentration in the weld region (including weld metal and HAZ) when the pipe strain reaches a certain target value. Option 1 is easier from test and qualification simplicity, but more difficult to meet than Option 2. To fully account for the distribution of the pipe strength, the qualification tests must be done on pipes with upper bound tensile properties and lowest hardenability. This may be difficult for large pipeline projects which may have multiple pipe and plate/coil suppliers with pipes delivered over a period of time. In addition, the variability of the manual welding processes must be considered. Regardless the options chosen, the qualification requirements are more likely to be successful if the strength of the pipe can be better measured and controlled.

Table 2 Factors affecting girth weld performance and their inclusion in API Standard 1104

Factors Affecting Girth Weld Performance		Factors Addressed in API 1104
Pipe wall thickness		Mostly Yes
Pipe strain hardening capacity		No
Weld strength mismatch	YS mismatch	No
	UTS mismatch	No
HAZ strength (Softening)		No
Weld profile	Cap reinforcement	Mostly No
	Misalignment	No
	Bevel geometry	Yes
Weld flaw	Flaw type	Yes
	Flaw dimensions	Yes
Toughness		No
Applied stress/strain		No

Quantifying HAZ Softening

For onshore pipelines, there are no tests or requirements to measure the materials' susceptibility to HAZ softening in either linepipe or girth welding standards. Development of a viable test to guard against excessive HAZ softening is necessary. Preferably this would be done in pre-production qualification, much like the API RP-2Z tests for offshore applications.

Bring Design and Field Practice into Alignment

One of the major gaps exposed by the girth weld incident is the gap or disconnect among the requirements in pipeline design, girth weld qualification, and actual field conditions. Most pipelines are designed by conventional stress-based design principles given in standards such as ASME B31.4 and B31.8. The pipelines and welds are required to be able to tolerate a longitudinal stress of 90% SYMS. This stress requirement is equivalent to requiring a strain tolerance of about 0.2% for X70 pipelines. Girth welding procedure qualification requirements, which permit failure in the girth weld provided the failure stress is greater than the specified tensile strength of pipe, effectively would lead to a lower end strain capacity of 0.25-0.35%.

In contrast to internal pressure, longitudinal stresses in pipelines are typically not managed. In fact, the level of longitudinal stresses is often not known. Recent more extensive use of IMU surveys indicate that there can be many locations in a pipeline with longitudinal strains exceeding the lower end strength capacity. The relatively rare occurrence of girth weld

Copyright © 2019 by ASME

incident without weld flaws in the past is probably due to a combination of higher strain capacity than that guaranteed by the qualification requirements and low strain demand in most girth weld locations. There have always been gaps and disconnects among the requirements in pipeline design, girth weld qualification, and actual field conditions; but they were not exposed until the strain capacity of the girth welds has been further reduced due to factors described in prior sections.

CONCLUDING REMARKS

The recent girth weld incidents at first appear to be unexpected. A more in-depth examination of the incidents provides reasonable and technically compelling explanations for most of these incidents. In this process, gaps in the customary approaches to linepipe manufacturing and girth weld qualification, pipeline design, and actual field conditions are revealed.

Misconceptions about standards could be a barrier towards reaching cost-effective solutions. Standards committees sometimes are not willing to specify restrictive requirements under the assumption that the standards are minimum requirements and project-specific requirements could be added by users when appropriate. In practice, these standards are often treated as providing sufficient conditions for sound pipelines.

The gaps between linepipe specifications and girth weld qualification requirements are quite telling. Welding procedures, including consumables, that are qualified for a grade are universally treated as good for the same and lower grades. On the other hand, the large range of strength permitted within a single grade in linepipe specifications, such as API 5L, effectively permit lower grades to have higher actual strength. Therefore, the notion that welds are good for lower grades than the qualified grade does not match with how linepipes are procured.

One should be clear that weld strength mismatch or HAZ softening does not come into the picture as long as the material is kept within its elastic limit. Satisfying the ASME longitudinal stress design requirement would ensure this is the case. In practice, longitudinal stresses are typically not actively managed once a pipeline is in service. Furthermore, stresses from construction can be difficult to control in practice. The current design requirements are satisfactory for a majority of onshore pipeline conditions. However, conditions exceeding the design limits are more common that most would realize.

Having robust girth welds that can provide satisfactory service under realistic field conditions should be a goal of our industry. Possible approaches to prevent such failures in new pipelines include (1) changes to the tensile properties of the linepipes and enhanced hardenability, (2) welding options aimed at increasing the weld strength and minimizing HAZ softening, and (3) reduction of stresses on girth welds through better management of construction practice and having excavation and backfill procedures that minimize localized post-backfill stresses of in-service pipelines.

ACKNOWLEDGMENT

A number of organizations and individuals supported the work that led to the content and background information covered in this paper, including US DOT PHMSA, PRCI, pipeline operators, linepipe manufacturers, and linepipe steel suppliers. The authors are grateful for the review and editing of Mr. Alex Wang and Dr. Dunji Yu at CRES. The views and opinions expressed here are those of authors and do not necessarily reflect the official position of the sponsoring organizations.

REFERENCES

1 Wang, Y.-Y., Horsley, D, and Rapp, S., "Evolution of Linepipe Manufacturing and its Implications on Weld Properties and Pipeline Service," Proceedings of the 11th International Pipeline Conference, Paper No. IPC2016-64632, September 26-September 30, 2016, Calgary, Alberta, Canada.

2 Wang, Y.-Y., Rapp, S., Horsley, D., Warman, D., and Gianetto, J. "Attributes of Modern Linepipes and their Implications on Girth Weld Strain Capacity," Proceedings of the 12th International Pipeline Conference, Paper No. IPC2018-78809, September 24-September 30, 2018, Calgary, Alberta, Canada.

3 Wang, Y.-Y., "Girth Welding in the Age of Evolving Steel Manufacturing Processes and under Realistic Field Conditions," *presentation at API 1104 and API 5L meetings*, January 22, 2019.

4 EPRG home page, materials and manufacturing – recent activities, http://www.eprg.net/about_eprg/our_activities/material

5 "Pipeline Safety: Potential Low and Variable Yield and Tensile Strength and Chemical Composition Properties in High Strength Line Pipe," Docket No. PHMSA-2009-0148, May 21, 2009.

6 Hillenbrand, H.G., et al., "Development of Large-Diameter Pipe in Grade X100," *Pipeline Technology*, Vol. 1, Rudi Denys, Eds., Elsevier Science B.V., 2000.

7 Wang, Y.-Y., "Development and Application of High-Strength and High-Toughness Line Pipes," *SATEC '01*, Beijing, China, 2001.

8 Walsh, W., Ehrhardt, B., Nast, S., Kompotiatis, L., and Huffman, T., "Pipe Yield Strength Measurement Using

Copyright © 2019 by ASME

Flattened Straps," *Materials Science and Technology (M &T 2015)*, October 4-8, 2015, Columbus, Ohio, USA.

9 Collins, L. and Rashid, M., "Best Practices for Yield Stress Determination Using the Flattened Strap Tensile Test," *Proceedings of the 10th International Pipeline Conference*, Paper No. IPC2014-33684, September 9-October 3, 2014, Calgary, Alberta, Canada.

10 Gould, M., Amend, W., Johnson, D., Rutherford, L, and Rapp, S., "Tensile Testing of Flattened-Strap Pipe Specimens," *28th International Pipeline Pigging & Integrity Management Conference*, February 8-11, 2016, Houston, Texas, USA.

11 Wang, Y.-Y., Chen, Y., and Salama, M., "Enhancing Tensile Strain Capacity through the Optimization of Weld Profiles," Proceedings of the 10th International Pipeline Conference, Paper No. IPC2014-33589, September 9-October 3, 2014, Calgary, Alberta, Canada

Copyright © 2019 by ASME

SECTION 8
IN-LINE STRAIN MONITORING

INTRODUCTION

By Jim Hart
(jim.hart@ssdinc.com)

The simplest approach for in-line monitoring of pipelines for the effects of ground movement is through the use of geometry (caliper) tools which are used primarily for the detection of localized pipe wall deformation features such as dents, wrinkles, buckles and ovality. Features of this sort are indirect indicators that the pipeline may have been affected by significant ground movement i.e., by the time pipe wall deformations of this sort have developed, the ground movement has most likely already resulted in significant bending and axial strains in the pipeline.

The most widely used and accepted ILI technology for measurement of pipeline bending strains are ILI tools equipped with high resolution (tactical grade) Inertial Measurement Units (IMUs) in a strapdown configuration on a wheel or cup-supported ILI tool body. The IMU data can be integrated between known tie-in coordinate locations to provide pipeline plan and elevation (mapping) profiles. The IMU data can also be differentiated to identify horizontally and vertically oriented pipeline curvature and bending strain features including those due to intentional bends (such as cold bends and elbows), those due to construction (e.g., pipe roping at tie-in locations or road crossings) and those due to transverse ground movement. High-resolution pipeline IMU surveys are being undertaken more frequently by pipeline operators in recent years because they provide an effective means of monitoring pipelines in geohazard areas (where pipelines may be exposed to hazards such landslides, lateral spreading, earthquake fault movements, etc.) and in arctic/permafrost environments (where pipelines may be exposed to ground movements such as frost heave and thaw settlement). Pipeline inertial survey data provides a reliable and repeatable basis for identifying locations in pipelines that have been affected by even low levels of transverse ground movement, for characterizing the attributes of the pipe bending deformation, and for prioritizing intervention locations. A major limitation of pipeline IMU surveys is that they cannot detect pure axial strains across the pipe cross section such as those caused by pipeline internal pressure or temperature differential or more importantly, ground movements that act parallel to the pipe axis.

A relatively new and emerging ILI technology is an axial strain tool which is based on magneto restriction principals. This tool is used to measure or infer the total elastic longitudinal strain including strains due to manufacturing, cold bending, backfilling and ground-movement induced deformation.

Under ideal conditions, all of the above ILI technologies (caliper, IMU and axial strain) could be combined into a single ILI tool to develop a clear picture of the current state of a pipeline including pipe wall deformations, curvature/bending strain signatures and axial strain profiles. Stand-alone or combined ILI technologies used to identify potential pipeline deformation features together with geotechnical and geological data (such as LiDAR) provides a multi-disciplinary approach for ranking and prioritizing geohazards. Many pipeline operators have adopted multi-tier, relative risk-based procedures for screening geohazards for implementation into the operator's pipeline integrity management program.

A brief overview of the papers presented in this Chapter is as follows:

The paper by Hart, Czyz, and Zulfiqar (AIMPIMG2019-1009) provides a detailed overview of pipeline inertial surveying, how the IMU data is processed including pipe position and bending strain calculations and their accuracies and describes how bending strain locations are identified in inertial data. It also includes recommendations for tables and plots to characterize bending strain features and recommended IMU survey deliverables and concludes with a range of possible responses to identified bending strain features.

The paper by Theriault, Hart, Mckenzie-Johnson, and Paulsen (AIMPIMG2019-1003) provides a discussion of how a bending strain data from database of over 4,600 IMU-identified bending strain features from single-run IMU surveys was utilized to examine relationships between total, vertical and horizontal bending strain at know geohazard location. The main finding from the work presented in this paper was that higher magnitudes of the horizontal component of bending strain are particularly likely to correlate with geohazard locations, with the correlation increasing from about 50% at 0.14% horizontal strain magnitude to 100% at 0.36% horizontal strain magnitude.

This paper by Goldfine, Dunford and Washabaugh (AIMPIMG2019-1001) addresses the ability to image and monitor strain (or stress), both residual and applied, in pipelines and other carbon steel structures such as risers and vessels. Strain (or stress) monitoring is needed to assess pipeline integrity after events such as ground motion, operating excesses, and prolonged operation in extreme environments. Three techniques are described: (1) an in-line-inspection (ILI) tool, (2) an

Copyright © 2019 by ASME

in-ditch or above ground imaging method, and (3) a permanently installed approach. Each of these methods uses MWM sensors and MWM-Arrays to make non-contact measurements of the steel's magnetic properties, which are then correlated to strain (or stress) in the material. Since the method uses magnetic fields, it can operate through non-conductive materials such as coatings and insulation.

The paper by Dawson, ElSeify and Murray (AIMPIMG2019-1021) discusses the benefits of integrating the assessment of pipelines based on both IMU bending strain and axial strain data streams. Utilizing both strain measurement technologies, AXISS and IMU, allows for a better understanding of the total axial strain condition in a pipeline, including: elimination of uncertainties when identifying pipeline strain induced by geological instabilities vs. pipeline construction practices; deviation between axial strain and IMU derived strain can be used to identify the presence of possible plastic or residual strain conditions; and assessment of the bi-axial stress/strain state of a pipeline section to determine severity and remediation needs.

Chapter 30 Review of Pipeline Inertial Surveying for Ground Movement-Induced Deformations (AIMPIMG2019-1009)
James D. Hart, SSD, Inc., USA; Jaroslaw A. Czyz, PipeNav Consulting, Ltd., Canada; and Nasir Zulfiqar, SSD, Inc, USA

Chapter 31 Correlation of Single-Run ILI IMU Bending Strain Features to Geohazard Locations (AIMPIMG2019-1003)
Bailey Theriault, Golder Associates; James D. Hart, SSD, Inc.; Alexander Mckenzie-Johnson, Consultant, and Sean Paulsen, Golder Associates, USA

Chapter 32 ILI, In-Ditch and Permanently Installed Tools for Stress/Strain Imaging and Monitoring (AIMPIMG2019-1001)
Neil J Goldfine, Todd M Dunford, Andrew P Washabaugh, JENTEK Sensors, Inc., USA

Chapter 33 Using ILI to Evaluate Pipeline Stain Conditions (AIMPIMG2019-1021)
Jane Dawson, Baker Hughes, a GE Company, UK; Mohamed ElSeify, Baker Hughes, a GE Company, Canada, and Ian Murray, Baker Hughes, a GE Company, UK

Copyright © 2019 by ASME

AIM-PIMG2019-1009

Chapter 30

REVIEW OF PIPELINE INERTIAL SURVEYING FOR GROUND MOVEMENT-INDUCED DEFORMATIONS

James D. Hart
SSD, Inc.
Reno, NV, USA

Jaroslaw A. Czyz
PipeNav Consulting, Ltd.
Calgary, Alberta, Canada

Nasir Zulfiqar
SSD, Inc.
Reno, NV, USA

ABSTRACT

This paper provides a state of practice review on the subject of in-line inspection (ILI) of pipelines using strapdown Inertial Measurement Unit (IMU) tools. The paper describes the IMU tool instrumentation (gyroscopes, accelerometers and odometers) and discusses processing of IMU data including basic processing of the raw data and the fundamentals of numerical integration of the gyroscope and odometer data into pipe position or out-of-straightness (OOS) coordinates and numerical differentiation of the gyroscope and odometer data into profiles of vertical, horizontal and total curvature/bending strain. A summary of typical IMU tool sensitivity and accuracy specifications is provided together with some commentary on pipe position accuracy and bending strain accuracy. The paper explains in detail how ground movement-induced or construction imperfection bending strain signatures are distinguished from curvature signatures corresponding to intentional bends in the pipeline (such as cold bends or elbow fittings). The paper also provides a detailed description of ways to best characterize pipeline bending strain features including tabulation of key parameters and spatial plots which graphically display pipeline geometry profiles at features of interest including plan and elevation profile (or out-of-straightness profiles), pipeline orientation (pitch and azimuth) profiles and profiles of vertical, horizontal, and total bending strain. The paper also outlines a framework for prioritization and integrity screening of identified bending strain features, possible responses to these features and a strategy for ongoing inertial survey monitoring of identified bending strain locations.

NOMENCLATURE

θ_{P2P}	Vertical angular misalignment between the IMU forward axis and pipe centerline axis
γ_{P2P}	Horizontal angular misalignment between the IMU forward axis and pipe centerline axis
θ_{IMU}	Raw pitch angle from the IMU
γ_{IMU}	Raw azimuth angle from the IMU
θ	Pipe centerline pitch angle
γ	Pipe centerline azimuth angle
ΔS	Increment of along the pipe distance between odometer data points
ΔN	Increment of northing between odometer data points
ΔE	Increment of easting between odometer data points
ΔH	Increment of elevation between odometer data points
L_{gage}	Gage length over which curvature is calculated
L_{pig}	Supported length of tool canister containing the IMU
Ψ_V	Vertical curvature
Ψ_H	Horizontal curvature
Ψ_R	Total (resultant) curvature
D	Outside diameter of pipe
ε_V	Vertical bending strain
ε_H	Horizontal bending strain
ε_R	Total (resultant) bending strain

1. OVERVIEW OF PIPELINE INERTIAL SURVEYING

Pipelines can be monitored for evidence of post-construction, ground movement-induced deformation signatures using ILI tools equipped with an IMU in a strapdown configuration. Several ILI vendors now provide high resolution inertial surveying services deploying a tactical grade IMU within a wheel or cup-supported tool cannister which travels with the pipe flow at optimal velocities in the range of about 1 to 4 m/s. The IMU includes two orthogonal sensor triads namely tri-axial accelerometers to measure linear motion and triaxial fiber optic gyroscopes to measure angular motion. The accelerometers and gyroscopes are aligned along common axes and are rigidly attached (i.e., "strapped down") to the tool with one axis aligned along the tool body axis running nominally parallel to the pipe centerline axis. Inertial survey tools also include odometers to measure pipeline chainage distance and instantaneous tool velocity. IMU data is typically sampled at a rate of at least 100 Hz.

Copyright © 2019 by ASME

The accelerometers in commonly used IMUs are solid-state Micro Electro-Mechanical Systems (MEMS) which output a velocity increment for a constant time increment Δt. The gyroscopes in commonly used IMUs are Fiber Optic Gyroscopes (FOGs) which output an angular increment for a constant time increment Δt. In recent years, MEMS based gyroscopes have also begun to be employed in ILI inertial surveys, particularly on small diameter tools due to the low power consumption and small size of these IMUs. The FOG-based IMUs commonly used in ILI surveys have a diameter of 3 inches so they can't fit into tools for inspections of pipelines of NPS smaller than 6. The performance of both the accelerometers and gyroscopes is characterized by their bias which is the instrument output in the absence of input stimulus (tactical grade IMUs have gyro and accelerometer biases of ≈1°/hour and ≈1 milli-g, respectively). Many IMU survey tools are deployed in combination tools which include magnetic flux leakage (MFL) and/or mechanical caliper instrumentation. Figure 1 presents an image of the Geopig inertial survey tool which includes spring-loaded caliper arms to measure the profile of the pipe cross-section.

FIGURE 1: Inertial Survey Tool [1]

For large diameter pipelines (typically ≥ NPS24), the inertial survey tool consists of a single supported canister body while for small diameter pipelines (typically < NPS24), the inertial survey tool is essentially a "train" of several supported canister bodies. Each canister is supported on a pair of flexible neoprene cups or support wheels which means that the orientation of the canister containing the IMU is always measured over a finite length – this length is typically in the range between 1 and 1.5 pipe diameters. The supported length of the canister containing the IMU is referred to herein as L_{pig} to acknowledge that the gyroscope-established orientation of the tool is not spatially instantaneous because it is always measured over this finite length.

2. PROCESSING OF INERTIAL DATA

The coordinate systems used in strapdown inertial navigation of pipelines are: the inertial frame, the local level frame and the body frame. For practical purpose the inertial frame can be considered to be a stationary frame in the solar system. The local level frame, shown on Figure 2, is the coordinate system with the origin moving on the earth surface together with the IMU, having three orthogonal axes pointing to the geodetic direction of east, north and zenith. The body frame is the coordinate system aligned with the IMU axes, with the selected roll axis pointing close to the direction of the movement of the cylindrical pig body along the pipeline. The earth-centered, earthed-fixed (ECEF) frame is also shown in Figure 2.

The gyroscopes measure the rate of rotation of the tool with respect to an inertial reference frame, including the earth rotation, the rotation of the local level frame with respect to the earth, and the rotation of the IMU with respect to the local level frame. The accelerometers measure both the gravity vector and the acceleration of the inertial system with respect to the inertial reference frame, including the tool vibrations, forward acceleration, the centripetal acceleration in bends and the Coriolis force. When the inertial system is stationary with respect to the earth, the accelerometers measure the gravity vector and the gyroscopes measure the earth rotation vector. This allows for determination of the initial orientation of the body frame with respect to the local level reference frame, i.e. pitch, yaw, and roll. The orientation of the tool during the survey is obtained primarily by integration of the angle rate gyroscope readings after subtraction of the gyro biases, the earth rotation and the rotation of the local level frame with respect to the earth.

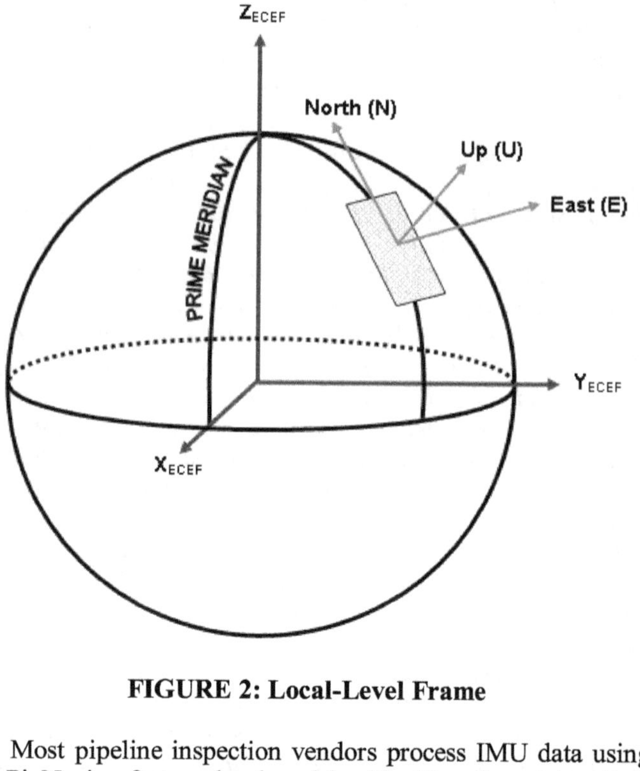

FIGURE 2: Local-Level Frame

Most pipeline inspection vendors process IMU data using the PigNavi software developed by PipeNav Consulting, Ltd. The software utilizes proprietary algorithms for calculation of IMU pitch and azimuth along the tool trajectory based on the gyroscope, accelerometer and odometer data as well as GPS data at selected points along the pipeline, typically taken at above ground marker (AGM) locations and pipeline installations, such as valves, tees and other features. The algorithms compensate for automatically calculated gyroscope and accelerometer biases as well as for gravitational acceleration and earth rotation effects and output the integrated navigation solution including the time,

Copyright © 2019 by ASME

distance, northing, easting, height, latitude, longitude, pitch, roll, azimuth, bending strain (horizontal, vertical, total) and tool velocity at a fixed distance or time interval.

As shown in Fig. 3, the pitch angle θ describes the inclination angle of the pig in the vertical plane (nose up is positive, nose down is negative) while the azimuth (or yaw) angle γ describes the angle between the pig travel direction/heading and north (azimuth is clockwise positive, yaw is clockwise negative).

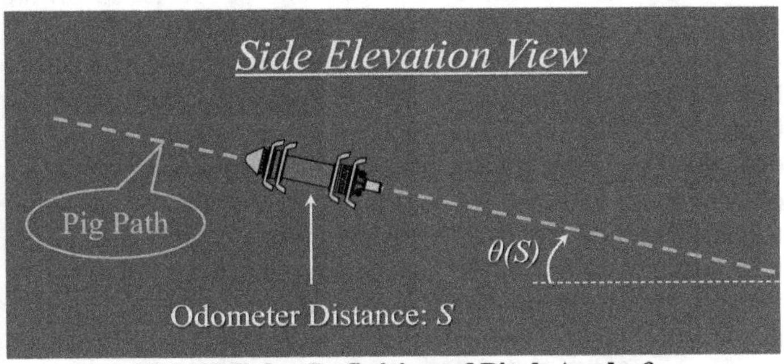

FIGURE 3a: Definition of Pitch Angle θ

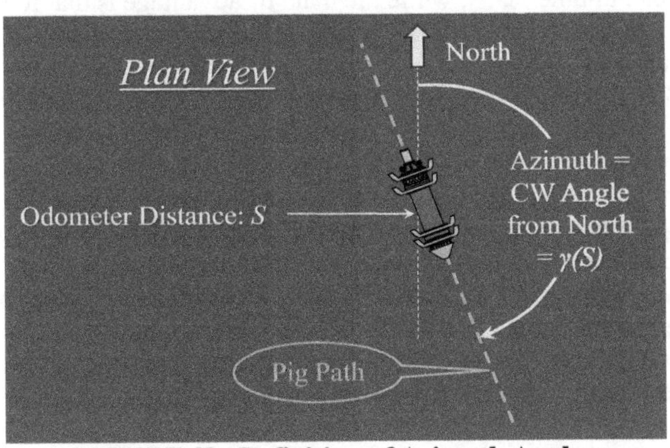

FIGURE 3b: Definition of Azimuth Angle γ

Note that the IMU forward axis usually is not perfectly aligned with the pipeline axis and there may be some angular misalignment between them in both the vertical plane (θ_{P2P}) and the horizontal plane (γ_{P2P}), therefore the pipe centerline pitch θ and azimuth γ are calculated based on the raw pitch (θ_{IMU}) and azimuth (γ_{IMU}) of the IMU as follows:

$$\theta = \theta_{IMU} + \theta_{P2P} \quad (1)$$
$$\gamma = \gamma_{IMU} + \gamma_{P2P} \quad (2)$$

The misalignment is commonly referred to as pig-to-pipe attitude, although a more precise name would be the IMU-to-pipe attitude as the IMU itself may not be perfectly aligned with the tool axis. This misalignment changes along the run, particularly with the tool roll angle, and is calculated based on several methods using proprietary algorithms.

2.1 POSITION CALCULATIONS

The pipeline position is calculated using numerical integration. Euler integration is a straight-forward and reliable integration method wherein the relative three-dimensional position of a profile point "k+1" is computed based on the pitch and azimuth (i.e., the slope of the profile) at a closely spaced adjacent point "k". In between the known coordinate tie-points, the incremental position changes of the pipe centerline path are computed using Euler integration of the pipeline centerline pitch and azimuth angles with the odometer distance data (S) as follows:

$$\Delta N = \Delta S \cdot cos\theta \cdot cos\gamma \quad (3)$$
$$\Delta E = \Delta S \cdot cos\theta \cdot sin\gamma \quad (4)$$
$$\Delta H = \Delta S \cdot sin\theta \quad (5)$$

where ΔN, ΔE and ΔH, are the increments of northing (N), easting (E) and elevation (H) along the pipe profile, ΔS is the along-the-pipe distance increment between odometer data points "k" and "k+1", and θ and γ are the pipe centerline pitch and azimuth angles corresponding to odometer data point "k" (where the integration direction is from "k" to "k+1"). The calculated centerline coordinate path is translated and rotated such that the starting and ending coordinates pass exactly through the coordinates at the known tie-points. The tie-points are typically established by GPS positioning at traps, valves, bends or wall thickness changes that are easily identified by the pig. Moreover, above ground markers (AGMs) are typically placed at the tie-point locations so that they can be easily identified in the field. Typical tie-point intervals range from roughly 1,500 to 4,000 m.

Out-of-Straightness (OOS) plots are often used to display the relative horizontal and vertical position profiles as a function of chainage over a region of interest. OOS plots present the relative position profiles as deviations from a straight line with zero OOS at the ends of the plot. OOS plots can be obtained by integrating locally between additional control points selected at the upstream and downstream ends of the feature of interest and performing a coordinate transformation from the northing-easting-elevation coordinates into the horizontal and vertical OOS deviations from a straight line connecting the feature end points.

2.2 POSITION ACCURACY

The specified accuracy of the pipe centerline position profile is typically 1/2,000[th] of the distance from the nearest tie-point. For a distance between tie points of say 4,000 m, the position error midway between the tie points for a single survey can reach ±1 m (i.e., 4,000/2/2,000), which corresponds to up to 2 m difference between two surveys (neglecting inaccuracies in the tie point coordinates). For tie point coordinates established using GPS, it should be noted that the elevation coordinate can be 2 to 3 times less accurate than the corresponding horizontal coordinates – this effect should be considered when the depth of the pipeline is important (e.g., depth of cover determination, or depth of pipe below crossings, etc.).

Copyright © 2019 by ASME

2.3 BENDING STRAIN CALCULATIONS

As discussed in Reference [2], the pipeline centerline pitch (θ) and azimuth (γ) angle data can be numerically differentiated with respect to the odometer distance data (S) to develop profiles of the pipe vertical and horizontal curvature.

As illustrated schematically in Figure 4, the vertical curvature (Ψ_V) and horizontal curvature (Ψ_H) are calculated as the change in pitch and azimuth angles ($\Delta\theta$ and $\Delta\gamma$), respectively, over a specified change in pipe distance ($\Delta S=L_{gage}$) along the path:

$$\Psi_V = \Delta\theta / L_{gage} \tag{6}$$
$$\Psi_H = -\Delta\gamma / L_{gage} \cdot \cos\theta \tag{7}$$

The total (resultant) curvature (Ψ_R) is:

$$\Psi_R = \sqrt{\Psi_V^2 + \Psi_H^2} \tag{8}$$

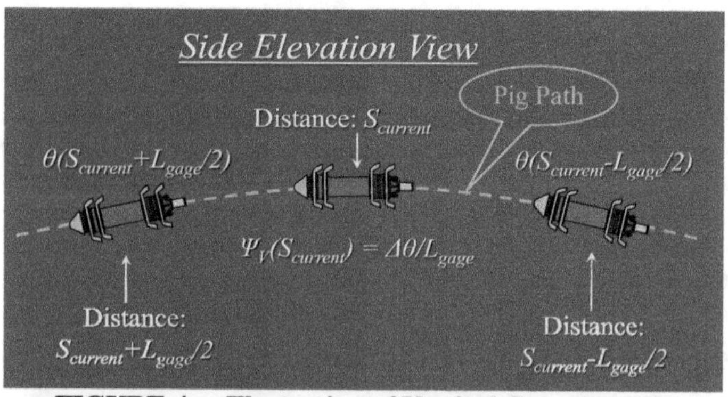

FIGURE 4a: Illustration of Vertical Curvature Ψ_V

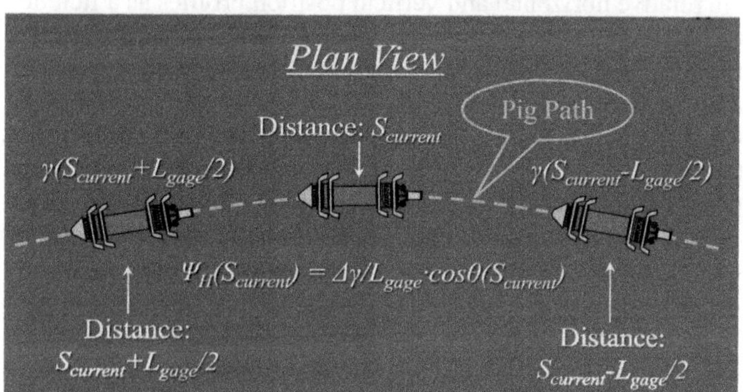

FIGURE 4b: Illustration of Horizontal Curvature Ψ_H

The vertical, horizontal and total bending strains are directly related to the corresponding curvature terms based on the assumption that plane sections remain plane as follows:

$$\varepsilon_V = \Psi_V \cdot D/2 \tag{9}$$
$$\varepsilon_H = \Psi_H \cdot D/2 \tag{10}$$
$$\varepsilon_R = \Psi_R \cdot D/2 \tag{11}$$

where D is the outside diameter of the pipe. In order to reduce noise (e.g., due to tool dynamics and minor imperfections along the pipeline) in the curvature signal, it is typically computed over lengths of 3 to 6 pipe diameters. The distance over which the curvature is calculated is referred to herein as the gage length L_{gage}. The sign convention used in Equations (6) and (7) is that sag bends and left turns correspond to positive vertical and horizontal curvatures, respectively. In other words, positive vertical bending strain corresponds to tension at the 6:00 orientation on the pipe cross section and positive horizontal bending strain corresponds to tension at the 3:00 orientation on the pipe cross section (looking downstream). The sign convention for curvature and bending strain is somewhat arbitrary and different vendors may use different definitions.

The most efficient curvature calculation is based on a simple finite difference approach. The $\Delta\theta$ and $\Delta\gamma$ numerators in Equations (6) and (7), respectively, are computed as the difference between the pitch or azimuth at distances of $L_{gage}/2$ downstream and $L_{gage}/2$ upstream of a given odometer distance coordinate. If the upstream or downstream location (at $\pm L_{gage}/2$) falls between two discrete pitch or azimuth values, the pitch or azimuth values used in the difference calculation are obtained by linear interpolation between the adjacent values. The main advantage of the finite difference approach is that it is very efficient numerically, while the main disadvantage is that it does not always sufficiently filter out noise that is present in the underlying data.

Significantly smoother and more accurate curvature profiles can be computed using a "regression line" method where the curvature is computed as the slope of a least-squares line fit through the profiles of θ vs. S and γ vs. S over the gage length L_{gage}. Figure 5 provides a schematic illustration of the regression line processing in θ vs. S space. At the current odometer distance coordinate, the least-squares fit lines on θ and γ as functions of S over the length L_{gage} centered on the current odometer distance coordinate are expressed (in the slope-intercept form $y=mx+b$ where m and b are the slope and intercept of the line) as follows:

$$\theta(S) = \Psi_V \cdot S + \theta_o \tag{12}$$
$$\gamma(S) = \Psi_H \cdot S + \gamma_o \tag{13}$$

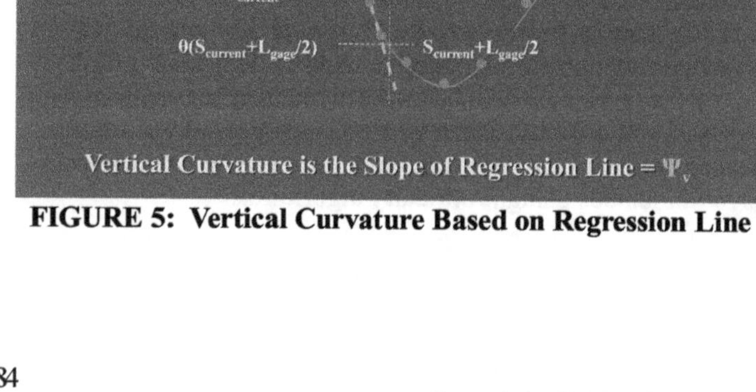

FIGURE 5: Vertical Curvature Based on Regression Line

Copyright © 2019 by ASME

The main advantages of the regression line approach are that it tends to smooth out noise that is present in the underlying data and that for a given gage length, it is more accurate than the finite difference approach. The main disadvantage of the regression line approach is that developing the regression line fit involves additional numerical (least-squares fitting) calculations that are not required for the finite difference approach.

2.4 BENDING STRAIN ACCURACY

Although some vendor reports or papers show results providing bending strain accuracies of ±0.005% to ±0.01% strain, the best available directly specified bending strain accuracy is ±0.02% strain (which is equivalent to a bend radius of ±2500*D*). The pipeline inspection industry has effectively adopted the ±0.02% bending strain accuracy specification which was originally established in the early 1990's based on the analysis of repeatability of strains from two Geopig runs [2]. This accuracy was further verified during a field test performed for SNAM in 1994 where a bending strain with an amplitude of 0.03% was induced into the pipeline between two ILI inspections, and the Geopig measured that change of strain with very high accuracy – about ±0.002% strain [3]. The same accuracy was reported in [4] for stain levels up to 0.1% based on test loop results. In the authors' opinion, the specification should be left at ±0.02% accuracy since it is not realistic to guarantee ±0.002% accuracy in all circumstances and for larger strain levels in actual pipeline inspections.

Theoretically, the bending strain measurement accuracy depends on two components: one due to the accuracy of the curvature of the calculated IMU trajectory and another due to the deviation of the IMU trajectory from the pipe centerline.

The accuracy of the curvature of the calculated IMU trajectory plays only a minor role in the bending strain accuracy outside of bends. The factors contributing to the accuracy of the curvature of the calculated IMU trajectory include gyro biases, odometer accuracy and the correction of IMU attitude using the accelerometer data. These factors have a significant effect on mapping accuracy, but not on bending strain measurements in originally straight sections of pipe. The contribution of gyro biases and odometer error to bending strain error is small because they affect the trajectory shape over long distances and have little effect on the local shape of the trajectory, which is the basis for calculation of bending strain. For this reason, the accuracy of bending strain measurements based on MEMS IMUs should be close to that obtained from FOG IMUs, although the latter produce more accurate absolute position data over longer distances due to their superior gyro bias stability which can be compensated for more accurately. Odometer error does not have a significant effect on bending strain measurements in initially straight sections of pipe; as discussed below, this effect can become significant at bends, particularly at fabricated elbows. The correction of IMU attitude using the accelerometer data won't normally result in significant bending strain error except when the correction is affected by large tool vibrations and accelerations.

The deviation of the IMU trajectory from the pipe centerline is the main cause of bending strain error. Numerous factors contribute to the deviation of the IMU trajectory from the pipe centerline including imperfections along the pipe wall within pipe joints such as expander marks as well as imperfections at girth welds such as angular or high-low misalignments. When the tool supports encounter these features, the pig-to-pipe attitude experiences short wave-length oscillations or abrupt "kicks" which are introduced in the IMU pitch and azimuth data. Fortunately, these features tend to be highly repeatable between surveys, particularly when the same tool is used for the IMU inspection. The tool's response to the short wave-length features does not represent true curvature of the pipeline centerline – it is basically "noise" in the pipe orientation data which is amplified upon differentiation. The effects of noise are reduced when longer curvature calculation gage lengths are used. When calculating bending strain in originally straight pipe, a gage length from say 1 to 5 m can be used in order to smooth out the noise. As previously described, regression line curvature calculations result in smoother curvature profiles or low-pass filtering can be used to smooth out the effects of noise.

The standard ±0.02% bending strain accuracy does not apply for intentional bends, particularly factory bends, but also field bends due to variations in run-to-run IMU trajectory when negotiating bends as well as due to local inaccuracy of the odometer measurements along such bends combined with the sensitivity of curvature calculations over short distances. The differences might be more pronounced when the tool is suspended on circumferentially distributed wheels rather than on cups. In order to measure the radius of curvature of intentional bends, a short gage length must be used such that the combined supported length of the IMU canister (L_{pig}) and the gage length (L_{gage}) doesn't exceed the bend length, however, this is impossible to do for short factory bends, for which the curvature is always underestimated. Instead of measuring change of curvature in bends between runs, particularly factory bends, it is preferable to measure the change of angle across the bend to look for evidence of pipeline movement. Using an inertial survey, the bend angle can be measured with an accuracy better than ±0.1° as long as the pig-to-pipe attitude is properly estimated. On the other hand, the corresponding change of curvature would be difficult to measure accurately. For example, a ±0.1° change of angle of a 3*D* bend would correspond to less than 0.02% change of strain, which is impossible to observe along such a tight bend. Actually, this change of strain would be even smaller if the combined length of IMU canister and the gage length ($L_{pig}+L_{gage}$) was shorter than the bend length.

Specialized software is required to produce the highest possible mapping and bending strain accuracies, particularly in case of significant tool accelerations, vibrations and pig-to-pipe attitude changes (as long as those changes are not random, but systematic and dependent only on the tool roll angle). The actual accuracy of bending strain measurements depends not only on the inertial processing software but also on the strain level and data quality, particularly the line conditions, the tool design and its ability to maintain a consistent pig-to-pipe attitude.

Copyright © 2019 by ASME

More comprehensive, statistically justified specifications could be developed by extensive analysis of repeatability of strain from multiple runs for several pipelines instead of analyzing data for a short pipeline section, probably in perfectly round pipe without significant pipe wall deformations, in a single test loop. In fact, the strain measurement accuracy is a function of strain feature magnitude and length, pig length and curvature calculation method, so these relationships could be further investigated to better establish accuracy specifications and correct interpretation of the results. In addition, the curvature of short strain features is inherently underestimated by inertial measurements due to the finite supported length of the IMU canister and the smoothing techniques employed in calculation of curvature.

3. IDENTIFICATION OF STRAIN LOCATIONS OF INTEREST BASED ON INERTIAL SURVEYS

Inertial surveys are utilized for "mapping" pipeline geometry which is computed based on numerical integration of the IMU and odometer data relative to specified tie-point coordinates. The mapped geometry can be used for locating key features and components along the pipeline alignment. However, the calculated pipe centerline coordinates are unlikely to identify areas of anomalous pipeline movement or strain since the accuracy of the position profiles is limited by the distance between tie points. In order to identify such areas, it is necessary to review the pipe orientation and curvature profiles.

For pipelines subjected to ground movement, the main interest is in screening the bending strain profiles, which are computed based on numerical differentiation of the centerline pitch and azimuth data with respect to odometer distance. In order to identify locations along the pipeline that have been deformed due to differential ground movement, the analyst must be able to distinguish pipeline geometry signatures associated with field bends and fabricated elbows (i.e., intentional bends) from pipeline geometry signatures associated with ground movement-induced transverse bending of the pipe.

All vertical and horizontal turns along a pipeline profile are recorded as "ramp-like" changes of the pipe centerline pitch and/or azimuth with odometer distance. Numerical differentiation of a ramp in the pitch or azimuth profiles will result in a lobe in the vertical or horizontal curvature/bending strain profiles. As explained below, ground movement-induced pipeline geometry signatures are distinguished from intentional pipeline bend signatures based primarily on three attributes: their pitch and azimuth signatures, their length, and their bending strain amplitude. Note that for convenience Figure 6, which will be discussed in more detail later in this paper, is used here to illustrate various pipeline geometry signatures (in this figure, the vertical geometry changes are the result of cold bends while the horizontal geometry changes are the result of ground motion-induced pipeline movement).

Intentional Bend Signatures. Intentional bends normally stand out as short angle change ramps in the pitch and/or azimuth profiles wherein the angle change always occurs within a single pipe joint between two girth welds. In the absence of ground movement-induced geometry changes or significant roping during construction, a pipeline will normally consist of a series of straight runs (flat sections in the pitch and/or azimuth profiles) and intentional bends (steep ramps in the pitch and/or azimuth profiles). This means that a section of pipeline consisting of only straight runs and intentional bends will exhibit pure "stair-stepped" pitch and azimuth profiles. Figure 6(c) illustrates the stair stepped pattern over a series of several pitch ramps corresponding to vertical cold bends. Numerical differentiation of a short ramp in the pitch and/or azimuth profiles at intentional bends results in a single, short curvature/bending strain lobe like those shown in the vertical bending strain profile in Figure 6(e). Note that these lobes *do not* have side-lobes. Depending on the bend length, the amplitude of the apparent bending strain at intentional bends tends to be relatively large, e.g., the B31.8 Code [5] minimum cold bend radii (18*D* for NPS12 and smaller and 30*D* for NPS20 and larger) correspond to apparent bending strains in the range of roughly 1.7% to 2.8%.

Ground Movement Signatures. Ground movement-induced orientation and curvature/bending strain signatures tend to develop over relatively long lengths (e.g., spanning over one or more girth welds sometimes up to several hundred feet) and are often characterized by longer wavelength features consisting of gentle/gradual ramps or one to several sinusoidal shaped "lobes" along the pipeline orientation and curvature/bending strain profiles e.g., like those shown in Figures 6(d) and 6(f). Sinusoidal orientation and curvature signatures are associated with sinusoidal ground movement shapes for example, within a long landslide where some sections of the slide will displace further downslope than adjacent sections. Ground movement-induced curvature signals associated with a fairly well-defined block or lobe of ground movement often exhibit a three-lobe "W-shape" variant of sinusoid corresponding to one dominant curvature lobe (nominally centered on the ground movement) in between two lesser side-lobes with curvature of the opposite sign e.g., as shown in Figure 6(f). Another important distinction is that ground movement-induced bending strain features are usually much less pronounced than intentional bends with smaller bending strain amplitudes with a typical "threshold of interest" in the range of 0.125% to 0.2%.

As part of a strain screening for a given inertial survey, profiles of vertical, horizontal and total bending strain will be computed for the entire length of the pipeline and the total bending strain profiles be screened for locations of interest. The inertial survey vendor has leeway in identifying or "flagging" anomalous bending strain features based on experience and/or using their standard feature screening procedures. In the absence of a vendor-defined approach for flagging anomalous bending strain features, the following approach is recommended:

- For the first inertial survey of a pipeline, a screening criterion that "flags" all locations along the pipeline with a peak total bending strain of 0.2% based on a curvature calculation gage length of 10 feet (3 m) which also have a feature signature length that is longer than one pipe joint (i.e., it extends over one or more girth welds) should be used. The screening threshold of 0.2% should be considered as an

Copyright © 2019 by ASME

adjustable parameter that can be specified for any given inertial survey (as explained later some vendors/operators use a lower threshold of 0.125%). Operators may choose to link the screening threshold to elastic stress-equivalent strain measure, e.g., a threshold strain value of 0.9·SMYS/E may be considered with reference to the B31.8 longitudinal stress limit of 90% SMYS [5].

- Secondary confirmation of the feature can be obtained by reviewing the pitch and/or azimuth profiles for the existence of long-wavelength ramps or sinusoidal features.

Based on only a single ILI inertial survey, it isn't usually possible to determine with absolute certainty if the bending strain feature was indeed the result of pipeline movement during operation or was induced during construction, e.g., by conforming the pipeline profile to an uneven profile while being laid in the trench. For pipelines that have been subjected to more than one inertial survey, the screening concepts described above should be supplemented to also look for specified survey-to-survey changes in the geometry profiles. Actually, comparing bending strain changes between two inertial inspections is a more reliable method of identifying areas of actual pipeline movement than screening a single run data for large bending strain anomalies, because a large percentage of such anomalies are introduced during construction and are not related to any pipeline movement during operation. Based on our experience with inertial survey data, a high degree of repeatability is typically observed in overlays of pipeline curvature/bending strain profiles obtained from different surveys at construction-induced bending features whereas the signals associated with real pipeline movement tend to exhibit curvature/bending strain profiles with increasing amplitudes. When running a subsequent survey with the same tool, subtle differences can be expected at girth welds and bends for example due to different tool velocities and slight differences in the pig-to-pipe attitude during the run. These differences can be more pronounced when the subsequent survey is run with a different tool, particularly if the lengths of the pig (L_{pig}) are significantly different. Various vendors have published papers that document survey repeatability and comparisons with discrete strain gage measurements (e.g., see References [3] and [4]). In order to compare the geometry data from different surveys at the same point along the pipeline, it is common to use the odometer distance at selected girth welds as a basis for aligning the two pipeline geometry profiles [6]. This is usually accomplished by scaling the odometer distance between selected girth welds from the new survey to match the odometer distance between the same selected girth welds from the old survey (or visa-versa). Note that these "odometer matching" adjustments aren't necessarily carried out for every girth weld but rather using a selected subset of girth welds that are separated by odometer distances on the order of 100 m or so. Once the odometer matching is carried out, profile plots of various geometric quantities (e.g., vertical and horizontal bending strain) can be developed directly overlaying the data from different surveys. As described in Reference [7], it is possible to calculate and plot *difference* profiles of selected geometric quantities by interpolating the data to a common odometer distance grid (e.g., typically from the survey with the finest data spacing). As a preliminary basis for screening using survey-to-survey changes, profiles of total bending strain differences (i.e., changes) with peak values in the range of ±0.04% is recommended.

As described above, the ILI vendor has leeway in identifying or "flagging" anomalous bending strain features based on experience and/or using their own feature screening procedures. As part of the screening process, it is expected that the vendor will experiment with a range of different curvature calculation lengths in order to establish and justify the gage length used to report strain features of interest on a case-by-case basis. The selected gage length should strike a balance between the tendency for short gage lengths to amplify noise in the data and the tendency for long gage lengths to underestimate curvature. Based on years of experience on pipelines ranging from NPS10 through NPS48, the authors have found that a gage length of 10 feet (or 3 m) will normally provide an acceptable balance and hence, this value is used as the "default" recommended curvature calculation gage length in the "flagging" procedure described above. The shortest gage length that would normally be considered is L_{pig}. For real (non-noise) strain features which develop over long lengths, large differences in the curvature profiles computed using gage lengths of L_{pig} vs. 10 feet would not be expected. However, the differences become more pronounced for short features such as bends, mitered welds, etc. Another consideration is that the curvature calculation gage length should be consistent with the gage length used to characterize the pipe strain capacity (e.g., from full-scale pipe tests).

References [8], [9] and [10] contain examples of bending strain and pipeline movement areas detected during single and multiple inertial surveys of pipelines susceptible to movement.

4. CHARACTERIZATION OF STRAIN LOCATIONS

As described above, a typical vendor-provided bending strain report should develop a list of apparent bending strain features of interest over the entire length of the survey based on identifying apparent post-construction movement signatures with peak total bending strains exceeding a specified threshold of interest e.g., 0.2%. Once the features are identified, they are typically characterized in some detail using tabulations and plots. Some vendors/pipeline operators utilize a lower threshold of interest e.g., 0.125% which may be beneficial for identifying pipeline movement locations associated with lower bending strain levels with the tradeoff that it is likely to flag a larger number of construction-induced features.

4.1 TABULATED ITEMS FOR FEATURES OF INTEREST

For each feature of interest, the inertial survey vendor should provide a unique feature identifier and tabulate the following information; the starting/ending odometer distance and estimated length of the feature; the plan coordinates (e.g., latitude-longitude) of the peak bending strain location, the values of the vertical, horizontal and total bending strain at the location of the peak total strain, the ID number and/or odometer distance of girth welds contained within the length of the strain feature

Copyright © 2019 by ASME

with their corresponding vertical, horizontal and total bending strains, the average tool velocity across the feature, a brief narrative description of the feature, and a listing of supplemental notes/comments (e.g., nearby corrosion, sleeves, ovality, dents, etc.). It will also be useful to include additional information from the pipeline database such as the pipe diameter, wall thickness, steel grade, MAOP, etc. Note that all references to bending strain should document the curvature calculation gage length.

4.2 PLOTS FOR FEATURES OF INTEREST

The vendor report should also provide detailed spatial plots of the feature geometry such as plan and elevation profile views, pitch and azimuth profile plots as well as along-the-pipe profile plots of vertical, horizontal and total bending strain. Plots of the vertical and horizontal out-of-straightness (OOS) profiles may be more useful than plan and elevation profile views. OOS profiles depict the vertical and horizontal distance between the pipe centerline and a straight line connecting two end points (usually at girth welds) which are located a short distance upstream and downstream of the pipeline movement feature of interest. The characterization plots should identify all girth welds using symbols and, in some cases, will denote the locations of intentional bends (e.g., using color coding).

Figure 6 provides an illustrative example of the type of multi-panel plot that should be used to characterize a pipeline strain feature. This example feature was identified by an inertial survey of a 26-inch diameter pipeline where it crosses a transverse horizontally-dominant landslide. In these plots, the pipeline geometry profiles are presented over a length of about 235 feet of relative odometer distance and the flow direction is from left to right. Figures 6(a) and 6(b) present profiles of the elevation and horizontal OOS across the feature; Figures 6(c) and 6(d) present the pipe centerline pitch and azimuth profiles, and Figures 6(e) and 6(f) present the profiles of the vertical and horizontal bending strain (for a 10-foot curvature calculation gage length). The locations of girth welds are identified by the "o" symbols. The plotted length includes a series of 7 fairly clear, intentional vertical cold bend (6 overbends, 1 sag bend) which are well-illustrated by the "stair-stepped" pattern in the pitch profile in Figure 6(c) and by the relatively sharp lobes (with no side lobes) in the vertical bending strain profile in Figure 6(e). As shown in the horizontal OOS profile in Figure 6(b), the landslide-induced pipeline movement has generated a horizontal OOS of about 3 feet with a main lobe of movement to the right (relative to the flow direction) with a peak horizontal OOS at a relative odometer distance of about 129 feet (about 4 feet downstream of a girth weld). The horizontal geometry is characterized by a clear, two-lobe sinusoidal undulation in the azimuth profile in Figure 6(d) extending over the distance range from about 65 feet to 215 feet with a total azimuth swing of ≈10°. The horizontal bending strain profile exhibits a fairly distinct "W-shape" with negative peak values of -0.78% and -0.43% at relative odometer distances of about 90 and 180 feet, respectively (at the upstream and downstream shoulders of the landslide) and a positive peak value of +0.78% at a relative odometer distance of 123 feet (near the middle of the landslide).

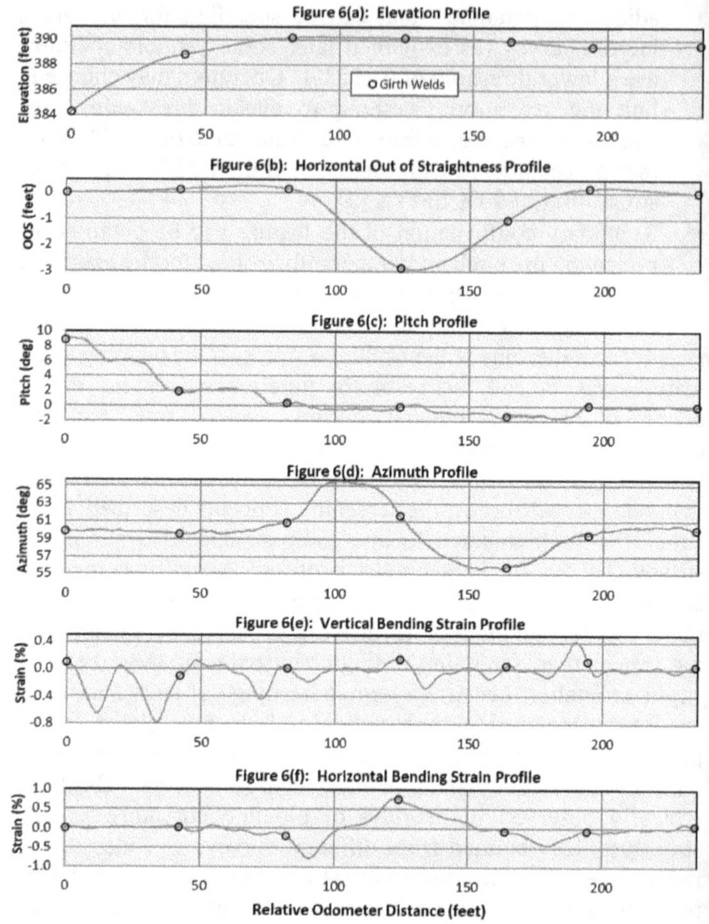

FIGURE 6: Pipeline Geometry Profiles from IMU Survey

4.3 RECOMMENDED DATA DELIVERABLES

In order to maximize the usefulness of a vendor-provided bending strain report, the authors recommend that pipeline owners/operators incorporate language into their IMU survey specification related to the report deliverables, for example to ensure that tables and plots similar to those outlined in Sections 4.1 and 4.2 above be provided for each bending strain feature. Additionally, it is recommended that the pipeline operators' IMU survey specification require that the vendor be prepared to provide processed digital data table exports for a selected subset of the strain features of interest or for the entire length of the survey based on an initial review of the bending strain report. The following processed inertial data should be exported at a spacing of 4 to 8 inches along the pipeline over a specified distance encompassing the length of the feature(s) of interest: odometer distance, northing, easting, elevation, azimuth, pitch, roll and tool velocity. A separate data table containing a listing of the girth welds and their odometer distance should also be provided. This additional data could be used to perform more detailed calculations and review of the selected features including gage length sensitivity studies, survey-to-survey comparisons, or other specialized processing steps (see Section 5 below).

Copyright © 2019 by ASME

Independent verification of the accuracy of the vendor-processed IMU data would require the raw data from the IMU together and the GPS tie-point coordinates as additional deliverables. The raw IMU data should consist of: distance, time, gyroscope (X, Y, Z), accelerometer (X, Y, Z) where X, Y, Z are the IMU axes defined in a right-handed coordinate system. The large volume of data collected at a sampling frequency above 100 Hz typically requires that the raw IMU data be split into several files, each below 100 MB. The GPS tie-point data should contain the following information: IMU passage time, longitude, latitude (or northing, easting) and elevation.

5. SPECIALIZED PROCESSING

The thrust of the reporting guidelines discussed herein is aimed at basic presentation of the pipeline geometry information needed to characterize strain locations of interest. However, there are certain situations where more advanced/specialized inertial data processing methods and techniques may be required. Examples of detailed processing applications that are not discussed in detail herein include the following:

(1) *Gage Length.* Gage length effects should be evaluated on a survey-by-survey basis in order to find a length that is long enough to minimize the effects of underlying noise in the data (which is amplified by differentiation) but not so long that it underestimates true curvature of a given feature. For the plots of bending strain vs. odometer distance, it may be useful to overlay the bending strain profiles computed using two different calculation gage lengths (e.g., a 10-foot (3 m) curvature calculation gage length and a gage length equal to the supported length of the pig canister containing the IMU). Plots of this sort can be used to delineate the beginning and ending locations of pipeline bends which may be useful for estimating strains at girth welds located in close proximity to a bend or in between two bends. Another consideration is that the curvature calculation gage length should be consistent with the gage length used to characterize the pipe strain capacity (e.g., from full-scale pipe tests).

(2) *Curvature Calculation Method.* Evaluation of curvature calculation method e.g., comparison of finite difference method vs. regression line method (or other methods). Different methods can have somewhat different "influence lengths" for the same curvature calculation gage length.

(3) *Noise.* Minor oscillations and vibration of the inertial survey tool is inevitable during a survey as the support wheels or cups encounter imperfections in the pipe wall, kicks at girth welds, and other features. These motions, which are referred to as noise, are exhibited in the pitch and azimuth profiles as short wave length features, much shorter than the wavelengths associated with real ground/pipe movement. The noise in the orientation signals tends to be amplified by numerical differentiation to obtain pipe curvature. As discussed in References [1], [7] and [11], it is possible to low-pass filter the pitch and azimuth data to rigorously remove this short wave length noise from the signal. The filter parameters must be carefully selected so that real features are not filtered out of the data.

(4) *Digital Pigging.* As described in Reference [11], digital pigging is the process of operating on a postulated (or measured) pipeline position coordinate profile in order to obtain a corresponding profile of orientation over a specified pig length (e.g., pigged pitch and/or azimuth) and the corresponding curvature profile computed for a specified gage length. In some situations, an improved understanding of a real pipeline geometry feature can be obtained using digital pigging techniques wherein a purely analytical geometry is postulated and digitally pigged and the results are compared to the measured profile.

6. PRIORITIZATION AND INTEGRITY SCREENING OF BENDING STRAIN FEATURES

Once the pipeline bending strain "demand" is established and characterized for a list of features, pipeline structural integrity assessments and/or screening and prioritization of the identified bending strain locations can be undertaken. It is important to understand that although pipeline inertial surveys are effective in capturing pipeline bending deformations, they do not capture the effects of axial strains which are uniform across the pipe cross-section such as those that would be induced by ground movements parallel to the axis of the pipeline or due to the restraint of the pipe's thermal expansion/contraction and/or due to restraint of Poisson contraction due to internal pressure. The simplest approach is to prioritize/rank the strain features based only on the bending strain demand with possible supplementary adjustments to include uniform axial strains on the pipe cross-section due to restraint of the pipe's thermal expansion/contraction or due to restraint of Poisson contraction possibly considering location-specific pipe operating pressures and temperature differentials. A more refined approach would be to prioritize/rank the strain features based on the ratio of the pipeline strain demand (including operating strains) to estimates of the pipe strain capacity i.e., using a strain demand/capacity ratio. This approach provides a framework that accounts for higher strain capacity at locations with increased pipe wall thickness and/or higher strength pipe, etc. The pipe strain capacity limits that are typically considered include the "critical" compressive strain capacity associated with wrinkling/local buckling of the pipe (a serviceability limit state) and the tension strain capacity at girth welds (an ultimate limit state). Many pipeline operators utilize a multi-level risk screening approach for example by applying low, medium and high resistance factors to the strain capacity estimates in order to provide different serviceability and safety margins.

7. POSSIBLE RESPONSES TO IDENTIFIED BENDING STRAIN FEATURES

Once a bending strain feature has been characterized and ranked, a range of possible response actions or interventions may be considered depending on the severity of the feature and it's ranking with respect to other bending strain features identified by the survey. Possible responses include:

- Review geological/geotechnical evidence at the location

Copyright © 2019 by ASME

- Review aerial reconnaissance and/or LiDAR for evidence of ground movement
- Field monitor the pipeline alignment and/or ground movement geometry for ongoing changes
- Use the inertial data in a buried pipe deformation analysis (i.e., finite element analysis) of the strain feature (considering the soil properties and cover depth) in order to more rigorously include axial force/uniform strain effects in the strain demand
- Review caliper data in/near the high strain zone for evidence of wrinkle formation, ripples or other anomalous wave forms that may be associated with pre-wrinkling or post-wrinkling conditions
- Review other sources (e.g., MFL data) for evidence of corrosion, gouges, girth weld high-low, etc. which could influence the pipe strain capacity
- Excavate the pipeline to inspect for evidence of local buckle formation or crack formation at/near the high strain zone
- Relieve stresses via soil removal allowing elastic spring back of the pipe
- Replace the affected section of the pipe
- Continue monitoring with periodic inertial surveys.

8. ONGOING INERTIAL SURVEY MONITORING OF BENDING STRAIN FEATURES

For locations that will be subject to ongoing periodic inertial survey monitoring, the bending strain feature identification and screening process can be supplemented to also look for specified survey-to-survey changes in the pipeline geometry profiles. As previously noted, in order to compare the geometry data from different inertial surveys at the same point along the pipeline, it is usually necessary to use the odometer matching at a subset of girth welds in order to align the surveys. Once the odometer matching is carried out, profile plots of various aligned geometric quantities (e.g., vertical and horizontal bending strain) can be developed by directly overlaying the data from different surveys. It is also possible to calculate and plot difference profiles of selected geometric quantities by interpolating the data sets to a common odometer distance grid (e.g., typically from the survey with the finest data spacing). As a preliminary basis for screening using survey-to-survey changes, profiles of bending strain differences (i.e., changes) with peak values in the range of ±0.04% may be considered, or even lower if a distinct strain change signature corresponding to movement can be identified. In cases where different inertial survey vendors or tools are used to perform the different surveys, it is imperative that both the same gage length and curvature calculation method be utilized in order to ensure an "apples-to-apples" comparison of the bending strain data sets.

9. SUMMARY

High-resolution pipeline inertial surveys are being undertaken more frequently by pipeline operators in recent years because they provide an extremely effective means of monitoring pipelines in geohazard areas (where pipelines may be exposed to hazards such landslides, lateral spreading, earthquake fault movements, etc.) and in arctic/permafrost environments (where pipelines may be exposed to ground movements such as frost heave and thaw settlement). Pipeline inertial survey data provides a very powerful basis for identifying locations in pipelines that have been affected by ground movement, for characterizing the attributes of the pipe bending deformation, and for prioritizing intervention locations. Many pipeline operators have adopted multi-level, risk-based procedures for ranking pipe bending strain locations which are also cross-referenced with respect to geotechnical/geological rankings to provide a multi-disciplinary screening approach that can be integrated into the operator's pipeline integrity management program.

Copyright © 2019 by ASME

REFERENCES

[1] Hart, J.D., Powell, G. H., Hackney, D., and Zulfiqar, N., "*Geometry Monitoring of the Trans-Alaska Pipeline*", Proceedings of the ASCE Cold Regions Conference, Anchorage, Alaska, May 20-22, 2002.

[2] Czyz, J.A. and Adams, J.R., "*Computation of Pipeline Bending Strains Based on Geopig Measurements*", Pipeline Pigging and Integrity Monitoring Conference, Houston, February 1994.

[3] Czyz, J.A., Fraccaroli, C. and Sergeant, A.P., "*Measuring Pipeline Movement in Geotechnically Unstable Areas Using an Inertial Geometry Pipeline Inspection Pig*", ASME 1-st International Pipeline Conference, Calgary, June 1996.

[4] Aue, H., Paeper, S., Brown, B., Humphreys, M. and Southerland, J., "*High Quality Geometry Module Data for Pipeline Strain Analysis*", Pigging Products and Services Association, 2007.

[5] American Society of Mechanical Engineers, B31.8 Gas Transmission and Distribution Piping Systems, ASME B31.8-2016.

[6] Clouston, S., Blair, G. and Hektner, D., "*Pipeline Out-of-Straightness Assessment Using Pipeline Inertial Geometry Survey Technology*", Alaska Pipeline Workshop, Anchorage, Alaska, November 8-9, 1999.

[7] Hart, J.D., Zulfiqar, N., and Erickson, T.H., "*3rd Party Review of Geometry Pig Inertial Survey Data at the Colville River HDD*", IPC08-64306, Proceedings of the International Pipeline Conference, Calgary, Alberta, Canada, September 29 – October 3, 2008.

[8] Czyz J. A., Wainselboim S., "*Monitoring Pipeline Movement and Its Effect on Pipe Integrity Using Inertial and Caliper In-line Inspection*", Rio Pipeline Conference and Exhibition 2003, Rio de Janeiro, Oct. 2003

[9] Czyz J. A, McClarty E., "*Prevention of Pipeline Failures in Geotechnically Unstable Areas by Monitoring with Inertial and Caliper In-line Inspection*", NACE Corrosion/2004 Conference, New Orleans, Mar. 2004

[10] Murray, Ian, et al. "*Pipeline Bending Strain Assessment using ILI Tools: Case Studies*", Pipeline Technology Conference 2015.

[11] Hart, J.D., Zulfiqar, N., Moore, D.H., and Swank, G.R., "*Digital Pigging as a Basis for Improved Pipeline Structural Integrity Evaluations*", IPC06-10349, Proceedings of the International Pipeline Conference, Calgary, Alberta, Canada, September 25-29, 2006.

Copyright © 2019 by ASME

AIM-PIMG2019-1003

Chapter 31

CORRELATION OF SINGLE-RUN ILI IMU BENDING STRAIN FEATURES TO GEOHAZARD LOCATIONS

Bailey Theriault
Golder Associates
Manchester, NH, USA

James D. Hart
SSD, Inc.
Reno, NV, USA

Alexander McKenzie-Johnson
Houston, TX, USA

Sean Paulsen
Golder Associates
Detroit, MI, USA

ABSTRACT

In-line-inspection (ILI) inertial measurement unit (IMU) data have been used to identify sections of pipelines bent or deflected by an outside force (e.g., a geohazard) for over 25 years, and particularly since the recent widespread adoption of the use of the IMU tool to more accurately map pipeline centerlines. In principle, the concept is simple: the ILI vendor reviews the IMU data and excludes construction related bends (e.g., sag bends, over bends, side bends). Bends that extend across one or more girth welds are inferred not to have been deliberately constructed and are reported by the vendor as "bending strain features," along with the magnitude of the bending strain and other attributes. Bending strain features exceeding a specified threshold magnitude are inferred to have formed post-construction (e.g., from an outside force such as a geohazard). As such, commissioning a bending strain report can be an attractive option to operators because the usage of an IMU has increasingly become a standard part of ILI data collections; thus, the additional cost of the bending strain report is low.

In practice, the use of IMU bending strain data is not nearly so simple. As multiple operators have experienced, the list of bending strain features provided by the ILI vendors often does not correlate to any apparent anomaly when excavated, nor any apparent geohazard. Absent context and supporting information, IMU bending strain data can appear to be essentially meaningless with little value added to real-World situations.

The goal of this paper is to provide context and examples for how IMU bending strain data have been used to successfully identify real-World impacts to pipelines from geohazards. The authors have access to a database containing over 4,600 IMU-identified bending features located in a variety of geographic settings, including areas with high densities of landslides. As described in the paper, the majority of reported bending strain features (over 90%) do not appear to correlate to a likely geohazard. Instead, most reported bending strain features are relatively low-magnitude (less than 0.2% strain) and are inferred to correlate with likely tie-in locations, such as at pipe replacement locations, road crossings, and river crossings, where low levels of bending may have been induced during construction. Although the number of bending strain features that correspond with a likely geohazard is a small percentage of the total, the authors have found statistical correlation between likely geohazard locations and relatively high magnitude bending strain features, as well as a correlation between geohazards and bending strain where the primary or significant component of strain is in the horizontal direction.

If IMU bending strain data can be put into context (e.g., overlain with known geohazard locations), the results can be used to aid in the prioritization of response, such as monitoring or mitigation of sites. Alternatively, through a close review of the IMU bending strain data and their locations, previously unidentified geohazards may be identified (e.g., the IMU data can be used to prioritize where to look for possible geohazards).

1 INTRODUCTION

The usage of inertial measurement unit (IMU) results to identify areas of bending strain is an established process offered by many in-line inspection (ILI) vendors. The purpose of a bending strain analysis is to identify sections of pipeline bent or deflected by an external force, such as a geohazard (e.g., landslide) or from a 3rd party (such as anchor drag at water crossings). It is typically assumed that bending strain features imposed by an external force formed or were induced post-construction. Collection of IMU data to map pipe centerlines is a standard service for most ILI runs and the additional costs for performance of bending strain analysis are relatively low,

Copyright © 2019 by ASME

making it an attractive option to evaluate areas where effects from an external force(s) are known or suspected.

In practice, the interpretation of bending strain results is not necessarily straightforward. The authors' experiences, as summarized in the sections that follow, demonstrate that the majority of reported bending strain features appear likely to be related to construction, with only a small subset likely originating from external forces. However, there are key statistical indicators that appear to separate the reported bending strain features that are most likely associated with geohazards. These statistical relationships demonstrate that bending strain analysis, if properly applied and understood, can be a valuable tool for identifying sections of pipe that have likely been strained by a geohazard.

This paper focuses solely on the results of single-run bending strain analysis. Run-to-run comparisons are not discussed herein, nor is the relationship between reported bending strain values and the actual strain experienced by a section of pipe. Both topics are relevant and could be considered in future papers.

2 BENDING STRAIN ANALYSIS

2.1 Background on Bending Strain

Bending strain can be thought of essentially as a measure of the curvature of the pipeline or its "out-of-straightness." The curvature of the pipeline is obtained by numerically differentiating the pitch and azimuth angles with respect to the odometer distance over a specified gage length (L_{gage}) typically in the range of 3 to 6 pipe diameters (i.e., 3D to 6D) centered on the current odometer distance. The curvature calculations are typically performed using a simple central difference approach, although some vendors utilize a more accurate regression line approach. The pipeline bending strain (ε_b) is calculated using the curvature (Ψ) and the pipe radius (r=D/2) based on the beam theory assumption that plane sections remain plane (i.e., $\varepsilon_b = \Psi \cdot r$). It is important to note that although pipeline inertial surveys are effective in capturing pipeline bending deformations, they do not capture the effects of axial strains which are: (1) uniform across the pipe cross-section (e.g., those that would be induced by ground movements parallel to the axis of the pipeline), (2) due to the restraint of the pipe's thermal expansion/contraction, and/or (3) due to restraint of Poisson contraction due to internal pressure.

2.2 Typical Contents of Bending Strain Report

A basic assumption of bending strain analysis is that bends that occur between two girth welds are cold bends or elbows, and thus only bends that cross one or more girth welds with strain magnitudes exceeding a specified threshold are reported. These "bending strain features" are reported over discrete lengths of the alignment with distinct start and end points along the surveyed pipeline.

For each bending strain feature, the ILI vendor typically provides profile plots that provide a graphical representation of the results and a table that summarizes key information for each bending strain feature (see example table of results in Figure 1). The focus of this paper is on the relationships between the tabular data and geohazard frequency, particularly regarding three key parameters:

1. Strain magnitude: The overall maximum calculated resultant bending strain for the bending strain feature (in percent).
2. Max Horizontal strain: The maximum calculated horizontal component of bending strain for the bending strain feature (in percent). Note that the location of the maximum horizontal component is not necessarily coincident with the location of overall maximum resultant bending strain within a given feature.

BAKER HUGHES a GE company

Client Name:	XXXXX	
Contract:	450564_30H	
Launch/Receive:	XXXXX	
Diameter:	30	in
Length:	66	miles
Strain Report Date:	23-Nov-18	
Reporting Threshold:	0.125	% Bending strain reporting threshold
	0.020	% Strain change reporting threshold

Go back to Table of contents

Confidential. Not to be copied, distributed, or reproduced without prior approval.
© 2017 Baker Hughes, a GE company, LLC - All rights reserved.

Bending Strain and Strain Change Assessment Report

Strain Reporting Threshold: Bending strain features and strain change features are greater or equal to the bending strain reporting threshold and length greater than 40ft (nominal pipe spool length)
The strain values summarized below represent axial bending strain only and do not include any additional axial strain due to other loading mechanisms

Notes:
Click on Strain feature no. in column A to view GW strains, list of bends, geometric anomalies & ILI features within strain event, loading type and stacked strain chart

Strain Feature or Strain Change ID	Abs Dist (ft)	Velocity (ft/s)	Strain Magnitude (%)	Peak Strain Distance (ft)	Strain Orientation	Horizontal Strain At Peak Strain (%)	Vertical Strain At Peak Strain (%)	Max Horizontal Strain (%)	Max Vertical Strain (%)	Strain Direction	Strain Change Magnitude (%)	Strain Change Orientation (hh:mm)	Strain Change Horizontal (%)	Strain Change Vertical (%)	U/S Weld
1	590.2	4.6	0.158	563.0	6:16	-0.023	-0.157	0.111	-0.157	Horizontal / Vertical					240
2	5881.5	5.8	0.161	5842.6	11:25	-0.048	0.154	-0.057	0.154	Vertical					1630
3	6299.4	5.4	0.145	6303.5	12:16	0.021	0.144	-0.076	0.144	Horizontal / Vertical					1770
4	10856.7	8.2	0.180	10844.2	5:46	0.020	-0.179	0.061	-0.179	Vertical					3030
5	33715.6	10.2	0.081	33689.0	11:54	-0.004	0.081	0.038	0.081	Horizontal / Vertical	0.054	12:09	0.033	0.054	9240
6	43941.1	10.5	0.143	43995.8	5:45	0.019	-0.142	0.104	-0.142	Horizontal / Vertical					11855
7	45951.5	11.7	0.153	45941.9	6:00	-0.001	-0.153	0.025	-0.153	Vertical					12420
8	61818.5	10.8	0.132	61807.7	6:34	-0.039	-0.126	-0.046	-0.126	Vertical					16590

Figure 1: An example of a portion of the tabular data generated by IMU vendor.

Copyright © 2019 by ASME

3. Max Vertical strain: The maximum calculated vertical component of bending strain for the bending strain feature (in percent). Note that the location of the maximum vertical component is not necessarily coincident with the location of overall maximum resultant bending strain within a given feature.

3 ANALYSIS OF BENDING STRAIN DATA

Bending strain analysis is often marketed as a tool to identify sections of pipe strained by external forces. In the remainder of this paper, we explore whether this claim is supported by a detailed geohazard review and statistical analysis of bending strain results.

To assess the possible relationship of bending strain features to possible geohazard locations, the authors manually reviewed the locations of 4,618 bending strain features, reported by three of the largest North American IMU survey vendors. The authors compared bending strain locations to a large dataset of previously identified locations of known or possible landslides, as well as available LiDAR data and aerial imagery to identify other possible geohazards (e.g., hydrotechnical, etc.). For each bending strain feature, the authors assigned the location as either a likely geohazard location (subdividing between "Landslide" and "River Channel") or a likely non-geohazard location (i.e., "Other"). Although the authors attempted to use as little subjectivity as possible, some subjective interpretation was needed. For instance, if a landslide was close to but not crossing a pipeline where a bending strain feature was located, the decision on whether the bending strain feature was associated with the landslide or not was made on a case-by-case basis. If there was evidence that the landslide had once crossed the pipeline, but that this evidence had been removed (such as the ROW cutting through the landslide toe), then the location was assigned as a landslide. Conversely, if it appeared likely that the landslide had never crossed the pipe centerline, then the location was not assigned as a landslide. River Channel was assigned to bending strain features that appeared to cross the flow channel of a stream or river, where repeated pipe exposure could occur due to scour; conversely, features that were adjacent to stream or river crossings, which may instead be related to construction, were assigned as other. Figure 2 provides an example of reviewed data where locations have been assigned.

3.1 Origin of Bending Strain Dataset

The bending strain data used in the analysis all originated from ILI runs conducted on natural gas transmission pipelines in the continental United States. Sixty-two percent originated from a single ILI vendor, with the remaining 38 percent divided between two other ILI vendors.

Approximately 90 percent of the reviewed bending strain reports were from ILI runs conducted within the eastern United States, in the Appalachian foothills, where landslides are common. Bending strain reports from ILI runs conducted in the western and southern US were reviewed as well.

FIGURE 2: Example of bending strain features and assigned locations. A hillshade generated from high-resolution LiDAR data is shown in the background, overlain by previously mapped landslides. Bending strain features are color-coded based on the assigned location type.

Pipe ages ranged from pipes installed in the 1940s to 2000s. Pipe diameters ranged from 8.625 inches to 42 inches. Grades ranged from X35 to X70. Wall thickness ranged from 0.219 inches to 0.72 inches.

Landslide locations were determined through geomorphic mapping by geologists and geotechnical engineers. Possible and confirmed landslides were identified by first reviewing high-resolution LiDAR topographic data (Figure 3) and/or aerial photographs and then confirmed (or refuted) by on-ground reconnaissance visits. Specific details of this landslide identification and evaluation program are outside of the scope of this paper, but additional details can be provided by the authors upon request.

The bending strain analysis was typically conducted to evaluate segments of pipe potentially affected by landslides. As such, the bending strain reports that the authors reviewed were for pipe segments previously identified to be potentially affected by one more landslides (i.e., through the landslide mapping program). Typically, the ILI vendors only reported bending strain features with a strain magnitude greater than or equal to 0.125%.

Copyright © 2019 by ASME

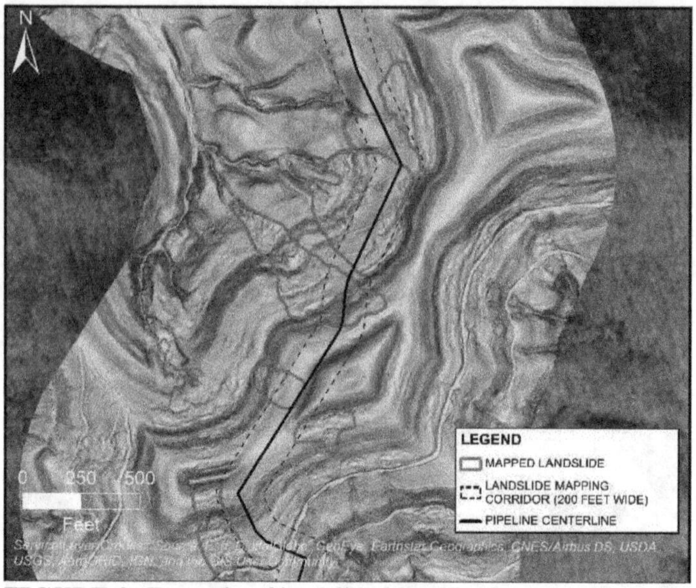

FIGURE 3: Example of landslide identification mapping using high-resolution LiDAR data, which is presented as a percent slope map and displayed in blue-scale. Landslides that cross into the mapping corridor are delineated in red.

3.2 Statistical Analysis of Bending Strain Data by Location

After completing the review of 4,618 bending strain locations and assigning a location type to each, the strain magnitude, horizonal strain, and vertical strain were compared between the three groups of locations (i.e., Landslide, River Channels, and Other).

The population of bending strain features was examined using scatter plots of the maximum vertical and horizontal strain components as functions of the maximum strain magnitude (which was used as the basis for selecting the bending strain features). As illustrated in Figure 4, scatter plots were developed for the three groups of locations. For each location group, linear regression trend lines were fit to the data and the goodness of fit was characterized based on the regression coefficient (R^2). A high R^2 value, which indicates a given strain component is becoming dominant (e.g., the horizontal strain is nearly equal to the strain magnitude), is interpreted as being indicative of more organized data. We note that the majority of bending strain features have at least some component of both horizontal and vertical strain; thus, the upper bound of the R^2 value is less than 1. Filtering of the data was accomplished by screening out data with strain magnitudes of <0.2%, and then further by screening out data with maximum vertical or horizontal strain components of <0.2%. Imposing the screening thresholds removed lower strain magnitude features and tended to improve the goodness of fit (i.e., R^2 values).

The R^2 results are summarized for the different location groups and screening thresholds using a bar chart in Figure 5. In addition to the three location groups shown on the scatter plots, for comparison, the bar chart also shows R^2 values for the entire dataset (i.e., All Data), and for a group which includes the combination of the landslide and river channel locations, which is intended to represent combined geohazard locations.

Lastly, exceedance plots are shown in Figure 6 for the strain magnitude, maximum horizonal strain, and maximum vertical strain, for the same five location groups that are shown on the bar charts. Each plot shows the number of features within a given location group that exceed increasing levels of strain or strain components. These plots are intended to show what strain levels may be most prominent within a given location group (if any).

4 RESULTS

Based on the analyses summarized in Section 3:

- Most vendor reported bending strain features have a magnitude of less than 0.2%, with only 1,046 out of the 4,618 analyzed features (~23%) having a magnitude equal to or greater than 0.2%. Vendor reported bending strain features of greater than 0.3% are very rare (only 116 out of the 4,618 features [~2.5%]). Bending strain features greater than 0.4% are exceedingly rare, at only 17 out of 4,618 (~0.4%).
- Overall correlation between likely geohazards (i.e., landslides and river channels) and reported bending strain features is relatively low at 298 features out of 4,618 (6%). However, this correlation rises sharply as strain magnitude increases. At and greater than 0.35% strain magnitude, the correlation between likely geohazards and strain magnitude rises to over 50%. At strain magnitude greater than 0.42%, the correlation between likely geohazards and strain magnitude rises to over 90%.
- The relationship between horizontal strain and likely geohazards is stronger than overall strain magnitude. At and greater than 0.14% horizontal strain, the correlation between likely geohazards and horizontal strain magnitude is greater than 50%, increasing to 100% at magnitudes greater than 0.36%.
- In contrast, the relationship between vertical strain and likely geohazards is relatively poor, with the correlation between vertical strain and geohazards rising to a maximum of 27% at a strain magnitude of 0.42%.
- The conclusion that high horizontal strain magnitude is predictive of likely geohazards is supported by the relatively high R^2 agreement shown on Figures 4 and 5.
- In contrast, the low R^2 agreement between vertical strain and likely geohazards supports the idea that vertical strain magnitude is not predictive of likely geohazard locations (for the reviewed dataset).

Copyright © 2019 by ASME

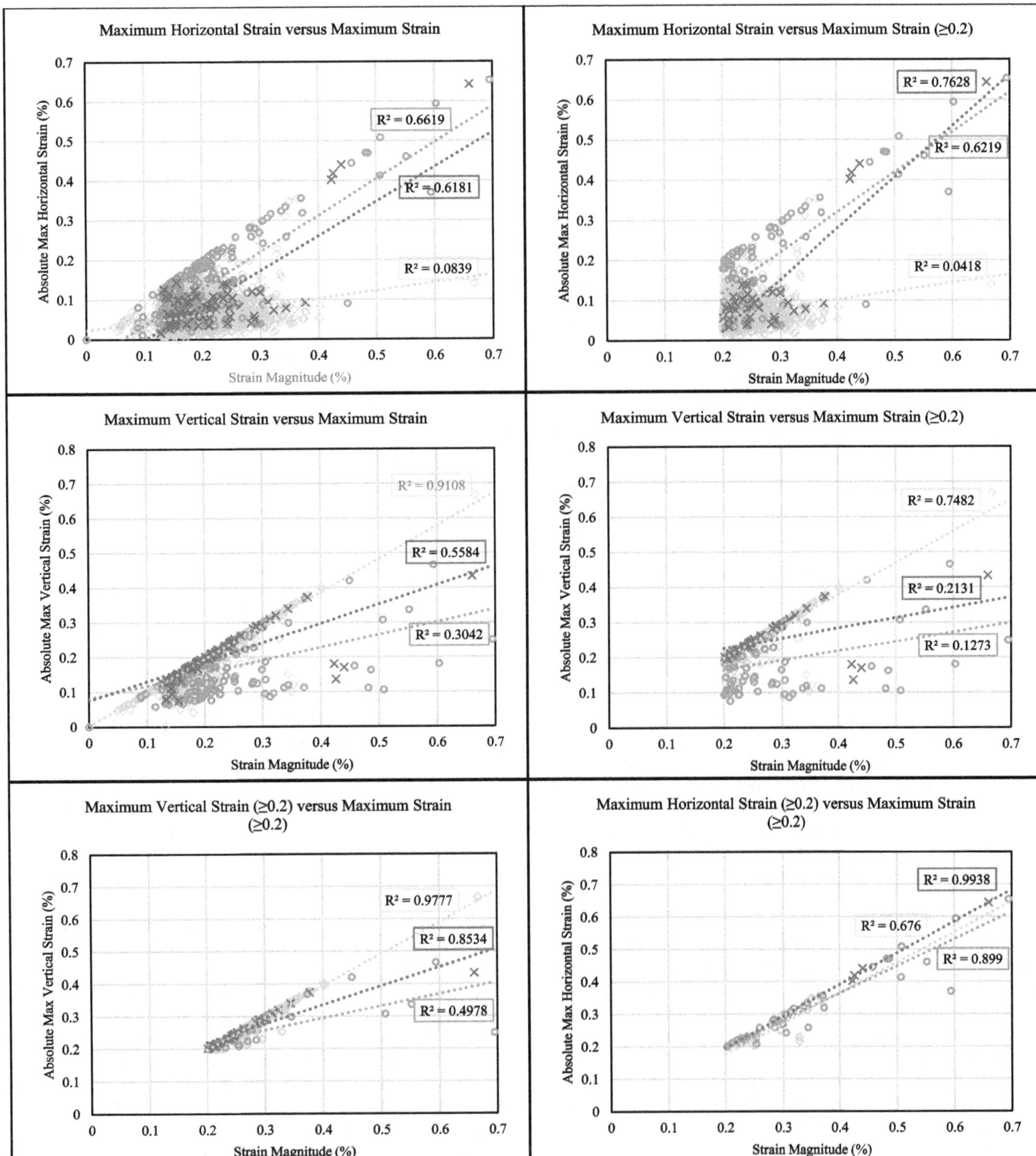

FIGURE 4: Scatter plots and linear regression lines for bending strain features with locations classified as landslides (shown in orange), river channels (shown in blue), or other (shown in yellow).

Copyright © 2019 by ASME

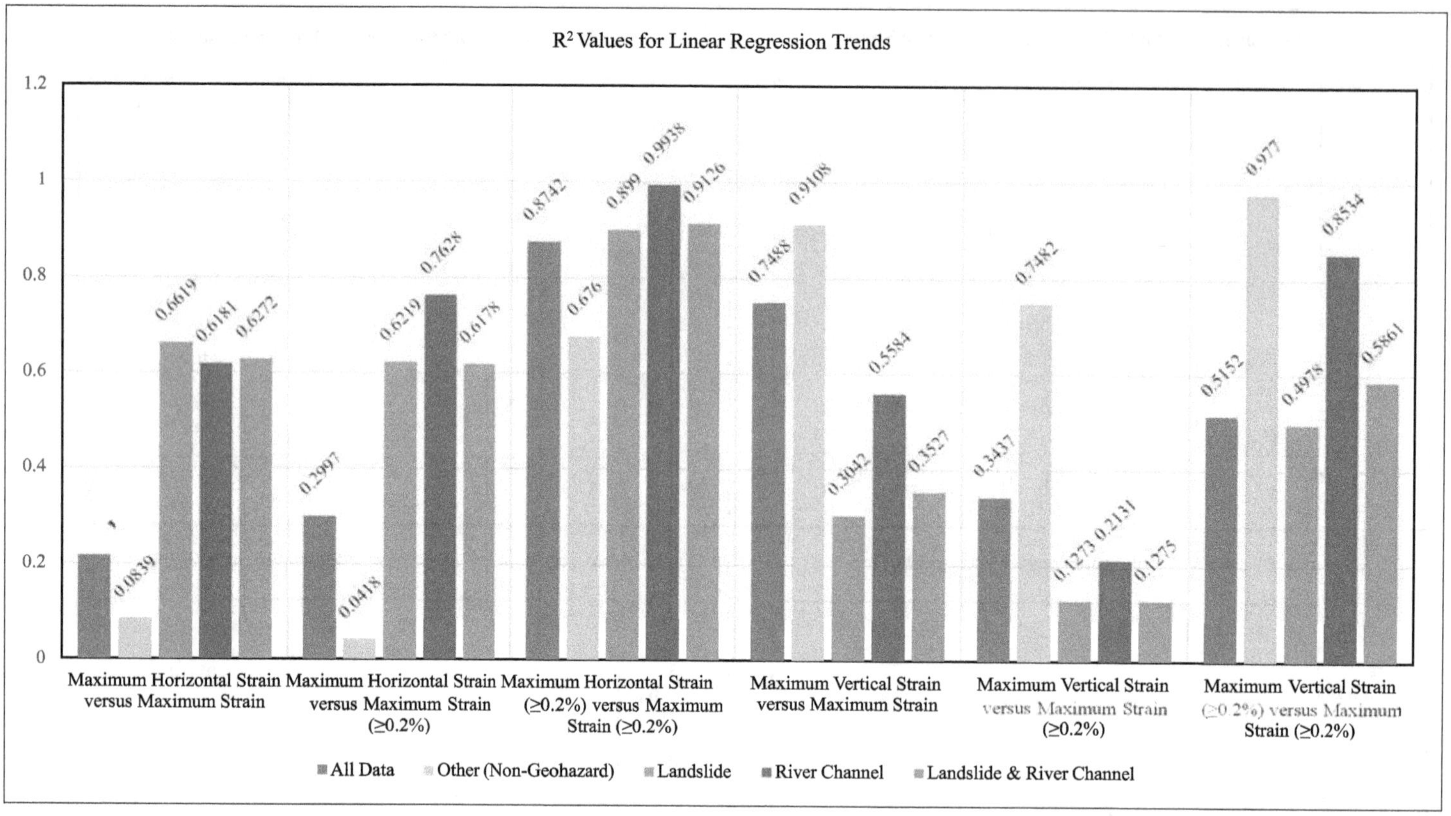

FIGURE 5: Bar chart showing R2 values for linear regression trends for various strain magnitude comparisons and data filtering scenarios, for different location groups.

- The bending strain features reported herein were generally selected based on screening for strain magnitudes ≥0.125%, although some features were reported at lower levels. Although the correlation between geohazards and bending strain features becomes stronger (e.g., higher R^2 values) at higher strain magnitudes, many low magnitude bending strain features (i.e., between 0.125 and 0.2%) also appear to be associated with geohazards.

5 DISCUSSION

Although the overall correlation between reported bending strain features and likely geohazards is low for the reviewed dataset, this correlation increases sharply at higher overall bending strain magnitudes. The correlation between increasing magnitude and likely geohazards is even more apparent with increases in horizontal magnitude. Based on the reviewed dataset, it appears that at strain magnitudes greater than approximately 0.35% and/or horizontal strain magnitudes greater than approximately 0.14%, it is likely that the bending strain may be caused by a geohazard. Furthermore, if the presence of a geohazard at such locations can be confirmed via other mechanisms (e.g., LiDAR data review, ground evaluation, etc.), confidence should be even higher that the geohazard is the cause of the detected strain. The observation that high magnitude bending strain features located within likely geohazards are likely caused by the geohazard is supported by the authors' experience of observing notable elastic rebound (ranging from about a foot to several feet) of pipelines exposed in locations where bending strain features intersect or are located within landslides.

In addition to the observed correlation between high magnitude total and horizontal bending strain and geohazards, a significant number of lower level strain sites appear to correspond to possible geohazards (i.e., 211 of the 298 bending strain features associated with landslides or river channels had strain magnitudes <0.2%). The association of such bending strain sites to a geohazard is based on the presence of a previously identified geohazard (i.e., identified through some other mechanism). As such, using a screening threshold of 0.125% to identify bending strain features is likely appropriate, and although a lower percentage of the low magnitude sites may correspond to geohazards, if the bending strain features can be overlain with previously identified geohazards, the data can still be efficiently reviewed and screened to identify possible locations where the pipe may be impacted by geohazards.

Based on the reviewed dataset, bending strain analysis appears to be a useful tool to evaluate whether a pipeline has been affected by a likely geohazard. Since the horizontal component of bending strain appears to be highly predictive of effects from geohazards, bending strain analysis is particularly

Copyright © 2019 by ASME

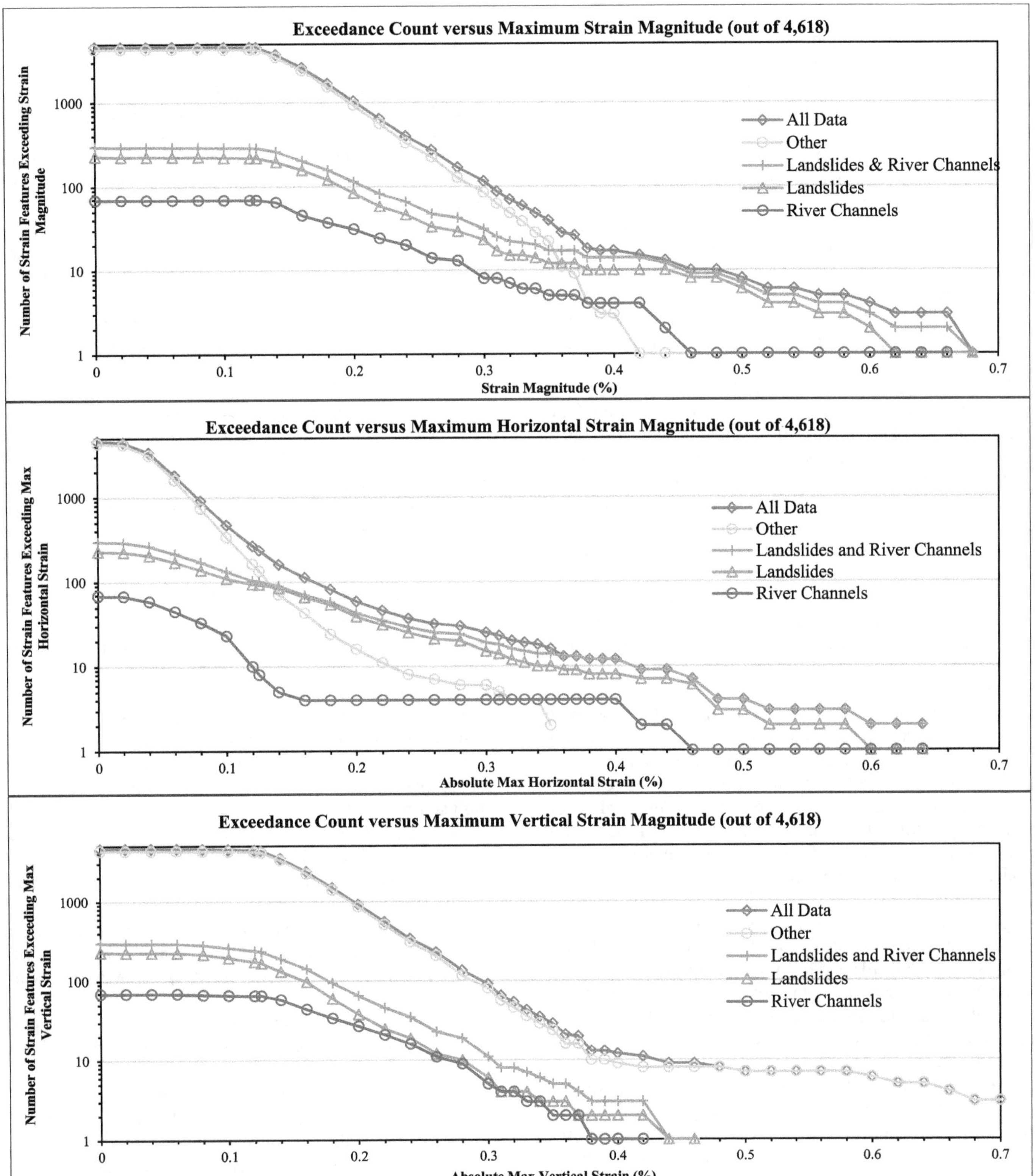

FIGURE 6: Exceedance plots for Strain Magnitude, Maximum Horizontal Strain, and Maximum Vertical Strain, for different location groups.

Copyright © 2019 by ASME

useful when the likely geohazards are oriented perpendicular or at an oblique angle to the analyzed pipeline. This type of analysis is also useful in regions with widespread geohazards, such as the landslide prone Appalachian foothills, because it allows for analysis of potential effects on a pipeline at many locations from many potential geohazards through a single assessment.

When analyzing for the effects of geohazards on pipelines, bending strain analysis provides two main benefits. The first benefit is when the location of a likely geohazard is known, it allows for an evaluation of whether the geohazard has already induced bending on the pipeline and the relative severity of the bending. The second benefit is that reported bending strain features in locations without known geohazards can prompt a geomorphic review of that area, to evaluate if there could be a previously unknown geohazard affecting the analyzed section of pipe.

While bending strain analysis is a useful tool in evaluating potential impacts on pipelines from geohazards, it should not be considered an absolute means of evaluation. For example, bending strain analysis will not detect pipeline deformation when the ground movement results in pure axial tension or compression with no associated bending (since transverse curvature of the pipe centerline is required for bending strain to develop). This situation (i.e., one where a geohazard is meaningfully impacting a pipeline without inducing bending) is likely to be rare, since even in situations where the direction of motion is purely axial to the pipeline, there will tend to be some indication of bending at the transition from the moving ground of the geohazard to stable ground (such as the head and toe of a landslide).

Although we have concluded (based on the analyzed dataset) that bending strain is a useful tool for evaluating whether a pipeline has been affected by a likely geohazard, the absence of a reported bending strain feature within a geohazard should not be construed as a guarantee that the geohazard is not affecting the pipeline. It is plausible there are mechanisms by which a geohazard could significantly affect a pipeline without being identified during bending strain analysis.

As noted previously, the vast majority (94%) of reported bending strain features do not correlate with a likely geohazard. While the focus of this paper has been on the correlation between likely geohazards and bending strain features, it is also important to understand the nature of bending strain features that do not appear to be caused by an external force. For example, do such features represent noise, or do they represent actual bending and is this bending injurious?

An informal analysis by the authors has shown that a large number (perhaps a majority) of these non-geohazard related features appear to be at likely tie-in locations, such as road and stream crossings, and pipes of different ages (i.e., replacement locations). We have inferred that these locations are reflecting bending induced by "roping" during construction, or the practice of transversely deflecting two disparate sections of pipe into alignment prior to welding. In the interest of keeping a focus on geohazards and the space limits allotted to this paper, we did not perform a detailed analysis of this hypothesis (i.e., that most bending strain features are at tie-in locations) but suggest that it would make for a worthwhile study to be published in another venue.

The significance of the observation that most bending strain features are associated with either likely geohazards or apparent construction related practices is that most reported bending strain features, even at relatively low magnitudes, appear to be identifying actual features, and thus should be understood to typically not constitute noise. While geohazard induced strain should be understood to be potentially dynamic (i.e., it could worsen over time), it is not clear if "static" bending strain induced by construction (i.e., it should not worsen over time) should be considered to be potentially injurious or not, and if so, the magnitude(s) at which the bending should be considered to be problematic.

6 CONCLUSIONS

Bending strain analysis is a useful and statistically valid tool for evaluating whether likely geohazards have affected a section of pipe. Based on the analyzed dataset, higher magnitudes of the horizontal component of bending strain are particularly likely to correlate with geohazard locations, with the correlation increasing from about 50% at 0.14% horizontal strain magnitude to 100% at 0.36% horizontal strain magnitude.

7 FURTHER RESEARCH

This paper focused on a limited evaluation of single-run bending strain analysis and its correlation to geohazard locations. It is the authors' understanding that while the mechanics of bending strain analysis are well understood ([see References [1] through [4]), there is a limited number of publications demonstrating an understanding of how these data pertain to real-world situations (which was part of the reason for preparing this paper). Ideas for further research and publication include:

- How would a larger dataset affect the conclusions summarized herein? Would the conclusions cited remain true for data collected in other locations and for other purposes?
- What is the correlation with strain change (i.e., run-to-run) analysis and likely geohazards? Does this analysis meaningfully identify bending changes between ILI runs?
- Is the hypothesis that most reported bending strain features are associated with likely tie-in locations correct?
- How does reported bending strain magnitude correspond to actual strain? Should this be understood to be an absolute or relative measure of strain?
- How do reported bending strain features compare between ILI vendors? For instance, is a reported bending strain of 0.4% from one vendor comparable to 0.4% reported from another vendor (i.e., are consistent gage lengths and curvature calculation methods being used)?

Copyright © 2019 by ASME

ACKNOWLEDGEMENTS

The authors would like to recognize Baker Hughes a GE company for permission to display an example table of bending strain results (Figure 1) from one of their Pipeline Strain Reports.

REFERENCES

[1] Czyz, J.A., and Adams, J.R., "*Computation of pipeline bending strains based on Geopig measurements,*" Pipeline Pigging and Integrity Monitoring Conference, Houston, February 1994.

[2] Czyz, J.A., Fraccaroli, C., and Sergeant, A.P., "*Measuring Pipeline Movement in Geotechnically Unstable Areas Using an Inertial Geometry Pipeline Inspection Pig,*" ASME 1-st International Pipeline Conference, Calgary, June 1996.

[3] Hart, J.D., Powell, G.H., Hackney, D., and Zulfiqar, N., "*Geometry Monitoring of the Trans-Alaska Pipeline,*" Proceedings of the ASCE Cold Regions Conference, Anchorage, Alaska, May 20-22, 2002.

[4] Hart, J.D., Zulfiqar, N., and Erickson, T.H., "*3rd Party Review of Geometry Pig Inertial Survey Data at the Colville River HDD,*" IPC08-64306, Proceedings of the International Pipeline Conference, Calgary, Alberta, Canada, September 29 – October 3, 2008.

Copyright © 2019 by ASME

Chapter 32

DRAFT

.I, IN-DITCH AND PERMANENTLY INSTALLED TOOLS FOR STRESS/STRAIN IMAGING AND MONITORING

Dr. Neil J Goldfine, Mr. Todd M Dunford, Dr. Andrew P Washabaugh JENTEK Sensors, Inc.
Marlborough, Massachusetts, USA

1 Abstract

This paper addresses the ability to image and monitor strain (or stress), both residual and applied, in pipelines and other carbon steel structures such as risers and vessels. Strain (or stress) monitoring is needed to assess pipeline integrity after events such as ground motion, operating excesses, and prolonged operation in extreme environments. Three techniques are described: (1) an in-line-inspection (ILI) tool, (2) an in-ditch or above ground imaging method, and (3) a permanently installed approach. Each of these methods uses MWM sensors and MWM-Arrays to make non-contact measurements of the steel's magnetic properties, which are then correlated to strain (or stress) in the material. Since the method uses magnetic fields, it can operate through non-conductive materials such as coatings and insulation. This work was funded in-part by JENTEK Sensors, Inc., the Department of Transportation (DOT), and end users. The opinions in this paper are those of the authors and not the sponsoring agencies.

2 Introduction

This paper first describes the MWM and MWM-Array sensor constructs. The MWM construct, originally conceived in the 1980s at the MIT Laboratory for Electromagenetic and Electronic Systems, initially had a meandering drive. Newer versions of the MWM (see FIGURE 1), although still providing a periodic magnetic field, use a novel interdigitated rectangle drive construct and are now referred to simply as MWM sensors, not Meandering Winding Magnetometers.

The MWM-Arrays are not typically periodic for scanning/imaging configurations, but can be periodic for permanently installed applications, as described later. Example one and two dimensional MWM-Arrays are also described in the following, including MR-MWM-Arrays that use magnetoresistive sensing elements to measure through the pipe wall thickness when needed. Finally, descriptions of the Quadri-Directional Magnetic Stress Gage (QD-MSG) and Bi-Directional Magnetic Stress Gage (BD-MSG) configurations used for simultaneous multidirectional magnetic permeability measurement at a single point are provided. The description of these sensors is followed by brief descriptions of the unique instrumentation used to acquire data and provide input to the model-based multivariate inverse methods (MIMs).

The JENTEK MIMs are called Grid Methods (for 2-unknown solutions) and HyperLattice® methods (for 3 or more unknowns) and are used to rapidly convert the MWM and MWM-Array impedance measurements into properties of interest such as magnetic permeability, pipe wall thickness, and liftoff (where liftoff is defined as the proximity of the sensor to the first conducting or magnetic layer). These methods are also described in the following along with specific applications to stress (strain) measurement. In addition, results are described for several static and dynamic tests demonstrating the stress monitoring capability. These descriptions included (1) stress monitoring with temperature correction in a permanently installed mode, (2) stress monitoring on a pressure cycled pipe, along with crack growth monitoring, also in a permanently installed mode, and (3) stress imaging at mechanical damage sites in a pipeline segment. Finally, an in-line-inspection tool configuration is described, for internal stress imaging (and corrosion and crack detection).

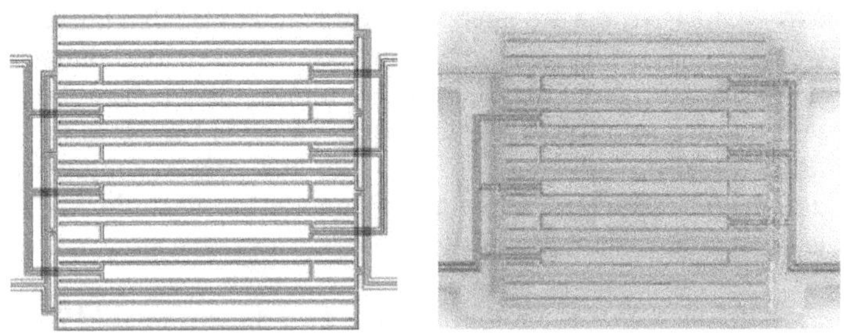

(a) *(b)*

FIGURE 1: MWM (A) SCHEMATIC, (B) SENSOR. [1]

In the following, both stress and strain monitoring is referred to as stress monitoring. Referring to stress instead of strain is often met with discomfort since some believe that only strain can be measured (through displacement) and stress is inferred. However, since the MWM measures the directional components of the magnetic permeability which are directly influenced by applied and residual stresses in multiple directions (i.e. stresses in one direction affect the magnetic permeability in other directions), it is preferred by the authors to call these sensors stress gages and not strain gages (this is not meant to imply anything conclusively about stress/strain causality, merely to highlight the issue. One possibility is that changes in magnetic permeability are related to historical work done on the magnetic material – represented by the product of stress and cumulative strain).

Copyright © 2019 by ASME

3 MWM Sensors

FIGURE 1(a) provides a schematic of an MWM sensor and FIGURE 1(b) shows an actual sensor. The MWM was designed for absolute property measurements (e.g., electrical conductivity and magnetic permeability for metals), without calibration standards, using a model-based inverse method [2-5]. The MWM sensor consists of a periodic primary winding with a modified, patented, design [6] for creating the magnetic field and series-connected secondary rectangles for sensing the response [7]. The MWM sensors were designed to permit the sensor response to be accurately modeled, dramatically reducing calibration requirements (as described in ASTM guidelines E2338 and E2884 [8,9].

4 MWM-Arrays and MR-MWM-Arrays

MWM-Arrays can be one dimensional or two dimensional. One dimensional MWM-Arrays typically have a single rectangle drive winding construct or a dual rectangle winding construct as shown in the sensing elements can be either inductive coils (for MWM-Arrays) or magnetoresistive (MR) sensing elements (for MR-MWM-Arrays). FIGURE 3 shows a large flexible array with a dual rectangle drive construct (hidden in the photograph) and two rows of MR sensing elements. The MR sensing elements provide improved signal-to-noise at lower frequencies, enabling measurement of properties deeper into the material under test. This is useful for both stress monitoring and for detection of damage such as corrosion through layers. MR-MWM-Arrays are needed if inspection through aluminum weather jacketing is required for stress or damage measurement. Note that stress monitoring through weather jacket and insolation is possible using methods similar to those used for corrosion imaging, but with the nominal thickness assumed constant to allow the magnetic permeability to be estimated instead.

FIGURE 4 shows a photograph of a two-dimensional MWM-Array with a periodic drive. Similar sensors are available with a single row of sensing elements in a one-dimensional array. These periodic drive constructs are typically used in a permanently installed configuration [6,7].

FIGURE 5 provides the depth of penetration chart for a range of sensor constructs for steel materials over a range of magnetic permeability (assuming an electrical conductivity of 1MS/m which is typical for steel alloys). The depth of penetration represents the depth of the applied field penetration into the material and should not be used alone to represent the depth of sensitivity for stress estimation or damage detection. Since the electrical conductivity does not vary significantly with stress, it can often be assumed to be constant. However, both the conductivity and magnetic permeability vary with temperature and, assuming the conductivity is constant, can produce errors in the stress estimates if care is not taken.

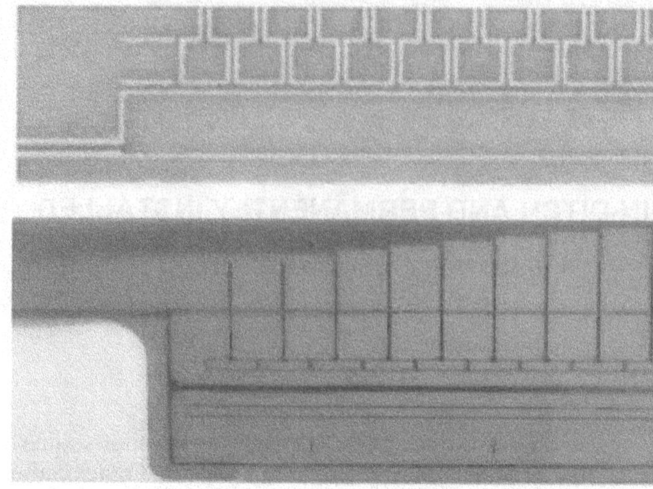

FIGURE 2: MWM-ARRAY (A) WITH A SINGLE RECTANGLE DRIVE WINDING CONSTRUCT, AND (B) WITH A DUAL RETANGLE DRIVE WINDING CONSTRUCT. [10, 11]

FIGURE 3: MR-MWM-ARRAY WITH A DUAL RECTANGLE DRIVE CONSTRUCT AND TWO ROWS OF MAGNETORESISTIVE (MR) SENSING ELEMENTS. [12]

Copyright © 2019 by ASME

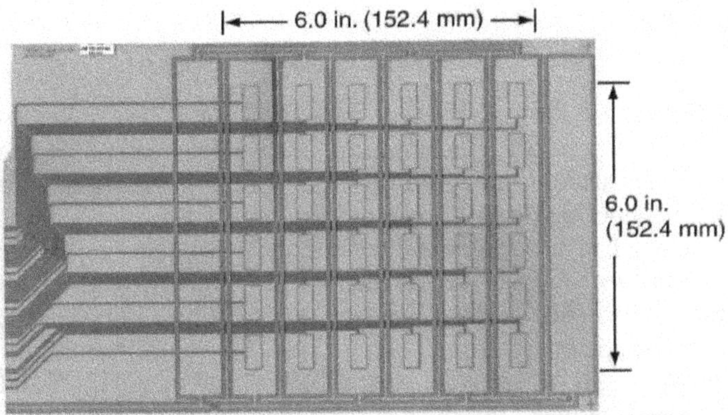

FIGURE 4: TWO DIMENSIONAL MWM-ARRAY WITH PERIODIC DRIVE WINDING CONSTRUCT AND 36 INDUCTIVE SENSING ELEMENTS. [13]

$$\Gamma_n = \sqrt{(2\pi n / \lambda)^2 + j2 / \delta^2}$$

$$\text{Skin depth: } \delta = \sqrt{\frac{1}{\pi f \mu \sigma}}$$

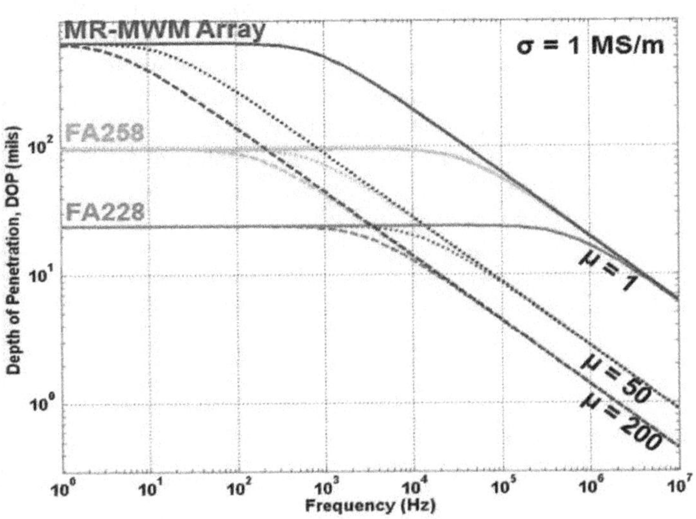

FIGURE 5: DEPTH OF PENETRATION CHART FOR VARIOUS DRIVE WINDING CONSTRUCTS, WHERE THE EFFECTIVE SPATIAL WAVELENGTH, λ, OF THE MWM OR MWM-ARRAY IS DERIVED FROM EITHER THE DRIVE TO SENSING ELEMENT GAP OR THE SPATIAL WAVELENGTH OF THE PERIODIC DRIVE CONSTRUCT OR A COMBINATION OF THESE DIMENSIONS. μ IN THE FIGURE IS THE RELATIVE MAGNETIC PERMEABILITY. 1 MIL= 0.001 INCHES. [13]

5 QD-MSGs

The MWM sensors or MWM-Arrays can provide a measurement of the magnetic permeability (independent of liftoff or other properties such as pipe wall thickness) under some practical circumstances. These measurements are directional in nature. In other words, the MWM or MWM-Array estimates the component of the magnetic permeability perpendicular to the longer drive winding segments within the sensor construct in the plane of the material under test. For practical applications, it is generally useful to normalize the measurement in one direction of interest by the measurement in another direction to remove material property variations that may not be of interest or to remove other stress components (e.g. from pressure in the pipe if that is known or not of concern). Furthermore, it is often useful to measure multidirectional or unidirectional magnetic permeability at multiple spatial locations (e.g. along a girth weld and in the axial direction away from the weld) to determine post weld heat treatment (PWHT) related residual stresses, before and after PWHT processing. For bending stresses and torque measurements, measurements at multiple locations are also of value to improve estimation accuracy and to reduce measurement noise.

The Quadri-Directional Magnetic Stress Gage (QD-MSG) (FIGURE 6) is a stack of four MWM sensors, with axes of sensitivity in four different directions (–45°, 0°, +45°, and 90°) [1,10]. A BD-MSG is simply a stack of two MWM sensors. The layout, orientation, secondary element size, and other geometrical properties are designed in a way that makes the sensing elements of one sensor insensitive to the magnetic fields generated by the primary windings of the other three sensors. This permits measurement of directional properties, such as conductivity or permeability, in four directions simultaneously and, although each sensor has a different off, the differences are known and can be accounted for by using the Grid Methods.

Arrays of these sensors (shown in FIGURE 7) have been used to demonstrate noncontact torque measurement capability on an unmodified main rotor shaft in a test cell at Boeing Rotorcraft Division [1]. These sensors might also be used for monitoring pipe or vessel stresses in either a permanently installed or scanning mode.

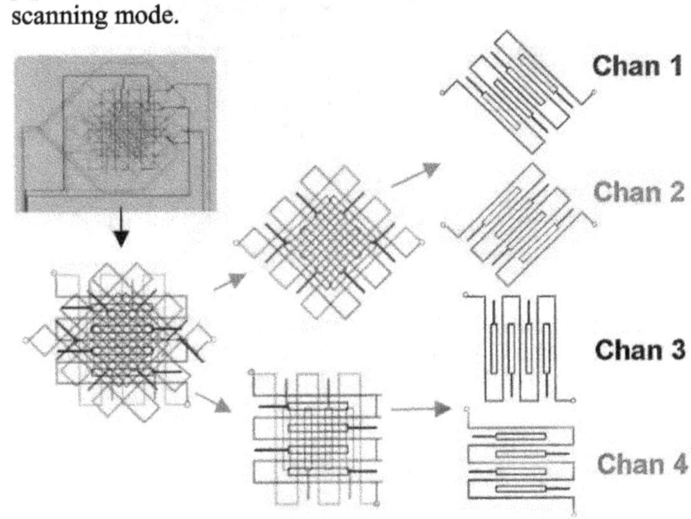

FIGURE 6: THE QUADRI-DIRECTIONAL MAGNETIC STRESS GAGE (QD-MSG™) IS A STACK OF FOUR DIRECTIONAL MWMS THAT PERMITS SIMULTANEOUS STRESS MEASUREMENT AT –45°, 0°, +45°, AND 90°. THIS PERMITS DETERMINATION OF AXIAL AND BENDING LOADS, AS WELL AS TORQUE. [1]

Copyright © 2019 by ASME

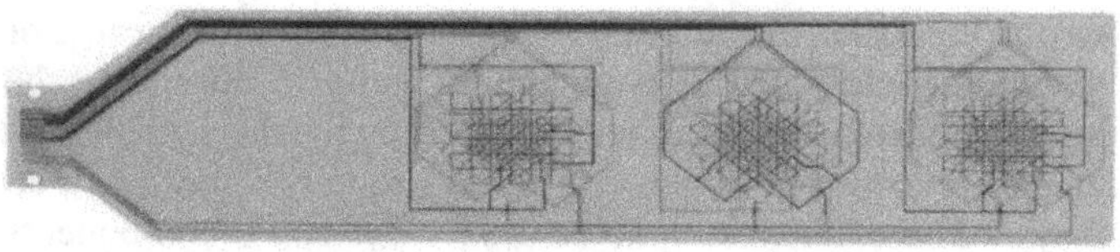

FIGURE 7: AN ARRAY OF THREE QD-MSG SENSORS USED FOR THE MAIN ROTOR SHAFT NONCONTACT TORQUE MEASUREMENT DEMONSTRATION. [1]

6 Parallel Architecture Instrumentation

FIGURE 8 and FIGURE 9 show JENTEK 7-channel and 39-channel parallel architecture instruments. Parallel architecture instruments are needed to enable simultaneous multidirectional sensing.

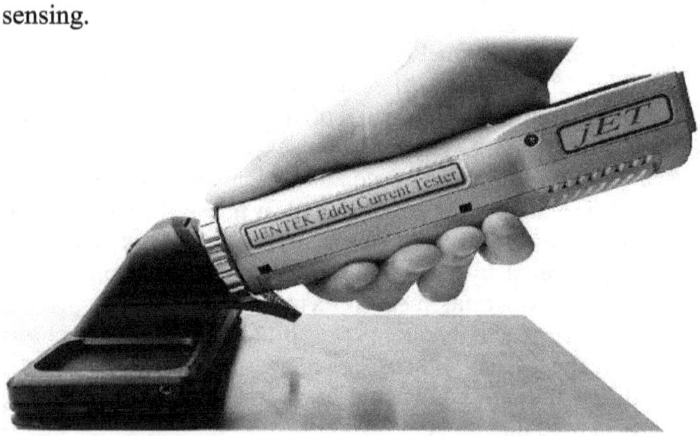

FIGURE 8: JENTEK 7-CHANNEL jET INSTRUMENT. [11]

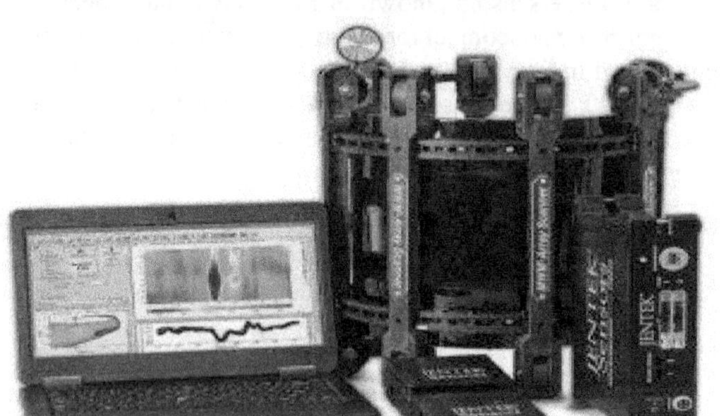

FIGURE 9: JENTEK 39-CHANNEL GRIDSTATION 8200 INSTRUMENT, WITH MR-MWM-ARRAY AND TWO MR PROBE ELECTRONICS UNITS. [11]

7 Grid and HyperLattice Methods

The Grid Methods use precomputed databases of sensor responses to represent the MWM or MWM-Array field interactions with the material under test. Grids are precomputed databases of sensor responses for estimation of two unknowns (such as magnetic permeability and liftoff). For three unknowns, the precomputed databases of sensor responses are called Lattices and for four or more unknowns, they are called HyperLattices [11].

FIGURE 10 shows a measurement grid for a two-unknown permeability/liftoff measurement. The measurement grid is generated using a model of the MWM field interactions with the neighboring material. The model used for this purpose was developed in the 1980s and refined over the years to enable extremely accurate representation of the MWM field interactions. The grid is generated once (off-line) and stored as a precomputed database for access by the GridStation® software. To generate the grid, all combinations of liftoff and magnetic permeability over the dynamic range of interest are input into the MWM models to compute the corresponding grid points. The visualization in FIGURE 10 includes lines of constant liftoff (h) and lines of constant magnetic permeability (μ). Calibration of the MWM and MWM-Arrays for all the data described in this paper was performed at the Air Point (Air Calibration) or at a single reference point on the part using the methods described in ASTM guidelines E2338 and E2884 [8,9].

To perform a permeability/liftoff measurement, first the real and imaginary parts of the complex transinductance (impedance/$j\omega$) are measured, at an instant in time, using JENTEK parallel architecture impedance instruments. It is important to make measurements at each sensing element simultaneously if the data is to be combined for dynamic stresses or if the sensor is moving relative to the material under test.

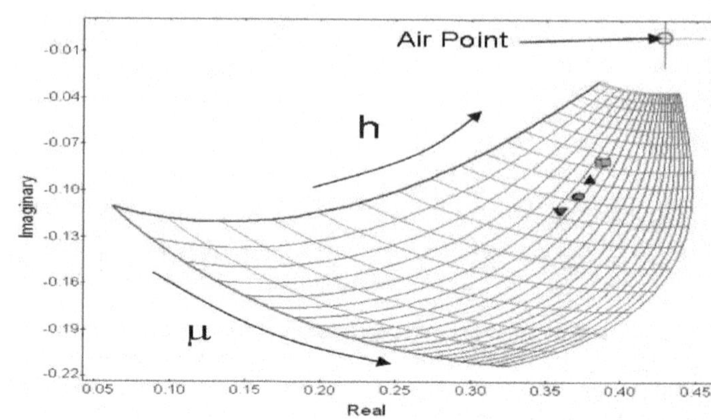

$$\text{Transinductance} = \frac{V_2}{j\omega i_1} = \text{Re}\left(\frac{V_2}{j\omega i_1}\right) + j\,\text{Im}\left(\frac{V_2}{j\omega i_1}\right)$$

FIGURE 10: MEASUREMENT GRID FOR LIFTOFF (h) AND MAGNETIC PERMEABILITY(μ) AT ONE APPLIED FREQUENCY F = ω/2π. NOTE THE TRANSINDUCTANCE REAL AND IMAGINARY PART ARE EASILY DERIVED FROM THE TRANSINDUCTANCE MAGNITUDE AND PHASE. THE COMPLEX REPRESENTATION IS PREFERRED FOR CALCULATIONS, BUT OFTEN THE MAGNITUDE AND PHASE PROVIDE A MORE INTUITIVE REPRESENTATION. EITHER CAN BE DISPLAYED IN THE JENTEK GRIDSTATION SOFTWARE. [1]

Copyright © 2019 by ASME

HyperLattice methods are for more than two unknowns and use more than one frequency or more than one sensor geometry or both. A typical application for stress measurement would be to determine pipe wall thickness, magnetic permeability, and liftoff as a three unknown method using a two or three frequency method and then estimating the magnetic permeability at multiple depths from the outer pipe surface using a second iteration on the MIM. FIGURE 11 shows a typical three unknown Lattice for permeability/liftoff/thickness estimation. Note that a HyperLattice is typically visualized as a series of Lattices (such as shown in FIGURE 11) that are plotted for different values of a fourth or fifth unknown, such as conductivity of the pipe wall.

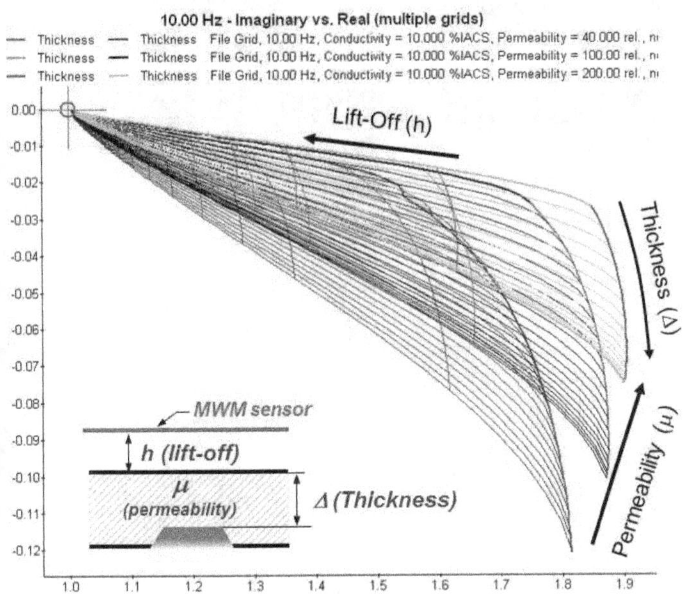

FIGURE 11: LATTICE FOR 3-UNKNOWN METHOD FOR PERMEABILITY/LIFTOFF/THICKNESS FOR A PIPE. [12]

8 Coupon Static Stress Testing

This section describes coupon tests at varied temperature along with correlation to stress (strain). FIGURE 12 provides a photograph of the setup with MWM-Arrays mounted to measure magnetic permeability in both the axial and transverse directions. FIGURE 13 shows the temperature sensitivity and the correlation plots between permeability and temperature. FIGURE 13 also shows that a monotonic response to stress can be obtained by combining the data in two orientations. This is needed since the response in the tensile direction peaks well before the stress reaches yield (see FIGURE 15). As the tensile stresses get higher, the transverse permeability measurement is used to obtain a monotonically increasing permeability response to stress variations.

FIGURE 12: PHOTOGRAPH OF TEST SETUP FOR OVEN TESTING OF BENDING COUPON WITH MOUNTED MWM-ARRAY FOR STRESS MONITORING. [13]

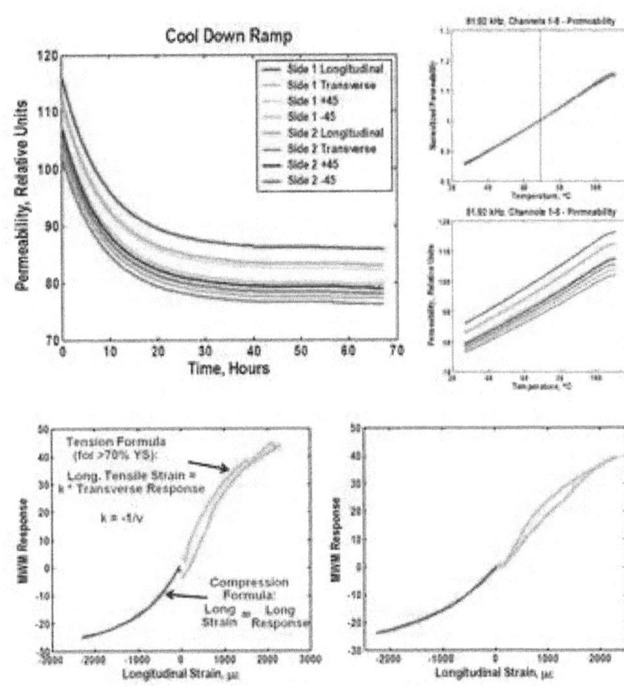

FIGURE 13:EXAMPLE RESULTS FROM OVEN TESTING OF BENDING COUPON WITH MOUNTED MWM-ARRAY FOR STRESS MONITORING. [13]

Copyright © 2019 by ASME

9 Dynamic Testing of a Pipe Segment

This section describes a stress monitoring capability demonstration on a pipe section where the pressure was varied using water pressure in a test at GDF Suez (now called Engie) under U.S. DOT funding. FIGURE 14 shows a photograph of the test setup for crack growth and stress monitoring at a mechanical damage site. The crack detection data has been presented previously [11]. FIGURE 15 provides a plot of the data taken using the installed MWM-Arrays at various locations along the axis of a gauged mechanical damage site.

FIGURE 14: PHOTOGRAPH GDF TEST SETUP FOR CRACK GROWTH AND STRESS MONITORING USING AN INSTALLED MWM-ARRAY AT A MECHANICAL DAMAGE SITE. [12]

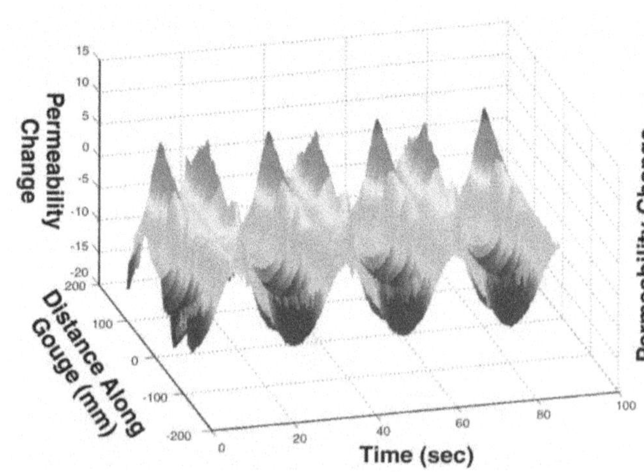

FIGURE 15: DYNAMIC STRESS DATA SHOWING VARIABLE PERMEABILITY AS THE PIPE SECTION PRESSURE IS VARIED CYCLICALLY OVER TIME. [12]

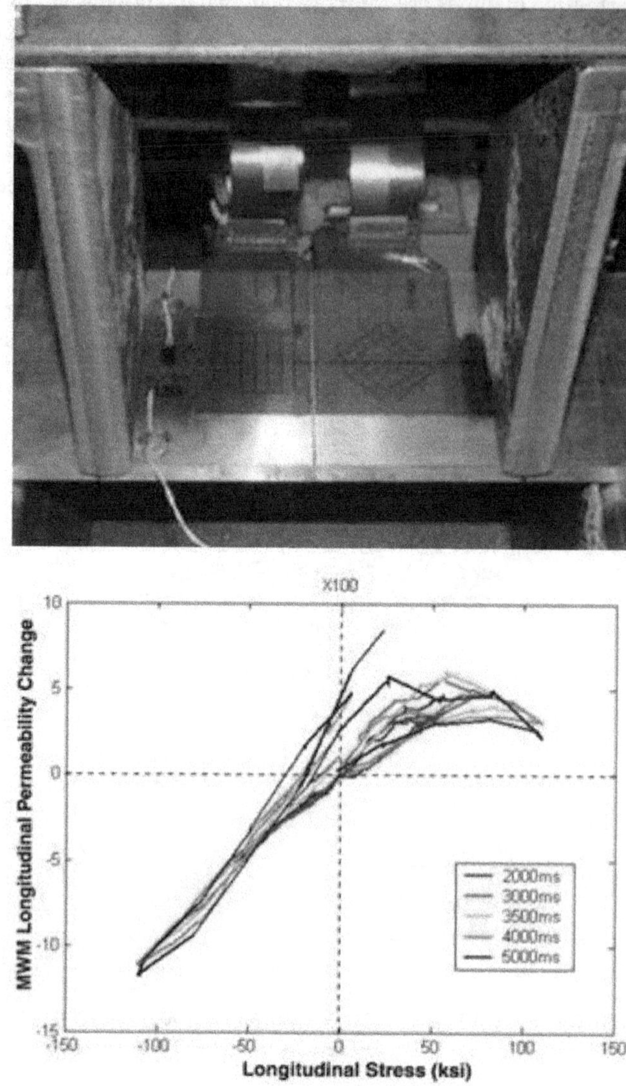

FIGURE 16: (TOP) STATIC TENSILE TEST COUPON DESIGN. (BOTTOM) COUPON TEST DATA SHOWING HOW THE PERMEABILITY VARIES WITH TENSILE AND COMPRESSIVE STRESSES. NOTE THAT THE RESPONSE PEAKS BEFORE YIELD IN THE TENSILE DIRECTION. THE VARIED COLORS REPRESENT REPEATED CYCLES OF THE STATIC (STEPPED) STRESS VARIATION. [12]

Note that using the method described in the previous section with bi-directional magnetic permeability measurements, the peak in the tensile stress response can be removed and a monotonic permeability vs stress relationship obtained (see FIGURE 13). It is valuable to note that the peak provides useful information about approaching tensile yield stresses that may be helpful in predicting sensitivity to stress corrosion cracking or overload conditions.

10 Residual Stress for Welds

FIGURE 17 provides the results of a measurement before and after post weld heat treatment for a sample with a nonvisible

Copyright © 2019 by ASME

mechanical damage site adjacent to the weld. The higher magnetic permeability region (dark blue to purple) indicates tensile stresses. This data was recorded with a 37-channel MWM-Array at a single frequency on an automated scanner. Note this data was presented by JENTEK previously [11].

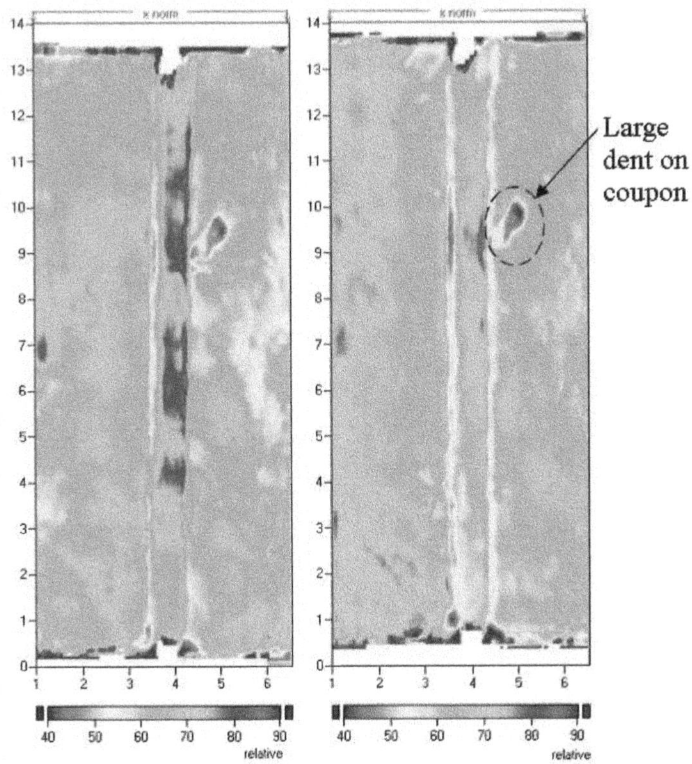

FIGURE 17: MAGNETIC PERMEABILITY MAP FOR A WELD WITH A MECHANICAL DAMAGE SITE, BEFORE AND AFTER POST WELD HEAT TREATMENT (PWHT). HIGHER PERMEABILITY CORRELATES TO HIGHER TENSILE STRESS. [12]

11 In-Line-Inspection (ILI) Tool

This section briefly describes an ILI tool for internal inspection of pipelines, with capability to map axial magnetic permeability and infer bending stress levels. FIGURE 18 shows the prototype tool. Two prototype tools where fabricated under DOT and other customer and JENTEK funding. An earlier version of these tools was successfully tested for imaging damage at the Pipeline Research Council International (PRCI) under DOT funding. Capability to image axial permeability was also demonstrated with suitability for stress imaging. This tool is not yet commercially available. Note that estimation of bending stresses is practical from axial permeability using apriori knowledge of spatial patterns in stress associated with bending loads. Estimating axial loads is more challenging, given typical variations in magnetic properties and the uniformity of the stress pattern.

FIGURE 18: A SCHEMATIC OF A PROTOTYPE ILI TOOL FOR INTERNAL CORROSION IMAGING, CRACK DETECTION, AND STRESS ESTIMATION USING MWM-ARRAYS. [11]

12 Summary

This paper has presented example configurations and methods for stress/strain monitoring using novel magnetic field-based methods. Multidirectional methods enable stress monitoring and model based MIMs with MWM-Arrays and QD-MSGs make stress monitoring practical. The data presented provides examples of capability demonstrations performed over many years by the authors and others at JENTEK Sensors, Inc.

REFERENCES

[1] Goldfine, Neil; Sheiretov, Yanko; Dunford, Todd; Denenberg, Scott; Grundy, David; Schlicker, Darrell; Zilberstein, Vladimir; Robuck, Mark; Parker, Cynthia; "Magnetic Stress Gages for Torque and Load Monitoring in Rotorcraft" AHS 64rd Annual Forum; Montreal, Canada; April 29-May 1, 2008.

[2] Melcher, J.R., "*Apparatus and Methods for Measuring Permeability and Conductivity in Materials Using Multiple Wavenumber Magnetic Interrogations*," US Patent Number 5,015,951, May 1991.

[3] Goldfine, N. J., "Uncalibrated, Absolute Property Estimation and Measurement Optimization for Conducting and Magnetic Media Using Imposed ω-k Magnetometry," Sc.D. thesis, Department of Mechanical Engineering, Massachusetts Institute of Technology, September, 1990.

[4] Goldfine, N., "*Magnetometers for Improved Materials Characterization in Aerospace Applications,*" *Materials Evaluation*, Volume 51, No. 3, pp. 396-405; March 1993.

[5] Melcher, James R.; "Apparatus and Methods for Measuring Permeability and Conductivity in Materials Using Multiple Wavenumber Magnetic Interrogations," U.S. Patent Number RE 36,986, December 12, 2000.

Copyright © 2019 by ASME

[6] Goldfine, N., Zilberstein, V., A. Washabaugh, D. Schlicker, I. Shay, D. Grundy, "Eddy Current Sensor Networks for Aircraft Fatigue Monitoring," *Materials Evaluation,* Volume 61, No. 7, July 2003, pp. 852-859.

[7] Goldfine, Neil J.; Schlicker, Darrell E.; Washabaugh, Andrew P.; Zilberstein, Vladimir A.; Tsukernik, Vladimir; "Surface Mounted and Scanning Spatially Periodic Eddy-Current Sensor Arrays," US Patent Number 6,952,095 B1, October 4, 2005.

[8] ASTM Standard Practice E2338-04, "Characterization of Coatings Using Conformable Eddy-Current Sensors without Coating Reference Standards," ASTM International, Book of Standards, Vol. 03, 2004.

[9] ASTM Standard Practice E2884-13, "Eddy Current Testing of Electrically Conducting Materials Using Conformable Sensor Arrays," ASTM International, Book of Standards Vol. 03.03.

[10] Sheiretov, Yanko; Goldfine, Neil; Dunford, Todd; Denenberg, Scott; Grundy, David; Schlicker, Darrell; Washabaugh, Andrew; Walrath, Karen; "Test Circuit with Sense Elements Having Associated and Unassociated Primary Windings," U.S. Patent Number 8,222,897, July 17, 2012.

[11] http://www.jenteksensors.com/

[12] Goldfine, Neil; Dunford, Todd; Denenberg, Scott; Sheiretov, Yanko; Washabaough, Andrew; Haque, Shayan; "Inspection of Refinery Piping and Vessels from the Outside for External and Internal Corrosion" API; Beijing, China; Septermber 9-13, 2013

[13] Goldfine, Neil; Dunford, Todd; Washabaugh, Andrew; "Tools for Stress/Strain Imaging and Monitoring" TP_2019_01_01, January 31, 2019

Copyright © 2019 by ASME

AIMPIMG2019-1021

Chapter 33

USING ILI TO EVALUATE PIPELINE STRAIN CONDITIONS

Jane Dawson
Baker Hughes, a GE company
Cramlington, Northumberland, UK

Mohamed ElSeify
Baker Hughes, a GE company
Calgary, Alberta, Canada

Ian Murray
Baker Hughes, a GE company
Cramlington, Northumberland, UK

ABSTRACT

Running an IMU (Inertial Mapping Unit) tool as part of an in-line inspection (ILI) survey has become a routine practice for many pipeline operators. Not only does it provide pipeline mapping information, but it also allows determination of the bending strain throughout the pipeline. Bending strain in a deformed pipeline can be derived from the change in curvature between the initial and the final pipeline shape. This provides the means to identify locations which may be affected by construction related strains and ground movement threats.

Whilst the IMU data provides reliable bending strain information about a pipeline; the more recent entrance of the axial strain measurement tool (AXISS™) fitted to standard magnetic flux leakage (MFL) tools provides the ability to measure pipeline axial strain. This delivers additional data that can be combined with the IMU inferred bending strain to have a fuller understanding of the total strain state of a pipeline. Historically, pure axial (tension or compression) threats would need to be monitored at a local level typically through periodic geotechnical assessments or strain gauges physically installed on the pipe in known threat areas, but now such threats can be detected and managed at the pipeline level.

This paper discusses the benefits of integrating the assessment of both the IMU bending strain and axial strain data streams. Utilizing both strain measurement technologies, AXISS and IMU, allows for a better understanding of the strain condition in a pipeline, for example:

- Elimination of uncertainties when identifying pipeline strain induced by geological instabilities vs pipeline construction practices
- Deviation between axial strain and IMU derived strain can be used to identify the presence of possible plastic or residual strain conditions
- Assessment of the bi-axial stress/strain state of a pipeline section to determine severity and remediation needs

A case study is provided in the paper to illustrate these benefits.

INTRODUCTION

The threat from axial strain to pipeline integrity can arise from several causes but is most often associated with geohazards and fabrication/construction residual stresses. Operators, faced with significant risk typically have engineering management programs involving geotechnical assessment, modelling and surveying to identify hazard areas requiring investigation and ongoing monitoring. Once a hazard is identified, monitoring systems (either geotechnical or strain gauges) are installed and monitored to ensure that pipeline strains are understood, and remediation work is completed proactively to relieve critical areas.

IMU's provide reliable and accurate bending strain data which is useful for identifying and monitoring pipeline movement; however, the IMU does not provide the total axial strain data which is needed to define the full strain state of the pipeline.
BHGE performed the first AXISS project with a prototype inspection tool in 2010. Since then, many successful inspections have been performed with ongoing experience and learning from each project being used to refine the quality of the inspection data collected, improve analysis techniques and the value of the deliverables. The overlap between the available measurement techniques (IMU and AXISS) allows the refinement of the strain assessment and better understanding of the strain condition of the pipeline.

The use of both AXISS and IMU technologies provides key advantages for axial strain monitoring:

Copyright © 2019 by ASME

- Enables cost-effective and continuous measurement of axial strain along the entire pipeline's length;
- Measures small changes in axial strain over the lifetime of the pipeline (fabrication phase through in-service condition);
- Reliable identification of the source of the axial strain, i.e., due to in-service geological instabilities vs pre-service pipeline construction and implementation practices.
- Provides proactive risk management by detecting critical strain conditions resulting from known and unknown geohazards or other causes of pipeline loading.

NOMENCLATURE

AXISS™Axial strain sensing technology (a trademark of Baker Hughes Incorporated)

ERW Electric Resistance Pipe
GPS Geographical Positioning System
HAZ Heat Affected Zone
HDD Horizontal Direct Drilling
ILI In-line Inspection
IMU Inertial Mapping Unit
INSAR Interferometric Synthetic Aperture Radar
MFL Magnetic Flux Leakage
ROW Right of Way
SDL Strain Demand Limit
SMYS Specified Minimum Yield Strength
TSC Tensile Strain Capacity
UOE Uing and Oing forming process; a method used for production of longitudinally welded large diameter pipes.

STRAIN MEASUREMENT INSPECTION TOOL

A key benefit of the axial strain sensing technology is that it can be retrofitted to standard Magnetic Flux Leakage (MFL) tools. This means that whilst dedicated inspections can be performed to determine the axial strain condition of the pipeline, in most cases, inspections can be conducted efficiently and cost effectively during scheduled metal loss, calliper and mapping inspection surveys. Moreover, this enables measurement of both axial and bending strains in the pipeline using both AXISS and IMU respectively.

As shown in Fig. 1, the axial strain probes are retrofitted to the ILI MFL combo tool. The location of the sensors is critical to optimize the data, both in terms of signal quality and effective analysis of the axial strain. Currently, axial strain modules are mounted behind Magnetic Flux Leakage (MFL) platforms and have either 4 or 8 probes that provide circumferential readings typically at 0.5 to 1 m intervals. This arrangement allows the MFL module to produce a homogenous magnetic state in the pipe surface which reduces the hysteresis effect resulting from previous loading cycles. There are two separate platforms with probe configurations as shown in Figures 1a and 1b:

- Large Diameter: 24" and above platforms. All have 4 probes except 30/36" sizes have 8 probes
- Small Diameter: 12" to 18" platforms. All have 8 probe

Figure 1a. Axial Strain Tool – Large Diameter with 4 Probes

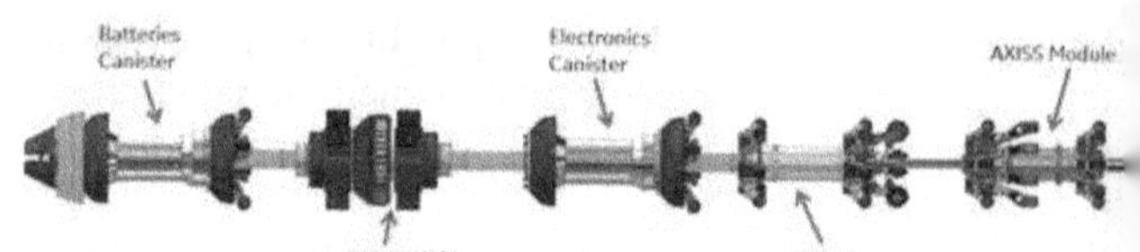

Figure 1b. Axial Strain Tool – Small Diameter with 8 Probes

Axial strain data sampling rate is set in order to record a minimum of one axial strain measurement per meter per probe for an overall average speed of 5m/s. Whilst this could be easily increased, it is more than adequate given the distances over which axial strain events typically occur.

The values reported are the axial strain variation about the axial strain mean across the entire pipeline. The variation refers to the amplitude of the change in axial strain reading compared to the mean. When using a calibration factor based on a representative pipeline sample of the inspected section, the axial strain will be absolute as opposed to relative. Sensitivity of the axial strain tool response to the differences in material properties between two pipes of the same grade but having different manufacturer has been observed. So, to obtain the most accurate results the operator needs to provide a material sample for each section where a change in the pipe grade or pipe manufacturer occurs.
The axial strain results are always expressed in terms of variations. If one accepts the data analysis assumptions, agrees on the strain model equations and representative pipeline

Copyright © 2019 by ASME

samples used during the calibration, the axial strain variations can be considered as absolute axial strain measurements. When no samples can be sourced, a standard calibration factor determined from laboratory tests and previous runs calibration factors library can be used [1].

When using a standard calibration factor, given the system's susceptibility to variations in steel permeability, the axial strain variation trend will be accurate allowing geotechnical and integrity engineers to identify areas of potential high axial strain and conduct further assessments and/or mitigation, if necessary.

ILI DATA ANALYSIS

Understanding the operating parameters of ILI strain measurement tools is fundamental when analysing loading conditions experienced by a pipeline.

The **IMU technology** provides geometry-based measurements of the centreline curvature. The data can be directly converted to strain using established definitions. Any elastic, plastic, or residual strain that changes the centreline curvature can be detected by the IMU. Detection capabilities for low level bending strain are limited (below 0.1% strain measurement or below 0.025% for changes in strain from one ILI run to the next run). This is due to tool vibration and the finite length of the IMU canister. However, IMU measurements alone cannot differentiate within which strain condition the bending strain exists. For example, using a subsequent inspection and run to run analysis is needed to determine whether a strain occurred during the operation of the pipeline. An increase/decrease in strain may be attributed to an active applied load.

The **AXISS technology** provides a measurement of a pipeline material's magnetic properties which can be converted to an absolute strain value based on the calibration factor for the pipe material. Axial strain sensors are designed and calibrated to operate in the elastic strain region of steel. Axial strain sensors have a small footprint and therefore represent a local elastic strain measurement at discrete locations around the pipe length and circumference.

UNDERSTANDING STRAIN HISTORY

When assessing any pipeline strain condition, the complete strain history must be considered to fully understand how the pipeline will respond to further stress. From the pipe manufacturing process to in-service loading conditions, the pipe experiences a series of strain events. Some may alter the pipe's response by permanent deformation while others may reduce the residual strain level by a combined aging process and strain relief through cyclic loading.

Pipe forming process: over the history of pipe manufacturing, a variety of pipe forming processes have been implemented (e.g. seamless, spiral welded, UOE, pyramid roll, mechanical expansion, cold drawing). Each forming process may influence the material properties such as the desired yield strength and the steel anisotropy. Pipe forming can also introduce significant residual strain distribution around the circumference. The residual strain may introduce local differences in the pipe strain response around the circumference especially in new pipes. Cyclic pressure loading and temperature variation (e.g. day/night) are expected to reduce residual strain levels over time by a work softening process.

Pipeline construction: during the pipeline construction and laying process, two kinds of strain are generally expected to be introduced, flexural elastic bending strain (e.g., by HDD) and residual strain (e.g. field bends). Additional strain inducing conditions may also be encountered by the pipeline during construction, e.g.:

- Tie-in alignment issues
- Interaction arising from ditch profile or soil support
- Topography
- Thermal expansion
- Unreported pipe weights
- Mishandling during pipeline roping

Roping pipe can potentially be classified as a field bend depending on the imposed curvature hence the response is similar to that of other fabrication bends. Roping pipe in this scenario would not reflect applied elastic strain, but purely a geometric change creating residual strain events. The IMU tool will measure the corresponding curvature change whereas axial strain inspection tool will measure a corresponding applied elastic axial strain, if any. Deviation between axial strain inspection tool and IMU measurements can identify the presence of plastic or residual strain conditions, i.e., problematic construction related conditions in pipe sections.

Pre-service hydrostatic testing: As part of the construction and testing processes, the pipeline will undergo hydrostatic pressure testing to a hoop stress level close to the yield strength. Hydrostatic pressure tests are based on the hoop strain alone and are designed to produce a strain range close to the plastic strain limit of the pipe without causing permanent deformation. However, the hydrostatic testing regime does not take into consideration any additional load which may have been introduced during the fabrication or the installation process, and thus producing a biaxial strain condition. One such example is referred to as expanded pipe [2] which often is the cause of unknown additional loads during the hydrostatic pressure test. Some expanded pipes result from the presence of pipe sections with lower yield strength than that required by the specified SMYS.

In-service loads: For buried pipes in-service loading is primarily related to the internal pressure and thermal expansion. Unless special design measures are implemented, the combined hoop and longitudinal stresses from operational loads shall be

Copyright © 2019 by ASME

limited to the elastic strain regime. However, additional external loading can produce a wide range of strain conditions ranging from one-off events such as a third-party strike to cyclic loading caused by frost heave or to continuous applied strains from a slow-moving landslide.

MATERIAL STRAIN STATE

The wall of a pipeline under operating conditions will be subjected to 3-dimensional strain state i.e., circumferential, longitudinal and radial strains. The radial direction may be neglected when appropriate, due to its low strain component. The resulting biaxial strain state of the pipe wall can be described using the von Mises theory of yield or the more conservative Tresca criterion [3]. The basic effect is illustrated in Fig. 2 and shows the tensile side of the yield ellipse. The compression side of the yield ellipse can generally be ignored for internally pressurized pipelines when the hoop stress component (y-axis) is positive. Figure 2 provides an illustration of a stress state that a pipe section may experience with a hoop stress at 90% of the yield stress. A calculated biaxial stress state resulting in the value falling inside the ellipse corresponds to an elastic stress state whereas one falling outside the ellipse corresponds to a plastic stress state.

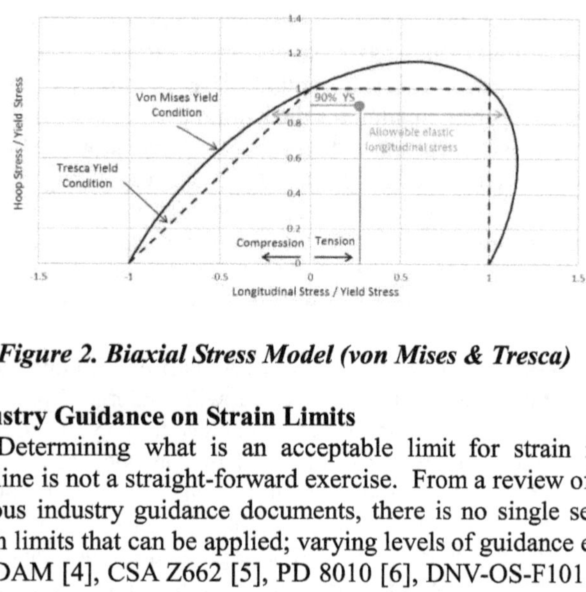

Figure 2. Biaxial Stress Model (von Mises & Tresca)

Industry Guidance on Strain Limits

Determining what is an acceptable limit for strain in a pipeline is not a straight-forward exercise. From a review of the various industry guidance documents, there is no single set of strain limits that can be applied; varying levels of guidance exist in PDAM [4], CSA Z662 [5], PD 8010 [6], DNV-OS-F101 [7], ASME B31.4 [8], ASME B31.8 [9], AS 2885 [10]. Also, the selection of an appropriate criterion is dependent on the loading that has caused the strain and specifically if the loading can be defined as being load controlled (also referred to as primary loading) or displacement controlled (secondary loading). Under load-controlled situations (e.g., internal pressure) stress limits will apply whereas with displacement-controlled situations (e.g., from thermal loading or ground movement induced loading) strain limits can be applied. However, in many cases the loading will be some combination of both mechanisms. In load-controlled situations the magnitude of the load is not affected by the amount of deformation or displacement of the structure. Whereas, in displacement-controlled situations, local yielding of the pipe will relieve the stresses placed on the pipeline. However, in practice it might not be possible to draw a clear distinction between load and displacement-controlled conditions. If there is any doubt regarding the loading condition load-controlled and stress-based criteria should be taken as the more conservative of the two options [4]. When developing or implementing a stress or strain-based assessment criteria the specific pipeline circumstances need to be explicitly considered.

Longitudinal Stress Limits (Load Controlled)

The tensile longitudinal stress limit in ASME B31.4 and B31.8 [8,9] is defined as 90% SMYS. The limits for compressive longitudinal stress are governed by the equivalent combined stress criterion. The highest allowable compressive longitudinal stress is approximately 29% SMYS under Von Mises criterion [11].

To convert a reported axial strain value to the corresponding stress the Ramberg-Osgood relationship (CSA-Z662 [12]) can be used, see Fig. 3 below. This enables the longitudinal stresses from the different loadings to be combined and compared with the allowable limits. It also allows for the presence of wall loss defects to be accounted for. There are several stress-based methods for assessing circumferentially orientated wall loss defects subject to axial stresses due to bending and internal pressure [13,14].

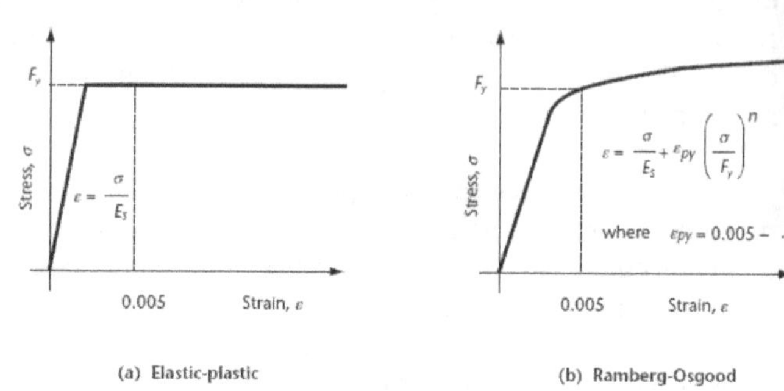

Figure 3. Stress-Strain relationships for steel pipe (taken from Fig. C.2 in CSA-Z662 [5])

Longitudinal Strain Limits (Displacement Controlled)

There is various published guidance for deriving strain acceptability limits. The CSA Z662 document offers relationships to calculate conservative compressive acceptability limits; these limits depend on pipe geometry, properties, defect dimensions and design or operating pressure. Tensile strain capacities as low as 0.2% and as high as 2 to 4% have been obtained from experimental tests [12]. The tensile strain capacity of pipe body material is expected to be relatively high; but at girth welds (even welds compliant with API 1104) it can be low [16].

Copyright © 2019 by ASME

It is noted that ASME B31.4 and B31.8 permit a tensile strain limit of 2% (significantly higher than the 90% SMYS stress limit) when certain conditions are met. However, these standards do require consideration of ductility and strain capacity of welds and pipe body material but the specific conditions by which the 2% strain limits are attainable are not fully defined.

DNV-OS-F101 allows a girth weld fracture (tensile strain) limit of 0.4% strain (good quality girth welds). The strain limit in AS 2885.1 applied to roped curvature during construction is 0.5% strain. Earlier versions of CSA-Z662 [12] did include (in the absence of pipeline specific data) a tensile strain limit of 0.5%. However, in more recent versions of the standard it is recommended that the tensile strain capacity of the pipe wall or weld should be determined by sound and proven methods that have been validated with physical tests. The methods shall take into consideration flaws, metallurgical damage, and mechanical property changes in the welds and HAZ. To obtain realistic strain limits, the strain capacity should be determined on a pipeline by pipeline basis using testing and/or modelling of the specific pipeline characteristics to predict the maximum amount of axial strain that the pipe can sustain before failing at a girth weld. Several tensile strain models have been published by PRCI/CRES [17-22]; this is an area of on-going research [11].

To add a degree of conservatism, a safety factor should be applied to the strain capacity to derive the strain demand limit (SDL); for example, CSA-Z662 requires tensile and compressive resistance factors of 0.7 and 0.8 to be applied to the tensile and compressive strain capacities.

CASE STUDY

This case study describes a specific HDD section of a 12" diameter ERW, X60 pipeline constructed in 2012. A highway and railway crossing location of the 12" pipeline was constructed using the HDD technique with a drill path located under the highway and adjacent railway crossing. The ROW is not shared with any other pipeline. The pipeline generally follows a straight path; however, the resulting drill path produces an oscillating horizontal and lower level vertical bending strains for the majority of the length.

Deviations in bending strain values reported by both IMU and the axial strain inspection tools were observed at the largest strain locations illustrated in Fig. 4. One of the sections located after the exit side matches a pipe roping section with vertical strain deviation.

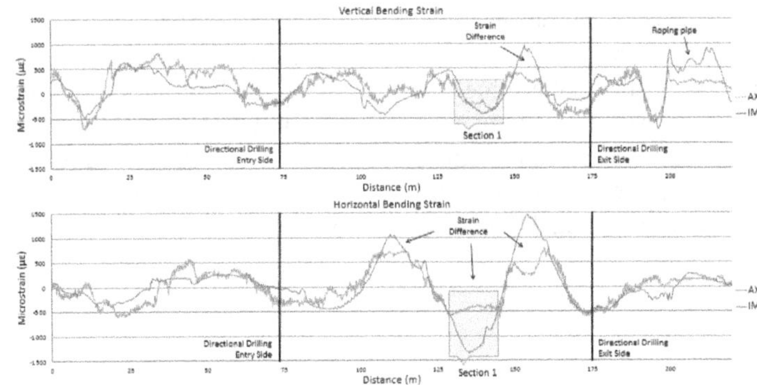

Figure 4. The bending strain components for the Highway & Railway Crossing as measured by IMU and Axial strain inspection tool

Section 1 is shown in Fig. 5; the longitudinal strain distribution pattern exhibits a distinct flat response in both the tension and compression sides and maintains the dominant orientation of the horizontal bending as shown in Fig. 5 with the max/min amplitude centered about the 90-270deg orientations (right-left). To fully understand the strain measurements, one must understand the pipe strain conditions. The anomalous readings toward the top of the pipe (330deg) correspond to one of the axial strain sensors riding on a seam weld for the duration of the plotted section. Seam weld locations are discarded during final analysis.

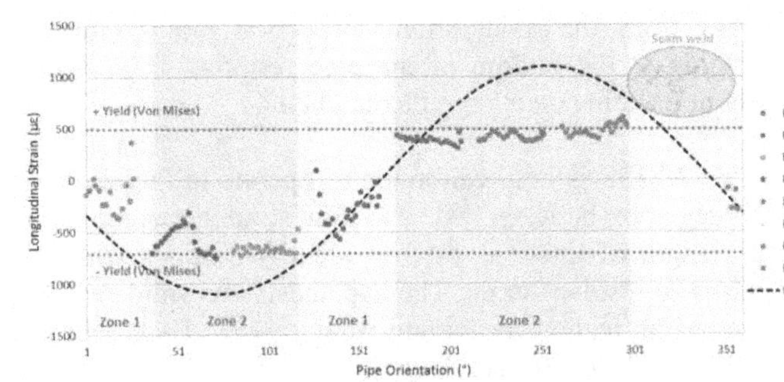

Figure 5. Longitudinal Strain Profile at Section 1

Knowing the hydrostatic pressure test parameters, the biaxial strain state can be assessed and compared with the yield ellipse for a defined set of pipe dimensions and material properties. The allowable longitudinal elastic strain before yield for both tension and compression can be estimated and compared to the measured longitudinal strain.

The von Mises yield was determined based on nominal yield stress defined by the pipe grade. The hoop stress ratio was calculated using the reported hoop stress level of 110% SMYS during hydrostatic testing. Because the pipe section was short (<100m) and capped (by pipeline end plug), the pipe was considered as unrestrained. Assuming hydrostatic testing did not introduce additional curvature to the HDD drill path, the initial

Copyright © 2019 by ASME

longitudinal stress prior to the hydrostatic testing is equivalent to the bending strain recorded by the IMU. The hardening rule was not applied due the limited information on loading during and prior to construction. Subsequently, the biaxial stress model was constructed as illustrated in Fig. 6 and the allowable elastic longitudinal strain limits were estimated for both tension and compression. These theoretical limits are plotted in Fig. 5.

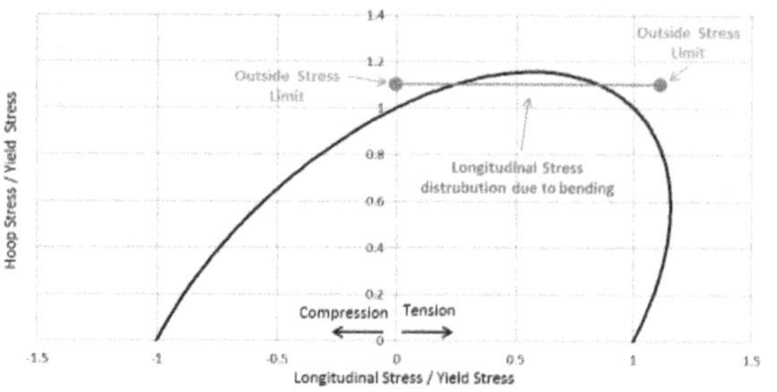

Figure 6. Biaxial Stress Model (von Mises) during Hydrostatic Testing at Section 1

According to the biaxial stress model in Fig. 6, the maxima and minima around the pipe circumference for the combined hoop and longitudinal stresses fall outside the yield stress limit as defined by von Mises yield ellipse. These theoretical limits for the hydrostatic test condition are added to Fig. 5.

The overlap between the theoretical yield (von Mises) in both tension and compression and the plateau in the longitudinal strain from the axial strain inspection tool confirms local yielding at the bottom of the pipe resulting from hydrostatic pressure testing on a large flexural bend.

Since AXISS is only calibrated to operate in the elastic range, the deviation between IMU and axial strain measurements along the length of line was used to identify sections with possible plastic or residual strain. The hypothesis was validated based on known fabrication bends and roping pipes. Extrapolating to a more local level, the deviation between both measurement techniques around the pipe circumference can also help to identify the presence of possible plastic or residual strain conditions in different pipe sections.

As shown clearly in Fig. 5, compared to the IMU data, the flatter axial strain distribution (identified as zone 2) around the circumference suggests that most of the stress was relieved by permanent deformation. In addition, the elastic strains (identified as zone 1) are concentrated in the transition zone from tension to compression. IMU tool will measure the curvature change whereas axial strain inspection tool will measure the corresponding applied elastic longitudinal strain which is reflected by the discrepancy in the bending strain amplitude.
In this case study, the hydrostatic testing is a time limited event in the pipeline timeline and occurs during the pipeline construction stage. The longitudinal strain measurement is conducted with the pipeline in-service; therefore, the biaxial stress model should be verified not only for any historical event, but also for pipeline in-service conditions with results compared to the relevant allowable stress level.

The in-service pipeline conditions are as follows, a restrained pipeline at hoop level of 27% SMYS, and a measured bending stress of +25% to -35% of SMYS (as measured by axial strain tool). The allowable stress limit is set to 90% SMYS for reference. The corresponding yield ellipse for the above condition is illustrated in Fig. 7.

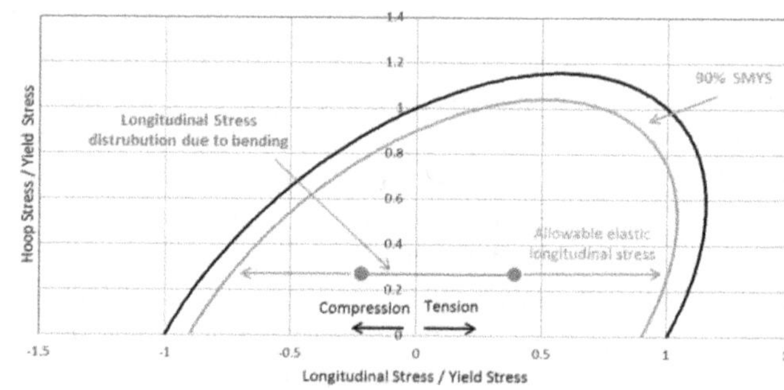

Figure 7. Biaxial Stress Model (von Mises) during In-Service Condition at Section 1

The combined stress effect from the calculated hoop and longitudinal stresses around the pipe circumference fall inside the 90% of SMYS criteria for the allowable stress level at Section 1.

No hardening rule was applied on the yield ellipse for the in-service pipeline stress assessment in Fig. 7. However, the largest biaxial stresses from hydrostatic testing in Fig. 8 are outside the yield limit. Translation or general expansion of the yield ellipse surface would have occurred according to the relevant hardening rules. In Fig. 8, the biaxial stress for in-service conditions was assessed using both the Isotropic and Kinematic Hardening rules based on the effects of hydrostatic testing during construction. For comparison, the hydrostatic testing condition is also illustrated to demonstrate the shift of the yield ellipse in response to the largest biaxial stress state experienced by the pipe at Section 1.

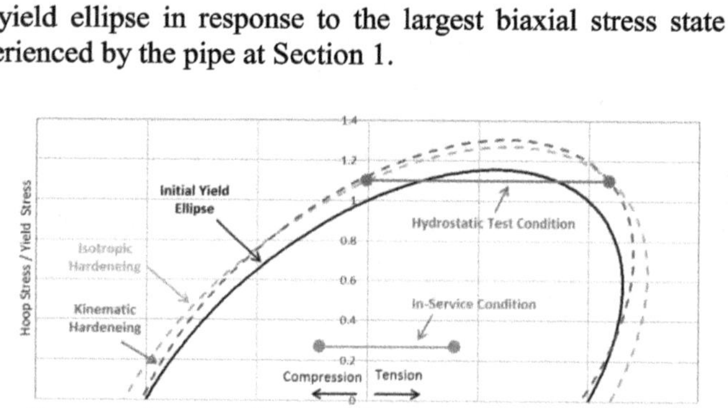

Copyright © 2019 by ASME

Figure 8. Biaxial Stress Model (von Mises) Hydrostatic test & In-Service Condition using Hardening rules at Section 1

Applying the hardening rules in Section 1 reinforces the previous assessment that the pipe is operating inside the allowable stress level criteria.

CONCLUSIONS

Utilizing both ILI strain measurement technologies, AXISS and IMU, allows for better understanding of an operating pipelines' stress and strain state.

- The IMU technology provides geometry-based measurements of the centreline curvature. The data can be directly converted to strain using established definitions. Any elastic, plastic, or residual strain that changes the centreline curvature can be detected by the IMU.
- The AXISS technology provides a measurement of a pipeline material's magnetic properties which can be converted to an absolute axial strain value based on the calibration factor for the pipe material.
- Active elastic longitudinal strains are detected by the axial strain inspection tool and bending strains are detected by IMU.
- Accurate measurements and correlation between both technologies in the elastic range is achieved using representative calibration for the axial strain inspection tool.
- Although there is no single set of acceptable strain limits that can be applied, there are various published guidance documents available. To determine what is an acceptable limit for strain in a pipeline, consideration should be given to the biaxial stress/strain state of the pipeline as well as to the line pipe material and weld properties, the presence and type of defects coincident with the loading and the need to incorporate a factor of safety to account for uncertainties
- Deviation between axial strain inspection tool and IMU measurements can help to identify the presence of possible plastic or residual strain conditions in the pipe sections studied.
- To assess biaxial stress/strain state of the pipeline, longitudinal strain is measured by axial strain inspection tool and hoop strain is pressure dependent and can be calculated. Understanding the yield ellipse for a defined set of material properties and using the pipeline operating parameters provides a full biaxial picture of the strain condition of the pipe section.
- It is important to determine the source and magnitude of the largest biaxial stress state encountered by the pipe during pipeline construction or operation. Comparison of the yield ellipse to the largest biaxial loading condition to assess possible historical permanent deformation should be carried out.
- To better understand the strain history, regular inspections provide the opportunity to measure and monitor the strain condition of the pipeline to distinguish active load and historical events.
- The use of IMU for pipeline bending strain assessment and AXISS for axial strain assessment is effective in detecting areas subject to external bending loads and fully complements any in-situ slope or strain monitoring being performed by operators.
- There are many advantages of combining multiple inspection technologies to effectively build a picture of the pipeline condition and additional loadings and how this changes over time. The approach of regular inspection and monitoring in this way, combined with effective in-field investigations can be utilised to address the threat from axial loadings to pipeline integrity.

ACKNOWLEDGMENTS

The authors would like to thank Baker Hughes, a GE company (BHGE) for the permission to publish this paper.

REFERENCES

1. Westwood, S., Jungwirth, D., Nickle, R., Dewar, D., Martens, M., 2014, "In-line Inspection of Geotechnical Hazards", Proceedings 10th International Pipeline Conference, Calgary, Alberta, Canada.
2. Rapp, S.C., Lamberth,W., Dimopoulos D., Gotsis V. and Rosenfeld, M.J., 2010, "Analysis of pipe expansion associated with field hydrostatic testing", Proceedings of International Pipeline Conference, IPC 2010-31666.
3. Hart, J.D., Powell, G.H., and Zulfiqar, N., 1996, "A Material Model for Pipeline Steels", ASME Offshore Mechanics and Offshore Engineering International Pipeline Conference, Calgary, Alberta, Canada.
4. Andrew Cosham et al, 2016, The Pipeline Defect Assessment Manual (PDAM), A Report to the PDAM Joint Industry Project.
5. The Canadian Standards Association (CSA), 2015, CAN/CSA-Z662-15, A National Standard of Canada, Oil and Gas Pipeline Systems, 2015.
6. Technical Committee PSE/17 & Subcommittee PSE/17/2, PD 8010-1:2004, Code of practice for pipelines - Part 1: Steel pipelines on land, British Standards Institution.
7. Det Norske Veritas (DNV), DNV-OS-F101, 2015, Offshore Standards, Submarine Pipeline Systems.
8. ASME B31.4, Pipeline Transportation Systems for Liquids and Slurries Code, 2016.

Copyright © 2019 by ASME

9. ASME B31.8, Gas Transmission & Distribution Piping Systems. 2016.
10. AS 2885.1. Pipelines - Gas and liquid petroleum. Part 1: Design and construction, 2012.
11. Guidance for Conducting Strain-Based Assessment of Buried Pipelines Subjected to Ground Movement, 2016, PRCI project SBD-1-2, draft report.
12. The Canadian Standards Association (CSA). CAN/CSA-Z662, A National Standard of Canada, Oil and Gas Pipeline Systems, 2011, 2007, 1999.
13. Review of Limit Loads of Structure Containing Defects, 1988, Int J Press Vess and Piping, 32, Nos.1-4, pp. 197-330. A G Miller.
14. Kastner, W., "Critical Crack Sizes in Ductile Piping", 1981, Int J Press Vessel and Piping, pp. 197-219.
15. Wang, Yong-Yi et al, 2015, "Integrity Management of Pipelines Subjected to Ground Movement Hazards", Paper 17, 20th Joint Technical Meeting, Paris, France.
16. API Standard 1104, Standard for Welding Pipelines and Related Facilities, 2013.
17. Wang, Yong-Yi et al, 2016, "Integrity Management of Ground Movement Hazards", International Pipeline Conference, Calgary, Alberta, Canada.
18. Wang, Y.-Y., Liu, M., Zhang, F., Horsley, D., and Nanney, S., 2012, "Multi-tier tensile strain design models for strain-based design Part I - fundamental basis," International Pipeline Conference, Calgary, Alberta, Canada.
19. Liu, M., Wang, Y.-Y., Song, Y., Horsley, D., and Nanney, S., 2012, "Multi-tier Tensile Strain Models for Strain- Based Design Part II - Development and Formulation of Tensile Strain Capacity Models," International Pipeline Conference, Calgary, Alberta, Canada.
20. Liu, M., Wang, Y.-Y., Horsley, D., and Nanney, S., 2012, "Multi-tier tensile strain design models for strain-based design Part III - model evaluation against experimental data," International Pipeline Conference, Calgary, Alberta, Canada.
21. Wang, Y.-Y. and Liu, M., 2013, "Status and Applications of Tensile Strain Capacity Models," Pipeline Technology Conference, Ostend, Belgium.
22. Kibey, S., Issa, J.A., Wang, X., and Minnaar, K., 2009, "A simplified, parametric equation for prediction of tensile strain capacity of welded pipelines", Pipeline Technology Conference, Ostend, Belgium.

Copyright © 2019 by ASME

SECTION 9
ABOVE GROUND STRAIN MONITORING

INTRODUCTION

By Mamdouh M. Salama
Mamdouh.m.salama@conocophillips.com

This chapter focuses on an interesting integrity assessment technology known as large standoff magnetometry (LSM). LSM is passive technique, it relies on the magnetic state of the material and knowledge of the ambient magnetic field. LSM detects changes in the magnetic properties of the pipe due to the presence of a defect such as dent, corrosion and cracking. Unlike MFL, LSM does not require applying magnetic field but it relies on the earth magnetic field. In the presence of an external magnetic field such as the Earth's magnetic field, changes in stress due to the presence of a defect can alter the magnetization of ferromagnetic materials as shown in Figure 1.

This technology is attractive because it operates at a distance from the pipe, hence the term 'large standoff and thus it can be used to inspect buried pipeline from above ground, i.e. without any contact with the pipeline as shown in Figure 2. The magnetometers used must be sensitive enough to detect small changes in magnetic field

LSM does not measure the dimensions of geometric defects but it measures the stress caused by these defects and identifies their character, location and orientation in accordance with the location and orientation of the area of stress. LSM offers several advantages including:

- It is often considered for the undiggable pipelines where in-line inspection method is inapplicable.
- It doesn't require any preparation of the pipeline for testing such as cleaning or stopping pipeline operation.
- It doesn't require magnetizing the pipe;
- It doesn't have limitation on the diameter, configuration, operating conditions, fluids transported and method of protection

The challenge with this technology is that the probability of false calls may be high in the range of 30-50 %, implying that one in three excavations will be unnecessary. The reason for high false calls is that the magnetic response is not only influenced by the change in the stress field, but it is also influenced by microstructure, metallurgical and physical changes in the material that may be caused by the defect. Therefore, it is expected that differ-

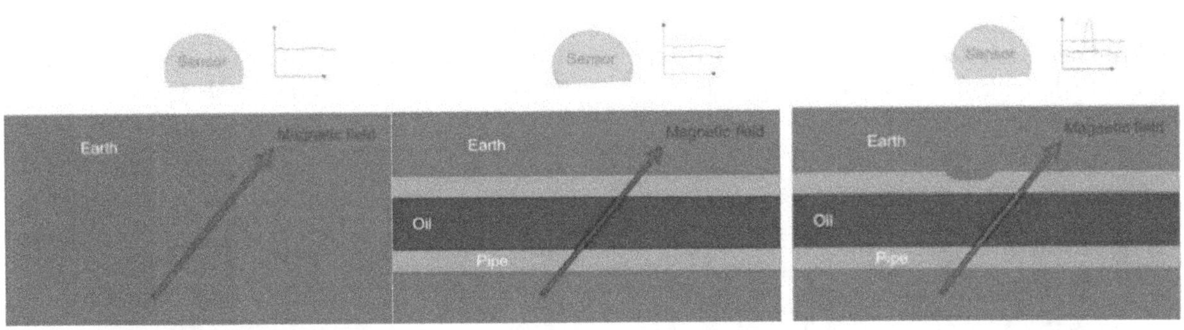

Figure 1. Change in the magnetic signal due to the presence on undamaged pipe (uniform change) and damaged pipe (non-uniform change).

Copyright © 2019 by ASME

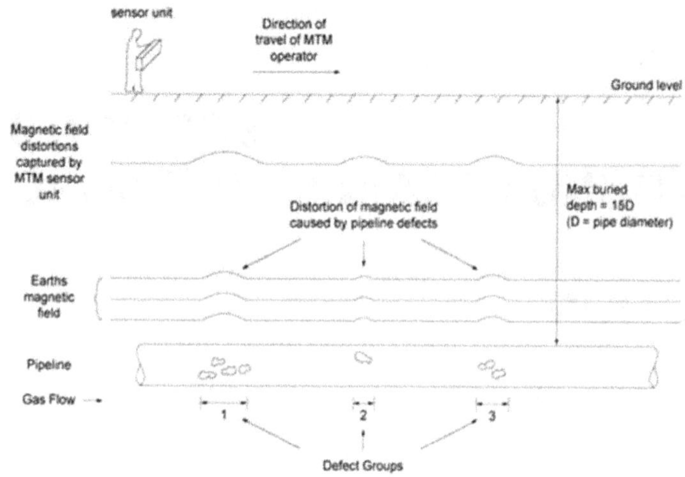

Figure 2. LSM Principle of Operation.

ences exist between magnetic field signatures caused by the different defects. This makes LSM good screening tool to identify areas where damage is likely that requires further inspection by other methods.

Commercially available LSM products utilize gradiometer technology based on fluxgate magnetometers. Fluxgate Magnetometers measure the magnetic field environment, the magnetic signature of metal objects, and the magnetic signature of non-metal objects. Fluxgate Magnetometers usually operate over the range of 100 pT to 1 mT. Gradiometers is generally used to monitor changes in magnetic field and thus are less susceptible to the ambient magnetic field. Gradiometers are suitable for detecting small, near-surface anomalies. While commercially available LSM systems have promise as complementary techniques to ILI and other methods, user needs to confirm several parameters such as: the probability of true call for the anomalies identified, the probability of identification, the probability of detection, the probability of false calls and locational accuracy.

Three papers in this chapter (AIM-PIMG 2019- 1007, 1019 and 1024) focus on LSM while the fourth paper (V1005) addresses the application of fiber optics in assessing permafrost stability thru temperature monitoring along the pipeline. The LSM papers addressed the important question of what the detected stress is based on as related to the existence of an anomaly. The papers suggest that the measured stress Is not the maximum stress but rather some form of an average stress. This means that the measured stress is likely to be an elevated elastic stress. This may be valuable because the magnetic sensors cannot clearly differentiate between elastic and plastic strains. Figure 3 presents the effect of stress on the magnetic signal showing that one can have the same signal intensity while the strain is either below or above the yield strain.

Chapter 34 Large Standoff Magnetometry Technology Advances to Assess Pipeline Integrity Under Geohazard Conditions and Approaches to Utilisation of It (AIMPIMG2019-1024)
Hamed Habibi, Chau Ky Vo, and Paul Jarram, Speir Hunter, UK

Chapter 35 Use of Large Standoff Magnetometry for Geohazard Pipeline Integrity Investigations (AIMPIMG2019-1019)
Chukwuma Onuoha, Eric Pozniak, Jerell Tolentino, Shamus McDonnell, Rob Zmud and Vignesh Shankar, PureHM, Canada

Chapter 36 Stress Characterization of Mechanical Damages in Gas Pipelines Using Large Standoff Magnetometry Technology (AIMPIMG2019-1007)
Xian-Kui Zhu, EWI, and Tianzong (David) Xu, Pacific Gas and Electrical Company, USA

Chapter 37 Monitoring and evaluation of pipeline stability at permafrost based on distributed optical fiber temperature measurement (AIMPIMG2019-1005)
Cai Yongjun, Bai Luyao, Tan Xianjun and Wu Zhangzhong, PetroChina Pipeline R&D Center, China

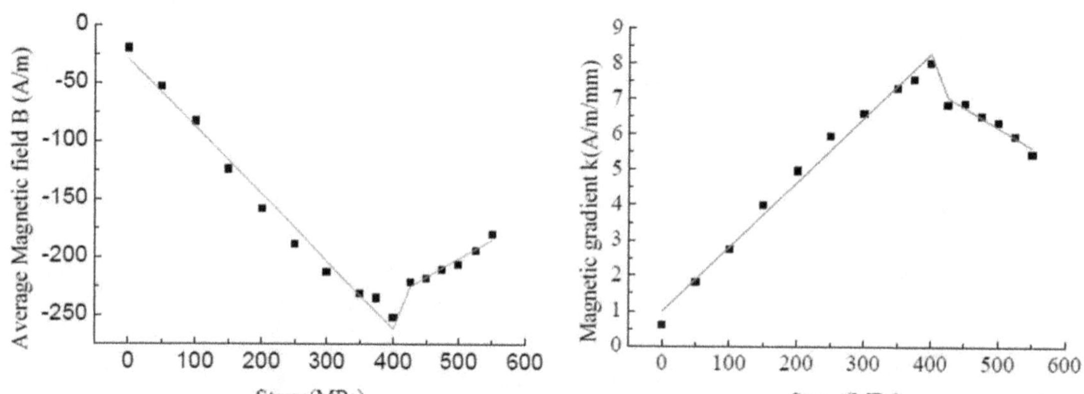

Figure 3. Magnetic signal as function of applied strain on steel specimen. (OMAE2015-41560, Characterization of stress-induced residual magnetic field in Ferromagnetic steels, by Sheng Bao, Li Lin, Da Zhang and Meili Fu)

Copyright © 2019 by ASME

AIM-PIMG2019-1024

Chapter 34

LARGE STANDOFF MAGNETOMETRY TECHNOLOGY ADVANCES TO ASSESS PIPELINE INTEGRITY UNDER GEOHAZARD CONDITIONS AND APPROACHES TO UTILISATION OF IT

Hamed Habibi[1], **Chau K Vo, Paul Jarram.**
Speir Hunter
Newark, United Kingdom

ABSTRACT

Large Standoff Magnetometry (LSM) has emerged as an innovative and non-intrusive inspection technology, able to detect localised mechanical stresses in pipeline walls as well as inferring the amount of stress at these sites. LSM has traditionally been applied to identify pipeline threats which are both time dependent (i.e. corrosion) and time independent (i.e. dents). However, continued development of this technology has opened the door to even more beneficial applications such as uncovering stress associated with earthquakes, subsidence and mud slides. LSM's unique ability to continuously log a pipeline's depth of cover as well as noting changes in the pipe diameter and wall thickness all contribute to the exporting of a three-dimensional (3D) pipeline map, which play a key role in assessing the severity of these geohazard conditions. To follow will be a summary of outcomes from numerous independent case studies of LSM performance on pipelines experiencing geohazardous conditions. The paper also illustrates the fundamental underlying science behind LSM technology and supplies insight into the original R&D rigor. Sources of detection uncertainty that need to be considered in assessing its applicability will also be discussed. LSM's specifications with relation to positional accuracy, pattern matching, and stress level estimation will be explored as well as lessons learned during field correlation exercises to verify overall system performance. The aim is to explore the current approaches and benefits of applying LSM in geohazardous conditions, with a realistic explanation of achievable end results. LSM results demonstrate the considerable benefits from its use in augmenting pipeline integrity management programs which are currently challenged by disruptive geohazardous conditions. LSM, as a complementary or screening technique, is capable of proactively identifying threats before they escalate to a more serious state, in turn enabling engineers to prioritize efforts and direct precious resources to other high risk areas.

Keywords: SCZ, LSM, ILI, GNSS, GPS, SMYS, FEA, DOC, Wrinkle bend, Landslide, Ground Movement, Sagging, Buckling, Weld Mismatch, Mechanical Stress, 3D Mapping, Dent, Plastic Deformation, SCC, External Corrosion, Pipeline Mapping, Displacement-Controlled Loading, Strain-Based Design.

1 INTRODUCTION

LSM is now known to all experts in the pipeline industry with ongoing debates around its competency to be used as a complementary or independent inspection technology with both applications showing a wide array of benefits. Sharing recent advances, evaluation results, case studies and utilization approaches adopted by some experts in the industry is therefore beneficial in order to open new windows to integrity departments working in exceptional conditions including pipelines suffering from geohazards.

1.1. Localised Stress Estimation

Magnetostriction was first established by Joule in 1842 using a bar of iron and mechanical levers to show that a bar expands in the direction of applied magnetisation.[1] In 1865, Villari showed that tensile stress on a bar of steel will alter the magnetic field around the bar which is known as Inverse-Magnetostriction, or Villari Effect. [2] Staples et. al. solved the inverse problem and defined the relationship between stress and magnetic fields produced by steel under localised stresses, focusing on steel pipe sections. [3] The problem is solved for carbon steels under elastic strain, which includes pipelines.

[1] Contact author: hamed.habibi@me.com

Copyright © 2019 by ASME

Unless failure of the pipeline is imminent, it is in the elastic stage. Geohazard conditions that cause local deformation usually affect the pipeline by way of displacement of ground around the pipe which may have nonhomogeneous stiffness due to large stone fragments. This displacement may stop after deformation of the pipeline or it may continue and cause failure. For pipelines that are in service and experiencing such conditions, the pipe material is in the elastic stage because the load will be released after permanent deformation and the material re-enters the elastic stage through a hysteresis behavior.

1.2. Three-Dimensional Mapping

FIGURE 1 illustrates a buried pipeline with an LSM scanner carried above it at an offset. The measured vector of the magnetic field from each magnetometer will point towards the centre of the pipelines.

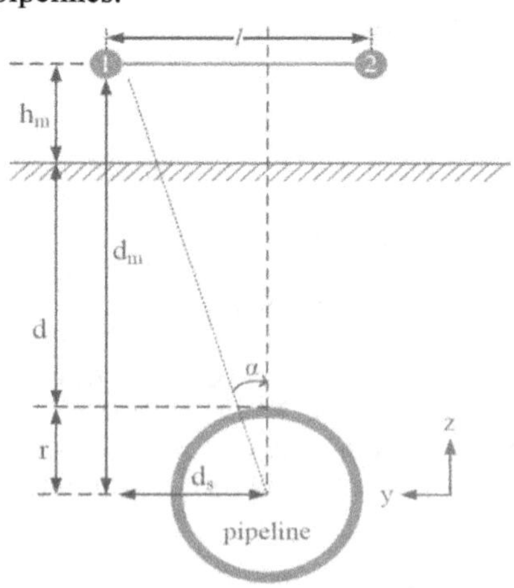

FIGURE 1: SCHEMATIC OF THE RELATIONSHIP BETWEEN TWO MEASUREMENTS LABELLED 1 AND 2, SEPARATED AT A DISTANCE *l*, AND THE BURIED DEPTH OF THE PIPELINE. [4]

The point at which all vectors converge is the centre of the pipeline and Depth of Cover (DOC) and Lateral Offset (LO) can be derived as follows: [4]

$$d_m = l/(\frac{B_{y1}}{B_{z1}} + \frac{B_{y2}}{B_{z2}}) \quad (1)$$

$$d_s = \frac{B_y}{B_z} d_m \quad (2)$$

$$d = d_m - h_m - r \quad (3)$$

So long as the LSM scanner is able to receive the magnetic field of the pipeline, it is possible to derive DOC and LO from the magnetic signal. This can then be added to the recorded pipeline route to form a Three-Dimensional map that includes the ground altitude. In equations (1) to (3), d is the Depth of Cover and $d_s - l/2$ will be the Lateral Offset.

2. EVALUATION APPROACHES

Lack of an internationally recognised standard for applying LSM has triggered several field trials in the industry, including pipelines operating under geohazardous conditions. Despite the fact that not all of these field trials have been ideal in all aspects, the objective is to attain as many promising results as possible to facilitate an understanding by integrity engineers of the LSM capabilities that might suit their needs. In this section three main approaches taken for the evaluation of LSM results are explained.

2.1. Field Verifications

Arguably the best, yet not always the most economical, approach to evaluate LSM is to apply it on a live pipeline where its condition is known or could be verified through excavations. It should be noted however, that there is no straightforward method to measure stress in the wall of an operational pipeline. Also, assumptions pertaining to the condition of the pipeline include many variables which could impact eventual results alignment with defined indications. For instance, Geo-coordinates projection transformation could add up to 10m of positional inaccuracy.[**5**] It is, therefore, crucial to use the same projection for both LSM and the reference survey. Alternatively, this additional inaccuracy must be considered during any evaluation.

Significant efforts have been made to minimize uncertainties and maximize evaluation reliability by selecting pipelines that are more likely to be in the technical scope of LSM. Pipelines with a diameter of larger than 8 inches, an average Depth of Cover (DOC) of less than 2m, nominal operating Hoop Stress (HS) of more than 65MPa and minimum length of survey of 100m are among such criteria. Pipelines that have been recently magnetized by a Magnetic Flux Leakage (MFL) tool were not selected because the impact of such a tool on LSM capabilities has not yet been thoroughly investigated.

For verification against known results, such as those from MFL ILI tools, it was considered that there are more SCZs than macroscopic defects due to the ability of LSM to detect all stress inducing conditions and the MFL limitation of detecting only metal loss. For verifications through excavation, there were only one to six excavations on inspected lengths ranging from 100m to 10km. On both occasions it was not possible to derive a Probability of Detection (POD) of LSM. In the first verification method, the definition of an indication varies between LSM and conventional methods, while the second method does not include enough actual data to provide statistical results.

2.2. Computational Methods

This approach is mainly used for the verification of reported stress levels by LSM. Very few methods exist that are capable of measuring the live stress in a pipeline wall, even when it is exposed. This is particularly important when the pipeline is suffering from geohazards but even strain gauges are not able to measure the actual stress in a pipeline wall. Rather, they are able to monitor changes in surface strain in particular orientations, and are usually effective in monitoring changes in strain along longer stretches of the pipeline.

Once a macroscopic defect is identified and accurately measured, one way to estimate its stress is to use a computational

Copyright © 2019 by ASME

method such as Finite Element Analysis (FEA). This still, however, includes a level of uncertainty resulting from the accuracy of observations coupled with the mathematical modeling of geometry and operating conditions. Some of these uncertainties exist when the data from strain-gauges is being studied.

The approach of developing the LSM technique by solving the inverse problem of magnetization-stress resolution has shown excellent agreement between stress and the measured magnetic field. [3] Stress-strain relationship is assumed to be elastic because some studies suggest the magnetisation-stress relationship follows a one-to-one function only is elastic mode. It is important to know the operating condition of the subject pipeline at the time of an LSM inspection as well as accurate dimensions of the known defect in order to have the best results. Modeling the geometry of a defect is, on the other hand, challenging. There is no 'ideal' defect, which could otherwise be easily modeled. However, modeling maximum width, length and depth of defect results in a simulated stress that is slightly higher than actual stress. For instance, when there is a large metal-loss that has a shallow depth for 90% of its area, and a greater depth for the remaining 10%, the resulting simulated stress is less than that of a modeled defect with its maximum depth distributed evenly across its entire area.

Knowing the mechanical properties of a pipeline material, for instance elastic modulus, also plays an important role. Forces acting on a pipeline could be from a combination of Load-Controlled (LC) and Displacement-Controlled (DC) loadings. LC loading examples are weights of the pipeline and surrounding earth. Ground movements around the pipeline will cause DC loadings. In computational analysis, it is always assumed that the stress-strain relationship is elastic. The reason behind this assumption is that under DC loadings the pipeline will only reach plastic mode during a deformation, or just before a rupture. This is also known as plastic strain. Weight of pipeline running on a slope is LC but it can be considered a DC loading due to presence of anchors and restrictions ground will make for movement of pipeline sections. Strain-based approach will only allow the pipeline to go to plastic mode only when strain demand is high. Displacement restriction will drop the stress when target displacement is made. The pipe will not settle in the plastic stage and once the deformation occurs, the load is removed, and it returns to the elastic stage with some residual strain that is the deformation, such as a dent or a wrinkle. An effort to estimate elevated stress from a wrinkle bend (dent) failed to achieve reasonable results due to an incorrect boundary condition and neglection of the fact that after formation of the dent, the load that makes it is removed and a residual stress will be dominated by operating stresses under new geometry.[6] In real world external forces acting on pipelines are due to displacements of ground and forces will settle after target displacement is made. The only forces that will remain are weights pipeline and surrounding earth. In this study a model of a full size pipe section of 12m length with an idealised defect is developed for FEA. To avoid any confusion in setting boundary conditions, an internal pressure and external atmospheric pressure is also applied to inner and outer surfaces as boundary conditions and displacements of one end of pipe is set to be zero.

In the case of subsidence, only pipeline deflection could be measured. Thus FEA was done in two stages. First, a load was applied in the middle of the pipeline until the calculated deflection reached the actual measured value. The required load to achieve such a situation was then used, in addition to operating pressures, to evaluate the stress at this point.

2.3. Data Fusion

The other approach taken to evaluate the performance of LSM was the fusion of all aspects of its output with all available data to assess the pipeline condition. For instance, in the case study for landslide assessment, a stretch of pipeline was selected where LSM reported a dramatic change in depth of cover, detection of a group of SCZs with high stress levels and deflection of the pipeline route from an imaginary straight line. Prior to inspection with LSM this area was considered to have a high risk of geohazard.

3. RESULTS

Evaluation results from a number of case-studies show the potential benefits of LSM in assessing pipelines under geohazard conditions. Subject to resources available at the time, the previously mentioned approaches have each been used to demonstrate LSM performance. The performance of LSM with regards to the detection and identification of stress levels has been discussed in other literature [7,8], with the focus of this study remaining on LSM capabilities in relation to geohazard monitoring.

3.1. Landslides

A LSM inspection was commissioned on a 20 inch diameter pipeline in Alberta, Canada. The subject pipeline shared a Right of Way (ROW) with a further 6 pipelines. TABLE 1 lists technical parameters of the subject pipeline. The pipeline was monitored using strain-gauges, slope inclinometers and water level sensors.

TABLE 1: TECHNICAL PARAMETERS OF THE PIPELINE

Diameter [in]	Wall Thickness [in]	Operating Pressure [MPa]	Pipeline Material Grade	SMYS of Pipeline Material [MPa]	Nominal Hoop Stress [MPa]
12.75	0.205	0.473	X56	386	14.7

FIGURE 2 shows a stress strip chart for a 200m length of this LSM survey. Considering the fact that operating pressure was only exerting a nominal hoop stress of 4% of Material SMYS in the pipeline wall, the detected SCZs showed a significant amount of increased localised stress. FIGURE 3 shows the reported DOC for the same section - also reported by LSM. The survey route is plotted in FIGURE 5 against an imaginary straight line that illustrates a 1.4m deflection. Offset

Copyright © 2019 by ASME

measurements by LSM shows the scanner as being within 20cm of the top dead center of the pipeline, as shown in FIGURE 4. From 1400m to 1480m of this survey, LSM showed an increase in density and stress levels of SCZ indications, along with a dramatic change in DOC and a 1.4m lateral deflection in the pipeline route – factors which may indicate that the pipeline is suffering from external forces caused by landslide. The deflection is shown in FIGURE 6.

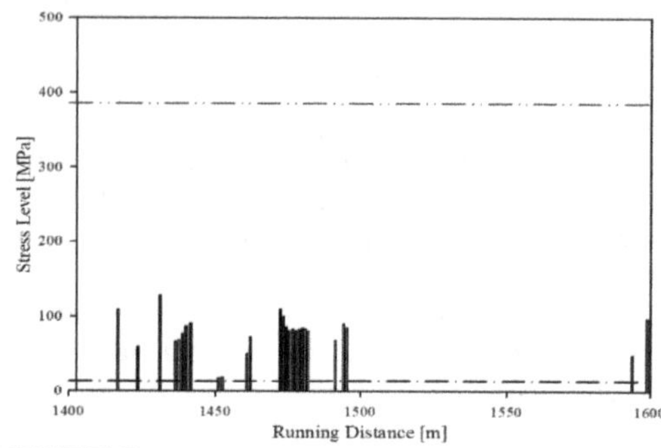

FIGURE 2: DISTRIBUTION OF SCZS AND THEIR STRESS LEVEL BETWEEN 1400M TO 1600M.

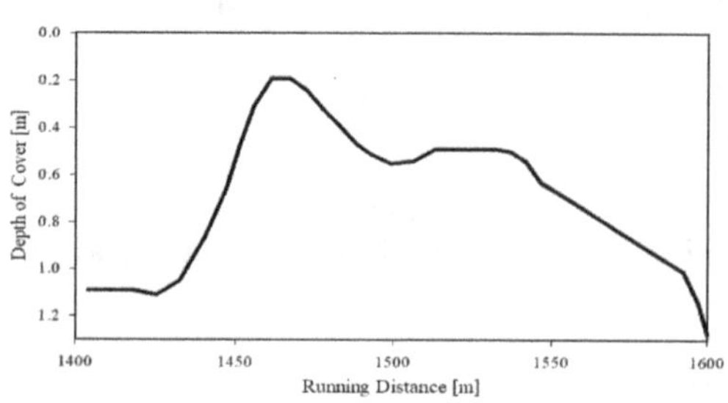

FIGURE 3: MEASURED DEPTH OF COVER BETWEEN 1400M TO 1600M

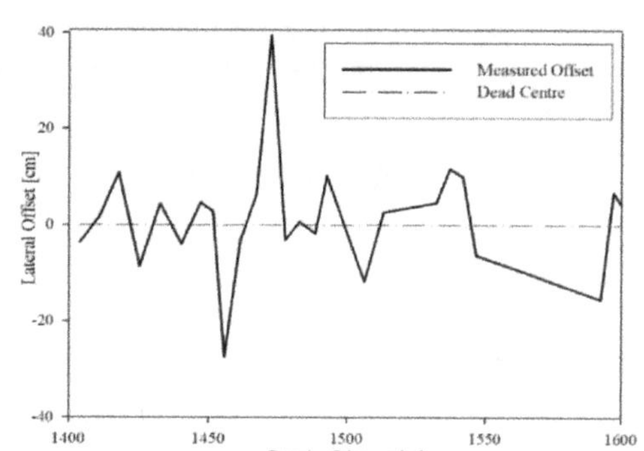

FIGURE 4: MEASURED LATERAL OFFSET BETWEEN 1400M TO 1600M

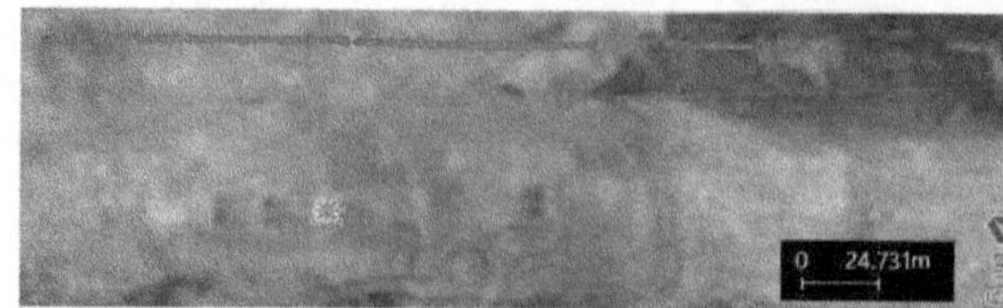

FIGURE 5: SURVEY ROUTE (RED) AND LOCATION OF FAILURE ON PARALLEL LINE (YELLOW)

FIGURE 6: SURVEY ROUTE (RED), IMAGINARY STRAIGHT LINE(WHITE) AND LOCATION OF FAILURE ON PARALLEL LINE (YELLOW)

The pipeline owner has since confirmed that this identified section of pipeline is located at an area of known geohazard conditions and is in the vicinity of a rupture on a parallel line which occurred two months after this LSM survey. FIGURE 5 shows the precise location of the aforementioned rupture, with safety fencing indicating the subsequent repair zone still visible in the satellite view image. The lateral distance from this repair zone to the area highlighted by LSM is 39m. Following an excavation in this area, the pipeline was shown to be sagging upward, a fact which was also in agreement with the LSM results.

3.2. Subsidence

LSM has been applied to an unpiggable pipeline in Europe with the aim of avoiding unnecessary repairs to zones that were marginally passing External Corrosion Direct Assessment (ECDA) thresholds. TABLE 2 lists the technical parameters of the pipeline.

TABLE 2: TECHNICAL PARAMETERS OF THE PIPELINE

Diameter [in]	Wall Thickness [in]	Operating Pressure [MPa]	Pipeline Material Grade	SMYS of Pipeline Material [MPa]	Nominal Hoop Stress [MPa]
6	0.189	5.6	X37	235	107

LSM predicted an SCZ with a stress level of 160Mpa (68% of SMYS) at a location that was 20cm from a collision point at which the pipeline was being pushed against a large rock, orientated to the 6 o'clock position. This caused a reversible

Copyright © 2019 by ASME

deflection on the exposed section of pipe. Evidence of this external force was obvious on the coating resulting in wrinkles and faults due to the collision. FIGURE 7 shows the condition of the coating at this collision point. A special machine was used to break the large rock and remove it, after which the pipeline was observed to deflect back to its original position by approximately 4cm over an exposed length of 3m.[9]

FIGURE 7: EVIDENCE OF A COLLISION ON COATING.

A finite element analysis using COMSOL Multi-Physics was carried out to estimate the stress level caused by this collision. A solid block was modeled to touch the pipeline, with a similar area to that which was observed from the coating damage. In an iterative analysis, the load on the solid block was increased until the deflection reached approximately 4cm under 5KN force. FIGURE 8 shows the exaggerated displacement of the analysis results. The whole pipeline length was modeled and was fixed at both ends at -5m and +5m.

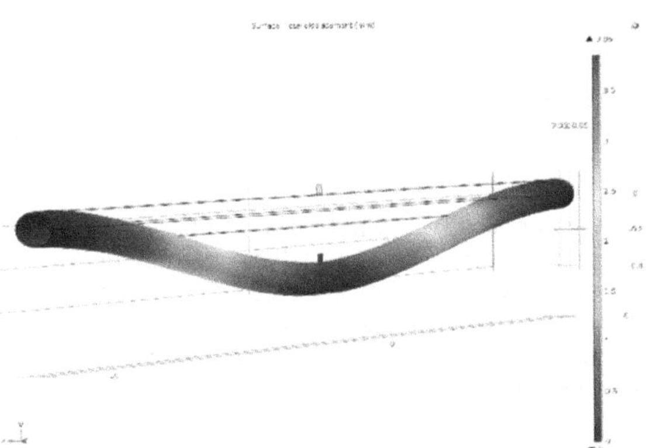

FIGURE 8: FEA RESULTS. EXAGERATED DEFORMATON OF PIPELINE UNDER EXTERNAL LOAD AT THE COLLISION POINT.

Under the same operating conditions, the FEA result for Von-Misses stress is shown in FIGURE 9. In this model, the stress level is 160MPa under a displacement of 3.85mm at the collision point. An effective force of 5KN has been assumed after taking out the soil weight. For a better representation, the block was modeled at 00:00 O'clock.

FIGURE 9: FEA RESULTS. VON-MISSESS STRESS DISTRIBUSTION AROUND THE COLLISION POINT.

A direct assessment of the SCZ using a handheld magnetometer showed that the magnetic level of the affected region reduced to a similar level as that of the remainder of the inspected pipeline length after 2 days of exposure following the removal of the external load.

3.3. Stress Corrosion Cracking

LSM reported another SCZ on the pipeline mentioned in section 4.2 that upon excavation showed evidence of external forces on the coating. Excavation was decided due to alignment between a severe DCVG reported coating fault and the LSM indication. FIGURE 10 shows the condition of the coating at a point 15cm from the reported SCZ which had a stress magnitude of 165MPa (70% of SMYS).

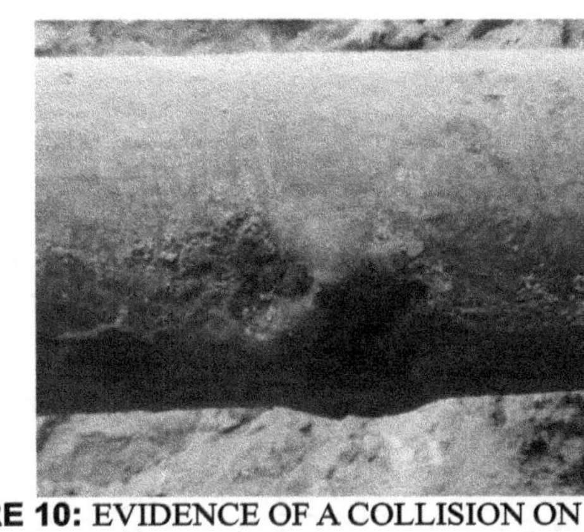

FIGURE 10: EVIDENCE OF A COLLISION ON COATING

In FIGURE 9 it can be seen that an external load point will cause symmetrical stress concentration zones. Likewise in this case, there were two areas of longitudinal linear indications at 04:00 and 08:00 O'clock that showed the beginnings of surface metal loss due to the presence of water after the coating failed, this is believed to be Stress Corrosion Cracking (SCC). FIGURE 11 shows an example of a linear indication measuring 140mm in length.

Copyright © 2019 by ASME

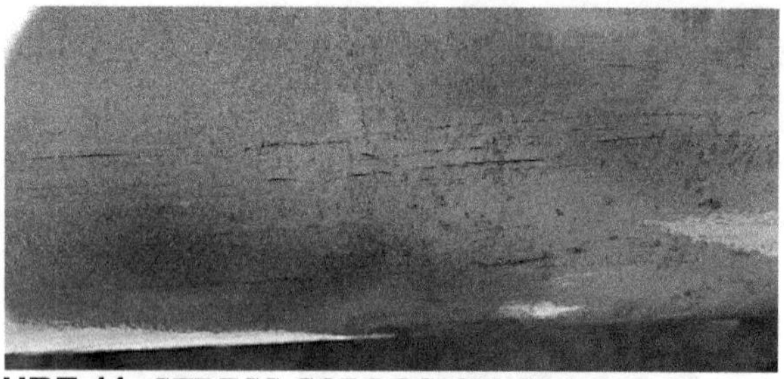

FIGURE 11: STRESS CORROSION CRACKING AT 04:00 O'CLOCK WITH LONGEST BEING 140MM.

A series of Magnetic Particle Inspections (MPI) following light grinding showed that the depth of cracks where about 18% of Wall Thickness (WT). A Two-Dimensional FEA using COMSOL was done for only one crack with 0.2mm opening and 18% WT. The boundary condition for this model was an internal pressure of 57bar and an external pressure of 1bar. FIGURE 12 shows the simulation result for Von-Misses stress are at 160MPa.

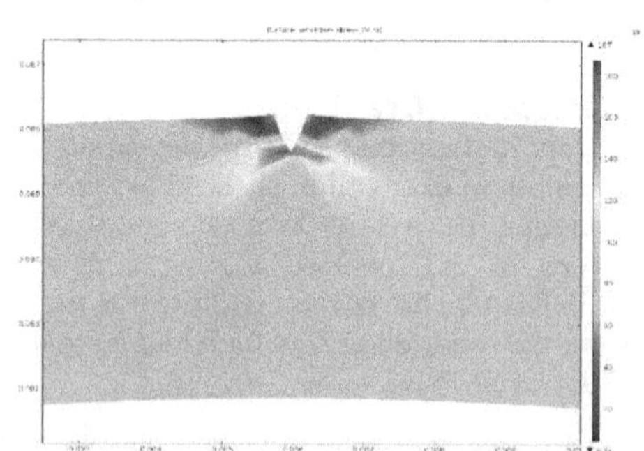

FIGURE 12: FEA RESULTS. VON-MISSESS STRESS DISTRIBUTION AROUND THE MICROCRACKS IN CROSS SECTION.

3.4. Wrinkle Bends

A 24-inch diameter pipeline whose technical parameters are listed in TABLE 3 was selected in a blind trial to evaluate LSM predictions against existing ILI and Direct Assessment (DA) results.

TABLE 3: TECHNICAL PARAMETERS OF THE PIPELINE

Diameter [in]	Wall Thickness [in]	Operating Pressure [MPa]	Pipeline Material Grade	SMYS of Pipeline Material [MPa]	Nominal Hoop Stress [MPa]
24	0.312	3.22	X60	413	124

LSM detected a SCZ whose localised stress magnitude was 162MPa (39% of SMYS) which was the highest along the survey length and was situated 20cm from a cluster of two wrinkle-bends. [6] The geometry of these wrinkle-bends have been modeled based on the dimensions shown in FIGURE 13.

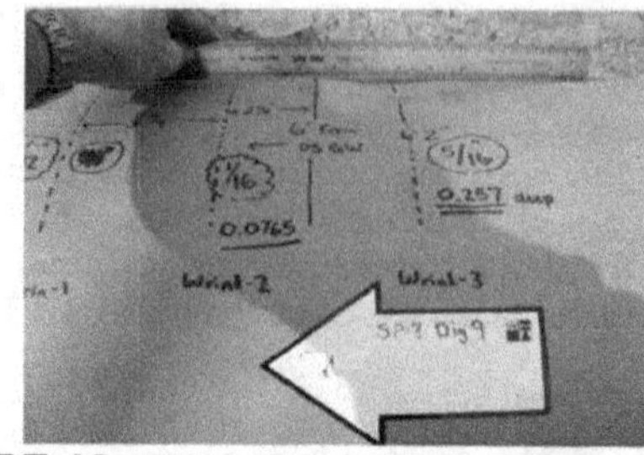

FIGURE 13: TWO WRINKLE-BENDS DUE TO GEOHAZARDOUS OPERATING CONDITIONS.

A FEA using COMSOL was carried out assuming an internal operating pressure of 3.32MPa and 0.1MPa of external pressure. The weight of soil was not computed and the pipeline material was considered to be in the elastic region. Although the wrinkles form when the material passes yield strength, the stress will drop and it enters an elastic stage again with some residual strain. This has been discussed in section 2.2. FIGURE 14 shows the FEA results for Von-Misses stress peak at 188MPa but directly above wrinkles at 165MPa which is in excellent agreement with LSM results.

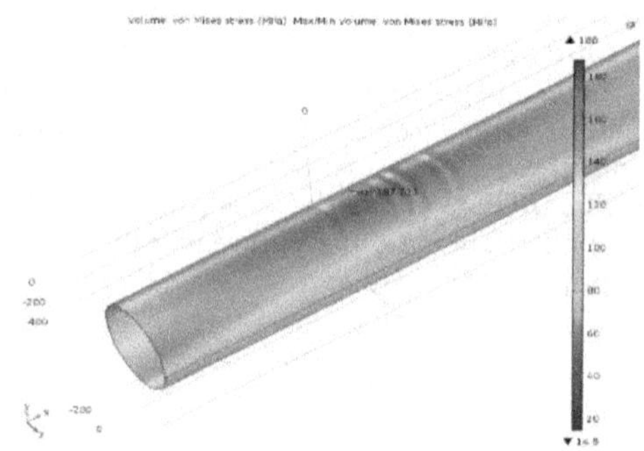

FIGURE 14: FEA RESULTS. VON-MISSESS STRESS DISTRIBUTION AROUND THE WRINKLE BENDS.

3.5. Weld Mismatch

A seven-year-old pipeline was surveyed by LSM due to construction problems. TABLE 4 lists the technical parameters of the pipeline. Among the top three SCZs identified by LSM was one with a 149MPa localised stress prediction (36% SMYS) which aligned with a DCVG fault. This location was selected for excavation and direct assessment.

TABLE 4: TECHNICAL PARAMETERS OF THE PIPELINE

Diameter [in]	Wall Thickness [in]	Operating Pressure [MPa]	Pipeline Material Grade	SMYS of Pipeline Material [MPa]	Nominal Hoop Stress [MPa]
30	0.340	1.0	X60	413	44

Direct assessment showed the damaged coating to be directly above a girth weld with the predicted SCZ 20cm away

Copyright © 2019 by ASME

from it. The pipe section was spirally welded and had a 4.3mm misalignment on the spiral weld at 12:00 O'clock. As a result of this there was a dent at 06:00 created on installation to facilitate joining of the two pipes. FIGURE 15 and FIGURE 16 show the extent of the misalignment and the dent.

FIGURE 15: 4.3MM MISALIGNMENT AT SPIRAL WELD DUE TO DIMENSIONAL MISMATCH OF TWO PIPES.

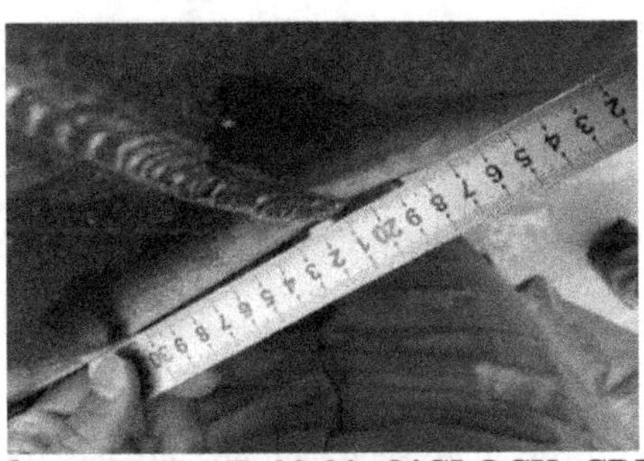

FIGURE 16: DENT AT 08:00 O'CLOCK CREATED ON ASSEMBLY AND INSTALLATION OF THE PIPELINE.

A Two-Dimensional FEA using COMSOL shows that a misalignment of 4.3mm will reduce the effective wall thickness at the weld which in turn increases the stress experienced at that location. FIGURE 17 shows the distribution of Von-Misses Stress rising to 160MPa peak stress on the outer surface.

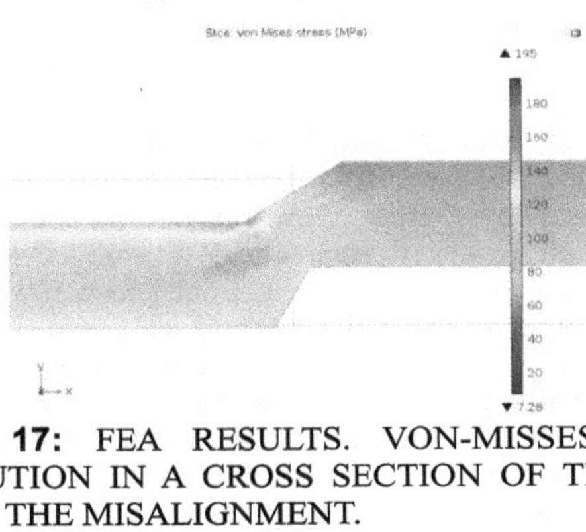

FIGURE 17: FEA RESULTS. VON-MISSES STRESS DISTRIBUTION IN A CROSS SECTION OF THE GIRTH WELD AT THE MISALIGNMENT.

3.6. Depth of Cover Measurement

The most accurate method of assessing the accuracy of LSM Depth of Cover (DOC) predictions is to use a plumb line on an excavated and exposed pipeline but cost in many cases renders this impractical. As-built DOC data cannot be used reliably due to the probability of soil erosion with time and accuracy of positioning device. An alternative method is to use readings from conventional methods as a reference. A 48inch pipeline was surveyed by LSM in the UK followed by two surveys using conventional pipeline locators with depth reading positions being recorded and mapped using an RTK GNSS system. FIGURE 18 compares the LSM prediction with results from two conventional locators with their industry standard error bars shown. Taking the same approach for lateral offset, FIGURE 19 shows the results of four individual LSM surveys on the same pipeline while the centre line was marked using a conventional pipe locator. [4]

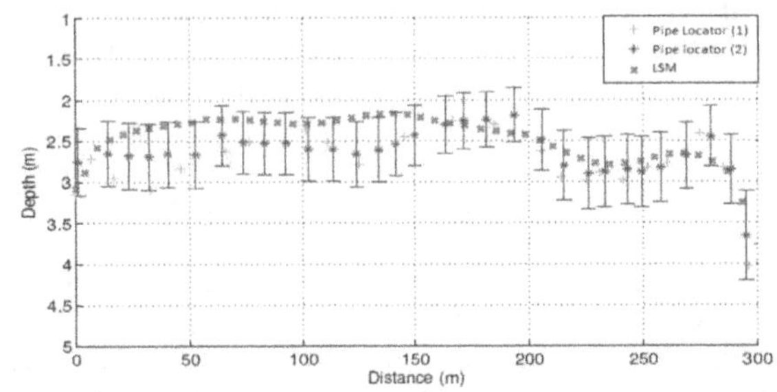

FIGURE 18: COMPARISON OF DEPTH FROM GROUND LEVEL TO CENTRE OF THE PIPE DETERMINED AS PREDICTED BY LSM AND TWO DIFFERENT CONVENTIONAL PIPE LOCATOR DEVICES. THE ERROR BAR OF THE PIPE LOCATOR MEASUREMENTS REPRESENTS A CONFIDENCE LEVEL OF 99.7%.

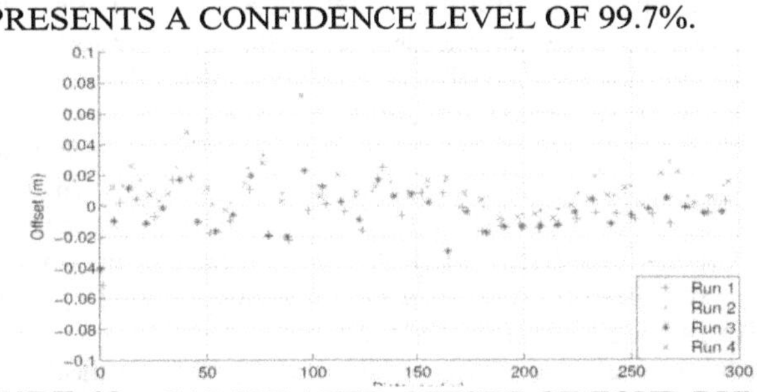

FIGURE 19: CALCULATED OFFSET OF FOUR RUNS IN TWO OPPOSITE DIRECTIONS. THE RESULTS INDICATED THAT THE SURVEYOR FOLLOWED THE REFERENCE POINTS AND STAYED WITHIN ±100 MM ON TOP OF THE PIPE.

4. DISCUSSION

Several case studies have been discussed in section 3 each reflecting accurate predictions made by LSM. However it is recognised that LSM is an emerging technology and that valid skepticism may exist in some members of the integrity community. Section 4 attempts to address these concerns.

Copyright © 2019 by ASME

4.1. Specifications of LSM

The main output from a LSM survey is detection of changes in pipe wall stress level based on the stress-magnetisation relationship. [2] This starts with detecting the presence of changes in measured magnetic field (Magnetic Anomalies) some but not all of which indicate the presence of an SCZ. The different mathematical algorithms used to discriminate between anomalies generated by an SCZ from those that do not will affect the performance accuracy of LSM providers. More trials are needed to confirm the accuracy of these algorithms and to define Probability of Detection (PoD) of LSM but an associated challenge is the definition of detection. Since LSM reports the location where the pipe wall experiences increased localised stress it will infer the presence of any type of defect ranging from strain caused by ground heave, through stress corrosion cracking to internal and external metal loss. However, no one conventional inspection technique can detect all of these defects itself. Consequently when comparisons are made between LSM and MFL results, for example, this should be taken into account. Also the positional errors, which can be up to +/- 10m caused by converting one system to another should also be taken into account. The one truly reliable method of assessing the accuracy of LSM predictions is direct assessment but this, in many cases, may be prohibitively expensive.

An equally significant output from LSM is its prediction of the magnitude of stress at each SCZ. This is independent from its performance in SCZ detection and again its accuracy is defendant upon the knowledge and specific algorithms developed by each vendor. Achievable accuracy is mainly determined by the amount of visible magnetic field of the pipeline which may be related to its distance from the sensors, mass of metal, material type and the magnetic behavior of pipeline material. In addition, potential sources of inaccuracy include movement of scanner during the survey which adds noise to the signal and sources of magnetic field other than the target pipeline which will mask the signal and add error. Factoring all sources of error, the claimed accuracy for LSM prediction of stress magnitude is ±25MPa which is only valid if a minimum of 65MPa Hoop Stress is exerted in the pipeline wall at-least once in its operational history.[2] A cyclic change in operating pressure will help to achieve a higher accuracy by establishing the Stress-Magnetisation relationship. Assessment of the accuracy of the predictions is limited by the methodologies available to convert the details of the features, as determined by direct assessment or ILI results, into a stress determination. [10]

Depth of Cover (DOC) predictions of pipeline as well as lateral offset measurement are keys to LSM implementation for pipelines located under geohazardous conditions. DOC from LSM is claimed to be as accurate as conventional pipeline locators i.e. ±5% of actual DOC.[11] LSM the pipeline route has to be marked prior to scanning and that to achieve this level of accuracy the scanner must be carried within 50cm of top dead-centre of the pipeline. Having a DOC reading every 2 to 6m is the advantage of LSM which does not need a stop and go survey. Lateral offset measurements are claimed to be as accurate as ±10cm when the signal from the pipeline is visible to the scanner.

4.2. Sources of Detection Uncertainty

LSM algorithms have been developed for application to naturally evolved features on ferromagnetic pipelines that are in service. Adding any unknown factor in will impact LSM performance.

Artificial defects, for example those fabricated by mechanical processes, may produce an easily detectable magnetic anomaly but their signatures may lead some LSM techniques to filter them out as noise or as external metallic objects.

An established Stress-Magnetisation relationship will improve the ability of LSM to find SCZs and quantify their severity level more precisely. Such a relationship is established by a combination of a minimum amount of stress in the ferromagnetic material usually between 40MPa and 62MPa with a cyclic stress which is usually the case for most pipelines under normal operating conditions.

Magnetic behavior of the pipeline material plays a key role and this factor becomes more potentially significant for older pipelines when industrial standards for chemical composition of pipe materials were not stringent or records have been lost as ownership changed with time. Although magnetic field strength can be examined by a short test survey, it remains uncertain what level of which alloying element affects LSM performance by leading to a reduced magnetic susceptibility.

A sufficiently strong magnetic signal received by the scanner is required to distinguish an SCZ from background noise and this can lead to a limitation in application. For example the vendors claim the maximum distance between scanner and the centre of the pipe to achieve the claimed performance is 12 to 15 times the pipeline diameter. This results in many 4 to 6 inch diameter pipelines rendered unsuitable for LSM if they are buried deeper than 0.5m.

Since LSM analysis techniques are based on statistical or signal processing methods a minimum amount of data is needed to achieve the claimed performance. For SCZ detection, a minimum of 30m survey length is needed while for DOC measurement, a minimum of 150m of survey length is required.

When the scanner is carried by a technician walking above the pipeline, preparation of ground surface can potentially impact the performance of LSM. If it is difficult to walk due to vegetation, excessive movement of the scanner will add more noise to the survey. This is also the case when obstacles prevent carrying the scanner on top dead centre of the pipeline.

4.3. Lessons Learned from Field Trials

When LSM results have been evaluated against ILI reports, there many sources of error in the alignment of the two datasets which makes it impractical to study Probability of Detection, for example conversion of one geospatial co-ordinate system to another, running distance and datum errors,. However, due to its simplicity, this is the preferred validation method. Many LSM validation exercises have been carried out this way with resulting POD figures ranging from 25% to 100%. The most accurate validation technique is direct assessment which is also the most expensive. However, when this has been applied one of

Copyright © 2019 by ASME

the vendors has demonstrated consistently a 100% POD for all of its deliverables with the exception of magnitude of stress which has not been validated mathematically. This vendor, though, can show results that have been validated by direct assessment consistently demonstrate the reported magnitude of stress reflects the level of severity of the defects present.

An alternative evaluation method has been suggested which proposes setting up a test pipeline for LSM. However it is a complicated process because it normally involves mechanical tasks that create false magnetic anomalies. It also factors out time variable pressures and operating history that play important roles in the establishment of Stress-Magnetisation.

If it is decided that evaluation of LSM must be by comparing results with MFL then there must either be a 12 month period between the application of the MFL inspection and the LSM survey. Alternatively if an MFL inspection is scheduled the LSM survey must be the first to take place. This is because MFL magnetisation saturates the pipeline wall thereby erasing the magnetic changes resulting from SCZs and leaves behind magnetised debris which might cause false calls. Magnetic anomalies will recover over time depending on the operating condition. All SCZs detected by LSM are usually reported by all vendors but this is not true of ILI reports which filter out defects that are considered within tolerance and not all ILI technologies are able to detect all types of defects. When evaluating correlation between the two techniques it must also be remembered a single SCZ may contain several defects and in view of this an evaluation based on a pattern alignment should be applied.

4.4. Criticisms to LSM

Integrity experts are familiar with many decision making protocols that are based on dimensions and characteristics of defects. Although all these protocols are driven from stress in the first place, they need to factor in many unknowns by considering safety factors that makes them very conservative and at the same time complicated to compare with direct stress measurement techniques which is the technique used by LSM. Numerous different shapes, sizes and types of defects at different clock positions may lead to the same magnitude of stress being calculated and ultimately the resulting calculated value is an estimate. LSM, on the other hand, exploits the direct relationship between magnitude of stress and the magnetic signal from the pipe and by doing so is potentially a more accurate integrity assessment method. It would not necessarily, though, lead to the same value as a value calculated from defect type, its geometry and clock position.

Another point of debate on LSM is the number of reported indications by LSM. It cannot distinguish between SCZs that are closer than 1m to one another but the number of indications detected and reported can be more than an MFL report. This is usually either because a threshold has been specified below which a metal loss report is not required by the pipeline operator or because defects undetectable by MFL, for example linear crack-like indications, are present and lead to increased localised stress in an SCZ. Note that if required some vendors have the capability to filter out from the report SCZs based on their severity level.

An additional criticism of LSM is that it cannot report on directional stresses alone but only volumetric changes in stress and the Von Misses stress. This is a limitation of LSM.

4.5. Underground Features

LSM can detect variations in magnetic field due to changes in mass of metal. This enables it to report on the location of ferromagnetic casings, sleeves and repair patches, fittings, changes in wall thickness and diameter and the presence of other pipelines. If, then, in a geohazardous area where a pipeline has been strained to the extent it touches a parallel or crossing pipeline, LSM is capable of reporting the situation.

5. APPROACHES TO APPLY LSM

LSM is an emerging technology that is still developing but already shows promising results. The main caveat in implementing LSM is to integrate its results into ongoing decision making protocols. In this section, some recommendations will be discussed.

5.1. Conventional Protocols

Internationally recognised standards are currently based on type, dimensions, number of defects and operating parameters. Extensive studies over many years led to an established practice that only recently have had their conservative approaches questioned. All above-mentioned factors are needed to estimate the stress in pipeline wall because a pipeline will only start to fail if stress at a specific point reaches its material SMYS and fails at SUT.

There are many combinations of dimensions of defects that will cause a certain level of stress. In other words, they will lead to the same recommendations by conventional protocols. Although it is impractical to find all possible combinations, estimating one is feasible by presumption of some and deriving the remainder. A lookup table between depth of defect and stress level can be made using FEA and assumption of type, clock position, length, width and operating condition. Stress level results from LSM can be associated with equivalent depth which can be input into conventional standards for decision making. An actual defect that is larger than assumed defect will be translated into a deeper defect that observed one but calculated ERF should remain the same because it is based on the raise in stress level.

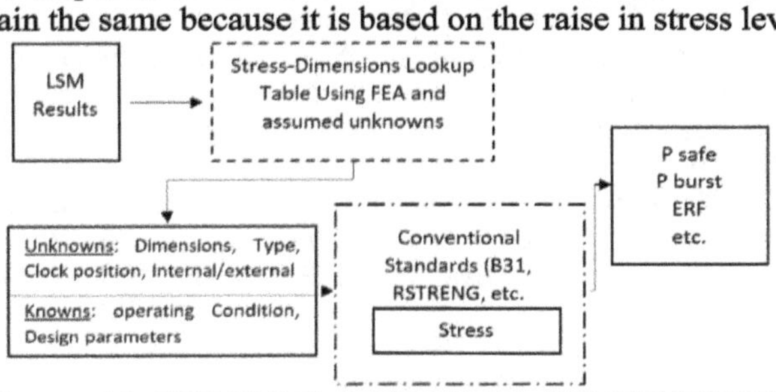

Figure 20: USING LSM REPORTED STRESS LEVELS WITH CONVENTIONAL PROTOCOLS.

Copyright © 2019 by ASME

A limitation of the abovementioned method is that it will not consider the growth rate of a defect due to its type. For example, a crack could be estimated as a deeper metal-loss at the time of inspection but a crack grows much faster than metal-loss. This can be overcome by repeating the LSM survey at regular intervals and then comparing the derived figures for the effective depth of the assumed defect. In this way it can be inferred if the source of the SCZ is a crack, metal-loss, an active external force due to landslide or inactive external force e.g. rock that dented the pipeline and is not increasing its pressure on the pipeline.

5.2. Screening Technique

LSM can be added to any inspection basket as a screening tool in order to increase efficiency regardless of the regulations that are in force provided that the follow up high resolution tools conform to currently applicable regulations. LSM will deliver an overall assessment of the condition of the pipeline regardless of type of defects and by doing so direct high resolution DA tools to locations where significant increases in magnitude of stress have been reported. Pipeline maintenance protocols define defects whose repair is mandatory but also defects that are marginally acceptable. An LSM survey can assist the repair point prioritisation following a DCVG coating assessment and a decision about marginal defects can be postponed or brought forward. For instance, an excavation to repair a coating fault can be postponed if LSM has not found any SCZ at in the vicinity.

A similar approach can be taken to monitor specific sections of the pipeline using LSM over time at regular intervals or to decide which parts of a pipeline should be monitored closely by alternative conventional methods.

6. CONCLUSIONS

It is not considered that LSM as an independent inspection technique, is a potential alternative to ILI. It does, however, improve efficiency and lower the cost of ECDA on unpiggable pipelines primarily by reducing the quantity of required excavations. It predicts if a damaged coating, reporte4d by DCVG inspection, has led to corrosion and therefore removes the need for an exploratory excavation with a resulting significant decrease in the overall ECDA cost.

For the condition monitoring of piggable pipelines located under geohazardous conditions, LSM can be deployed to measure the magnitude of stress that exists even if the pipeline wall is not deformed. It can also be applied to short sections of the pipeline to monitor the effect of ground movement, its impact on the pipeline in terms of increased stress and by deploying the three dimensional mapping function which can detect the extent of strain. Three-Dimensional mapping using LSM will also help to monitor geohazards that can alter the position of the pipeline in longer stretches. A repeated LSM survey at intervals of less than 12 month is a key to significantly increase the benefits of LSM by reducing the risk of pipeline rupture in geohazardous regions. Combined with high resolution satellite imaging of ground movement LSM can be used to focus on lengths of pipeline that are buried in known areas of ground movement to assess and monitor the impact that movement has on the pipeline in terms both of measured stress and strain.

ACKNOWLEDGEMENTS

The case study results presented in this paper have been sourced from validation reports made available by companies worldwide that have evaluated or are in the process of evaluating LSM. The authors appreciate the support of these companies in pioneering the implementation of LSM.

REFERENCES

[1] Joule, J. P., Sturgeons Annals of Electricity, 8, 359. 1842.

[2] Villari, E., Ann. Phys., Lpz., 126, 87. 1865.

[3] Staples, S. G. H., Vo, C., Cowell, D. M. J., Freear, S., Ives, C., Varcoe, B. T. H. *Solving the Inverse Problem of Magnetization–Stress Resolution,* Journal of Applied Physics Vol.113 (2013): pp.133905.

[4] Vo, C., Staples, S. G. H, Cowell, D. M. J., Cookson, C., Varcoe, B. T. H., Freear, S., *Determining the Depth and Location of Buried Pipeline by Magnetometer Survey,* Preprint submitted to Elsevier, 2015.

[5] Ordnance Survey. *A guide to Coordinate Systems in Great Britain*, 2015.

[6] Xu, Tianzong. *Understanding Quantitative Performance of Large Standoff Magnetometry in Detecting Live Gas Pipeline Anomalies with Stress Estimation.* Proceedings of IPC2018, 2018.

[7] Habibi, Hamed, *Stress Concentration Tomography. A New Tool to Remotely Monitor Pipeline Wall Condition, Depth of Cover and Stress Concurrently,* Journal of Corrosion Management 2018(142):13-16, 2018.

[8] Habibi, Hamed, *Developments in Remote Magnetic Monitoring of Carbon Steel Pipelines to Locate and Measure Abnormal Stress*, Proceedings of CEOCOR2018, 2018.

[9] Habibi, Hamed, Jarram, Paul, Vo, Chau, *Field Verification of Remote Magnetic Monitoring of Stress for buried unpiggable pipelines.* Proceedings of UPSF2017, 2017.

[10] Jiles, D., Atherton, D., *Theory of Ferromagnetic Hysteresis* , Journal of Magnetism and Magnetic Materials, Vol. 61 No.1-2 (1986): pp.48-60, 1982.

[11] Radiodetection, *RD8000 Universal Precision Cable, Pipe And Rf Marker Locator,* 2014

Copyright © 2019 by ASME

Chapter 35

Use of Large Standoff Magnetometry for Geohazard Pipeline Integrity Investigations

Chukwuma Onuoha[1]
PureHM
Edmonton, AB, Canada

Eric Pozniak
PureHM
Edmonton, AB, Canada

Jerell Tolentino
PureHM
Edmonton, AB, Canada

Shamus McDonnell
PureHM
Edmonton, AB, Canada

Rob Zmud
PureHM
Edmonton, AB, Canada

Vignesh Shankar
PureHM
Edmonton, AB, Canada

ABSTRACT

Buried steel pipelines are subjected to mechanical stress emanating from either internal or external forces, resulting from geohazards, shear or external loading, and hoop stress. These conditions are important factors that can be detrimental to the integrity of a pipeline and lead to potential ruptures and failures if not properly and proactively managed with applicable monitoring and / or inspection tools. Some of the failures have resulted from the development of concentrations of stress that have led to dents, ruptures, cracks, buckles, leaks, etc. Pipeline operators have proactively installed ground monitoring devices or frequently performed inspections in the areas that present concern to mitigate problems prior to an incident. These conditions become even more challenging on difficult to pig pipelines driving the need for use of non-intrusive, and non-destructive indirect inspection tools.

Using large standoff magnetometry (LSM), a novel screening technology, has shown strong industry relevance in the identification of stress concentrations for geohazard applications. LSM detects inverse magnetostriction (also known as the Villari effect) "which is the change of the magnetic susceptibility of a material when subjected to mechanical stress". This innovative technology can detect changes in the magnetic field of a pipeline which can indicate the presence of stress on the pipe wall. Geometric anomalies (ovalities, dent, wrinkle), hoop stress, ground / slope movement, bend strain, thermal expansion, cracks, and material defects are potential sources of stress that LSM can detect from aboveground.

LSM technology has been deployed in several slope monitoring and ground displacement projects to proactively monitor the development of stress concentration zones in areas of concern. The technology is used to perform a baseline inspection in these areas of concern followed by periodic inspections that monitor the change to the magnitude of the stress concentrations that were initially measured with the intent to prevent future failures and limit downtime. In addition, LSM has been used to verify and prioritize stress concentration call-outs by ILI tools in sections that have developed dents, buckles, cracks, SCC, leaks, etc., due to environmental conditions.

This paper summarizes the use of LSM in several pipeline integrity geohazard assessments, that were conducted on different oil and gas pipelines, to identify dents, cracks, buckle, slope movement, and other geotechnical anomalies. Lessons learnt from practical real-life projects and validations of the technology are presented to demonstrate the effectiveness of LSM for geohazard applications.

Keywords - LSM, Inline inspection, Stress Concentration, Geometric Anomalies, Hoop Stress, Ground / Slope Movement, Cracks, Casing, Leak, Bend Strain, Buckle, SCCDA, ECDA, Digs.

INTRODUCTION

Because buried steel pipelines are subjected to mechanical stress emanating from either internal or external forces resulting from geohazards, shear or external loading, and hoop stress, pipeline operators have proactively installed ground monitoring devices or frequently performed inspections in areas that present concern to mitigate problems prior to an incident. These threats can be detrimental to the integrity of a pipeline and lead to potential ruptures and failures if not properly and proactively managed with applicable monitoring and / or inspection tools. These conditions become even more challenging on difficult-to-pig pipelines, driving the need for the use of non-intrusive, non-destructive indirect inspection tools such as Large Standoff Magnetometry (LSM).

[1]Contact author: Chukwuma.onuoha@xyleminc.com

Copyright © 2019 by ASME

LSM is an emerging aboveground, non-intrusive inspection technique that is becoming very attractive because of its operational simplicity and economic advantages, especially on difficult-to-pig pipelines that can be challenging for other robust inspection methods. LSM is a non-contact, passive, geo-magnetization flux leakage measurement technology for identifying and locating elevated levels of stress through the measurement of the magnetic field surrounding steel pipelines, regardless of the ability to pig or whether the pipeline is buried, exposed, or elevated. A change in magnetic field can indicate the presence of stress in the pipe wall. A change in material size, volume, type, age, etc., of a ferromagnetic material during magnetization is characterized as magnetostriction. Inverse magnetostriction, known as the Villari effect, characterizes the change of the magnetic susceptibility of a material when mechanical stresses are applied. This change in the magnetic field due to mechanical stress is what LSM technology can detect [1 – 2].

This emerging technology has been successfully used to identify unknown buried structures, such as valves, taps, etc., by measuring changes in the magnetic field across the buried structure. The inspection is conducted at an offset from the area of interest, then data is continuously collected and logged by walking along the buried pipeline with the LSM sensor. A magnetic field is sampled by a tool operator while walking along the centerline of a pipeline using an LSM sensor, recording data points hundreds of times per second. The recorded data is analyzed using a mathematical approach that identifies indications with signal characteristics similar to a dipole. Figure 1 depicts LSM scanning of an underground pipeline with anomalies [3].

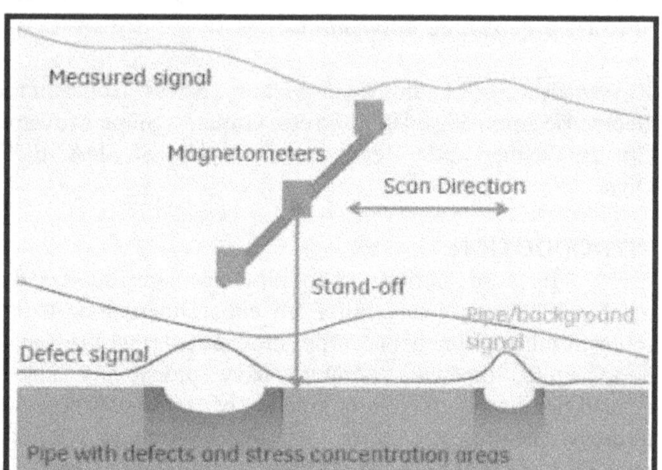

FIGURE 1: LSM SCANNING OF AN UNDERGROUND PIPELINE WITH ANOMALIES [3]

After the field inspection, the signal response is carefully interpreted by a data analyst to report changes in the magnetic field, signal quality, location of interest, repeatability and relative size of the indication. The dipole signature represents an area of possible stress concentration. Figure 2 depicts a dipole signature demonstrating area of stress concentration. Please note that there is no definite length of dipole but is generally in the range of ~10 – 50m depending the defect type and specific circumstances.

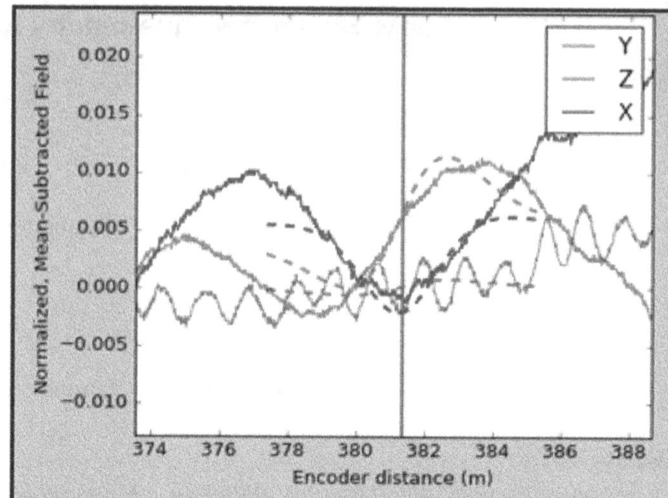

FIGURE 2: LSM DIPOLE INDICATING STRESS CONCENTRATION AREA

These possible stress concentration areas are recommended to be further investigated using high-definition tools and/or correlation using other data sets that may be available to assist in making a sound integrity decision [1-2]. However, it should be noted that other non-stress-related factors such as metal heterogeneity, extra metal objects and magnetic interference from the surrounding environment, etc., can affect changes in the magnetic field response, and therefore, measurement data analysis and interpretation are crucial to the proper discrimination of all the factors.

LSM technology is used as a screening tool to identify areas that require further examination, help target inspections and prioritize excavations. It may also help make informed decisions for further investigations on potential excavations based on ILI results, geo-technical surveys, or as part of a direct assessment program. The use of LSM has been shown to produce significant outcomes when correlated with additional data. For instance, LSM technology correlated with Inline Inspection (ILI) results and other pipeline construction information has aided in the accurate selection of digs to mitigate failures due to stress concentration. Successfully identifying digs to mitigate stress concentration is vital, as it substantially reduces costs due to potential failures and avoiding unnecessary digs.

LSM technology has recently found strong relevance in geotechnical applications. As underground buried pipelines pass through geographically diverse areas including slopes and other terrain, it becomes critical to manage these areas by identifying, inspecting, monitoring and, where necessary, remediating slopes where ground movement could impact the

[1]Contact author: Chukwuma.onuoha@xyleminc.com

Copyright © 2019 by ASME

pipeline. Slope inspection and monitoring are used to assess ground movement in key locations along pipelines. Slopes identified as having the potential for movement or known to have movement are carefully managed through aerial, ground and / or in-line inspections if the pipeline is piggable. Since this technology can be applied on difficult-to-pig pipelines, it has been deployed in several slope monitoring and ground displacement projects to proactively monitor the development of stress concentration zones in areas of concern. This technology is used to perform a baseline inspection in these areas of concern followed by periodic inspections that monitor changes in the magnitude of stress concentrations that were initially measured with the intent of preventing future failures and limiting downtime [1].

In addition, LSM has been used to verify and prioritize stress concentration call-outs by ILI tools in sections that have developed dents, buckles, cracks, SCC, leaks, etc., due to environmental conditions.

SOME PRACTICAL APPLICATIONS AND VALIDATIONS OF LSM TECHNOLOGY

LSM has been successfully deployed in an SCCDA project to accurately pinpoint where near-neutral SCC was occurring as shown in Figure 3 [4].

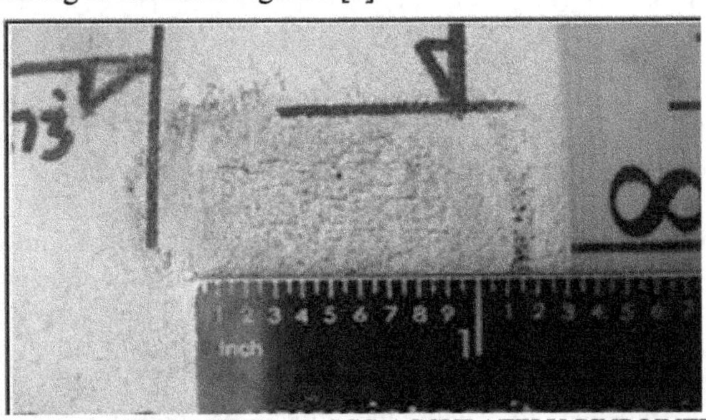

FIGURE 3: NEAR-NEUTRAL SCC ACCURATELY PINPOINTED DURING SCCDA DIRECT EXAMINATION WITH THE HELP OF LSM [4].

In this project, LSM was used as one of the integrated indirect inspection tools to help confidently pinpoint crack locations. Cathodic protection close interval survey (CP CIPS) was used to ascertain the level of cathodic protection. Direct current and alternating current (DCVG & ACVG) were employed to identify coating anomaly locations, whereas LSM identified areas of stress concentrations. In addition, ILI results (identification of metal loss locations), environmental data, and soil data were used. All these datasets in conjunction with LSM were combined, classified, and prioritized to accurately pinpoint crack locations.

LSM has also been used in conjunction with the above-mentioned integrated indirect inspection tools to validate dent locations identified by ILI in another DA project. CP CIPS was employed to ascertain the level of cathodic protection. DCVG and ACVG were utilized with the integrated technology to pinpoint coating anomaly locations, whereas LSM identified areas of stress concentrations. In addition, ILI results environmental data, and soil data were also used. All these datasets in conjunction with LSM were combined, classified, and prioritized to accurately pinpoint dent location.

The direct examination findings of the predicted dent location depicted in Figure 4 confirmed the presence of a dent as identified by ILI and LSM [5], demonstrating growing confidence in the use of this technology as a secondary screening tool.

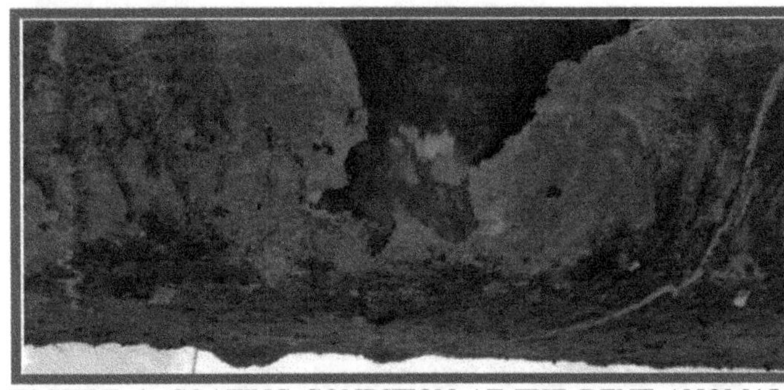

FIGURE 4: COATING CONDITION AT THE DENT ANOMALY AREA IDENTIFIED BY ILI AND VALIDATED BY LSM [5].

LSM has also been employed in another project to confirm the success in identifying casing ends previously identified by ILI as depicted in Figures 5 and 6 [1-2].

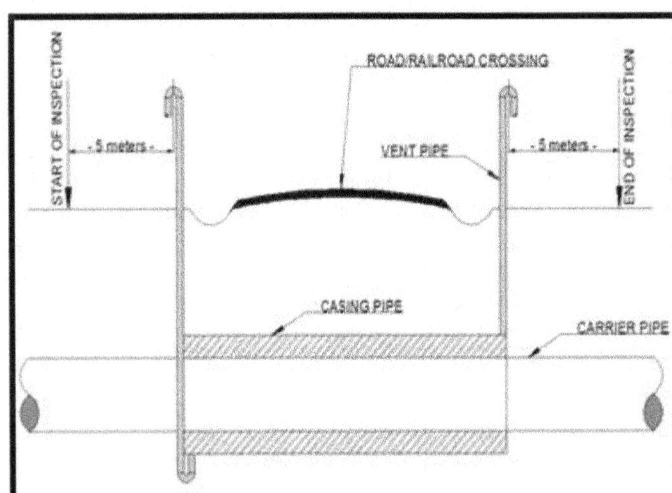

FIGURE 5: LSM VALIDATION TESTING INSPECTION SITE [1-2]

[1]Contact author: Chukwuma.onuoha@xyleminc.com

Copyright © 2019 by ASME

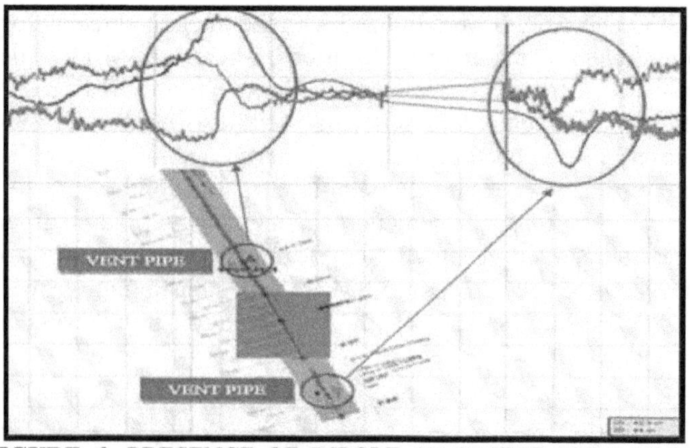

FIGURE 6: PRESENCE OF DIPOLE AT CASING ENDS FROM LSM [1 – 2]

It is evident from Figure 6 that changes in the magnetic field were observed in all three axes of the LSM sensor for both ends of the casings being inspected. The consistency of the dipole was verified by an analyst during post-inspection data analysis. The collected LSM data was correlated with ILI data and the results were a perfect match. The correlation and validation of LSM and ILI gives strong confidence in applying LSM technology for finding unknown casing ends and buried ferromagnetic materials [2].

LSM has also been deployed on an operating 16-inch pipeline with known ILI defects (call-outs) to validate the technology as part of the holistic continuous improvement approach. This validation project was conducted on two pipeline segments that were inspected by the LSM technology. The areas of interest were inspected, and the tool was able to find changes in the magnetic field caused by stress developed on the pipeline in both sites. Figures 7 and 8 depict the changes in the magnetic field seen in Sites 1 and 2 obtained by the LSM technology. The LSM tool identified one indication at Site 1 and two indications at Site 2, indicating possible stress concentrations based on the changes in the magnetic field observed and the consistency of the captured dipole.

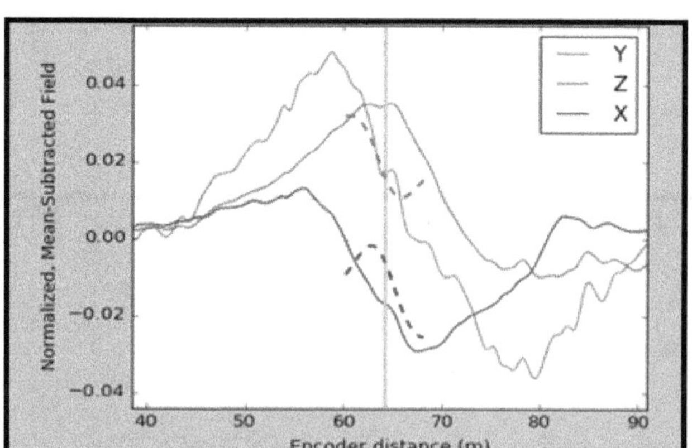

FIGURE 7: LSM INDICATION AT DIRECT EXAMINATION LOCATIONS DEPICTED IN TABLE 1 AT SITE 1[2]

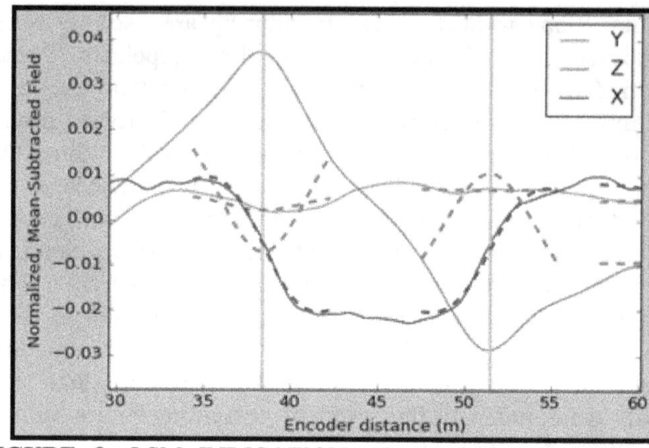

FIGURE 8: LSM INDICATION AT DIRECT EXAMINATION LOCATIONS DEPICTED IN TABLE 1 AT SITE 2 [2]

Site No.	No. of LSM Indications	Defect Category	Defect Dimension			Defect Morphology
			Circ. Width (mm)	Axial Length (mm)	Max Depth (mm)	
1	1	Axial Slotting	1	14	1.24	Hook Flaw
2	2		1	22	6.22	
			1	10	2.30	

TABLE 1: DIRECT EXAMINATION CONFIRMATION OF LSM & ILI INDICATIONS [2].

The outcome of the direct examination presented in Table 1 was correlated with the LSM technology and ILI findings. This correlation provided further validation for the usage of LSM technology for screening purposes to identify stress concentration zones. The result of the non-destructive examinations (NDE) classified the defect as axial slotting and the defect morphology as a hook flaw at all three LSM indications [2].

The last validation was conducted with LSM on an 8-inch pipeline to study the effect of pressure on the reproducibility and identification of indications. The methodology involved creating a dent on the pipe (Figure 9) and conducting LSM on the dent before and after a pressure of 175 PSI was applied (Figure 10).

FIGURE 9: DENT LOCATION

[1]Contact author: Chukwuma.onuoha@xyleminc.com

Copyright © 2019 by ASME

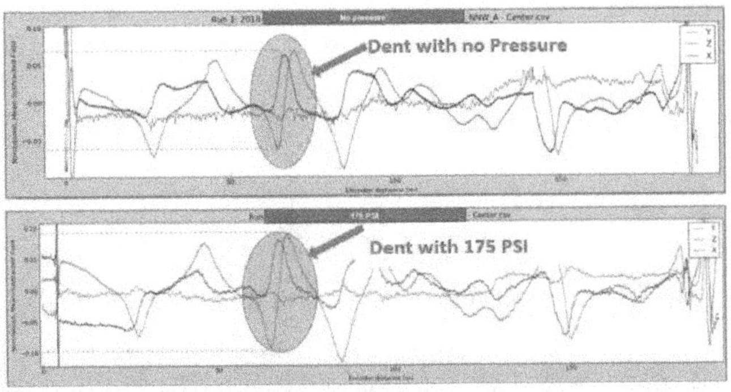

FIGURE 10: COMPARISON OF LSM INDICATIONS ON A DENT WITH NO PRESSURE & APPLIED PRESSURE AT 175 PSI

The result of this validation is promising and showed that the identification of dents with LSM can be successfully reproduced, and the detection sensitivity of the dipole increased at a higher pressure.

LSM IN GEOHAZARD APPLICATIONS

Normally, pipelines that are above ground can be flexibly supported, but most pipelines are below ground due to environmental, safety, or logistic reasons. Pipelines that are buried are confined by surrounding native soil to varying degrees depending on the compaction of the soil. The rigidity of the soil restricts the free movement of the pipe due to soil-pipe interaction. Previous studies have shown that if the surrounding soil is firm, it will take a shorter distance for a pipeline segment to absorb fault movements, with a higher level of strain induced on the pipe [6]. This strain is further exaggerated when pipelines are on slopes or the ground in the region has shown evidence of periodic movement due to environmental conditions. There are recorded incidents where soil movement has caused pipeline failures [6 -7].

Some of the failures have resulted from the development of concentrations of stress that led to dents, ruptures, cracks, buckles, leaks, etc. Pipeline operators have proactively installed ground monitoring devices or frequently performed inspections in these areas of concern to mitigate problems prior to an incident [2, 6 -7].

LSM has been deployed in several slope monitoring and ground displacement projects to proactively monitor the development of stress concentration zones in areas of concern. LSM is used to perform a baseline inspection in areas of concern followed by periodic inspections that monitor changes to the magnitude of the stress concentrations that were initially measured with the intent of preventing future failures and limiting downtime.

CASE STUDY 1

For the first case study, LSM was employed as an additional screening tool to support geohazard risk assessment of a 12-inch pipeline where slope movement was a concern. LSM was used to perform the baseline inspection and periodic inspections to ascertain and refine locations prone to slope movement and prevent failure. Figure 11 and 12 depict the LSM indication at a location of concern and the outcome of the direct examination.

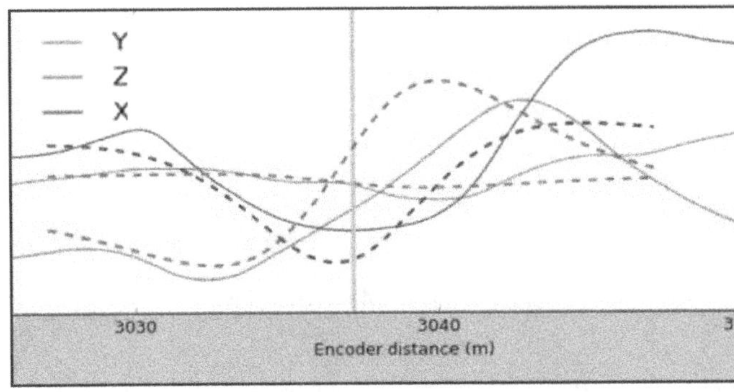

FIGURE 11: LSM INDICATION

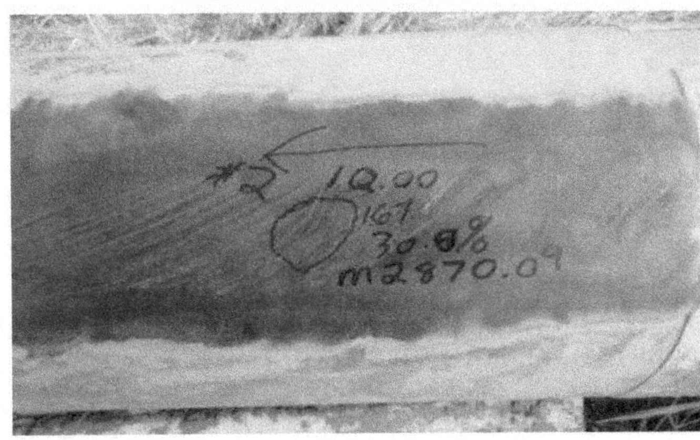

FIGURE 12: CONFIRMED DIRECT EXAMINATION FINDING AT THE LSM PREDICTED INDICATION

The LSM tool identified this location, indicating possible stress concentration based on the changes in the magnetic field observed and the consistency of the captured dipole (Figure 11). The result of NDE classified the defect as pitting and a pipeline external defect.

In another similar project on a 12-inch pipeline, LSM was used as a screening tool in a geohazard risk assessment to precisely identify locations prone to slope movement. A baseline inspection and periodic inspections were conducted prior to selecting one of the locations for direct examination as noted in Figures 13 and 14.

[1]Contact author: Chukwuma.onuoha@xyleminc.com

Copyright © 2019 by ASME

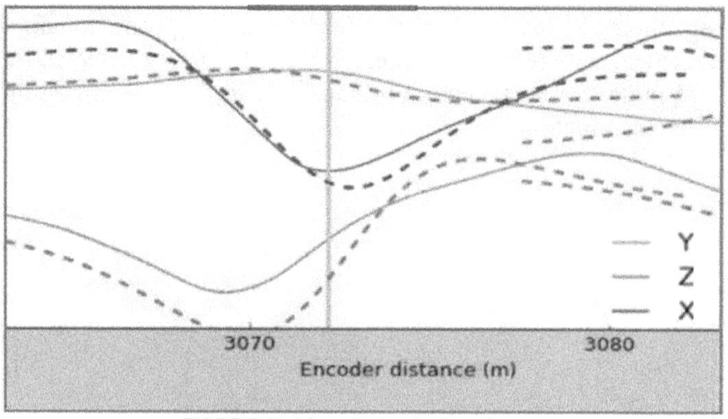

FIGURE 13: LSM INDICATION

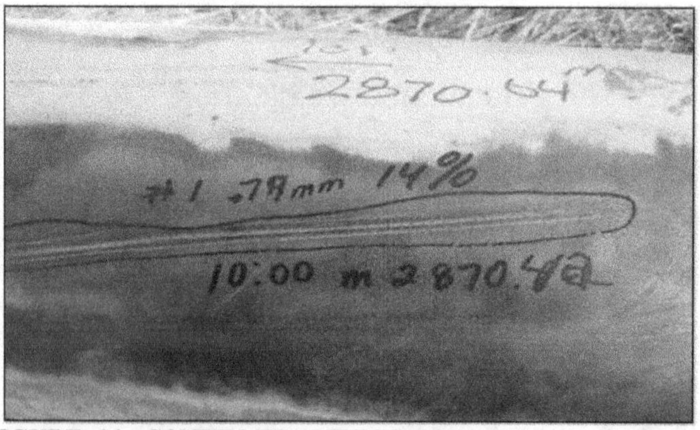

FIGURE 14: CONFIRMED DIRECT EXAMINATION FINDING AT THE LSM PREDICTED INDICATION

The result of the NDE classified the defect as axial slotting, demonstrating the relevance of this technology in geohazard investigations.

CASE STUDY 2

The next case study was conducted on a 12-inch pipeline in a location where slope movement was a concern. LSM was initially used to perform a baseline inspection in these areas of concern followed by periodic inspections that monitor the changes to the magnitude of the stress concentrations that were initially measured with the intent of preventing future failures and limiting downtime. In addition to LSM, ILI results were used to validate LSM indications, and both inspection tools identified an indication in the area.

The direct examination findings at one of the locations, predicted by LSM and validated by ILI (Figures 15 and 16), showed that the pipeline buckled due to significant slope movements.

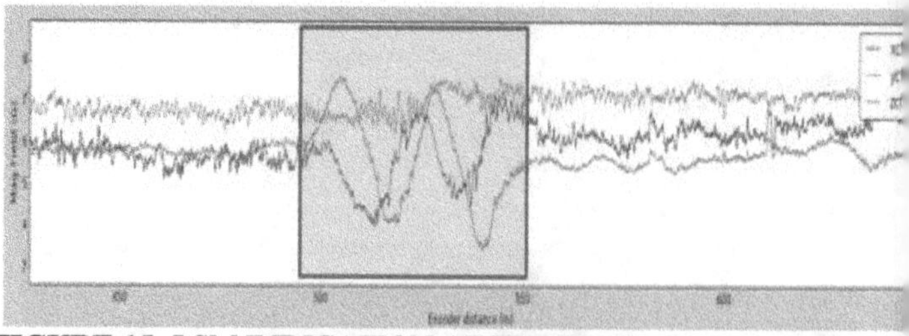

FIGURE 15: LSM INDICATION AT THE BUCKLE LOCATION

FIGURE 16: CONFIRMED DIRECT EXAMINATION FINDING (BUCKLE) AT THE LSM PREDICTED & ILI VALIDATED INDICATION LOCATION

CASE STUDY 3

Generally, geohazard conditions that cause deformations can affect the pipeline by way of displacement and this displacement may stop after deformation of the pipeline and / or may continue and cause failure. In another related project, LSM was utilized on a 6-inch pipeline in a location where slope movement was a major concern. LSM detected continuous stress concentration indications on submerged section of the pipe believed to be curved and bent due to geohazard operating conditions. The direct examination finding indicated that external and localized corrosion were reported on the curved sections of the pipe where coating was damaged. The external corrosion reported on the bend may have been caused by synergistic effects of failed coating due to stress at the bend, ingress of corrosive species to the pipe due to exposed coating, inadequate cathodic protection and / or shielding of cathodic protection exacerbated by stress concentration.

The predicted LSM indication location (Figure 17) and outcome of the direct examination (Figure 18) correlated perfectly. Stress concentration zones on the submerged, curved and / or bent sections of the pipe (Figure 18), presumably due to geohazard impact was successfully detected by LSM.

[1]Contact author: Chukwuma.onuoha@xyleminc.com

Copyright © 2019 by ASME

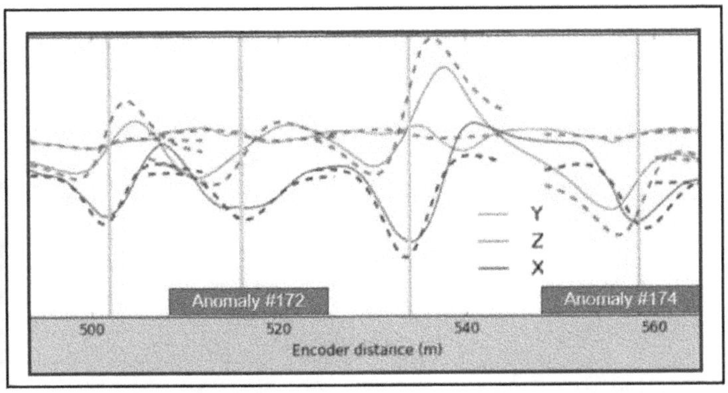

FIGURE 17: LSM INDICATIONS AT BENDS DUE TO GEOHAZARD OPERATING CONDISTIONS

FIGURE 18: CONFIRMED DIRECT EXAMINATION FINDING AT THE LSM PREDICTED INDICATION

This shows strong confidence in the use of LSM to predict areas of concern based on the identification of stress concentration zones demonstrating the relevance of this technology in geohazard investigations.

The lessons learnt in the use of LSM is that it is an emerging technology that can be used with other robust inspection technologies to make informed pipeline integrity decisions in geohazard applications. For condition monitoring under geohazards conditions, LSM stands out because it can be used to measure areas of stress concentration that could exist even if the pipeline wall is not deformed.

Intensive and full-scale validation exercises are currently being done to aid in developing the technology and elevating it to a level where it could be a standalone inspection technology in the future.

CONCLUSIONS

Large standoff magnetometry (LSM), a novel screening technology, has shown strong industry relevance in the identification of stress concentrations for geohazard applications. It has been successfully used to identify external corrosion and stress corrosion cracking locations in direct assessment programs in conjunction with other DA tools. In addition, it has become a screening tool for refining and prioritizing indications from ILI to make informed pipeline integrity decisions.

For difficult-to-pig pipelines, where no other inspection technologies can be applied, LSM may serve as a qualitative inspection solution due to its operational simplicity and economic advantages.

Continuous improvement efforts and validation exercises are currently occurring to further refine the process of determining stress concentration indications and potentially discover additional valuable applications of the LSM technology in the oil and gas industry.

LSM results show the considerable benefits from its use in augmenting pipeline integrity management programs which are currently challenged by disruptive geohazard conditions and / or impacts.

ACKNOWLEDGMENTS

Authors would like to thank Jerell Tolentino, Project Manager, PureHM, Canada for coordinating with pipeline operators to obtain permission in publishing results from recently completed inspections. In addition, we would also like to thank Rob Zmud in providing support in analyzing the collected LSM tool's data for this paper. It is a pleasure to acknowledge their support.

REFERENCES

1. S. McDonnell, C. Onuoha, E. Pozniak, V. Shankar, and et al. "Improved Methodology for Identification of Buried Casings using Indirect Inspection Method", NACE CORROSION 2017, Paper No.C2017-9400, Houston, TX, March 26-30, 2017.

2. S. McDonnell, C. Onuoha, E. Pozniak, V. Shankar, "Identifying Stress Concentrations on Buried Steel Pipelines using Large Standoff Magnetometry Technology", IPC 2018, Paper No. IPC2018-78162, Calgary Canada, September 24-28, 2018.

3. T. D Xu. "Understanding Quantitative Performance of Large Standoff Magnetometry in Detecting Live Gas Pipeline Anomalies with Stress Estimation". Proceedings of the 2018 12th International Pipeline Conference. IPC2018-78655. Calgary, Alberta, Canada, September 24 – 28, 2018.

4. Onuoha, Chukwuma, McDonnell, Shamus, Eric, Pozniak and Et Al., "Advancements in Stress Corrosion Cracking Direct Assessment using an Integrated Approach", *NACE Corrosion 2018*, Paper No.C2018-11194, Phoenix, AZ, April 15-19, 2018.

[1]Contact author: Chukwuma.onuoha@xyleminc.com

Copyright © 2019 by ASME

5. Onuoha, Chukwuma, McDonnell, Shamus, Eric, Pozniak and Et Al., "Interpretation and Selection of Direct Examination Locations with respect to SCCDA Methodology", *NACE Corrosion 2019*, Paper No.C2019-12812, Nashville, TN, March 24-28, 2019.

6. Lee, James P. and Bohinsky, Joseph A, "Design of Buried Pipeline Subjected to Large Fault Movements", Eleventh World Conference on Earthquake Engineering, Paper No. 1600, Acapulco, Mexico, June 23-28, 1996.

7. Transport Safety Board Canada, "Pipeline Occurrence Report-Natural Gas Pipeline Rupture", TCPL, Report No. P96H0012, April 15, 1996, St. Norbert, Manitoba.

[1]Contact author: Chukwuma.onuoha@xyleminc.com

Copyright © 2019 by ASME

AIM-PIMG2019-1007

Chapter 36

STRESS CHARACTERIZATION OF MECHANICAL DAMAGES IN GAS PIPELINES USING LARGE STANDOFF MAGNETOMETRY TECHNOLOGY

Xian-Kui Zhu
EWI
1250 Arthur E Adams Drive
Columbus, Ohio 43221, USA

Tianzong (David) Xu
Pacific Gas and Electrical Company
6111 Bollinger Canyon Road
San Ramon, CA 94583, USA

ABSTRACT

Large standoff magnetometry (LSM) is an emerging non-intrusive, above ground, passive geo-magnetization flux leakage inspection technology used to detect pipeline features or anomalies associated with elevated local stresses. The stress estimation and characterization are the important part of the LSM technology and need to be understood properly so that the LSM output stresses are correctly analyzed and assessed for the anomalies in the pipeline. For this purpose, the present paper briefly reviews the Stress Concentration Tomography (SCT) – one version of the LSM technology, including its advantages, capacities, limitations, challenges and practical uses as well. Theoretical basis of SCT technology and its equivalent stress definitions are discussed. Followed are two field trial examples of LSM inspection results of mechanical dents in operating gas pipelines. With the dent profiles measured by ILI or in-ditch NDE, the finite element analyses (FEA) are performed to determine stresses of the dents in the gas pipelines in three-dimensional conditions. Then the FEA stresses and the LSM estimated stresses local to stress concentration zones or dents are comparatively analyzed to show their differences for the selected plain dents found in the two field trials.

Keywords: LSM, SCT, ILI, FEA, mechanical damage, pipeline dent, equivalent stress, magnetic field

INTRODUCTION

Pipeline integrity management plans require an effective inspection technology to detect existing anomalies in pipelines. A non-intrusive, above-ground inspection of underground pipelines is very attractive to the pipeline operators due to its operation simplicity and economic advantages, particularly for unpiggable or difficult-to-inspect pipelines because no other effective above-ground inspection techniques are available today. With commercially available, high-sensitive magnetic sensors capable of detecting weak magnetization signals at the nano-Tesla level in the earth magnetic field, various versions of passive geo-magnetic detection techniques were developed and have gained broad attentions over the past decade. Large Standoff Magnetometry (LSM) is an emerging and promising inspection technology commercially available for field trials with quite sound outcomes in recent years. McDonnell et al. [1] stated that the LSM technology is used to identify and locate the elevated stress zones through measuring the magnetic field surrounding stressed pipelines, regardless of the ability to pig or whether the pipeline is buried, exposed or elevated. A change in the magnetic field indicates the presence of stress concentration in the pipe wall due to anomalies or features, and a shape change of magnetization in a ferromagnetic steel pipe is characterized as magnetostriction. The LSM technology was developed to measure such changes.

Recently, Xu [2] pointed out that the basic principle of the LSM technology is the inverse magnetostriction, that was known as Villari Effect, first discovered by Italian Physicist E. Villari in 1865, where applied stresses cause a change of magnetization in a ferromagnetic material. Changes in the earth magnetic field surrounding a steel pipeline by magnetostriction from applied stresses can be measured at a large standoff distance from the pipeline using a magnetometer with highly sensitive magnetic sensors, as shown schematically by Navalgund et al. [3] in Fig. 1, where a magnetic field is sampled with magnetic data recorded at a rate of hundreds per second, and the LSM tool is hand carried by an operator walking along the pipe centerline. After a LSM scan, data recorded by the LSM tool are analyzed to identify the signal characteristics that are correlated to anomalies. Typically, those anomalies in a pipeline are the main source of elevated stresses in the pipe wall. The location of elevated stresses or stress concentration zone (SCZ) can be thus captured and determined through a correlation between the magnetic change and stress estimated from the LSM measured signals. As a result,

Copyright © 2019 by ASME

this technology allows determination of the SCZ location and stress amplitude associated with a pipeline anomaly or feature.

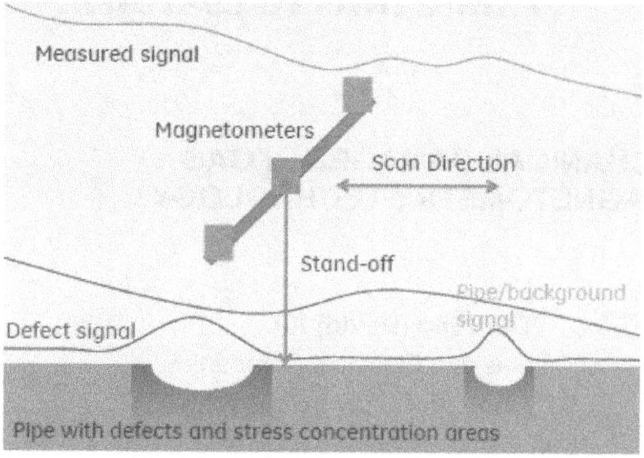

Figure 1. LSM scanning underground pipeline with anomalies, taken from Reference [3]

However, other factors such as metal heterogeneity, extra metal objects and magnetic interference from surrounding environment can also attribute to changes in the magnetic field, and thus the measured data analysis and interpretation are critical to proper discrimination of all factors. Different versions of the LSM Technology utilized different proprietary techniques and methodologies for their own tools. Therefore, the exact mathematical models built in the LSM tools to define the stress-magnetization relation and to output stresses are not accessible today by public or pipeline operators.

Xu [2] briefly reviewed the historical development of the LSM technology and pointed out that University of Leeds [4-6] in UK made a major contribution to the fundamental modeling as well as system development and commercialization of LSM with a private partner – Speir Hunter Ltd. At the current stage, all LSM tools aim to detect features or anomalies in pipelines using the output stresses from the LSM tool, rather than geometric shapes. Currently there are three active LSM technology providers in the pipeline industry, and they benchmarked their own LSM technology with some noticeable differentiations in terms of equipment, claimed capabilities and limitations. However, the pipeline operators have limited knowledge of the LSM technology, its fundamental physics and mathematical models used to estimate stresses in the SCZs associated with pipeline anomalies.

To evaluate the LSM technology and to understand the LSM outcome of stresses, the three active LSM technology providers were hired by PG&E [2] to perform a small-scale field trial evaluation on selected natural gas pipelines in 2016 to 2017. Xu [2] detailed the results of the LSM field trials and discussed the fundamental physics of the LSM technology with a preliminary understanding of its current performance on selected live gas pipelines that have anomalies characterized by in-line inspection (ILI) or in-ditch non-destructive evaluation (NDE) methods. Through comparisons, some differences of the three LSM systems were better understood, including their capabilities and limitations.

This paper continues to evaluate the LSM technology with use of the field trial outcomes, as reported by Xu [2]. The focus here is on one LSM technology, its fundamental physics model and possible equivalent stress definitions. The FEA simulations are performed for two mechanical dents selected from the field trials, and then the numerical stresses are used to evaluate the LSM output stresses. On this basis, suggestions are given on how to use and how to improve the LSM technology.

SCT TECHNOLOGY

The stress concentration tomography (SCT) is one version of the LSM technology, invented by University of Leeds [7] through a joint research project funded by a private partner - Speir Hunter Ltd. Recently in 2018, the SCT technology with a tool design and data processing [7] was patented in USA. The early project was initiated to characterize signatures of the magnetic field of anomalies associated with elevated local stresses or a SCZ based on a mathematical model [4-6] that was suggested to describe the relationship between one-dimensional (1D) stress and change in magnetic field for a steel bar. This model was then expanded into a three-dimensional (3D) stress condition and verified successfully by testing pressurized pipes with capped ends and an external force in the pipe middle. On this basis, a fully automated software algorithm was developed for the patented SCT tool [7] and used to interpolate magnetic data and to calculate elevated stresses. Because of proprietary nature, explanation of the stress-magnetization relation seems still hypothetical to public or the pipeline operators.

There are some advantages of SCT as an inspection tool for existing pipelines, as claimed by Cowell et al. [8], Jarram [9], and Habibi [10-12]. SCT is a remote inspection technology that requires no contact with the target, nor the input of energy into it, and thus with not really many limitations on pipeline configurations or operational conditions. It can detect all causes of elevated stresses local to corrosion defects, cracks, dents, wrinkles, girth welds, and part penetration in welds. Some field verifications of the SCT system were presented in References [1-2, 8-12] for various types of pipeline features or anomalies. The attractions, claimed benefits, capacities, specifications, limitations and challenges of SCT are outlined below.

Attractions to Pipeline Operators

- Ideal tool for aging and non-piggable pipelines
- Non-invasive, remote pipeline integrity assessment
- No interruption of pipeline service
- No change required to operating service conditions
- No dependence on flow rate of transported medium
- Suitable for all ferromagnetic pipelines

Copyright © 2019 by ASME

Benefits of SCT Inspection

- Detect stresses at all metal defects irrespective of types and orientations
- Detect stress corrosion cracking (SCC)
- Detect stresses caused by geological hazards
- Detect stress local to a defect, not its geometry
- Can confirm if damaged coating led to metal corrosion

Claimed Capabilities of SCT Tool

- Measure changes of the magnetic field along the pipeline center for shape, strength and direction
- Report (1) locations of elevated stresses, (2) maximum allowable operating pressure, (3) metallurgical and mechanical conditions, and (4) safe operation term
- Reported stresses are associated with anomalies of corrosion defects, cracks, dents, wrinkles, SCCs, metallurgical defects, soil slippage and subsidence
- Reported stress is at the geometric center of a SCZ caused by one or multiple anomalies
- Magnetic field generated by pipeline stress decays with distance from the pipeline

Claimed Specifications of SCT Tool

- Probability of detection of SCZs is larger than 80%
- Location accuracy of SCZs is less than one meter in the axial direction, and ±10 cm in the lateral directions
- Estimated stress has an accuracy of 25 MPa (3.6 ksi) regardless of its resource
- 10 km of pipeline typically inspected per team per day on terrain

Limitations of SCT Technology

- An area of multiple SCZs within a range of one meter will not be individually distinguishable
- The highest level of stress is reported in a SCZ range, including all defects within one meter
- Collected magnetic data are the sum of three individual magnetic signatures, and its maximum amplitude is associated with the highest level of stress at each SCZ
- SCT cannot measure geometric size of any defect

Challenges of SCT Technology

Although magnetic signatures are of a minute intensity, they are detectable with advanced sensor and instrumentation technology commercially available today. Still detecting weak magnetic signatures in a pipeline was one of major challenges but understanding them to correlate magnetic change with stress was even more challenging in developing the SCT tool.

In general, verification of the accuracy of SCT reported stress level is hard to achieve in real-world conditions, because SCT reports one SCZ which reflects the highest localized stress in an area with the geometric center of one or more defects without considering number and type of defects. Specifications of the patented SCT tool are clear, but interpretations of the SCT outcomes remain unclear and causes most of issues.

Because the SCT reports elevated stresses and SCZs at or near an anomaly, and ILI reports the geometric profile of the anomaly, the numbers of SCZs in the pipeline reported by SCT may be much more than the number of indications reported by ILI. Increasing accuracy of SCT reported stress level may help get better correlation between the two inspection technologies at least on those medium-to-significant indicators.

In summary, the math model of stress-magnetization adopted in the SCT tool [7] needs a proper understanding so as to solve the challenges and to correctly analyze the SCT outcomes to assess the pipeline integrity. Thus, fundamental basis of the math model for SCT is briefly reviewed.

THEORETICAL BASIS OF LSM

The magnetostriction and magnetization caused by stress in ferromagnetic steels has long been known, but only partially understood. When stress is applied to a ferrous structure, such as a carbon steel bar, magnetization in the bar changes as a function of stress, leading to a change in the magnetic field around the structure.

A Stress-Magnetization Model

Magnetostriction has been theoretically modeled by Jiles and his coworkers [13-18], and a basic magnetization-stress model was developed for ferrous materials in conjunction with early experimental data and the elastic energy:

$$\frac{dM}{d\sigma} = \frac{1}{\epsilon^2}(\sigma \pm \eta E)(1 - c)(M_{an} - M_{irr}) + c\frac{dM_{an}}{d\sigma} \tag{1}$$

where M is magnetization, σ is an applied uniaxial stress, ϵ is a material property, E is the Young's modulus, η and c are constants reflecting the ability of magnetic domains to become magnetized, M_{an} is the anhysteretic magnetization, and M_{irr} the irreversible part of magnetization. This equation is also referred to as the Modified Jiles-Atherton-Sablik model that has been improved by Viana et al. [18] and Li and Xu [19].

Because elastic energy was used during the derivation of Eq. (1), the stress-magnetization is only applicable to an elastic stress condition, but not to a plastic condition. Equation (1) can be solved numerically to determine a relationship between stress and magnetization. Figure 2 shows a typical solution using test data from Jiles [15] for a carbon steel bar, as reported in Staples et al. [4], where the x-axis denotes compressive or tensile stress in the bar, and the y-axis denotes the normalized magnetization (M/Ms). The solution shows that irreversible magnetization or magnetic memory is expected for a steel bar in cycled stresses. For complicated structures with 3D stresses, the available theory is difficult to use directly, and further understanding is required on the magnetic field surrounding an object. Wang et al. [21-22] have developed theoretical and experimental work in this direction.

Copyright © 2019 by ASME

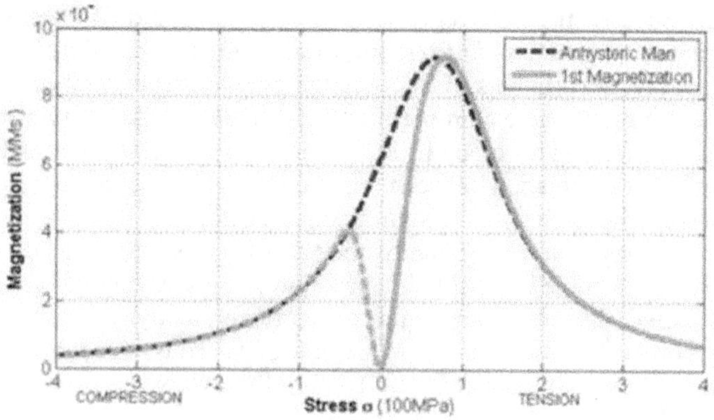

Figure 2. A solution of stress-magnetization in carbon steel

To understand the magnetic field surrounding a stressed steel object, a simple relation between stress and magnetization needs to develop for use in a multiphysics modeling tool to predict stress from a detected magnetic field. To this end, Staples et al. [4-6] conducted different experimental tests using 1D tensile sheet bars and 3D cylindrical pipes and observed the similar variations between stress and magnetization, as described by Eq. (1). However, a reverse solution of Eq. (1) to express the stress as a function of magnetization was not yet available to public.

Definitions of Equivalent Stress

A uniaxial stress model cannot be directly applied to multi-axial stress models, and an equivalent stress needs to describe the multiaxial stresses as a fictitious but equivalent uniaxial stress. Different definitions of an equivalent stress were proposed for the stress-magnetization analysis.

Schneider and Richardson [21] proposed a very simple equivalent stress:

$$\sigma_{eq} = \sigma_1 - \sigma_2 \tag{2}$$

where σ_1 and σ_2 are the principal stresses in the sheet plane, and the σ_1-axis is parallel to the applied magnetic field.

Kashiwaya [22] proposed an equivalent stress:

$$\sigma_{eq} = K(\sigma_i - \sigma_{max}) \tag{3}$$

where K is a constant, σ_i is the principal stress coaxial with the magnetic field, and the σ_{max} is the maximum principal stress.

Sablik et al. [23] proposed another equivalent stress:

$$\sigma_{eq} = \begin{cases} \frac{1}{3}(2\sigma_1 - \sigma_2), & for\ \sigma_1 < 0 \\ \frac{1}{3}(\sigma_1 - 2\sigma_2), & for\ \sigma_1 \geq 0 \end{cases} \tag{4}$$

where the principal stress σ_1 is coaxial with the magnetic field. However, this definition is discontinuous for $\sigma_2 = 0$ and the equivalent stress is not equal to the applied stress in 1D loading.

All those three definitions of equivalent stress are simple, but they have strong limitations: the mechanical loading is restricted to biaxial stresses, and the magnetic field must be applied along the eigendirection of the stress tensors. A more general equivalent stress is needed to consider a 3D stress sate or a more complicated range of combined magnetic field and mechanical loading in practical applications.

For a live pipeline, the operating conditions introduce a uniform stress in the longitudinal, circumferential and radial directions of the pipeline without any anomaly. This creates a baseline for the magnetic field that can be measured at a distance from the pipeline. Pipeline features (like girth weld, casing end), defects (such as corrosion/metal loss, crack, dent and wrinkles) and external forces (such as landslip movements) will cause an increase of localized stresses in comparison to undamaged sections of the pipeline. This will change the orientation of the magnetic field in the defect-free pipeline, which in turn disturbs the baseline magnetic field. This is the magnetic response to increased local stresses or SCZ associated with a defect in the pipeline.

To describe the relationship between triaxial stresses and magnetic signals for a pressurized pipe or pressure vessel in the 3D deformation conditions, the von Mises equivalent stress was used by different investigators of Rosakosz and Bienick [24] and Yi et al. [25].

$$\sigma_{eq} = \frac{1}{\sqrt{2}}\sqrt{(\sigma_{11} - \sigma_{22})^2 + (\sigma_{22} - \sigma_{33})^2 + (\sigma_{33} - \sigma_{11})^2 + 6[\tau_{12} + \tau_{23} + \tau_{31}]} \tag{5}$$

where $\sigma_{11}, \sigma_{22}, \sigma_{33}$ are the normal stresses, and $\tau_{12}, \tau_{23}, \tau_{31}$ are the shear stresses. It has been demonstrated that the trends of the von Mises equivalent stress and the magnetic signal along the measurement line is primarily the same [24-25].

Li et al. [26] investigated the magneto-mechanical effect on the geomagnetic field of ferromagnetic pipelines and proposed a theoretical model that matches well with actual test data. For an infinite long pipeline, these authors showed that pipeline diameter, wall thickness, operating pressure as well as cover of soil depth all have impacts on the magnitude of the measured magnetic field, but not the pipe length.

Bar Magnetic Model for LSM

Staples et al. [6] presented their modelling of a pipeline using a bar magnet model, in which the pipeline was represented as a series of magnetic dipoles. Using the theory proposed for the bar magnet field, various situations can be considered as combinations of bar magnets, and the magnetic field is computed using the Python program VectorFieldPlot.

Using this approach, the magnetic flux leakage can be measured and analyzed using the LSM technology that operates in the near magnetic field at a significant distance from the object. Both modelling and experimental/field work are presented to justify this approach. Figure 3 shows three pipeline sections with two girth weld joints represented as bar magnets, where the magnetic field lines have been plotted using the VectorFieldPlot program, and the weld is strongly magnetized with the same orientation as the pipeline [6].

Copyright © 2019 by ASME

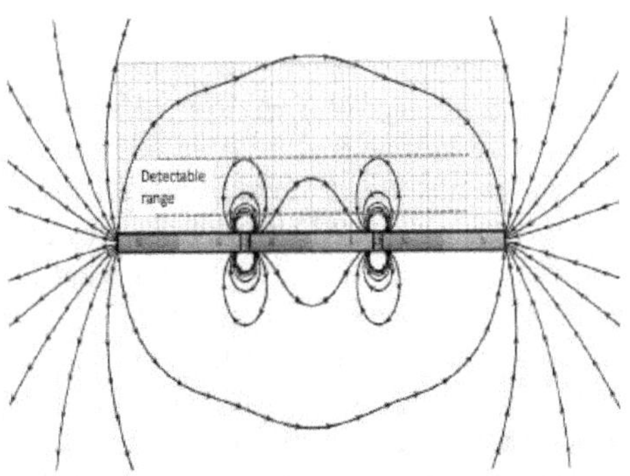

Figure 3. Bar magnet models for three pipeline sections with two jointed welds, each represented as a bar model [6]

Figure 3 illustrates the possible variation of magnetic field with distance from an underground pipeline. A grid has been imposed over the pipeline model, with the vertical scale set to measure distance above the pipe sections, each major tick on the scale being one pipe diameter. The weld magnetic field variations are detectable in the near field, but not the far field.

As an example, a field survey is discussed here. A 175-m section of an existing underground gas pipeline was scanned, and the pipeline girth welds were located using the dipole model [6] that is similar to the bar magnet arrangement. The survey was selected as there was confirmed information on the girth weld locations, obtained from ILI together with a planned excavation to locate and repair a pipeline defect.

Figure 4 shows the field survey results of the underground pipeline, where the top part shows the magnetic field and the predicted weld locations along the survey distance compared to the weld positions located by ILI, and the bottom part shows the gradient modulus function of the magnetic field located as the function of distance along the pipeline. The locations of SCZs and ILI welds/features are also plotted in both parts of the figure. It shows that the highest peak in the modulus function coincides with the pipeline defect that was located by ILI or the position confirmed by a later excavation.

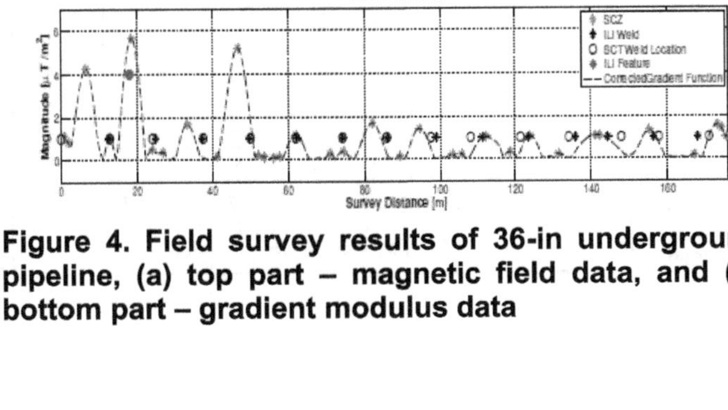

Figure 4. Field survey results of 36-in underground pipeline, (a) top part – magnetic field data, and (b) bottom part – gradient modulus data

This field work shows that the dipole model is viable on above-ground survey of underground pipeline, where SCZ such as welds can be located. The results also confirm that the SCT is a viable inspection technology for steel pipelines. However, the inverse solution of stress-magnetization equation and the equivalent stress were not given by Staples et al. [4-6]. In recent two LSM projects funded by PRCI [29-30], the LSM output stress was claimed as the von Mises stress, but not the hoop stress of the inspected pipeline.

FIELD TRIALS OF LSM INSPECTION

To evaluate the LSM technology and to understand its outcomes, PG&E recently hired three active LSM technology providers to conduct small-scale field trials on selected live natural gas pipelines containing known anomalies or feature types, as detailed by Xu [2]. The LSM inspection results were analyzed for quantitative evaluations of the anomaly location error and elevated stress estimation.

Anomaly or Feature Types for Evaluation

Girth weld and casing end are the two unique features used in field trials due to relatively easy detection. Landslide damage is an anomaly type selected due to interests at PG&E. Dent is a common mechanical damage associated with high elevated stresses and thus serves as a good candidate in a stress quantification study. For simplicity, only plain dents were discussed here for the field trial study. Data obtained from the planned excavation per the integrity dig program were used as the accurate reference. In case there is no planned excavation, ILI data or other survey data were used as the reference. No dig was initiated solely for evaluation of the LSM indications.

Selection of Field Trial Sites

Six sites with 150 ~ 1000 feet segment each were selected after consideration of the LSM limitation. All the selected sites have previous ILI records as a preliminary reference plus excavation data before or after LSM scan as the main reference. The selected sites were intended to cover representative field conditions, including single pipeline at an open hill terrain as well as typical industrial-residential area with some sorts of interference. Noted that most detected dent anomalies are minors without a requirement to repair, and their magnetic signals are typically weak, which challenges the LSM detection sensitivity.

To avoid the impact of residual magnetization, pipelines exposed to ILI MFL (magnetic flux leakage) run over 2 years ago were first selected per one LSM technology provider's suggestion. In addition, a pipeline with MFL run just 6 months ago was also selected as a typical field case. But a quick check at the live pipe OD surface indicates negligible magnetic field strength (< 1-3 gauss per a simple gauss gauge). So far only one vendor has defined a specification of residual magnetization level (≤ 100 A/m) per past experience. A fundamental study is needed to help define a specification because this can pose a limitation to LSM field applications.

Copyright © 2019 by ASME

Quantitative Evaluation of Location Accuracy

All LSM reported SCZs were used to match ILI indications of pipeline feature or anomaly location with a focus in this work on the targeted girth welds, casting ends and plain dents.

For verification of girth weld location, one or two girth welds exposed in the ditch were used as the benchmark to correct ILI girth weld GPS coordinates within LSM scanning range. Since the LSM scanning distance per site is relatively short, the drift of ILI's positioning devices is negligible and just a linear offset correction per dig reference is accurate enough.

For the targeted landslide damage, there is no excavation, and so only ILI and location survey data were available to serve as a reference. Therefore, some errors from the LSM measured data may be expected up to three feet (one meter).

For the targeted casing end and other anomalies, excavation NDE data were used as the accurate reference. It is noted that there could be human errors or GPS positioning errors due to poor wireless communication conditions in the deep ditch. In the study of LSM field trials, only one large GPS positioning error was identified by Xu [2].

LSM Inspection Results

Basic observation

There are no large differences in field operation among the three providers except one asked for field calibration through pressure adjustment (≥10%) or a dig to have higher confidence. The major field observations were focused on the accurately pre-marked pipe centerline, reliable magnetometry equipment, and accurate GNSS (GPS) system. Two vendors claimed good repeatability of their inspection results but did not provide data for proof. In terms of post-scan data processing and analysis, one vendor did provide preliminary data processed by their automatic software in 24 hours, and the other two provided results two-three weeks later due to manual data analysis.

Girth weld detection by Vendor-A

Two rural sites of Site #1 and Site #2 were scanned by Vendor A who has the state-of-the-art GNSS system. One site is on an open cow-grazing hill with no interference or blockage. The other is on a flat open space surface terrain with another pipeline in parallel at ~15 feet away. Table 1 summarized the LSM outcomes after implementing correction technique as mentioned above. It is clear that the claimed capability of ±2 m errors is not even close to what is actually achieved. However, the large error includes a consistent shift (~11-12 feet). By subtraction of this shift, the error goes down to the claimed range, as shown in Table 1. Per the vendor, the magnetic signature of girth welds is very unique with a large width that can add some errors in determining the location, and girth weld location error does not affect location accuracy of other features or anomalies. This is not really convincing.

Casing end detection by Vendor-C

Casing end detection was considered as a fairly easy application not associated with SCZs. Since LSM cannot really identify such feature type yet, detection has to be based on a known approximate presence range. The vendor with the advanced LSM technology was chosen for the field work on two targeted casing ends in an urban environment (Site#3). One of the two LSM scans was matched with an error of 2.5 feet even though the scan area has some surrounding bushes and a steel wire fence along the pipeline. Although this error amount is still larger than ILI specification (±0.5 foot) and what is practically desired (±1.0 foot), it is still a good starting point especially in case no other better technique is applicable. Surely more field validations are needed especially as this vendor's GPS unit sometimes may produce very large location error.

Xu [2] reported the detailed analysis of LSM measured location errors for girth weld detection by Vendor A, casing end detection by Vender B, and landslide damage detection by Vendors B and C. This paper will focus on mechanical dents detected in the field trials on gas pipelines, as discussed next.

Table 1. LSM girth weld and casing end location error

Locations	LSM Providers	ILI Girth Weld ID	LSM locating error			
			As it is		substraction of the average offset	
			(m)	(feet)	(m)	(feet)
Site #1 (24", 0.312") (X60) (~410 psi)	A	36410	3.2	10.5	**-0.1**	**0.1**
		36420	3.3	10.7	**-0.1**	**0.3**
		36430	3.0	9.8	**-0.3**	**-0.7**
		36440	3.0	9.7	**-0.3**	**-0.7**
		36450	3.2	10.4	**-0.2**	**-0.1**
		36460	3.2	10.4	**-0.2**	**0.0**
		36470	3.0	9.9	**-0.3**	**-0.5**
		36480	3.4	11.0	**0.0**	**0.6**
		36490	3.7	12.0	**0.3**	**1.6**
		36500	3.9	12.9	**0.6**	**2.5**
		36510	4.1	13.5	**0.8**	**3.1**
		36520	4.3	14.1	**1.0**	**3.7**
		36530	4.4	14.4	**1.1**	**4.0**
		36540	4.3	14.2	**1.0**	**3.8**
		36550	0.9	2.8	**-2.5**	**-7.6**
		36560	2.3	7.7	**-1.0**	**-2.7**
		Average	**3.3**	**10.9**	**0.0**	**0.5**
		Std	**0.9**	**2.9**	**0.9**	**2.9**
Site #2 (24", 0.312) (X60) (~410 psi)	A	41560	10.1	33.2	poor data quality at ends per Provider-A	
		41570	7.5	24.5		
		41580	4.6	15.2	**0.9**	**2.9**
		41590	3.1	10.2	**-0.6**	**-2.0**
		41600	3.7	12.2	**0.0**	**-0.1**
		41610	3.2	10.6	**-0.5**	**-1.7**
		41620	3.3	10.9	**-0.4**	**-1.4**
		41630	3.2	10.7	**-0.5**	**-1.6**
		41640	3.9	12.7	**0.1**	**0.4**
		41650	4.0	13.1	**0.2**	**0.8**
		41660	4.6	15.1	**0.9**	**2.8**
		Average	**3.7**	**12.3**	**0.0**	**0.0**
		Std	**0.6**	**1.9**	**0.0**	**0.0**
Site #3 (12", 0.281") (X42) (~330 psi)	B	Casing end #1	**0.8**	**2.5**	N/A	
		Casing end #2	Not accessible:under highway wall			

Copyright © 2019 by ASME

Dent Location Errors

Table 2 shows the LSM outcomes of dent location errors and estimated stresses obtained by three providers for plain dents at different sites. As evident in this table, the detected location errors have a range from 0.5 m to 3.7 m with a maximum error of 5 m, but the stress has no error to give.

Table 2. LSM measured dent location error and stress

Locations	LSM Providers	ILI Dent ID	LSM location error & stress			
			(m)	(feet)	Stress (MPa)	% of SMYS
Site #1 (24", 0.312") (X60) (~410 psi)	A	#1	0.6	2.0	162	39%
	B		5.0	16.3	small signal	
	C		N/A		missed	
Site#4 (24", 0.312") (X42) (~235 psi)	C	#2	1.9	6.2	91	32%
Site #5 (24", 0.271") (X60) (~235 psi)	B	#8	1.1	3.6	medium signal	
		#9	0.5	1.9	large signal	
		#10	3.2	10.6	large signal	
		#11	2.2	7.1	medium signal	
		#12	2.8	9.1	medium signal	
		#13	N/A		missed	
	C	#7	2.8	9.2	117	40%
		#8	2.8	9.2	142	49%
		#9	2.7	8.7	102	35%
		#10	3.7	12.3	158	54%
		#11	N/A		wrong pipeline marking	
		#12				
		#13				

(a)

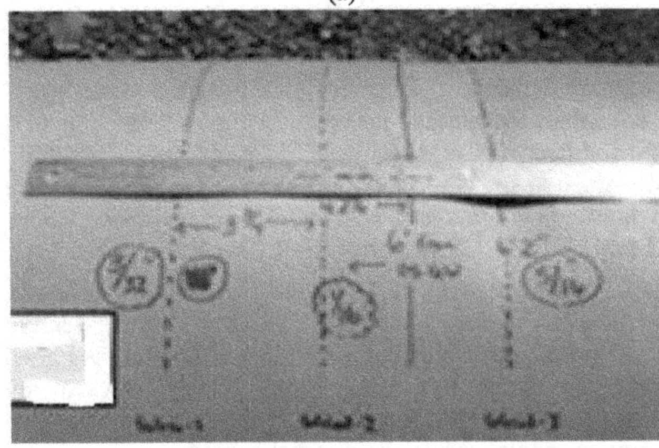

(b)

Figure 5. A cluster of 2 wrinkle-like dents marked as (a) Dent #1-1 (wrinkle 1 + wrinkle 2) and (b) Dent #1-2 (wrinkle 3 + wrinkle 4)

At an open space hill Site #1, there is a cluster of two wrinkle-like dents that are close to each other at pipe top surface (12:00 o'clock position), as shown in Fig. 5. From the ILI and excavation results obtained in 2008, no repair was required. The most recent ILI run in 2014 did not show any noticeable change from the inspection data in 2008. The small wrinkle-like dent is a challenge to detect using the LSM tool due to its weak magnetic signatures. Vender A detected its location with a small location error of 0.6 m and reported the maximum stress of 162 MPa or 39% of SMYS for the X60 gas pipeline at the SCZ associated with the wrinkle-like dents. However, Vendor C was unable to detect these dents, and Vendor B can detect, but gave a large location error of 5 m.

Figure 6 shows an example of the LSM outcomes of the pipeline section near the two wrinkle-like dents provided by Vendor A, where the upper plot indicates the pipeline altitude, the upper-middle plot shows the estimated stress together with SMYS and operating hoop stress, the lower-middle plot shows speed of the LSM tool operator, and the lower plot records the identification number of pipeline features. The red arrow in the figure shows the dent location at a hill valley and the LSM estimated stress of 162 MPa at the SCZ. However, it's unknown that the LSM stress is a hoop stress or von Mises stress.

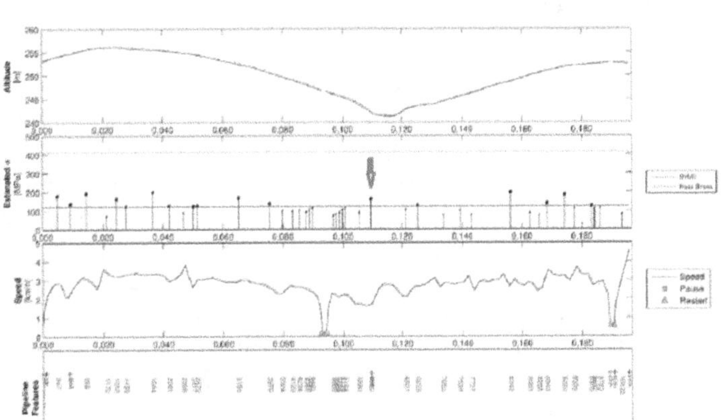

Figure 6. LSM outcomes of the pipeline section near the two wrinkle-like dents provided by Vendor A with the upper plot indicating the pipeline altitude, the upper-middle plot showing SMYS, operating hoop stress and estimated stress, the lower-middle plot recording speed of the LSM tool operator, and the lower plot numbering the pipeline features detected.

Copyright © 2019 by ASME

During the LSM scanning, the X60 pipeline has an internal pressure of 410 psi, and the operating hoop stress is 108.7 MPa or 26.3% of SMYS, as marked in Fig. 6. The lower estimated stresses by LSM tool should correspond to the pipeline sections without any defects, in which the operating von Mises stress is 94.2 MPa that is lower than the operating hoop stress. Due to this reason and from the LSM results in Fig. 6, one may conclude the LSM estimated stress is the von Mises stress. This is consistent with the statement that von Mises stress was estimated from the LSM tool in the PRCR Reports [29-30].

At Site #4, there is a large dented area, shallow Dent#2, as shown in Fig. 7 after excavation. The LSM scan by Vendor C detected this shallow but long dent (39" x 10", depth<1% OD) near the pipe bottom OD surface (5:15 o'clock) with a location error of 1.9 m. At Site #5, most dents detected are surrounded by very minor anomalies such as external corrosion pits, tiny dents or gouges that do not require any repairs. Those minor anomalies are beyond the ILI detection capability although they can be seen at field excavation sites. As such, dents at Site #1 and Site #4 detected from the LSM tool are selected here for further analyses.

Figure 7. A large very minor Dent #2 at Site #4.

Dent stress estimated from LSM

As shown in Table 2, all LSM detected dents have a low stress level in comparison to SMYS of the pipeline steels (≤50% SYMS). This seems consistent with the fact that none of those dents require a repair in the actual integrity assessment. Such low LSM stress levels may indicate plastic deformation at the dents was not considered. To further understand and quantify those LSM stresses, an accurate stress for each dent need to be calculated for each pipe using the FEA method. The FEA stresses can be used as a reference to evaluate the LSM output stresses. Otherwise, there is no other viable way to measure actual stresses in a dent area of live pipelines.

DENT STRESS DETERIMNATION BY FEA

Two sets of dents shown in Figs 5 and 7 are selected for the FEA calculations to determine the stress distribution along the axially central line of the pipeline for each dent. The FEA models of dents, simulations and result analyses were detailed in the EWI Report [31]. A brief introduction is given as follows.

FEA Models of SCT Inspected Dents

PRCI Project MD-4-4 validated FEA models for plain dents [32-33] are adopted here to simulate the two dents. The elastic and/or elastic-plastic FEA calculations are performed using the commercial FEA software ABAQUS for the two dents in the 3D conditions.

In the FEA model for a single dent, one quarter of the pipe were used to consider the pipe symmetry and to reduce the FEA calculation costs. Four-node shell elements are adopted in the FEA meshes. The length of pipe model is 10 times the pipe OD, and the remote pipe ends are fixed to reflect the infinitely long pipeline and to reduce the pip end effect. Only internal pressure is applied to the end capped pipe after a dent formed. A dent is created by a quasi-static contact between an assumed rigid indenter and the pipe body. During indentation, the soil restrain on the pipe is modeled by rigid radial supports at the pipe bottom over an arc of 120°. For each pipeline, the true stress-strain data leveraged from the yield stress (YS) and the ultimate tensile stress (UTS) of the pipeline steel are utilized in the FEA simulations.

For the wrinkle-like dents at Site #1, the pipe has an actual OD of 24.18 in and actual wall thickness of 0.324 in, and the pipeline grade is X60. During the LSM inspection, the operating pressure in this X60 pipeline is 410 psi, leading to the operating stress of 26.3% of SMYS. For the long and shallow dent at Site #4, the pipe has an OD of 24 in and a wall thickness of 0.313 in, and the pipeline grade is X42. During the LSM scan, the operating pressure in this X42 pipeline is 235 psi, leading to the operating stress of 21.5% of SMYS. The geometric sizes of those dents can be found in Report [31].

To simulate the dents, two FEA modeling approaches are adopted. One is the idealized dent approach [32], and another is the pre-formed dent approach [34]. In the former approach, the profile of an accurate dent can be approximately created with indentation history, residual stresses, large plastic deformation, and the actual stresses local to the dent can be calculated under the operating pressure. In the latter approach, the profile of an actual dent measured at in-ditch is mapped to create a purely geometric profile of the dent without residual stresses or plastic deformation, and elastic stresses of the purely geometric dent can be numerically calculated under small operating pressures.

It should be noted the FEA outcomes are ABAQUS-based numerical results that may be different from the LSM outputs. Since this FEA work just aims to evaluate LSM stress, both LSM and FEA results are not intentionally used to make a judgement on the integrity decision for the dented pipelines.

Numerical Results of Dent Stresses

Using the dent modelling approaches noted above, two sets of the plain dents matched to LSM indications were simulated using ABAQUS in the 3D conditions to define a reference stress. One is the small cluster of wrinkle-like Dent #1-1 and Dent #1-2 at Site #1 in a gas pipeline constructed in 1991. The other is Dent #2 at Site #4 in a gas pipeline constructed in 1970.

The idealized dent approach was first utilized to calculate actual stresses for the pipelines with the wrinkle-like dents at Site

Copyright © 2019 by ASME

#1 and the single dent at Site #4 at their operating pressure conditions. Figure 8 shows the contours of von Mises stress on the pipe OD surface of the two wrinkle-like dents (dent #1-1 and dent #1-2 at Site #1) in the X60 pipeline for the loading condition of (a) zero pressure at the residual stress status where the indenter was removed from the pipe, and (b) operating pressure of 410 psi. One half of pipeline symmetry was used in the FEA simulation. Figure 9 shows the stress distributions on the pipe top OD surface along the axial direction of the X60 pipeline at the operating pressure, where the axial stress, hoop stress, shear stress and von Mises stress are included. Figures 8 and 9 reveal that the stresses in the two wrinkle-like dents are comparable and have very high values in comparison to the SMYS of 60 ksi. As such, the interaction between the two wrinkle-like dents is small and can be neglected in the FEA simulation. The deeper wrinkle-like dent #1-2 is then selected as the representative of the two dents.

As noted, the von Mises stress in the dents is very high and close to the level of true UTS of the X60 pipeline steel even in the low operating stress condition. This contributes to the high residual stress and large plastic deformation generated from a large indenter force during the dent formation at the zero pressure. This result is consistent with other numerical results for dents, such as observed by Baek et al. [35].

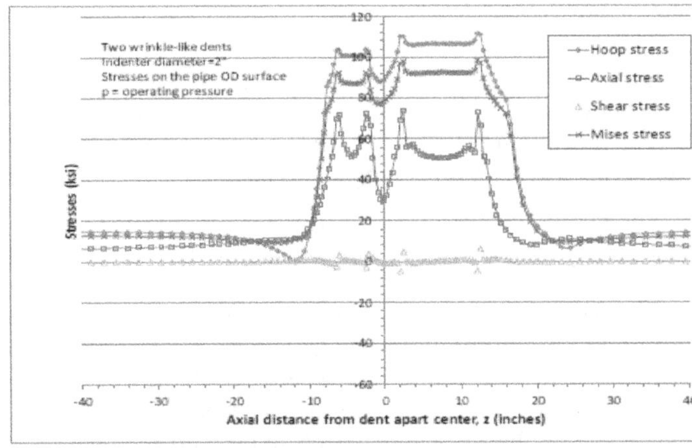

Figure 9. Distributions of axial stress, hoop stress, shear stress and Mises on the pipe top OD surface in the axial direction for two wrinkle-like dents at operating pressure.

Figure 10 shows the contours of von Mises stress on the pipe top OD surface of the single dent (dent #2 at Site #4) in the X42 pipeline for loading condition of (a) zero pressure at the residual stress status, and (b) operating pressure of 235 psi,

(a)

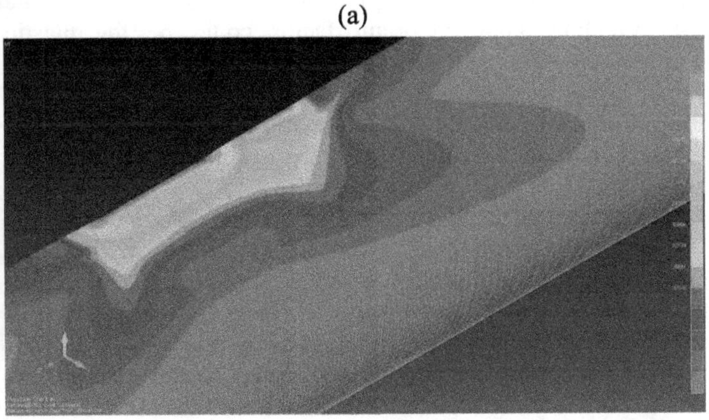

(b)

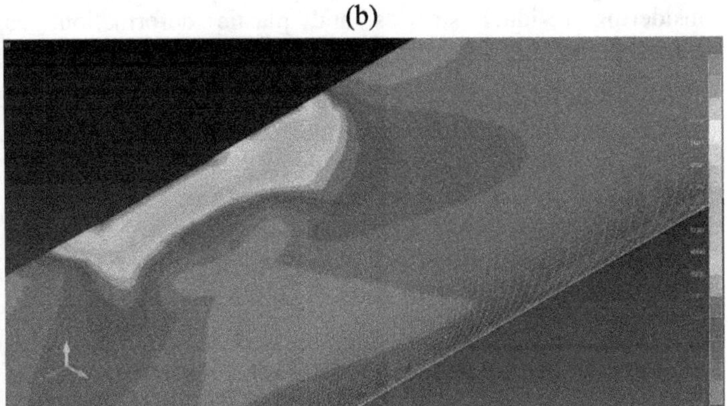

Figure 8. Contours of von Mises Stress on pipe OD surface for two wrinkle-like dents (Dent #1-1 and #1-2) for loading conditions at (a) residual stress and (b) operating pressure

(a)

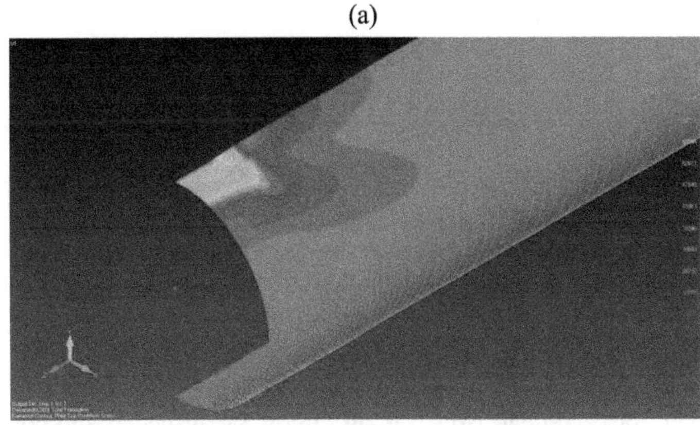

(b)

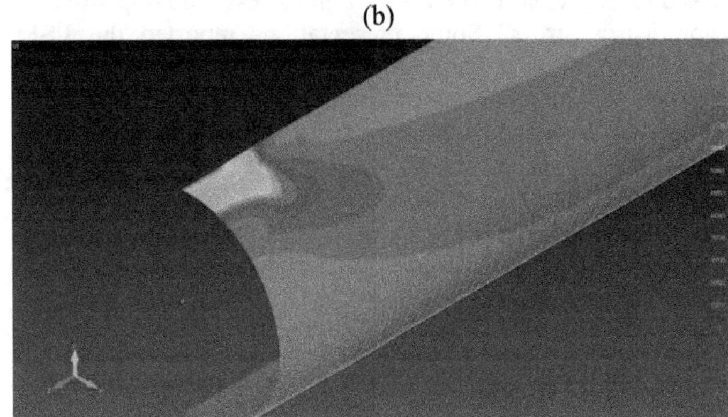

Figure 10. Contours of the Von Mises Stress on pipe OD surface for Dent #2 for loading conditions at (a) residual stress and (b) operating pressure

Copyright © 2019 by ASME

where a quarter of the pipe model was used in the FEA simulation due to the two pipeline symmetries. As evident in Fig. 10, the large plastic deformation and the maximum Mises stress are generated in the vicinity of the single dent, but other areas of the X42 pipeline remain in the elastic state with small elastic deformation and lower operating Mises stress.

Figure 11 shows the stress distributions on the pipe top OD surface along the axial direction of the X42 pipeline at operating pressure, where the axial stress, hoop stress, shear stress and von Mises stress are included. Again, the very high von Mises stress close to the level of true UTS and large plastic deformation are generated in the dent of the X42 pipeline at the low operating pressure. This observation is consistent with the results shown in Figs 8 and 9 for the wrinkle-like dents in the X60 pipeline.

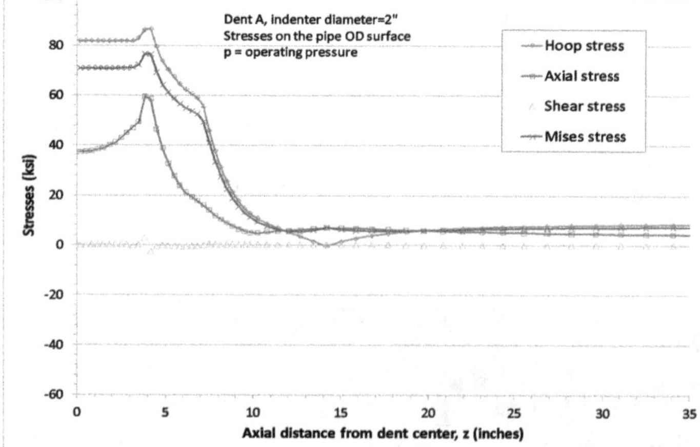

Figure 11. Distributions of axial stress, hoop stress, shear stress and von Mises on the pipe top OD surface in the axial direction for the single dent at the operating pressure.

Evaluation of LSM Output Stress

In Table 2, the stress estimated from the LSM scan was given for each dent. However, all three LSM providers did not marked clearly what is the LSM output stress – a hoop stress or a von Mises stress? Some presentations reported the LSM estimated stress is a hoop stress, but it may be not a good candidate of equivalent stress in comparison of the previously reviewed definitions. On the other hand, as claimed in the PRCI Reports [29-30] and the analysis of Fig. 6, the LSM output stress could be the von Mises stress. Thus, it is assumed in this work that the LSM stress given in Table 2 is the maximum von Mises stress at the center of the SCZ near a single defect or a cluster of defects within a range of one-meter area in the axial direction of the pipeline.

Comparison of LSM stress and the FEA stress

For the wrinkle-like dents at Site #1 as listed in Table 2, the maximum von Mises stress estimated from the LSM tool is 162 MPa or 39% of SMYS for the X60 pipeline at the operating pressure of 410 psi that corresponds to an operating hoop stress of 26.3% of SMYS or an operating von Mises stress of 22.8% of SMYS. As a result, the LSM estimated Mises stress at the center of the wrinkle-like dent #1-2 is 1.71 times the operating Mises stress. From Fig. 9, the numerical Mises stress at the center of wrinkle-like dent #1-2 is about 92.4 ksi or 154% of SMYS for the X60 pipeline that is 6.75 times the operating Mises stress, or 3.95 times the LSM estimated Mises stress. Accordingly, the LSM estimated Mises stress is pretty low, and only about 25% of the FEA or actual Mises stress at the center of the wrinkle-like dent.

For the single dent at Site #4, the maximum von Mises stress estimated from the LSM scan is 92 MPa or 32% of SMYS for the X42 pipeline at the operating pressure of 235 psi that corresponds to an operating hoop stress of 21.5% of SMYS or an operating Mises stress of 18.6% of SMYS. As a result, the LSM estimated Mises stress at the center of the single dent #2 is 1.72 times the operating Mises stress. From Fig. 11, the numerical Mises stress at the center of the single dent #2 is about 71 ksi or 169% of SMYS for the X42 pipeline that is 9.09 times of the Mises operating stress, or 5.28 times of the LSM estimated Mises stress. In this case, the LSM estimated Mises stress is less than 20% of the FEA or actual Mises stress.

In summary, the LSM estimated stress is significantly lower than the actual stress determined by the FEA calculations for the wrinkle-like dent in the X60 pipeline and the single dent in the X42 pipeline. One important factor could be the plastic deformation effect. In the current LSM technology, only the elastic stress was considered. In the idealized dent approach of the FEA model, large plastic deformation was generated during the dent creation and considered in the loading history.

Comparison of LSM stress and other FEA stresses

To exclude the plastic deformation effect in the FEA simulations, two additional approaches are utilized. One is a net stress approach that determines a net stress as the difference of the operating stress and the corresponding residual stress from the idealized dent approach. The second one is the purely geometrical stress approach that determines stresses from a pure geometric profile of a dent at the operating pressure without considering residual stresses and plastic deformation, as modeled using the pre-formed dent approach.

Figure 12 compares the von Mises stresses on the X60 pipe top OD surface obtained by (1) the idealized dent approach at the operating pressure, (2) the idealized dent approach at the residual stress state with zero pressure, (3) the net stress as described above, and (4) the pre-formed dent approach with a purely geometric profile of the dent at the operating pressure. As evident in Fig. 12 for the wrinkle-like dent #1-2, one has:

1). The actual Mises stresses at the residual stress state with zero pressure and at operating pressure are comparable, and thus actual Mises stresses at both loading conditions are considerably higher than the LSM estimated stress.

Copyright © 2019 by ASME

2). The net Mises stress at the center of wrinkle-like dent #1-2 is 14.7 ksi or 24.1% of SMYS that is 0.62 times of the LSM estimated Mises stress in the X60 pipeline.

3). The purely geometric Mises stress at the center of the wrinkle-like dent #1-2 is 58 ksi or 97% of SMYS that is 2.5 times the LSM estimated Mises stress in the X60 pipeline.

As a result, the net stress approach may be inappropriate to use because the simple superposition principle is theoretically incorrect for a plastic deformation case. In contrast, the pre-formed dent approach is technically correct to determine the purely geometrical stress in the elastic conditions, and thus the determined Mises stress is closer to the LSM estimate stress, but the difference remains large.

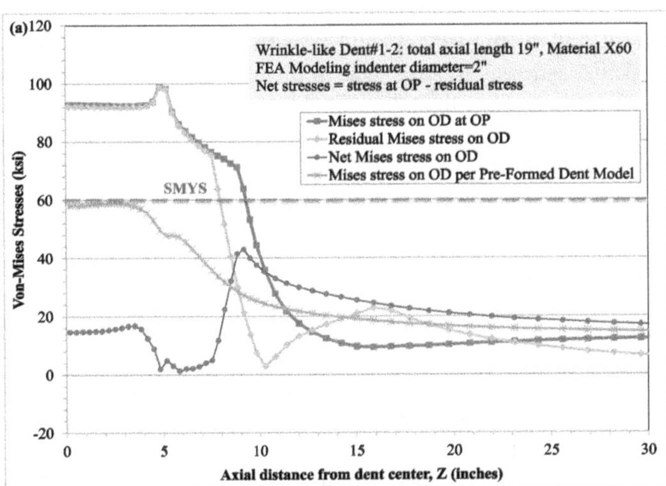

Figure 12. Comparison of von Mises stress obtained by different methods on X60 pipe top OD surface in the axial direction for wrinkle-like Dent #1-2 at Site #1

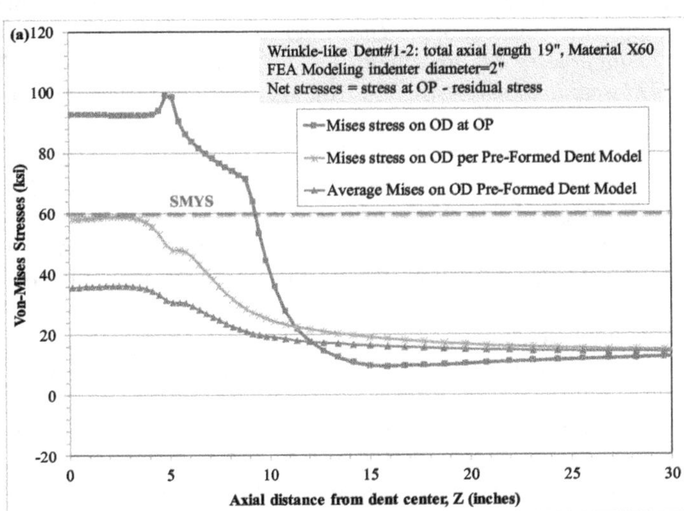

Figure 13. An averaged von Mises on the X60 pipe top OD surface for Wrinkle Dent #1-2

As previously described, all LSM providers did not share the details of stress-magnetization model and the definition of equivalent stress used in their math models. As such, it is difficult to quantitatively evaluate the LSM output stress in comparison to the FEA stress. Xu [2] pointed out that the LSM stress is likely a volume-averaged result, while the FEA stress is distributed along the axial direction and varies with the pipe wall. Let's do a simple average of the Mises stresses obtained from the pre-formed dent approach. An averaged Mises stress is defined as the average of the maximum Mises stress at the dent center and the operating Mises stress at the opposite pipe wall. Figure 13 shows the distribution of such averaged Mises stress on the X60 pipe top OD surface for the wrinkle dent#1-2. Also included in Fig. 13 are two FEA results of von Mises stress on the X60 pipe OD surface for the wrinkle dent#1-2 at the operating pressure that obtained numerically using the idealized dent approach and the pre-deformed dent approach.

Figure 13 shows that the averaged von Mises stress at the dent center is significantly lower than the Mise stress obtained using the idealized dent approach and the pre-deformed dent approach. At the dent center, the averaged Mises stress is 35.8 ksi, while the Mises stress is 92.4 ksi for the idealized dent approach and 58 ksi for the pre-deformed dent approach. As a result, the averaged Mises stress is only 1.54 times the LSM estimated stress 0f 162 MPa (23.5 ksi) for the X60 pipeline dent. In contrast, the maximum Mises stresses obtained by the idealized dent approach and the pre-deformed dent approach are 3.95 and 2.5 times the LSM estimated stress for the wrinkle dent. Therefore, the averaged Mises stress matches better with the LSM estimated stress for the X60 pipeline dent.

Figure 14 shows von Mises stresses obtained using different approaches for the single dent in the X42 pipeline. Comparison of Figs 12 and 14 shows that the observation and implication for the dent in the X42 pipeline are similar to those for the wrinkle-like dents in the X60 pipeline.

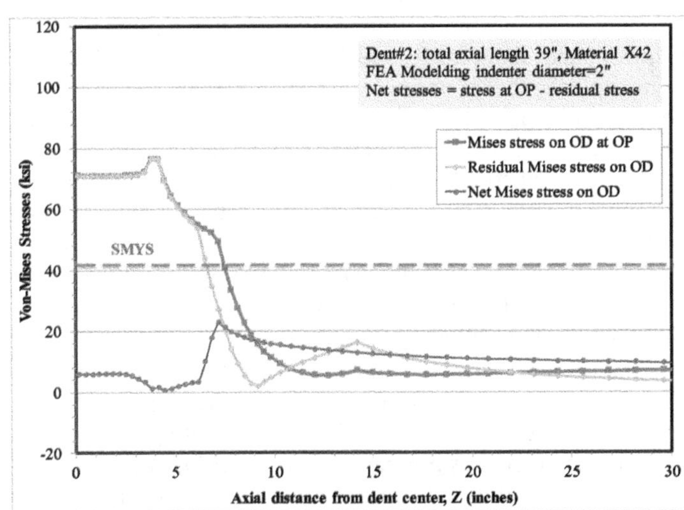

Figure 14. Comparison of von Mises stress obtained by different methods on X42 pipe top OD surface in the axial direction for the Dent #2 at Site #4

Copyright © 2019 by ASME

LSM stress quantification

From the analyses of FEA numerical stress and LSM estimated stress made above, one may conclude that (1) the LSM output stress could be the von Mises stress, but not fully sure without confirmation by the LSM providers; (2) the LSM estimated stress is significantly smaller than the actual Mises stress for the dents in the X60 or X42 pipeline, and thus the large residual stresses generated during the dent creation were not considered in the LSM tools; (3) the LSM estimated stress is closer to the Mises stress obtained using the pre-formed dent approach in the elastic conditions; (4) the LSM estimated stress is not the actual maximum Mises stress at the center of SCZ or dent, but a nominal or volume-averaged stress in the SCZ, as demonstrated by a simple averaged Mises stress using the pre-deformation dent approach, where the residual stress and plastic deformation are not considered in the FEA simulation; and (5) the LSM output stress is not an actual Mises stress but a nominal result, and thus the elevated stress can be used as an indication of possible anomaly presence to serve as a screen tool in comparison to the operating Mises stress (not hoop stress) to define severity of SCZ. As a result, the LSM estimated stress should not be directly used in an integrity assessment of a dent or any other anomaly in a pipeline.

As well known, the classical magnetoelastic theory only considers an elastic stress in the sheet bar, as evident in Eq. (1) to quantify the stress-magnetization relationship. Accordingly, the plastic deformation or residual stress generated in an actual dent is for sure not be considered in the LSM tools. At the present stage, a plastic stress-magnetization model is still too complicated to develop. As shown by Bao et al. [36], the averaged magnetic field B monotonically increases with the increasing stress for different ductile steels when the applied stress is less than the yield strength. However, when the applied stress is larger than the yield strength, plastic deformation occurs, and the averaged magnetic field B can increase or decrease with the increasing stress until the fracture occurs.

Due to the extreme complexity of stress-magnetization relation once the plastic yielding occurs in the object, a magnetoplastic theory has not been developed yet. Therefore, at the present stage the LSM technology can only use the magnetoelastic theory to correlate a simple relationship between stress and magnetization. As a result, the LSM tools cannot determine the accurate stress for pipeline dents because of large plastic residual stress effect, as shown previously in this work, but may be able to determine more accurate stress for corrosion defects or cracks in a pipeline because there is no residual stress generated in such anomalies.

The LSM tools can be more reliable if the math model used to correlate stress-magnetization can be further improved. This may depend the LSM providers if they are willing to share the mathematical model as programed in their LSM tools. If so, the corresponding equivalent stress can be properly calculated from the FEA simulations, and the LSM estimated stress can be correctly analyzed in the assessment of a pipeline containing dents or other anomalies.

CONCLUSIONS

The LSM inspection is an emerging technology to scan stress in a pipeline and report the elevated stress and SCZs. The new inspection tool showed many advantages over traditional or conventional methods to inspect anomalies or defects existing in operating pipelines. To better understand the LSM inspection technology, this paper made a simple review on the LSM technology, including its development, advantages, limitations, challenges and its practical uses. Its theoretical basis and equivalent stress definitions were also discussed.

Two field trial examples of LSM inspection were outlined with a focus on the plain dents found by the LSM scan on two gas pipelines. In general, the LSM determined location of SCZs was acceptable as compared to ILI or evacuation NDE results. However, the LSM output stress was not properly understood. For this reason, two kinds of dents were simulated using the FEA method, and the numerical von Mises stress was used to evaluate the LSM stress. Results showed that the LSM stress is significantly smaller than the FEA Mises stress for the two dents. Thus, LSM tool can serve as a screening tool only but not an assessment tool. Further study of LSM technology is needed to solve unclear issues and to improve its reliability.

REFERENCES

[1] McDonnell S, et al., "Identifying stress concentrations on buried steel pipelines using large standoff magnetometry technology", *Proceedings of International Pipeline Conference*, September 24-28, 2018, Calgary, Alberta, Canada, IPC2018-78162.

[2] Xu T, "Understanding quantitative performance of large standoff magnetometry in determining live gas pipeline anomalies with stress estimation", *Proceedings of International Pipeline Conference*, September 24-28, 2018, Calgary, Alberta, Canada, IPC2018-78655.

[3] Navalgund M, et al. "Large standoff magnetometry for pipeline inspection." *The International Conference on Diagnostics of Structures and Components Using Metal Magnetic Memory Method*, Budapest, Hungary, May 19-20, 2016.

[4] Staples, SGH, et al., "Solving the inverse problem of magnetization–stress resolution." *Journal of Applied Physics*, Vol.113, 2013, 133905.

[5] Staples SGH, et al., *Dipole Model for Stress Concertation Zones in Earth's Field*, Research Paper. University of Leeds, Leeds, United Kingdom, 2014.

[6] Staples SGH, et al., *Bar Magnet Model for Large Stand Off Magnetometry*, Research Paper. University of Leeds, Leeds, United Kingdom, 2017.

[7] Freear S, et al., *Fault Detection for Pipelines*, US Patent Publication No. US 2018/018827 A1, July 5, 2018.

[8] Cowell D, et al., "Detection of Stress due to Corrosion through Passive Measurement and Analysis of Magnetic Fields", NACE *Corrosion 2013 Conference & Expo*, March 17-21, 2013, Orlando, Florida, USA.

Copyright © 2019 by ASME

[9] Jarram P, "Remote Measurement of Stress in Carbon Steel Pipelines–Developments in Remote Magnetic Monitoring", NACE *Corrosion 2016 Conference & Expo*, March 6-10, 2016, Vancouver, BC, Canada.

[10] Habibi H, et a., "Field Verification of Remote Magnetic Monitoring of Stress for Buried unpiggable Pipelines", *Unpiggable Pipeline Solutions Forum*, June 14-15, 2017, Houston, USA.

[11] Habibi H, "Developments in remote magnetic monitoring of carbon steel pipelines to locate and measure abnormal stress, *The CEOCOR 2018 Congress*, May 15-18, 2018, Stratford-upon-Avon, UK.

[12] Habibi H, "Stress Concentration Tomography – A New Tool to Remotely Monitor Pipeline Wall Condition, Depth of Cover and Stresses, Concurrently", *Corrosion Management*, Issue 142, March/April, 2018.

[13] Jiles D and Atherton D, "Theory of the magnetization process in ferromagnets and its application to the magnetomechanical effect." *Journal Physics D: Applied Physics* Vo.17, 1984, pp.1265-1282.

[14] Jiles D and Atherton D, "Theory of ferromagnetic hysteresis." *Journal of Magnetism and Magnetic Materials* Vol. 61, 1986, pp. 48-60.

[15] Jiles D, "Theory of the magnetomechanical effect." *Journal of Physics D- Applied Physics,* Vol.28, 1995, pp.1537-1546.

[16] Sablik M, "A model for asymmetry in magnetic property behavior under tensile and compressive stress in steel." *IEEE Transactions on Magnetics* Vol.33, 1997, p.2958.

[17] Jiles D and Li L, "Modified law of approach for magneto-mechanical model - Application of Rayleigh law to stress." *IEEE Transaction on Magnetics* Vol. 39, 2003, p.3037.

[18] Jiles D and Li L, "A new approach to modeling the magnetomechanical effect." *Journal of Applied Physics* Vol. 95, 2004, p.7058.

[19] Viana A, Coulomb JL, Rouve LL, and Cauffet G, "numerical resolution of the modified Langevin equation using a differential expression: application to the Jiles magnetostriction law of approach." *Journal of Magnetism and Magnetic Materials,* Vol.322, 2010, pp. 186-189.

[20] Wang ZD, Yao K, Deng B, and Ding KQ, "Quantitative study of metal magnetic memory signal versus local stress concentration", *NDT&E International*, Vol. 43, 2010, pp. 513-518.

[21] Wang ZD, Deng B and Yao K, "Physical model of plastic deformation on magnetization in ferromagnetic materials", *Journal of Applied Physics*, Vol. 109, 2011, 083928.

[22] Li J and Xu M, "Modified Jiles-Atherton-Sablik Model for asymmetry in magnetomechanical effect under tensile and compressive stress." *Journal of Applied Physics* Vol.110, 2011, 063918.

[23] Schneider CS and Richardson JM, "Biaxial magneto-elasticity in steels", *Journal of Applied Physics*, Vol. 53, 1982, p.8136

[24] Kashiwaya K, "Fundamentals of nondestructive measurement of biaxial stress in steel utilizing magneto-elastic effect under low magnetic field", *Japanese Journal of Applied Physics*, Vol. 30, 1991, p.2932

[25] Sablik MJ, et al., "Micromagnetic model for biaxial stress effects on magnetic properties", *Journal of Magnetic Materials*, Vol. 132, 1994, pp. 131-148.

[26] Roskosz M and Bieniek M, "Evaluation of residual stress in ferromagnetic steels based on residual magnetic field measurements", *NDT&E International*, Vol. 45, 2012, pp. 55-62.

[27] Yi S, Wang W and Su S, "Being Experimental Study on Metal Magnetic Memory Signal Based on von Mises Yield Criterion", *International Journal of Applied Electro-Magnetics and Mechanics*, Vol. 49, 2015, pp. 547-556.

[28] Li C, et al., "Theoretical research on the characteristics of the self-magnetic leakage field induced by ferromagnetic pipelines", *Electomagnetics*, Vol. 58, 2016, pp. 601- 608.

[29] Wall M and Lee R, *Evaluation of Large Standoff Magnetometry Techniques for Above Pipe Integrity Assessment (NDE-3-1)*, Final Report No. PR-398-133725. Pipeline Research Council International, August 2015.

[30] Habibi H, Jarram P, Gan J, Keogh P and Jarram R, *Evaluation of Large Standoff Magnetometry Techniques NDE-3-2*, Final Report No. PR-410-163767, Pipeline Research Council International, 2017.

[31] Zhu XK, *Stress Simulation of Mechanical Dents in 24-inch Gas Pipelines for PG&E*, Technical Report No. 57597CSP. Edison Welding Institute, Columbus, Ohio, 2018.

[32] Leis BN and Zhu XK, *Quantifying Re-Rounding in Pipeline Damage Severity Models (MD-4-4)*, Technical Report No. PR-185-133739, Pipeline Research Council International, 2017.

[33] Zhu XK and Leis BN, "Finite element modeling and quantification of mechanical damage severity in pipelines." *Proceedings of the 11th International Pipeline Conference*, September 26-30, 2016, Calgary, Alberta, Canada, IPC2016-64680.

[34] Bratton JP, Alexander T, Bubenik TA, Finneran S and Heggen HO, "An approach for evaluating the integrity of plain dents reported by in-line inspection tools", *Proceedings of the 9th International Pipeline Conference*, Calgary, Alberta, Canada, September 24-28, 2012. IPC2012-90643.

[35] Baek JH, et al., "Load bearing capacity of API X65 pipe with dent defect under internal pressure and in-plane bending", *Materials Science and Engineering A*, Vol. 540, 2012, 70-82.

[36] Bao S, et al., "Characterization of stress induced residual magnetic field in ferromagnetic steels," *Proceedings of the ASME 2015 34th International Conference on Ocean, Offshore and Arctic Engineering*, May 31 – June 5, 2015, St. John's, Newfoundland, Canada.

Copyright © 2019 by ASME

AIM-PIMG2019-1005

Chapter 37

EVALUATION OF THERMAL PROBE COOLING EFFECT BASED ON PIPELINE PARALLEL OPTICAL CABLE

Cai Yongjun[1], Bai Luyao, Tan Xianjun
PetroChina Pipeline R&D Center
Langfang, Hebei, China

Wu Zhangzhong
PetroChina Pipeline R&D Center
Langfang, Hebei, China

[1] Contact author: yjcai@petrochina.com.cn

Copyright © 2019 by ASME

ABSTRACT

In this paper, Distributed temperature measurement by Raman scattering distributed optical fiber temperature measurement technology is used to monitoring of soil temperature along pipeline, which can realize the stability evaluation of pipeline at permafrost. This is a low-cost realization to solve the long-distance temperature measurements at the same time, to avoid the heavy work of establishing temperature monitoring section along the pipeline. It's a simple realization of freezing monitoring and evaluation along the pipeline. Experiments on Mohe-Daqing pipeline show that the technology can identify the thawing and freezing zones along the pipeline in seasonal frozen soil area, evaluate the effect of frozen soil protection measures, such as active cooling measures. It will help pipeline operators to quickly evaluate the safety of pipelines at permafrost.

Keywords: Raman scattering; Distributed Temperature System; Pipeline at permafrost; Thermal Probes; Distributed optical fiber temperature measurement

1. SURVEY OF MOHE-DAQING PIPELINE

The Mohe-Daqing(Mo-Da) pipeline is an important pipeline, which is built for China's import crude oil from Russia. It starts from the Heilongjiang River, which is the boundary between China and Russia, to Linyuan terminal station at Daqing City. It runs across the Daxinganling Mountains in the north, and crosses the hinterland of Songnen Plain in the south, with a total length of 933 Km. The geological environment around the pipeline is complex, the environment is harsh, and the extreme minimum temperature is -52.3 ℃. It passes through a large area of continuous permafrost, island permafrost, scattered permafrost and deep (>1.5 m) seasonal permafrost to reach Daqing (Fig. 1). The permafrost area is about 441 km, the seasonal permafrost area is about 512 km, and the high temperature and ice-rich soil area is about 119 km[1-3].

The pipeline is laid by traditional trench burial method. The design buried depth at permafrost section is between 1.6 m and 2 m, which is close to the permafrost table. The oil temperature in Mohe station from 2011 to 2013 ranges from 4.8 to 16.1 ℃, at the same time the designed oil temperature ranges from -6 to 10 ℃. The actual oil temperature of exceeds the design temperature usually, which play an adverse effects on stability of frozen soil. Since the pipeline was put into operation in January 2011, the surface of the high ice rich section along the pipeline has undergone significant thawing settlement, and seasonal frost heaving mounds have appeared in some areas [1-2]. The annual precipitation along the pipeline is large, the groundwater level is high, the distribution of permafrost marsh wetland is about 50 km. Through a large area of paddy fields, the surface runoff water system develops (Fig. 1). In the warm season, the subsidence zone collects surface water, active layer melting water and other water bodies, forming a large-scale water accumulation ditch, which further intensifies the development rate of the melting layer around the pipeline. In the cold season, the melting layer around the pipeline develops rapidly. Under the action of pipeline thermal release, water migrates upward and freezes to form an ice ditch, which makes backfilling difficult in some sections of the ditch.

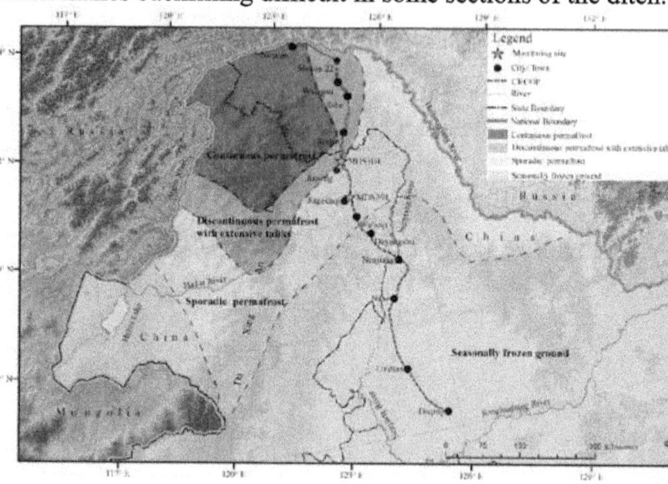

FIGURE 1: Trend of Mohe-Daqing Crude Oil Pipeline and Its Main Research Areas

2. RAMAN SCATTERING PRINCIPLE

The laser pulses propagating in optical fibers will be scattered. These scattered lights can be divided into Rayleigh scattering light, Raman scattering light and Brillouin scattering light. The frequency of Rayleigh scattering light does not change. At the same time, the frequency of Raman scattering and Brillouin scattering will change, and their spectral distributions are shown in Fig. 2. Brillouin scattering is related to temperature and stress. The change of Raman scattering signal is only related to temperature. At the same time, Raman scattering demodulation is easier. Distributed temperature system (DTS) mainly uses Raman scattering demodulation[4-9].

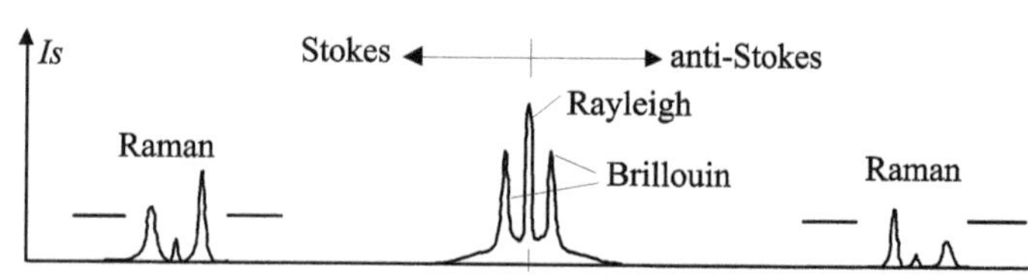

FIGURE 2: Light scattering spectrum schematic diagram

Copyright © 2019 by ASME

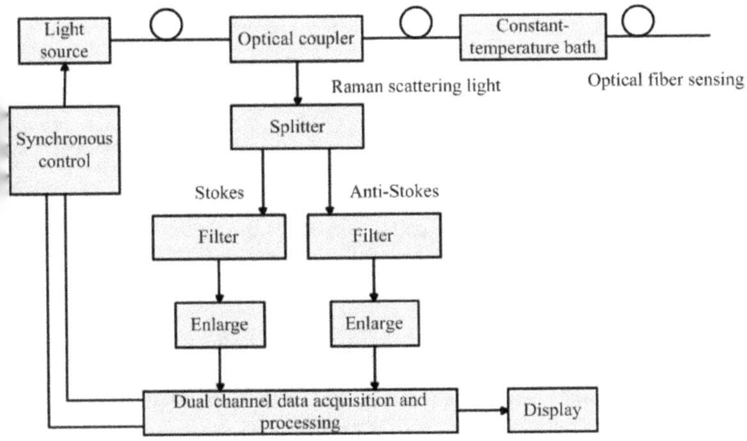

FIGURE 3:. Schematic diagram of distributed temperature measurement system

Inelastic collision between incidents pulsed light and fiber molecules results in the exchange of energy between photons and fiber molecules. The energy of photons may be transferred to molecules, and the energy of molecules may also be transferred to photons. Thus, the frequency of photons changes. The scattered light is Stokes light and anti-Stokes light respectively. This phenomenon is called Raman scattering. The intensity of anti-Stokes has a strong correlation with temperature. The temperature can be obtained by calculating the signal intensity ratio of anti-Stokes peak to Stokes peak. The location can be achieved by the time difference between incidents light and reflected light. Fig. 3 is the schematic diagram of distributed optical fiber temperature sensing monitoring system.

When a certain energy laser pulse is injected into the optical fiber, Raman scattering occurs at the distance from the injection point L. The relationship between the temperature in the optical fiber and the propagation characteristics of Raman scattering light can be expressed as follows.

$$T = \frac{h\Delta\upsilon T_0}{h\Delta\upsilon - kT_0\left[\ln\frac{P_{AS}(L)/P_S(L)}{P_{AS}(L')/P_S(L')} - (\alpha_S - \alpha_{AS})(L - L')\right]} \quad (1)$$

In the formula (1): P_S and P_{AS} are Stokes Raman scattering light and Anti-Stokes Raman scattering light power respectively; α_S and α_{AS} are Stokes Raman scattering light and Anti-Stokes Raman scattering light loss coefficient respectively; L is the length of optical fibers from backscattering point to detection section; L' is the length of calibrated optical fibers placed at constant temperature T_0; h is Planck constant; k is Boltzmann constant.

According to formula (1), the temperature coefficient of the distributed optical fiber temperature measurement system based on Raman scatting is related to the properties of the optical fiber itself. Without knowing the precise value of refractive index and loss of the optical fiber itself, only the temperature trend along the optical fiber can be obtained. The temperature, which is calibrated, is the real environmental temperature. T_0 in formula (1) is used to calibrate the temperature of the optical fibers to be measured by a section of optical fibers with certain parameters inside the instrument. Because it is difficult to measure the parameters of the optical fibers in used, and it is impossible to be consistent with the calibrated optical fibers in the instrument, there is a big error in this method. Another calibration method is to use $T+T_0$ to provide a standard temperature by external platinum resistance.

When a high-power pulsed light is injected into the sensing fiber, assuming that the time at which the backward Raman scattering light returns to the input end is t, the distance between the point and the incident end is as follows

$$L = \frac{c}{n} \bullet \frac{t}{2} \quad (2)$$

In the formula, n is the limited refractive index of the optical fiber and c is the speed of light.

3. FIELD TEST

3.1 Test conditions

Field tests were carried out at Jiagedaqi pumping station, the intermediate station of Moda crude oil pipeline. The pipeline transportation volume was 2100 m^3/h and the transportation temperature was 7.1°C. The cable layed in the same trench with oil pipeline is GYTA53-24D. The total length of the cable is 10.12 km. The DTS device used in the field has a sampling spatial resolution of 2m and a temperature resolution of 1°C.

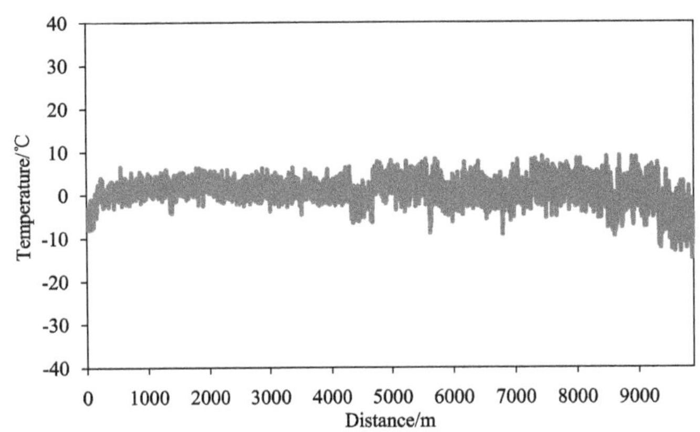

FIGURE 4:. Temperature along Optical Fiber

3.2 Cooling effect of the thermal probes

Figure 4 presents temperature measured by optical cable at different positions. A test area for active cooling of thermal probes was constructed at Jiagedaqi pump station. Three groups of thermal probes were set up at 20 m intervals. At the same time, a set of temperature reference points were set at the same interval. T1 temperature hole was set near the pipeline 20 m from the thermal probes area, T2 temperature hole was set 15 m from the T1 temperature hole of the pipeline, and a T3 temperature hole was set near the thermal probes, as shown in Figure 4 (circle is thermal probes, diamond is Temperature hole). Over the top 5m, the temperature is measured every 0.5m. And, between 5m and 15m, it is measured every

Copyright © 2019 by ASME

1.0m.The temperature sensor is a platinum resistance with a precision of <0.05°C. The data acquisition unit includes the CR3000 digital acquisition instrument manufactured by Campbell Scientific in the United States.

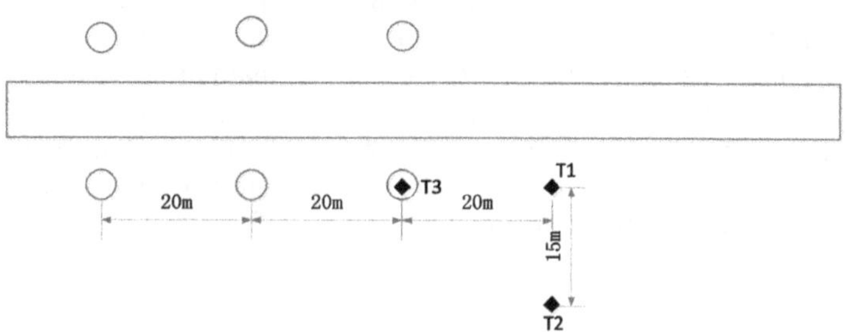

FIGURE 5: Diagram of Temperature Calibration Cross Section

From Figure 5, it can be seen that the temperature where the thermal probes is not installed in T1 is higher than the temperature where the thermal probes is installed in T3 at a depth of about 2 m.

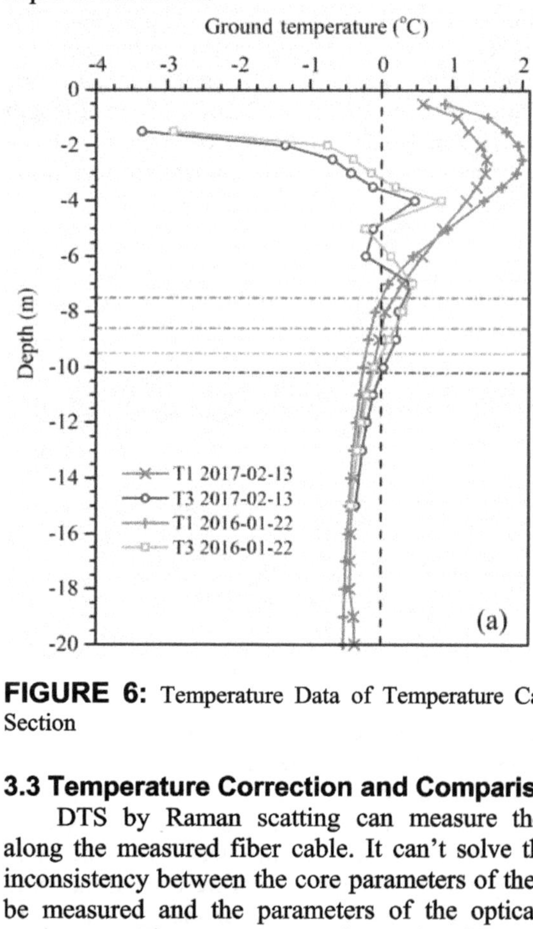

FIGURE 6: Temperature Data of Temperature Calibration Cross Section

3.3 Temperature Correction and Comparison

DTS by Raman scatting can measure the temperature along the measured fiber cable. It can't solve the problem of inconsistency between the core parameters of the fiber cable to be measured and the parameters of the optical fiber in the equipment. The temperature shown in Fig. 4 is the relative temperature of each point at the fiber optical cable. If the true temperature is to be obtained, two points of accurate temperature on the cable is need to correct it.

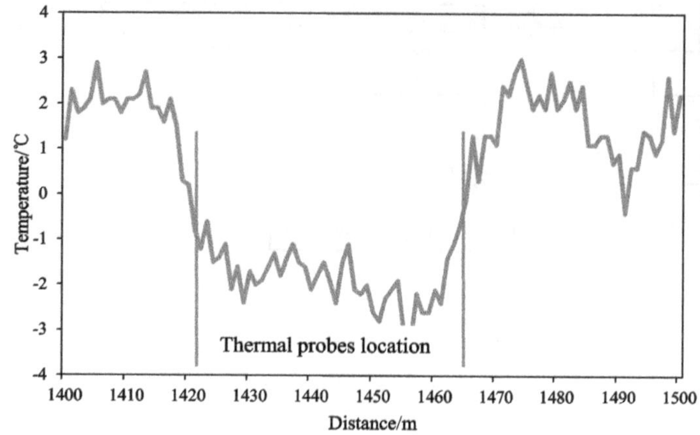

FIGURE 7: Temperature Variation of Optical Fiber Cable in the Area Affected by thermal probes

Fig. 7 is segmented and analyzed. The temperature of optical fiber cables in the range of 1410-1470m is lower than another, as showed in Fig8. The core temperature drop zone is 1420-1460m, which is lower than the surrounding temperature of 4-5 ℃. The length of this zone is about 40 m is the same with the length of three sets of thermal probes. The two zones of 1410-1420m and 1460-1470m are the temperature drop affected zone. The temperature calibration section is used to calibrate the temperature of DTS. The calibration results are shown in Fig. 8. At the same time, it can be concluded that a group of thermal probes can affect the temperature field in the region of about 10 m each, and that setting a group of thermal probes at 20 meters interval can effectively control the temperature of pipeline trench.

Hot rod is a kind of refrigeration device without additional power. It uses liquid-gas two-phase convection circulation to realize the automatic heat transfer. The data of the hot rod temperature calibration area show that the temperature difference between the thermal probe and the non- thermal probes area is about 5°C, and the measurement results are consistent with the optical fiber temperature measurement. The temperature data of T1 and T3 in two temperature sections of hot rod working area are real temperature data. The temperature data of T1 and T3 in two temperature sections of hot rod working area are used to correct the temperature of optical fiber temperature measurement. The relative temperature can be corrected to real temperature by uniformly reducing the temperature data of optical fiber temperature measurement by 8 degrees centigrade. After correction, the basic temperature along the pipeline is 1-2 °C, and the local temperature is higher than 3 °C.

3.3 Optical Cable and Pipeline Position Measurement

The temperature measured by optical fiber is affected by the temperature field and ground temperature field of the pipeline at the same time because of the temperature measurement experiment with the pipeline accompanying

Copyright © 2019 by ASME

optical cable. Because the distance between the optical cable and the pipeline can not be guaranteed during the laying process, there is a certain fluctuation in the temperature measurement data of optical fiber. RD8000 pipe detector and 3M2273E cable routing detector were used to measure the relative position and buried depth of the cable and pipeline. As shown in Figures 8 and 9.

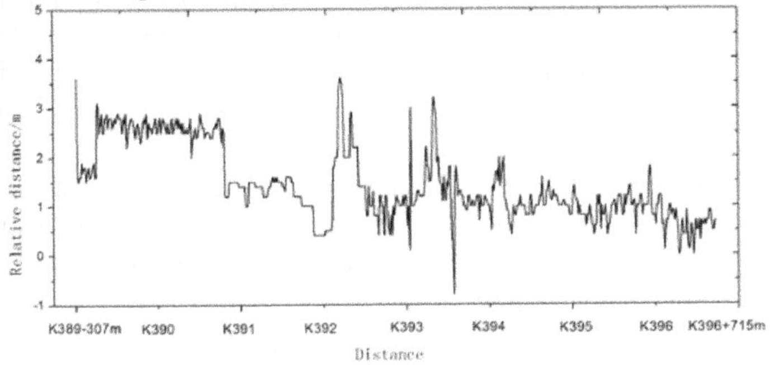

FIGURE 8: Horizontal Position of Optical Cable and Pipeline

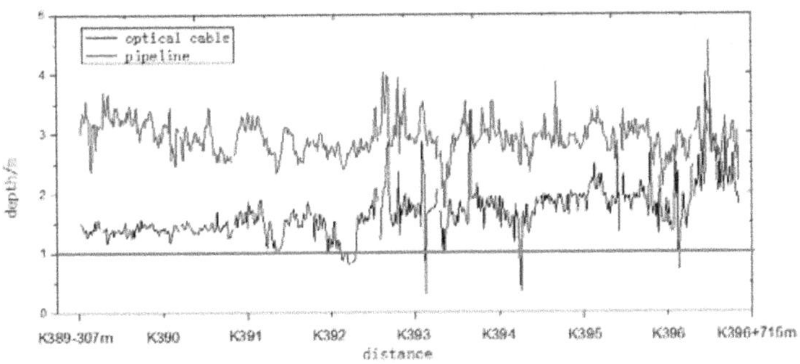

FIGURE 9 Buried Depth of Optical Cable and Pipeline

Comparing the temperature monitored by optical fiber with the buried depth of optical cable, it is not difficult to see that the trend of the temperature change along the buried depth of optical fiber is basically the same. As shown in Fig. 10, the temperature measured by optical cable is significantly lower than the surrounding temperature in the range of 4 500-4 800 m from the exit of Gagdage Daqi oil station. This section of optical cable is located near the pipeline mileage K392+200 m and belongs to shallow buried section. The buried depth is less than 1 m, which is greatly affected by the ground temperature. Therefore, the temperature measured in the shallow buried section of the optical cable is lower.

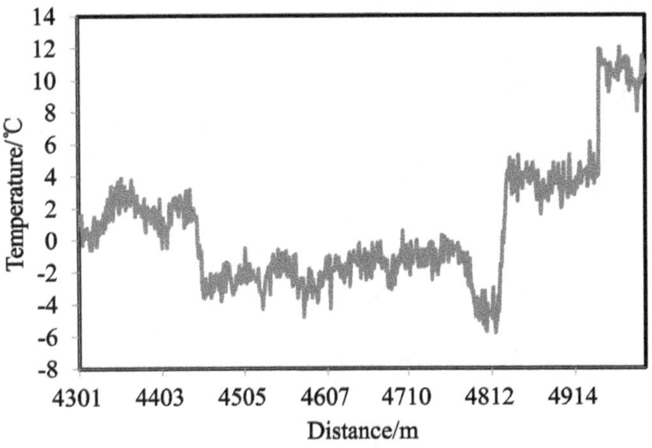

FIGURE10: Temperature Data of Shallow Buried Section of Optical Cable

Fig. 10 shows the temperature measured by the cable. The buried depth of the cable is too shallow in 4450-4750m section. The temperature is about - 4 °C, 4750-4820m section. The cable deviates from the pipeline by about 3M and the soil temperature is about - 6°C. From figs. 9 and 10, it can be seen that although the optical cables and pipelines in this area are close to each other and deviate from the horizontal direction about 0.8 m, they are buried about 2 m above the pipeline, and the distance between the optical cables and the pipeline is about 4 M. Fiber optic cable is more affected by the surface temperature field, and the surface temperature during the test period is about - 25°C.

4. RESULTS AND DISCUSSIONS

The distributed measurement of soil temperature along the pipeline can be realized by using the fiber optic sensing system based on Raman scattering combined with the correction of cross-section temperature measurement. The measurement results show that:

1) The heat-affected zone of the hot rod can affect the pipeline's axial 10 m range, and a group of Thermal Probes with a distance of 20 m can stabilize the soil around the pipeline.

2) Thermal Probe in winter can reduce the soil around the pipeline by 4-5 °C.

3) The long-term positive temperature operation of the pipeline, the soil temperature field around the pipeline is normal temperature, there is a melting circle.

4) The temperature field of the pipeline is less than 2 m above the pipeline, and the surface is more affected by the ambient temperature.

ACKNOWLEDGEMENTS

This research acknowledges the financial support provided by National Key Research and Development Program of China

Copyright © 2019 by ASME

(No. 2016YFC0802103).

REFERENCES

[1] Hao Jiaqian, He Ruixia. Discussion of the permafrost environment and pipeline construction technique along the China-Russia pipeline[J]. Journal of Glaciology and Geocryology,2013,35(5):1224-1231.

[2] JIN Hui—jun , YU W en—bing , CHEN You—chang,et al. (Differentia1)FrostHeaveand Thaw Settlem entin theEngineering Design and Construction ofOilPipelinesin Perm afrostRegions : A Review[J],JOURNAI OF GIACIOIOGY AND GE0CRYOIOGY,2005,27(3):454-464

[3] Ingrid Pederson, Millan Sen, Andrew Bidwell,et al. ENBRIDGE NORTHERN PIPELINE: 25 YEARS OF OPERATIONS, SUCCESSES AND CHALLENGES[C], Proceedings of the 8th International Pipeline Conference, IPC2010-31611.

[4] PANDIAN Chelliah, KASINATHAN Murgesan, SOSAMMA Samvel, et al. Looped back fiber mode for reduction of false alarm in leak detection using distributed optical fiber sensor[J].Applied Optics , 2010，49(20)：3869-3874.

[5] GUNES Yilmaz , SAITESER Karlik.A distributed optical fiber sensor for temperature detection in power cables [J].Sensors and Actuators，2006，125：148-155.

[6] Yang Rui, Li Xiaoyan, Gao Xiang. Correct the measurement error of distributed Raman optical fiber temperature measurement system [J]. Acta Photonica Sinica, 2015,44(10):1006006

[7] LI Guo-yu1, MA Wei1, WANG Xue-li, et al, Frost hazards and mitigative measures following operation of Mohe-Daqing pipeline of China-Russia crude oil pipeline[J], Rock and Soil Mechanics,2015,36（10）:2963-2973

[8] ZOU Jiang，WANG Zhong． Temperature resolution of distributed optical fiber temperature measurement system[J]． Laser & Infrared, 2000,30(5) : 304－306．(in Chinese).

[9] HUANG Xiang , GAN Xiaoqing , LI Qiang , et al . Application of distributed optical fiber temperature measurement system based on Raman scattering［J］. Journal of Yangtze River Scientific Research Institute， 2013， 30(2) : 92 － 96. (in Chinese)

Copyright © 2019 by ASME

SECTION 10
STRUCTURAL RELIABILITY AND RISK ASSESSMENT

INTRODUCTION

By Bernt Johan Leira
(bernt.leira@ntnu.no)

This introduction presents summary of the papers covered by this section and discusses the main steps of the reliability analysis and risk assessment procedure. The role of this procedure in connection with risk management is also briefly touched upon. Subsequently, the contents of the relevant papers are briefly reviewed and their profiles in relation to the assessment procure are summarized by means of a table.

Steps of the structural reliability analysis and risk assessment procedure:

I. Define the pipeline system to be analyzed as well as the system boundaries (e.g. decide which sub-stations such as end-terminals and process equipment that should be included)

II. Subdivide the pipeline system into a number of characteristic "segments". For each of the segments the "conditions" as exemplified in item III should be as constant (i.e. homogenous) as possible within the segment.

III. For each segment, identify relevant hazards and critical sites within the segment (e.g. Earthquakes, fault movement, soil sliding)

IV. Perform a mapping of different sources of geo-hazards (e.g. earthquakes, ground movement)

V. Estimate probabilities associated with different levels of the different hazards at the critical sites

VI. Identify critical cross-sections and relevant mechanical limit states for the pipeline segment under investigation (e.g. tensile and compressive failure, fracture e.g. in girth welds, rupture, burst)

VII. Establish probability distributions for load and resistance variables for the critical cross-sections

VIII. Compute failure probabilities for critical cross-sections. Update analysis results based on inspection data.

IX. Consider system effects, e.g. correlation between random load and resistance variables at the different cross-sections and limit states. Compute system reliability (or equivalently: probability of failure). Common-cause failure and network vulnerability are also topics which are highly relevant in this connection.

X. Quantify consequences of failure associated with the relevant limit states at the critical cross-sections. Make distinction between monetary, ecological and fatality consequences. Compute system risk by combining consequences with system failure probabilities which are obtained from item 9.

XI. Having completed these steps, a necessary tool for proper risk management is available. The effects of various management decisions and viability of different scenarios can then be analyzed by repeating the assessment procedure as many times as required. If required, comparison between application of different decision-making procedures can also be made.

The contents of the papers covered by this section are briefly reviewed and their respective profiles in relation to the steps of the assessment procedure are subsequently summarized in the form of a table. The paper by Keprate, Eldevik, Avice, Tornqvist, and Erling Østby (AIMPIMG2019-1022) presents a Structural Reliability Assessment (SRA) of pipeline girth welds by the use of a probabilistic machine learning approach coupled with Finite Element Analysis (FEA). The assessment has been used to evaluate the probability of fracture for pipelines subjected to large deformations. Samples of input variables (wall thickness, outer diameter, initial crack length and height, weld material) are first generated, and these are used to perform corresponding nonlinear FEA in order to derive the crack driving force (CDF) as a function of applied load (i.e. bending strain). The CDF is characterized in terms of the crack tip opening displacement (CTOD) for girth welds. Results from different machine learning (ML) surrogate models (SMs) have been compared to select the most adequate method. Monte Carlo Simulations (MCS) are applied in order to evaluate the probability of failure. The advantage of the methodology is its ability to assess uncertainty based on relatively small sample sizes where analytical limit state functions are not available.

The paper by Abdolrazaghi and Hassanien (AIMPIMG2019-1047) is concerned with Integrated Risk Informed Decision Making (IRIDM) for pipelines in moving slopes. This decision making is based on four pillars: the definition of the decision, the role of stakeholders involved in this process, the assessment methodology, and the timeline to complete the projects. Slope remediation projects fall into non-programmed decision making to ensure the safety of pipeline and environment and optimizing the cost expenditure. To effectively initiate and plan the remediation, Input roles are required to complete multiple assessments other than stress and strain-based analysis. In addition, a number of considerations including cost benefit analysis, risk reductions, and regulatory and political factors need to be utilized. A deliberation process

Copyright © 2019 by ASME

among the stakeholders is developed to account for the uncertainties around the decision-making process.

The paper by Brewer, Bowlin, West and Johnson (AIMPIMG2019-1056) assessed the significant hazard for buried pipelines due to permanent ground displacement (PGD) caused by coseismic fault displacement in earthquakes. Possible large displacements can result in pipeline buckling, rupture, and failure. In order to evaluate the relative likelihood of failure for multiple pipeline fault crossings, a natural gas pipeline operator performed a simplified Probabilistic Fault Displacement Hazard Analysis (PFDHA) for a number of potentially high hazard fault crossings in Southern California. Quantitative estimates of the displacement magnitude and probability of ground movement were defined for nine faults. These were combined with non-linear elastic-plastic finite element analysis (FEA) to determine the probability of varying levels of strain in the pipeline at each crossing location. This enables the operator to screen and rank the relative criticality of locations in the pipeline and/or determine if fault displacement mitigation is necessary.

The paper by *Wan Ismail, Ali Napiah, Tuselim and Zabidi* (AIMPIMG2019-1069) is concerned with the Sabah-Sarawak Gas Pipeline (SSGP) that traverses hilly and mountainous terrain with significant rain fall in East Malaysia i.e. in the state of Sabah &Sarawak. The pipeline experienced leak in 2014 and 2018. It was found that both pipeline failures were due to combination of external force (from geohazards), girth weld defects and under-matching of girth welds. Following the incidents in 2014 and 2018, PETRONAS implemented strategies for managing pipeline integrity i.e. risk assessments, inspection and monitoring, integrity assessments, and repairs. The focus of the paper is on the usage of Structural Reliability Assessment (SRA) and Quantitative Risk Assessment (QRA) as part of risk and integrity assessments. Pipeline threats considered for the SRA and QRA are (i) girth weld anomalies (including under-matched effect), (ii) bend stress/strain (representing external force/geohazards), (iii) pipe buoyancy/stability at exposed river/stream crossings, (iv) dent/gouge (from third party), and (v) internal/external corrosion. The resultant reliability and risk numbers were compared to target reliability based on procedures from ISO 16708 and target risk level from PETRONAS Technical Standard 16.71.04, respectively.

The paper by Abdolrazaghi and Hassanien (AIMPIMG2019-1048) presented formulation of the integrity decisions for oil pipelines near water bodies against stresses and strains beyond their intended designs. The estimated risk was presented in terms of potential loss of containment and adverse consequences is a critical task for safe operations. In practical integrity situations, the epistemic uncertainty (e.g. due to model uncertainties) in the estimated risk is equally significant and relevant as compared to aleatory uncertainties (which are due to inherent randomness). This paper discusses a process of practical analysis to evaluate the risk distribution associated with imperfect information (i.e. aleatory uncertainties) and epistemic uncertainties.

The paper by Koduru and Nessim (AIMPIMG2019-1050) addresses limit state design checks that have been developed for geotechnical loads using a strain-based design format. The application of these checks requires that the user is able to develop probability distributions for the maximum strain demand and for the minimum strain capacity. These design checks are provided for both local buckling and for tensile rupture. The design formats and safety factors are calibrated to meet specified risk-consistent reliability targets. The paper presents a detailed description of the safety factors which are developed as part of the limit states design checks, and the implications of their use in the assessment of pipelines in moving slopes.

A tentative summary of the paper profiles in relation to the steps of the assessment procedure which was described in this introduction is provided by means of the table below.

The structural reliability and risk assessment procedure which is outlined in this section comprises a wide range of specialist topics. This includes e.g., geophysics, material science, structural reliability analysis, accident consequence evaluation and risk assessment. This indicates that a written guideline providing further details in relation to the various steps of the assessment procedure could well be developed. The papers covered by this section have (according to the opinion of the present reviewer) served in an excellent manner to highlight the essential features of the various steps involved in the assessment and decision process. This is also illustrated by Table 1 where all the steps are addressed by at least one of the papers.

Table 1. Tentative classification of paper profiles in relation to the steps of the assessment procedure

Paper number	I.	II.	III.	IV.	V.	VI.	VII.	VIII.	IX.	X.	XI.
1	x						x	x			x
2			x								x
3	x	x			x						x
4	x	x	x	x		x		x	x	x	x
5							x	x		x	x
6	x						x	x			x

Copyright © 2019 by ASME

Chapter 38 Structural Reliability Assessment (SRA) of Pipeline Girth Welds using Gaussian Process Regression (AIMPIMG2019-1022)
Arvind Keprate, Simen Eldevik, Marc Avice, Rikard Tornqvist, and Erling Østby, DNV GL, Norway

Chapter 39 Assessment of Pipelines in Moving Slopes: A risk-informed decision-making approach (AIMPIMG2019-1047)
Mona Abdolrazaghi and Sherif Hassanien, Enbridge, Canada

Chapter 40 Probabilistic Fault Displacement Analysis for a Pipeline (AIMPIMG2019-1056)
Jonathan Brewer and Elizabeth Bowlin, Stress Engineering Services, Inc.; Don West and Clayton Johnson, Golder Associates Inc., USA

Chapter 41 Application of Structural Reliability Assessment (SRA) & Quantitative Risk Assessment (QRA) in managing Risk (AIMPIMG2019-1069)
Wan M Marzuki Wan Ismail, Mohd Nazmi Mohd Ali Napiah, Ahmad Sirwan M Tuselim and Mohd Zazli Mohd Zabidi, Petronas, Malaysia

Chapter 42 A Practical Assessment of Pipeline Risk under Epistemic Uncertainty: A hydro-technical threat (AIMPIMG2019-1048)
Mona Abdolrazaghi and Sherif Hassanien, Enbridge, Canada

Chapter 43 Implications of the Limit State Design Checks for Geotechnical Loads (AIMPIMG2019-1050)
Smitha D. Koduru and Maher A. Nessim, C-FER Technologies, Canada

Copyright © 2019 by ASME

AIM-PIMG2019-1022

Chapter 38

STRUCTURAL RELIABILITY ASSESSMENT OF PIPELINE GIRTH WELDS USING GAUSSIAN PROCESS REGRESSION

Arvind Keprate[1], Simen Eldevik, Marc Avice, Rikard Tornqvist, Erling Østby

DNV GL, Høvik, Norway

ABSTRACT

A Structural Reliability Assessment (SRA) of pipeline girth welds by the use of probabilistic machine learning approach coupled with Finite Element Analysis (FEA) have been used to evaluate the probability of fracture for pipelines subjected to large deformations. Firstly, Design of Experiment (DoE) is performed (using two different methods namely One (factor) at a time (OAT) and Latin Hypercube Sampling (LHS)) to generate samples of input variables (wall thickness, outer diameter, initial crack length and height, weld material), which are used to perform FEA. Based on the DOE, a set of non-linear FEA have been performed to derive the crack driving force (CDF) as a function of applied load (i.e. bending strain). The CDF is characterized in terms of the crack tip opening displacement (CTOD) for girth welds with different combinations of input. Different machine learning (ML) surrogate models (SMs) such as Linear regression, Lasso, Ridge, k-Nearest Neighbor, Random Forest, Support Vector Machine, Neural Network and Gaussian Process regression (GPR) have been compared to select the method which best generalize the CTOD response. Based on this comparison the SM GPR_$M_{v=5/2}$ have been trained and tested on the generated FEA samples, and then used to predict values of CTOD for input combinations that have not explicitly been simulated by FEA. Then, a Monte Carlo Simulations (MCS) have been used to evaluate the probability of failure by comparing CTOD predictions based on the fitted response surface with a representative failure criteria. The advantage of the methodology is its ability to assess uncertainty based on relatively small sample sizes where analytical limit states functions do not exist.

Keywords: Design of Experiment, Gaussian Process Regression, Structural Reliability Analysis, Finite Element Analysis, Probability of Failure

NOMENCLATURE

a (mm)	crack height
2c (mm)	crack length
g(X)	limit state function
$CTOD_{Cap}$ (mm)	CTOD capacity
$CTOD_{Pred}$ (mm)	CTOD predicted
$M_{v=5/2}$	Matern kernel of GP
σ_{ys} (MPa)	specified material yield stress
ε_{ys}	specified material yield strain
σ_{UTS} (MPa)	ultimate tensile strength
E (MPa)	Young's modulus
ν	Poisson's ratio
n_f	number of samples failed
N	total number of samples

1.0 INTRODUCTION

The fracture mechanics evaluation of the pipe girth weld defects is subject to scatter in the input parameters. In order to cater for the uncertainty in the input parameters, it is a common industry practice to evaluate the probability of failure (PoF) using the Structural reliability analysis (SRA). The SRA estimates the failure probability using the limit state function (LSF; i.e. g(X)=0) which separates the design domain into failure (g(X)≤0) and safety (g(X)>0) regions [1]. To address the problem, several efficient approximate techniques such as the First Order Reliability Methods (FORM), Second Order Reliability Methods (SORM), and the Monte Carlo Simulation

[1] Contact author: Arvind.Keprate@dnvgl.com

Copyright © 2019 by ASME

(MCS) have been widely adopted to achieve efficient reliability evaluations [2]. Although widely used, all of the methods have various drawbacks. The FORM/SORM methods may provide unstable or unreasonable failure probabilities for highly nonlinear limit function (see e.g. the case of corroded pipelines in [3][4]). While the MCS, the most straightforward and intuitive method, might suffer from high computational cost, especially for low failure probabilities. Moreover, the outcome of the SRA depends both on the applicability of the variability in input parameters defined and accuracy of the model applied (i.e. the model uncertainty). Use of non-linear FEA represents a possible way of arriving at better crack driving force relations. However, full 3D analysis of pipes with welds is still a computationally expensive approach. Due to this it is of interest to try to optimize the number of FEA runs necessary to perform. In this paper a methodology to apply such an optimization process is explored, with the output of the analyses being used as input to a probabilistic fracture assessment of pipes with weld defects.

While performing the SRA, some of the quintessential requirements are:

1. The formulation of the LSF (g (X) = 0, where X=R-S, represent the difference between the system's R – structural resistance and S – structural load).
2. The variables and their corresponding probability distribution and parameters.
3. The joint probability distribution of the random variables.

During one of the projects of estimating the PoF for a gas pipeline (where R was the $CTOD_{Cap}$ value obtained from the SENB tests), authors were faced with the challenge of unavailability of the mathematical equation for $CTOD_{Pred}$ (S of the LSF). Although the probability distributions of the random variables (i.e. $CTOD_{Cap}$, wall thickness (WT), outer diameter (OD), yield strength of weld (Weld_R_0.5), crack length (2c), crack height (a), and strain) were available, nevertheless the joint probability distribution of the aforementioned variables which are used to estimate $CTOD_{Pred}$ (i. e. S of LSF)), was also not available in the literature. Consequently, the traditional approaches of FORM, SORM, MCS were unsuitable for performing SRA. As an alternative a surrogate model (SM) was utilized to estimate the value of $CTOD_{Pred}$. Thereafter, the limit state function coupled with Monte Carlo Simulation (MCS) was used to estimate PoF at various strain levels.

Surrogate models (SMs), are data-driven models that try to emulate the complex input/output (I/O) behavior of the underlying system, trained on a limited set of computational expensive simulations (CES) [5]. The two basic steps while constructing a SM are training and testing/validation. The first one corresponds to fitting a model to the training points, while the second step involves comparing the predictions of a SM to the actual response [5]. A surrogate model is not a simplified version (with less fidelity) of the CES, rather, it is a statistical method to emulate the data generating process of the CES as accurately as possible based on a set of observed data from the CES. Furthermore, it should have a lower computational cost than the CES to be applicable [5].

In the SRA domain, to date, SMs have been developed for several applications. Yuvraj et al. [6] used SM to predict critical Stress Intensity Factor (SIF) for concrete beams, Keprate et al. [7] utilized Gaussian Process to predict SIF in offshore piping, Moreover, API 579 [8], uses a third order polynomial equation (which is a type of SM) along with an influence coefficient to predict SIF. Building on the successful implementation of the SMs to predict SIF, it seems logical to utilize SM to estimate the value of $CTOD_{Pred}$.

In this manuscript Gaussian Process (GP) or Kriging has been used as the SM to predict $CTOD_{Pred}$ for a given input combination. GPs have successfully been used for SRA applications [9]. Two mature approaches of this kind of method are EGRA (efficient global reliability analysis) [10] and AK-MCS (Active Learning Reliability Method combining Kriging and MCS) [11]. In these two approaches, the Kriging or GP model is used as a SM to substitute for the expensive response function. The key advantage of using GP is that it is much simpler since analyst does not have to choose any topology (as in Artificial Neural Networks (ANN)). Furthermore, the hyperparameters of the GP (more discussed in the next section), can be chosen based on the prior knowledge of the problem at hand.

The most superior feature of GP (in comparison to other SMs) is the ability to provide an uncertainty measure associated with its predictions, which can then be used to adaptively train the GP regression model for improving prediction accuracy. This paper does not use the uncertainty in the GP predictions to adaptively train the GP, however it is used to estimate uncertainty bounds on the PoF curve.

The paper is structured as follows: Section 2 discusses the theory and fundamentals of the Gaussian Process Regression. Section 3 describes the SRA methodology and the FEA used to describe the underlying physical process, while Section 4, concludes the methodology used.

2. Gaussian Process Regression Model

Gaussian process regression (GPR) is a non-parametric kernel-based stochastic technique, built on the fundamentals of spatial statistics [12]. The core idea of a GPR model is that the output values, i.e. Y ($CTOD_{Pred}$, for our case), assessed at varying values of the input variables, i.e. X (WT, OD, Weld_R_0.5, 2c, a, and strain) are modeled as a Gaussian random field, with a known mean and covariance (or kernel) function. A common choice is to assume a zero-mean prior (i.e. that the observations of the GP model has a distribution with mean equal to zero), and that the covariance between two observations decrease with increasing distance between the two observations (i.e. the predictions from a GP model are more dependent on the observations of near neighbors than more distant neighbors). The latter is achieved by applying a covariance (or kernel) function which takes in two observation points and outputs the covariance between them. There are a

Copyright © 2019 by ASME

wide range of possible kernel functions, and apart from the assumption on the mean, the selection of a kernel function is the core of a specific GP with respect to modeling how similar two observations are. Based on this, an appropriate kernel function needs to be considered and selected for the specific data under investigation. The complexity of the GP functions is largely determined by the covariance function, ranging from the very smooth, infinitely differentiable, functions generated by the radial basis function (RBF), to the "spiky", non-differentiable functions generated by the exponential (EXP) (or Matérn $M_{v=1/2}$) kernel. Samples from the distribution of prior GP functions are shown for four kernels with different "smoothness" properties in Figure 1.

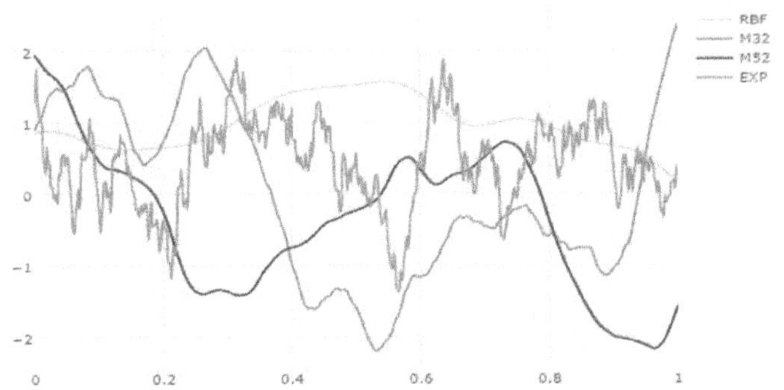

Figure 1 Sample of prior GP functions for zero mean and RBF, M32, M52 and EXP kernel

When both a mean function and a kernel function have been selected, the next step is to fit a GP model to the data. During this process, the hyper parameters of the GP are also determined (e.g. the parameters of the mean and covariance function). This is done by optimizing the GP hyper parameters to maximize the likelihood of the observed data. Once the hyper-parameters are estimated, the GPR can then be used for making predictions. The detailed mathematical theory of the GPR can be found in [12], As an illustrative example, Figure 2 depicts the function in Equation 1, which has been evaluated at six random (1D) points x, on the interval [0,1], and then a GP has been fitted and used to make predictions for the whole interval.

$$G(x) = 0.6\sin(7x) + 2x \quad (1)$$

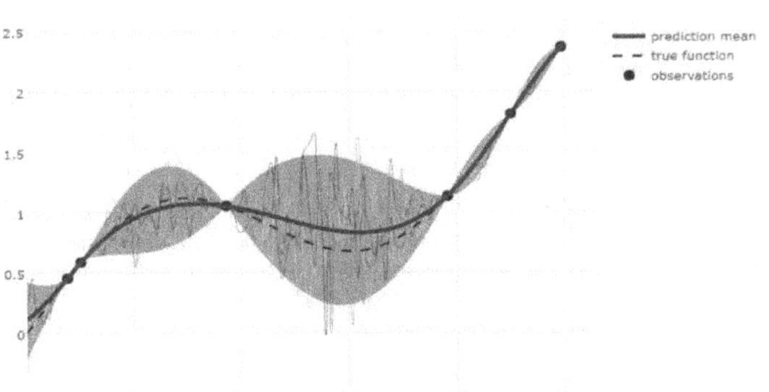

Figure 2 GP predictions of mean ±2std including 10 sample functions

3. SRA METHODOLOGY

The methodology used to perform SRA consists of the following steps:

1. Generate Design Space
2. Perform FEA to generate training and validation data
3. Training and Testing of GPR model
4. Sample space generation
5. Prediction of $CTOD_{Pred}$ using GPR model
6. PoF evaluation using MCS

3.1 Generate design space

To explore an unknown response space, modellers use a method called design-of-experiments (DOE). However, modellers have a tendency to prefer non-optimal designs like one-at-a-time (OAT) approaches due to aspects like [13]:

- The baseline being a "safe" starting point where the model properties are well known (i.e., the best estimate).
- Changing one factor at the time ensures that whatever effect is observed on the output can be attributed to this factor alone.
- A non-zero effect implies influence, i.e., OAT do not make Type I errors (false positives)

This makes OAT approaches intuitive, but to be a valid approach, it must be assumed that all the probability density functions for the uncertain factors have sharp peaks on this multidimensional best-estimate point. This is not usually the case. [13] also show how the curse-of-dimensionality proves that an OAT approach is inefficient with respect to exploring only a tiny fraction of a multidimensional input space.

A more "space-filling" approach can be constructed by using Latin hypercube sampling (LHS). LHS is a statistical method for generating a near-random sample of parameter values from a multidimensional distribution. LHS ensures that the samples are spread out over the sampling space in a way that avoids clustering of sample points. This is achieved by dividing the range of each variable into N equally probable intervals, and N sample points are the placed in this hyper-grid under the requirement that each sample point should be the only one in each hyper-plane. In 2D this is similar to placing N rooks on a N x N chessboard without threatening each other. If no information on the distribution of the variability of the variables is available, it is common to assume a uniform distribution so that sub-intervals of individual variables are of the same size. Note that LHS do not solve the dimensionality problem, but that it covers the sampling space in a better way than OAT for a given set of samples. This means that a SM based on GPs will have less uncertainty associated with the distance from samples for a LHS set than for a OAT sample set. Note also that the efficacy of a SM is dependent on how many influencing parameters affect the relevant response, and whether it is possible to explore a sufficiently large portion of the input space using the expensive model.

Copyright © 2019 by ASME

3.2 Perform FEA to generate training and validation data

Nonlinear Finite Element analyses were performed on pipeline with girth weld strength mismatch as well as pipe geometry mismatch (misalignment). The objective of the analyses was to evaluate the integrity of the pipeline for a large set of flaw sizes, pipe misalignment and weld materials properties. As shown in Figure 3, the pipe dimensions are kept constant on one side whereas the other side is scalable to create a misalignment. Figure 4 presents an example of a typical FE model. In this example, the dimensions of the scalable pipe are 356.2mm (OD) and 19.7mm (WT) and the flaw dimensions are 2.1 x 36.2 mm. All the flaws in the FE analyses are modelled with a radius of 0.05mm and are located on the fusion line at the weld root.

The finite element results are then used to develop response surfaces at different strain levels for how the crack driving force (CTOD) varies with the flaw sizes and misalignments using machine learning. The analyses were performed considering various flaw sizes, misalignments and weld material properties. The nonlinear FE analyses were performed using ABAQUS/CAE software version 2017. In total, 26 Finite Element analyses were performed combining different OD, WT, weld material properties and flaw sizes (11 OAT samples, 11 LHS samples and 4 LHS test samples).

Table 1: Material parameters for pipeline

Parameter	Value
Pipe material, OD	352.4 mm
Wall thickness to cater for design pressure and corrosion allowance	19.2 mm
Specified Material Yield Stress, σ_{ys}	502 MPa
Specified Material Yield Strain, ε_{ys}	0.005 (0.5% or 5‰)
Ultimate Tensile Strength, σ_{UTS}	572 MPa
Young's Modulus, E	207,000 MPa
Poisson's Ratio, ν	0.3

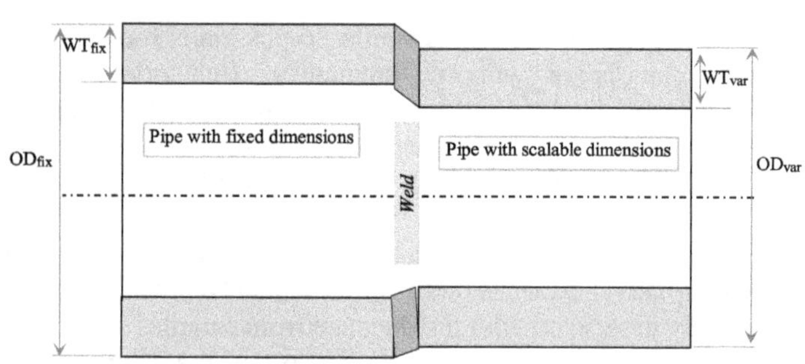

Figure 3 Schematic of the Pipeline with Weld Misalignment

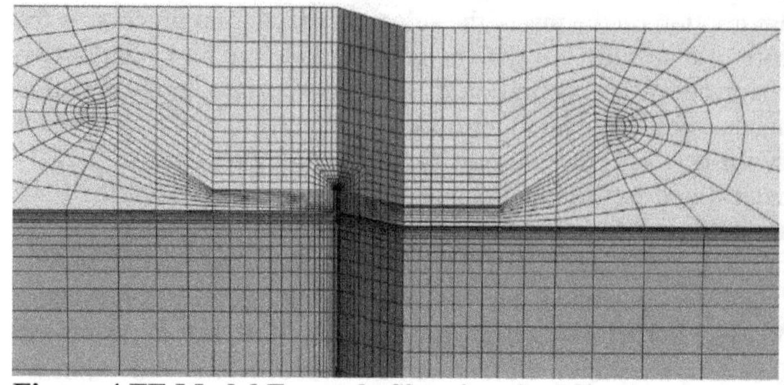

Figure 4 FE Model Example Showing the Misalignment and Flaw

The pipeline models were meshed using 3D solid elements of type C3D8R (eight-node brick element with reduced integration). The pipeline models were meshed using 3D solid elements of type C3D8R (eight-node brick element with reduced integration). In the finite element model symmetry conditions along the pipe is applied to save modelling and computational time, Figure 5.

In addition, one end of the pipe is constrained from displacements and rotations in all degrees of freedom. The other end is constrained from displacement in X and Y directions as well as in rotation in all directions. The analyses were carried out in two static steps:

- 1st step: apply internal pressure and endcap effect
- 2nd step: Apply global bending of the pipeline by applying rotation to each end of the pipe

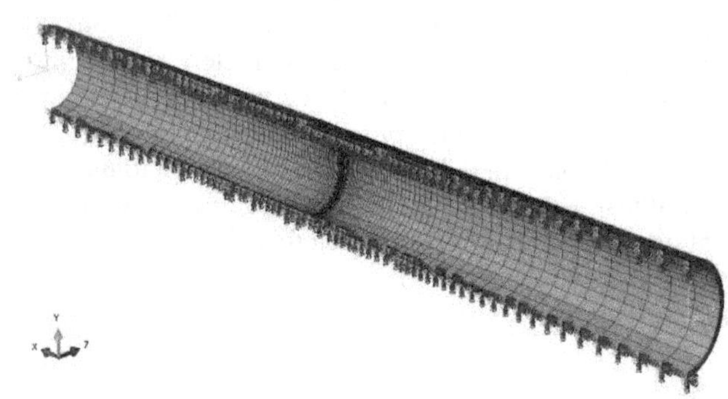

Figure 5 Symmetry Boundary Condition

An internal pressure of 400 bar is applied to the inner surface of the pipe as well as inside the crack. Endcap effect due to pressure loading is applied to one end of the pipe surface. It was clearly seen that the plastic strain in the shear band increases as the joint misalignment increases. FE Analyses for other flaw sizes show similar behavior. An example of the plastic strain in the shear band is shown in Figure 6.

Copyright © 2019 by ASME

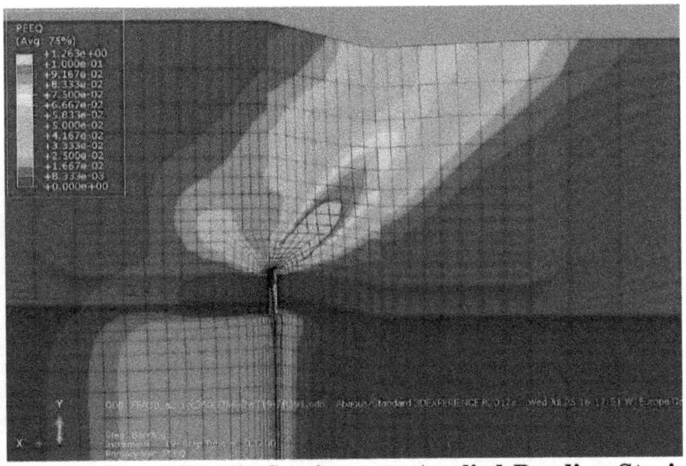

Figure 6 Local Plastic Strain at an Applied Bending Strain (i.e. Elastic + Plastic) of About 0.5% for a 2.1X36.2mm Root Flaw

The main result from the FEA is to evaluate how the crack driving force increases with the bending strain for the weld materials, flaw dimensions and misalignments. It must be mentioned here that misalignment is a function of WT and OD. The crack driving force is typically measured by the crack tip opening displacement (CTOD), which is extracted at nodal pairs that closest measures the 90° opening angle. Figure 7 shows the location of simulations in the flaw size, OD, WT, Weld R0.5 inputs space. Notice how 5 of the green (OAT) simulations are located at the center point, and 6 other simulations creates perpendicular planes of exploration, leaving large volumes unexplored. The blue (LHS) simulations are spread more stochastically around the entire input volume. Figure 8, 9 and 10 represent a typical CTOD versus the applied rotational strain (elastic + plastic) for 2 different input sets (OAT and LHS) vs. WT. From Figure 9 and 10 it is seen that the CTOD vs strain increases as the misalignment increases as well as with the flaw length.

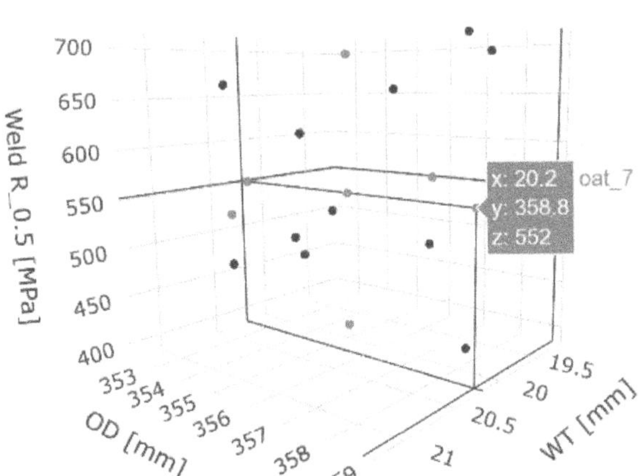

Figure 7 Simulation inputs for an OAT (green) and LHS (blue) approach plotted for different sets of input variables

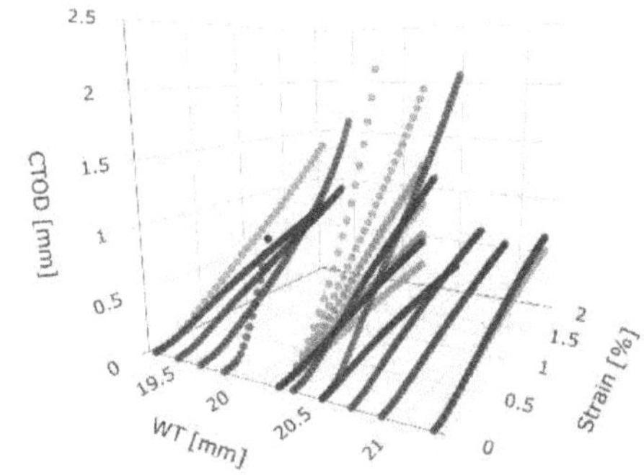

Figure 8 CTOD response curves for an OAT (green) and LHS (blue) input set plotted against WT and Strain

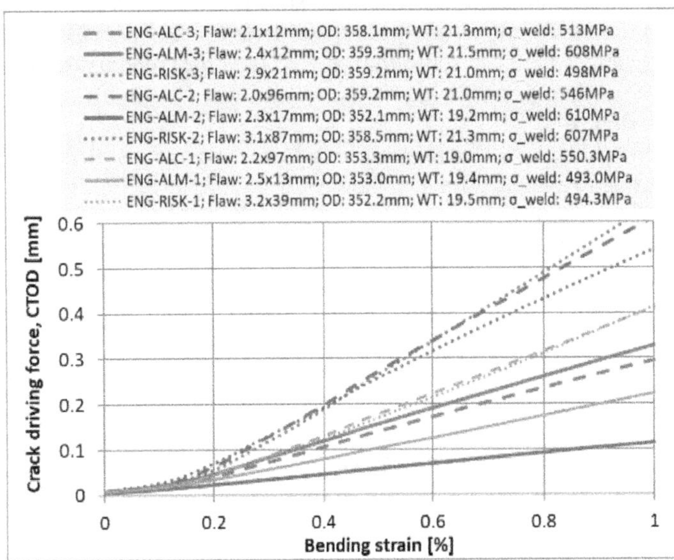

Figure 9 CTOD response curves for OAT input set plotted against Strain

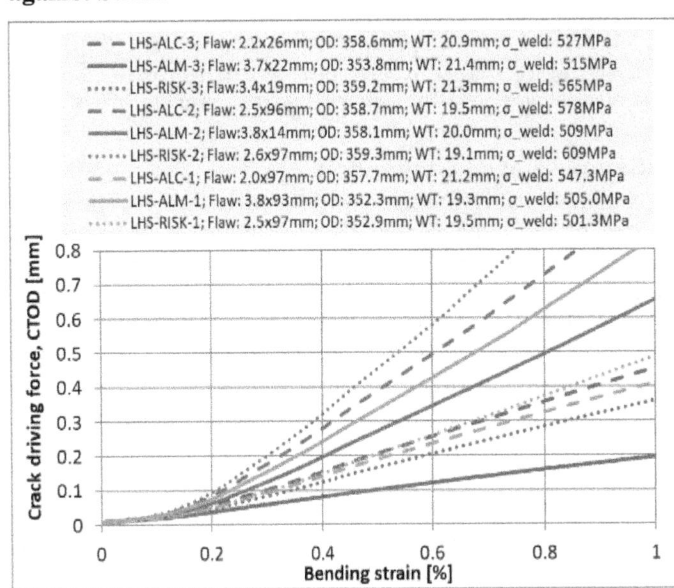

Figure 10 CTOD response curves for LHS input set plotted against Strain

Copyright © 2019 by ASME

3.3 Training and Testing of GPR Model

To be able to perform SRA, assessing the probability of a strain induced CTOD larger than the pipeline's capacity to withstand the crack driving force, we need to be able to estimate CTOD response values across the potential variations of input parameters that have not explicitly been simulated by FEA. The number of potential analysis runs is limited, and we need to be able to select a SM based on its efficacy to predict out of sample. A good screening measure is to calculate the r^2 score based on a a model trained on an initial set of simulations and predicting on a smaller set of randomized simulations used as test set. In Table 2 below, the r^2 score have been calculated based on two training sets, denoted OAT and LHS, and used to predict the result of 6 random test simulations. The Gaussian Process regressor with Matérn kernel (with smoothness parameter $v = 3/2$ or $5/2$) gives consistently better r^2 score than any of the other models out-of-the-box (that is, no tuning of the scikit-learn [14] models have been done).

In addition to a screening based on the r^2 score, it is also relevant to visually look at how well a sample of predictions across the entire input space compares with the simulations. Figure 12 shows CTOD predictions based on MC samples (see section) from the GP with Matérn kernels and smoothness parameter $v =$ a) 3/2 and b) 5/2 for the OAT set of simulations. As can be seen, $v = 5/2$ is more consistent with the simulations at lower strain values. In the following, the GP M52 is used as basis for the surrogate model.

Table 2: Surrogate model screening – r^2 score

DOE	SM type[2]											
	GP M32	GP M52	GP RBF	KNN	KNN 2	KNN 4	Lasso	Linear	Ridge	RF	MLP	SVM
OAT	0.63	0.74	-1.1	0.36	0.35	0.32	-0.04	0.52	0.52	0.42	-681	-0.38
LHS	0.77	0.56	-0.67	0.46	0.45	0.43	-0.11	0.55	0.55	0.19	-761	-0.46

Note that this result can be heavily dependent on how the random test simulations are located in the input space compared to the set of training simulations.

3.4 Sample Space Generation

A simple and intuitive way to visualize the potential response of the pipeline under consideration is to perform a Monte Carlo (MC) simulation based on the input distributions given in Table 3 and predict the corresponding CTOD response from the GP regressor. Figure 11 shows such a MC simulation plotted together with the FEA results, and is a useful way to assess consistency between the FEA and the SM.

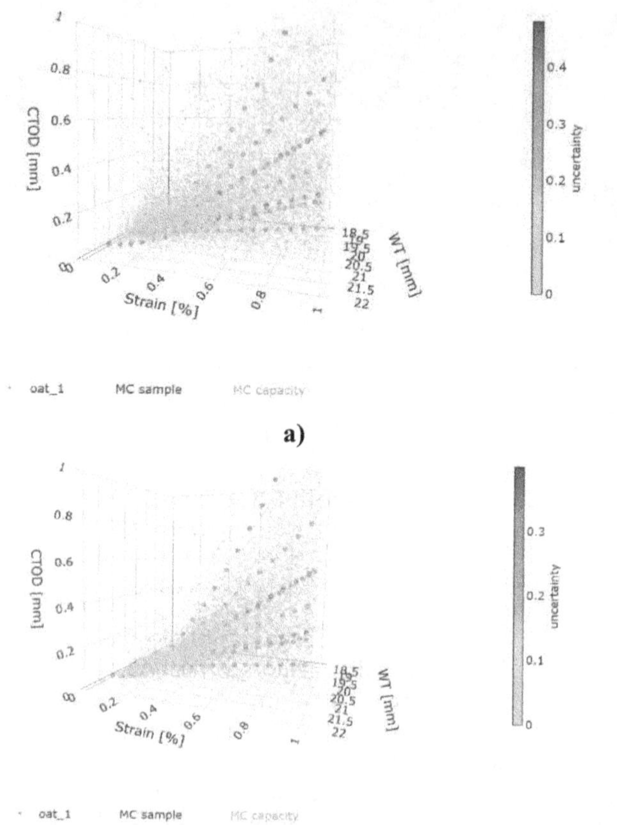

Figure 11 Example of comparing CTOD predictions with simulation results a) GP M32 b) GP M52

Table 3 Distributions of input and capacity variables

	Parameter Name	Distribution Type	Mean	StdDev
Input	*WT (mm)*	Normal	20.2	0.5
	OD(mm)	Normal	355.6	1.6
	Weld R_0.5 (MPa)	Normal	552	70
			Lower Bound	**Upper Bound**
	a (mm)	Uniform	2	4
	2c (mm)	Uniform	10	100
	Strain (%)	Uniform	0	3
			LogMean	**LogStdDev**
Capacity	*CTOD_cap*	Lognormal	-1.84	0.14

3.5 Prediction of CTOD response using GP regression

To compare the different DOE approaches, three sets of simulations where performed. One set of 11 simulations was based on the OAT approach, while two sets where generated based on 11 LHS simulations. The third set was used as a test set, and consisted of 6 simulations based on LHS. A GP was fitted to

[2] Surrogate model types: GP – Gaussian process, MX2 – Matérn kernels with $v = x/2$, RBF – Radial basis function kernel, KNN – K nearest neighbours, RF – Random forrest, MLP – Multi layer perceptron (Neural Network), SVM – Support vector machines. See scikit-learn.org documentation for details.

Copyright © 2019 by ASME

each of the two sets of 12 simulations, and to the combined set of 22 simulations, and a MC simulation of predicted CTOD response and CTOD capacity was performed. Figure 11 shows a MC simulation where the grey-to-orange dots are the predicted CTOD response (the color represent the uncertainty in the GP prediction), while the red dots is the MC sample of CTOD capacity of the pipeline. By visualizing the multidimensional MC response together with the FE results, helps the modeler assess the goodness of fit between the response surface and the FEA response.

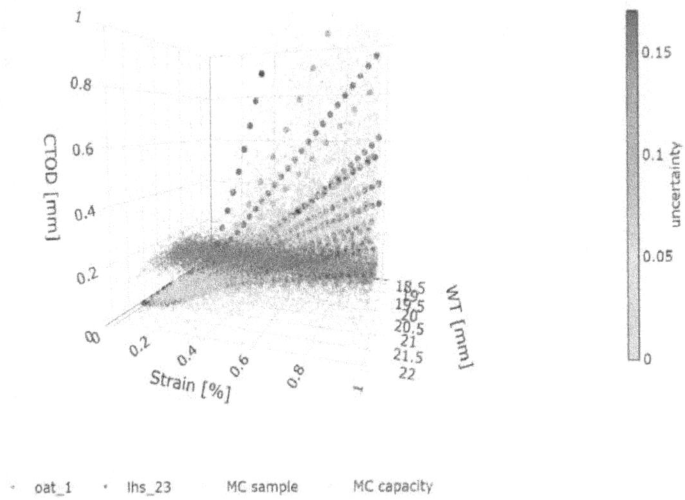

Figure 12: Example of GP prediction of CTOD from MC simulation based on 11 OAT and 11 LHS expensive FE simulations

3.6 PoF evaluation using Monte Carlo

The objective of this paper is to assess a critical level of applied bending strain with respect to the resulting CTOD and the pipeline's capacity to withstand this CTOD level. To do this a series of MC estimates of the probability-of-failure at consecutive strain levels was performed based on the input distribution in Table 3, corresponding CTOD predictions using the GP regressor and comparing it with samples from the capacity distribution, also in Table 3. The PoF level has been estimated by comparing the predicted responses and corresponding samples from the capacity distribution and calculating the fraction of samples where the response exceeds the capacity. To obtain sufficient confidence in the PoF level, the MC sample size was set to at least 100 times the level of individual failures (e.g. for a PoF level of 10^{-3}, 10^5 samples where generated from the relevant distributions of the input variables and capacity).

Figure 12 shows a MC simulation of pairs of responses and capacities. The grey-to-orange dots are the predicted CTOD response (gray signify low uncertainty in the GP prediction, while orange to red signify high uncertainty in the GP prediction). The bright red dots are samples from the capacity distribution. Thus, each pair where the grey response prediction is higher that its corresponding red capacity sample, counts as a failed realization (e.g. where the limit state function given by Equation 2:

$$g = CTOD_{Cap} - CTOD_{Pred} < 0 \quad (2)$$

where $CTOD_{Cap}$ is the capacity and $CTOD_{Pred}$ is the CTOD response predicted by the GP for a given input realization. The PoF is calculated at consecutive strain levels as:

$$\text{PoF} = \frac{n_f}{N} \quad (3)$$

where n_f is the number of samples that failed (i.e. for which g <0) and N is the total number of samples (e.g. 10^5). In Figure 13 the PoF is plotted on a log scale as a function of strain. The solid lines are the PoF calculated based on the mean prediction of the GP, whil the dashed lines signify the PoF calculated by the GP predicted mean ± one standard deviation. Figure 13 shows clearly that the uncertainty in the PoF estimate based on the 11 simulations in the OAT set is higher than for the 11 simulations in the LHS set. Combining the two set gives the PoF level plotted on a log scale in Figure 14. Here it is clear that the added simulations have reduced the uncertainty in the PoF estimate considerably across the whole strain range.

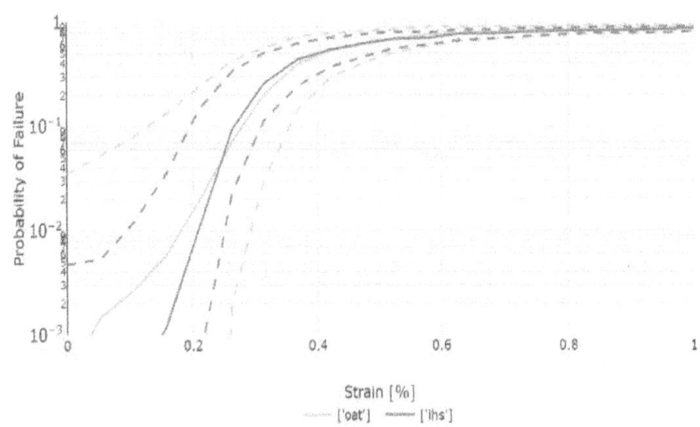

Figure 13 PoF vs. Strain for OAT and LHS simulation sets

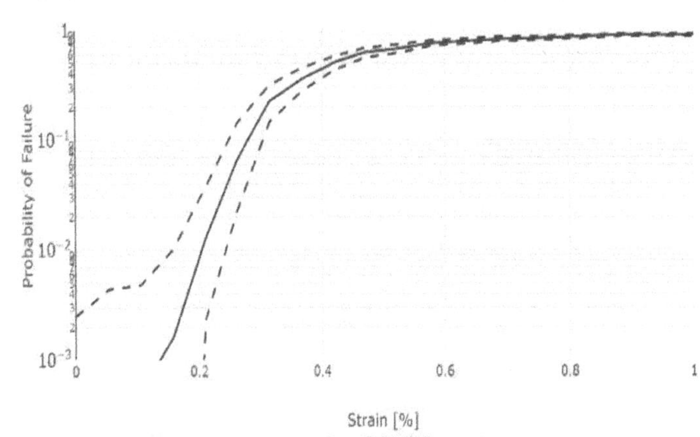

Figure 14 PoF vs. Strain for the combination of OAT and LHS simulation sets

Copyright © 2019 by ASME

Authors wish to state that for the assumed distribution of $CTOD_{Cap}$, failure is most likely predicted to happen when the pipe reaches global plasticity. This is believed to be due to the accelerated growth of CTOD as the crack ligament becomes fully plastic resulting in applied CTOD values of similar magnitude as the critical CTOD values assumed. Thus, for the case considered, a significantly higher fracture toughness would be required to ensure that the pipe could go into global plasticity with low probability of the weld failing in presence of significant weld defects.

4.0 CONCLUSIONS

This paper has outlined a method for consistent use of GPR surrogate models to estimate PoF levels for an engineering system with a non-linear and complex response.We have shown that the often used OAT approach, although intuitive and simple, do not explore a multidimensional input space sufficiently, which result in large uncertainties both in point predictions and in the PoF estimates. Although the LHS explores the input space in a more space-filling way, it may still not capture the input combinations with the most critical response. The PoF level estimated by the 11 LHS simulations had considerable associated uncertainty.

By combining the two sets, totalling 22 simulations, the uncertainty in the estimated PoF level was reduced significantly (although the mean prediction did not change significantly between those based on the 11 LHS and the 22 OAT+LHS). The reduced uncertainty is expected as more information is used in the latter case. However, the approach shown here does not guarantee that the input variation that influence the PoF the most has been explored. A natural continuation of this work would be to explore ways to adaptivly explore the critical responses based on a risk reducing approach as suggested in [15] and possibly incorporate physics-based constraints as suggested by [15] and explored in [16].

5.0 ACKNOWLEDGEMENTS

The authors acknowledge our colleagues at DNVGL for their assistance in performing the work upon which this paper has been prepared and the permission by management to publish this paper.

6.0 REFERENCES

[1] Keshtegar, B., Miri, M., "Reliability analysis of corroded pipes using conjugate HL-RF algorithm based on average shear stress yield criterion", Eng. Fail. Anal. 46, 104–117, 2014.

[2] Seghier, M. El. A. B., Bettayeb, M., Correia, J., Jesus, A. D., Calçada, R., "Structural reliability of corroded pipeline using the so-called separable Monte Carlo method", J. Strain Anal. Eng. Des, 730-737, 2018.

[3] Seghier, M. El. A. B., Keshtegar, B., Elahmoune, B., "Reliability analysis of low, mid and high-grade strength corroded pipes based on plastic flow theory using adaptive nonlinear conjugate map", Eng. Fail. Anal. 90, 245-261, 2018.

[4] Witek, M., Batura, A., Orynyak, I., Borodii, M., "An integrated risk assessment of onshore gas transmission pipelines based on defect population", Eng. Struct. 150–165, 2018.

[5] Deschrijver, D., and Dhaene, T., "Surrogate Modelling Lab," University of Gent, Belgium, 2012. http://www.sumo.intec.ugent.be/research.Accessed January 2019.

[6] Yuvraj, P., Murthy, A. R., Iyer, N. R., Samui, P., and Sekar, S. K., "Prediction of Critical Stress Intensity Factor for High Strength and Ultra-High Strength Concrete Beams Using Support Vector Regression", Journal of Structural Engineering, vol. 40(3), pp. 224-233, 2014.

[7] Keprate, A., Ratnayake, R.M.C. and Sankararaman, S., "Adaptive Gaussian Process Regression as an Alternative to FEM for Prediction of Stress Intensity Factor to Assess Fatigue Degradation in Offshore Piping," Int J Pres Ves Pip, 153, pp. 45-58, 2017.

[8] API 579-1/ASME FFS-1. Fitness-for-service. Second ed. Washington DC: American Petroleum Institute;

[9] Huang, C.W., Hami, A., El., and Radi, B., "Overview of Structural Reliability Analysis Methods-Part 3: Global Reliability Methods", ISTE Open Science, 2017.

[10] Bichon, B.J., Eldred, M.S., Swiler, L.P., Mahadevan, S., and McFarland, J.M., "Multimodal Reliability Assessment for Complex Engineering Applications Using Efficient Global Optimization", AIAA paper No. 2007-1946, 2007.

[11] Echard, B., Gayton N., and Lemaire, M., "AK-MCS: An Active Learning Reliability Method Combining Kriging and Mote Carlo Simulation", Structural safety, V33(2), 145-154, 2011.

[12] Rasmussen C.E., Williams C.K.I., "Gaussian processes in machine learning", ISBN 026218253X, MIT Press 2006, USA.

[13] Saltelli A. and Annoni P., "How to avoid a perfunctory sensitivity analysis", Environmental Modelling & Software, Vol. 25, Issue 12, pp. 1508-1517 2010.

[14] scikit-learn.org, "Open source Machine Learning in Python", Accessed January 2019.

[15] Eldevik S., Hafver A., Agrell C., Pedersen F.B., "AI+Safety, Safety implications for artificial intelligence", www.ai-and-safety.dnvgl.com, DNV GL, 2018

[16] Agrell C., "Gaussian processes with linear operator inequality constraintst", https://arxiv.org/abs/1901.03134 arXiv, 2019

Copyright © 2019 by ASME

AIMPIMG2019-1047

Chapter 39

ASSESSMENT OF PIPELINES IN MOVING SLOPES: A RISK-INFORMED DECISION MAKING APPROACH

Mona Abdolrazaghi
Enbridge
Edmonton, Alberta, Canada

Sherif Hassanien
Enbridge
Edmonton, Alberta, Canada

ABSTRACT

Integrated Risk Informed Decision Making (IRIDM) is a concept that can be used for typically unique high impact complex situations. There are four pillars of the effective decision making: the definition of the decision, the role of stakeholders involved in this process, the assessment methodology, and the timeline to complete the projects. There are five main roles defined within IRIDM: Recommend, Agree, Perform, Input, and Decide (RAPID). Slope remediation projects fall into non-programmed decision making to ensure the safety of pipeline and environment and optimizing the cost expenditure. To effectively initiate and plan the remediation, Input roles are required to complete multiple assessments other than stress and strain based analysis. In addition, a number of considerations including cost benefit analysis, risk reductions, and regulatory and political factors are required to be utilized by Input through IRIDM decision making. A deliberation process among the stakeholders is developed to account for the uncertainties around the decision making process.

INTRODUCTION

Safety critical industries are considered as operating systems which any failures can result in loss of life, significant property, and/or environmental damage such as nuclear, aerospace, pipeline hazardous materials. The safety measures in these industries are exceptionally high. In millennium era, the multi-level protection, so called Defense-In-Depth (DID) has been developed to effectively maintain the safety in operations. In DID approach, multiple hierarchy physical barriers are in placed that in the event of failing every hurdle, there will be a prevention. The intention of this approach is firstly, prevention and secondly, predictions of the effective mitigation approach. If one level fails, the successive level will be still active to prevent the failure [1-3].

Probabilistic Safety Assessment (PSA) is another affective reliability management approach that uses analytic evaluation to account of associated uncertainties.

And, lastly, Risk based versus Risk-Informed decision making approach includes the uncertainty and risk in decision making process. Risk-based approach uses a risk metrics with narrow mindset of interpretations [4, 5].

Integrated Risk-Informed decision making (IRIDM) approach includes various ranges of parameter s incorporated in decision making such as cost-benefit analysis, expert judgement, stakeholders concerns. In this approach, the limitations of the project along with an optimization approach are utilized to devise a unique solution. IRIDM provides a single solution for an individual problem through an intensive deliberation approach.

Assessing the safety of pipelines crossing moving slopes and remediation projects are one of the most challenging integrity problems that face decision makers. IRIDM can be an effective approach which can be used for integrity problems with unique/high impacts on pipelines business. This paper presents a slope remediation case study using IRIDM approach. Generally, the four pillars of the effective decision making: defining decision statement, assigning roles of stakeholders in the decision making process, understanding technical methodologies and their limitations/uncertainties, and meeting timelines constraint to complete the project. The five main roles within an IRIDM are: Recommend, Agree, Perform, Input, and Decide (RAPID). Moreover, the integrity decision making can be classified in two categories; namely: programmed (e.g. regulations, integrity rules, best practice), and non-programmed which falls beyond systematic integrity processes. Remediating moving slopes and maintaining the safety of crossing pipelines fall into the non-programmed decision making category since it requires balancing between safety of pipeline/environment and cost expenditure. In order to effectively initiate and plan a slope remediation project, several inputs are required to complete a fulsome assessment in addition to stress and strain-based analysis. A number of considerations for such a project including cost benefit analyses, anticipated risk reductions, and political factors are essential to achieve an optimum integrity

Copyright © 2019 by ASME

decision. A deliberation process among stakeholders needs to take place in order to account for various uncertainties associated with decision making process. The Slope remediation project typically looks for devising a unique solution to obtain the most cost benefit remediation approach while maintaining the safety of the pipeline. Multiple slope remediation alternatives are proposed with a given constrains; next, a comprehensive cost benefit analysis for each case is completed. In the recommendation process, multiple factors including risk, regulatory, logistics, and budgetary constraints are embedded. A decision based on the given recommendation and other relevant contributing attributes is accomplished.

INTEGRATED RISK INFORMED DECISION MAKING (RAPID)

There are four elements in the effective decision making behavior through IRIDM approach: "what", "when", "who", "how". "What" refers as the definition and framework of the decision; "who" is referred to the roles play that decision," "when" is about identifying the timeline and project milestones, and "how" is about applying the best practice in decision process (shown in **Figure 1**) [6-8].

The role of people involved in the project has been proposed using so called RAPID model. "R" stands for the role of recommender, "A" for the person whom agrees, "P" for performer," I" for person (s) 's inputs, and lastly, "D" for the decider [15].

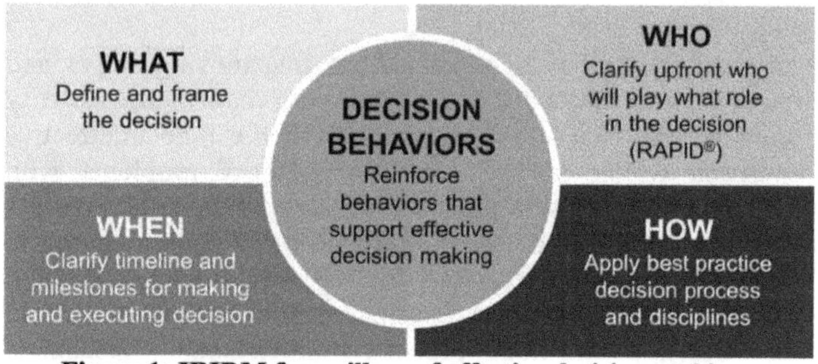

Figure 1: IRIDM four pillars of effective decision making

The deliberation process Using IRIDM approach starts with defining the project and the scope of implementations followed by the proposed alternatives to the project among "R", "A", and "D". Then, the inputs are all gathered; these inputs are not limited to the deterministic and probabilistic, it includes regulatory, budgetary or any other forecasted limitations on the project implementations. Next, the risks for each identified alternatives for the decisions are weighted by "R" and "A". Next, "R" recommends the proposed decision to "D". Once, "D" decides on the recommendations, "P" performs the project. The effectiveness of the implementation is monitored by "R" and "A". Lastly, the learnings from the decision is recorded by "R".

Flowchart in **Figure 2** shows the details of the process of IRIDM along with the roles of individuals involved in [9].

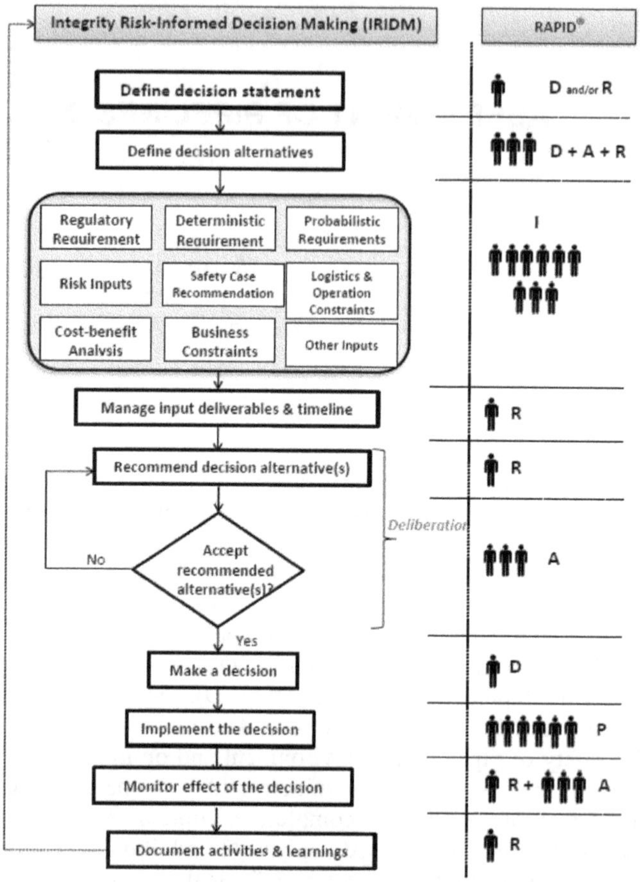

Figure 2: IRIDM framework flowcharts

To ensure the effectiveness of the project as well as for future continuous improvement, it is essential to track the effectiveness of the decision after completion of every project.

IRIDM approach is designed for projects with unique circumstances. This is not recommended for any routine and defined action plans. The unique projects are divided into two categories: structured and non-structured. The non-structured risk informed is for the projects require an immediate decision and the timeline for the full IRIDM approach is not permitted. In this case, a simplified version of IRIDM is proposed with the roles of "R","D", and "P" only. In any case, the documentation and the record of effectiveness of the project are required.

CASE STUDY

In this paper, a geotechnical site remediation with the application of IRIDM and thorough assessment is discussed.

In this hypothetical scenario, five lines are crossing 1.4 Km slope with the average annual movement of 50 mm/ year. There has been multiple failures on the slope historically that were resulted in straining the pipeline at toe of the slope.

Copyright © 2019 by ASME

The mitigation decision for this case is based on short term and long term plans. The short term plan includes running In line inspection (ILI)s annually and repair the high strain or areas with the reported wrinkles exceeding the permissible threshold. The long term plan includes several options and alternatives for the mitigation.

IRIDM approach has been implemented in this cased as follows:

- **Definition:** To identify the most efficient and effective long term plan on risk mitigation on this slope
- **Timeline:** Assessment time line
- **Roles and responsibilities:** Identified through RAPID
- **Methodologies:** Probabilistic, Deterministic, cost benefit analysis, budgetary and regulatory restriction

Once, the definitions and roles and responsibilities were established, the alternatives for this project are discussed as follows:

1) **Option 1:** To relocate two stress lines and replace the rest with the heavy wall
2) **Option 2:** To replace all line to above grade
3) **Option 3:** To continue with the existing plan (running ILI and repair the areas with the high strain)

Next, the inputs for these scenarios are all collected including ground measurements strain measurements, ILI and Geopig data, pipeline operating condition (temperature and pressure), and pipe properties. A finite element model of the pipe is developed and a limit state approach is implemented to calculate the probability of failure (existing and predicted for future).

FEA is the most preferred method to assess pipeline integrity due to ground movement and pipe-soil interaction. FEA model can be as much detailed as possible and all the non-linearity in pipe-soil interaction or ground movement to be considered to have more realistic understanding of the geohazard system. The most common practice in FEA model of pipe-soil interaction is to model the pipeline with the beam element and represent the soil loading on the pipeline using discrete soil springs (axial, horizontal, or vertical). The beam element can account for biaxial stress due to internal pressure and temperature as well as ground movement. To accommodate for large ground movement, nonlinear pipe-soil interaction and stress-strain can be conducted in FEA model. It should be noted that the beam element is an appropriate representative of the pipeline up to the point that pipe wrinkles.

Alternatively, the pipe-soil can be modeled with three-dimensional (3D) shell model to represent the continuity of the pipe and soil interaction. However, analyzing shell model is typically computationally expensive and sometimes inability of continuum model to capture non-continuum behaviours.

FEA provides the strain at every point on the pipeline along the pipe profile.

Compressive capacity model was based on Dorey's model [10]. And, the tensile capacity model was based on CRES' model [11] which is a modification of CSA-Z662 model [12].

The probability of failure (PoF) is calculated at the highest strain demand using limit state approach (**Figure 3**).

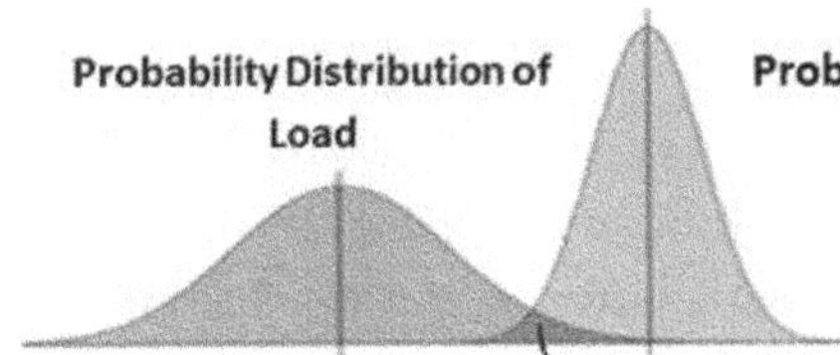

Figure 3: Limit state approach

PoF is calculated for all six pipelines and shown in Table 1. The existing PoF is calculated to be 6E-3 for the worst case of the lines within right of way. The permissible target for geotechnical site is set as to be 1E-3. So, the geotechnical sites within the current mitigation plan are exceeding the permissible target. Safety factor was calculated using the mean capacity over mean demand values. This safety factor is shown in Figure 4.

Table 1: Safety factor and PoF for the lines within the slope

Line	Aggregated PoF	Average Mean Capacity/Mean demand
Line 1	1E-03	6
Line 2	1E-04	11
Line 3	6E-03	8
Line 4	3E-03	21
Line 5	4E-03	7
Line 6	2E-03	8

The relationship between safety factor and PoF can be shown as follows [13]:

Figure 4 shows the relationship of mean capacity over mean demand versus allowable probability of failure for different coefficient of variation (COV) in demand model. As Figure 4 shows, in order to achieve to the permissible PoF 1E-3, a safety factor of 10 is at least required. However, that is not a case for these pipelines.

Copyright © 2019 by ASME

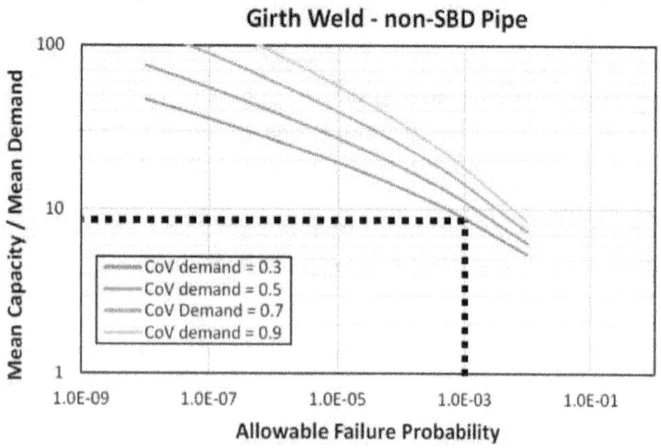

Figure 4: Design safety factor and PoF relationship

A cost benefit analysis using decision tree and utilizing Precision Tree© has completed for every scenario **Figure 5**. It should be noted that all the cost numbers on this assessment is hypothetical. Expected monetary value is the utility function that accounts for the probability, consequence and the cost of the option.

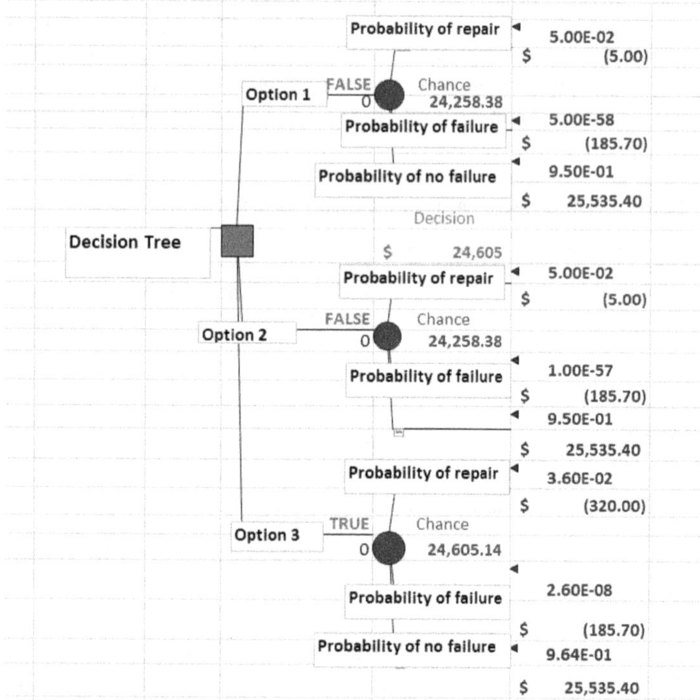

Figure 5: Decision tree and alternatives for 20 year interval

Figure 5 shows the cost of the option in order to mitigate a risk associated with implementing of the option. Decision tree identifies the optimum option with the maximum monetary value. Decision tree shows option 3 which is continue with the current plan have the highest expected monetary value for the 20 year interval. In addition, it is also shown that the difference between each option is not significant in every approach. Budgetary or regulatory constraints do not pose any limitation in this case study.

"R" summarized these findings and presented to the decision maker, "D". D decides to vouch for Option 1 given the expected monetary values among options are not significant and option 1 will ensure the ground movement poses no impact on the pipeline. In this decision, "D" accounted not only the risk and budget but also political and economic factors. This case study showed a complete implementation of IRIDM approach within a geotechnical project.

CONCLUSION

In this paper, we have shown a full implementation of IRIDM approach in geotechnical reliability assessment. IRIDM approach included the deterministic, probabilistic, cost-benefit analysis, budgetary and regulatory constraints.

The probabilistic assessment showed that the pipeline PoF is exceeding the permissible target currently. The deterministic safety factor is below the target safety factor (here 10) for the probabilistic target of 1E-3.

The cost-benefit analysis compares the multiple alternatives against each other. It is shown in the cost benefit analysis that the existing management plan that includes repair of the pipeline every few years is more cost effective than any pipeline replacements.

There were no regulatory or budgetary constraints assumed in this case study.

Including all above mentioned assessment, the decision maker decides on replacing the pipelines by first reroute line 2 and 3 and replaces the rest sequentially. The thought behind this decision was not only the risk, deterministic, or cost analysis but also additional non-integrity factors were involved in. The decision maker decided that a pipeline with minimum risk of soil movement and reducing the PoF to the value well below the target is the most desirable option in this case study.

REFERENCES

[1] INSAG-3. 1988."*Basic Safety Principles for Nuclear Power Plants*". International Nuclear Safety Group. International Atomic Energy Agency (IAEA).

[2] INSAG-10. 1996. "*Defense in Depth in Nuclear Safety*". International Nuclear Safety Group. International Atomic Energy Agency (IAEA).

[3] INSAG-12. 1999."*Basic Safety Principles for Nuclear Power Plants 75 INSAG-3 Rev. 1*". International Nuclear

Copyright © 2019 by ASME

Safety Group. International Atomic Energy Agency (IAEA).

[4] Aven, T. 2003. "*Foundations of Risk Analysis: A Knowledge and Decision-Oriented Perspective*". John Wiley & Sons.

[5] Zio, E. & Pedroni, N. 2012. "*Risk Informed Decision Making Processes*". Foundation for an Industrial Safety Culture (FONCSI).

[6] Blenko, M.W., Mankins, M.C., and Rogers, P. 2010. "*Who Has the D?: How Clear Decision Roles Enhance Organizational Performance*". Harvard Business Review. January Issue.

[7] Ersdal, G. & Aven, T. 2008. "*Risk Informed Decision-Making and its Ethical Basis*". Journal of Reliability Engineering and System Safety. 93, pp. 197-205.

[8] Rogers, P. & Blenko, M.W. 2006. "*Who Has the D?: How Clear Decision Roles Enhance Organizational Performance*". Harvard Business Review. January Issue.

[9] S. Hassanien, D. Langer, and M. Abdolrazaghi, "Integrity Risk-Informed Decision Making", IPC2018-78149.

[10] Dorey, A. (2001) Critical buckling strains in energy pipelines, Ph.D., thesis, Department of Civil and Environmental Engineering, University of Alberta, Edmonton, Alberta, Canada.

[11] Liu, Wang, Zhang, and Kotian, *Realistic Strain Capacity Models for Pipeline Construction and Maintenance*, Center for Reliability Energy Systems (CRES), 2013.

[12] Steel Pipe, CSA-Z 662, March 2014.

[13] M. Nessim and S. Koduru, "Limit States Design for pipeline in Moving Slopes", CFER Technologies, M154, January 2015.

[14] Precision Tree6, Palisade, Copyright 2014.

[15] RAPID: Bain's tool to clarify decision accountability –Bain & Company, August 11, 2011.

Copyright © 2019 by ASME

AIM-PIMG2019-1056

Chapter 40

PROBABILISTIC FAULT DISPLACEMENT ANALYSIS FOR A PIPELINE SYSTEM

Jonathan Brewer, PE, Elizabeth Bowlin, PE
Stress Engineering Services, Inc.
Houston, Texas

Don West, LEG, LG, Clayton Johnson, LG
Golder Associates Inc.
Redmond, Washington

ABSTRACT

Permanent ground displacement (PGD) caused by coseismic fault displacement in earthquakes presents a significant hazard for buried pipelines. Although large displacements are not frequent, they can result in pipeline buckling, rupture, and failure. In order to evaluate the relative likelihood of failure for multiple pipeline fault crossings, a natural gas pipeline operator performed a simplified Probabilistic Fault Displacement Hazard Analysis (PFDHA) for a number of potentially high hazard fault crossings in Southern California.

As a part of the PFDHA, quantitative estimates of the displacement magnitude and probability of ground movement were defined for nine faults. These were combined with non-linear elastic-plastic finite element analysis (FEA) to determine the probability of varying levels of strain in the pipeline at each crossing location. Based on the analysis presented, the expected compressive/tensile strain to failure can be incorporated to evaluate the fault rupture likelihood associated with each pipeline-fault crossing. By reviewing the analysis results of this and similar analyses, the operator can screen and rank the relative criticality of locations in the pipeline and/or determine if fault displacement mitigation is necessary.

INTRODUCTION

A probabilistic fault displacement analysis (PFDA) was carried out in 2018 for a confidential pipeline client with pipeline assets exposed to potential coseismic surface faulting from nine active tectonic faults crossing its pipelines in southern California. Permanent ground displacement (PGD) caused by coseismic fault displacement presents a significant hazard for buried pipelines and although large displacements are infrequent, they can result in pipeline buckling, rupture, and/or subsequent failure.

The PFDA is comprised of three primary steps or elements (Figure 1). In the first step, an initial simplified probabilistic fault displacement hazard analysis (PFDHA), completed by Golder Associates Inc. (Golder), was used to develop quantitative estimates of the nature, amount, and probability of surface fault rupture for the nine fault zones. In the second step, the results and data from the PFDHA were used by Stress Engineering Services Inc. (SES) as inputs in their finite element analysis (FEA) to determine the pipe response to potential fault displacements. The information from the combined, overall PFDA (Step 3) can allow the pipeline operator to evaluate:

- Locations of interest for further analysis,
- Potential exposure to expected surface fault rupture,
- Potential exposure to pipeline damage/failure, and/or
- Potential mitigation measures, if appropriate.

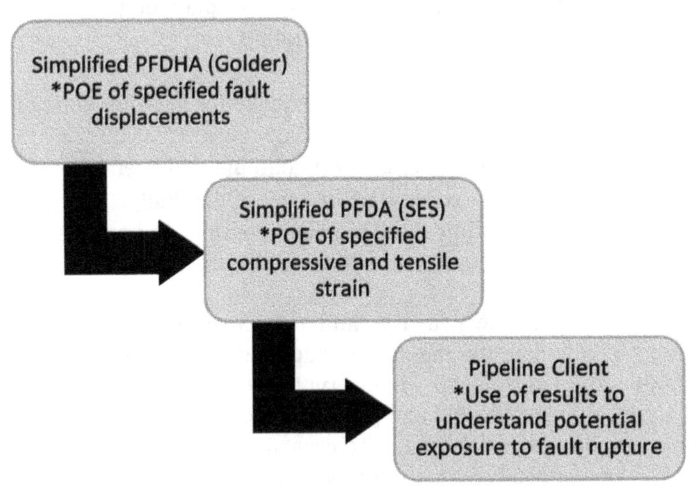

Note: POE = Probability of Exceedance

FIGURE 1. OVERALL PROBIBALISTIC FAULT DISPLACEMENT ANALYSIS (PFDA) PROCESS

Copyright © 2019 by ASME

DEVELOPMENT OF THE SIMPLIFIED PROBABILISTIC FAULT DISPLACEMENT HAZARD ANALYSIS (PFDHA)

Background and Approach

The simplified PFDHA was performed as a follow-on, desktop study to an initial office-based, regional-scale fault rupture hazards assessment completed by Golder in 2014 in which the nine potentially active fault zones (historic or Holocene movement) were identified crossing the pipeline system. The nine fault zones addressed in the PFDHA are listed by number and fault type below, and as classified in the USGS Quaternary Fault and Fold Database of the United States [8].

- Fault 1; strike-slip
- Fault 2; normal-slip
- Fault 3; strike-slip
- Fault 4; strike-slip
- Fault 5; strike-slip
- Fault 6; strike-slip
- Fault 7; strike-slip
- Fault 8; strike-slip
- Fault 9; reverse/oblique-slip

The approach to the PFDHA consisted of five steps, which are summarized in Figure 2. Step 1 included the collection of several primary geologic and geometric characteristics of the nine active faults. These characteristics were derived from several sources of data including the available literature pertaining to the specific individual faults, the fault zone-specific data from [8], and data from [3]. The pertinent fault characteristics data used in the development of the PFDHA included, but were not limited to:

- Fault type and sense of displacement
- Fault geometry (strike and dip)
- Total fault length and segment or splay lengths
- Fault subsurface rupture width
- Fault rupture area
- Maximum magnitude earthquake for the fault zone or system
- Displacement per event (maximum and average)
- Slip rate or recurrence interval of the fault
- Fault location at the pipeline crossing, and orientation with respect the pipeline (the crossing angle)

In Step 2, the primary fault rupture characteristics were used to calculate the maximum magnitude earthquake for each fault zone using the fault rupture-earthquake magnitude relationships of [9], [1] and [6]. Once the maximum earthquake magnitude was established for each fault zone, the maximum and average displacements per event were back-calculated from the earthquake magnitude using the fault displacement relationships of [9].

In Step 3, the estimated maximum and average displacements per event, as well as estimated slip rate were used in running a simplified PFDHA approach of [10].

In Steps 4 and 5, the results of the PFDHA were assembled and submitted to SES. In addition, estimated geotechnical soil properties for each of the fault crossings were submitted as part of Step 5.

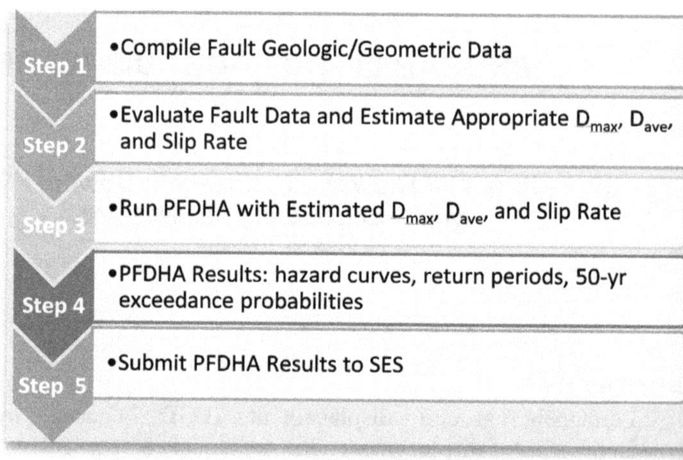

FIGURE 2. SIMPLIFIED PROBILISTIC FAULT DISPLACEMENT HAZARD ANALYSIS (PFDHA) PROCESS

Table 1 lists the estimated specific fault parameters, for each of the nine faults, implemented in the PFDHA. Note that Faults 1, 3, and 9 have high combined maximum and average displacements/event, along with the highest slip rates. Although Faults 5 and 6 have high maximum and average displacements/event, they have very low slip rates (< 1 mm/yr).

TABLE 1. ESTIMATED SITE-SPECIFIC CHARACTERISTICS OF THE NINE FAULTS

Fault (Type)[1]	Fault Dip (deg.)	Approx. Crossing Angle[2] (deg.)	Est. Max. Displ./Event (m)[3]	Est. Ave. Displ./Event (m)[3]	Est. Slip Rate (mm/yr)[3]
1 (SS)	90	55	4.19	1.86	1.6
2 (N)	70E	74	0.26	0.15	1.4
3 (SS)	90	19	7.0	4.4	7.0
4 (SS)	90	55	2.2	0.8	0.6
5 (SS)	90	59	5.25	2.45	0.6
6 (SS)	90	10	5.72	2.46	0.6
7 (SS)	90	44	2.2	0.8	0.8
8 (SS)	90	64	0.136	0.084	0.2
9 (R, SS)	65SE	65	5.72	2.46	6.0

Notes: [1] SS = strike-slip fault; N = normal-slip fault; R = reverse-slip fault.
[2] Difference between fault strike azimuth at the pipeline, and pipeline orientation azimuth.
[3] Maximum and average displacement, and slip rate are inputs to the simplified PFDHA.

Copyright © 2019 by ASME

Youngs et al. [10] identify and describe two basic approaches to PFDHA. The first approach is termed the "earthquake approach," and it is derived from probabilistic seismic hazard analysis (PSHA) for the development of earthquake design ground motions. The approach relates the occurrence of fault displacement at the surface explicitly to the occurrence of earthquakes (i.e., fault slip at depth) in the region of the site. The second approach is termed the "displacement approach," which utilizes the site-specific characteristics of surface fault displacement to quantify the displacement hazard without invoking a special mechanism (e.g., an earthquake occurrence) for the cause.

We employed a simplified displacement approach of [10] following data requirements guidance for a PFDHA provided by [2]. By simplified, we mean that potential aleatory (physical, inherent, and random) and potential epistemic (data, model and data completeness) uncertainties in the tectonic environment, fault geometric and geologic data for each of the nine faults were not explicitly addressed, and multiple potential scenarios for fault displacement, arising from the potential uncertainties were not accommodated in the simplified approach. Rather, geologic professional judgment was used to identify the appropriate, estimated geologic and geometric data to be used in the PFDHA.

In the displacement approach, the annual frequency of exceeding a specified level of maximum or average displacement is calculated, thus producing hazard curves for the displacements. The annual rate of exceedance is converted to a return period for the specified displacements, and a percent probability of exceedance in 50-years, is also calculated for the specified displacements.

Results of the PFDHA

Selected results of the PFDHA, for the nine faults considered, are summarized in Table 2. Table 2 provides the return period (years) and percent probability of exceedance in 50 years for the selected average displacements of 100 cm (3.28 ft), 200 cm (6.56 ft) and 300 cm (9.84 ft), for the nine faults, although the probability of displacements as low as 1 cm were also produced in the simplified PFDHA.

From Table 2, it appears that the nine faults can be roughly grouped into three categories based on return period and probability of exceedance in 50 years, for average displacements of 100 cm to 300 cm. The first group appears to be the most active, with the shortest return periods and highest probabilities of exceedance in 50 years and includes only Fault 9. Fault 9 has return periods between about 600 years and 1,000 years for average displacements from 100 cm to 300 cm, while the probabilities of exceedance in 50 years for displacements from 100 cm to 300 cm range from 8.3 percent to 4.9 percent, respectively.

The second group of faults includes Faults 1 and 3. These faults are characterized by moderately long return periods ranging from about 1,200 years to 3,600 years for displacements from 100 cm to 300 cm (Table 3), with Fault 3 having slightly shorter return periods. They have probabilities

TABLE 2. SUMMARY OF PFDHA RESULTS FOR SELECTED AVERAGE DISPLACEMENTS/EVENT

Fault	Return Period and Percent Probability of Exceedance in 50 Years for Select Average Displacements					
	100 cm (39.4 in)		200 cm (78.7 in)		300 cm (118.1 in)	
	Return Period (Years)	% Prob. Exceed. In 50 Years	Return Period (Years)	% Prob. Exceed. In 50 Years	Return Period (Years)	% Prob. Exceed. In 50 Years
1	1,792	2.8	2,570	1.9	3,626	1.4
2	6,965	0.7	335,000	< 0.05	--	--
3	1,230	4.0	1,454	3.4	1,702	2.9
4	3,300	1.5	7,215	0.7	15,390	0.3
5	5,743	0.9	7,610	0.7	9,943	0.5
6	13,390	0.4	17,730	0.3	23,140	0.2
7	2,475	2.0	5,411	0.9	11,540	0.4
8	575,400	< 0.05	--	--	--	--
9	576	8.3	762	6.4	995	4.9

of exceedance in 50 years that range from 4.0 percent to 1.4 percent, with Fault 3 having consistently slightly higher probabilities of exceedance for displacements from 100 cm to 300 cm.

The third group of faults includes the remaining six faults (Faults 2, 4, 5, 6, 7, and 8). These faults are characterized by moderately long to very long return periods, and probabilities of exceedance in 50 years generally less than 1 percent (Table 2).

Geotechnical Soil Properties

The identification of the estimated soil properties at the nine fault/pipeline crossings was based on a request by SES that stated the soil properties needed for the PFDA should include the following:

- For undrained conditions: the soil undrained shear strength, and
- For drained conditions:
  - the effective unit weight
  - the coefficient of lateral soil pressure at the pipe centerline (K_0)
  - the soil friction angle

Table 3 is a summary of the soil properties at each of the fault/pipeline crossings. These properties were based on accessing the online soils data from [14] for the specific mapped soil at each fault/pipeline crossing. The soil descriptions of [14], along with their listed various percentages of clay, silt and sand for each mapped soil were used to estimate a specific Unified Soil Classification System (USCS) description for each soil. The USCS information was then used to estimate typical unit weights, and internal friction angles from publications regarding the geotechnical properties of

Copyright © 2019 by ASME

geologic materials (e.g., [4], [5], and [7]). The at-rest lateral soil pressure coefficient (K_0) was calculated using the angle of internal friction (ϕ) where $K_0 = 1 - \sin \phi$.

Note that all of the fault/pipeline crossings are located in arid to semi-arid environments, and the soil descriptions of [14] indicate that the soils are all relatively granular and well-drained. Thus, the soil properties listed in Table 3 are for drained, unsaturated conditions.

TABLE 3. ESTIMATED GEOTECHNICAL SOIL PROPERTIES AT EACH FAULT CROSSING

Fault Crossing	Estimated Soil Properties		
	Unit Weight (pcf)	Coeff. of Lat. Press. at rest (K_0)	Internal Friction Angle (deg.)
1	115	0.46	33
2	115	0.46	33
3	115	0.47	32
4	115	0.44	34
5	115	0.46	33
6	120	0.44	34
7	115	0.46	33
8	115	0.46	33
9	115	0.46	33
Note: Values for unit weight, friction angle, and K_0 are expected to be within ±15 pcf, ±5 degrees, and ±0.08, respectively.			

DEVELOPMENT OF THE SIMPLIFIED PROBABILISTIC FAULT DISPLACEMENT ANALYSIS (PFDA)

Background and Approach

The above described PFDHA provides the quantitative estimates of the nature, amount, and probability of soil movement for the nine fault zones. SES then prepared to perform the PFDA using finite element analysis (FEA) based on the guidance of PRCI 268-9823 [11]. The approach to the PFDA consisted of five steps, which are summarized in Figure 3.

Step 1 included collection of the primary pipeline design, geometric, and material data. The pipeline properties for each fault crossing are shown in Table 4. The pipeline route information was provided by the client in the form of Universal Transverse Mercator (UTM) coordinates for each bend in the pipeline and the associated radius of curvature. At the time of the analysis, the pipeline depth of cover at each pipeline/fault crossing was unknown. Therefore, for each crossing, the pipeline was assessed for depths of cover (DOC) of 3 feet and 7 feet.

In Step 2, the provided pipeline geometric data were used to generate finite element models (FEM) for each pipeline/fault crossing location. Finite element modeling for this study was performed using the Abaqus general purpose finite element code version 6.14-1. The models of each pipeline/fault crossing were constructed using:

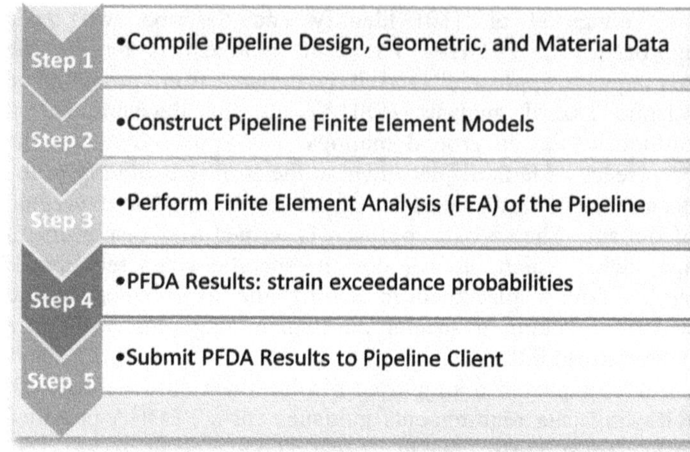

FIGURE 3. SIMPLIFIED PROBILISTIC FAULT DISPLACEMENT ANALYSIS (PFDA) PROCESS

- First order, three-dimensional pipe elements (Abaqus type PIPE31) to model the pipeline, and
- Three-dimensional, 4-node, pipe-soil interaction elements (Abaqus type PSI34) to capture the interaction with the soil.

In the vicinity of the fault crossings, an element size of approximately 6 inches was used for the pipeline model. In regions away from the fault, an element size of approximately 12 inches was used. A minimum of 650 feet of pipeline was modeled on each side of the fault crossing. During analysis, the results were reviewed to verify that a sufficient length of pipeline was modeled such that the applied boundary condition did not affect the magnitude of the computed longitudinal strains.

The pipeline steel was modeled using elastic-plastic material properties, with the stress-strain curve defined using the specified minimum material properties. The monotonic stress-strain curve is developed using the guidance of the ASME BPVC [12]. The axial, lateral, and vertical soil loading on the pipeline were represented using bi-linear soil springs computed based on the soil properties shown in Table 2 and references [11] and [13].

TABLE 4. PIPELINE GEOMETRIC AND MATERIAL DATA

Faults	Pipeline Geometric and Material Data			
	Outside Diameter (in)	Wall Thickness (in)	API 5L Material Grade	Maximum Allowable Operating Pressure (psi)
1, 2, 5, 7, 8	30	0.358	X70	1,200
3, 4, 6, 9	42	0.5	X70	1,200

Copyright © 2019 by ASME

In Step 3, finite element analysis was performed for the pipeline with associated soil movement. The pipeline was initially modeled in its undeformed configuration. Loading was then applied in several steps, which are listed below:

Step 1: Apply gravity and internal pressure to the pipeline
Steps 2-10: Apply soil movement

In the first step of the analysis, gravity was applied to the pipeline along with internal pressure equivalent to the MAOP. In the following steps, the fault movement was applied using enforced displacements on the base of the soil spring elements. Figure 4 is a schematic showing the applied soil movement. On one side of the fault, the base of the soil spring elements and the end of the pipeline are fixed in all six degrees of freedom. On the opposite side of the fault, the base of the soil spring elements and the end of the pipeline model were displaced in accordance with the direction and magnitudes from the PFDHA. The overall soil movement was ramped up in 9 steps:

- 10, 20, 30, 50, 100, 200, and 300 cm,
- The average displacement per event, and
- The maximum displacement per event.

The average and maximum displacements per event were inserted between the 7 fixed displacement steps as necessary to result in monotonically increasing displacement magnitudes. The analyses were post-processed to determine the maximum (tensile) and minimum (compressive) strain in the pipeline as a function of magnitude of soil movement.

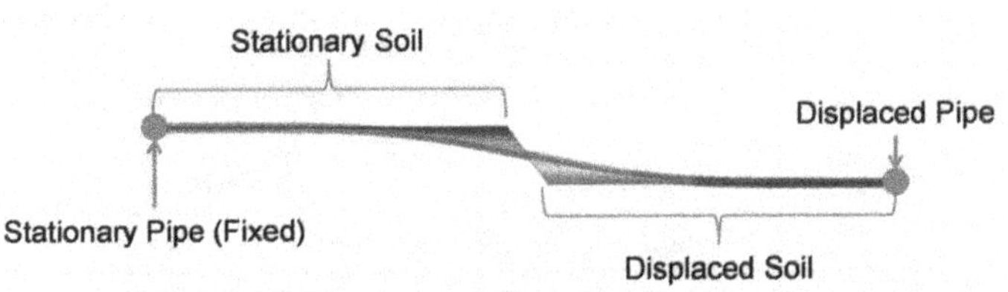

FIGURE 4. SCHEMATIC SHOWING SOIL AND PIPELINE BOUNDARY CONDITIONS

In Step 4, plastic strain vs. soil movement results were combined with the probabilities of exceedance from Golder, culminating in the curves showing strain exceedance probabilities.

Although analysis was performed for all nine (9) fault crossings presented in the previous sections, detailed analysis results are presented here only for faults 2, 3, and 9.

Analysis Results – Fault 2

The finite element model for Fault 2 is shown in Figure 5. The pipeline in this case is relatively straight and subject to a normal-slip fault, where the east side of the pipeline shifts down relative to the west side.

Figures 6 and 7 show the tensile and compressive longitudinal strains as a function of fault movement. As the fault shifts, the pipeline is subject to high tensile and bending stresses near the fault crossing. As a result of the tensile force in the pipeline from the fault movement, the tensile strains are higher in magnitude than the compressive strains. In addition, for this fault, the strains in the pipeline with 7 ft DOC are over twice the magnitude of the strains in the pipeline with 3 ft DOC. This is due to the significant increase in the maximum uplift soil spring force with increasing DOC. This results in higher bending stresses and strains.

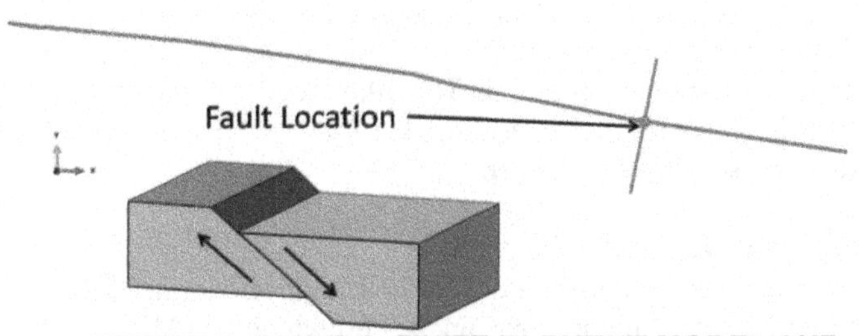

FIGURE 5. FAULT 2: FINITE ELEMENT MODEL AND FAULT LOCATION

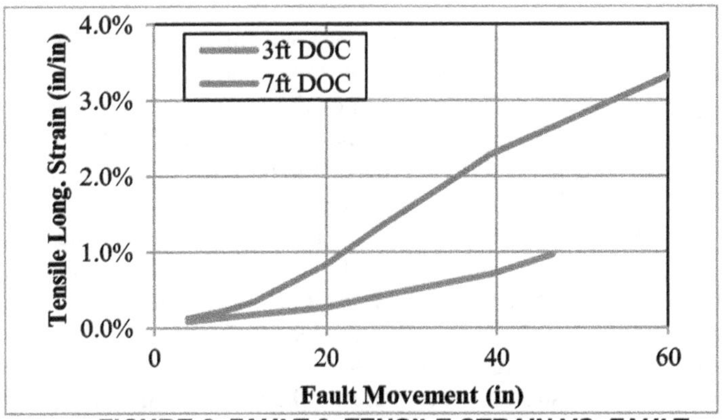

FIGURE 6. FAULT 2: TENSILE STRAIN VS. FAULT MOVEMENT

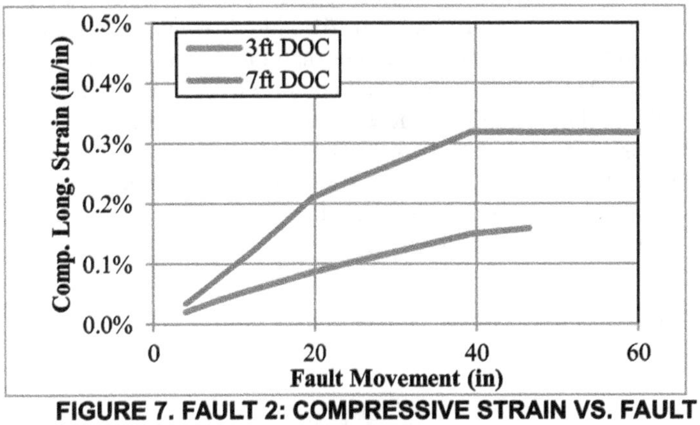

FIGURE 7. FAULT 2: COMPRESSIVE STRAIN VS. FAULT MOVEMENT

Copyright © 2019 by ASME

Once the relationships between strain and fault movement are defined, they are combined with the fault displacement probability of exceedance (POE) values from the PFDHA (shown in Table 2). This results in tensile and compressive strain POE curves, which are shown in Figures 8 and 9. As a result of the larger maximum vertical uplift soil spring force for the 7 ft DOC, the POE curves are shifted up (higher strain) and to the right (higher probability of exceedance). This indicates that the likelihood of failure is higher for deeper depths of cover. These curves are used with defined strain limits to define the likelihood of the pipeline exceeding allowable strain limits.

For the purposes of this study, a 1% allowable strain was arbitrarily chosen. The POE of 1% tensile and compressive strain is shown in Table 5. The likelihood of exceeding 1% strain in 50 years increases from 0.57% to 4.55% as a result of the increased depth of cover.

Analysis Results – Fault 3

The finite element model for Fault 3 is shown in Figure 10. In this case, the pipeline has an S-bend which straddles the left-lateral strike-slip fault. Since the exact location of the soil movement is not known, three analysis cases were performed by shifting the location of the soil movement relative to the S-bend.

Using a similar process as described for Fault 2, the tensile and compressive POE curves are developed and are shown in Figures 11 and 12. The direction of the fault movement results in tensile loading of the pipeline between the bends of the pipeline. This leads to much higher magnitudes of tensile strain than compressive strain for this case.

Similar to Fault 2, increasing the DOC increases the soil stiffness. This increases the tensile loading between the bends, which shifts the tensile POE curves up (higher strain) and to the right (higher probability of exceedance). However, this also results in a tension-dominated configuration and the compressive strains are below 0.2% for all analyzed configurations. However, for the smaller 3-foot DOC, the bending strain is a larger percentage of the total strain. This results in higher compressive strains. A comparison of the curves also shows the effect of the fault location with the expected strains. Fault locations #1 and #3, which are closer to the bends, results in higher strains than Fault location #2.

The POE of 1% tensile and compressive strain is shown in Table 5. The likelihood of exceeding 1% strain in 50 years increases from 3.75% to 4.21% as a result of the increased depth of cover. Fault location #3 results in the highest tensile strains (Figure 11) and fault location #1 results in the highest compressive strains (Figure 12).

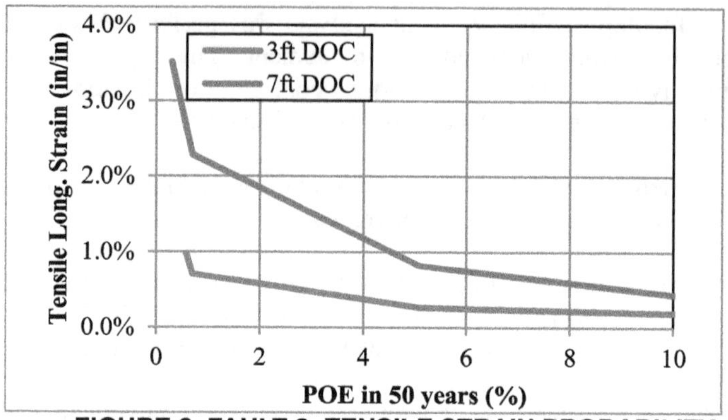

FIGURE 8. FAULT 2: TENSILE STRAIN PROBABILITY OF EXCEEDANCE

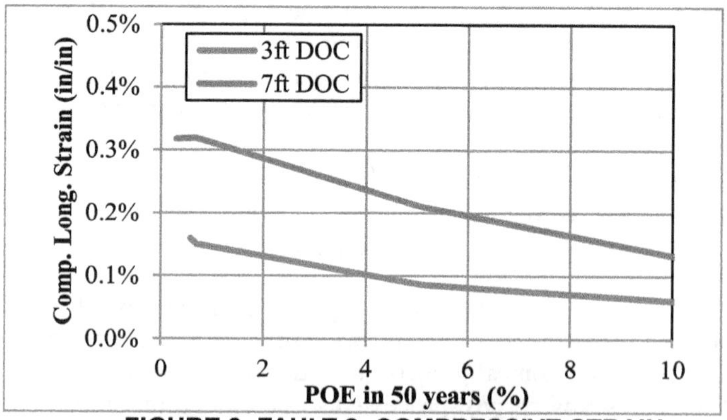

FIGURE 9. FAULT 2: COMPRESSIVE STRAIN PROBABILITY OF EXCEEDANCE

TABLE 5. ANALYSIS SUMMARY OF SHOWING 1% STRAIN PROBABILITY OF EXCEEDANCE

Fault Crossing	Probability of Exceedance (POE) in 50 years of 1% Strain			
	3ft DOC		7ft DOC	
	Comp.	Tensile	Comp.	Tensile
2	0.00% [3]	0.57% [2]	0.00% [3]	4.55%
3[1]	2.64% [2]	3.75%	0.00% [3]	4.21%
9	5.50% [2]	5.53% [2]	8.76%	8.74%

Notes: [1] The POE for fault crossing 3 is taken from the worst case crossing location
[2] The FEA did not converge to 1% strain. The tabulated POE is computed by extrapolation
[3] The analysis did not predict that 1% compressive strain would occur

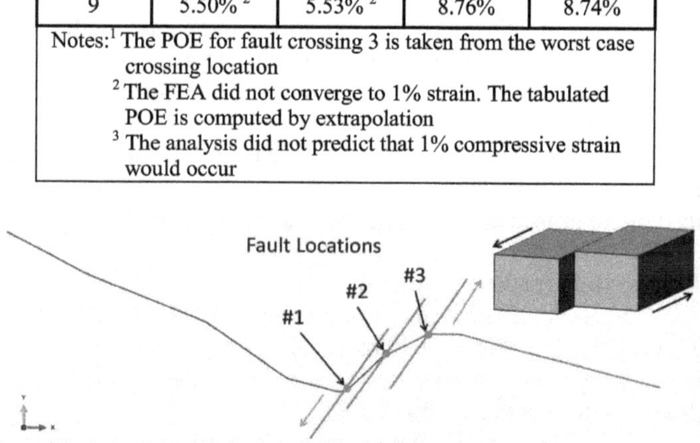

FIGURE 10. FAULT 3: FINITE ELEMENT MODEL AND FAULT LOCATION.

Copyright © 2019 by ASME

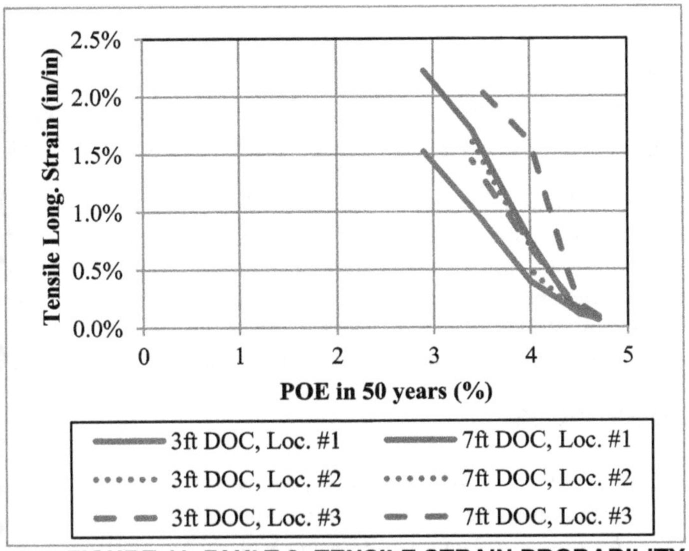

FIGURE 11. FAULT 3: TENSILE STRAIN PROBABILITY OF EXCEEDANCE

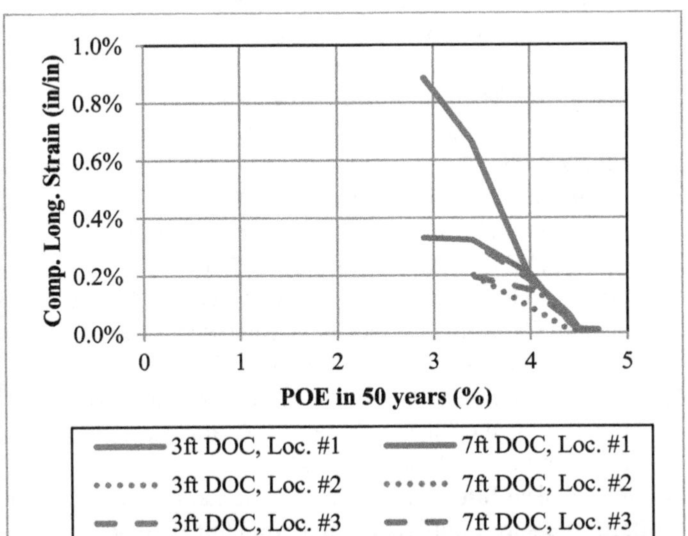

FIGURE 12. FAULT 3: COMPRESSIVE STRAIN PROBABILITY OF EXCEEDANCE

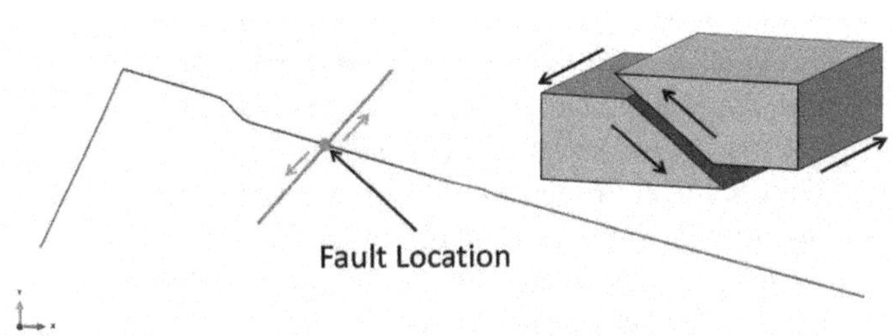

FIGURE 13. FAULT 9: FINITE ELEMENT MODEL AND FAULT LOCATION

Analysis Results – Fault 9

The finite element model for Fault 9 is shown in Figure 13. In this case, the pipeline is relatively straight at the fault location. The fault in this case is a combination reverse-slip, left lateral strike-slip fault. The magnitude of the left-lateral strike-slip movement is 77% of the reverse-slip movement.

Using a similar process as described for Faults 2 and 3, the tensile and compressive POE curves are developed and are shown in Figures 14 and 15. The direction of the fault movement results in nearly pure bending at the critical location. This results in very similar compressive and tensile POE curves.

Since a significant portion of the soil movement for Fault 9 is in the vertical direction, the large increase in the maximum uplift soil spring force with increasing DOC results in significantly higher strains and POE for the 7-foot DOC case

The POE of 1% tensile and compressive strain is shown in Table 5. The likelihood of exceeding 1% strain increases from 5.53% in 50 years to 8.76% in 50 years as a result of the increased depth of cover.

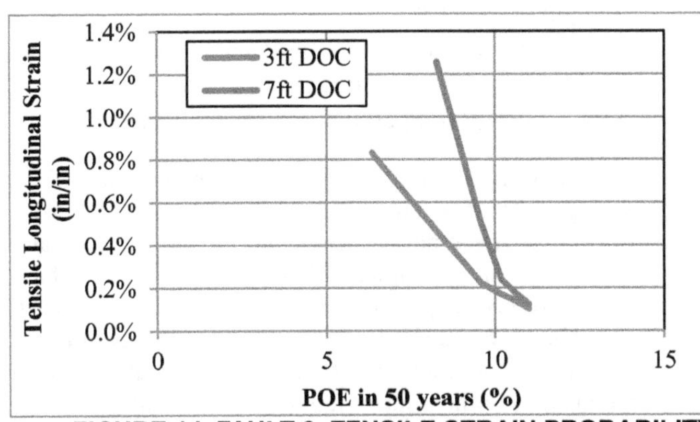

FIGURE 14. FAULT 9: TENSILE STRAIN PROBABILITY OF EXCEEDANCE

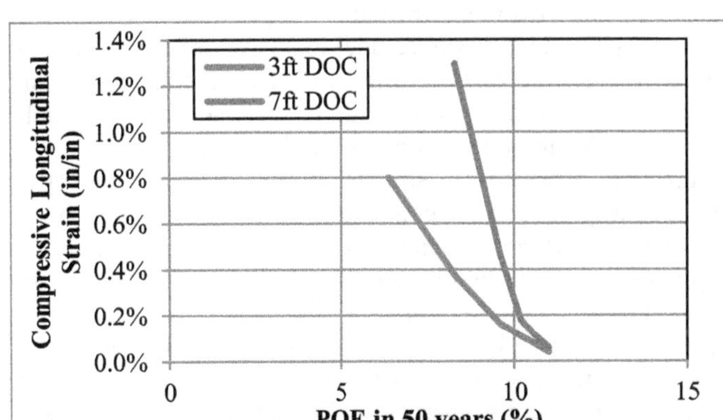

FIGURE 15. FAULT 9: COMPRESSIVE STRAIN PROBABILITY OF EXCEEDANCE

Copyright © 2019 by ASME

DISCUSSION

The results of the simplified PFDHA and PFDA described in this paper represent the starting point for a pipeline operator to determine the risks associated with fault-generated permanent ground displacement (PGD) on a buried pipeline system. These results can be used as a screening tool to identify the relative likelihood of failure for different locations on the pipeline and identify where additional detailed analysis would be most beneficial.

Although the magnitude and probability of soil movement computed in the PFDHA alone can be a good indicator of the most critical locations in a pipeline, changes in the soil properties and/or the presence other pipeline geometric features can result in regions of the pipeline with higher strain demand with smaller soil movement. Determining the strain demand requires the use of analysis tools, such as FEA.

For the higher hazard faults which exhibit higher strain demand, follow-up studies would be sensible and could include:

- Verifying site-specific fault locations and geometries,
- Quantifying site-specific soil properties with testing,
- Verifying the analysis inputs, such as depth of cover and pipe geometry,
- Performing additional sensitivity analyses,
- Defining the pipe-specific strain capacity, and/or
- Designing risk mitigation for critical locations.

One of the primary inputs for the finite element analysis is the soil spring stiffnesses. The effect of changing the properties of the soil and/or the pipeline depth can be very significant. Using Fault 2 as an example, although the predicted fault displacements are low, depending on the soil stiffness, the POE of 1% strain increased by an order of magnitude. Understanding the soil behavior at this location could result in significant changes to the relative ranking of the fault.

Additional sensitivities would also help quantify the strain demand for the pipeline design. Some of these are discussed in reference [11]. These sensitivities include: upper bound (as opposed to specified minimum) material properties for the pipe, different operating pressures, additional modifications in the fault location and/or transition, etc. In several cases, the associated probabilities associated with the sensitivities could also be included to better define the strain demand.

For this study, the 1% strain POE is presented. This value was arbitrarily chosen for comparative purposes only. In order to truly define the consequences of PGD, the strain capacity for the pipeline needs to be defined. The first step in defining the strain capacity of the pipeline is to collect additional information on the pipeline system. This includes determining pipeline geometric anomalies (within and between pipe joints), pipe properties, weld properties, weld quality specification, and weld inspection results. The second step is to define the meaning of "failure" for the pipeline. Depending on the location of the pipeline and the pipeline operator's risk-tolerance, this could be to maintain pressure integrity and/or to maintain normal operability. Often, operators may select different allowable probabilities of exceedance for these two "failure" modes.

Finally, once the overall likelihood of failure has been defined, the operator can also determine how, or if the risk associated with pipeline damage due to PGD needs to be mitigated. There are many options available to mitigate the risk, which are described in detail in reference [11]. However, most of the mitigations for existing pipelines fall into one or more of three categories: (1) decrease the strain demand on the pipeline due to soil movement, (2) increase the strain capacity of the pipeline, or (3) address the consequences of failure by planning.

REFERENCES

[1] Anderson, J.G., Wesnousky, S., and Stirling, M.W. "Earthquake size as a function of fault slip." *Bulletin of the Seismological Society of America* Vol. 86 No. 3 (1996): pp. 683-690.

[2] Coppersmith, K.J., and Youngs, R.R. "Data needs for probabilistic fault displacement hazard analysis." *Journal of Geodynamics* Vol. 29 (2000): pp. 329-343.

[3] Field, E.H., Biasi, G.P., Bird, P., Dawson, T.E., Felzer, K.R., Jackson, D.D., Johnson, K.M., Jordan, T.H., Madden, C., Michael, A.J., Milner, K.R., Page, M.T., Parsons, T., Powers, P.M., Shaw, B.E., Thatcher, W.R., Weldon, R.J., II, and Zeng, Y. "Uniform California earthquake rupture forecast, version 3 (UCERF3)—The time-independent model." U.S. Geological Survey Open-File Report 2013–1165, California Geological Survey Special Report 228, and Southern California Earthquake Center Publication 1792, 2013. http://pubs.usgs.gov/of/2013/1165/.

[4] Geotechdata.info. (2013). "Soil Friction Angle." Retrieved from http://geotechdata.info/parameter/angle-of-friction.html on February 15, 2017.

[5] Geotechnical Info.com. "Angle of Internal Friction, Lateral Earth Pressure Coefficients, and Soil Unit Weights." Retrieved from http://www.geotechnicalinfo.com/soil_unit_weight.html on February 15, 2017.

[6] Hanks, T.C., and Bakun, W.H. "A Bilinear Source-Scaling Model for M-log A Observations of Continental Earthquakes." *Bulletin of the Seismological Society of America* Vol. 92 No. 5 (2002): pp. 1841-1846.

[7] Koloski, J.W., Schwarz, S.D., and Tubbs, D.W. "Geotechnical Properties of Geologic Materials." *Engineering Geology in Washington, Volume 1, Washington Division of Geology and Earth Resources, Bulletin 78* (1989): pp. 19-26.

[8] U.S. Geological Survey and California Geologic Survey (USGS-CGS). (2006). "Quaternary fault and fold database

Copyright © 2019 by ASME

for the United States." Retrieved from http://earthquakes.usgs.gov/hazards/qfaults/ on December 1, 2016.

[9] Wells, D.L. and Coppersmith, K.J. 1994. "New empirical relationships among magnitude, rupture length, rupture width, rupture area, and surface displacement." *Bulletin of the Seismological Society of America* Vol. 84 No. 4 (1994): pp. 974-1002.

[10] Youngs, R.R., Arabasz, W.J., Anderson, R.E., Ramelli, A.R., Ake, J.P., Slemmons, D.B., McCalpin, J.P., Doser, D.I., Fridrich, C.J., Swan, F.H., III, Rogers, A.M., Yount, J.C., Anderson, L.W., Smith, K.D., Bruhn, R.L., Knuepfer, P.L.K., Smith, R.B., dePolo, C.M., O'Leary, D.W., Coppersmith, K.J., Pezzopane, S.K., Schwartz, D.P., Whitney, J.W., Olig, S.S., and Toro, G.B. "A Methodology for Probabilistic Fault Displacement Hazard Analysis (PFDHA)." *Earthquake Spectra* Vol. 19 No. 1 (2003): pp. 191-219.

[11] Honegger, D.G., and Nyman, D.J., "Guidelines for the Seismic Design and Assessment of Natural Gas and Liquid Hydrocarbon Pipelines." PRCI Contract PR-268-9823, Catalog No. L51927. 2004.

[12] American Lifelines Alliance (ALA). (2001 with Addenda through February 2005). "Guidelines for the Design of Buried Steel Pipe." Retrieved from https://www.americanlifelinesalliance.com/pdf/Update061305.pdf.

[13] American Society of Mechanical Engineers (ASME). "ASME Boiler and Pressure Vessel Code: Division 2, Alternative Rules." ASME BPVC.VIII.2-2017. ASME, New York, NY (2017).

[14] Natural Resources Conservation Service (NRCS). (2017). "Web Soil Survey." Retrieved from https://websoilsurvey.sc.egov.usda.gov/App/WebSoilSurvey.aspx.

Copyright © 2019 by ASME

Chapter 41

Managing Risk: Effective Use of Structural Reliability Assessment (SRA) and Quantitative Risk Assessment (QRA) for Sabah-Sarawak Gas Pipeline (SSGP)

Wan M Marzuki Wan Ismail
PETRONAS
Kuala Lumpur, Malaysia

Mohd Nazmi Mohd Ali Napiah
PETRONAS
Kuala Lumpur, Malaysia

Mohd Zazli Mohd Zabidi
PETRONAS
Kuala Lumpur, Malaysia

Ahmad Sirwan M Tuselim
PETRONAS
Kuala Lumpur, Malaysia

ABSTRACT

PETRONAS is operating Sabah-Sarawak Gas Pipeline (SSGP) that traverses hilly and mountainous terrain with significant rain fall in East Malaysia i.e. in state of Sabah & Sarawak. SSGP is an important pipeline asset to PETRONAS flowing around 700-800 MMSFCD of sales gas to PETRONAS LNG complex in Bintulu, Sarawak. The pipeline experienced leak in 2014 and 2018. From root cause failure studies, both pipeline failures were due to combination of external force (from geohazards), girth weld defects and under-matched of girth weld. Post incidents in 2014 and 2018, PETRONAS implements strategies in-line with industry practices in managing pipeline integrity i.e. risk assessments, inspection and monitoring, integrity assessments, and repairs. As such, this paper will focus on the usage of Structural Reliability Assessment (SRA) and Quantitative Risk Assessment (QRA) as part of risk and integrity assessments portion of the strategies. PETRONAS selected SRA and QRA due to availability of inspection, monitoring and operating data and also the fact that both methods consider uncertainties of data/parameters. Pipeline threats considered for the SRA and QRA are (i) girth weld anomalies (including under-matched effect), (ii) bend stress/strain (representing external force/geohazards), (iii) pipe buoyancy/stability at exposed river/stream crossings, (iv) dent/gouge (from third party), and (v) internal/external corrosion. This paper will dwell onto the evaluation and selection of appropriate equation of states for the threats taking into account the correct 'load' and 'resistance' factors. The resultant reliability and risk numbers were compared to target reliability based on procedures from ISO 16708 and target risk level from PETRONAS Technical Standard 16.71.04, respectively. At the end of the paper will discuss and elaborate the actions that need to be taken by pipeline operator consist of short, medium and long term measures aim to sustain the integrity and reliability of the pipeline.

1.0 INTRODUCTION

Sabah Sarawak Gas Pipeline (SSGP) consists of 512 km NPS 36 onshore pipeline with 22 block valves, metering stations and compressor facility and was designed with flow capacity of 760 MMscfd of natural gas. It connects the Sabah Oil and Gas Terminal (SOGT) in Kimanis, Sabah to the PETRONAS LNG Complex in Bintulu, Sarawak, refer to Figure 1.

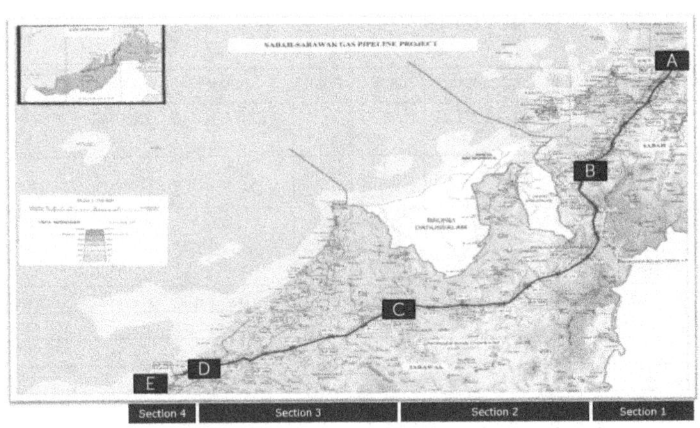

Figure 1 – SSGP route running through two Malaysia's states i.e. Sabah (Kimanis) and Sarawak (Bintulu)

On the 10th June 2014, the first incident of the SSGP occurred at KP135+766 whereby the gas from the pipeline was ignited and caused explosion. On 10th January 2018, the second incident occurred nearly 45 km from the previous failure joint

Copyright © 2019 by ASME

in which a gas leak was detected, no explosion and no injuries were reported.

As part of the integrity management program, PETRONAS employs Structural Reliability (SRA) and Quantitative Risk Assessment (QRA) QRA for SSGP.

NOMENCLATURE

KP	kilometer point
POF	probability of failure
COF	consequence of failure
M	total mass
t	defect depth
F	force
σ_m	membrane stress
σ_h	bending stress
PTG	PETRONAS Technical Guideline
PTS	PETRONAS Technical Standard
p	Pressure
D	Diameter
Kr	Material toughness
Sr	Stress ratio
M	Folias factor

2.0 SSGP DESIGN INTENT

SSGP's design, construction and testing was performed according to ASME B31.8 [1] whereby the pipeline intent was to transport dry gas. Subsequently, post commissioning of SSGP in February 2014, the pipeline's integrity is managed as per PTG 11.36.01 [2]/ASME B31.8S [3] focusing on risk-based approach.

The pipeline transverse 512 kilometers through Borneo tropical forest. Predominantly a mountainous region with highest elevation of 1,050 meters above mean sea level and over 5,000 river crossings.

Parameters	Section 1	Section 2	Section 3	Section 4
KP	KP0 – KP121	KP121 – KP343	KP343-KP481	KP481-KP503
Distance (502km)	121 km	222 km	138 km	22 km
Launcher/ Receiver Location	SOGT/Lawas (BV106)	Lawas/Long Lama (BV115)	Long Lama/BCS (BV121)	BCS/MLNG
BV stations	6 BVs	9 BVs	6 BVs	1 BV
Diameter	36in.			
Design Pressure	96 barg			
MAOP	96 barg			
Operating Pressure	66 – 83 barg			
Operating Temperature	28 deg C			
Material	API 5L Grade X70 PSL 2			
Location Class	Class 1, 2 & 3			
Wall Thickness	Class 1 = 14.23mm; Class 2 = 16.8mm; Class 3 = 19.8mm			

Table 1- SSGP's parameters

Wall thickness calculation as per ASME B31.8 [1] is influenced by the location class. SSGP was identified to have 3 different location classes i.e. 1, 2 and 3. The weld design is in accordance to API 1104 [4].

3.0 INSPECTION AND MONITORING PROGRAM

SSGP embarks on pipeline inspection using ILI tool. An MFL and IMU combo tool was selected to capture and record pipeline anomalies and suspicious bends as well as high strain pipes. As part of the established SSGP Inspection and Maintenance Plan (IMP), MFL and IMU tool shall be conducted on annually and 6-monthly basis; respectively. The anomalies detected by MFL can either be metal loss, girth weld anomalies, mill defects or mechanical defects.

As for IMU, exposure of SSGP towards geohazard makes the curvature inspection data vital in estimating the local strain value in which are used to calculate the probability of failure of girth welds. The radius of curvature reported from IMU are categorized into 3 categories:

- Radius less than 250D categorize as 'a'; require immediate check and control
- Radius more than 250D and less than 500D categorize as 'b'
- Radius more than 500D; categorize as 'c'

Figure 2 – IMU and EMFL Tool (courtesy of VDT)

Apart from pipeline inspection, SSGP also embark on Inspection Monitoring Maintenance Repair (IMMR) whereby walk down of the pipeline ROW is consistently done to identify any change of geohazard towards pipeline. This is supplemented by LiDAR survey and aerial surveillance. This geotechnical inspection information are then consolidated to be reflected in the geohazard mapping depicting the severity of the Damage Potential (DP) and Terrain Type (TT).

4.0 STRUCTURAL RELIABIITY ANALYSIS (SRA)

Probabilistic models are developed to determine the risk of leak or rupture due to threats prevalent in SSGP. The assessment is performed using Monte Carlo simulation by substituting a range of values with a probability distribution for any factor that has inherent uncertainty.

Copyright © 2019 by ASME

4.1 Target Reliability (TR)

Target Reliability (TR) is used as "acceptance criteria" for the calculated probabilities of failure to ensure that the likelihood of any failure mode occurring is comparable to general failure statistics for onshore pipeline. The guidance described in ISO 16708 [5] for determining an appropriate target reliability for an onshore pipeline, given its assigned safety class, has been adopted. The appropriate reliability target is estimated based on the following equations:

Safety class 1(low) $$P_{f,target} = \frac{5\times10^{-3}}{P\cdot D^3} \quad (1)$$

Safety class 3 (high) $$P_{f,target} = \frac{5\times10^{-5}}{P\cdot D^3} \quad (2)$$

For SSGP, the target reliability (TR) of weldment and other anomalies should be 1x10^{-6} per km-year at populated areas (Class 3) and 1x10^{-4} per km-year at other areas (Class 1). The recommended reliability target of 1x10^{-4} will be considered if the QRA shows that risk to human life is ALARP.

4.2 Limit State Equation

The SRA used for SSGP is under probabilistic approach. The limit state equation used for each threat is identified and explained as follows.

4.2.1 Corrosion

Corrosion threat is assessed based on limit state equation from ASME B31G [6].

$$\sqrt{1+0.6275\left(\frac{l}{\sqrt{Dt}}\right)^2-0.003375\left(\frac{l}{\sqrt{Dt}}\right)^4} \quad (3)$$

From ASME B31G, it specifies that user may choose to conduct the analysis based on a Level from 0 to 3. The Level to choose is depending on the quantity and quality of the data available. In this assessment, Level 1 is chosen.

4.2.2 External Interference

The failure modes considered in the case study stemmed from damage due to external interference. Such external interference may result in either a puncture of the pipe wall or in a dent and/or gouge in the pipe wall. The metal loss, and associated stress concentration and intensification, corresponding to a gouge may result in failure of the pipe wall ligament under internal pressure loading.

External interference to buried onshore pipelines is most usually caused by excavators which the damage may result in:

- puncture due to penetration of the pipe by an excavator bucket tooth
- a gouge and/or dent in the pipe wall

For the above damage scenarios, the model extracted from UK HSE titled Investigation of numerical anomalies associated with Transco Pipeline Upgrade [7] were used.

a. Puncture Model

The probability of puncture by an excavator bucket tooth is assessed as follows.

The force applied to the pipeline is assumed to be related to the mass of the excavator by an empirical equation :

$$F = 5.625\ M \quad (4)$$

Where M is the total mass of the excavator in tonnes and F is the force applied to the pipeline in kN.

The force that the pipeline can resist is calculated using the following limit state function given as,

$$r = \left[1.17 - 0.0058\cdot\frac{R}{w}\right]\cdot(L_t + w_t)\cdot w\cdot\sigma_u \quad (2)$$

Equation above is then modified as,

$$r = \left[1.17 - 0.0058\cdot\frac{R}{w}\right]\cdot(L_t + w_t)\cdot w\cdot\sigma_u + C \quad (3)$$

Where C is an additive error term to be described by an appropriate probability distribution. The limit state function for puncture is then defined as,

$$F\text{-}r = 0 \quad (4)$$

b. Gauge/Dent Model

Gouge/dent defects are represented by gouge depth and length, and dent depth. The dent depth could be zero when a gouge occurs without a dent.

These consequences resulted as presence of a gouge defect:

- The gouge depth may be of a sufficient to cause a leak occurred.
- If the gouge length exceeds a critical value, then rupture will occur.
- Neither leak nor rupture will occur if the gouge depth and length are less than their respective critical values (although there may be fatigue implications, not considered here, resulting from pressure cycling).

Copyright © 2019 by ASME

To solve the problem following two failures functions are required,

- Critical gouge depth and
- Critical gouge length

A. Critical gouge depth

Gouges situated in dents are assessed using a fracture mechanics approach that assumes that a gouge behaves as a crack. The failure function is given by:

$$K_r = S_r\left[\frac{8}{\pi^2}\ln\left\{\sec\left(\frac{\pi}{2}S_r\right)\right\}\right]^{-1/2} \tag{5}$$

Where Kr and Sr are dimensionless material toughness and stress ratio parameters, respectively. Their compositions are explained below. Before that, it is more convenient to rearrange this function and note that the depth of a gouge exceeds a critical value if:

$$\cos\left(\frac{\pi}{2}S_r\right) - \exp\left[-\frac{\pi^2}{8}\cdot\frac{S_r^2}{K_r^2}\right] < 0 \tag{6}$$

Kr and Sr are given by:

$$K_r = \frac{[\sigma_m Y_m(a,w) + \sigma_b Y_b(a,w)]\sqrt{\pi a}}{K_{IC}} \tag{7}$$

$$S_r = \frac{\sigma_m\left(1-\frac{a}{Mw}\right)}{\sigma_f\left(1-\frac{a}{w}\right)} \tag{8}$$

In these equations σm is a membrane stress given by:

$$\sigma_m = \sigma_h\left(1-1.8\frac{D}{2R}\right) \tag{9}$$

σ_h is a bending stress, present due to dent of depth D, given by:

$$\sigma_b = 10.2\sigma_h\frac{D}{2w} \tag{10}$$

And σ_b is the hoop stress in the pipe wall, given by:

$$\sigma_h = \frac{PR}{w} \tag{11}$$

Since the dent depth D is assumed to be zero here, σm = σh, and σb=0.

The quantities Ym and Yb are functions of gouge depth and pipe wall thickness and are normalised stress intensity factors for an edge-cracked strip in tension and bending, respectively. (The expression for Yb is given for completeness but is not used since σb= 0). They are given by:

$$Y_m = 1.12 - 0.23\left(\frac{a}{w}\right) + 10.6\left(\frac{a}{w}\right)^2 - 21.7\left(\frac{a}{w}\right)^3 + 30.4\left(\frac{a}{w}\right)^4 \tag{12}$$

$$Y_b = 1.12 - 1.39\left(\frac{a}{w}\right) + 7.3\left(\frac{a}{w}\right)^2 - 13.0\left(\frac{a}{w}\right)^3 + 14.0\left(\frac{a}{w}\right)^4 \tag{13}$$

The Folias factor M is given by:

$$M = \left[1 + 0.26\frac{L^2}{Rw}\right]^{1/2} \tag{14}$$

Where L is the gouge length. The material flow stress σfis expressed in terms of the material yield and ultimate strengths of σy and σu as:

$$\sigma_f = \alpha(\sigma_y + \sigma_u) \tag{15}$$

Where a is a flow stress parameter.

The material fracture toughness, K, is found from the Charpy energy following correlation:

$$K_{IC} = \left(\frac{EC_{v0}}{A}\right)^{1/2}\left(\frac{C_v}{C_{v0}}\right)^{1/2b} \tag{16}$$

Where E is Young's modulus of the material, A is the area of the Charpy test specimen, b is a dimensionless parameter, and Cv0is a reference Charpy energy.

Copyright © 2019 by ASME

B. Critical Gouge Length

If the depth of a gouge exceeds a critical value, then a through-wall defect results. If the length of such a through-wall defect exceeds a critical value, then rupture will occur. Such a situation obtains if:

$$\sqrt{\left[\left(\frac{\sigma_h}{1.150\sigma_y}\right)^{-2}-1\right]\cdot\frac{Rw}{0.4}}-L<0 \qquad (17)$$

W

The terms within the squared-root sign can be interpreted as representing a critical length Lc.

It is worthwhile noting that, whilst the failure function for gouge depth incorporates both plastic collapse and fracture (i.e. incorporates both flow stress and fracture toughness), the failure function for gouge length appears to only involve collapse.

4.2.3 Girth Weld Anomalies with Geohazard

The methodology for geohazard POF calculation is based on fracture mechanics approach from ASME FFS-1/API 579-1 Fitness for Service [8]. The limit state considered pipeline characteristic including wall thickness, girth weld anomalies and external forces in the assessment.

The inspection data used in the assessment include MFL ILI and IMU inspection. All locations with bend stress reported by IMU indicate probability that the pipeline have moved or forced aligned during construction fit-up. For location not detected by IMU, bend stress is assumed as negligible unless reported in other assessment report i.e. pipe to soil interaction study by Geotech. Geohazard Mapping (DPTT updated as of 27 February 2018 is referred for estimation of damage potential onto the pipelines.

These assessment procedures are based on the Failure Assessment Diagram (FAD) method. Crack-like flaws are planar flaws that are predominantly characterized by a length and depth, with a sharp root radius. Crack-like flaws may either be surface breaking, embedded, or through-wall. Examples of crack-like flaws include planar cracks, lack of fusion and lack of penetration in welds, sharp groove-like localized corrosion, and branch type cracks associated with environmental cracking [6].

As the conservative approach, it is assumed that the crack-like flaw is subject to loading conditions and/or an environment that will not result in crack growth. The assessment considered the material fracture toughness, yield strength, wall thickness, defect height and the applied loading on to the pipeline.

The FAD is defined using the following equation:

$$K_r=\left(1-0.14\left(L_r^P\right)^2\right)\left(0.3+0.7\exp\left[-0.65\left(L_r^P\right)^6\right]\right) \quad for \quad L_r^P \le L_{r(max)}^P \qquad (18)$$

Lr(max) is determined as below:

$$L_{r(max)}^P=\frac{\sigma_f}{\sigma_{ys}} \qquad (19)$$

The parameter Kr and Lr is defined by:

$$K_r=\frac{K_I^P}{K_J} \qquad (20)$$

$$L_r=\frac{\sigma_{ref}}{\sigma_{ys}} \qquad (21)$$

The stress intensity factor, K_I and reference stress, σ_{ref} solutions are for surface crack at circumferential direction due to internal pressure and external axial force. This is obtained from API 579-1/ASME FFS-1 [8]. The equation proposed as below:

$$K_I=G_o\left(\frac{pR_i^2}{R_o^2-R_i^2}+\frac{F}{\pi\left(R_o^2-R_i^2\right)}\right)\sqrt{\frac{\pi a}{Q}} \qquad (22)$$

, and

$$\sigma_{ref}=\frac{P_b+\left(P_b^2+9\left\{Z\cdot P_m\cdot(1-\alpha)^2\right\}^2\right)^{0.5}}{3(1-\alpha)^2} \qquad (23)$$

The equation for the ratio of induced bending stress to applied membrane stress for the circumferential joints of pipelines with centerline offset (misalignment) is defined as:

$$R_b^{cajc}=\mathbf{abs}\left[\frac{12}{R_1t_{1c}}\left(0.25672R_2t_{2c}\left\{\frac{C_1}{C_3}\right\}+\frac{eR_a}{2}\left\{\frac{C_2}{C_3}\right\}\right)\right] \qquad (24)$$

Hence, the bending stress due to misalignment can be calculated as:

$$\sigma_b=\sigma_m\,(Rb) \qquad (25)$$

4.2.4 Pipeline Instability (Buoyancy)

The methodology for pipeline instability frequency calculation is based design criteria for pipeline stability concept in which total pipe weight must be more than 1.2 times buoyancy forces. Based on basic buoyancy calculation in reference with PTS 11.30.07 [9], a general limit state function as per following paragraphs describe the relationship between

Copyright © 2019 by ASME

buoyancy force (Fbuoyancy1,2), pipe weight (Woverall1,2) to the pipeline stability.

Calculation for the buoyancy force:

$$F_{buoyancy1,2} = \rho_{water} g A_{overall1,2} \quad (26)$$

$$F_{buoyancy1} = \rho_{water} g A_{overall1} \quad (27)$$

$$F_{buoyancy2} = \rho_{water} g A_{overall2} \quad (28)$$

$$A_{overall1} = \frac{\pi(OD_{cc}^{2})}{4} \quad (29)$$

$$A_{overall2} = \frac{\pi(OD_{cwc}^{2})}{4} \quad (30)$$

$$OD_{cc} - OD_{pipe} + 2t_{cc} \quad (31)$$

$$OD_{cwc} = OD_{pipe} + 2t_{cc} + 2t_{cwc} \quad (32)$$

Calculation for the pipe weight:

$$W_{overall2} - W_{product} + W_{pipe} + W_{cc} + W_{cwc} \quad (33)$$

$$W_{overall2} = \rho_{product} g \left[\frac{\pi(OD_{pipe} - 2t_{wt})^{2}}{4}\right] + \rho_{pipe} g \left\{\frac{\pi\left[OD_{pipe}^{2} - (OD_{pipe} - 2t_{wt})^{2}\right]}{4}\right\} + \rho_{cc} g \left\{\frac{\pi\left[OD_{cc}^{2} - OD_{pipe}^{2}\right]}{4}\right\} + \rho_{cwc} g \left\{\frac{\pi\left[OD_{cwc}^{2} - OD_{cc}^{2}\right]}{4}\right\} \quad (34)$$

$$W_{overall1} = W_{product} + W_{pipe} + W_{cc} \quad (35)$$

$$W_{overall1} = \rho_{product} g \left[\frac{\pi(OD_{pipe} - 2t_{wt})^{2}}{4}\right] + \rho_{pipe} g \left\{\frac{\pi\left[OD_{pipe}^{2} - (OD_{pipe} - 2t_{wt})^{2}\right]}{4}\right\} + \rho_{cc} g \left\{\frac{\pi\left[OD_{cc}^{2} - OD_{pipe}^{2}\right]}{4}\right\} \quad (36)$$

Calculation for pipeline stability:

$$W_{soil} L_{buried} + W_{overall1} L_{buried} + W_{overall2} L_{expose} \geq 1.2\left(F_{buoyancy1} L_{buried} + F_{buoyancy2} L_{expose}\right) \quad (37)$$

For the SSGP buoyancy calculation, it is assumed that one spool length of pipe is exposed and submerged in water, with additional half spool on each end is submerged and buried in soil.

5.0 QUANTITATIVE RISK ASSESSMENT (QRA)

Risk is considered as the likelihood or chance of individual being harmed by a hazard. The level of risk is based on a combination of the likelihood of the event happening and the consequence of the event which expressed as frequency or probability. Results from the Structural Reliability Analysis in Section 5.0 is used for the QRA.

For SSGP, the failure frequencies according to the following prevalent threats were considered;

1. Corrosion (Internal/External)
2. External Interference
3. Girt Weld Anomalies with Geohazard
4. Pipeline Instability (Buoyancy)

The Methodology is based on PTS 16.71.04 [10] as described in Figure 3.

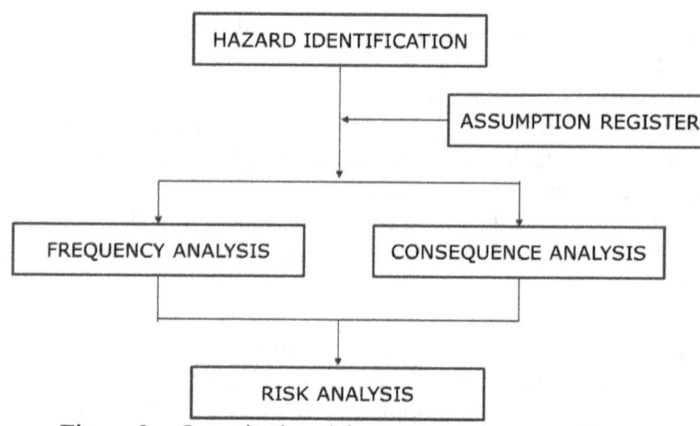

Figure 3 – Quantitative risk assessment process flow

Population data along the ROW was gathered through observation of aerial imaging together with field reports from operations team of any buildings with potential human occupancy of up to 1 mile away from pipeline ROW.

5.1 Frequency Analysis

The frequency values for the QRA are taken at the design conditions. The values for specific onshore pipeline hazards identified in Section 4.2 are compared to historical data of UK Onshore Pipeline Operators' Association (UKOPA) [11] and SSGP Failure History.

Product Loss Cause	No. of Incidents	Incident rate
Girth Weld Defect	36	4.00E-05
External Interference	43	4.78E-05
Internal Corrosion	2	2.22E-06
External Corrosion	42	4.67E-05
Unknown	7	7.78E-06

Copyright © 2019 by ASME

Other	43	4.78E-05
Pipe Defect	13	1.44E-05
Ground Movement	7	7.78E-06
Seam Weld Defect	3	3.33E-06
Total	**196**	2.18E-04

Table 2 – Incident rates used for SSGP

By multiplying with the respective probability of failure of each Location Class with the associated incident rate, then the risk or failure frequency per km-year is calculated as following equation:

$$Failure\ Frequency = Probability\ of\ failure \times number\ of\ incidents\ per\ km.yr \quad (38)$$

5.2 Consequence Analysis

Consequence analysis or also known as consequence modeling evaluates the impact to human, equipment, structure and environment if an event occurs.

The consequence analysis covers the following:

- Orientation of the release.
- A standard release rate equation with coefficient of discharge, 0.8. For rupture case, a pipeline model is used to determine the release rate of the inventory in the pipeline.
- Determination of consequence distance such as jet fire, flash fire, gas dispersion etc. Calculation of thermal radiation which affect the surrounding population.

In analyzing the consequence analysis, PETRONAS use the software PHAST by DNVGL

5.3 Location Specific Individual Risk (LSIR)

LSIR is used to indicate the risk at a particular location. It is the risk for a hypothetical individual who is positioned in the specified location for 24 hours per day, 365 days per year. In SSGP, the geographical variation of LSIR is represented by iso-risk contour plot. In determining LSIR, appropriate failure frequency is calculated based on the Location Class hence it varies according to the location of concern.

Onshore facility in Malaysia shall comply with the Malaysia's Department of Environment's risk acceptance criteria [13]:

- The 1E-05 per year LSIR contour must not extend beyond the site boundary.
- The 1E-06 per year LSIR contour must not reach residential area.

The risk can be rephrased as:

For residential area, a fatality in a residential area must not exceed 1 chance in a million per person per year. Meanwhile for the industrial area, the risk must not exceed 1 chance in 100,000 per person per year.

Figure 4 – Sample of risk contour for SSGP

6.4 Societal Risk

Societal risk is the relationship between the frequency of an event and the number of people affected. SR thus gives the number of individuals at risk. SR from a major hazard facility can thus be represented by F-N curve, which is plot of the cumulative frequency (F) of various accident scenarios against the number (N) of casualties associated with the modeled incidents.

The plot is cumulative in the sense that, for each frequency, N is the number of casualties that could be equaled or exceeded. Often 'casualties' are defined in a risk assessment as fatal injuries, in which case N is the number of people that could be killed by the incidents. Example of F-N curve is given below:

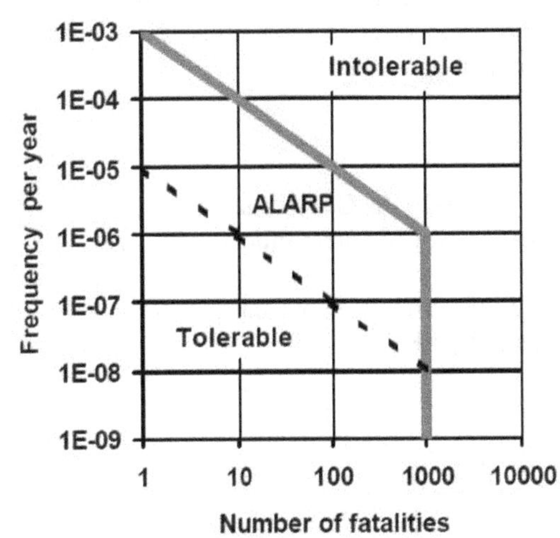

Figure 5 – Societal risk acceptance criteria

Copyright © 2019 by ASME

6.0 DISCUSSION

SSGP has over 42,000 joints installed. With the challenging terrain it traverses, and ever changing geohazard condition, a comprehensive inspection program shall be required to complement the risk and pipeline assessment.

Prevalent multiple threats acting on SSGP require accurate assessment methodology to ensure all resources are focus at the correct locations and appropriate risk reduction measures are correctly identified and applied. The limit state equations are identified from established references whereby probabilistic approach is used to ensure uncertainties are accounted for while Monte Carlo simulation to calculate the probability of failure.

The methodology as explained in this paper have been successfully implemented for qualitative risk assessment of SSGP which is exposed to multiple threats from geohazard, corrosion, external interference, hydrology and girth weld anomalies.

7.0 CONCLUSIONS

The learnings from SSGP failure incidents have shown that SSGP Quantitative Risk Assessment needs to consider the failure probability of multiple threats. Assessing and to a certain extent predicting SSGP risk proven to be a challenging task as shown by the two incidents that had already experienced, however with a better understanding of the root cause and development of a structured SRA and QRA methodology tailored made for SSGP and complemented by relevant inspection program, the risk management of SSGP will ensure continuous safe operations.

8.0 ACKNOWLEDGMENTS

The authors acknowledge our colleagues at PETRONAS for their assistance in performing the work upon which this paper has been prepared and the permission by management to publish this paper.

REFERENCES

[1] ASME B31.8-2010 Gas Transmission and Distribution Piping System by ASME International.

[2] PTG 11.36.01-2015 Pipeline Integrity Management System (PIMS) Manual by PETRONAS.

[3] ASME B31.8S-2018 Managing System considering the integrity of Gas Pipelines standard by ASME International.

[4] API 1104-2013 Welding of Pipelines and Related Facilities by API.

[5] ISO 16708:2006 Petroleum and natural gas industries – Pipeline transportations system – Reliability-based limit state methods by ISO.

[6] ASME B31G-2012 Manual for Determining the Remaining Strength of Corroded Pipelines by ASME.

[7] Turner, Richard. Investigation of numerical anomalies associated with Transco Pipeline Upgrade. s.l. : UK HSE, 2002. RR 021.

[8] ASME FFS-1/API 579-1:2016 Fitness for Service by API.

[9] PTS 11.30.07-2017 Pipeline and Riser Engineering Design by PETRONAS.

[10] PTS 16.71.04-2017 Quantitative Risk Assessment (QRA) by PETRONAS.

[11] UK Onshore Pipeline Operators' Association (UKOPA)-2018 UKOPA Pipeline Product Loss Incidents and Faults Report (1962 – 2016) by UKOPA.

Copyright © 2019 by ASME

AIMPIMG2019-1048

Chapter 42

A PRACTICAL ASSESSMENT OF PIPELINE RISK UNDER EPISTEMIC UNCERTAINTY: A HYDRO-TECHNICAL THREAT

Mona Abdolrazaghi
Enbridge
Edmonton, Alberta, Canada

Sherif Hassanien
Enbridge
Edmonton, Alberta, Canada

ABSTRACT
In formulating integrity decisions for oil pipelines near water bodies against stresses and strains beyond their intended designs, the estimated risk in terms of potential loss of containment and adverse consequences is a critical task for safe operations. In practical integrity situations, the epistemic uncertainty in the estimated risk is equally significant and relevant as aleatory uncertainties. It is the integrity engineers' responsibility to provide a proper reliability and risk technical information to the decision maker and stakeholders in order to achieve the highest level of safety. This paper discusses a process of practical analysis to evaluate the risk distribution associated with imperfect information and epistemic uncertainties.

INTRODUCTION

There are two main categories of the uncertainties: epistemic (or knowledge-reliant) and aleatory (random) uncertainty [1, 2]. Aleatory uncertainty represents the random variability in a model. This type of uncertainty may never be eliminated because of its random nature. Examples of aleatory uncertainty include variation in pipe manufacturing processes, signal fluctuations in sensor data collected from inline inspections, localized pipeline operating pressure variations and precise voltage levels in cathodic protection (CP) systems. Aleatory uncertainty has been the main focus in pipeline reliability analysis where the use of probabilistic distributions and analyses (either through simulation or analytical means), have been the main strategy for incorporating uncertainties into decision making processes. However, many instances and recent development into the quantification of uncertainties for the purpose of probabilistic modeling fall into the category of epistemic uncertainty. The basis of epistemic uncertainty is a lack of knowledge; in contrast to aleatory uncertainty it is not random and may be reduced as more knowledge is gained. Examples of epistemic uncertainty include model error, depth-dependent biases in ILI measurements, human factor risks, and the reliance on SME experience [3, 4]. In this paper, a number of geohazard uncertainties in the probabilistic assessment are considered.

Epistemic uncertainty process is expected to increase the confidence in integrity decisions and support risk-averse measures. Specifically, hypothetical onshore pipeline is assumed to be exposed to water due to erosion and a lake natural expansion. Accordingly, the pipeline deflected in both horizontal and vertical directions with no downward support. This unplanned situation excreted stress and strains on the pipe that might be beyond its design limitations. A deterministic stress-based and/or strain-based analysis can be conducted in order to assess the pipe response, address any integrity concerns, and plan for remediation. Moreover, a reliability analysis is necessary to evaluate the risk of exceeding allowable stresses/strains and potential loss of containment under both aleatory and epistemic uncertainties. The epistemity arises from the lack of knowledge and/or imperfect information about: evidence of presence of several factors including circumferential cracks in/at girth welds and other manufacturing anomalies, potential material depletion due to ageing/cycling and high/low weld misalignment, weld properties mismatch, and potential progressive erosion over time. The integrity reliability was evaluated by simulating such uncertainties using renewal stochastic processes. The distribution of the probability of failure (leak and burst failure modes) are then multiplied by the consequence factor that include multiple factors such as anticipated environmental, health and safety, operation disruption, operator's reputation consequences in order to estimate the risk distribution. Risk information could be then the essence of integrity/remediation plans for this site.

The integrated risk-informed decision-making (IRIDM) [5, 6, 7] is a deliberative process that uses a set of performance measures, together with other considerations, to "inform" the decision maker. In this paper, a probabilistic assessment is completed to be utilized as part of IRIDM process [8].

Copyright © 2019 by ASME

CASE STUDY

In this paper, a hypothetical pipeline within a hydro-technical site remediation is discussed. This pipeline (Spiral X52, Dia. 48", WT 0.406", 1972 construction) is routed near the shore of a lake. During caliper In-line inspection (ILI), it was noticed that the lake has naturally expanded and a section of the pipeline (approximately 50ft) is currently passes through the lake; the pipeline has been deflected both vertically and laterally reported by IMU 5ft, 25ft; respectively.

The site analysis reported: (1) no pipeline movement was observed in the most recent most recent ILI inspections within a span of three year, (2) field survey indicated that the pipeline is stable at the lake area, and (3) the pipeline may have settled in place for quite some time. However, the soil has a potential of being unstable due to ice jacking, frost-heave, and/or thaw settlement. Hence, there could be future pipeline movement.

A finite element analysis (FEA) is conducted and reports longitudinal stress, Von Mises stress, and longitudinal strain. Based on the FEA results, three modes of failure (limit state) are chosen for the probabilistic analysis; namely:

- **Strain Based Rupture**: Longitudinal Strain Demand and Longitudinal Strain Capacity
- **Stress Based Rupture**: Longitudinal Stress Demand and Longitudinal Stress Capacity (Ultimate Tensile Strength)
- **Yielding:** Von Mises Stress Demand and Yield Strength Capacity

The stress based rupture indicates that the base metal has reached (or exceeded) the ultimate tensile strength of the parent material. Yielding limit state only signifies that the base metal has yielded and there is permanent deformation, and the corresponding *PoF* is essentially a *PoE* (Probability of Exceedance). The tensile strain limit state is reached when the strain demand exceeding the pipe strain capacity at a girth weld flaw. The failure modes and corresponding limit states are summarized in Table 1. Of these three modes, the strain based rupture is considered the most appropriate limit state for integrity reliability assessment. The tensile capacity model was based on CRES' model [9] which is a modification of CSA-Z662 model [10].

Table 1: Failure modes and limit states considered in the analysis

Failure Mode	Demand Parameter	Capacity Parameter	Limit State	*PoF* Designation
Strain Based Rupture	Tensile Strain (ε_T)	CRES Strain Capacity (ε_{CRES})	$LS_{\varepsilon_T}=\varepsilon_T - \varepsilon_{CRES}$	$PoF^{\varepsilon}_{Rupture}$
Stress Based Rupture	Tensile Stress (σ_T)	Ultimate Tensile Strength (σ_{UTS})	$LS_{\sigma_T}=\sigma_T - \sigma_{UTS}$	$PoF^{\sigma}_{Rupture}$
Yielding	Von Mises Stress (σ_{VM})	Yield Strength (σ_{YS})	$LS_{\sigma_{Yield}}=\sigma_{VM} - \sigma_{YS}$	$PoF_{Yielding}$

The probability of failure is calculated at the highest strain demand using limit state approach (**Figure 1**).

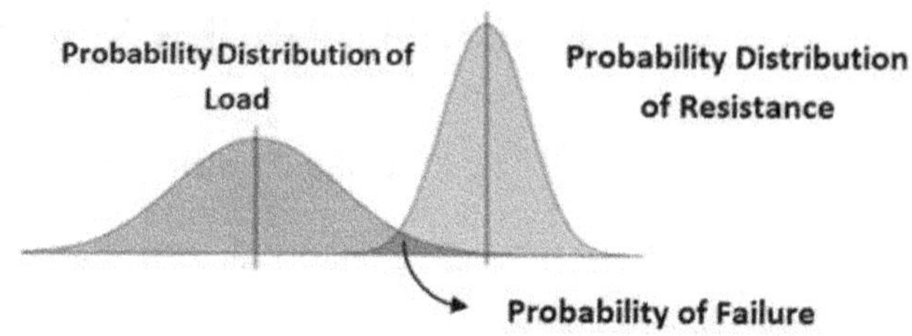

Figure 1: Limit state approach

The input parameters for strain capacity are based on CRES report and summarized in **Error! Reference source not found.**. The statistical uncertainties associated with these parameters are listed in **Error! Reference source not found.**.

Table 2: Strain capacity input parameters

Parameter	Value	Reference
Weld Strength under-matching	10%-20%	[3]
Crack flaw height	3mm	API 1104 (girth weld inspection)
Crack flaw depth	25mm	API 1104 (girth weld inspection)
High-Low misalignment	0.8mm and 1.6mm	Vintage girthweld
CTOD	Average-1STDEV=0.61mm	Mean= 0.98mm and STDEV=0.37mm

Table 3: Statistical uncertainties of demand and capacity variables

	Description	Distribution Type	Mean (μ)	COV	Truncation Limits
ε_{CRES}	Strain Capacity	Log Normal	0.65%, 0.83%, 1.16%, 1.49%	5%	μ*CoV
ε_T	Strain demand	Log Normal	CRES Analysis	10%	μ*CoV
σ_{UTS}	Stress rupture capacity	Normal	80.5 Ksi	N/A	4.5 Ksi
σ_T	Stress rupture demand	Log Normal	CRES Analysis	10%	μ*CoV
σ_{YS}	Stress yield capacity	Normal	56 Ksi	N/A	3.14 Ksi
σ_{VM}	Stress yield demand	Log Normal	CRES Analysis	5%	μ*CoV

Copyright © 2019 by ASME

The maximum PoF in three failure modes was observed at the pipe bend (Table 4); the stress-based rupture designed showed the maximum PoF value of 3.2E-2 per girth weld; the Yielding PoF showed the maximum PoF value of 1.80E-1 per girth weld (i.e. the probability that the pipeline metallurgy structure passes the elastic mode and enter the plastic mode); The strain-based PoF which is the most appropriate PoF for this calculation has shown the highest PoF value of 1.7E-2 per girth weld; The proposed permissible PoF for tensile strain rupture limit state is 1E-5 per girth weld.

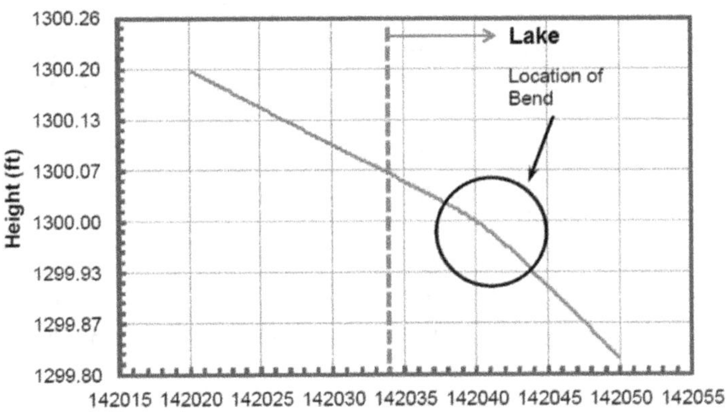

Figure 2: Location of pipe bend on the pipeline

Deterministic analysis has shown that the pipelines are meeting the target except at the pipe 1

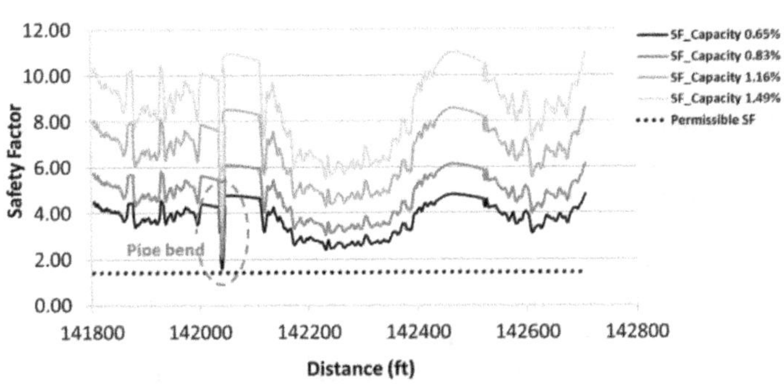

Figure 3: Deterministic Safety Factor Analysis

Probabilistic analysis has shown the high probability of failure at the pipe bend. The preliminary results are updated based on the frequency of circumferential crack, high-low misalignment weld, and weld strength under-match occurrence. PI Diagnostics [5] collected available NDE results and construction repair data based on SME input as summarized in **Error! Reference source not found.**.

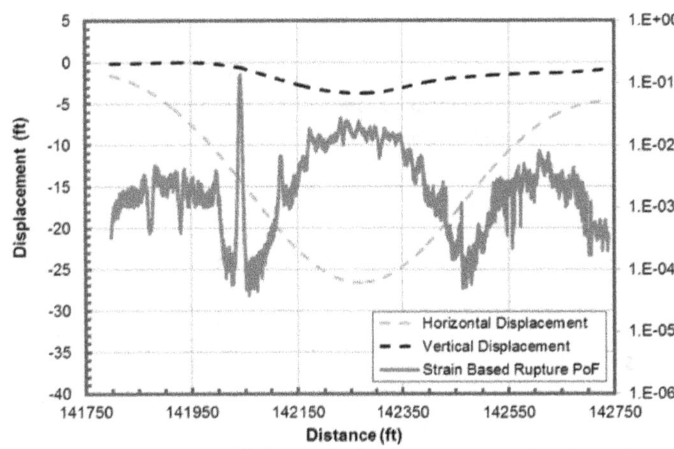

Figure 4: Probabilistic analysis of rupture strain along the displacement

Table 4: Frequency of flaw hazard occurrence per GW [5]

Flaw Hazard	Frequency
Existing a crack size with depth 3 mm and length 25 mm	3%
High/Low misalignment (o.8 mm and 1.6 mm)	3%
Under-matched 10% and 20%	10%

In general, there is no enough statistically evidences about the frequency of manufacturing anomalies; hence, the MP SME opinion was taken into consideration in order to form the lower bound of probability of failure. Accordingly, a design of experiment was conducted to address all possible scenarios of flaw hazards combination. **Error! Reference source not found.** provides the list of the design of experiment, inputs, deterministic demand and capacity, probability of failure, and safety factor of each scenario.

The updated PoF is evaluated using a renewal process system reliability given multiple hazards [11]. If the multiple hazards (circumferential crack, high low misalignment, and weld under-match) are randomly modeled using Poisson process, the updated PoF is given by:

$$PoF_{Pipe-tenisle\ rupture} = \sum_{i=1}^{n} \lambda_i \times PoF_i + \sum_{i=1}^{n}\sum_{j=1}^{n} \lambda_{ij} \times PoF_{ij} + \sum_{i=1}^{n}\sum_{j=1}^{n}\sum_{k=1}^{n} \lambda_{ijk} \times PoF_{ijk} + \cdots$$

where, λ_i and PoF_i are the rate of occurrence and probability of failure of scenario i; respectively. The updated PoF results are summarized in Table 6.

Copyright © 2019 by ASME

Table 5: Pipeline failure for possible cases

Scenario	Circumferential crack with depth 3 mm and length 25 mm exists	High-Low Misalignment	Weld mismatched	Demand (%)	Capacity (%)	PoF	SF
1	Y	0	100%		4.83	2.8E-12	11.92
2	Y	0.8	100%		1.87	6.7E-09	4.62
3	Y	1.6	100%		1.6	4.7E-08	3.95
4	Y	0.8	20%		0.83	1.2E-03	2.05
5	Y	0.8	10%		1.49	1.0E-06	3.68
6	Y	1.6	10%		1.16	1.6E-05	2.86
7	**Y**	**1.6**	**20%**	**0.41**	**0.65**	**1.7E-02**	**1.60**
8	Y	0	10%		1.63	3.8E-08	4.02
9	Y	0	20%		0.94	2.9E-04	2.32
10	N	0.8	10%		5.05	2.4E-12	12.46
11	N	0.8	20%		2.91	8.9E-11	7.18
12	N	1.6	10%		3.73	1.2E-11	9.21
13	N	1.6	20%		2.15	1.3E-09	5.31
14	N	0	10%		6.96	5.5E-13	17.18
15	N	0	20%		4.01	8.8E-12	9.90
16	N	0.8	100%		6.77	6.4E-13	16.71
17	N	1.6	100%		5.00	2.4E-12	12.34

Table 6: Tensile Strain Rupture PoF results

	PoF per girth weld
Lower bound	2.50×10^{-6}
Most Likely (Reliability SME)	$\sim 1.00 \times 10^{-4}$
Upper bound	1.70×10^{-2}

Strain analysis conducted by the external consultant indicates the maximum strain is ~0.41% at the deflected pipeline section where it is below the minimum calculated strain capacity of 0.65% with the safety factor of 1.60 barely above an acceptable safety factor of 1.5. This analysis assumes the existence of a circumferential crack flaw with 3.00 mm depth and 25 mm length at the girth weld along with 20% weld strength under-match and 1.6 mm high-low misalignment. The probability of failure changes between lower bound 2.5E-6 and 1.7E-2 with the maximum likelihood of 1E-4. Assuming the maximum consequence value of $1B at the lake, the risk of failure for this line is approximately $100 K. Using integrated risk informed decision making [8], accounting for deterministic safety factor, probability of failure, risk of pipeline failure, and any other budgetary or other constrains, a sound integrity decision can rise from this assessment.

CONCLUSION

The reliability analysis is completed for three modes of failure (limit states): Stress Rupture, Yielding and Tensile Strain exceedance.

Strain analysis conducted by the external consultant indicates the maximum strain is ~0.41% at the deflected pipeline section where it is below the minimum calculated strain capacity of 0.65% with the **safety factor of 1.60** barely above an acceptable safety factor of 1.5. This analysis assumes the existence of a circumferential crack flaw with 3.00 mm depth and 25 mm length at the girth weld along with 20% weld strength under-match and 1.6 mm high-low misalignment.

Herein, the most representative limit state is the tensile strain rupture. The analysis has indicated that probability of failure is between 2.5E-6 and 1.7E-2 with most likely value of ~1E-4 at the highest strain demand near a girth weld. The main reasons behind the wide range of PoF are:

- anecdotal evidence of the presence/rate of circumferential cracks in/at girth welds and other manufacturing anomalies;

Copyright © 2019 by ASME

- potential material depletion due to ageing/cycling;
- Relatively wide uncertainties of pipe strain capacity.

The PoF's wide range can be reduced to a single value if it can be proven that no circumferential cracks in/at girth welds and any other manufacturing anomalies; with a confidence equals to the confidence of the method of proving (e.g. inspection technology).

This assessment provides a valid basis for integrated risk informed decision making for remediation approach and timeline.

REFERENCES

[1] Haldar A., Recent Developments in Reliability-Based Civil Engineering, World Scientific Publishing, 2006.

[2] Koduru, S., “Strategies for Separation of Aleatory and Epistemic Uncertainties”, 12th International Conference on Applications of Statistics and Probability in Civil Engineering (ICASP12), 2015, Paper No. 472.

[3] Nessim, M., Zhou, J., Fuglem, M.,, 2008, "Dealing with Knowledge Uncertainties in Pipeline Reliability Based Design and Assessment," Proceedings of the International Pipeline Conference, ASME, Calgary, Alberta, Canada, Paper No. IPC2008-64478.

[4] Gelman, A., Carlin, J. B., Stern, H. S., Rubin, D. B., “Bayesian Data Analysis”, 2nd Edition, Chapman & Hall/CRC, 2004.

[5] Aven, T. 2003. “Foundations of Risk Analysis: A Knowledge and Decision-Oriented Perspective”. John Wiley & Sons.

[6] Zio, E. & Pedroni, N. 2012. “Risk Informed Decision Making Processes”. Foundation for an Industrial Safety Culture (FONCSI).

[7] Blenko, M.W., Mankins, M.C., and Rogers, P. 2010. “Who Has the D?: How Clear Decision Roles Enhance Organizational Performance”. Harvard Business Review. January Issue.

[8] S. Hassanien, D. Langer, and M. Abdolrazaghi, “Integrity Risk-Informed Decision Making”, IPC2018-78149.

[9] Liu, Wang, Zhang, and Kotian, Realistic Strain Capacity Models for Pipeline Construction and Maintenance, Center for Reliability Energy Systems (CRES), 2013.

[10] Steel Pipe, CSA-Z 662, March 2014.

[11] S. Hassanien, “Life Cycle Utility Optimization for Structural System (LCUO)”, 2003.

Copyright © 2019 by ASME

AIM-PIMG2019-1050

Chapter 43

IMPLICATIONS OF LIMIT STATE DESIGN CHECKS FOR GEOTECHNICAL LOADS

Smitha D. Koduru[1], Maher A. Nessim[2]
C-FER Technologies (1999) Inc.
Edmonton, Alberta, Canada

ABSTRACT

Limit states design checks have been developed for geotechnical loads using a strain-based design format. The application of these checks requires the user to develop probability distributions for the maximum strain demand and minimum strain capacity. These checks are provided for both local buckling and tensile rupture, and they are calibrated to meet specified risk-consistent reliability targets. This paper presents a detailed description of the safety factors developed as part of the limit states design checks, and the implications of their use in the assessment of pipelines in moving slopes.

Keywords: limit states design, pipe-soil interaction, strain-based design, girth-weld rupture, local buckling, tensile strain, compressive strain

NOMENCLATURE

D	diameter (mm)
L	sliding length in the longitudinal movement direction
SC	safety class
W	sliding width in the transverse movement direction
c	coefficient for allowable probability corresponding to a safety class
m	number of loading locations in a 1.6 km distance for a location-specific loading process
t	wall thickness (mm)
α	product factor
γ	safety factor
θ	movement direction angle (degrees)
τ	time
V_c	strain capacity CoV
V_d	strain demand CoV
r_m	ratio of the mean capacity and mean demand
$pf_{a\text{-}ULS}$	allowable failure probability for ULS
$\Phi^{-1}(.)$	inverse of cumulative standard Normal distribution
ε_c	characteristic strain capacity
ε_{cc}	characteristic compressive strain capacity
ε_{ct}	characteristic tensile strain capacity
ε_d	characteristic strain demand
ε_{dc}	characteristic compressive strain demand
ε_{dt}	characteristic tensile strain demand
γ_c	safety factor for compressive strain
γ_t	safety factor for tensile strain

INTRODUCTION

The limit states design approach is well established as the basis for civil structural design codes and has been used over the last two decades in offshore pipeline standards such as DNV-OS-F101 [1] and API RP 1111 [2]. By contrast, application of the approach to onshore pipelines has been limited.

In the past two decades, however, significant progress has been made on the application of reliability-based and limit states (LS) design to onshore pipelines [3, 4, 5, 6, 7]. The approach has been used as a basis for decision making under uncertainty for defect management based on in-line inspection (ILI) data, mechanical damage prevention for class location changes, and other integrity management issues. The benefits of the approach, which are now recognized by many in the industry, include achieving quantified and consistent safety levels, and identifying the most cost-effective ways to reach a specified safety target.

The LS design approach for onshore pipelines has been developed as part of a multi-year joint industry project (JIP), and a portion of the JIP results were published by Nessim et al. [3], Nessim and Adianto [4], and Adianto and Nessim [5]. As part of the JIP, reliability-based design rules were developed for geotechnical loads. The loading processes explicitly considered in the development included slope movement (slow or sudden), frost heave and thaw settlement. Under these geotechnical loads, a buried onshore pipeline may experience local buckling due to

[1] Contact author: s.koduru@cfertech.com
[2] Chief Engineer and C-FER Fellow: m.nessim@cfertech.com

Copyright © 2019 by ASME

excessive compressive strains and excessive tensile strains that can cause pipeline rupture of the pipe body or a circumferential girth weld flaw.

Based on the LS categories defined in Nessim et al. [3], LS design checks were developed for the following:

1. *Local buckling*. A serviceability limit state (SLS) that applies to the compression side of the pipe. The LS is typically defined at buckle initiation, which is assumed to coincide with the peak of the moment-curvature relationship for a pipeline subjected to combined bending and axial forces. Treatment of local buckling as an SLS is contingent on demonstrating, through detailed analysis or the use of a monitoring and intervention program, that local buckling will not progress to a condition that causes loss of containment.
2. *Rupture*. An ultimate limit state (ULS) that applies to the tension side of the pipe.

In consideration of the modelling complexity of soil movement mechanisms and pipe-soil interaction, and to allow for flexibility to incorporate future model improvements, the design rule formulation developed is dependent only on the distribution parameters of the strain demand and capacity of the pipeline. The format of the LS checks is as follows:

$$\varepsilon_d(\tau) < \gamma\varepsilon_c \quad (1)$$

where ε_d is the characteristic strain demand, ε_c is the characteristic strain capacity, γ is a safety factor and τ is time. It is recognized that most limit state design implementations involve using separate load and resistance factors, however, a single factor format is used here. While the separate load and resistance factor format is effective where multiple checks that utilize different combinations of the same basic loads apply, that format offers no advantage for the current case where two independent ULS and SLS checks are being developed.

Characteristic values of the strain demand and strain capacity are a function of the probability distributions of the strain demand and capacity, pipeline allowable probability of failure and the LS type. The safety factor is defined as a function of the allowable failure probability, the 'coefficients of variation' (CoV) of the strain demand and capacity probability distributions, and the probability distribution type of the strain demand. Separate ULS checks were developed for the pipe body and girth weld to account for differences related to local geometry, material properties and the potential for weld flaws.

The objective of this paper is to present the implications of using the LS design checks for integrity management of pipelines designed for internal pressure according to ASME B31.8 and buried in moving slopes. The paper begins with a description of the LS design checks developed for SLS and ULS, followed by a description of a set of hypothetical pipeline cases used to demonstrate application of the approach. Implications of the LS checks and the practicality of meeting them under different conditions is discussed.

LIMIT STATE DESIGN CHECKS

The LS checks developed are applicable to both the design of new pipelines and the assessment of existing pipelines. The main difference between the two applications is the time period over which the maximum strain demand distribution is defined (referred to here as the "evaluation period") for time-dependent loading processes. This evaluation period equals the future time period over which the pipeline can be operated under a time-dependent loading process, such as slope creep, based on meeting the requirements of the LS check. The evaluation period should be no less than the design life for a new pipeline or the time to the next assessment for an existing pipeline.

The parameter ranges over which the LS checks can be applied, which include the allowable probability and the demand and capacity probability distribution types and CoVs, were selected to cover a wide range of pipeline design and assessment cases. The basis for the distribution types and range of CoVs is explained in greater detail in Nessim and Koduru [8] and Koduru and Nessim [9].

The basis for the allowable probability values and the safety classes used to assign them are provided in Table 1 from Nessim et al. [3]. An approach has also been developed by Nessim and Koduru [8] to convert the per km-year allowable failure probabilities in the table into event-specific allowable probabilities based on the spatial and temporal variabilities characterizing geotechnical loads. For this purpose, each geotechnical loading processes is classified as:

- Random-in-time and Random-location (e.g., landslides along the mountainside);
- Random-in-time and Location-specific (e.g., landslides at a known slope);
- Time-dependent and Random-location (e.g., thaw settlement along frozen-unfrozen interfaces in discontinuous permafrost); or
- Time-dependent and Location-specific (e.g., slope creep).

The event-specific allowable probability is defined 'per event' for random-in-time loads, and 'per location' for random-location loads.

TABLE 1: SAFETY CLASS AND ALLOWABLE FAILURE PROBABILITY (pfa)

Safety Class	Description	Allowable Failure Probability, pf_a (per km-year)
1	Very Low	0.5×10^{-3}
2	Low	1.5×10^{-4}
3	Medium	1.5×10^{-5}
4	High	1.5×10^{-6}
5	Very High	1.5×10^{-7}
6	Extreme	1.5×10^{-8}

Copyright © 2019 by ASME

Serviceability Limit State

The SLS checks are applicable for the following parameter ranges [8]:

- Strain demand: Gamma, Weibull, or Lognormal distribution with CoVs of 0.1 to 1.4;
- Strain capacity: Lognormal distribution with CoV of 0.3; and
- Allowable failure probability: 10^{-1} per event and/or location.

Because of the single strain capacity CoV and the high value of allowable probability, it was possible to calibrate a single LS check for all distribution types and for both the pipe body and girth weld, with the safety factor being defined as a function of the strain demand CoV. The resulting check is as follows:

$$\varepsilon_{dc} < \gamma_c \varepsilon_{cc} \quad (2)$$

$$\gamma_c = 0.865 - 0.272 \times V_d \quad (3)$$

where

ε_{dc} is the characteristic compressive strain demand, defined as the value associated with a probability of exceedance (PoE) of 0.3 on the strain demand distribution;

ε_{cc} is the characteristic compressive strain capacity, defined as the value associated with a PoE of 0.7 on the compressive strain capacity distribution;

γ_c is the safety factor for compressive strain; and

V_d is the strain demand CoV.

Ultimate Limit State

The ULS checks are applicable for the following parameter ranges [8]:

- Strain demand: Gamma, Weibull or Lognormal distribution with CoVs of 0.3 to 1.0;
- Strain capacity: Lognormal distribution with CoVs of 0.15 to 0.75; and
- Allowable failure probability between 10^{-8} and 10^{-2} per event and/or location based on the safety class and the loading process characteristics, as explained in Nessim and Koduru [9].

The checks were developed separately for three subsets of the tensile strain capacity CoVs representing the pipe body (0.15 CoV), girth welds in strain-based design (SBD) pipe[3] (CoV = 0.35 to 0.55) and girth welds in non-SBD pipe (CoV = 0.55 to 0.75). The rationale behind these CoV ranges is explained in Koduru and Nessim [9]. They are provided for guidance only as the LS checks are calibrated as a direct function of the CoV and their applicability for a given CoV value does not depend on the type of pipe or weld. The calibration cases consist of all combinations of the values summarized in Table 2.

[3] This is normally defined as pipe with a minimum wall thickness of 13 mm and "round-house" stress-strain curve behaviour of the material in the longitudinal direction (axial tension).

TABLE 2: PARAMETER RANGES FOR ULS CHECK

Parameter	Values	
Allowable failure probability (per event or location)	10^{-8} to 10^{-2} (7 values at an interval of 10)	
Strain demand CoV	0.3 to 1.0 (8 values at an interval of 0.1)	
Strain capacity CoV	Pipe Body	0.15
	Girth Weld	0.35 to 0.75 (nine values at an interval of 0.05)

The resulting check is as follows:

$$\varepsilon_{dt} < \gamma_t \varepsilon_{ct} \quad (4)$$

For the Gamma or Weibull strain demand distribution types, γ_t is given by

$$\gamma_t = 10^a \quad (5)$$

$$a = a_1 V_d + a_2 V_c + \log 10(pf_{atl-ULS} \times \frac{a_3 V_c^2}{V_d^2} + a_4 V_d) \quad (6)$$

For the Lognormal strain demand distribution type, γ_t is given by

$$\gamma_t = e^b \quad (7)$$

$$b = \phi^{-1}(p_d)\sqrt{\ln(1+V_d^2)} - \phi^{-1}(p_c)\sqrt{\ln(1+V_c^2)} + \phi^{-1}(pf_{atl-ULS}) + \sqrt{(\ln(1+V_d^2) + \ln(1+V_c^2)} \quad (8)$$

$$p_d = 1 - (pf_{a-ULS})^{0.3} \quad (9)$$

$$p_c = (pf_{a-ULS})^{0.3} \quad (10)$$

where

ε_{dt} is the characteristic tensile strain demand, defined as the value associated with a PoE of $(pf_{a-ULS})^{0.3}$ on the strain demand distribution;

ε_{ct} is the characteristic tensile strain capacity, defined as the value associated with a PoE of $1 - (pf_{a-ULS})^{0.3}$ on the strain capacity distribution;

γ_t is the safety factor for tensile strain;

V_d is the strain demand CoV;

V_c is the strain capacity CoV;

a_1 to a_4 are coefficients defined in Table 3;

$\Phi^{-1}(.)$ is the inverse of cumulative standard Normal distribution; and

pf_{a-ULS} is the ULS allowable failure probability estimated as a function of time and location of geotechnical load, which is estimated as follows for slope creep:

Copyright © 2019 by ASME

Time-dependent – Location-specific

$$pf_{a-ULS} = 1.6c \times 10^{-[SC+2]}/(\tau/m) \quad (11)$$

where

- c is a coefficient determined by the safety class (SC); 0.5 for SC = 1 and 1.5 for SC ≥ 2;
- SC is the safety class value from 1 to 6 as described in Nessim and Koduru [8];
- m is the number of loading locations in a 1.6 km distance for a location-specific loading process; and
- τ is the evaluation period (years).

TABLE 3: TENSILE STRAIN SAFETY FACTOR COEFFICIENTS – EQUATION (6)

Coefficient		a_1	a_2	a_3	a_4
Pipe Body ($V_c \approx 0.15$)		-0.4680	-0.2800	0.0135	-0.0095
Girth Weld	$0.35 \le V_c \le 0.55$	-0.2410	-0.5420	0.0070	0.0037
	$0.55 < V_c \le 0.75$	-0.0845	-0.6610	0.0041	0.0130

The safety factor expression used for the Lognormal strain demand distribution is different from the one used for the Gamma and Weibull distributions because the Lognormal distribution has an extended upper tail, making it difficult to find a satisfactory common solution for the low probability levels considered. While the Normal distribution for strain demand was not considered in the calibration because it did not provide the best fit for any of the cases considered, a separate analysis demonstrated that the SLS or ULS checks developed for Gamma and Weibull distributions are conservative for the Normal distribution.

EXAMPLE APPLICATIONS

The LS checks were applied to hypothetical pipeline cases in moving slopes described in Table 4. The safety classes used for each case correspond to gas pipelines in class locations 1 and 2, as well as low vapour pressure (LVP) pipelines in sensitive and non-sensitive environments. Class locations 3 and 4 for gas pipelines were not included because it is unlikely that an unstable slope will exist in these highly populated areas. The safety classes were obtained using the approach described in Nessim and Koduru [8].

Descriptions for the pipeline and soil parameters used are provided in Table 5. Each pipeline and soil combination was assumed to be subject to soil movement in the longitudinal and transverse directions (see Fig. 1). It was found that small changes in movement angle from the idealized cases of longitudinal movement and transverse movement can result in significant changes to pipe response. Therefore, limited deviations in movement direction were accounted for in estimating the strain demand distributions by treating the movement direction (angle, θ, in Fig. 1) as a variable between 0° and 10°.

TABLE 4: BASIC CASES USED IN THE EXAMPLE APPLICATIONS

Case	Pipeline	Soil	Movement Direction	D/t	Safety Classes
1	NPS 30	Clay	Longitudinal	80	1, 2, 3
2	NPS 30	Clay	Transverse	80	1, 2, 3
3	NPS 30	Sand	Longitudinal	80	1, 2, 3
4	NPS 30	Sand	Transverse	80	1, 2, 3
5	NPS 12	Clay	Longitudinal	80	1, 2
6	NPS 12	Clay	Transverse	80	1, 2
7	NPS 12	Sand	Longitudinal	80	1, 2
8	NPS 12	Sand	Transverse	80	1, 2
9	NPS 30	Clay	Longitudinal	40	1, 2, 3

TABLE 5: PIPE AND SOIL PARAMETERS

Parameter	Case Descriptions
Pipeline	NPS 12 / 5.9 MPa / Grade 359
	NPS 30 / 7.9 MPa / Grade 448
Soil	Loose Sand (effective unit weight = 17 kN/m^3; friction angle = 25°)
	Stiff Clay (undrained shear strength = 100 kPa)

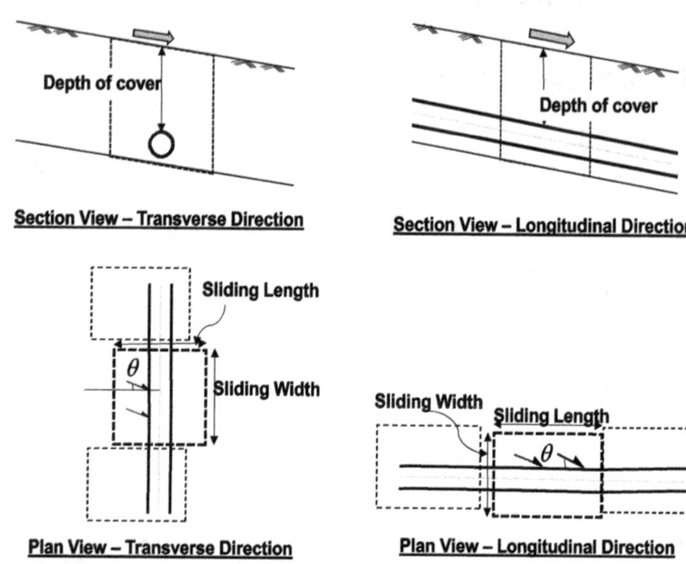

FIGURE 1: SOIL MOVEMENT GEOMETRY AND ORIENTATION

The example cases were designed for internal pressure using ASME B31.8 without considering slope movement and the LS checks were assumed to be applied to the existing pipelines in moving slopes. The maximum strain demand distributions (see Table 3) were then calculated using the methodology described in Koduru and Nessim [9] for a mean slope movement rate of 50 mm per year and a 10-year assessment period. Assuming that

Copyright © 2019 by ASME

the sliding length, sliding width and the movement angle are known for an existing moving slope, specific values were selected for these parameters. For the sliding length in the longitudinal movement direction, a single value of L = 100 m was used, and for the sliding width in the transverse movement direction, a single value of W = 200 m was used. Three movement angles of 0°, 5° and 10° from the principal movement direction were used for both the longitudinal and transverse movement directions. The Lognormal distribution was used to model strain demand.

The results are shown in Figure 2 for the compressive strain (SLS) check and in Figures 3 and 4 for the tensile (ULS) check. The results marked as SBD pipe in Figures 3 and 4 utilize CoVs that are representative of SBD pipe, but do not necessarily have wall thicknesses exceeding 13 mm, per the definition provided previously. These results were included for illustration. It can be seen from these figures that the required capacities increase with movement angle. This is consistent with the findings of the deterministic finite element analyses [9], which demonstrated that small changes in the movement angle from the idealized pure longitudinal or pure transverse directions result in compounding the longitudinal and transverse strains and change the pipe response significantly.

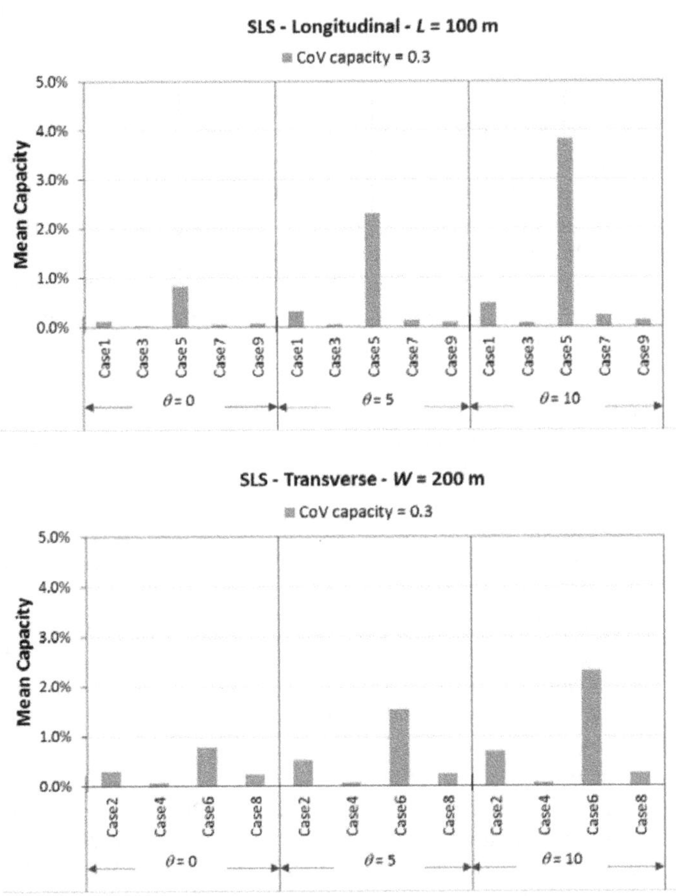

FIGURE 2: MEAN CAPACITY REQUIRED FOR THE COMPRESSIVE STRAIN CHECK

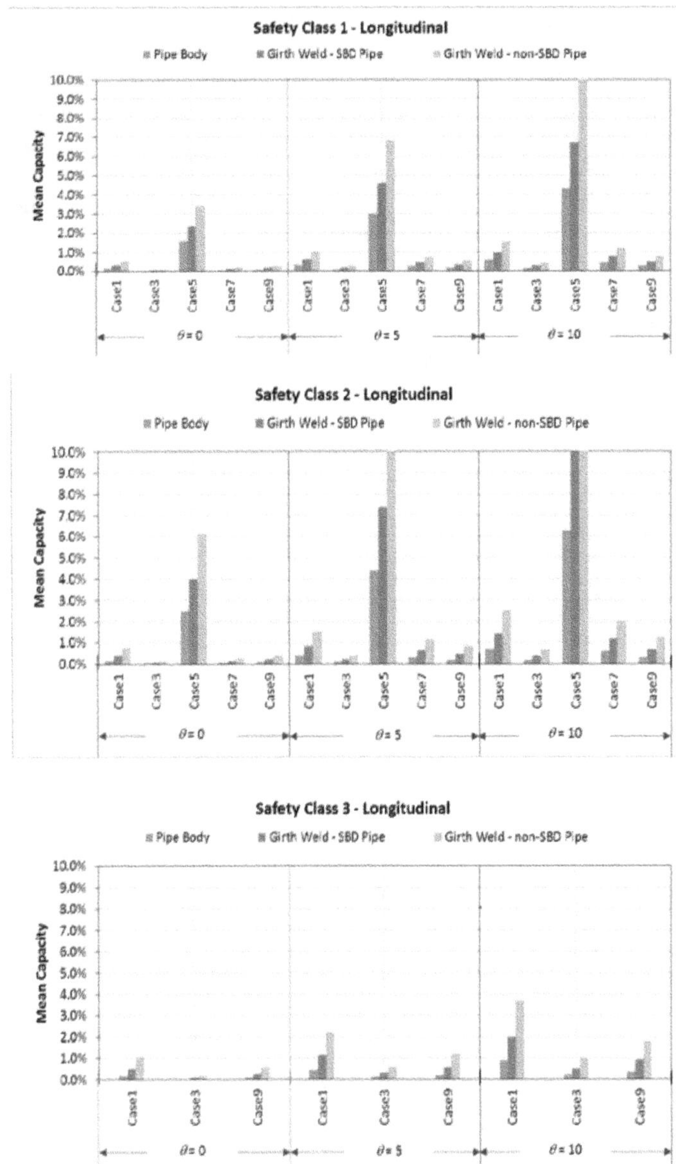

FIGURE 3: MEAN CAPACITY REQUIRED FOR THE TENSILE STRAIN CHECK FOR LONGITUDINAL MOVEMENT

Assuming the mean compressive strain capacity of an existing pipe is less than 0.5%, Figure 2 shows that the LS checks are met in all cases except when the movement angle is 10° and for the case of small diameter pipeline in clay soil. Assuming that mean tensile strain capacity of the pipe body is 7%, LS checks are met for all cases shown in Figures 3 and 4. Assuming that the mean tensile strain capacity of girth welds is 0.5%, LS checks were met mostly for the large diameter pipes in sparsely populated Safety Class 1.

As full-scale experimental tests have shown that girth weld tensile strain capacities greater than 0.5% can be achieved [10, 11], assuming an upper bound tensile strain capacity of 4%, the LS checks can be met for most cases shown in Figure 3. Similar

Copyright © 2019 by ASME

to the observations for the SLS check for compressive strain in Figure 2, ULS checks for small diameter pipeline in clay soil are not met for the tensile strains shown in Figure 3. A similar observation for small diameter in clay soil applies to the ULS checks for the tensile strains shown in Figure 4. In Figure 4, the mean tensile capacity required to meet the tensile strain check exceeds 4% for the large diameter pipe in the clay soil only when the CoV of strain capacity is greater than 0.55. Reducing the variability in girth weld tensile strain capacity through monitoring weld quality and reducing girth weld flaws would reduce the mean tensile strain capacity required to meet the tensile strain checks.

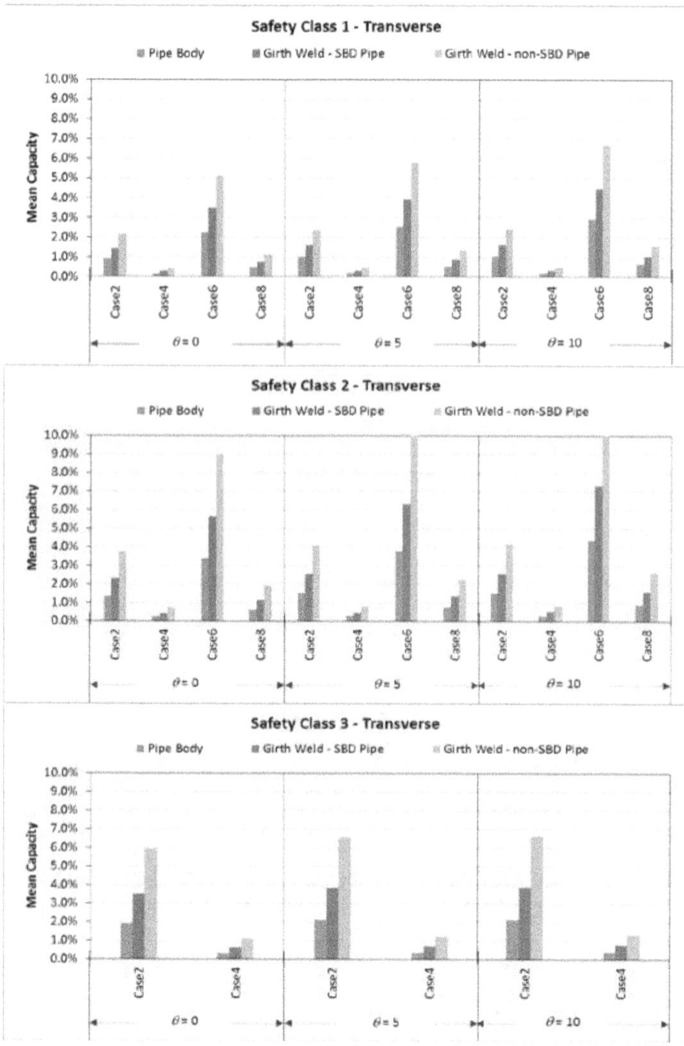

FIGURE 4: MEAN CAPACITY REQUIRED TO MEET THE TENSILE STRAIN CHECK FOR TRANSVERSE MOVEMENT

CONCLUSIONS

Geotechnical LS design checks that were developed in previous work [8, 9] were applied to the assessment of a set of hypothetical existing pipeline cases in moving slopes. For an existing pipeline, it was assumed that the direction of slope movement, rate of soil movement, and length or width of the moving soil block are known.

To assess the practicality of the approach, the mean strain capacities required to meet the checks were calculated for selected slope creep cases. The cases represented combinations of large and small diameter pipelines buried in stiff clay and loose sand. The results show reasonable trends in all cases, with the required mean strain capacity increasing with decreasing pipe size and increasing soil strength, D/t, safety class and capacity CoV. The required mean capacity also increases with increasing deviation of the moving angle from the idealized longitudinal and transverse movement directions. The required mean capacities are generally higher for ULS than for SLS, and for transverse movement than for longitudinal movement. The absolute values of the required mean strain capacity were found to be achievable in many cases, even under conservative assumptions for the strain capacities. This implies that the LS check requirements can be met for many practical cases.

The implication analyses also demonstrate that there is an incentive to lower the uncertainties associated with strain demand and capacity estimates as this significantly improves the ability to meet the requirements of the ULS check. These uncertainties can be reduced through site data acquisition, pipe capacity tests and the development of better predictive models. It is noted, however, that LS checks are based on the assumption that global buckling will not govern and this should be verified independently from the LS checks.

ACKNOWLEDGEMENTS

The work described here was performed under a JIP sponsored by Alliance Pipeline Ltd., China Petroleum Pipeline Engineering Corporation, Enbridge Employee Services Inc., ExxonMobil Development Company and TransCanada Pipelines Limited. The authors wish to thank the sponsor organization representatives on the project steering committee for their input and guidance throughout the work.

REFERENCES

[1] Det Norske Veritas AS. "Submarine Pipeline Systems – Offshore Standard." DNV-OS-F101 (2013).

[2] American Petroleum Institute. "Design, Construction, Operation, and Maintenance of Offshore Hydrocarbon Pipelines (Limit State Design)." API Recommended Practice 1111 (2015).

[3] Nessim, M., Adianto, R. and Stephens, M. "Limit States Design for Onshore Pipelines – Methodology and Implications." *Proceedings of 10th International Pipeline Conference*. IPC2014-33434. Calgary, Alberta, Canada, September 29-October 3, 2014.

[4] Nessim, M. and Adianto, R. "Limit States Design for Onshore Pipelines – Design Rules for Operating Pressure and Equipment Impact Loads." *Proceedings of the 2014 10th International Pipeline Conference.* IPC2014-33436. Calgary, Alberta, Canada, September 29-October 3, 2014.

[5] Adianto, R. and Nessim, M. "Limit States Design for Onshore Pipelines – Designing Hydrostatic Test Pressure

Copyright © 2019 by ASME

and Restrained Thermal Expansion." *Proceedings of the 2014 10th International Pipeline Conference.* IPC2014-33437. Calgary, Alberta, Canada, September 29-October 3, 2014.

[6] Nessim, M., Zhou, W., Zhou, J. and Rothwell, B. "Reliability Based Design and Assessment for Location-Specific Failure Threats with Application to Natural Gas Pipelines." *Journal of Pressure Vessel Technology* Vol. 131 (2009).

[7] Nessim, M., Zhou, W., Zhou, J. and Rothwell, B. "Target Reliability Levels for Design and Assessment of Onshore Natural Gas Pipelines." *Journal of Pressure Vessel Technology* Vol. 131 No. 6 (2009).

[8] Nessim, M.A. and Koduru, S.D. "Limit States Design Checks for Geotechnical Loads". *Proceedings of the 12th International Pipeline Conference.* IPC208-78569. Calgary, Alberta, Canada, September 24-28, 2018.

[9] Koduru, S.D. and Nessim, M.A. "Development of Strain Demand and Capacity Distributions to Use in Limit States Design." *Proceedings of the 12th International Pipeline Conference.* IPC208-78784. Calgary, Alberta, Canada, September 24-28, 2018.

[10] Tang, H., Fairchild, D., Panico, M., Crapps, J. and Cheng, W. "Strain Capacity Prediction of Strain-based Pipelines." *Proceedings of the 10th International Pipeline Conference.* IPC2014-33749. Calgary, Alberta, Canada, September 29-October 3, 2014

[11] Cakiroglu, C., Adeeb, S., Cheng, J.J.R. and Sen, M. "Design of the full scale experiments for the test of the tensile strain capacity of X52 pipes with girth weld flaws under internal pressure and tensile displacement." *Proceedings of the 10th International Pipeline Conference.* IPC2014-33225. Calgary, Alberta, Canada, September 29-October 3, 2014

Copyright © 2019 by ASME

www.ingramcontent.com/pod-product-compliance
Lightning Source LLC
LaVergne TN
LVHW062117300926
849161LV00010B/2478
* 9 7 8 0 7 9 1 8 6 1 9 9 8 *